Antenna Handbook

Antenna Handbook

VOLUME II ANTENNA THEORY

Edited by
Y. T. Lo
Electromagnetics Laboratory
Department of Electrical and Computer Engineering
University of Illinois–Urbana

S. W. Lee
Electromagnetics Laboratory
Department of Electrical and Computer Engineering
University of Illinois–Urbana

VAN NOSTRAND REINHOLD
New York

Copyright © 1993 by Van Nostrand Reinhold
Published by Van Nostrand Reinhold

I(T)P™ A division of International Thomson Publishing, Inc.
The ITP logo is a trademark under license

Printed in the United States of America

For more information, contact:

Van Nostrand Reinhold
115 Fifth Avenue
New York, NY 10003

International Thomson Publishing GmbH
Königswinterer Strasse 418
53227 Bonn
Germany

International Thomson Publishing Europe
Berkshire House 168-173
High Holborn
London WC1V 7AA
England

International Thomson Publishing Asia
221 Henderson Road #05-10
Henderson Building
Singapore 0315

Thomas Nelson Australia
102 Dodds Street
South Melbourne, 3205
Victoria, Australia

International Thomson Publishing Japan
Hirakawacho Kyowa Building, 3F
2-2-1 Hirakawacho
Chiyoda-ku, 102 Tokyo
Japan

Nelson Canada
1120 Birchmount Road
Scarborough, Ontario
Canada M1K 5G4

International Thomson Editores
Campos Eliseos 385, Piso 7
Col. Polanco
11560 Mexico D.F. Mexico

All rights reserved. No part of this work covered by the copyright hereon may be reproduced or used in any form or by any means—graphic, electronic, or mechanical, including photocopying, recording, taping, or information storage and retrieval systems—without the written permission of the publisher.

ARCFF 95 96 97 98 99 . . . 8 7 6 5 4 3 2

Library of Congress Cataloging-in-Publication Data

The antenna handbook/edited by Y.T. Lo and S.W. Lee
    p.   cm.
    Includes bibliographical references and indexes.
    Contents: v. 1. Fundamentals and mathematical techniques—v. 2. Antenna theory—v. 3. Applications—v. 4. Related topics.
    ISBN 0-442-01592-5 (v. 1).—ISBN 0-442-01593-3 (v. 2).—ISBN 0-442-01594-1 (v. 3).—ISBN 0-442-01596 (v. 4).
    1. Antennas (Electronics)   I. Lo, Y.T. II. Lee, S. W.
TK7871.6.A496   1993
621.382'4—dc20

93-6502
CIP

Contents

Volume II ANTENNA THEORY

5. **Radiation From Apertures** — 5-3
   E. V. Jull

6. **Receiving Antennas** — 6-1
   P. K. Park and C. T. Tai

7. **Wire and Loop Antennas** — 7-1
   L. W. Rispin and D. C. Chang

8. **Horn Antennas** — 8-1
   Constantine A. Balanis

9. **Frequency-Independent Antennas** — 9-1
   Paul E. Mayes

10. **Microstrip Antennas** — 10-1
    William F. Richards

11. **Array Theory** — 11-1
    Y. T. Lo

12. **The Design of Waveguide-Fed Slot Arrays** — 12-1
    Robert S. Elliott

13. **Periodic Arrays** — 13-1
    R. J. Mailloux

14. **Aperiodic Arrays** — 14-1
    Y. T. Lo

15. **Reflector Antennas** — 15-1
    Y. Rahmat-Samii

16. **Lens Antennas** — 16-1
    J. J. Lee

Appendices

A. Physical Constants, International Units, Conversion of Units, and Metric Prefixes — A-3
B. The Frequency Spectrum — B-1
C. Electromagnetic Properties of Materials — C-1
D. Vector Analysis — D-1
E. VSWR Versus Reflection Coefficient and Mismatch Loss — E-1
F. Decibels Versus Voltage and Power Ratios — F-1

Index — I-1

Preface

During the past decades, new demands for sophisticated space-age communication and remote sensing systems prompted a surge of R & D activities in the antenna field. There has been an awareness, in the professional community, of the need for a systematic and critical review of the progress made in those activities. This is evidenced by the sudden appearance of many excellent books on the subject after a long dormant period in the sixties and seventies. The goal of this book is to compile a reference to *complement* those books. We believe that this has been achieved to a great degree.

A book of this magnitude cannot be completed without difficulties. We are indebted to many for their dedication and patience and, in particular, to the forty-two contributing authors. Our first thanks go to Mr. Charlie Dresser and Dr. Edward C. Jordan, who initiated the project and persuaded us to make it a reality. After smooth sailing in the first period, the original sponsoring publisher had some unexpected financial problems which delayed its publication three years. In 1988, Van Nostrand Reinhold took over the publication tasks. There were many unsung heroes who devoted their talents to the perfection of the volume. In particular, Mr. Jack Davis spent many arduous hours editing the entire manuscript. Mr. Thomas R. Emrick redrew practically all of the figures with extraordinary precision and professionalism. Ms. Linda Venator, the last publication editor, tied up all of the loose ends at the final stage, including the preparation of the Index. Without their dedication and professionalism, the publication of this book would not have been possible.

Finally, we would like to express our appreciation to our teachers, students, and colleagues for their interest and comments. We are particularly indebted to Professor Edward C. Jordan and Professor George A. Deschamps for their encouragement and teaching, which have had a profound influence on our careers and on our ways of thinking about the matured field of electromagnetics and antennas.

This Preface was originally prepared for the first printing in 1988. Unfortunately, it was omitted at that time due to a change in the publication schedule. Since many readers questioned the lack of a Preface, we are pleased to include it here, and in all future printings.

Preface to the Second Printing

Since the publication of the first printing, we have received many constructive comments from the readers. The foremost was the bulkiness of a single volume for this massive book. The issue of dividing the book into multivolumes had been debated many times. Many users are interested in specific topics and not necessarily the entire book. To meet both needs, the publisher decided to reprint the book in multivolumes. We received this news with great joy, because we now have the opportunity to correct the typos and to insert the original Preface, which includes a heartfelt acknowledgment to all who contributed to this work.

We regret to announce the death of Professor Edward C. Jordan on October 18, 1991.

PART B
Antenna Theory

Chapter 5

Radiation from Apertures

E. V. Jull
University of British Columbia

CONTENTS

1. Alternative Formulations for Radiation Fields . . . 5-5
   Plane-Wave Spectra   5-5
   Equivalent Currents   5-8
2. Radiation Patterns of Planar Aperture Distributions . . . 5-10
   Approximations   5-10
   Rectangular Apertures   5-11
   Circular Apertures   5-20
   Near-Field Patterns   5-24
3. Aperture Gain . . . 5-26
4. Effective Area and Aperture Efficiency . . . 5-27
5. Near-Field Axial Gain and Power Density . . . 5-29
6. References . . . 5-34

Edward V. Jull was born in Calgary, Alberta, Canada. He received a BSc degree in engineering physics from Queen's University, Kingston, Ontario, in 1956, a PhD in electrical engineering in 1960, and a DSc (Eng.) in 1979, both from the University of London, England.

In 1956–57 and 1961–72 he was a research officer in the microwave section and later the antenna engineering section of the Division of Electrical Engineering of the National Research Council of Canada Laboratories in Ottawa. During 1963–65 he was a guest worker in the Electromagnetics Institute of the Technical University of Denmark and the Microwave Institute of the Royal Institute of Technology, Stockholm, Sweden. In 1972 he joined the University of British Columbia, Vancouver, Canada, where he is now a professor in the Department of Electrical Engineering.

In 1964 Dr. Jull was a joint winner of the IEEE Antennas and Propagation Society Best Paper Award. He has been chairman of Canadian Commission VI for the International Union of Radio Science (URSI), chairman of the Canadian National Committee for URSI, an associate editor of *Radio Science*, and an international director of the Electromagnetics Society. He is currently a vice president of URSI and is the author of *Aperture Antennas and Diffraction Theory* (Peter Peregrinus, 1981).

1. Alternative Formulations for Radiation Fields

An aperture antenna is an opening in a surface designed to radiate. Examples are radiating slots, horns, and reflectors. It is usually more convenient to calculate aperture radiation patterns from the electromagnetic fields of the aperture rather than from the currents on the antenna. There are now basically two methods for doing this. Traditionally the pattern has been derived from the tangential electric and magnetic fields in the aperture. This aperture field method is an electromagnetic formulation of the Huygens-Kirchhoff method of optical diffraction. In application it is convenient and accurate for the forward pattern of large apertures. More recently the pattern has also been derived from fields associated with rays which pass through the aperture and rays diffracted by the aperture edges. Its origins can be traced to the early ideas on optical diffraction of Young as more recently formulated by Keller [1] in his geometrical theory of diffraction. It is particularly useful in deriving the radiation pattern in the lateral and rear directions and is described in Chapter 4.

This chapter deals only with the derivation of radiation patterns from the tangential fields in the aperture. Two methods of formulating the radiation integrals are given in this section. They lead to the same result but differ in their concepts of radiation from apertures.

Plane-Wave Spectra

This approach has the advantages of conceptual simplicity for the radiative fields and completeness in its inclusion of the reactive fields of the aperture [2–5]. It uses the fact that any radiating field can be represented by a superposition of plane waves in different directions. The amplitude of the plane waves in the various directions of propagation, or the spectrum function, is determined from the tangential fields in the aperture. This spectrum function is the far-field radiation pattern of the aperture for radiation in real direction angles. The reactive aperture fields are represented by the complex directions of propagation in the total spectrum function.

The method is most appropriate for planar apertures. If the aperture lies in the $z = 0$ plane of Fig. 1 and radiates into $z > 0$, components E_x and E_y of the electric field in $z \geq 0$ can be written in terms of their corresponding spectrum functions P_x and P_y as

$$\begin{Bmatrix} E_x(x,y,z) \\ E_y(x,y,z) \end{Bmatrix} = \frac{1}{(2\pi)^2} \int_{-\infty}^{\infty} \int_{-\infty}^{\infty} \begin{Bmatrix} P_x(k_x,k_y) \\ P_y(k_x,k_y) \end{Bmatrix} e^{-j(k_x x + k_y y + k_z z)} \, dk_x \, dk_y \quad (1)$$

If α, β are the directions of propagation of each plane wave in Fig. 1, then $k_x = k \sin \alpha \cos \beta$, $k_y = k \sin \alpha \sin \beta$, and $k_z = k \cos \alpha$ are the components of the propagation vector $\mathbf{k}$ for each plane wave. It is necessary to specify

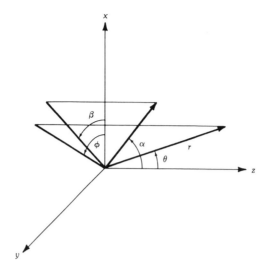

Fig. 1. Radiation from an aperture in the $z = 0$ plane: a plane wave radiates in a direction defined by the angles α and β, and the coordinates of a far-field point are (r, θ, ϕ).

$$k_z = \begin{cases} \sqrt{k^2 - (k_x^2 + k_y^2)} & \text{when } k_x^2 + k_y^2 \leq k^2 \\ -j\sqrt{(k_x^2 + k_y^2) - k^2} & \text{when } k_x^2 + k_y^2 > k^2 \end{cases} \quad (2)$$

so for real angles α, β (with $k_x^2 + k_y^2 \leq k^2$) there is radiation, and for complex angles α, β (with $k_x^2 + k_y^2 > k^2$), the fields decay exponentially outwards from the aperture.

Putting $z = 0$ in (1) and inverting the transforms shows the relation between the spectrum functions and the aperture fields:

$$\begin{Bmatrix} P_x(k_x, k_y) \\ P_y(k_x, k_y) \end{Bmatrix} = \int_{-\infty}^{\infty} \int_{-\infty}^{\infty} \begin{Bmatrix} E_x(x, y, 0) \\ E_y(x, y, 0) \end{Bmatrix} e^{j(k_x x + k_y y)} \, dx \, dy \quad (3)$$

Each plane wave has a z component associated with its x and y components in (1) and their relative magnitudes follow from $\mathbf{k} \cdot \mathbf{E} = 0$ for each plane wave. Thus the total field is

$$\mathbf{E}(x, y, z) = \frac{1}{(2\pi)^2} \int_{-\infty}^{\infty} \int_{-\infty}^{\infty} [(\hat{\mathbf{x}} k_z - \hat{\mathbf{z}} k_x) P_x(k_x, k_y) \\ + (\hat{\mathbf{y}} k_z - \hat{\mathbf{z}} k_y) P_y(k_x, k_y)] e^{-j(k_x x + k_y y + k_z z)} k_z^{-1} \, dk_x \, dk_y \quad (4)$$

To evaluate (4) at large distances from the aperture it is convenient to first convert to spherical coordinates, then apply stationary phase integration to the two double integrals. The stationary point is $\alpha = \theta$, $\beta = \phi$, and the final result for $kr \gg 1$ is

Radiation from Apertures

$$\mathbf{E}(r,\theta,\phi) \simeq \frac{je^{-jkr}}{\lambda r}[\hat{\boldsymbol{\theta}}(P_x\cos\phi + P_y\sin\phi) - \hat{\boldsymbol{\phi}}\cos\theta(P_x\sin\phi - P_y\cos\phi)] \quad (5)$$

where

$$\begin{Bmatrix} P_x \\ P_y \end{Bmatrix} = \int_{-\infty}^{\infty}\int_{-\infty}^{\infty} \begin{Bmatrix} E_x(x,y,0) \\ E_y(x,y,0) \end{Bmatrix} e^{jk(x\sin\theta\cos\phi + y\sin\theta\sin\phi)} \, dx\, dy \quad (6)$$

The far magnetic field follows from

$$\mathbf{H}(r,\theta,\phi) = Y_0 \hat{\mathbf{r}} \times \mathbf{E}(r,\theta,\phi) \quad (7)$$

where $Y_0 = \sqrt{\epsilon_0/\mu_0}$ is the free-space wave admittance.

Equation 5 provides the complete far-field radiation pattern from the Fourier transforms (6) of the tangential electric field in the aperture. It is also possible to use instead the tangential magnetic field in the aperture. Then, in terms of the spectrum functions

$$\begin{Bmatrix} Q_x \\ Q_y \end{Bmatrix} = \int_{-\infty}^{\infty}\int_{-\infty}^{\infty} \begin{Bmatrix} H_x(x,y,0) \\ H_y(x,y,0) \end{Bmatrix} e^{jk(x\sin\theta\cos\phi + y\sin\theta\sin\phi)} \, dx\, dy \quad (8)$$

the total electric field for $kr \gg 1$ is

$$\mathbf{E}(r,\theta,\phi) \simeq \frac{-jZ_0 e^{-jkr}}{\lambda r}[\hat{\boldsymbol{\theta}}(Q_x\sin\phi - Q_y\cos\phi)\cos\theta + \hat{\boldsymbol{\phi}}(Q_x\cos\phi + Q_y\sin\phi)] \quad (9)$$

where $\simeq$ means asymptotically equal and $Z_0 = \sqrt{\mu_0/\epsilon_0}$ is the free-space wave impedance.

An equivalent result in terms of both electric and magnetic tangential components of the aperture field is half the sum of (5) and (9), i.e.,

$$\mathbf{E}(r,\theta,\phi) \simeq \frac{je^{-jkr}}{2\lambda r}\{\hat{\boldsymbol{\theta}}[P_x\cos\phi + P_y\sin\phi) - Z_0\cos\theta(Q_x\sin\phi - Q_y\cos\phi)]$$
$$- \hat{\boldsymbol{\phi}}[(P_x\sin\phi - P_y\cos\phi)\cos\theta + Z_0(Q_x\cos\phi + Q_y\sin\phi)]\} \quad (10)$$

in which P_x, P_y and Q_x, Q_y are defined by (6) and (8), respectively. This superposition of electric and magnetic current sources provides a field which satisfies Huygens' principle in that radiation into $z < 0$ is suppressed.

The three expressions (5), (9), and (10) all yield the exact far-field pattern from the exact aperture field integrated over the entire aperture plane. Usually electric fields are more convenient to measure and calculate than magnetic fields, so (9) is rarely used. The choice between (5) and (10) should depend on how well the true boundary conditions are satisfied by whichever approximations are used. For example, it is convenient to assume that the field vanishes in the aperture plane outside the aperture. Then (5) should be used for apertures mounted in a large conducting plane as the boundary conditions on the conductor are rigorously satisfied. For apertures not in a conducting plane, however, (5) with this assumption generally yields less accuracy near the aperture plane than (10).

For apertures which are large in wavelengths the pattern is well predicted away from the aperture plane by all of these expressions. Then the aperture electric and magnetic fields are related by essentially free-space conditions, i.e., $E_x(x,y,0) = Z_0 H_y(x,y,0)$, $E_y(x,y,0) = -Z_0 H_x(x,y,0)$ and $Q_x = -Z_0^{-1} P_y$, $Q_y = Z_0^{-1} P_x$. Equations 9 and 10 become, respectively,

$$\mathbf{E}(r,\theta,\phi) \simeq \frac{je^{-jkr}}{\lambda r} [\hat{\boldsymbol{\theta}}(P_x \cos\phi + P_y \sin\phi)\cos\theta - \hat{\boldsymbol{\phi}}(P_x \sin\phi - P_y \cos\phi)] \quad (11)$$

and

$$\mathbf{E}(r,\theta,\phi) \simeq \frac{je^{-jkr}}{2\lambda r} (1 + \cos\theta)[\hat{\boldsymbol{\theta}}(P_x \cos\phi + P_y \sin\phi) - \hat{\boldsymbol{\phi}}(P_x \sin\phi - P_y \cos\phi)] \quad (12)$$

For small angles of θ, $\cos\theta \cong 1$ and the three expressions (5), (11), and (12) yield essentially identical results in that region of the pattern where their accuracy is highest. All have less precision at angles far off the beam axis, where (12) yields values which are the average of (5) and (11). As (12) is in terms of an aperture electric field which vanishes in the rear ($\theta = \pi$) direction, it is most commonly used for larger apertures, such as horns or reflectors, which are not in a conducting screen.

Equivalent Currents

The fields of sources within a surface S of Fig. 2 can be calculated from the tangential electric and magnetic fields $\mathbf{E}_s$ and $\mathbf{H}_s$ of the sources on S. Alternatively, these surface fields may be replaced by equivalent currents [3,6,7]. An electric surface current density $\mathbf{J}_s = \hat{\mathbf{n}} \times \mathbf{H}_s$, where $\hat{\mathbf{n}}$ is a unit vector normally outward from S, represents the tangential magnetic fields. Their contribution to the magnetic vector potential

$$\mathbf{A} = \frac{1}{4\pi} \int_S \mathbf{J}_s \frac{e^{-jkR}}{R} dS \quad (13)$$

gives a magnetic field $\mathbf{H} = \nabla \times \mathbf{A}$ and an electric field

$$\mathbf{E} = \frac{1}{j\omega\epsilon_0} \nabla \times \nabla \times \mathbf{A} \quad (14)$$

Similarly, a magnetic surface current density $\mathbf{K}_s = \mathbf{E}_s \times \hat{\mathbf{n}}$ represents the tangential electric fields on S and the resulting electric vector potential

$$\mathbf{F} = \frac{1}{4\pi} \int_S \mathbf{K}_s \frac{e^{-jkR}}{R} dS \quad (15)$$

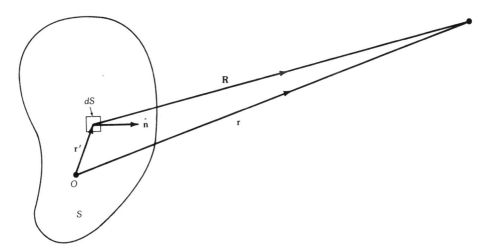

Fig. 2. Coordinates for calculation of radiation from a surface S.

provides an electric field

$$\mathbf{E} = -\nabla \times \mathbf{F} \tag{16}$$

and a magnetic field $\mathbf{H} = \nabla \times \nabla \times \mathbf{F}/(j\omega\mu_0)$.

Contributions from both electric and magnetic current sources give the total electric and magnetic fields outside S:

$$\mathbf{E} = -\nabla \times \mathbf{F} + \frac{1}{j\omega\epsilon_0} \nabla \times \nabla \times \mathbf{A} \tag{17}$$

$$\mathbf{H} = \nabla \times \mathbf{A} + \frac{1}{j\omega\mu_0} \nabla \times \nabla \times \mathbf{F} \tag{18}$$

With this superposition of the fields of electric and magnetic surface current densities radiation into the region enclosed by S is suppressed and Huygens' principle satisfied. There will also be no fields inside S if that region is a perfect conductor. Then $\mathbf{K}_s = 0$ on S and the total electric field outside S can be calculated from (14) with $2\mathbf{J}_s$ in (13).

As discussed previously, under "Plane-Wave Spectra," the fields can be calculated from the tangential electric or magnetic fields on S or from a superposition of the two sets of fields and all three methods yield the same result if the boundary conditions are rigorously satisfied. Usually $\mathbf{E}_s$ and $\mathbf{H}_s$ are approximated by incident fields, and scattered fields on S are neglected. The choice between the three methods should depend on the convenience of accurately approximating the true boundary conditions.

At distances r much larger than the maximum dimension of S in Fig. 2, $R \cong r - \hat{\mathbf{r}} \cdot \mathbf{r}'$ and (13) and (15) become

$$\begin{Bmatrix} \mathbf{A} \\ \mathbf{F} \end{Bmatrix} = \frac{e^{-jkr}}{4\pi r} \int_S \begin{Bmatrix} \mathbf{J}_s \\ \mathbf{K}_s \end{Bmatrix} e^{jk\hat{\mathbf{r}}\cdot\mathbf{r}'} dS \qquad (19)$$

where $\hat{\mathbf{r}} = \mathbf{r}/r$ and $\mathbf{r}'$ is the vector distance from the origin to dS. Also (14) and (16) become, in the far field in spherical coordinates, respectively,

$$\mathbf{E}(r,\theta,\phi) = -j\omega\mu_0(\hat{\boldsymbol{\theta}}A_\theta + \hat{\boldsymbol{\phi}}A_\phi) \qquad (20)$$
$$= -jk\mathbf{F} \times \hat{\mathbf{r}} \qquad (21)$$

Consequently the electric field of electric and magnetic current sources in the far field is

$$\mathbf{E}(r,\theta,\phi) = -j\omega\mu_0(\hat{\boldsymbol{\theta}}A_\theta + \hat{\boldsymbol{\phi}}A_\phi) - jk\mathbf{F} \times \hat{\mathbf{r}} \qquad (22)$$

and the corresponding magnetic fields are given by (7). Again, if electric or magnetic currents alone are used to calculate the total fields from (20) or (21), a factor of 2 must be included in the right side of (19).

With the surface S in the plane $z = 0$ of Fig. 2 and radiation into $z > 0$, $\hat{\mathbf{n}} = \hat{\mathbf{z}}$ in (19), which becomes

$$\mathbf{A} = \frac{e^{-jkr}}{4\pi r}(-\hat{\mathbf{x}}Q_y + \hat{\mathbf{y}}Q_x) \qquad (23a)$$

$$\mathbf{F} = \frac{-e^{-jkr}}{4\pi r}(-\hat{\mathbf{x}}P_y + \hat{\mathbf{y}}P_x) \qquad (23b)$$

in which P_x, P_y and Q_x, Q_y are defined by (6) and (8). Using (23a) in (20) with a factor of 2 gives (9). Equation 23b in (21) with a factor of 2 gives (5), and (23a) and (23b) in (22) yields (10). Thus the equivalent-current and plane-wave spectrum formulations provide identical results for the radiation fields of an aperture. The equivalent current method is simpler mathematically, but does not account for the reactive fields of the aperture.

2. Radiation Patterns of Planar Aperture Distributions

The expressions (5), (8), and (10) for the radiating far fields of an aperture in terms of the Fourier transforms of the tangential electric and magnetic aperture fields (6) and (8) are exact, but approximations are required in their application.

Approximations

In obtaining the approximations the usual assumptions are the following:

(a) The integration limits in (6) and (8) are the antenna aperture dimensions, i.e., fields in the aperture plane outside the aperture are assumed negligible.

Radiation from Apertures

(b) The aperture field is assumed to be the incident field from the antenna feed, i.e., scattered fields in the aperture are assumed negligible.

(c) Aperture electric and magnetic fields are assumed related by free-space conditions, i.e., the aperture is assumed large in wavelengths. This is so in (11) and (12) but not in (5), (9), and (10).

(d) Aperture fields are assumed separable in the coordinates of the aperture. Fortunately this assumption applies in many antenna designs; otherwise numerical integration of double integrals is usually required.

Clearly, the accuracy of the final result will depend on the degree to which all of the above assumptions are satisfied.

Rectangular Apertures

The Uniform Aperture Distribution—If the rectangular aperture of Fig. 3 is large and not in a conducting plane, its far field is conveniently calculated from (12) (assumption c above). Each component of aperture electric field can be dealt with separately. From (12) with $P_y = 0$, the x component of aperture field produces the far field

$$\mathbf{E}(r, \theta, \phi) = \mathbf{A} \int_{-a/2}^{a/2} \int_{-b/2}^{b/2} E_x(x, y, 0) e^{j(k_1 x + k_2 y)} \, dx \, dy \qquad (24)$$

where

$$\mathbf{A} = \mathbf{A}(r, \theta, \phi) = j \frac{e^{-jkr}}{2\lambda r} (1 + \cos\theta)(\hat{\boldsymbol{\theta}} \cos\phi - \hat{\boldsymbol{\phi}} \sin\phi) \qquad (25)$$

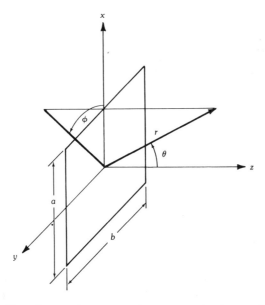

Fig. 3. Coordinates of a rectangular aperture in the $z = 0$ plane.

and

$$k_1 = k \sin\theta \cos\phi$$
$$k_2 = k \sin\theta \sin\phi \qquad (26)$$

Assumption a has been used in (24). For separable aperture fields (assumption d) $E_x(x,y,0) = E_0 E_1(x) E_2(y)$, where $E_1(x)$ and $E_2(y)$ are the field distributions normalized to the field E_0. Then (24) becomes

$$\mathbf{E}(r,\theta,\phi) = \mathbf{A} E_0 F_1(k_1) F_2(k_2) \qquad (27)$$

where

$$F_1(k_1) = \int_{-a/2}^{a/2} E_1(x) e^{jk_1 x} dx \qquad (28)$$

$$F_2(k_2) = \int_{-b/2}^{b/2} E_2(y) e^{jk_2 y} dy \qquad (29)$$

It is now necessary to choose an aperture field (assumption b). For the ideal case of a field uniform in amplitude and phase across the aperture, $E_1(x) = E_2(y) = 1$ and

$$F_1(k_1) = a \frac{\sin(k_1 a/2)}{k_1 a/2} \qquad (30)$$

$$F_2(k_2) = b \frac{\sin(k_2 b/2)}{k_2 b/2} \qquad (31)$$

This specifies the complete three-dimensional radiation pattern. For practical reasons it is usually the pattern in the two principal planes which are of major interest. These are the E-plane pattern in the (x,z) or $\phi = 0$ plane,

$$\mathbf{E}(r,\theta,0) = \frac{\hat{\boldsymbol{\theta}} j e^{-jkr}}{2\lambda r}(1+\cos\theta) ab \frac{\sin[(\pi a/\lambda)\sin\theta]}{(\pi a/\lambda)\sin\theta} \qquad (32)$$

and the H-plane pattern in the (y,z) or $\phi = \pi/2$ plane

$$\mathbf{E}(r,\theta,\pi/2) = -\frac{\hat{\boldsymbol{\phi}} e^{-jkr}}{2\lambda r}(1+\cos\theta) ab \frac{\sin[(\pi b/\lambda)\sin\theta]}{(\pi b/\lambda)\sin\theta} \qquad (33)$$

These patterns are of the same form but scaled in θ according to the aperture dimensions in their respective planes. If the aperture is large ($a,b \gg \lambda$), the main beam and first side lobes are contained in a small angle θ, so $1+\cos\theta \cong 2$ and the function $(\sin u)/u$, where $u = (\pi a/\lambda) \sin \theta$, determines the pattern. This function is plotted in Fig. 4.

The first null in the pattern occurs at $u = \pi$ or $\theta = \sin^{-1}(\lambda/a)$. Hence the full width of the main beam is $2 \sin^{-1}(\lambda/a) \cong 2\lambda/a$ radians for $a \gg \lambda$. At $u = 1.39$ the field is 0.707 its peak value. Thus the half-power beamwidth in radians is

$$\Delta\theta_{HP} = 2 \sin^{-1}\left(\frac{1.39\lambda}{\pi a}\right) \cong 0.88\lambda/a \qquad (34)$$

for $a \gg \lambda$. The first side lobes are at $u = \pm 1.43\pi$ and are $20 \log_{10}(0.217) = -13.3$ dB below the peak value of the main beam.

Simple Distributions—Radiation patterns are usually characterized by their principal plane half-power beamwidths and first side lobe levels. These parameters are given for several simple symmetrical aperture distributions in Table 1. The pattern functions there are derived from the Fourier cosine transforms of the aperture distributions, i.e., if $E_1(-x) = E_1(x)$, then (28) becomes

$$F_1(k_1) = 2 \int_0^{a/2} E_1(x) \cos k_1 x \, dx \qquad (35)$$

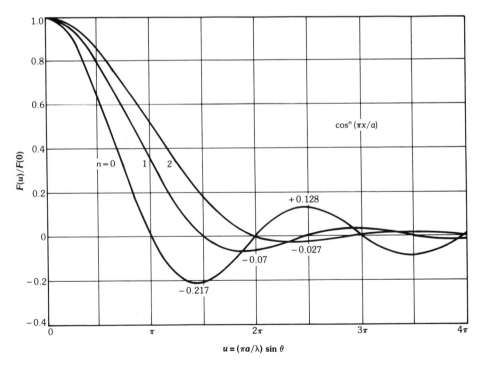

Fig. 4. Pattern functions of in-phase symmetrical field distributions in a rectangular aperture: uniform ($n = 0$), cosinusoidal ($n = 1$), and cosine-squared ($n = 2$).

The results in Table 1 are arranged in order of increasing beamwidths and decreasing first side lobe levels. For in-phase distributions the uniform aperture field has the highest gain but it also has a high side lobe level. The more the distribution decreases toward the aperture edges, the broader is the main beam and the lower the side lobe levels.

If the aperture distribution is an odd function, i.e., if $E_1(-x) = -E_1(x)$ in (28), its pattern is a Fourier sine transform,

$$F(k_1) = 2j \int_0^{a/2} E_1(x) \sin k_1 x \, dx \tag{36}$$

and is itself an odd function $[F(-k_1) = -F(k_1)]$. Several examples are shown in Table 2. These patterns have nulls at $\theta = 0$ and consequently their beamwidths and side lobe levels are unspecified. Instead, the angle at which the first and main lobe of the pattern appears is given, and the examples are arranged in order of increasing values of this angle. Such patterns may arise from the cross-polarized fields of a paraboloidal reflector, or they may be used for tracking on the pattern null, but otherwise they are rarely encountered.

The two-dimensional patterns of Tables 1 and 2 combine in (27) to give three-dimensional patterns of rectangular apertures. For example, an open-ended rectangular waveguide with the TE_{01} mode has an electric-field distribution in the aperture of Fig. 3

$$E_x(x, y, 0) = E_0 \cos(\pi y/b) \tag{37}$$

Table 1. Radiation Patterns of Even Distributions in a Rectangular Aperture

Aperture Distribution	$E(x)$ ($\|x\| < a/2$)	$F(u)$ $u = (\pi a/\lambda)\sin\theta$	First Null (rad)	3-dB Beamwidth (rad)	Relative Gain	Side Lobe Level (dB)
	$\frac{2}{a}\|x\|$	$a\left(\dfrac{\sin u}{u} - \dfrac{1-\cos u}{u^2}\right)$	$\sin^{-1}(0.75\lambda/a)$	$0.796\lambda/a$	0.742	-4.6
	1	$a\dfrac{\sin u}{u}$	$\sin^{-1}(\lambda/a)$	$0.886\lambda/a$	1.0	-13.3
	$1 - (2x/a)^2$	$\dfrac{2a}{u^2}\left(\dfrac{\sin u}{u} - \cos u\right)$	$\sin^{-1}(1.43\lambda/a)$	$1.179\lambda/a$	0.833	-21.3
	$\cos(\pi x/a)$	$2\pi a\dfrac{\cos u}{\pi^2 - (2u)^2}$	$\sin^{-1}(1.5\lambda/a)$	$1.189\lambda/a$	0.810	-23.1
	$1 - 2\|x\|/a$	$\dfrac{a}{2}\dfrac{\sin^2(u/2)}{(u/2)^2}$	$\sin^{-1}(2\lambda/a)$	$1.273\lambda/a$	0.742	-26.5
	$\cos^2(\pi x/a)$	$\dfrac{a}{2}\dfrac{\sin u}{u}\dfrac{\pi^2}{\pi^2 - u^2}$	$\sin^{-1}(2\lambda/a)$	$1.441\lambda/a$	0.667	-31.5

Table 2. Radiation Patterns of Odd Distributions in a Rectangular Aperture

Aperture Distribution	$E(x)$ ($\|x\| < a/2$)	$F(u)$ $u = (\pi a/\lambda)\sin\theta$	First Peak (rad)
RECTILINEAR	$\dfrac{2x}{a}$	$-j\dfrac{a}{u^2}(u\cos u - \sin u)$	$\sin^{-1}(0.6\lambda/a)$
ANTIPHASE CONSTANT	± 1 ($x \neq 0$)	$-ja\dfrac{1-\cos u}{u}$	$\sin^{-1}(0.7\lambda/a)$
SINE	$\sin(2\pi x/a)$	$-ja\dfrac{\pi \sin u}{\pi^2 - u^2}$	$\sin^{-1}(0.81\lambda/a)$
ANTIPHASE SINE-SQUARED	$\pm\sin^2(2\pi x/a)$ ($x \neq 0$)	$-j\dfrac{a}{2}\dfrac{1-\cos u}{u}\dfrac{4\pi^2}{4\pi^2 - u^2}$	$\sin^{-1}(0.9\lambda/a)$
ANTIPHASE TRIANGULAR	$\pm 1 - 2x/a$ ($x \neq 0$)	$-ja\dfrac{u - \sin u}{u^2}$	$\sin^{-1}(\lambda/a)$

and a radiation pattern, from (27) and Table 1,

$$\mathbf{E}(r,\theta,\phi) = j\frac{e^{-jkr}}{2\lambda r}(1+\cos\theta)(\hat{\boldsymbol{\theta}}\cos\phi - \hat{\boldsymbol{\phi}}\sin\phi)2\pi ab\, E_0 \frac{\sin[(\pi a/\lambda)\sin\theta\cos\phi]}{(\pi a/\lambda)\sin\theta\cos\phi}$$

$$\times \frac{\cos[(\pi b/\lambda)\sin\theta\sin\phi]}{\pi^2 - [(2\pi b/\lambda)\sin\theta\sin\phi]^2}, \quad \text{for } -\frac{\pi}{2} < \theta < \frac{\pi}{2} \tag{38}$$

This is not a very accurate representation of the pattern of an open-ended rectangular waveguide which supports only the dominant mode since the dimensions a, b are less than $\lambda/2$ and the assumptions above, under "Approximations," are invalid. The inaccuracy is largest near the aperture plane.

For the same rectangular waveguide aperture set in a conducting plane in $z = 0$ of Fig. 3, the far field follows from (5) with $P_y = 0$, i.e.,

$$\mathbf{E}(r,\theta,\phi) = j\frac{e^{-jkr}}{\lambda r}(\hat{\boldsymbol{\theta}}\cos\phi - \hat{\boldsymbol{\phi}}\cos\theta\sin\phi)2\pi ab\, E_0 \frac{\sin[(\pi a/\lambda)\sin\theta\cos\phi]}{(\pi a/\lambda)\sin\theta\cos\phi}$$

$$\times \frac{\cos[(\pi b/\lambda)\sin\theta\sin\phi]}{\pi^2 - [(2\pi b/\lambda)\sin\theta\sin\phi]^2} \tag{39}$$

This result satisfies the aperture plane boundary conditions and so is accurate even for narrow rectangular slots in a conducting plane.

Compound Distributions—The patterns of aperture distributions which are linear combinations of the simple distributions of Tables 1 and 2 are the same linear combinations of their patterns. That is, if

$$\int_{-\infty}^{\infty} E_n(x) e^{jk_1 x} dx = F_n(k_1), \qquad n = 1, 2, \ldots m \tag{40}$$

are the patterns of m simple aperture distributions, the pattern of a *compound distribution* is

$$\int_{-\infty}^{\infty} \sum_{n=1}^{m} a_n E_n(x) e^{jk_1 x} dx = \sum_{n=1}^{m} a_n F_n(k_1) \tag{41}$$

For example, the compound distribution

$$E(x) = \begin{cases} C + (1 - C) \cos^2(\pi x/a) & \text{for } |x| < a/2 \\ 0 & \text{for } |x| > a/2 \end{cases} \tag{42}$$

has the radiation pattern

$$F(u) = a \frac{\sin u}{u} \left[C + \left(\frac{1 - C}{2} \right) \frac{\pi^2}{\pi^2 - u^2} \right] \tag{43}$$

in which $u = k_1 a/2$. The half-power beamwidth of (43) lies between those of uniform and cosine-squared distributions, according to the value of C, as indicated by the solid curve in Fig. 5. Also shown are the half-power beamwidths of compound uniform and cosinusoidal and uniform and parabolic distributions. The side lobe levels of these compound distributions are also between those of the component parts.

Another practical example is that of aperture blockage by a reflector feed or subreflector of width $\delta \ll a$. The blocked aperture distribution is then

$$E'(x) = \begin{cases} E(x), & \delta/2 < |x| < a/2 \\ 0, & |x| < \delta/2, \ |x| > a/2 \end{cases} \tag{44}$$

and the resulting pattern is

$$F'(k_1) = F(k_1) - \delta \frac{\sin(k_1 \delta/2)}{k_1 \delta/2} \tag{45}$$

Here $F(k_1)$ is the radiation pattern of the unblocked distribution $E(x)$. If $k_1 \delta \ll 1$, then $F'(k_1) \cong F(k_1) - \delta$. The blocked pattern is uniformly reduced by δ. The effect is a narrower main beam and higher side lobe levels than the unblocked pattern.

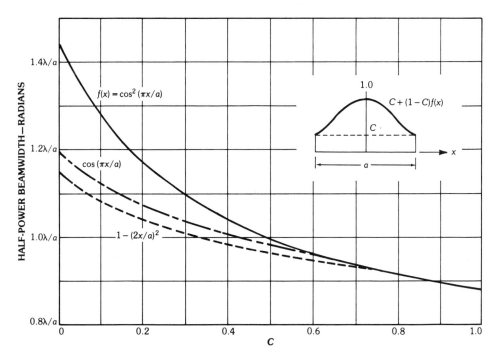

Fig. 5. Half-power beamwidths of compound symmetrical distributions in a rectangular aperture.

Displaced and Phase-Shifted Distributions—If the aperture distributions $E_n(x)$ of (41) are laterally displaced by x_n in the aperture plane, the radiation pattern is

$$\int_{-\infty}^{\infty} \sum_{n=1}^{m} a_n E_n(x - x_n) e^{jk_1 x}\, dx = \sum_{n=1}^{m} a_n F_n(k_1) e^{jk_1 x_n} \qquad (46)$$

For example, the radiation pattern of an array of $2m + 1$ uniform, in-phase distributions of symmetrical amplitudes $a_n = a_{-n}$ and widths w_n, symmetrically placed about the array center as in Fig. 6, is

$$F(k_1) = a_0 w_0 \frac{\sin(k_1 w_0/2)}{k_1 w_0/2} + 2 \sum_{n=1}^{m} a_n w_n \frac{\sin(k_1 w_n/2)}{k_1 w_n/2} \cos(k_1 x_n) \qquad (47)$$

Thus the patterns of arbitrary distributions can be obtained by summing contributions from segments of the aperture of essentially uniform amplitude and phase.

If the aperture distributions of (41) have a progressive linear phase shift exponent $(-j2\pi nx/a)$ across the aperture of width a, the pattern is

$$\int_{-\infty}^{\infty} \sum_{n=1}^{m} a_n E_n(x) e^{j[k_1 - (2\pi n/a)]x}\, dx = \sum_{n=1}^{m} a_n F_n[k_1 - (2\pi n/a)] \qquad (48)$$

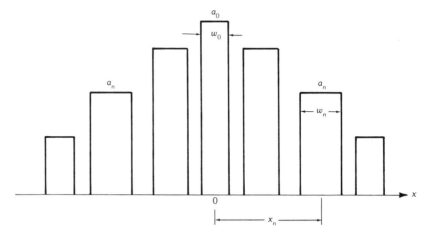

Fig. 6. Uniform apertures symmetrical about the array center.

This is a linear superposition of radiation patterns, each with an angular shift of the main beam $\theta_n = \sin^{-1}(n\lambda/a)$ from the original beam axis.

As an example, consider $2m + 1$ distributions with uniform amplitude b_n but with a linear phase variation of $2\pi n$ radians across the aperture. The radiation pattern is

$$F(k_1) = \int_{-a/2}^{a/2} \sum_{n=-m}^{m} b_n e^{j[k_1 - (2\pi n/a)]x} \, dx = a \sum_{n=-m}^{m} b_n \frac{\sin[(k_1 a/2) - n\pi]}{(k_1 a/2) - n\pi} \quad (49)$$

Equation 49 can be used to synthesize an antenna pattern $F_1(k_1)$ by equating b_n with $F_1(2\pi n/a)$. The accuracy of this approximation will be limited by the aperture width a since the largest value of m for real angles $\theta = \sin^{-1}(n\lambda/a)$ is a/λ.

Optimum Pattern Distributions—Generally antenna radiation patterns with narrow beams and low side lobe levels are sought. This involves a compromise, for in reducing side lobe levels by symmetrically tapering the aperture distribution the main beam broadens. An *optimum distribution* is one which has the lowest side lobe levels for a given width of main beam, or the narrowest main beam for a given side lobe level. Optimizing in this way results in a pattern with a maximum number of side lobes, all of the same level and width. The Chebyshev polynomials conveniently provide this as pattern functions for arrays of a finite number of elements (see Chapter 11) and the corresponding current distributions are rather similar to those of a cosine-squared function on a pedestal.

For a continuous field distribution in an aperture of width a, an ideal pattern function, which may be derived from an approximation to the Chebyshev polynomials of large order, is

$$F(u) = \begin{cases} \cosh \sqrt{(\pi A)^2 - u^2} & \text{for } u < \pi A \\ \cos \sqrt{u^2 - (\pi A)^2} & \text{for } u > \pi A \end{cases} \quad (50)$$

Radiation from Apertures

with $u = k_1 a/2 = (\pi a/\lambda) \sin \theta$.

As the side lobe levels are all unity the main beam to side lobe ratio is

$$R = F(0) = \cosh \pi A \tag{51}$$

Pattern nulls occur for

$$\begin{aligned} u_n &= \pm \pi \sqrt{A^2 + (n - 1/2)^2}, \quad n = 1, 2, \ldots \\ &\cong \pm (n - 1/2)\pi, \quad u \gg \pi A \end{aligned} \tag{52}$$

Hence the main beamwidth is determined by the side lobe levels and the far side lobes are equispaced. This ideal pattern is impractical, however, because the aperture distribution required to produce it is infinite at the aperture edges.

For a realizable pattern the far side lobe levels must decrease. This occurs in the normalized pattern [9]

$$F(u, A, \bar{n}) = \frac{\sin u}{u} \prod_{n=1}^{\bar{n}-1} \frac{1 - (u/u_n)^2}{1 - (u/n\pi)^2} \tag{53}$$

in which the nulls are

$$u_n = \begin{cases} \pm \pi\sigma\sqrt{A^2 + (n - 1/2)^2}, & 1 \leq n < \bar{n} \\ \pm n\pi & , \bar{n} \leq n < \infty \end{cases} \tag{54}$$

with

$$\sigma = \frac{\bar{n}}{\sqrt{A^2 + (\bar{n} - 1/2)^2}} \tag{55}$$

This pattern has a first side lobe level of $1/R$ and gradually decreasing near side lobes with null positions determined by the product in (53). Beyond the nth null the pattern is approximately that of the uniform distribution $(\sin u)/u$ depressed by the factor $(\pi/u_1)^2$, with side lobes of equal width and decreasing height. The scaling parameter σ ensures that the nulls of the two patterns coincide for $n = \bar{n}$.

This optimum pattern (53) may be synthesized by the method indicated by (49). The normalized aperture distribution is expressed as a Fourier series,

$$E(x) = 1 + 2 \sum_{p=1}^{\bar{n}} f(p, A, \bar{n}) \cos(2\pi p x/a) \tag{56}$$

in which the coefficients are samples of the pattern (53) for $u = m\pi$ and $n < \bar{n}$ or

$$f(p, A, \bar{n}) = \frac{[(\bar{n} - 1)!]^2}{(\bar{n} - 1 + p)!(\bar{n} - 1 - p)!} \prod_{m=1}^{\bar{n}-1} [1 - (n\pi/u_m)^2] \tag{57}$$

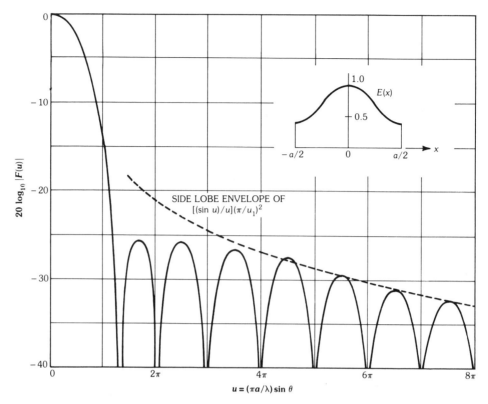

Fig. 7. Optimum aperture distribution and radiation pattern for first side lobe levels down 25 dB and $\bar{n} = 5$.

Tables of these coefficients are available [10]. The half-power beamwidth of the pattern is approximately ([10], p. 56)

$$\Delta\theta_{HP} \cong 2\sin^{-1}\left\{\frac{\lambda\sigma}{\pi a}[(\cosh^{-1}R)^2 - (\cosh^{-1}R/\sqrt{2})^2]^{1/2}\right\} \quad (58)$$

with σ given by (55). The half-power beamwidth of the ideal pattern (50) is slightly smaller and given precisely by (58) with $\sigma = 1$.

In the numerical example of Fig. 7 the first side lobe of the pattern is 25 dB below the main beam ($R = 17.78$) and the pattern nulls are those of $(\sin u)/u$ for $u \geq 5\pi$ ($\bar{n} = 5$). Hence from (51), (54), and (55), $A = 1.1365$, $\sigma = 1.0773$, and the first four nulls are at $u/\pi = 1.34, 2.03, 2.96,$ and 3.96.

Circular Apertures

For a circular aperture of diameter a as in Fig. 8 the far field of the x component of aperture field $E_x(\varrho', \phi')$ is, from (24) with $x = \varrho'\cos\phi'$ and $y = \varrho'\sin\phi'$,

$$\mathbf{E}(r, \theta, \phi) = \mathbf{A}\int_0^{2\pi}\int_0^{a/2} E_x(\varrho', \phi') e^{jk\varrho'\sin\theta\cos(\phi - \phi')}\varrho'\, d\varrho'\, d\phi' \quad (59)$$

Radiation from Apertures

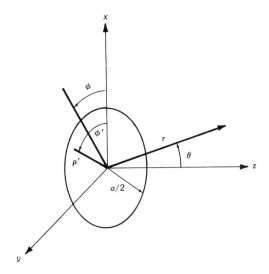

Fig. 8. Coordinates for radiation from a circular aperture.

where

$$\mathbf{A} = \frac{je^{-jkr}}{2\lambda r}(1 + \cos\theta)(\hat{\boldsymbol{\theta}}\cos\phi - \hat{\boldsymbol{\phi}}\sin\phi) \tag{60}$$

for an aperture not in a conducting plane and

$$\mathbf{A} = \frac{je^{-jkr}}{2\lambda r}(\hat{\boldsymbol{\theta}}\cos\phi - \hat{\boldsymbol{\phi}}\cos\theta\sin\phi) \tag{61}$$

for a conducting plane in $z = 0$, $\varrho' > a/2$.

If the aperture distribution is independent of ϕ' then $E_x(\varrho', \phi') = E_x(\varrho')$. Integration of (59) in ϕ' gives the Bessel function J_0 and

$$\mathbf{E}(r, \theta, \phi) = \mathbf{A}2\pi \int_0^{a/2} E_x(\varrho') J_0(k\varrho' \sin\theta) \varrho' \, d\varrho' \tag{62}$$

The family of aperture distributions

$$E_x(\varrho') = E_0[1 - (2\varrho'/a)^2]^n, \quad n = 0, 1, 2, \ldots \tag{63}$$

in (62) yield the radiation patterns

$$\mathbf{E}(r, \theta, \phi) = \mathbf{A}E_0 \frac{\pi a^2}{2} 2^n n! \frac{J_{n+1}[(\pi a/\lambda)\sin\theta]}{[(\pi a/\lambda)\sin\theta]^{n+1}} \tag{64}$$

The normalized patterns are illustrated in Fig. 9 for the uniform distribution ($n = 0$), the parabolic distribution ($n = 1$), and the parabolic-squared distribution

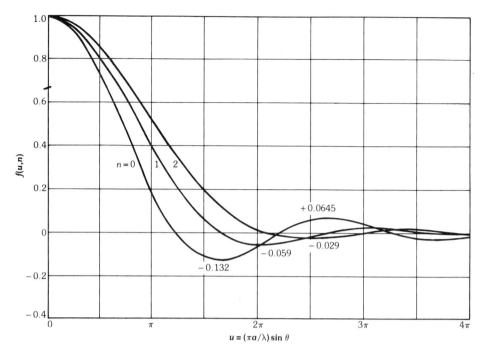

Fig. 9. Pattern functions of in-phase symmetrical field distributions in a circular aperture: uniform ($n = 0$), parabolic ($n = 1$), and parabolic-squared ($n = 2$).

($n = 2$). Values of the angular position off the beam axis of the first pattern null, the half-power beamwidth, and first side lobe levels are given in Table 3. Also given are corresponding values for the H-plane pattern for the dominant (TE_{11}) mode in the aperture of an open-ended circular waveguide. The E-plane pattern characteristics in this situation are those of a uniform aperture distribution in a circular aperture.

Compound distributions in a circular aperture are analyzed as above, under "Compound Distributions." For the distributions

$$E(\varrho') = C + (1 - C)[1 - (2\varrho'/a)^2]^n, \quad n = 0, 1, 2 \tag{65}$$

in a circular aperture, the normalized patterns are

$$\frac{Cf(u,0) - (1 - C)/(n + 1) f(u,n)}{C + (1 - C)/(n + 1)} \tag{66}$$

where

$$f(u,n) = 2^{n+1}(n + 1)! \frac{J_{n+1}(u)}{u^{n+1}} \tag{67}$$

are the normalized patterns of (64) in which $u = (\pi a/\lambda) \sin \theta$. The half-power beamwidths of patterns of a parabolic taper on a pedestal distribution for $n = 1$ and

Table 3. Radiation Patterns of Simple Distributions in a Circular Aperture

Aperture Distribution	$E(r)$, $\varrho' < a/2$	$F(u)$, $u = (\pi a/\lambda)\sin\theta$	First Null (rad)	3-dB Beamwidth (rad)	Relative Gain	First Side Lobe Level (dB)
UNIFORM	1	$\dfrac{\pi a^2}{2}\dfrac{J_1(u)}{u}$	$\sin^{-1}(1.22\lambda/a)$	$1.016\lambda/a$	1.0	−17.6
PARABOLA	$1-(2\varrho'/a)^2$	$\pi a^2 \dfrac{J_2(u)}{u^2}$	$\sin^{-1}(1.63\lambda/a)$	$1.267\lambda/a$	0.75	−24.6
PARABOLA-SQUARED	$[1-(2\varrho'/a)^2]^2$	$4\pi a^2 \dfrac{J_3(u)}{u^3}$	$\sin^{-1}(2.03\lambda/a)$	$1.47\lambda/a$	0.55	−30.6
H-PLANE PATTERN OF TE_{11} MODE	$\dfrac{\partial}{\partial\varrho'}[J_1(3.682\varrho'/a)]$	$\dfrac{\pi a^2}{2}\dfrac{J_0(u)-J_1(u)/u}{1-(u/1.841)^2}$	$\sin^{-1}(1.71\lambda/a)$	$1.29\lambda/a$		−26.2

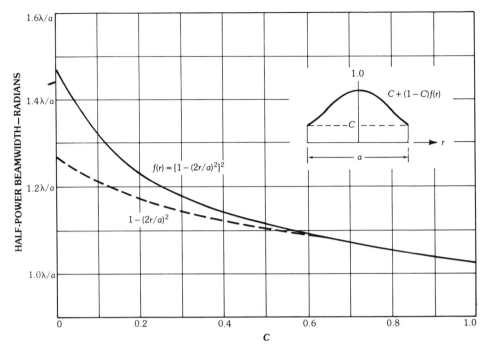

Fig. 10. Half-power beamwidths of compound symmetrical distributions in a circular aperture.

$n = 2$ in the above expressions are plotted in Fig. 10 for various values of C.

Data for the design of optimum pattern distributions in circular apertures is also available [11, 12].

Near-Field Patterns

The pattern expressions preceding are far field, i.e., the distance is very large compared with the aperture dimensions. There are linear phase differences in the radiation reaching the field point from different parts of the aperture. The first-order *near-field* effect is an additional quadratic phase difference in the radiation from the aperture. Thus for an x-polarized separable field in the aperture of Fig. 3,

$$\mathbf{E}(r, \theta, \phi) = \mathbf{A} E_0 D_1(k_1) D_2(k_2) \tag{68}$$

where

$$D_1(k_1) = \int_{-a/2}^{a/2} E_1(x) e^{-j(\beta x^2 - k_1 x)} dx \tag{69}$$

$$D_2(k_2) = \int_{-b/2}^{b/2} E_2(y) e^{-j(\beta y^2 - k_2 u)} dy \tag{70}$$

with $\beta = \pi/r\lambda$ and k_1 and k_2 defined by (26). As $r \to \infty$, $\beta \to 0$ and the Fresnel transforms (69) and (70) become the Fourier transforms (28) and (29).

For a uniform distribution $E_1(x) = 1$, so

$$D_1(k_1) = \sqrt{\frac{r\lambda}{2}} e^{jr\lambda k_1^2/4\pi} \{C(u_2) - C(u_1) - j[S(u_2) - S(u_1)]\} \tag{71}$$

where the *Fresnel integrals** $C(u)$ and $S(u)$ are defined by

$$C(u) - jS(u) = \int_0^u e^{-j\pi v^2/2} dv \tag{72}$$

and have arguments

$$\begin{Bmatrix} u_2 \\ u_1 \end{Bmatrix} = \pm \frac{a}{\sqrt{2r\lambda}} - \frac{k_1}{2\pi}\sqrt{2r\lambda} \tag{73}$$

For the cosinusoidal distribution $E_2(y) = \cos(\pi y/b)$, so

$$D_2(k_2) = \frac{1}{2}\sqrt{\frac{r\lambda}{2}} (e^{j(r\lambda/4\pi)(k_2+\pi/b)^2}\{C(v_2) - C(v_1) - j[S(v_2) - S(v_1)]\}$$
$$+ e^{j(r\lambda/4\pi)(k_2-\pi/b)^2}\{C(w_2) - C(w_1) - j[S(w_2) - S(w_1)]\}) \tag{74}$$

where

$$\begin{Bmatrix} v_2 \\ v_1 \end{Bmatrix} = \pm \frac{b}{\sqrt{2r\lambda}} - \frac{k_2}{2\pi}\sqrt{2r\lambda} - \frac{1}{b}\sqrt{\frac{r\lambda}{2}}$$
$$\begin{Bmatrix} w_2 \\ w_1 \end{Bmatrix} = \pm \frac{b}{\sqrt{2r\lambda}} - \frac{k_2}{2\pi}\sqrt{2r\lambda} + \frac{1}{b}\sqrt{\frac{r\lambda}{2}} \tag{75}$$

The above expressions together give the near-field pattern of an open-ended rectangular waveguide and reduce to the far-field pattern expressions (38) and (39) as $r \to \infty$.

Some near-field patterns in the $\phi = \pi/2$ plane ($k_1 = k\sin\theta$, $k_2 = 0$) for a cosinusoidal distribution $E_2(y) = \cos(\pi y/b)$ in a rectangular aperture are shown in Fig. 11. The near-field effects on the pattern are a broadening of the main beam, a filling in of the nulls, and a raising of the side lobe levels. Similar effects are observed in the near-field patterns of circular apertures [14].

---

*Computer subroutines and tabulated values are available. For example, see [13].

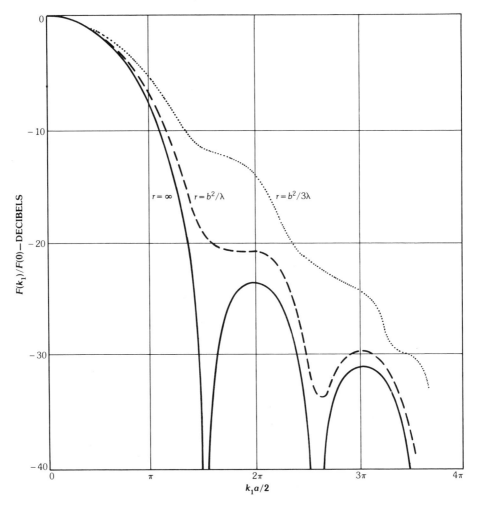

Fig. 11. Normalized radiation patterns of a cosinusoidal distribution. (*After Jull [5], © 1981 Peter Peregrinus Ltd.*)

3. Aperture Gain

The transmitting properties of an antenna are usually characterized by its gain. The *directive gain* $G(\theta, \phi)$ of an antenna in the (θ, ϕ) direction is the ratio of the radiated power density from it at (r, θ, ϕ) to its average radiated power density about the antenna at the range r. Thus

$$G(\theta, \phi) = \frac{\tfrac{1}{2} Y_0 \{|E_\theta(r, \theta, \phi)|^2 + |E_\phi(r, \theta, \phi)|^2\}}{P_r / 4\pi r^2} \qquad (76)$$

Here P_r is the total power radiated and can be calculated by integrating the numerator of (76) over a sphere of radius r surrounding the antenna. For planar

Radiation from Apertures

antennas it is usually simpler to integrate the time-averaged complex Poynting vector over the aperture. For a rectangular aperture in the $z = 0$ plane

$$P_r = \frac{1}{2} \text{Re} \left\{ \int_{\text{ap.}} (E_x H_y^* - E_y H_x^*) \, dx \, dy \right\}$$

$$= \frac{1}{2} Y_0 \int_{\text{ap.}} (|E_x|^2 + |E_y|^2) \, dx \, dy \qquad (77)$$

where the integration is over the aperture with the aperture dimensions sufficiently large in wavelengths that electric and magnetic fields in it are related by essentially free-space conditions.

For in-phase symmetrical aperture distributions the maximum directive gain, or directivity, is in the $\theta = 0$ direction. Then, from (12), (76), and (77), this is

$$G = \frac{4\pi}{\lambda^2} \frac{\left| \int E_x \, dx \, dy \right|^2 + \left| \int E_y \, dx \, dy \right|^2}{\int (|E_x|^2 + |E_y|^2) \, dx \, dy} \qquad (78)$$

For a uniform in-phase distribution the aperture fields $E_x = E_y$ are a constant and

$$G = 4\pi A/\lambda^2 \qquad (79)$$

where A is the aperture area. It can readily be shown that for in-phase distributions this uniform distribution yields the highest gain, i.e., $G \leq 4\pi A/\lambda^2$. "Supergain" antennas with nonuniform phase distributions can in principle have gains higher than (79) but are almost invariably impractical for apertures large in wavelengths.

4. Effective Area and Aperture Efficiency

The receiving properties of an antenna are characterized by its *effective area*. This is defined as the ratio of the power available from an antenna to the power density incident on it which is polarization matched to it. Simple reciprocity arguments show that the ratio of gain to effective area is a constant for all antennas. The value of this constant is, from (79), $4\pi/\lambda^2$. Hence the effective area A_{eff} of an antenna is related to the antenna gain G by

$$A_{\text{eff}} = \lambda^2 G/4\pi \quad \text{or} \quad G = 4\pi A_{\text{eff}}/\lambda^2 \qquad (80)$$

From (78) and (80) for an in-phase distribution in a large rectangular aperture, the effective area can be calculated from

$$A_{\text{eff}} = \frac{\left|\int\int E_x \, dx \, dy\right|^2 + \left|\int\int E_y \, dx \, dy\right|^2}{\int(|E_x|^2 + |E_y|^2) \, dx \, dy} \tag{81}$$

For a uniform aperture distribution with $E_x = E_y$, a constant, (81) gives an effective area equal to the physical area of the aperture. This is the maximum effective area for in-phase distributions. If the aperture field is polarized in the x direction ($E_y = 0$) and separable in the aperture coordinates $E_x = E_1(x)E_2(y)$, (81) may be written as

$$A_{\text{eff}} = (A_{\text{eff}})_1 (A_{\text{eff}})_2 \tag{82}$$

where

$$(A_{\text{eff}})_1 = \left|\int E_x \, dx\right|^2 \bigg/ \int |E_x|^2 \, dx$$
$$(A_{\text{eff}})_2 = \left|\int E_x \, dy\right|^2 \bigg/ \int |E_x|^2 \, dy \tag{83}$$

The *aperture efficiency* is the ratio of the effective area to the physical area of the aperture. For separable distributions in a rectangular aperture, as above,

$$\varepsilon = \frac{A_{\text{eff}}}{ab} = \varepsilon_1 \varepsilon_2 = \frac{(A_{\text{eff}})_1}{a} \frac{(A_{\text{eff}})_2}{b} \tag{84}$$

The two-dimensional aperture efficiency A_{eff}/a for simple aperture distributions is given under "Relative Gain" in Table 1. For compound distributions in a rectangular aperture the two-dimensional aperture efficiencies, or gain relative to a uniform distribution, are as follows:

For a parabola on a pedestal of height C, as in Fig. 12, the one-dimensional aperture efficiency is

$$\varepsilon_1 = \frac{(2 + C)^2}{9[1 - (2/3)(1 - C) + (1/5)(1 - C)^2]} \tag{85}$$

for a cosine on a pedestal,

$$\varepsilon_1 = \frac{[C + (2/\pi)(1 - C)]^2}{C^2 + (1/2)(1 - C) + (4C/\pi)(1 - C)} \tag{86}$$

for a cosine-squared on a pedestal,

$$\varepsilon_1 = \frac{[(C + 1)/2]^2}{C^2\{1 + (1 - C)/C + (3/8)[(1 - C)/C]^2\}} \tag{87}$$

Radiation from Apertures

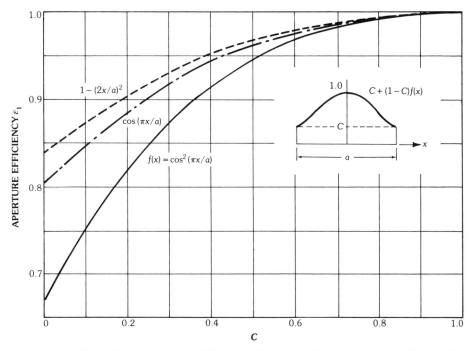

Fig. 12. One-dimensional aperture efficiencies of compound symmetrical distributions in a rectangular aperture.

These are plotted as dashed, broken, and solid curves, respectively, in Fig. 12. The gain of a rectangular aperture with separable distributions is then

$$G = \frac{4\pi}{\lambda^2} \varepsilon_1 \varepsilon_2 ab \tag{88}$$

Circular apertures with simple distributions have the aperture efficiency or relative gain given in Table 3. For the compound distributions of a parabola and a parabola-squared on a pedestal, the aperture efficiencies are plotted as the dashed and solid curves, respectively, in Fig. 13.

5. Near-Field Axial Gain and Power Density

On the beam axis ($\theta = 0$) the gain of an x-polarized aperture distribution is, from (76),

$$G = \frac{\tfrac{1}{2} Y_0 |E_\theta(r,0)|^2}{P_r/4\pi r^2} \tag{89}$$

For a uniform x-polarized distribution in the aperture of Fig. 3 the axial near-field follows from (68) and (71):

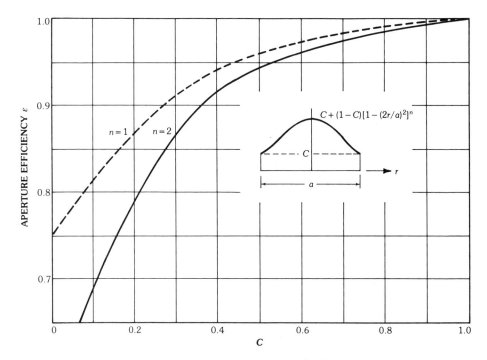

Fig. 13. Aperture efficiencies of compound symmetrical distributions in a circular aperture.

$$\mathbf{E}(r,0) = \hat{\mathbf{\theta}} 2j E_0 e^{-jkr}[C(u) - jS(u)][C(u') - jS(u')] \qquad (90)$$

where $C(u)$ and $S(u)$ are the Fresnel integrals defined by (72), $u = a/\sqrt{2r\lambda}$, and $u' = b/\sqrt{2r\lambda}$. The power radiated is $P_r = \tfrac{1}{2} Y_0 E_0^2 ab$ and the axial near-field gain can be written as

$$G = \frac{4\pi ab}{\lambda^2} R_E(u) R_E(u') \qquad (91)$$

where

$$R_E(u) = \frac{C^2(u) + S^2(u)}{u^2} \qquad (92)$$

is the near-field gain reduction factor for a one-dimensional uniform distribution. This factor is plotted in Fig. 14.

For a uniform and cosinusoidal distribution linearly polarized in the aperture of Fig. 4, the axial near-field is, from (68) with (71) and (74),

$$\mathbf{E}(r,0) = \hat{\mathbf{\theta}} j E_0 e^{-jkr+j(\pi r\lambda/4b^2)}[C(u) - jS(u)]\{C(v) - C(w) - j[S(v) - S(w)]\} \qquad (93)$$

where, from (75),

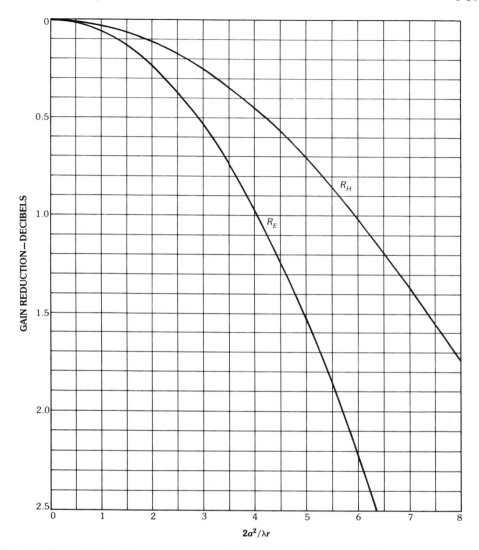

Fig. 14. Near-field axial gain reduction factors in decibels: R_E, (92), for a uniform distribution, and R_H, (96), for a cosinusoidal distribution in a rectangular aperture. (*After Jull [5], © 1981 Peter Peregrinus Ltd.*)

$$\begin{Bmatrix} v \\ w \end{Bmatrix} = \pm \frac{b}{\sqrt{2r\lambda}} + \frac{1}{b}\sqrt{\frac{r\lambda}{2}} \tag{94}$$

The power radiated is now $P_r = \frac{1}{4} Y_0 E_0^2 ab$ and the axial gain can be written as

$$G = \frac{32ab}{\pi\lambda^2} R_E(u) R_H(v, w) \tag{95}$$

in which

$$R_H(v, w) = \frac{\pi^2}{4} \frac{[C(u) - C(w)]^2 + [S(v) - S(w)]^2}{(v - w)^2} \quad (96)$$

is the gain reduction factor for the cosinusoidal distribution. It is plotted in Fig. 14. The factors R_E and R_H also appear in expressions for the axial gain of pyramidal and sectoral horns and are tabulated [15].

For the circular aperture of Fig. 8 with an *x*-polarized uniform distribution the axial field at a distance r is

$$\mathbf{E}(r, \theta) = \hat{\boldsymbol{\theta}} 2jE_0 e^{-j(kr+t)} \sin t \quad (97)$$

with $t = \pi a^2/(8\lambda r)$. As the power radiated is $P_r = \frac{1}{2} Y_0 E_0^2 \pi a^2/4$ the axial near-field gain is

$$G = \frac{4\pi}{\lambda^2} \frac{\pi a^2}{4} \frac{\sin^2 t}{t^2} \quad (98)$$

It is evident from (98) that at very close ranges r, this axial gain has maxima at

$$r = \frac{a^2}{4(2n + 1)\lambda}, \quad n = 0, 1, 2, \ldots \quad (99)$$

and is zero at

$$r = \frac{a^2}{8n\lambda}, \quad n = 1, 2, \ldots \quad (100)$$

These correspond to ranges at which the aperture contains even and odd numbers of Fresnel zones, respectively. The situation is illustrated in Fig. 15a, where the axial near-field power density of a uniform circular aperture of diameter a,

$$2 Y_0 E_0^2 \sin^2 t \quad (101)$$

is plotted.

Fig. 15b shows corresponding results for a uniform square aperture of side a. The axial power density

$$2 Y_0 E_0^2 [C^2(u) + S^2(u)] \quad (102)$$

is not completely canceled at any range because the Fresnel zones are annular and incomplete in a square aperture.

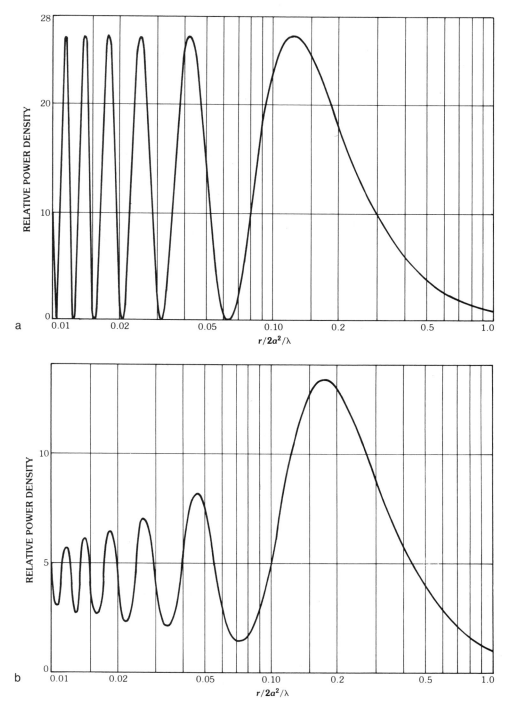

Fig. 15. Axial power density in the near field of uniform distributions. (*a*) Circular aperture of diameter *a*; (*b*) Square aperture of side *a*.(*After Jull [5], © 1981 Peter Peregrinus Ltd.*)

6. References

[1] J. B. Keller, "Diffraction by an aperture," *J. Appl. Phys.*, vol. 28, pp. 426–444, 1957.
[2] P. C. Clemmow, *The Plane Wave Spectrum Representation of Electromagnetic Fields*, New York: Pergamon Press, 1966, pp. 11–37.
[3] R. E. Collin and F. J. Zucker, *Antenna Theory, Part 1*, New York: McGraw-Hill Book Co., 1969, pp. 62–74.
[4] D. R. Rhodes, *Synthesis of Planar Antenna Sources*, Oxford: Oxford University Press, 1974, pp. 9–42.
[5] E. V. Jull, *Aperture Antennas and Diffraction Theory*, Stevenage, Herts., England: Peter Peregrinus, 1981, pp. 7–17.
[6] W. L. Stutzman and G. A. Thiele, *Antenna Theory and Design*, New York: John Wiley & Sons, 1981, pp. 375–384.
[7] C. A. Balanis, *Antenna Theory: Analysis and Design*, New York: Harper and Row, 1982, pp. 446–456.
[8] S. Silver, ed., *Microwave Antenna Theory and Design*, New York: McGraw-Hill Book Co., 1946, pp. 159–168.
[9] T. T. Taylor, "Design of line-source antennas for narrow beamwidth and low side lobe levels," *IRE Trans. Antennas Propag.*, vol. AP-3, pp. 16–28, January 1955.
[10] R. C. Hansen, ed., *Microwave Scanning Antennas, Volume 1*, New York: Academic Press, 1964, pp. 419, 427.
[11] T. T. Taylor, "Design of circular apertures for narrow beamwidth and low side lobes," *IRE Trans. Antennas Propag.*, vol. AP-8, pp. 17–22, January 1960.
[12] R. C. Hansen, "Table of Taylor distributions for circular aperture antennas," *IRE Trans. Antennas Propag.*, vol. AP-8, pp. 23–26, January 1960.
[13] M. Abramowitz and I. A. Stegun, *Handbook of Mathematical Functions*, NBS, US Dept. of Commerce, 1964, pp. 321–322 (reprinted by Dover Publications, 1965).
[14] R. C. Hansen and L. L. Baillin, "A new method of near-field analysis," *IRE Trans. Antennas Propag.*, vol. AP-7, pp. 458–467, 1959.
[15] E. V. Jull, "Finite range gain of sectoral and pyramidal horns," *Electron. Lett.*, vol. 6, pp. 680–681, 1970.

Chapter 6

Receiving Antennas

P. K. Park
Hughes Aircraft Company

C. T. Tai
University of Michigan

CONTENTS

1. Equivalent Circuit of a Receiving Antenna — 6-3
2. Vector Effective Height of an Antenna — 6-3
3. Receiving Cross Section, Impedance-Matching Factor, and Polarization-Matching Factor — 6-6
4. Generalized Friis Transmission Formula — 6-8
5. Mutual Impedance between Distant Antennas — 6-9
6. Small Antennas — 6-9
    The Short Dipole 6-10
    The Small Loop 6-13
7. Ferrite Loop Antennas — 6-18
8. Bandwidth and Efficiency — 6-22
9. Noise — 6-24
10. Satellite TV Earth Station Receiving Antenna — 6-25
11. References — 6-32

Pyong Kiel Park was born in Pyong-An-Do, Korea. He received his BSEE degree from In-Ha University, Korea, in 1962, the MSEE degree from Yon Sei University, Korea, in 1965, and the PhD in electrical engineering from UCLA in 1979. He is a senior staff engineer at Hughes Aircraft Company, Missile Systems Division, which he joined in 1979. Currently he is responsible for designing a monopulse concurrent low side-lobe antenna, a coherent side lobe canceller antenna of the AIM-54, and their monopulse feed networks. His prior experience includes lecturer at Yon Sei University, assistant professor at Kwang Woon University, member of the technical staff at TRW, and senior engineer at Ford Aeronatronics. His main interests are in electromagnetic theory, antennas, and microwave circuits. He is a Senior Member of IEEE.

Chen-To Tai received his BS degree in physics from Tsing-Hua University in 1937 and his DSc degree in communication engineering from Harvard University in 1947. He has been a professor of electrical engineering at the University of Michigan since 1964. He has been a visiting professor at several universities in the United States, Europe, and the Far East, and held an honorary professorship from the Shanghai Normal University. A Life Fellow of IEEE and a past chairman of the Antennas and Propagation Society (1971), he received a Distinguished Faculty Award from the University of Michigan in 1975, and several awards from the Department of Electrical and Computer Engineering and the College of Engineering at that university.

1. Equivalent Circuit of a Receiving Antenna

By applying either Thevenin's theorem or Norton's theorem to a receiving antenna placed in a harmonically oscillating incident electromagnetic field, the load current can be calculated from the corresponding equivalent circuit (Fig. 1), i.e.,

$$I_L = \frac{V_{oc}}{Z_{in} + Z_L} \quad (1)$$

and

$$I_L = \frac{Z_{in}}{Z_{in} + Z_L} I_{sc} \quad (2)$$

where

V_{oc} = the open-circuit voltage measured at the receiving terminals
Z_{in} = the input impedance of the antenna when it is operating in its transmitting mode
Z_L = the load impedance
I_{sc} = the short-circuit current through the receiving terminals

The open-circuit voltage and the short-circuit current are related by

$$V_{oc} = Z_{in} I_{sc} \quad (3)$$

2. Vector Effective Height of an Antenna

When an antenna is operating in its transmitting mode the far-zone electric and magnetic fields can be written in the form

$$\mathbf{E} = \frac{-jkZ_0 \mathbf{N} e^{-jkr}}{4\pi r} \quad (4)$$

and

$$\mathbf{H} = \frac{\hat{\mathbf{r}} \times \mathbf{E}}{Z_0} \quad (5)$$

where

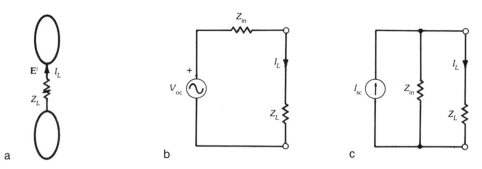

Fig. 1. Illustration for calculating load current with Thevenin's and Norton's theorems. (*a*) Receiving antenna in incident electromagnetic field. (*b*) Thevenin's equivalent circuit for load current. (*c*) Norton's equivalent circuit for load current.

$k = 2\pi/\lambda = \omega(\mu_0 \epsilon_0)^{1/2}$

$Z_0 = (\mu_0/\epsilon_0)^{1/2}$

$\hat{r}$ = the unit vector along **r**, the distance vector from antenna terminal to point of observation

r = the spherical radial distance from the antenna terminal to the point of observation of the field

The function **N** in (4) is related to the amplitude vector **A** defined by (39) in Chapter 1, namely,

$$\mathbf{N} = j\frac{4\pi}{k\sqrt{Z_0}}\mathbf{A} \qquad (6)$$

It is understood that the medium under consideration is free space with constitutive constants μ_0 and ϵ_0. The physical dimension of the vector **N** is measured in ampere-meters while that of **A** is measured in (watts)$^{1/2}$.

One of the most useful parameters in antenna theory was introduced by G. Sinclair [1]. It is designated as the *vector effective height function* or simply the *vector effective height*. It is defined by

$$\mathbf{h} = \frac{\mathbf{N}}{I_{in}} \qquad (7)$$

where I_{in} denotes the input current to the antenna when it is operating in its transmitting mode.* The dimension of **h** is measured in meters, hence it is identified as the "height" function. The function **h**, like the function **N** or **A**, is a function of the spherical angular variables (θ, ϕ) only, being independent of the radial variable r. A plot of $|\mathbf{h}|$ yields the field pattern and that of $|\mathbf{h}|^2$ yields the power pattern.

---

*Sinclair's original definition includes a negative sign, i.e., $\mathbf{h} = -\mathbf{N}/I_{in}$.

Receiving Antennas

The vector effective height is particularly useful in relating the open-circuit voltage or the short-circuit current in the theory of receiving antennas. By means of the reciprocity theorem one finds

$$V_{oc} = Z_{in} I_{sc} = \mathbf{h} \cdot \mathbf{E}^i \qquad (8)$$

where $\mathbf{E}^i$ denotes the incident electric field measured at the site of the antenna terminals. The expressions for the effective height of several simple antennas are listed in Table 1. It is assumed that the dipoles are pointed in the z direction and θ denotes the polar angle measured from the z axis. For the small loop, its axis is assumed to coincide with the z axis and $\hat{\mathbf{t}}$ denotes the unit tangent vector to the loop. Its shape, however, could be arbitrary as long as its largest linear dimension is electrically small.

For loop type antennas it is convenient to introduce a parameter designated as the *vector effective area*, which is defined by

$$\mathbf{S} = \frac{j}{k}(\hat{\mathbf{r}} \times \mathbf{h}) \qquad (9)$$

where $\mathbf{r}$ is the unit vector pointed in the radial direction. In terms of $\mathbf{S}$ the open-circuit voltage V_{oc} given by (8) can be written in the form

$$\begin{aligned} V_{oc} &= \mathbf{h} \cdot \mathbf{E}^i = \mathbf{h} \cdot (-\hat{\mathbf{r}} \times \mathbf{H}^i) Z_0 \\ &= \frac{-\mathbf{h} \cdot (\hat{\mathbf{r}} \times \mathbf{B}^i) Z_0}{\mu_0} = -j\omega \left[\frac{j}{k} \mathbf{B}^i \cdot (\hat{\mathbf{r}} \times \mathbf{h}) \right] \\ &= -j\omega (\mathbf{S} \cdot \mathbf{B}^i) \end{aligned} \qquad (10)$$

where $\mathbf{H}^i$ and $\mathbf{B}^i$ denote, respectively, the incident magnetic field and the incident magnetic induction. Thus the open-circuit voltage can be calculated using the vector effective area and the incident magnetic induction field, instead of the vector effective height and the incident electric field. This alternative formula conforms to the expression of the induced emf in a loop deduced from Faraday's law. For a small loop the vector effective area is given by

Table 1. Vector Effective Height

Antenna Type	Current Distribution	h
Hertzian dipole	$\mathbf{I} = I\hat{\mathbf{z}}, \quad \|z\| \leq \ell$	$-2\ell \sin\theta \, \hat{\boldsymbol{\theta}}$
Abraham dipole	$\mathbf{I} = I\left(1 - \frac{\|z\|}{\ell}\right)\hat{\mathbf{z}}, \quad \|z\| \leq \ell$	$-\ell \sin\theta \, \hat{\boldsymbol{\theta}}$
Half-wave dipole	$\mathbf{I} = I\cos\left(\frac{2\pi}{\lambda}z\right)\hat{\mathbf{z}}, \quad \|z\| \leq \lambda/4$	$-\dfrac{\lambda}{\pi} \dfrac{\cos[(\pi/2)\cos\theta]}{\sin\theta} \hat{\boldsymbol{\theta}}$
Small loop of area A	$\mathbf{I} = I\hat{\mathbf{t}}$	$jkA \sin\theta \, \hat{\boldsymbol{\phi}}$

$$\mathbf{S} = \frac{j}{k}(\hat{\mathbf{r}} \times \mathbf{h}) = A \sin\theta\, \hat{\boldsymbol{\theta}} \tag{11}$$

where θ is the angle between $\hat{\mathbf{r}}$ and $\hat{\mathbf{h}}$, and A is the area of the loop.

3. Receiving Cross Section, Impedance-Matching Factor, and Polarization-Matching Factor

The *receiving cross section* of an antenna is defined by

$$\sigma_R = \frac{\text{power received by an antenna}}{\text{incident power density}}$$
$$= \frac{1}{2}|I_L|^2 R_L \bigg/ \frac{1}{2}\frac{|\mathbf{E}^i|^2}{Z_0} \tag{12}$$

Using (1), (3), and (8) we have

$$\sigma_R = \frac{|\mathbf{h}\cdot\mathbf{E}^i|^2 R_L Z_0}{|Z_L + Z_{\text{in}}|^2 |\mathbf{E}^i|^2} \tag{13}$$

The receiving cross section is a measure of the effective "capture" area of an antenna with respect to the incident power density. As far as the dependence on the load impedance is concerned, the maximum value of σ_R occurs when Z_L is equal to Z_{in}^*, the conjugate of Z_{in}. Thus it is convenient to introduce an impedance-matching factor defined by

$$q = 1 - \left|\frac{Z_L - Z_{\text{in}}^*}{Z_L + Z_{\text{in}}}\right|^2 = \frac{4 R_L R_{\text{in}}}{|Z_L + Z_{\text{in}}|^2} \tag{14}$$

whose value lies between zero and unity. Equation 13 can then be written as

$$\sigma_R = \frac{q Z_0 |\mathbf{h}\cdot\mathbf{E}^i|^2}{4 R_{\text{in}} |\mathbf{E}^i|^2} \tag{15}$$

The quantity Z_0/R_{in} can be expressed in terms of the effective height and the directivity of the antenna.

We start with the expression for the power input to an antenna:

$$P_{\text{in}} = 1/2\, |I_{\text{in}}|^2 R_{\text{in}} \tag{16}$$

The radiated power P_r is related to the input power by

$$P_r = \eta P_{\text{in}} \tag{17}$$

where η denotes the efficiency of radiation to account for the ohmic loss of the antenna. Furthermore, the *directivity* or the *directive gain* of the antenna in an arbitrary direction (θ, ϕ) is defined by

$$D(\theta, \phi) = \frac{4\pi r^2 S(\theta, \phi)}{P_r} \tag{18}$$

where $S(\theta, \phi)$ denotes the power density in the far-zone region in the (θ, ϕ) direction, i.e.,

$$S(\theta, \phi) = \frac{|\mathbf{E}(\theta, \phi)|^2}{2Z_0} = \frac{k^2 Z_0 |I_{in} \mathbf{h}|^2}{32\pi^2 r^2} \tag{19}$$

in view of (4) and (7). Hence

$$\begin{aligned}D(\theta, \phi) &= \frac{k^2 Z_0 |I_{in}|^2 |\mathbf{h}|^2}{8\pi P_r} \\ &= \frac{k^2 Z_0 |\mathbf{h}|^2}{4\pi \eta R_{in}}\end{aligned} \tag{20}$$

on account of (16) and (17).

Eliminating R_{in}/Z_0 between (15) and (20), we obtain

$$\sigma_R = \frac{\lambda^2}{4\pi} \eta q p D(\theta, \phi) \tag{21}$$

where

$$p = \frac{|\mathbf{h} \cdot \mathbf{E}^i|^2}{|\mathbf{h}|^2 |\mathbf{E}^i|^2} \tag{22}$$

is designated as the *polarization-matching factor*, and its value lies between zero and unity. Equation 21 is the most general formula to characterize the receiving capability of an antenna [2].

To compute the value of p, it is convenient to introduce the complex polarization ratios for $\mathbf{h}$ and $\mathbf{E}^i$ defined by

$$\frac{h_\theta}{h_\phi} = t e^{j\beta} \tag{23}$$

$$\frac{E^i_\theta}{E^i_\phi} = s e^{j\alpha} \tag{24}$$

Then

$$p = \frac{1 + 2st\cos(\alpha + \beta) + s^2 t^2}{(1 + s^2)(1 + t^2)} \qquad (25)$$

The maximum value of p, equal to unity, occurs when $s = t$ and $\alpha = -\beta$, and corresponds to two identical elliptically polarized states with opposite sense of rotation. When $\alpha = \beta = 0$ both $\mathbf{h}$ and $\mathbf{E}^i$ are linearly polarized. A circularly polarized incident field corresponds to $s = 1$ and $\alpha = \pm\pi/2$. An antenna which radiates a circularly polarized far-zone field corresponds to $t = 1$ and $\beta = \pm\pi/2$. The polarization-matching factor between a circularly polarized incident field and a linearly polarized antenna is 1/2. The matching factor can also be displayed graphically on the Poincaré sphere or alternatively on a Smith chart [3].

4. Generalized Friis Transmission Formula

When we are dealing with two distant antennas, one transmitting, designated as no. 1, and another receiving, designated as no. 2, the pertinent parameters involved would be the power transfer ratio and the mutual impedance between the two antennas. Both parameters can most conveniently be expressed in terms of the vector effective heights of these two antennas. The power transfer ratio between the two antennas can be written in the form

$$\frac{P_2}{P_1} = \frac{\sigma_2 \eta_1 D_1(\theta, \phi)}{4\pi r^2} \qquad (26)$$

where

P_2 = the power received by antenna no. 2
P_1 = the input power to antenna no. 1
σ_2 = the receiving cross section of antenna no. 2
η_1 = the radiation efficiency of antenna no. 1
D_1 = the directivity of antenna no. 1
r = the distance between the terminals of the two antennas

Using the expression for σ_2 given by (21) we obtain

$$\frac{P_2}{P_1} = \left(\frac{\lambda}{4\pi r}\right)^2 \eta_1 \eta_2 q_2 p D_1(\theta, \phi) D_2(\theta, \phi) \qquad (27)$$

where

η_2 = the radiation efficiency of antenna no. 2 when it is operating in its transmitting mode
q_2 = the impedance-matching factor of antenna no. 2

Receiving Antennas

$$= 1 - \left|\frac{Z_L - Z_{in,2}^*}{Z_L + Z_{in,2}}\right|^2$$

$Z_{in,2}$ = the input impedance of antenna no. 2
Z_L = the load impedance connected to the terminals of antenna no. 2
$D_2(\theta, \phi)$ = the directivity of antenna no. 2 in the direction (θ, ϕ)
$p = |\mathbf{h}_1 \cdot \mathbf{h}_2|^2 / |\mathbf{h}_1|^2 |\mathbf{h}_2|^2$ = polarization-matching factor of the two antennas
$\mathbf{h}_1, \mathbf{h}_2$ = the vector effective heights of the two antennas

Equation 27 is the generalized version of Friis' transmission formula [4] originally formulated for two linear antennas without considering the radiation efficiency, the impedance-matching factor, or the polarization-matching factor.

5. Mutual Impedance between Distant Antennas

The mutual impedance between two antennas, one of which lies in the far-zone field of the other, can be calculated by using the vector effective height of these two antennas. By considering the relation

$$Z_{12} = \frac{(V_2)_{oc}}{I_1} \tag{28}$$

where

$(V_2)_{oc}$ = the open-circuit voltage excited at the terminals of antenna no. 2
I_1 = the input current to the terminals of antenna no. 1

one finds

$$Z_{12} = \frac{jkZ_0 e^{-jkr}}{4\pi r} (\mathbf{h}_1 \cdot \mathbf{h}_2) \tag{29}$$

This expression has been tested for two half-wave dipoles. The result shows that the expression is quite accurate where r, the distance between the two antennas, is greater than half of a wavelength. The formula could be used to estimate the coupling between antennas since the vector effective heights of most antennas can be calculated using some reasonable assumption of their current distributions or aperture field distribution.

6. Small Antennas

Most receiving antennas in practical use at low frequencies are small in wavelength. An antenna is said to be *small* if its physical dimensions do not exceed approximately one eighth of the wavelength. The small transmitting antenna is never satisfactory from the standpoint of efficiency because much of the generated power is wasted in heating the ohmic resistance. On the other hand, small receiving antennas often are very satisfactory. Sensitivity rather than power efficiency is the important factor in reception, and sensitivity is limited by noise. Schelkunoff and Friis ([5], Chap. 10) are suggested for a complete discussion of small antennas. There are two types of small antennas: the short dipole and the small loop. In the limiting case of vanishingly low frequency the short dipole is just a capacitor and the small loop is an inductor.

The Short Dipole

Radiation Pattern and Gain—The far fields of a short dipole oriented in the $\hat{z}$ direction (see Fig. 2a) are given by

$$E_\theta = j \frac{60\pi \sin\theta}{r\lambda} M \tag{30}$$

$$H_\phi = \frac{E_\theta}{120\pi} \tag{31}$$

where

M = the moment of current distribution defined as $M = \int_{-\ell}^{\ell} I(z)\,dz$

r = distance in meters to the observation point

λ = wavelength in meters

k = wave number $(2\pi/\lambda)$

The radiation pattern of a short dipole ($\sin\theta$) is shown in solid lines in Fig. 3. For comparison purposes the radiation pattern of a half-wave dipole is shown in dotted lines. The directivity of a short dipole with respect to the isotropic radiation is

$$D(\theta,\phi) = \frac{4\pi}{\int_0^{2\pi}\int_0^\pi \Psi \sin\theta\,d\theta\,d\phi} = 1.5 \tag{32}$$

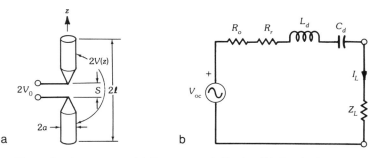

Fig. 2. Short dipole antenna. (*a*) Incremental dipole. (*b*) Equivalent circuit of (*a*).

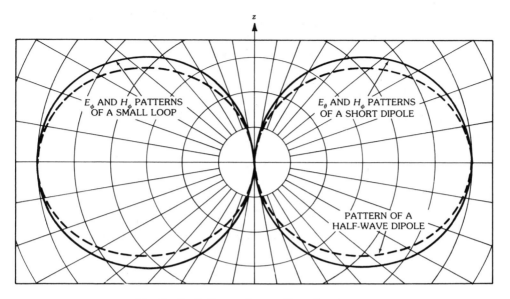

Fig. 3. Radiation patterns of small antennas.

where $\theta = 90°$ and the radiated power pattern Ψ is proportional to $\sin^2\theta$. Note that its directivity is only slightly less than that (1.64) of a half-wave dipole ([6], p. 54).

Impedance—The input impedance of a short dipole as shown in Fig. 2b is composed of the series inductance, the capacitance, the radiation resistance, and the ohmic resistance, which are expressed as

$$L_d = \frac{1}{|I_0|^2} \int_0^\ell L(z) |I(z)|^2 \, dz \tag{33}$$

where

$L(z) = (\mu_0/\pi) \ln(2z/a)$ = the inductance per unit length

$$C_d = \frac{1}{|V_0|^2} \int_{S/2}^\ell C(z) |V(z)|^2 \, dz$$
$$= \int_{S/2}^\ell C(z) \, dz, \qquad V_0 \cong V(z) \tag{34}$$

$C(z) = \pi\epsilon_0/\ln(2z/a)$ = the capacitance per unit length

$$R_r = \frac{80\pi^2}{I_0^2} \left(\frac{M}{\lambda}\right)^2 \tag{35}$$

$$R_0 = \frac{2\ell}{2\pi a} R_s \tag{36}$$

and where I_0 is the input current at the center, S is the input gap separation, and R_s is the surface resistance $(\omega\mu_0/2\sigma)^{1/2} = 4.5567 \times 10^{-3}/\sqrt{\lambda}\,\Omega$. Note that for a short monopole of length ℓ above a perfect ground the capacitance is twice and the inductance and the radiation resistance are half of those given by the above equations.

To illustrate the input impedance calculation for a short dipole, consider a short dipole 0.1λ long with a radius of 0.001λ. Assuming the linear current distribution $I(z) = I_0(1 - |z|/\ell)$, we find that ([5], p. 312)

$$L_d = \frac{\mu_0 \ell}{3\pi}\left(\ln\left(\frac{2\ell}{a}\right) - \frac{11}{6}\right) = 0.0147\mu_0\lambda \quad \text{(H)}$$

$$C_d = \pi\epsilon_0\ell/\ln(2\ell/a) = 0.034\epsilon_0\lambda \quad \text{(F)}$$

$$R_r = 80\pi^2(\ell/\lambda)^2 = 1.974 \quad (\Omega)$$

$$R_o = 7.2522 \times 10^{-2}/\sqrt{\lambda} \quad (\Omega)$$

The capacitive reactance ($X_{C_d} = 1/j\omega C_d = -j1763.5\,\Omega$) for this short dipole is much greater than the inductive reactance ($X_{L_d} = j\omega L_d = j34.769\,\Omega$). This example demonstrates that a short dipole is just a capacitor in the low-frequency limit.

Maximum Effective Height—The magnitude of the maximum effective height of a straight linear receiving antenna in (8) is equal to the effective length of a straight linear transmitting antenna, which is defined as the moment of its current distribution divided by the input current ([5], p. 301)

$$h_{\max} = \frac{M}{I_0} = \int_{-\ell}^{\ell} I(z)\,dz/I_0 \tag{37}$$

Maximum Receiving Cross Section $(\sigma_R)_{\max}$—From (15) the maximum receiving cross section $(\sigma_R)_{\max}$ occurs when the matching factor q becomes 1, and the vector effective height $\mathbf{h}$ is at its maximum (at $\theta = 90°$):

$$(\sigma_R)_{\max} = \frac{Z_0(h_{\max})^2 (E^i)^2}{4R_r(E^i)^2} = \frac{3}{8\pi}\lambda^2 \tag{38}$$

where $Z_0 = 120\pi$ is the free-space impedance and R_r is the radiation resistance of the antenna. In Table 2 the radiation resistance, the maximum effective height, and the maximum receiving cross section are listed for various short dipoles. Note that the half-wave dipole is also included in Table 2 for comparison. The receiving cross section of a half-wave dipole is $0.13\lambda^2$ ([6], p. 51), which is only slightly greater than the cross section of the short dipole $0.119\lambda^2$. (See Table 2.) This interesting condition exists because the radiation resistance of the antenna decreases with antenna size as rapidly as the square of the induced voltage, so the power available

Receiving Antennas

from the antenna remains constant. In practice the available power cannot be obtained because the antenna and its matching network are generally lossy, and also the high antenna reactance generally limits the bandwidth over which an effective match can be made with a simple network.

To induce resonance an inductance may be added in series with the antenna and the generator. The loading inductor keeps the current distribution nearly constant from the feed to the load point, with a linear decrease from the load to the end as shown in row 2 of Table 2. The inductive loading is advantageous since the loading increases the current moment, and the receiving parameter—effective length—varies as the current moment. The location of the inductor in the radiating structure affects the value of the radiation resistance in the sense that the current distribution along the antenna is modified.

The optimum location for this inductance with regard to efficiency is about four-tenths of the length of the antenna above the ground plane [7]. The radiation efficiency of this inductively loaded monopole is still small (in practice, on the order of 10 percent at the lower end of the band).

Top loading as shown in row 3 of Table 2 is another traditional approach toward achieving a fairly high efficiency in a relatively small size antenna. Note that the current distribution in these structures is almost constant. Fig. 4 shows plots of theoretical radiation resistance R_r of a top-loaded monopole antenna for an assumed linear current distribution. Further improvement of radiation efficiency can be achieved by combining the inductive loading and the top loading as shown in row 4 of Table 2.

The Small Loop

Radiation Pattern and Gain—The radiation pattern of a small loop is identical with that of a short dipole oriented normal to the plane of the loop with **E** and **H** fields interchanged. If the normal direction to the plane of the loop is the z axis (see Fig. 5a), the radiated electromagnetic fields are given by

$$E_\phi = 120\pi \frac{\pi N}{r} \frac{A}{\lambda^2} I \sin\theta \qquad (39)$$

and

$$H_\theta = -\frac{\pi N}{r} \frac{A}{\lambda^2} I \sin\theta = -\frac{E_\phi}{120\pi} \qquad (40)$$

where

- r = distance from the antenna
- I = antenna current
- N = number of turns of the loop
- A = area of loop
- λ = wavelength

Table 2. Short Dipole Parameters

Dipole	Current Distribution	Current Moment $M = \int_{-L}^{L} I(z)\,dz$	Radiation Resistance $R_r = 80\pi^2 \dfrac{(M/\lambda)^2}{I_0^2}$	Maximum Effective Height $h_{max} = \dfrac{M}{I_0}$	Maximum Receiving Cross Section $\sigma R_{max} = \dfrac{Z_0 h^2}{4R_r}$
Short (or Abraham)	(triangular current distribution, $2L$ wide, I_0 at base)	$I_0 L$	$80\pi^2 \dfrac{L^2}{\lambda^2}$	L	$\dfrac{3}{8\pi}\lambda^2 = 0.119\lambda^2$
Inductively loaded short	(trapezoidal current distribution, $2L$ wide, L_1, I_0 with inductors)	$I_0(2L - L_1)$	$80\pi^2 \dfrac{(2L - L_1)^2}{\lambda^2}$	$2L - L_1$	$\dfrac{3}{8\pi}\lambda^2$

Type	Diagram				
Top-loaded short	(diagram: inverted-V top load, width $2L$, height l_0, stubs l_1)	$(I_0 + I_1)L$	$80\pi^2 \dfrac{(I_0 + I_1)^2}{\lambda^2 I_0^2}$	$\left(1 + \dfrac{I_1}{I_C}\right) L$	$\dfrac{3}{8\pi}\lambda^2$
Top-loaded and inductively loaded short (or Hertzian)	(diagram: rectangular top load width $2L$, height l_0, with inductors)	$2I_0 L$	$320\pi^2 \dfrac{L^2}{\lambda^2}$	$2L$	$\dfrac{3}{8\pi}\lambda^2$
Half-wave	(diagram: semicircular element, width $\lambda/2$, height l_0)	$I_0 \dfrac{\lambda}{\pi}$	$73\,\Omega$	$\dfrac{\lambda}{\pi}$	$\dfrac{30}{73\pi}\lambda^2 = 0.13\lambda^2$

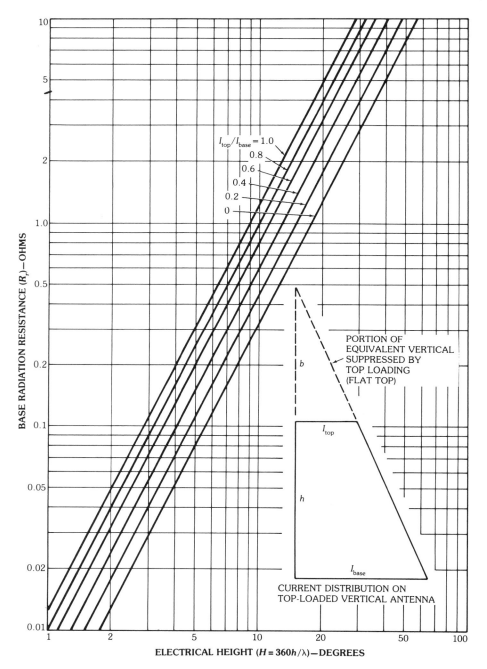

Fig. 4. Theoretical radiation resistance of vertical antenna for assumed linear current distribution. (*After Laport [15]; reprinted with permission of McGraw-Hill Book Company*)

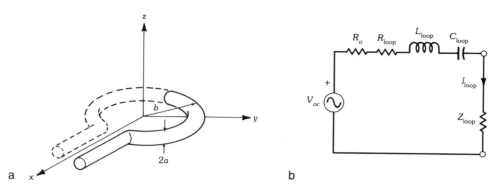

Fig. 5. Magnetic dipole. (*a*) Incremental magnetic dipole. (*b*) Equivalent circuit of (*a*).

The above equations are valid for arbitrary loop cross sections, provided that the loop diameter is small (i.e., loop radius $a < \lambda/2$) and uniform so that the loop current is uniform. The radiation pattern is shown in Fig. 3. The gain of a small loop with respect to the isotropic radiator is again 1.5 (= 1.76 dB).

Impedance—The input impedance of a small loop of one turn shown in Fig. 5a is composed of the ohmic resistance R_o, the series inductance L_{loop}, and the radiation resistance R_{loop}, which may be expressed as

$$R_o = \frac{\text{loop length} \times R_s}{\text{perimeter of wire cross section}} = \frac{b}{a} R_s \tag{41}$$

$$L_{\text{loop}} = \mu_0 b \ln(b/a) \tag{42}$$

$$R_{\text{loop}} = 20(k^2 S)^2 = 20[(h_{\text{max}})^2/S]^2 = 320\pi^4 \frac{S^2}{\lambda^4} \tag{43}$$

where

- b = the loop radius
- a = the wire radius
- S = the loop area
- R_s = the surface resistance = $\sqrt{\omega\mu/2\sigma}$, with σ the conductivity of the metal that the loop is made of

A satisfactory formula for the series capacitance of a coil is not available. However, since the capacitive reactance is much smaller than the inductive reactance of a multiple-turn small loop, one may ignore the series capacitance. If there are N turns, the total circulating current is the current per turn multiplied by N; hence R_{loop} and L_{loop} are to be multiplied by N^2. Note that while the radiation resistance of a dipole varies as the second power of the frequency, the radiation resistance of a small loop varies as the fourth power. To illustrate the input impedance calculation for a small loop, consider a circular loop of diameter 0.1λ and wire radius of 0.001λ. Then

$$R_o = \frac{b}{a}\sqrt{\frac{\mu_0 \omega}{2\sigma}} = 0.227\,835/\sqrt{\lambda} \quad (\Omega)$$

$$R_{\text{loop}} = 320\pi^6 \left(\frac{b}{\lambda}\right)^4 = 1.9227 \quad (\Omega)$$

$$L_{\text{loop}} = \mu_0 b \ln\left(\frac{b}{a}\right) = 0.1956 \mu_0 \lambda \quad (\text{H})$$

The inductive reactance ($X_L = j\omega L = j463\,\Omega$) for this small loop is much greater than R_o and R_{loop}, indicating that a small loop is just an inductor in the low-frequency limit.

Maximum Effective Height—For a loop placed in a medium of its relative permeability μ_r where a uniform field exists, the maximum induced voltage is given by

$$V_{\max} = -\frac{\partial \phi}{\partial t} = -j\omega\mu_0\mu_r HS \tag{44}$$

where ϕ is the magnetic flux ($\phi = \int_{\text{loop}} \mathbf{B}\cdot d\mathbf{S} = \mu_r\mu_0 HS$) and S is the area of the loop. Since the magnetic field in free space is related to the electric field as $H = \sqrt{\epsilon_0/\mu_0}\,E$, the maximum induced voltage can be expressed in terms of the incident electric-field vector $\mathbf{E}^i$ as

$$V_{\max} = -j\mu_r kSE^i, \quad k = \omega\sqrt{\mu_0\epsilon_0} \tag{45}$$

or the magnitude of the maximum vector effective height $h_{\max}$ becomes $h_{\max} = -j\mu_r kS$ (see Table 1).

Maximum Receiving Cross Section $(\sigma_R)_{\max}$—The maximum receiving cross-section for a small loop placed in a free space ($\mu_r = 1$) is identical with that for a short dipole, that is,

$$(\sigma_R)_{\max} = \frac{Z_0 h_{\max}}{4R_r} \frac{120\pi k^2 S^2}{4 \times 320\pi^4 S^2/\lambda^2} = \frac{3}{8\pi}\lambda^2 \tag{46}$$

Table 3 lists the receiving parameters for various small loops.

7. Ferrite Loop Antennas

For a given current a loop wound on a magnetic core produces a stronger field than the loop alone. If the loop and its core are small, the directive pattern is still of

Receiving Antennas

Table 3. Radiation Resistance of Various Small Loops

Loop	Shape	Area	Radiation Resistance $31\,000\ S^2/\lambda^4$
Circular	(circle, diameter $2A$)	πA^2	$31\,000 \left(\dfrac{\pi^2 A^4}{\lambda^4}\right)$
Ellipse	(ellipse, $2A \times 2B$)	πAB	$31\,000 \left(\dfrac{\pi^2 A^2 B^2}{\lambda^4}\right)$
Rectangular	(rectangle, $A \times B$)	AB	$31\,000 \left(\dfrac{A^2 B^2}{\lambda^4}\right)$
N-turn loop	(coil)	$N\pi A^2$	$N^2 \left(31\,000\ \dfrac{\pi^2 A^4}{\lambda^4}\right)$
Ferrite-core N-turn coil	(coil on rod)	$N\pi A^2$	$N^2 (\bar{\mu}_{\text{eff}})^2 \left(31\,000\ \dfrac{\pi^2 A^4}{\lambda^4}\right)$

the figure-8 shape; hence the directivity of the loop is not affected by the core. Since for the given current the field and hence the radiated power are increased by the core, the radiation resistance must become larger.

Mean Effective Permeability of a Ferrite Rod—The permeability of a core $\mu_c(x)$ of any configuration (see Fig. 6), with the exception of an ellipsoid, is a function of the core cross-section shape and the location of the coil in the core's longitudinal axis. Fig. 7 shows the curve μ_c as a function of the axial coordinate of the cross section.

The distribution of μ_c along the rod may be approximated by a trinomial [8]

$$\mu_c(x) = \mu_{cs}(1 + 0.106\bar{x} - 0.988\bar{x}^2) \quad (47)$$

where

$$\bar{x} = \frac{|x|}{\ell/2}$$

$$\mu_{cs} = \begin{cases} \dfrac{\mu}{1 + (\mu-1)(D/\ell)^2(\ln(\ell/D)\{0.5 + 0.7[1 - \exp(-\mu \times 10^{-3})]\} - 1)} & \text{for cylindrical cross section} \\ \dfrac{\mu}{1 + (\mu-1)(4ab/\pi\ell^2)\{\ln[k\ell/(a+b)] - 1\}} & \text{for rectangular cross section} \end{cases}$$

$k = 4 - 0.732[1 - \exp(-5.5b/a)] - 1.23\exp(-\mu \times 10^{-3})$

μ = initial or reversible permeability of the core material

ℓ = the core length

D = the cylinder diameter

a = the height of the rectangular cross section

b = the width of the rectangular cross section

The empirical equation (47) has very little error for permeabilities of from 50 to 1000 and for relative core lengths (core length divided by core diameter) of from 10 to 40. These ranges embrace practically all ferrite antenna construction. The mean

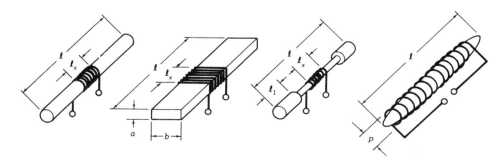

Fig. 6. Ferrite antennas.

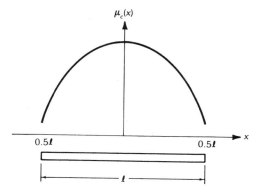

Fig. 7. Variation of permeability along ferrite rod.

effective core permeability $\bar{\mu}_{\text{eff}}$ for a coil length ℓ_{coil} enclosed between coordinates x_1 and x_2 is defined as

$$\bar{\mu}_{\text{eff}} = \frac{1}{\ell_{\text{coil}}} \int_{x_1}^{x_2} \mu_c(x)\, dx \qquad (48)$$

It was reported that the mean effective permeability $\bar{\mu}_{\text{eff}}$ for a hyperboloid core was about 1.8 times larger than that for a cylindrical core having the same weight. With the dumbbell shape, however, the reduction of weight was accomplished with a factor less than 1.2.

Impedance—An equivalent circuit of a ferrite antenna (which is a coil of N turns around a ferrite core) consists of an antenna coil inductance L_{coil}, radiation resistance R_r, an ohmic resistance R_o, an equivalent resistance of losses in the ferrite R_{loss}, a parasitic antenna capacitance C_p, and an equivalent emf V_{oc}, the effective value of which equals

$$V_{\text{oc}} = \mathbf{h} \cdot \mathbf{E}^i \qquad (49)$$

where

$$\mathbf{h} = \bar{\mu}_{\text{eff}} k A \sin\theta\, \boldsymbol{\theta}$$
$$R_o = N(b/a) R_s$$
$$L_{\text{coil}} = N^2 \bar{\mu}_{\text{eff}} \mu_0 b \ln(b/a) = \bar{\mu}_{\text{eff}} L_{\text{loop}} = \bar{\mu}_{\text{eff}} N^2 \mu_0 b \ln(b/a)$$

The radiation resistance of a small loop around the ferrite core can be found by applying the reciprocity theorem, given in the form

$$\int \mathbf{B}^b \cdot \mathbf{K}_m^a \, dV = \int \mathbf{B}^a \cdot \mathbf{K}_m^b \, dV \qquad (50)$$

To see how this can be accomplished, let the situation a be an air loop and the situation b be a loop around a ferrite core where $\mathbf{B}$ and $\mathbf{K}_m$ are the magnetic flux density inside of the loop and the magnetic dipole moment of the loop, respectively.

Then from (50) the magnetic dipole current J_m^b for the ferrite-core loop becomes $J_m^b = (H_{\text{ferrite}}/H_{\text{air}}) J_m^a$. Therefore the radiation resistance of the ferrite loop becomes [9]

$$R_{\text{loop}} = \frac{320\pi^4}{\lambda^4} \left| \iint_S \frac{H_{\text{ferrite}}}{H_{\text{air}}} dS \right|^2 = 320\pi^4 (\mu_{\text{eff}})^2 \frac{S^2}{\lambda^4} \qquad (51)$$

It is noted that since no satisfactory formulas for the inductance and the capacitance of a ferrite antenna are available, the quality factor of the ferrite antenna cannot be determined theoretically. It was observed that the quality factor increases with a shifting of the coil toward the end of the core, and in addition by making the coil diameter larger than the diameter of the core (the optimum diameter ratio is 4/3).

It is easy to verify that the maximum receiving cross section for a ferrite loop antenna remains the same as for a small loop antenna.

8. Bandwidth and Efficiency

A certain relationship exists between antenna size and its theoretical maximum bandwidth and efficiency. Chu [10] obtained the quality factor Q for the electrically small antenna as

$$Q = \frac{1 + 3k^2a^2}{k^3a^3(1 + k^2a^2)} \qquad (52)$$

where a is the radius of the smallest sphere which encloses the longest antenna dimension and k is the wave number in free space.

This Q is based on the lowest TM mode. When both a TM mode and a TE mode are excited, the value of Q is halved. Note that for $ka \ll 1$ the Q varies inversely as the cube of sphere radius in wavelengths. The importance of the Chu result is that it relates the lowest achievable Q to the largest dimension of an electrically small antenna, and the result is independent of the art that is used to construct the antenna within the hypothetical sphere, except in determining whether a pure TE or TM, or both modes, is excited. The above equation is for a lossless antenna. If the antenna is lossy, its ohmic resistance can be added in series with the radiation resistance, so the effect on Q is apparent. Fig. 8 plots single-mode Q for various efficiencies. The radiation efficiency is defined as

$$\eta = \frac{R_r}{R_r + R_L} \qquad (53)$$

where R_r is the radiation resistance of the small antenna and R_L is all the losses in the circuit. As Fig. 8 shows, improvement of the bandwidth for an electrically small

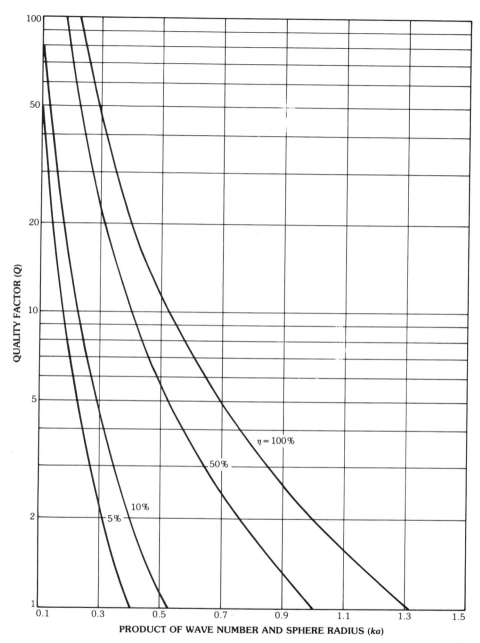

Fig. 8. Chu-Harrington fundamental limitations for single-mode antenna versus efficiency. (*After Hansen [16], © 1981 IEEE*)

antenna is only possible by fully utilizing the volume in establishing a TM and TE mode or by reducing the efficiency. A summary of bandwidth characteristics of various small antennas [11] is given in Table 4. In this table the bandwidth is defined such that the input vswr is less than 3 after matching with the L network.

9. Noise

Antenna noises can be classified, according to their origins, into two kinds: (1) internal and (2) external. The former consist of those due to the ohmic loss of the antenna itself, the transmission line, the feeding networks, and the amplifiers. Clearly these contributions depend only on their ambient temperature and amplifier noise figure, not the antenna radiation resistance and patterns. In contrast the latter depend on the antenna directive property and its ability to receive the noise emitted by external sources such as heavenly bodies, cosmic gases, ionosphere, atmosphere, the Earth, and many human-made sources. For small receiving antennas, because of their lack of directivity, the received noise is in general essentially constant. Fig. 9 shows the median values of average noise power

Table 4. Summary of Bandwidth Characteristics of Various "Small" Antennas (*After Desantis [11]*)

Number	Antenna Type or Technique	Impedance and Pattern Bandwidth for Input VSWR $\leq$ 3	Size (Height × Diameter)
1	Stub plus L network	1.16	$0.1\lambda \times 0.005\lambda$
2	Loop plus L network	1.05	$0.1\lambda \times 0.05\lambda$
3	Top-loaded stub plus L network	$\cong 1.24$	$\lambda/8 \times \lambda/8$
4	Top-loaded, folded, plus L network	$\cong 1.22$	$0.07\lambda \times 0.1\lambda$
5	Electrically thick monopole	$\cong 1.8$	$\lambda/2 \times \lambda/4$
6	Monopole-slot	1.3	$\lambda/4 \times 3/8\lambda$
7	Parasite-loading	1.8	$\lambda/2 \times 0.05\lambda$
8	Goubau antenna	2	$0.05\lambda \times 0.2\lambda$
9	Electrically small, complementary pair	>2.5	$\lambda/9 \times \lambda/4$
10	Slotted-cone antenna	>3	$\lambda/8 \times 0.44\lambda$
11	Hallen	>3	$\lambda/2 \times 0.03\lambda$

Receiving Antennas

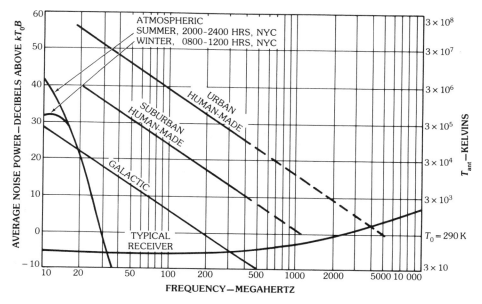

Fig. 9. Median values of average antenna noise temperature for an omnidirectional antenna near the Earth surface. (*Reprinted with permission of Howard W. Sams*)

expected from various sources. In this figure $k = 1.38 \times 10^{-23}$ J/B, $T_0 = 290$ K, and B is the receiver bandwidth in hertz. External human-made noise levels decrease with frequency but can increase rapidly from a suburban to an urban area. Furthermore, for frequencies below the gigahertz range, the sky noise increases sharply due to contributions from what is called atmospheric noise and galactic noise, and it is desirable to maximize the desired signal-to-noise ratio. This problem is considered in Chapter 11. For frequencies above the gigahertz range, the internal noise may become significant. It has been shown that for the internal noise a proper amount of mismatch between the antenna and the amplifier may actually improve the signal-to-noise ratio [12, 13, 14].

10. Satellite TV Earth Station Receiving Antenna*

The design of an Earth station antenna (which is not electrically small) requires a broad variety of system considerations which include microwave antenna, satellite link, and receiver characteristics. There are a number of factors that contribute to the overall performance of any satellite system. A typical tvro (television receiving only) system is shown in Fig. 10. The function of the uplink transmitter is to send a frequency modulated video signal and an fm audio signal to the satellite using a carrier in the 5.9- to 6.5-GHz band. Then the satellite retransmits those signals back to Earth on a lower carrier in the 3.7- to 4.2-GHz

---

*The material contained in this section is based mainly on the private note "Earth station considerations," dated June 1983, by James F. Corum of West Virginia University and Basil F. Pinzone, Jr., president of Pinzone Communication Products, Inc. Their assistance is duly acknowledged.

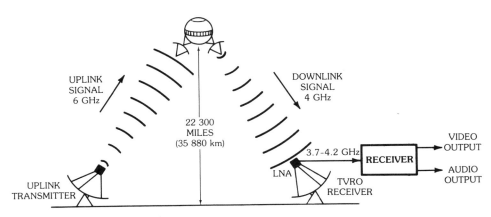

Fig. 10. Typical television-receiving-only system. (*After Decker [17], copyright 1982, Gernsback Publications; reprinted with permission from May 1982* Radio Electronics)

band. Using two separate frequencies allows simultaneous reception and transmission. Both the uplink and the downlink paths have a 500-MHz-wide frequency range (5.9 to 6.4 GHz and 3.7 to 4.2 GHz), which means that there is room for 12 channels, each 40 MHz wide. However, 24 channels are allocated in the same space by overlapping the channels with cross-polarized signals as shown in Fig. 11. The downlink signal transmitted by the satellite has an output power of 36 dBW, including the onboard antenna gain (approximately 30 dB). Fig. 12 shows an antenna pattern "footprint" on the Earth. This signal travels 22 300 miles to Earth, losing 196 dB of its initial strength. Therefore the ground receiving station is not left with much to work with. Once the signal is received by the receiving antenna, the lna (low-noise amplifier) amplifies the 4-GHz signal by about 40 dB while adding only 1 or 2 dB of noise. It is important to introduce as little noise at this stage as possible, since any that is introduced will be carried and amplified through the rest of the system. Then it is fed to a receiver where the 4-GHz rf signal transforms into standard composite video and audio. A summary of the frequency allocations for each television delivery technology is shown in Fig. 13.

Satellite Location

Communications satellites have geosynchronous orbits, which are directly above the equator and have an angular velocity the same as the angular velocity of the Earth's rotation. An observer on Earth sees the satellite as a stationary point in the sky and wants to locate the satellite. Figs. 14 and 15 are self-explanatory. Given the satellite point F (the point along with Earth's equator directly below the satellite) and the Earth station's coordinates (latitude and longitude), the standard spherical trigonometric quantities provide the observer with the azimuth and elevation angles of the desired satellite.

Satellite Link Calculation

Since the satellite is line of sight for the Earth station the received power P_R at the Earth receiving antenna can be expressed in a form of the Friis transmission formula

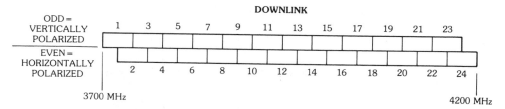

Fig. 11. How to fit 24 channels in a 12-channel bandwidth. (*After Decker [17], copyright 1982, Gernsback Publications; reprinted with permission from May 1982* Radio Electronics)

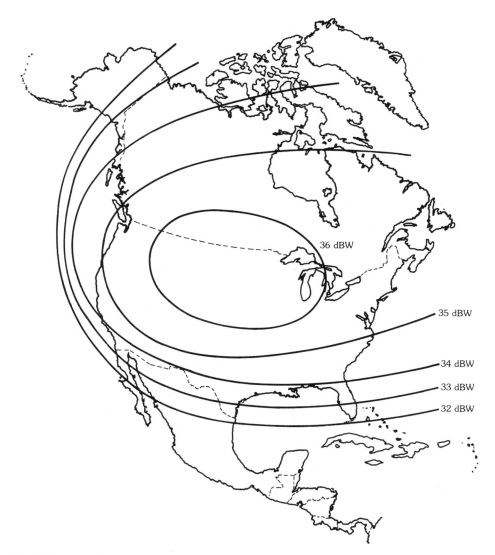

Fig. 12. Pattern footprint (contour of EIRP) on North America. (*Courtesy James Corum and Basil Pinzone*)

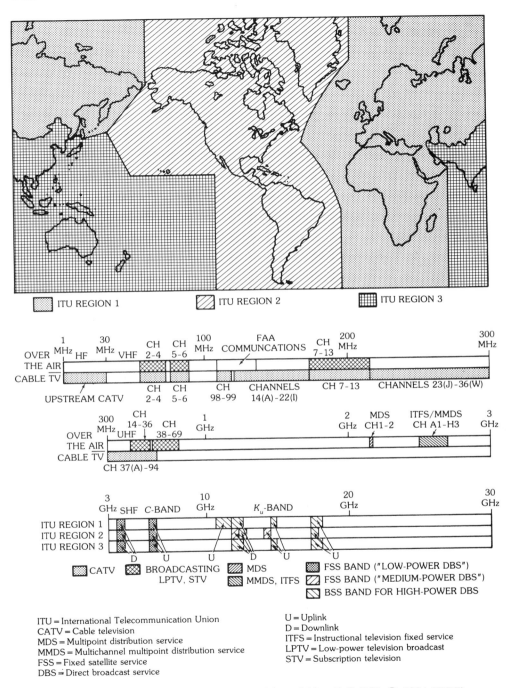

Fig. 13. Frequency allocations for television. (*After Bell [18], © 1984 IEEE*)

Receiving Antennas

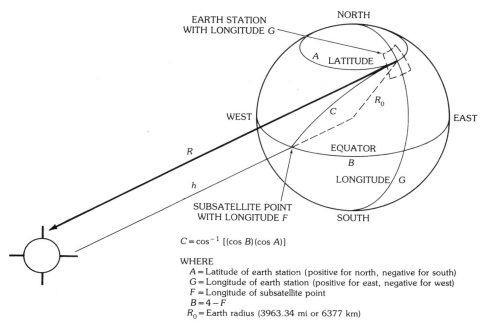

$$C = \cos^{-1}[(\cos B)(\cos A)]$$

WHERE
- A = Latitude of earth station (positive for north, negative for south)
- G = Longitude of earth station (positive for east, negative for west)
- F = Longitude of subsatellite point
- $B = 4 - F$
- R_0 = Earth radius (3963.34 mi or 6377 km)

Fig. 14. Standard spherical trigonometric quantities required for locating a desired satellite from a specified Earth station. (*Courtesy James Corum and Basil Pinzone*)

$$a_N = 180° + \tan^{-1}(\tan B / \sin A)$$
$$R = \sqrt{R_0^2 + (R_0 + h)^2 - 2(R_0 + h)R_0 \cos C}$$
$$e = \tan^{-1}\{[\cos C - (R_0/R)]/\sqrt{1 - \cos^2 C}\}$$

EXAMPLE:
Site: Morgantown ($A = 39°$, $G = -79°$)
Satellite: Statcom I ($0°$, $-135°$, $h = 22\,282$ mi or $35\,852$ km)
Great Circle Arc: $C = 64.24°$
Azimuth: $a_N = 247° = 113°$ west of north
Elevation: $e = 16.81°$
Slant Range: $R = 24\,396$ mi ($39\,253$ km)

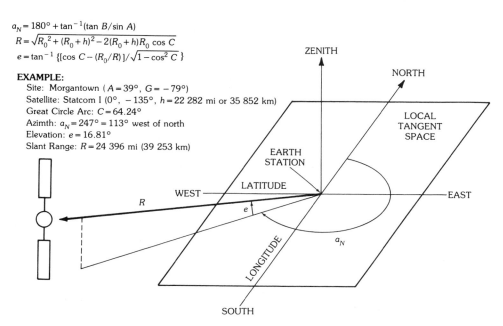

Fig. 15. Earth station look angles for satellite antennas. (*Courtesy James Corum and Basil Pinzone*)

$$P_R = \left(\frac{\lambda}{4\pi r^2}\right)^2 \frac{P_T G_T G_R}{L_0} \tag{54}$$

where

P_T = the satellite transmitter output power
G_T = the satellite transmitter antenna gain
G_R = the Earth receiving antenna gain
L_0 = the atmospheric attenuation loss
$(\lambda/4\pi r^2)^2$ = the free-space path loss factor

The Earth receiving antenna received the signal from the satellite as well as noise. If an effective antenna noise temperature is T in kelvins, then the available noise power at the receiving port N in watts is given by

$$P_n = kT\Delta f \tag{55}$$

where k is Boltzmann's constant (1.38×10^{-23} J/K) and Δf is the bandwidth of the receiving antenna system in hertz. The ratio of signal power to the noise power appearing at the antenna's terminal is given by

$$\text{snr} = \frac{P_R}{P_n} = \frac{(\lambda/4\pi r^2)^2 (P_T G_T G_R / L_0)}{kT\Delta f}$$

or

$$\text{snr} = \left(\frac{\lambda}{4\pi r^2}\right)^2 \frac{P_T G_T}{L_0} \left(\frac{G_R}{T}\right) \frac{1}{k\Delta f} \tag{56}$$

This equation can be written as

$$(G_R/T)_{\text{dB}} = (\text{snr})_{\text{dB}} - P_L - \text{EIRP} + (L_0)_{\text{dB}} + (\Delta f)_{\text{dB}} + k_{\text{dB}} \tag{57}$$

where

$(G_R/T)_{\text{dB}}$ = ratio of ground receiver antenna gain to noise temperature
$(\text{snr})_{\text{dB}}$ = signal-to-noise ratio in the ground receiver
$P_L = 10\log(\lambda/4\pi r^2)^2$, the path loss from satellite to Earth
EIRP = $10\log(P_T G_T)$, the effective isotropically radiated power in decibels above 1 W (dBW)
$(L_0)_{\text{dB}}$ = atmospheric loss in decibels
$(\Delta f)_{\text{dB}} = 10\log(\Delta f)$, channel bandwidth in decibels
$k_{\text{dB}} = 10\log(1.38 \times 10^{-23}) = -228.6$ dB

Receiving Antennas

The G_R/T quantity above is a figure of merit for the Earth station. Typically the sky temperature is on the order of 3 to 20 K. The contribution to the antenna temperature due to the side lobes facing the Earth will decrease as the antenna elevation is increased. Typically the antenna temperature will be on the order of 25 K for elevation angle above 45° and will rise to perhaps 90 K as the antenna elevation is lowered to about 5° above the horizon as shown in Fig. 16. Consequently one would provide a high antenna gain and a vanishingly small side lobe in order to increase the G_R/T figure of merit for the Earth station.

Earth Station Receiving Antenna—From (57) one can determine the antenna type and size. As an example, let us consider low-power, direct-broadcast, satellite services which are those available in the C band between 3.7 to 4.2 GHz; the satellite transmits an output power of 36 dBW (an output power P_T on the order of 5 to 10 W and the onboard antenna gain G_T of 30 dB), namely, the EIRP of 36 dBW. The path loss of −196 dB (22 300 mi or 35 880 km from satellite to Earth at 4 GHz), an atmospheric attenuation of 2 dB, and the channel bandwidth of 74.7 dB (for assuming a bandwidth of 30 MHz) are estimated.

To ensure reasonably good picture quality the signal-to-noise ratio $(snr)_{dB}$, which depends on the receiver quality, needs to be about 12 dB. The ground receiver antenna gain-to-noise ratio $(G_R/T)_{dB}$ becomes 20.1 dB by substituting all the estimated values into the above equation. Assuming the effective antenna temperature of 40 K, the minimum required receiving antenna gain G_R becomes 36 dB. Once the antenna gain is defined, the aperture size, aperture taper, and antenna type can be selected. The design procedure of the antenna for a specified

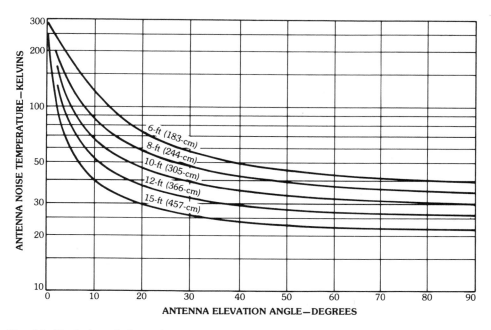

Fig. 16. Typical variation of antenna noise temperature with elevation angle, for five popular antenna sizes. (*Courtesy James Corum and Basil Pinzone*)

gain, side lobe level, and bandwidth are described in the chapters on reflector antennas and arrays.

11. References

[1] G. Sinclair, "The transmission and the reception of elliptically polarized waves," *Proc. IRE*, vol. 38, pp. 148–151, February 1950.
[2] C. T. Tai, "On the definition of the effective aperture of antennas," *IEEE Trans. Antennas Propag.*, vol. AP-9, pp. 224–225, March 1961.
[3] G. A. Deschamps, "Geometrical representation of the plane electromagnetic wave," *Proc. IRE*, vol. 39, pp. 540–544, 1961.
[4] H. T. Friis, "A note on a simple transmission formula," *Proc. IRE*, vol. 34, pp. 254–256, 1946; see also [5], pp. 388–394.
[5] S. A. Schelkunoff and H. T. Friis, *Antenna Theory and Practice*, New York: John Wiley & Sons, 1952.
[6] J. D. Kraus, *Antennas*, New York: McGraw-Hill Book Co., 1950.
[7] R. C. Hansen, "Efficiency and matching trade-offs for inductively loaded short antennas," *IEEE Trans. Commun.*, vol. COM-23, no. 4, April 1975.
[8] E. A. Mel'nikov and L. N. Mel'nikova, "Receiving induction ferrite antennas," *Izvestiya Vysshikh Uchebnykh Zavedeniy: Radioelectronica*, tr. Leo Kanner Associates, vol. 17, no. 10, 1974. Obtainable from the Department of the Army, US Army Foreign Science and Technology Center, 230 Seventh St. NE, Charlottesville, Va 22901.
[9] V. H. Rumsey and W. L. Weeks, "Electrically small loaded long antennas," *IRE Natl. Con. Rec.*, vol. 4, pt. 1, pp. 165–170, 1956.
[10] L. J. Chu, "Physical limitations of omnidirectional antennas," *J. Appl. Phys.*, vol. 19, p. 1163, December 1948.
[11] C. M. Desantis, "Low profile antenna performance study, pt. III: bibliography," *R&D Tech. Rep ECOM-4547*, November 1977.
[12] J. J. Sweeny and G. A. Deschamps, "A wave amplitude approach to noise analysis with application to a new method of noise measurement," *Tech. Rep. AFAL-TR-70-20*, Air Force Avionics Laboratory, Wright-Patterson AFB, Ohio 45433, March 1970.
[13] E. W. Harold, "An analysis of the signal-to-noise ratio of ultrahigh-frequency receivers," *RCA Rev.*, vol. 6, pp. 302–322, January 1942.
[14] H. T. Friis, "Noise figure of radio receivers," *Proc. IRE*, vol. 32, pp. 419–422, July 1944.
[15] E. A. Laport, *Radio Antenna Engineering*, chapter 1, New York: McGraw-Hill Book Co., 1952.
[16] R. C. Hansen, "Fundamental limitations in antennas," *Proc. IEEE*, vol. 69, no. 2, pp. 170–173, February 1981.
[17] D. Decker, "Satellite tv receiver," *Radio Electronics*, vol. 53, no. 5, p. 51, May 1982.
[18] T. E. Bell, "The new television: looking behind the tube," *IEEE Spectrum*, pp. 52–53, September 1984.

Chapter 7

Wire and Loop Antennas

L. W. Rispin
MIT Lincoln Laboratory

D. C. Chang
University of Colorado

CONTENTS

1. Introduction — 7-5
   - The Thin-Wire Antenna   7-5
   - Input Admittance or Impedance   7-7
   - Far-Field Radiation from a Thin-Wire Transmitting Antenna   7-8
   - The Receiving Antenna   7-8
   - Loaded Thin-Wire Antennas   7-9
   - Transient Response   7-10
   - Equivalent Radius for Noncircular Cylindrical Thin-Wire Conductors   7-10
   - Solid Thin-Wire Antenna Conductors   7-10
   - Antenna Parameters   7-11
2. The Linear Dipole Antenna — 7-11
   - Unloaded Transmitting Antennas   7-12
   - Input Admittance or Impedance   7-14
   - Far-Field Radiation from a Linear Thin-Wire Transmitting Antenna   7-15
   - Unloaded Receiving Antennas   7-18
   - Impedance-Loaded Antennas   7-19
   - An Impedance-Loaded Monopole Antenna   7-20
   - Transient Behavior of a Dipole Antenna   7-21
3. The Sleeve Antenna — 7-23
   - The Junction Effect   7-23
   - The Sleeve Dipole Antenna   7-26
   - The Sleeve Monopole Antenna   7-29

Lawrence W. Rispin received the BSEE degree from Akron University, in Akron, Ohio, in 1971, and an MSE degree from Arizona State University, in Tempe, Arizona, in 1973. In 1982 Dr. Rispin was awarded a PhD in electrical engineering from the University of Colorado, Boulder, Colorado. His doctoral dissertation involved research in cylindrical antenna theory.

His industrial experience includes cooperative student employment with Diebold, Inc., Canton, Ohio, and staff positions with Hughes Aircraft Company, Fullerton, California, and Motorola, Inc., Phoenix, Arizona, where he helped develop microwave power transistors. Currently he is in the Communication Antennas Group at MIT Lincoln Laboratory, in Lexington, Massachusetts, where he has been engaged in building an earth terminal for experimental EHF satellite communications.

He is a member of the IEEE, Eta Kappa Nu, and Sigma Tau. His major interests lie in analytical methods in antenna theory.

David C. Chang received his PhD in applied physics from Harvard University, Cambridge, Massachusetts, in 1967. He joined the faculty in the Department of Electrical and Computer Engineering, University of Colorado, Boulder, Colorado, in September 1967 and has been a professor of electrical and computer engineering since 1975, and Chairman of the Department since 1982.

Dr. Chang has been active in electromagnetic theory, antennas, and microwave circuits research. He has served, on various occasions, as the associate editor (1980–82), coordinator for the Distinguished Lecturers Program (1982–85), Chair of the Ad Hoc Committee for Basic Research (1985–present), and a member of the Administrative Committee (1985–present) of the IEEE Professional Society on Antennas and Propagation; as a member of the Technical Subcommittee on Microwave Field Theory of the IEEE Professional Society on Microwave Theory and Techniques (1975–85); as a member-at-large, US National Committee (1982–85), Chair of Technical Program Committee, Commission B on Fields and Waves (1983–86) of the URSI (International Union of Radio Science).

Dr. Chang is a Fellow of the IEEE.

  The Coaxial Sleeve Antenna with a Decoupling Choke 7-31
4. The Folded Dipole Antenna 7-37
5. The Thin-Wire Loop Antenna 7-42
  Far-Field Radiation from a Circular Loop Antenna 7-44
  The Electrically Small Receiving Loop Antenna 7-46
  Loaded Loop Antennas 7-47
6. Concluding Remarks 7-48
7. References 7-49

1. Introduction

Wire and loop antennas are widely used in communication systems from low to ultra-high frequencies, either in the form of individual elements or arranged with other similar elements to form phased arrays. They are also frequently used as probes to sense unknown environments or as bases for modeling more complex systems and structures. In this chapter we shall be concerned with the important properties of wire and loop antennas as isolated elements. Our emphasis will be to develop simple expressions with sufficient accuracy for some basic antenna forms so that readers can generalize them to more complicated, composite structures pertaining to their particular needs without excessive reformulation and computing. For this reason our approach will be substantially different from those computationally more demanding methods reported in the literature. It is important for readers to recognize the physical description of our solution process in the later sessions in order to fully benefit from this approach.

The Thin-Wire Antenna

A general thin-wire antenna structure having the radius a and of length $2h$ is shown in Fig. 1a. The wire is assumed to be perfectly conducting and satisfies the following conditions:

$$a \ll \lambda_0 \quad \text{and} \quad a \ll h \tag{1}$$

where λ_0 is the free-space wavelength of a plane wave at an angular operating frequency ω in radians per second. The electrical properties of this antenna can be described by the axial current on its surface. Choosing the thin-wire conductor axis to coincide with the curvilinear coordinate s and assuming the axially directed current $I(s)$ to be uniformly distributed about the conductor surface, the average induced tangential electric field $\langle E_s(s) \rangle$ at the conductor surface generated by this current can be written in the general manner [1]

$$\langle E_s(s) \rangle = \int_C I(s') K(s;s') \, ds' \tag{2}$$

where C is the line contour along the antenna from $s = -h$ to h. The kernel $K(s;s')$ represents the tangential electric field produced by an elemental dipole moment, $I(s') \, ds'$. The specific form of the kernel $K(s;s')$ depends on the particular antenna structure at hand. For example, in the case of a linear dipole antenna, i.e., a straight wire as shown in Fig. 1b where the natural coordinate system is a cylindrical coordinate system, it is given as [2], [3],

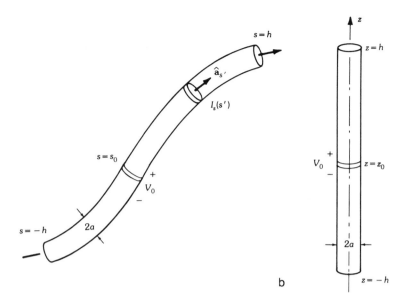

Fig. 1. The thin-wire antenna. (*a*) General form. (*b*) Straight wire.

$$K(z;z') = -j\frac{\xi_0}{4\pi}\left[\frac{d^2}{dz^2} + k_0^2\right] G(z;z') \tag{3}$$

where

$$G(z;z') = \frac{1}{2\pi}\int_{-\pi}^{\pi} \frac{e^{-jk_0 R}}{k_0 R} d\phi, \quad R = [(z-z')^2 + 4a^2 \sin^2(\phi/2)]^{1/2} \tag{4}$$

and $\xi_0 = 120\pi$ ohms is the characteristic impedance in free space, k_0 is the wave number, and λ_0 is the free-space wavelength.

For a transmitting antenna we usually assume the antenna is excited by a source potential V_0 maintained across an infinitesimally small gap (i.e., a so-called delta-function voltage generator) in the thin-wire conductor at $s = s_0$ so that

$$\langle E_s(s) \rangle = -V_0 \delta(s - s_0), \quad -h < s < h \tag{5}$$

Equating (5) with (2) yields an integral equation for finding the transmitting current.

In the receiving situation the induced electric field on the conductor must cancel the axially directed electric-field component associated with an incident plane wave, i.e.,

$$\langle E_s(s) \rangle = -\hat{\mathbf{a}}_s \cdot \mathbf{E}^i e^{-jk_0 \hat{\mathbf{a}}_i \cdot \mathbf{x}}, \quad \mathbf{x} \in C \tag{6}$$

where $\mathbf{E}^i$ in volts per meter is the amplitude of the plane-wave field and $\hat{\mathbf{a}}_i$ is the unit vector in the direction of propagation of the plane wave; $\mathbf{x}$ is the position vector at an observation point s along the antenna and $\hat{\mathbf{a}}_s$ is the axially directed unit vector at the same point. Equating (6) with (2) gives us the desired integral equation for finding the receiving current distribution.

Various approximate methods have been employed to solve the appropriate form of the integral equation in (2). Many of these methods [2–4] rely on the fact that the value of the kernel function $K(s;s')$ becomes extremely large at $s = s'$. Numerical techniques may also be employed to obtain numerically "exact" solutions for specific thin-wire structures. These methods will be consulted for the purposes of comparison in the sections to follow. The particular method we have adopted for our discussion, except in the case of loop antennas, is derived from the observation that the current distribution on a finite-length thin-wire antenna is essentially that of a standing wave resulting from multiply reflected currents which "bounce" back and forth between discontinuities (such as the ends, conductor radius changes, lumped impedance elements, etc.) in the thin-wire structures. Thus, by characterizing the reflected and transmitted current waves for each type of discontinuity, we can construct the standing-wave current distributions by superposing these effects.

Before we proceed with this method of constructing a solution for any particular thin-wire antenna, we need to first address some of the general issues regarding thin-wire antenna structures.

Input Admittance or Impedance

As mentioned earlier, the excitation of a thin-wire transmitting antenna is usually modeled mathematically by a finite voltage source V_0 maintained across an infinitesimal gap along the thin-wire structure. In practice, however, the antenna must be fed by some sort of transmission line, such as coaxial cable or open two-wire line. The mathematical model is therefore only a convenient way to separate the antenna problem from the circuit analysis of the transmission line. We now designate $J^T(s;s_0)$ as the current distribution of a transmitting antenna due to a *unit* voltage source at a feed point s_0, i.e., $I^T(s) = V_0 J^T(s;s_0)$. The superscript T refers to a transmitting antenna. The input admittance in siemens is then given by

$$Y_{in} = I^T(s_0)/V_0 \quad \text{or} \quad J^T(s_0;s_0) \tag{7}$$

and the input admittance in ohms is

$$Z_{in} = Y_{in}^{-1} \tag{8}$$

which acts as a load terminating the connecting transmission line. We should caution the readers, however, that because the electromagnetic coupling between the antenna and the transmission line usually cannot be completely avoided, an "end-correction" network usually has to be added if a high degree of accuracy is desired. The specific form of this network obviously depends on the particular way the antenna is fed. This topic is beyond the scope of this chapter and readers are referred to [2] for more detail.

An additional complication which often arises in thin-wire antenna analyses concerns the gap capacitance associated with the highly idealized delta-function voltage source. Because this capacitance is an integral part of the antenna admittance, we usually cannot separate one from the other. This means that the input susceptance calculation has an inherent error which can be avoided by detailed modeling of the actual source region, which is usually impractical.

Far-Field Radiation from a Thin-Wire Transmitting Antenna

The radiation field at an observation point (r, θ, ϕ) in the far zone or the Fresnel zone, i.e., $r \gg h$ and λ_0, can be expressed approximately in terms of a vector potential function $\mathbf{A}(r, \theta, \phi)$ as follows:

$$\mathbf{H} = -jk_0 \hat{\mathbf{a}}_r \times \mathbf{A} \quad \text{and} \quad \mathbf{E} = \xi_0 \mathbf{H} \times \hat{\mathbf{a}}_r \tag{9}$$

where $\hat{\mathbf{a}}_r$ is a radially directed unit vector at the observation point. Provided the normalized current distribution $J^T(s; s_0)$ is known, we can write the expression for $\mathbf{A}$ approximately as

$$\mathbf{A}(r, \theta, \phi) = V_0 \mathbf{F}(\theta, \phi; s_0) \frac{e^{-jk_0 r}}{4\pi r} \tag{10}$$

where

$$\mathbf{F}(\theta, \phi; s_0) = \int_C \hat{\mathbf{a}}_s J^T(s'; s_0) e^{jk_0 \hat{\mathbf{a}}_r \cdot \mathbf{x}'} \, ds' \tag{11}$$

in siemens-meters and $\mathbf{x}'$ again is the position vector at the source point s' and $\hat{\mathbf{a}}_s$ is the axial unit vector at that point. The function $\mathbf{F}$, which varies only angularly, is the so-called vector far-field pattern. In the case of a linear dipole antenna, $\mathbf{F}$ readily reduces to the more familiar form,

$$\mathbf{F}(\theta, \phi; z_0) = \hat{\mathbf{a}}_z F(\theta; z_0) \tag{12}$$

where

$$F(\theta; z_0) = \int_{-h}^{h} J^T(z'; z_0) e^{jk_0 z' \cos \theta} \, dz' \tag{13}$$

The Receiving Antenna

It is well known that the far-field pattern of a transmitting antenna has the same angular dependence as the current of an unloaded receiving antenna at the location corresponding to the feed point of the transmitting case. This can be readily observed when we compare (5) with (6), and conclude that the receiving case is actually equivalent to a transmitting antenna, excited by a distributed voltage source of amplitude $\hat{\mathbf{a}}_{s'} \cdot \mathbf{E}^i \, ds'$ at point s' on the antenna. The receiving current can thus be obtained by integrating over all the elementary sources on the antenna:

Wire and Loop Antennas

$$I^R(s) = \int_C (\mathbf{E}^i \cdot \hat{\mathbf{a}}_{s'}) J^T(s;s') e^{-jk_0 \hat{\mathbf{a}}_i \cdot \mathbf{x}'} \, ds' \tag{14}$$

where $\mathbf{E}^i = \hat{\mathbf{a}}_e E^i$ is the field intensity vector of an incident plane wave with an amplitude E^i volts per meter and polarized in the direction of the unit vector $\hat{\mathbf{a}}_e$; on the other hand, $\hat{\mathbf{a}}_i$ is the direction of propagation of the plane wave. To relate this current with the far-field pattern of a transmitting antenna fed at s, we first have to recognize that the reciprocity theorem requires that $J^T(s;s') = J^T(s';s)$. Now since the amplitude E^i of a plane-wave field is constant everywhere and the direction of the incident plane wave is opposite to the observation direction in the transmitting case, i.e., $\hat{\mathbf{a}}_r = -\hat{\mathbf{a}}_i$, we have from (11) and (14) the following result:

$$I^R(s) = \mathbf{E}^i \cdot \mathbf{F}(\pi - \theta, \phi; s) \tag{15}$$

The superscript R refers to the receiving current in this case. A normalized receiving current can be similarly defined as

$$J^R(\theta, \phi; s) = I^R(s)/E^i = \hat{\mathbf{a}}_e \cdot \mathbf{F}(\pi - \theta, \phi; s) \tag{16}$$

Loaded Thin-Wire Antennas

The electrical properties of a thin-wire antenna can be drastically altered by introducing lumped impedance elements along the antenna. The effects of such loading can be modeled mathematically by equivalent voltage sources corresponding to the actual voltage drops across individual loads. The overall current distribution on a thin-wire transmitting antenna having N voltage sources $V_1, V_2, \ldots, V_N$ in series with impedances $Z_1, Z_2, \ldots, Z_N$ located at $s = s_1, s_2, \ldots, s_N$ along the antenna can be written as

$$I^T(s) = \sum_{n=1}^{N} [V_n - Z_n I^T(s_n)] J^T(s;s_n), \quad -h < s < h \tag{17}$$

The values of $I^T(s_n)$ can be found from the N linear equations obtained by setting $s = s_1, s_2, \ldots, s_N$ on both sides of (17). If any of the Z_n's are simple passive impedance elements not associated with a voltage source, the corresponding V_n's in (17) are simply set to zero.

The first term in the square bracket of (17) represents the overall unloaded current distribution resulting from the N voltage sources. In the receiving situation this term would be replaced by one representing the unloaded receiving current distribution excited by any number (say M) of incident plane-wave fields. The loaded receiving current distribution is then given by

$$I^R(s) = \sum_{m=1}^{M} E_m^i J^R(\theta_m, \phi_m; s) - \sum_{n=1}^{N} Z_n I^R(s_n) J^T(s;s_n) \tag{18}$$

Again, we can determine $I^R(s_n)$ for $n = 1, 2, \ldots, N$ by setting $s = s_n$ on both sides of (18) and solving the set of N linear equations.

Transient Response

Study of the transient response on thin-wire structures has many contemporary applications, among them the use of these structures as electromagnetic pulse (EMP) simulators. Basically, the homogeneous integral equation [5, 6] associated with the expression in (2) possesses nontrivial solutions, $I(s; \omega_\alpha)$ for some complex frequencies $\omega = \omega_\alpha$, where $\alpha = 1, 2, \ldots$, so that

$$\int_C I(s'; \omega_\alpha) K(s, s'; \omega_\alpha) \, ds' = 0 \tag{19}$$

Each of these solutions can be identified as a resonance of the structure, very much in the same way as a waveguide cavity, where the electromagnetic waves excited by the current distribution on the cavity wall constructively interact with each other in phase. A thin-wire antenna, in fact, can be considered as an open resonator with complex natural frequency ω_α and associated natural mode currents $G_\alpha(s; \omega_\alpha)$, with $\alpha = 1, 2, \ldots$. Transient response of a thin-wire structure is then given by the so-called SEM method [5, 6]:

$$I(t, t_0; s, s_0) = \text{Re}\left\{ \sum_{\alpha=1}^{\infty} A_\alpha G_\alpha(s; \omega_\alpha) G_\alpha(s_0; \omega_\alpha) e^{j\omega_\alpha(t-t_0)} \right\} \tag{20}$$

where s_0 and t_0 are, respectively, the source location and turn-on time, and A_α is the excitation factor, which, of course, depends on the particular pulse shape of the voltage source. According to the time causality principle, such an expression can be used only after the arrival time, which is determined by the observation distance divided by the speed of light.

Equivalent Radius for Noncircular Cylindrical Thin-Wire Conductors

Should a thin-wire antenna be constructed from a noncircular cylindrical conductor, an equivalent radius [7] can be assigned to the antenna, provided that the current can be assumed to be fairly uniform around the noncircular conductor. For the cross section shown in Fig. 2, an approximate equivalent radius can be written as [7]

$$a_e = \exp\left\{ \frac{1}{S^2} \oint_\ell \oint_\ell \ln|\mathbf{w} - \mathbf{w}'| \, dw \, dw' \right\} \tag{21}$$

where S is the peripheral length around the cross section and $|\mathbf{w} - \mathbf{w}'|$ is the distance between two points on the contour ℓ bounding the peripheral surface that are located by the position vectors $\mathbf{w}$ and $\mathbf{w}'$. The equivalent radii for a few common noncircular thin-wire antenna-conductor cross sections are given in [7].

Solid Thin-Wire Antenna Conductors

For the most part the solutions given in this chapter are based on analyses of tubular thin-wire antennas. In practice, however, thin-wire antennas are usually

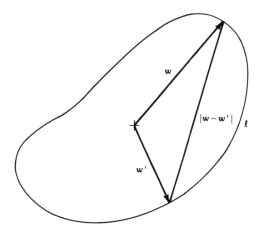

Fig. 2. General noncircular cylindrical conductor.

constructed from solid conductors. The essential effect is that an antenna made from a solid thin-wire conductor appears to be electrically slightly longer. The apparent increase in overall length has been shown to be approximately $0.26a$, which is indeed negligible for practical thin-wire structures [8].

Antenna Parameters

Before we conclude this section we should consider two important parameters frequently used to characterize a thin-wire antenna:

$$\Omega = 2\ln(2h/a), \qquad C_a = -\ln(k_0 a) - \gamma \tag{22}$$

where $\gamma = 0.577$ is the Euler constant. For thin-wire antennas that satisfy the conditions in (1), either Ω or C_a can be used as the large expansion parameter for developing asymptotic solutions in antenna problems.

2. The Linear Dipole Antenna

The linear dipole antenna depicted in Fig. 3 is one of the most basic antenna forms. By assuming the current is concentrated along the axis of the antenna, the integral equation given in (2) becomes more readily tractable. In fact, based on the observation that the kernel function $K(z;z')$ is at its peak when $z = z'$, iterative solutions and a few trigonometric-term solutions for the current distribution have been successfully developed and a large amount of data compiled [2, 3, 9, 10, 11]. Accurate numerical computation of this current can also be achieved using moment methods [4, 12, 13], sometimes without the constraint of the thin-wire approximation [14–15]. The approach we will adopt in this chapter, however, involves the construction of a current standing wave on the linear thin-wire antenna based upon a hollow cylindrical model [16–22]. Although this approach was generally regarded as a long-antenna theory in the past, recent improvements in the analysis [23–24] have removed many of the restrictions. The formulas to be presented in the

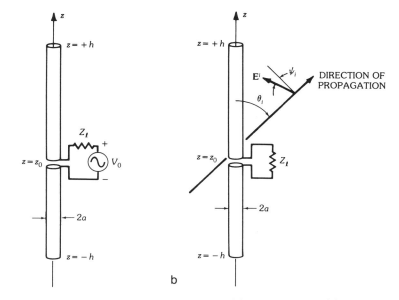

Fig. 3. Linear thin-wire dipole antenna. (*a*) Transmitting. (*b*) Receiving.

later sections of this chapter are well within the acceptable accuracy in engineering practice provided that

$$k_0 a \leqq 0.1 \quad \text{and} \quad \Omega \geqq 10 \tag{23}$$

which, of course, encompasses almost all the thin-wire antenna structures commonly used. As we mentioned earlier, our emphasis is to develop a simple procedure which will allow readers to extend the results in this chapter to more complicated, composite structures.

Unloaded Transmitting Antennas

Consider the linear thin-wire transmitting antenna illustrated in Fig. 3a. The voltage source V_0 applied across the electrically small gap at z_0 excites currents which travel (i.e., progress in phase) outward away from the source along the thin wire in a manner analogous to the propagation of current along a transmission line. Unlike the transmission-line current, however, the antenna current is subject to attenuation due to radiation. This initial, or "primary," current can be obtained from the current distribution existing between $-h < z < h$ on a similarly excited, infinitely long, linear, thin-wire antenna. An approximate expression for this current, here normalized to V_0, is given by [20]

$$U_a^s(|z-z_0|) = \frac{-j}{\xi_0} e^{-jk_0|z-z_0|}$$

$$\times \ln\left\{1 + \frac{j2\pi}{2C_a + \gamma - j3\pi/2 + \ln[k_0|z-z_0| + (k_0^2|z-z_0|^2 + e^{-2\gamma})^{1/2}]}\right\} \tag{24}$$

where $\xi_0 = 120\pi$ ohms and $\gamma = 0.577$. Originally derived for $k_0|z - z_0| > 1$, the above expression gives an accurate value for the real part of the input current even when it is evaluated at the feed point [11].

As depicted in Fig. 4a, the primary current emanating from the source at z_0 impinges on the ends of the antenna and is reflected back in the opposite direction. The form of these initial "secondary" currents can be obtained from a semi-infinitely long antenna of the same radius, excited by waves incident at the angle π with respect to a particular end. Determined from a Wiener-Hopf analysis, the initial secondary currents shown in Fig. 4b from both ends are denoted by the terms $-V_0 U_a^s(h - z_0) R_a U_a(h - z)$ and $-V_0 U_a^s(h + z_0) R_a U_a(h + z)$, respectively, where $V_0 U_a^s(h \mp z_0)$ is the value of the primary current at $z = \pm h$. The *reflection coefficient* R_a is given approximately by

$$R_a = \frac{\xi_0}{2\pi}(2C_a - j\pi)/(1 + \delta_a)^2 \tag{25}$$

where

$$\delta_a = \operatorname{Re}\left\{\frac{-j}{2\pi}\ln\left(1 + \frac{j2\pi}{2C_a - j3\pi/2}\right) - \frac{1}{2C_a - j\pi + \ln 2}\right\} \tag{26}$$

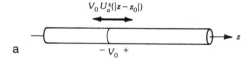

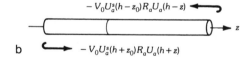

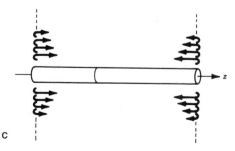

Fig. 4. The multiple reflection concept as applied to a finite-length thin-wire transmitting antenna. (*a*) Primary current emanating from the source. (*b*) Initial reflected currents from both antenna ends. (*c*) Subsequent multiple reflected currents from antenna ends.

The expression for the current distribution of the reflected wave is given by

$$U_a(z) = \frac{2\pi}{\xi_0} \frac{e^{-jk_0 z}}{2C_a + \gamma - j\pi/2 + \ln(2k_0 z) + e^{j2k_0 z} E_1(j2k_0 z)} \quad (27)$$

where E_1 is the exponential integral of the first kind ([25], Chapter 5). We note that the primary current U_a^s and the reflected distribution U_a are essentially the same for $k_0 z \gtrsim 1$. As mentioned earlier, the small kz behavior of the former expression was manipulated in order to achieve an accurate real component of input current.

Continuing, the two secondary currents travel (progress in phase and attenuate) on the antenna in opposite directions until they impinge on the other ends, whereupon another secondary current is generated. This process continues on and on as depicted in Fig. 4c, leading to two sets of infinite, though summable, series of multiply reflected waves. Assembling the constituent components gives the total current distribution per unit volt on an unloaded thin-wire transmitting antenna as

$$J_a^T(z; z_0) = U_a^s(|z - z_0|) - Q_a^T(h) R_a U_a(h - z) - Q_a^T(-h) R_a U_a(h + z),$$
$$\text{for } -h < z < h \quad (28)$$

where

$$Q_a^T(\pm h) = [U_a^s(h \mp z_0) - U_a^s(h \pm z_0) R_a U_a(2h)]/\Delta(h) \quad (29)$$

$$\Delta(h) = 1 - R_a^2 U_a^2(2h) \quad (30)$$

The tacit equivalence between the primary transmitting current $U_a^s(z)$ and the reflected current $U_a(z)$ means that the total current distribution may be thought of as the superimposed distributions caused by an independent unit voltage source and two dependent voltage sources one at each end of the antenna. To obtain the actual current due to a voltage source V_0, one obviously needs only to multiply by V_0 to obtain $I_a^T(z) = V_0 J_a^T(z; z_0)$ in amperes.

The current distribution on a center-driven, half-wave ($k_0 h = \pi/2$) linear antenna where $\Omega = 2 \ln(2h/a) = 10$ is shown in Fig. 5. Corresponding data from the three-term theory of King and Wu [3] and from the approximate second-order iteration procedure of King and Middleton ([2], Chapter 1, Section 22) are also shown in the figure.

Input Admittance or Impedance

The input admittance of a linear thin-wire antenna at the position z_0 is given in siemens by

$$Y_{\text{in}} = U_a^s(0) - Q_a^T(h) R_a U_a(h - z_0) - Q_a^T(-h) R_a U_a(h + z_0) \quad (31)$$

Figs. 6a and 6b show the input conductance and susceptance, respectively, of a center-fed antenna as a function of half-length $k_0 h$. Agreement with the results in

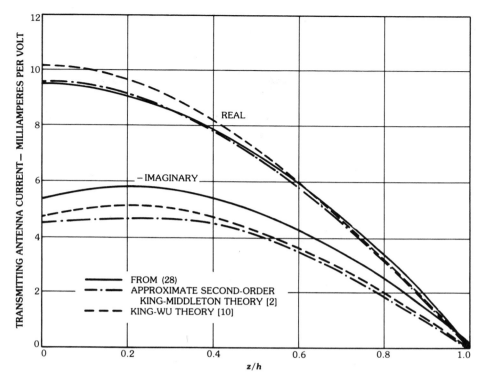

Fig. 5. Current distribution on a linear thin-wire transmitting antenna with $kh = \pi/2$, $\Omega = 10$.

[2] and [3] is indeed very good in the case of input conductance for all lengths. The shift in the input susceptance is closely related to the issue of how to determine the gap capacitance inherent in the idealized voltage source as we addressed earlier [26].

A close look at the input conductance near the first resonance $k_0 h = \pi/2$ for several values of the antenna parameter Ω is given in Fig. 7. Note that as the antenna becomes thinner (Ω increases), the bandwidth as well as the maximum value of G decreases and the resonance peak approaches $k_0 h = \pi/2$.

Far-Field Radiation from a Linear Thin-Wire Transmitting Antenna

Using the current distribution in (28), the far-field pattern function of a thin-wire antenna can be obtained approximately from (11) as

$$F(\theta; z_0) = \left(\frac{j2\pi}{k_0 \xi_0}\right) \Big\{ e^{jk_0 z_0 \cos\theta} [W_a(h + z_0; \theta) + W_a(h - z_0; \pi - \theta)] \\ - R_a [e^{jk_0 h \cos\theta} Q_a^T(h) W_a(2h; \theta) \\ + e^{-jk_0 h \cos\theta} Q_a^T(-h) W_a(2h; \pi - \theta)] \Big\} \quad (32)$$

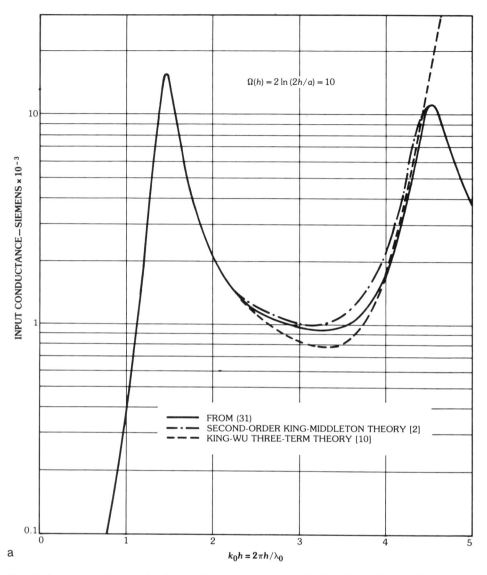

Fig. 6. Input impedance of a linear thin-wire antenna with $0 < kh < 5$, $\Omega = 10$. (*a*) Input conductance. (*b*) Input susceptance.

where, based on a similar evaluation in [27],

$$W_a(z;\theta) = \frac{1}{1+\cos\theta}\left[\frac{e^{-jk_0 z(1+\cos\theta)}}{2C_a - j\pi/2 + \gamma + P(\theta;z)} - \frac{1}{2C_a - j\pi - 2\csc^2(\theta/2)\ln\cos(\theta/2)}\right] \quad (33a)$$

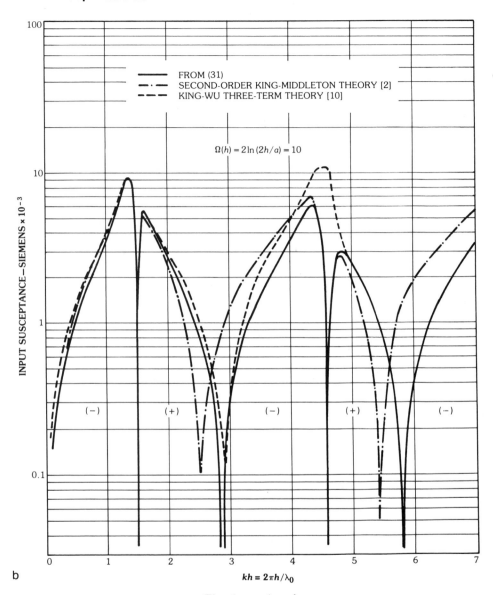

Fig. 6, *continued.*

$$P(\theta; z) = \ln(2k_0 z) + \csc^2(\theta/2) e^{jk_0 z(1+\cos\theta)} E_1(jk_0 z[1 + \cos\theta])$$
$$- \cot^2(\theta/2) e^{j2k_0 z} E_1(j2k_0 z) \qquad (33b)$$

Here $\gamma = 0.577$ is the Euler constant and E_1 again is the exponential integral of the first kind. We note that for an infinitely thin antenna, i.e., $C_a \to \infty$, the pattern function readily reduces to the well-known expression for a sinusoidal current given earlier in Chapter 1.

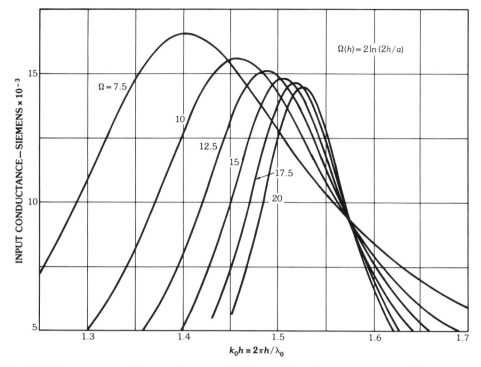

Fig. 7. Close-up view of the conductance of a linear thin-wire antenna near $\lambda_0/2$ resonance.

Unloaded Receiving Antennas

A pictorial description of a linear thin-wire receiving antenna is given by Fig. 3b. The uniform plane-wave field illuminating the antenna is polarized at the angle ψ_i and incident at the angle θ_i with respect to the antenna axis. Based on (14) and (16), we can write the expression for the receiving current as

$$J_a^R(\theta_i; z) = \cos\psi_i \sin\theta_i F(\pi - \theta_i; z), \quad 0 < \theta_i < \pi \tag{34}$$

with $F(\theta; z)$ given in (32). Also, we note that $J_a^R(\theta_i; z) = J_a^R(\pi - \theta_i; -z)$. The actual receiving current is $I^R(z) = E^i J_a^R(\theta_i; z)$, where E^i is the complex amplitude of the incident plane wave in volts per meter.

The current distributions on a half-wave receiving antenna for several incident angles are shown in Fig. 8. Corresponding data for the normal incidence ($\theta_i = \pi/2$) case as determined from the first-order King-Middleton theory ([2], Section IV.7) and the three-term theory of King [9] are also shown.

The receiving current antenna pattern at three different positions is shown in Fig. 9. Note the almost sinusoidal (as it is often assumed) behavior with respect to θ_i. This figure also corresponds to the far-field pattern (when θ_i is replaced by $\pi - \theta_i$) of a linear thin-wire transmitting antenna fed at the same points indicated.

Wire and Loop Antennas

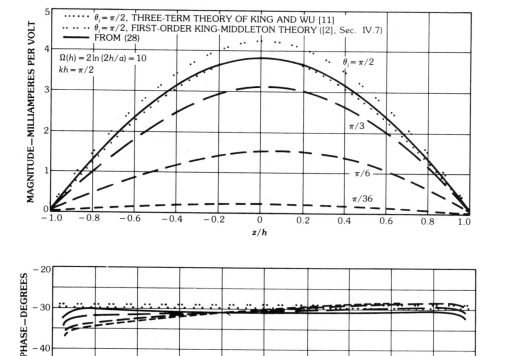

Fig. 8. Receiving current distribution (normalized to the incident electric field and the wavelength, i.e., $I_a^R(\theta_i;z)/\lambda_0 E^i$) with $kh = \pi/2$, $\Omega = 10$.

Impedance-Loaded Antennas

Expressions for the total current on a dipole antenna with an impedance loading of Z_ℓ at the location z_0 is immediately available from (17) and (18):

$$I_a^T(z) = V_0\left(\frac{Y_\ell}{Y_{in} + Y_\ell}\right) J_a^T(z;z_0) \qquad (35)$$

for a transmitting antenna, and

$$I_a^R(\theta_i;z) = E^i\left\{J_a^R(\theta_i;z) - \frac{J_a^R(\theta_i;z_0)}{Y_{in} + Y_\ell} J_a^T(z;z_0)\right\} \qquad (36)$$

for a receiving antenna, where $Y_\ell = Z_\ell^{-1}$. A multiple impedance loading can be handled by following a similar procedure as in (17) and (18).

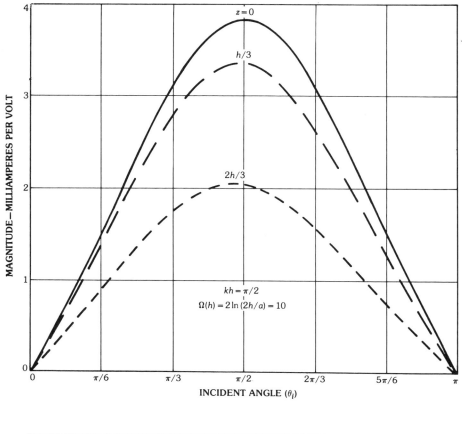
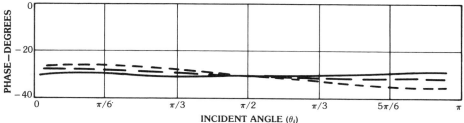

Fig. 9. Receiving current (normalized to the incident electric field and the wavelength, i.e., $I_a^R(\theta_i;z)/\lambda_0 E^i$) with $0 < \theta_i < \pi$, $kh = \pi/2$, $\Omega = 10$.

An Impedance-Loaded Monopole Antenna

A monopole antenna is a thin-wire antenna mounted vertically above a large conducting ground plane and driven at or near its base. Use of the image theorem allows us to replace it by a dipole antenna with two voltage sources at $\pm z_0$ and the associated load impedances. We can therefore write, with the help of (17) and (18),

Wire and Loop Antennas

$$I^T(z) = \frac{Y_\ell V_0}{Y_\ell + Y_{\text{in}} + J_a^T(z_0; -z_0)} [J_a^T(z; z_0) + J_a^T(z; -z_0)] \quad (37)$$

for the transmitting antenna and

$$I^R(z) = E^i \left\{ J_a^R(\theta_i; z) + J_a^R(\pi - \theta_i; z) \right. \\ \left. - \frac{[J_a^R(\theta_i; z_0) + J_a^R(\pi - \theta_i; z_0)]}{Y_\ell + Y_{\text{in}} + J_a^T(z_0; -z_0)} [J_a^T(z; z_0) + J_a^T(z; -z_0)] \right\} \quad (38)$$

for a receiving antenna. Here the input admittance $Y_{\text{in}} = J_a^T(z_0; z_0)$, and the currents J_a^R and J_a^T refer to those on a linear antenna having an overall length of $2h$. Allowing the feed point to approach the conducting plane, i.e., $z_0 \to 0$, yields the well-known relationships between monopole and dipole antennas.

Transient Behavior of a Dipole Antenna

As we mentioned in the general discussion, transient behavior of a thin-wire dipole antenna acting as an open resonator can be described by a set of natural frequencies and natural modes. According to (19), these modes can be determined by seeking the nontrivial solution of the current distribution in the absence of a voltage source. From the expression that $I^T(z) = V_0 J_a^T(z; z_0)$ with the normalized current J_a^T given by (28) through (30), we can readily establish the resonant condition is

$$\Delta(h; \omega_\alpha) \equiv 1 - R_\alpha^2(\omega_\alpha) U_\alpha^2(2h; \omega_\alpha) = 0, \quad \alpha = 1, 2, \ldots \quad (39)$$

where $R_\alpha(\omega_\alpha)$, $U_\alpha(2h; \omega_\alpha)$ and $\Delta(h; \omega_\alpha)$ are given by (25), (27), and (30), with k_0 replaced everywhere by ω_α/c.

Now since R_a and U_a in (25) through (27) are given by simple functions, searching for the roots of (39) in the complex plane is a relatively simple task. The transient current response of a transmitting antenna with a step-function voltage source, $V(t) = V_0$ for $t > t_0$ and 0 for $t < t_0$, is derived in [27] as

$$I^T(t, t_0; z, z_0) = 2 V_0 \operatorname{Re} \left\{ \sum_{\alpha=1}^{\infty} \frac{R_\alpha(\omega_\alpha)}{\omega_\alpha \dfrac{\partial \Delta(h; \omega_\alpha)}{\partial \omega_\alpha}} G_\alpha(z; \omega_\alpha) G_\alpha(z_0; \omega_\alpha) \right. \\ \left. \times \exp[j\omega_\alpha(t - t_0 - |z - z_0|/c)] \right\} \quad (40a)$$

after the arrival time $t > t_0 + |z - z_0|/c$, where c is the speed of light, and z_0 and t_0 are respectively the location and the turn-on time of the source. In a similar manner the transient current response of a receiving antenna due to a step-function uniform plane wave $\mathbf{E}^i(t:\mathbf{r}) = \mathbf{E}^i \exp[j\omega(t - z\cos\theta_i/c)]$ for $t > [(h+z)/c]\cos\theta_i$, and 0 for $t < [(h+z)/c]\cos\theta_i$ can be written as [27]

$$-I^R(t:\theta_i;z) = 2E^i \cos\psi_i \sin\theta_i \operatorname{Re}\left\{\sum_{a=1}^{\infty} \frac{R_a(\omega_a)}{\omega_a \dfrac{\partial\Delta(h;\omega_a)}{\partial\omega_a}} G_a(z;\omega_a)\right.$$
$$\times \left(\frac{j2\pi}{\xi_0 \omega_a/c}\right)\{\exp(j\omega_a h \cos\theta_i/c)[W_a(z_2;\theta_i;\omega_a)$$
$$- W_a(z_1;\theta_i;\omega_a)] - (-1)^a \exp(-j\omega_a h \cos\theta_i/c)$$
$$\times [W_a(2h - z_2;\pi - \theta_i;\omega_a)$$
$$\left. - W_a(2h - z_1;\pi - \theta_i;\omega_a)]\}\exp(j\omega_a t)\right\} \quad (40b)$$

after the arrival time $t > (h+z)\cos\theta_i/c$ where $W_a(z;\theta;\omega_a)$ is based on (33) with k_0 replaced everywhere by ω_a/c and

$$z_1 = \begin{cases} \left(\dfrac{h + z - ct}{1 - \cos\theta_i}\right) & \text{for } (h+z)\cos\theta_i/c < t < (h+z)/c \\ 0, & \text{otherwise} \end{cases}$$

$$z_2 = \begin{cases} 0 & \text{for } t < (h+z)\cos\theta_i/c \\ \left(\dfrac{h + z + ct}{1 + \cos\theta_i}\right) & \text{for } (h+z)\cos\theta_i/c < t < (2h\cos\theta_i + h - z)/c \\ 2h & \text{for } t > (2h\cos\theta_i + h - z)/c \end{cases}$$

In both transmitting and receiving cases the natural-mode current $G_a(z;\omega_a)$ is simply

$$G_a(z;\omega_a) = U_a(h + z;\omega_a) - (-1)^a U_a(h - z;\omega_a) \quad (41)$$

and has either a basically sinusoidal (a even) or a cosinusoidal (a odd) distribution. Here, $U_a(z;\omega_a)$ is obtained from (27) for $U_a(z)$ with k_0 everywhere replaced by ω_a/c.

Transient response of the feed-point current of a center-fed dipole antenna with $\Omega = 2\ln(2h/a) = 10$ due to a step-function voltage source is shown in Fig. 10 and compared to the numerical results in [29] and as previously derived as a closed-form result in [28]. The transient response for the midpoint current of a receiving antenna ($\Omega = 10$) illuminated by a plane wave at an angle of 90° with respect to the antenna, and with a step-function electric field in the plane of incidence, is shown in Fig. 11 and again compared with other numerical [29] and analytical [28] results.

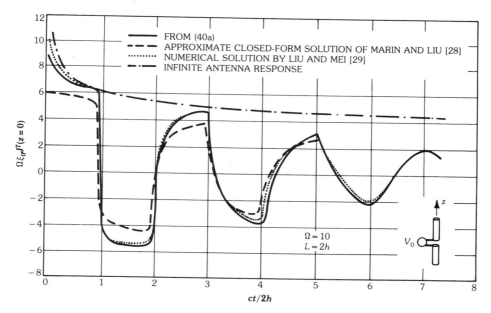

Fig. 10. Time-domain response of the driving-point current on a linear antenna excited by a step-function voltage applied at $z_0 = 0$ at $t_0 = 0$. (*After Hoorfar and Chang [27]*, © 1982 IEEE)

3. The Sleeve Antenna

A thin-wire sleeve, or, as it is sometimes called, a "coaxial" antenna, basically consists of two coaxial tubular cylinders, the thinner, longer one fitting partially within the larger (in radius) shorter one, such as depicted in Fig. 12. In normal applications this type of antenna is vertically oriented and is quite similar in appearance as well as electrical characteristics to the linear thin-wire antenna described in the previous section. An important practical advantage of the sleeve antenna is the coaxial line contained within the structure, which can be used as a means of feeding the antenna. Utilization of this type of feed, however, requires that the feed line exiting the antenna be decoupled from the antenna itself.

Approximate simulations [2, 30] of the sleeve antenna have been formulated using the linear thin-wire antenna as a basis. A direct approach [31] based on the same concepts of constructing a standing-wave solution as in the last section will be discussed here. Our method is similar to the one pursued by Hurd [32].

The Junction Effect

We need to first examine the effect of the junction created by the truncation of the outer conductor of the thin-wire coaxial line shown in Figs. 13a through 13c. There are three possible current waves that can impinge on this junction: an antenna current $U_a(z_a - z)$ emanating from a source at z_a on the extended portion of the inner conductor of radius a which acts as an antenna, an antenna current $U_b(z - z_b)$ emanating from a source at z_b on the outer surface $\varrho = b^+$ of the outer

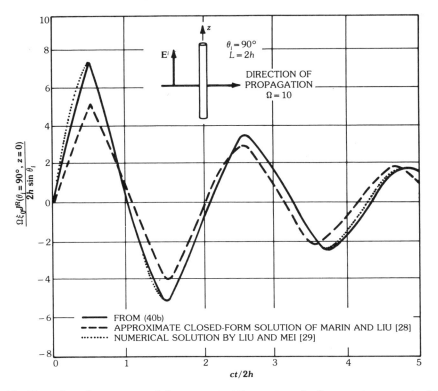

Fig. 11. Time-domain response of the current at the center of a linear antenna excited by a normally incident uniform plane wave arriving at $t_0 = 0$. (*After Hoorfar and Chang [27]*, © *1982 IEEE*)

conductor of radius b which also acts as an antenna, and finally a transmission-line current $Z_0^{-1} \exp[-jk_0(z - z_c)]$ amperes from a source at z_c inside the coaxial region with a characteristic impedance $Z_0 = (\xi_0/2\pi) \ln(b/a)$ ohms where $\xi_0 = 120\pi$ ohms. For convenience we should only consider a "one-sided" voltage source [33] for the excitation of the antenna currents U_b so that we can ignore any other current excitation on the inner surface of the same conductor. Reflections and transmission of these current waves have been previously determined by [31] using a Wiener-Hopf method. The results given in [31] are summarized as follows:

1. The junction has essentially no effect on an antenna current incident from the left. The current wave $U_a(z_a - z)$ in this case continues unperturbed from conductor a to the outer surface of conductor b. In doing so, transmission-line currents $U_a(z_a) \exp(jk_0 z)$ on conductor a and $-U_a(z_a) \exp(jk_0 z)$ on the inner surface of conductor b are transmitted into the coaxial region, thereby providing a smooth transition between the antenna current $U_a(z_a - z)$ and the transmission-line current $U_a(z_a) \exp(jk_0 z)$ at the junction.

2. When an antenna current $U_b(z - z_b)$ is incident from the right, it also continues to propagate across the junction from conductor b to conductor a. How-

Wire and Loop Antennas

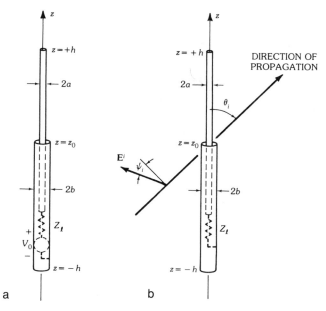

Fig. 12. Sleeve antenna illustration. (*a*) Transmitting. (*b*) Receiving.

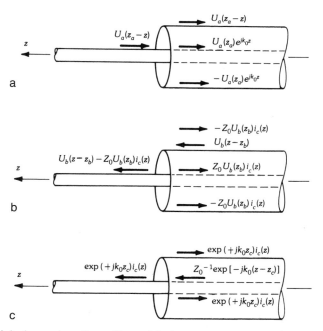

Fig. 13. Coaxial sleeve junction effect. (*a*) Antenna current $U_a(z_a - z)$ impinging on junction. (*b*) Antenna current $U_b(z - z_b)$ impinging on junction. (*c*) Transmission-line current $Z_0^{-1} \exp[-jK_0(z - z_c)]$ impinging on junction. Here, junction location is taken to be $z = 0$.

ever, the junction discontinuity now acts like a secondary voltage source of amplitude $Z_0 U_b(z_b)$, exciting currents in all three regions which have the form

$$i_c(z) = \begin{cases} 2U_a^s(z) & \text{for } \varrho = a, z \geqq 0 \\ 2\nu U_b^s(-z) & \text{for } \varrho = b, z \leqq 0 \\ -\nu Z_0^{-1} \exp(jk_0 z) & \text{for } \varrho = a, z \leqq 0 \end{cases} \quad (42)$$

where $\nu = U_a^s(0)/U_b^s(0)$. The antenna current is then given by $U_b(z + z_b) - [U_b(z_b) Z_0] i_c(z)$ in both antenna regions and simply $-[U_b(z_b) Z_0)] i_c(z)$ in the coaxial region ($z < 0$).

3. The same junction effect exists when the TEM current $Z_0^{-1} \exp[jk_0(z + z_c)]$ due to unit voltage excitation on conductor a in the coaxial region is incident from the right, except for a reversal in sign so that the total current in the coaxial region is $Z_0^{-1} \exp[-jk_0(z - z_c)] + \exp(jk_0 z_c) i_c(z)$ and $\exp(jk_0 z_c) i_c(z)$ in the antenna regions.

It should be noted that the definition of $i_c(z)$ in (42) in terms of the functions $U_a^s(z)$ and $U_b^s(z)$ from (24) is somewhat an approximation whose intention is to yield (in a manner analogous to the case of a linear thin-wire antenna) a more accurate value for the real part of the input current at the junction. As a result of this approximation one can determine that the sum of the currents entering the junction at $z = 0$ is only approximately equal to the sum of the currents leaving the junction. If the expressions for the $U_a(z)$ and $U_b(z)$ from (27) are used in the definition of $i_c(z)$ in (42), the sum of the currents entering and leaving the junction is zero. But this is at the expense of the accuracy of the real part of the input current and the accuracy of any subsequent input impedance calculations.

The Sleeve Dipole Antenna

A transmitting sleeve dipole antenna is depicted in Fig. 12a with built-in voltage source V_0. Such an antenna can be used for remote sensing and telemetry purposes. To determine the input impedance (or admittance) defined at the coaxial aperture we need first to write all the possible current components on the two halves of the antenna due to an incident current wave of $Z_0^{-1} \exp[-jk_0(z - z_0)]$ as

$$J^T(z) = i_c(z - z_0) - R_a Q_a^T(h) U_a(h - z) - R_b Q_b^T(-h) \\ \times [U_b(h + z) - Z_0 U_b(h + z_0) i_c(z - z_0)] \quad (43)$$

for $z > z_0$, $\varrho = a$, and $z < z_0$, $\varrho = b$, where Q_a^T and Q_b^T are yet undetermined constants and $i_c(z - z_0)$ is a junction-induced current given in (42) for each region of interest. Since the junction effect at the coaxial opening has now been fully incorporated, we only need to use the end conditions $J^T(h) = J^T(-h) = 0$ to determine the two unknown constants. At $z = +h$, the reflected current is $U_a(h_1 - z)$ and the incident currents are $i_c(z)$ and $U_b(h + z)$. Hence

$$Q_a^T(h) = i_c(h - z_0) - R_b Q_b^T(-h)[U_b(2h) - Z_0 U_b(h + z_0) i_c(h - z_0)] \quad (44a)$$

Wire and Loop Antennas

Likewise, at the other end, $z = -h$, the reflected current is $U_b(h + z)$ and the incident current is $i_c(z - z_0)$, and $U_a(h - z)$ so that

$$Q_b^T(-h) = i_c(-h - z_0) - R_a Q_a^T(h) U_a(2h) \\ + R_b Q_b^T(-h) Z_0 U_b(h + z_0) i_c(-h - z_0) \quad (44b)$$

The two equations now allow us to obtain explicit expressions for Q_a^T and Q_b^T as follows:

$$Q_a^T(h) = [i_c(h - z_0) - R_b U_b(2h) i_c(-h - z_0)]/\Delta \quad (45a)$$

$$Q_b^T(-h) = [i_c(-h - z_0) - R_a U_a(2h) i_c(h - z_0)]/\Delta \quad (45b)$$

where

$$\Delta = 1 - R_a R_b U_a(2h) U_b(2h) - Z_0 R_b U_b(h + z_0) \\ \times [i_c(-h - z_0) - R_a U_a(2h) i_c(h - z_0)] \quad (46)$$

and R_a and R_b are given in (25) for radii a and b. We note that, interestingly enough, the formulas for a sleeve dipole antenna as given in (43) through (46) reduce immediately to those of a simple dipole antenna in (28) through (30) when we set $a = b$ and replace the incident current $i_c(z)$ with the transmitting antenna current $U_a^s(|z|)$.

To find the input impedance we only need to know that each component current in (43) has to continue into the coaxial region so that the total coaxial current for an incident voltage amplitude V_0 is given by

$$I_c = \left(\frac{V_0}{Z_0}\right)[e^{-jk_0(z-z_0)} - \Gamma_c e^{jk_0(z-z_0)}] \quad (47)$$

where

$$\Gamma_c = v + Z_0 R_a Q_a^T(h) U_a(h - z_0) + v Z_0 R_b Q_b^T(-h_2) U_b(h + z_0) \quad (48)$$

is actually the (voltage) reflection coefficient for the reflected TEM wave. The input impedance is then defined as

$$Z_{in} = Z_0(1 + \Gamma_c)/(1 - \Gamma_c) \quad (49)$$

The far-field pattern of such an antenna can be obtained in exactly the same manner as in the simple dipole case. Using the current expression in (43) and the far-field formula in (11) we obtain

$$\left(\frac{j2\pi}{k_0\xi_0}\right)^{-1} F(\theta; z_0) = 2[1 + Z_0 R_b U_b(h + z_0) Q_b^T(-h)] e^{jk_0 z_0 \cos\theta}$$
$$\times [W_a(h - z_0; \pi - \theta) + \nu W_b(h + z_0; \theta)]$$
$$- R_a Q_a^T(h) e^{jk_0 h \cos\theta} W_a(2h; \theta) - R_b Q_b^T(-h)$$
$$\times e^{-jk_0 h \cos\theta} W_b(2h; \pi - \theta) \tag{50}$$

and W_a (W_b) is given in (33) for radius a (b).

Fig. 14 shows the input impedance as determined by using (48) and (49) for an

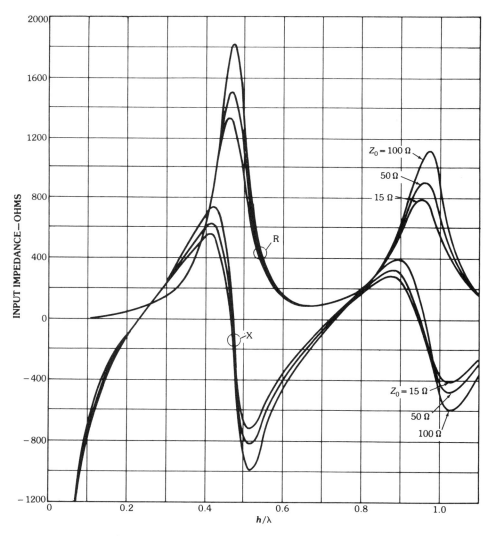

Fig. 14. Input impedance of isolated center-fed sleeve dipoles, where $\Omega = 2\ln(2h/b) = 10$ and $Z_0 = (\xi_0/2\pi)\ln(b/a)$.

isolated thin-wire $\Omega = 2\ln(2h/b) = 10$ sleeve dipole antenna as a function of its half-length kh for several ratios of b/a. These data show the input impedance to be fairly insensitive to the characteristic impedance of the internal coaxial line. In fact, the nominal input impedance in all the cases is quite similar to that of a linear thin-wire antenna having a constant radius (where $\Omega = 2\ln(2h/\sqrt{ab})$) such as discussed in Section 2. Furthermore, the current distributions on both transmitting and receiving sleeve dipoles would be essentially quite similar to those on equivalently sized linear thin-wire dipoles.

The Sleeve Monopole Antenna

Fig. 15 depicts a transmitting sleeve antenna mounted on a ground plane, which is often used to experimentally study the effect of the location of feed point on the excitation of a monopole antenna [2, 30]. To construct a solution for such an antenna we first have to use the image theorem to determine the kind of current components that can exist on the structure, as shown in Fig. 15. We then incorporate the required conditions at the junctions into our solution to obtain

$$J^T(z) = i_c(z - z_0) - R_a Q_a^T U_a(h - z) + A i_c(-z - z_0) + B U_a(h + z)$$
$$- Z_0[A i_c(-2z_0) + B U_a(h + z_0)] i_c(z - z_0), \quad \text{for } 0 < z < h \quad (51)$$

Here $i_c(z - z_0)$ as given in (42) represents the primary current, due to an incident current wave of $Z_0^{-1} \exp[-jk_0(z - z_0)]$ inside the coaxial region; $U_a(h - z)$ is the reflected current from the end at $z = h$, and, as we know, the junction has no effect on this current; $i_c(-z - z_0)$ and $U_a(h + z)$ are the corresponding image currents originating from the "junction" at $z = -z_0$ and the other "end" at $-h$,

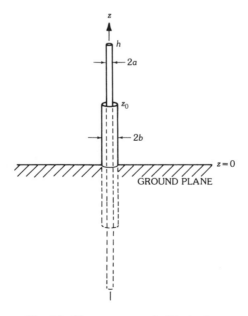

Fig. 15. Sleeve monopole illustration.

respectively. Together these two image currents impinge onto the junction at $z = z_0$ and produce yet another scattered current represented by the last term in the square bracket. In order to determine the unknown constants Q_a^T, A, and B, we invoke the symmetry requirement with respect to the ground plane $J^T(z) = J^T(-z)$ to obtain

$$B = -R_a Q_a^T$$
$$A = 1 - Z_0[Ai_c(-2z_0) + BU_a(h + z_0)]$$

and because of the end condition at $J^T(h) = 0$, we know the reflected current amplitude is related to the (total) incident current amplitude by $-R_a$ so that

$$Q_a^T = i_c(h - z_0) + Ai_c(-h - z_0) + BU_a(2h)$$
$$- Z_0[Ai_c(-2z_0) + BU_a(h + z_0)i_c(h - z_0)]$$

These equations can then be used to obtain an explicit expression for the current distribution:

$$J^T(z) = A[i_c(z - z_0) + i_c(-z - z_0)]$$
$$- R_a Q_a^T[U_a(h - z) + U_a(h + z)], \quad 0 \leq z < h \quad (52)$$

where

$$A = [1 + R_a U_a(2h)]/\Delta \quad (53a)$$

$$Q_a^T = [i_c(h - z_0) + i_c(-h - z_0)]/\Delta \quad (53b)$$

$$\Delta = [1 + R_a U_a(2h)][1 + Z_0 i_c(-2z_0)] - Z_0 R_a U_a(h + z_0)[i_c(h - z_0) + i_c(-h - z_0)] \quad (53c)$$

The reader can show that the current expression reduces immediately to the result of a simple monopole when we set $b \to a$ and $z_0 \to 0$.

The input impedance again can be found once we recognize that each component at the junction has to be continuous into the coaxial region. Thus, for an incident voltage wave of amplitude V_0 at $z = z_0$, the current in the coaxial region is given from (51) as

$$I_c = \left(\frac{V_0}{Z_0}\right)\{e^{-jk_0(z-z_0)} + i_c(z - z_0) - Z_0 R_a Q_a^T U_a(h - z_0) e^{-jk_0(z_0-z)}$$
$$- Z_0[Ai_c(-2z_0) + BU_a(h + z_0)]i_c(z - z_0)\} \quad (54)$$

which, on the substitution of all the relevant constants, provides us an explicit expression for the reflection coefficient Γ_c:

$$I_c = \frac{V_0}{Z_0}[e^{-jk_0(z-z_0)} - \Gamma_c e^{jk_0(z-z_0)}] \tag{55}$$

where $\Gamma_c = \nu A + Z_0 R_a Q_a^T U_a(h - z_0)$, with A and Q_a^T given in (53). The input impedance is then given by the expression $Z_0(1 + \Gamma_c)/(1 - \Gamma_c)$. Likewise, using the far-field formula in (11) and the current expression in (52), we can obtain the far-field pattern function

$$\begin{aligned}\left(\frac{j2\pi}{k_0\xi_0}\right)^{-1} F(\theta; z_0) = {} & 2A e^{jk_0 z_0 \cos\theta}[W_a(h - z_0; \pi - \theta) \\ & + \nu W_b(h + z_0; \theta)] + 2A e^{-jk_0 z_0 \cos\theta} \\ & \times [\nu W_b(h + z_0; \pi - \theta) + W_a(h - z_0; \theta)] \\ & - R_a Q_a^T [e^{jk_0 h \cos\theta} W_a(2h; \theta) \\ & + e^{-jk_0 h \cos\theta} W_a(2h; \pi - \theta)] \end{aligned} \tag{56}$$

Figs. 16 and 17 show the current distributions on two quite different monopole sleeve antennas. In Fig. 16 the currents on the larger, *b* conductor are near resonance while those on the smaller, *a* conductor are not. In Fig. 17 the opposite situation is the case. The current distributions on the antennas depend not only on the overall length (including image) but on the position of the coaxial junction(s) as well, the currents emanating from the junction(s) being excited by the internal source or by the effect of the junction on the external currents. The data for these figures were calculated by using (52) and assuming an incident current of $1/Z_0 = 1/90 = 11.1$ mA. Such a current could be generated by a matched (source impedance $Z_\ell = Z_0$) source having an open circuit voltage of 2 V. Furthermore, these figures are consistent with the experimentally determined distributions obtained by Taylor [30] and readily available in the book *The Theory of Linear Antennas* ([2], Section III.30) by King.

The input impedance to a monopole sleeve antenna for several different overall lengths is shown in Fig. 18 as a function of the position of the coaxial junction. The same conductor radii ($ka = 0.02$ and $kb = 0.09$) were chosen here as they were used in the preceding figures. Note that the input resistance becomes very large in all cases when $kz_0 = 2\pi z_0/\lambda_0 \cong \pi/4$ and $3\pi/4$, corresponding to situations in which the junctions are $\lambda_0/2$ and $3\lambda_0/2$ apart. At these positions the source and its image are opposing one another. For the present theory to yield results comparable to the experimental data measured by Taylor (see [2], Section III.30) we found that it was necessary to include a shunt susceptance of $j2 \times 10^{-3}$ siemens in Fig. 18 in order to account for the finite thickness of the outer conductor in the experiment.

The Coaxial Sleeve Antenna with a Decoupling Choke

In most practical applications, sleeve antennas are fed via the inherent coaxial line within them with the feed line exciting the larger conductor in the manner illustrated in Fig. 19. Normally of coaxial construction itself, the outer sheath of the

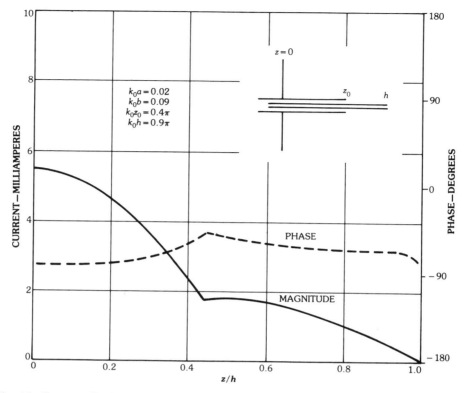

Fig. 16. Current distribution on a monopole sleeve antenna due to an incident coaxial current of $Z_0^{-1}\exp(-jk_0z)$.

feed line and the hollow larger conductor form another sleeve-type junction. With the proper choice of terminating impedance $Z_t = 1/Y_t$ for the b-c coaxial line indicated in the figure, the sleeve antenna itself may be virtually isolated from its feed line at a single frequency. In the most practical situation, Z_t is chosen so that the equivalent of an open circuit appears at the lower truncation of conductor b. If physically permissible, this can be accomplished by shorting the b-c coaxial line at a distance of $\lambda_0/4$ from the opening. Obviously this type of decoupling has a strong dependence on the operating frequency. In the forthcoming analysis, current waves traveling up the c conductor toward the b and a conductors, which form the intended antenna structure, are not considered. Such waves could result if the antenna is not mounted sufficiently high enough above the surrounding environment (especially the ground) and operated too far from the choke resonance point.

To determine the input impedance of such an antenna we again first write all the component currents existing on the two halves of the antenna, i.e., conductors a and b, due to an incident current wave in the coaxial region:

$$J^T(z;z_0) = i_c(z - z_0) - R_a Q_a^T U_a(h - z) + A[i_{ch}(-h - z) \\ - Z_0 i_{ch}(-h - z_0) i_c(z - z_0)], \quad \text{for } -h < z < +h \quad (57)$$

Wire and Loop Antennas

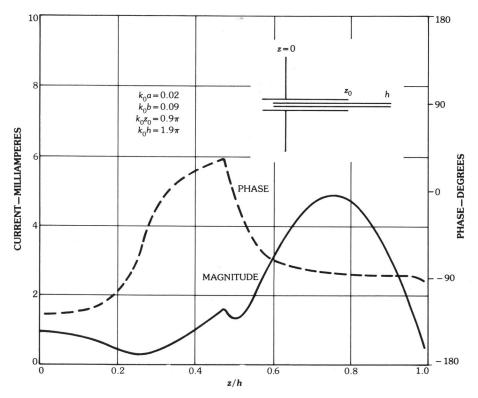

Fig. 17. Current distribution on a monopole sleeve antenna due to an incident coaxial current of $Z_0^{-1}\exp(-jk_0z)$.

Here i_{ch} is the junction current due to the choke at $z = -h$. The expression for i_{ch} is the same as i_c in (42) except that radii (a, b) are replaced by (c, b). Here we note that the choke current $i_{ch}(-h - z)$ is a result of multiple bounces of current waves in the choke region $-h < z < z_{ch}$ back to conductor b. Since for each subsequent bounce a factor of $-v_{ch}\exp(-j2k_0\ell)$ is introduced, where ℓ is the length of the choke, i.e., $\ell = z_{ch} + h$ and $v_{ch} = U_c^s(0)/U_b^s(0)$, the amplitude of the total current wave scattered back from the choke at $z = -h$ can be summed together as

$$-Z_{ch} + Z_{ch}v_{ch}e^{-j2k_0\ell}[1 - v_{ch}e^{-j2k_0\ell} + v_{ch}^2 e^{-j4k_0\ell} - \cdots]$$
$$= -Z_{ch}(1 + v_{ch}e^{-j2k_0\ell})^{-1} \tag{58}$$

for an incident current wave of unity on conductor b. Here $Z_{ch} = (\xi_0/2\pi)\ln(b/c)$ is the characteristic impedance of the coaxial choke section. Using the above equation to relate the incident wave and reflected wave at $z = -h$, we have

$$A = -Z_{ch}(1 + v_{ch}e^{-j2k_0\ell})^{-1}$$
$$\times [i_c(-h - z_0) - R_a Q_a^T U_a(2h) - A Z_0 i_{ch}(-h - z_0) i_c(-z_0 - h)]$$

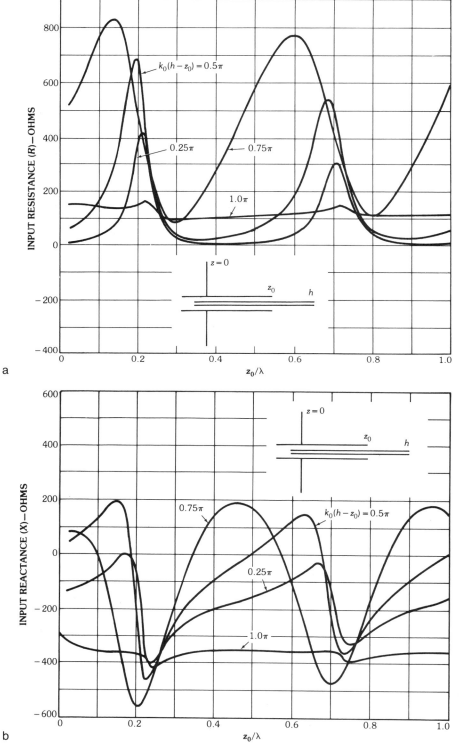

Fig. 18. Input impedance of a monopole sleeve antenna. (*a*) Input resistance. (*b*) Input reactance.

Wire and Loop Antennas

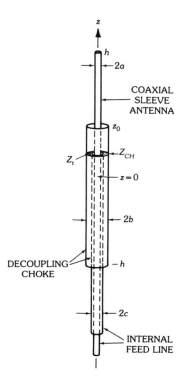

Fig. 19. Decoupled sleeve antenna.

We note that the sum of the incident and the reflected current at the choke is proportional to

$$1 - Z_{ch}(1 + v_{ch}e^{-j2k_0\ell})^{-1}i_{ch}(0)$$

which can be shown to vanish when the length of the choke is a quarter-wave long, i.e., $[1 - Z_{ch}i_{ch}(0)/(1 - v_{ch})] \to 0$. Thus the current is forced to zero at $z = h$ by the choke and remains vanishingly small beyond that point, along the conductor c. In any event, if we now employ the condition at the end of the upper half of the antenna at $J^T(h; z_0) = 0$

$$Q_a^T = i_c(h - z_0) + A[i_{ch}(-2h) - Z_0 i_{ch}(-h - z_0)i_c(h - z_0)]$$

we can solve for the two unknown constants A and Q_a^T:

$$A = [-i_c(-h - z_0) + R_a U_a(2h) i_c(h - z_0)]/\Delta \qquad (59)$$

$$Q_a^T = [-i_c(-h - z_0) i_{ch}(-2h) + (1 + v_{ch}e^{-j2k_0\ell}) i_c(h - z_0)/Z_{ch}]/\Delta \qquad (60)$$

where

$$\Delta = (1 + v_{ch}e^{-j2k_0\ell})/Z_{ch} - R_a U_a(2h) i_{ch}(-2h) - Z_0 i_{ch}(-h - z_0)$$
$$\times [i_c(-h - z_0) - R_a U_a(2h) i_c(h - z_0)] \tag{61}$$

For easy reference we will repeat some of the known expressions here:

$$i_c(-h - z_0) = 2v U_b(h + z_0), \qquad i_c(h - z_0) = 2U_a(h - z_0)$$
$$i_{ch}(-2h) = 2v_{ch} U_b(2h), \qquad i_{ch}(-h - z_0) = 2v_{ch} U_b(h + z_0)$$
$$v = U_a^s(0)/U_b^s(0), \qquad v_{ch} = U_c^s(0)/U_b^s(0)$$

and $U_a(z)$ and $U_a^s(z)$ were previously defined in (27) and (24). These expressions, when substituted into (57), now give the explicit expression for the current.

The reflection coefficient and hence the input impedance of the antenna can be obtained in the manner described before. For an incident voltage wave of amplitude V_0 in the coaxial region, we have from (55)

$$I_c = \frac{V_0}{Z_0}[e^{-jk_0(z-z_0)} - \Gamma_c e^{jk_0(z-z_0)}]$$

where

$$\Gamma_c = v - vAZ_0 i_{ch}(-h - z_0) + Z_0 R_a Q_a^T U_a(h - z_0) \tag{62}$$

Fig. 20 shows the input impedance to a decoupled sleeve dipole antenna as a function of its electrical half-length $k_0 h$, which is proportional to frequency, i.e., $k_0 = 2\pi f/c$. The electrical length of the choke is taken to be the same as the antenna half-length, $k\ell = kh$. A somewhat hypothetical case, to be sure (unless the b-c coaxial line is filled with dielectric so that the physical choke length and half-length are appreciably different, with $\ell < h$), but one which illustrates the frequency dependence of this type of antenna quite well. At the resonances of $kh = \pi/2$ and $3\pi/2$, the choke makes the antenna behave essentially like an isolated sleeve dipole. The bandwidths about these resonant points, however, are seen to be quite narrow. For single-frequency operation, though, this does not present a problem.

4. The Folded Dipole Antenna

Folded dipole antennas, such as the ones depicted in Fig. 21, offer performance similar to that of the linear thin-wire antenna (discussed in Section 2) with the added advantage of a certain measure of control over their input admittance near resonance. A simplified analysis [34] of this type of antenna is possible through the consideration of a length of uniform two-conductor transmission line shorted at both ends (which the folded dipole resembles) with the aid of the equivalent radius concept mentioned in Section 1.

The currents excited on each arm of an unloaded, folded dipole antenna can be approximately determined through a decomposition of the source voltage into a

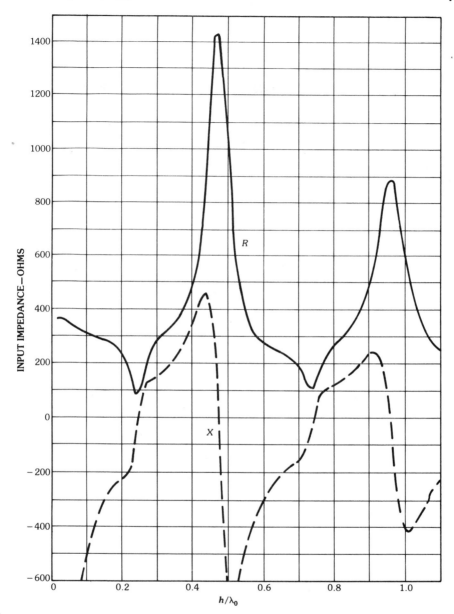

Fig. 20. Input impedance to a decoupled sleeve dipole antenna, where $\Omega = 2 \ln (2h/b) = 10$, $Z_c = (\xi_0/2\pi) \ln (b/a) = 50$ ohms, $Z_{ch} = (\xi_0/2\pi) \ln (c/b) = 72$ ohms, and $k_0 \ell = \pi/2$.

symmetrical and an antisymmetrical arrangement [2] as shown in Figs. 22a and 22b. Requiring that $d \ll \lambda_0$, the strong mutual coupling between the two closely spaced parallel conductors allows the symmetrical current (sometimes called *antenna current*) to be approximated by one corresponding to a single antenna of the same overall length $2h$ and an equivalent radius (see the discussion in Section 1) equal to

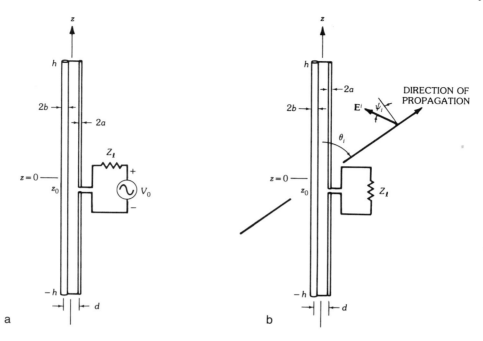

Fig. 21. Folded dipole antenna. (*a*) Transmitting. (*b*) Receiving.

$$a_e \approx \exp\left\{\frac{1}{(a+b)^2}[a^2 \ln a + b^2 \ln b + 2ab \ln d]\right\} \qquad (63)$$

The antenna current is divided between two parallel conductors according to the "current division factor" [34] which, when a and b are much less than d, is approximately given by

$$\eta = \frac{J_b^T}{J_a^T} \approx \frac{\ln(d/a)}{\ln(d/b)} \qquad (64)$$

The antisymmetrical current on the two conductors, on the other hand, can be determined from a two-wire transmission line short-circuited at both ends. The overall current distribution on each conductor of the unloaded, folded-dipole transmitting antenna can then be written as

$$I_{a,b}^T(z;z_0) = V_0 \begin{cases} \frac{1}{2}\left(\frac{\eta}{1+\eta}\right)J_{a_e}^T(z;z_0) - J_{tl}(z;z_0), & b \text{ conductor} \\ \frac{1}{2}\left(\frac{1}{1+\eta}\right)J_{a_e}^T(z;z_0) + J_{tl}(z;z_0), & a \text{ conductor} \end{cases} \qquad (65)$$

where $J_{a_e}^T$ is given by (28). Now since the transmission-line current also can be formulated in terms of bouncing waves, we can still use (28) for J_{tl} with the

Wire and Loop Antennas

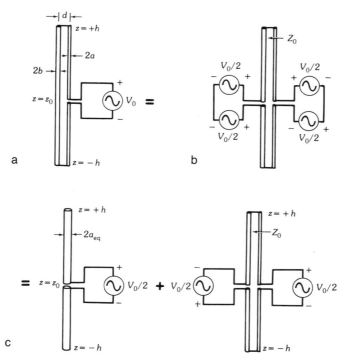

Fig. 22. Approximate model for a folded dipole transmitting antenna. (*a*) Folded dipole transmitting antenna. (*b*) Decomposition of voltage source into symmetric and unsymmetric parts. (*c*) Equivalent symmetrical and antisymmetrical problems.

following replacements:

$$U_a^s(z - z_0) \to (2Z_0)^{-1} e^{-jk_0|z-z_0|}$$

$$U_a(z) \to (2Z_0)^{-1} e^{-jk_0 z}$$

$$R_a \to -2Z_0$$

and Z_0 is the characteristic impedance of the two-wire line, i.e.,

$$Z_0 = (\xi_0/\pi) \cosh^{-1}\left(\frac{d}{2\sqrt{ab}}\right), \qquad \xi_0 = 120\pi \text{ ohms}$$

The input admittance at the feed point $z = z_0$ in conductor a is easily deduced from the input current in (65), i.e.,

$$Y_{\text{in}}(z_0) = \frac{1}{2} \frac{1}{(1+\eta)} J_{a_c}^T(z_0; z_0) + J_{tl}(z_0; z_0), \qquad -h < z_0 < +h$$

where

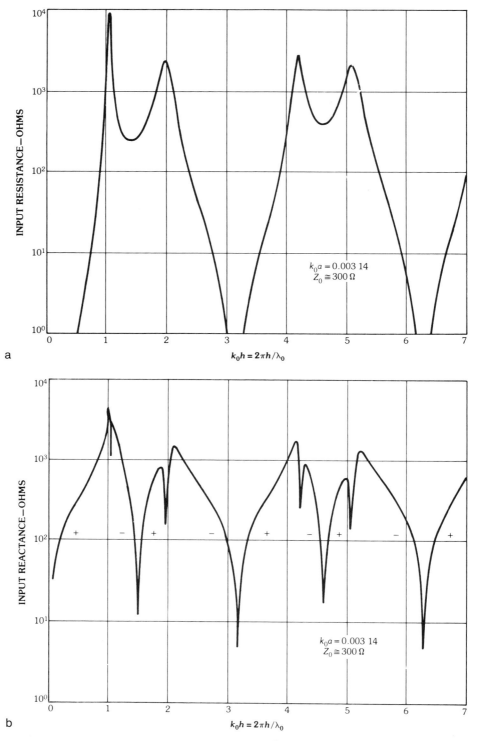

Fig. 23. Input resistance and reactance of a folded dipole antenna, with $a = b$ and $d = 12.5a$. (*a*) Input resistance. (*b*) Input reactance.

$$J_{tl}(z_0; z_0) = Y_{tl}(z_0) = \frac{1}{jZ_0\{\tan[k_0(h+z_0)] + \tan[k_0(h-z_0)]\}}$$

is the input admittance of the transmission-line circuit.

The far-field radiation of a folded dipole, however, comes only from the in-phase antenna current on the two conductors. Again, if we use the equivalent radius concept, the far-field pattern function is then given by (32) and (33) in Section 2, with the exception that a be replaced by a_e and that the actual field strength should be multiplied by $V_0/2$ instead of V_0.

Figs. 23a and 23b show the input resistance and reactance, respectively, to a center-fed folded dipole antenna as a function of its electrical half-length, $k_0 h = 2\pi h/\lambda_0$. The electrical radius of both conductors is constant at $k_0 a = 0.001\pi$ and their separation remains fixed at $d = 12.5a$, thereby yielding a nominal transmission-line impedance of $Z_0 = 300$ ohms.

Fig. 24 affords a close-up view of the input resistance to a center-fed folded dipole antenna near half-wave resonance. The conductors have the electrical radii $k_0 a = 0.001\pi$ and $k_0 b = k_0 a, 2k_0 a, 3k_0 a,$ and $4k_0 a$. The separation of the conductors has been appropriately chosen so that the nominal transmission-line characteristic impedance is $Z_0 \cong 300$ ohms in each case, i.e., $d = 12.5\sqrt{ab}$.

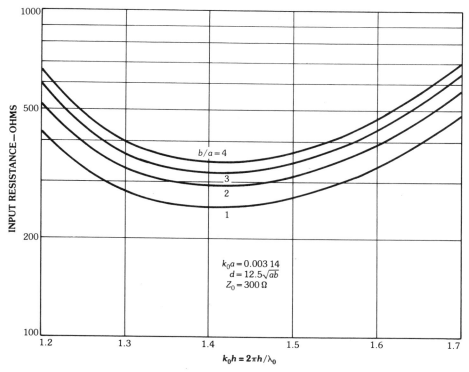

Fig. 24. Close-up view of the input resistance of thin-wire folded dipoles near $\lambda_0/2$ resonance.

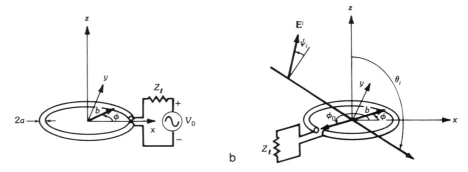

Fig. 25. Circular thin-wire loop antennas. (*a*) Transmitting. (*b*) Receiving.

5. The Thin-Wire Loop Antenna

The thin-wire loop antenna discussed in this section is constructed from a single turn of electrically thin wire (as shown in Fig. 25). Such antennas find application in direction-finding systems and uhf communications, and serve as probes for magnetic intensity measurement.

Following the analysis given by King [35], the basic characteristics and properties of a transmitting loop antenna can be inferred from solving the integral equation given as

$$-V_0 \delta(\phi) = -j \frac{\xi_0}{4\pi} \int_{-\pi}^{+\pi} \left[k_0 b \cos(\phi - \phi') + \frac{1}{k_0 b} \frac{\partial^2}{\partial \phi^2} \right] \\ \times G(\phi - \phi') I^T(\phi') d\phi' \qquad (66)$$

where the kernel is given by

$$G(\phi - \phi') = \frac{1}{2\pi} \int_{-\pi}^{+\pi} \frac{e^{-jk_0 b R}}{R} d\psi \qquad (67a)$$

with

$$R = \left[4\sin^2\left(\frac{\phi - \phi'}{2}\right) + \left(\frac{2a}{b}\right)^2 \sin^2\left(\frac{\psi}{2}\right) \right]^{1/2} \qquad (67b)$$

and $\xi_0 = 120\pi$ ohms. However, because the antenna does not possess any abrupt discontinuities as in the case of a dipole antenna, the physical picture of current waves bouncing between the two ends is no longer viable. Instead, we invoke the Fourier series expansion method to obtain

$$I^T(\phi) = V_0 J^T(\phi) = -j \frac{V_0}{\pi \xi_0} \left\{ \frac{1}{A_0} + 2 \sum_{n=1}^{\infty} \frac{\cos n\phi}{A_n} \right\} \qquad (68)$$

and likewise, for the kernel function G,

$$G(\phi - \phi') = K_0 + 2 \sum_{n=1}^{\infty} K_n \cos n(\phi - \phi') \qquad (69)$$

Substitution of these expressions into the integral equation in (66) permits the solution of the A_n coefficients in terms of

$$A_n = \frac{k_0 b}{2}(K_{n+1} + K_{n-1}) - \frac{n^2}{k_0 b} K_n, \qquad K_{-n} = K_{+n} \qquad (70)$$

The K_n coefficients above are expressible in terms of integrals involving Bessel and Lommel-Weber (Anger) functions [36].

In most practical situations, the electrical radius $k_0 b = 2\pi b/\lambda_0$ of the loop is seldom very much greater than unity. Keeping only the most important terms, the coefficients A_0, A_1, and A_2 can be approximated as

$$A_0 = \frac{k_0 b}{\pi}\left[\ln\left(\frac{8b}{a}\right) - 2\right] + \frac{1}{\pi}[0.667(k_0 b)^3 - 0.267(k_0 b)^5]$$
$$- j[0.167(k_0 b)^4 - 0.033(k_0 b)^6] \qquad (71a)$$

$$A_1 = \left(k_0 b - \frac{1}{k_0 b}\right)\frac{1}{\pi}\left[\ln\left(\frac{8b}{a}\right) - 2\right] + \frac{1}{\pi}[-0.667(k_0 b)^3 + 0.207(k_0 b)^5]$$
$$- j[0.333(k_0 b)^2 - 0.133(k_0 b)^4 + 0.026(k_0 b)^6] \qquad (71b)$$

$$A_2 = \left(k_0 b - \frac{4}{k_0 b}\right)\frac{1}{\pi}\left[\ln\left(\frac{8b}{a}\right) - 2.667\right]$$
$$+ \frac{1}{\pi}[-0.40(k_0 b) + 0.21(k_0 b)^3 - 0.086(k_0 b)^5]$$
$$- j[0.050(k_0 b)^4 - 0.012(k_0 b)^6] \qquad (71c)$$

Above $k_0 b = 1.3$, the accuracy of the above expressions rapidly deteriorates. Figs. 26a and 26b show the real and imaginary components of the inverses of the above approximate Fourier coefficients for loops where $\Omega = 8$, 9, 10, 11, and 12 as a function of the electrical radius $k_0 b$. Over the limited range of $k_0 b$ shown, these approximate coefficients are consistent with the more exact numerically evaluated data given by King [35]. The higher-order ($n \geq 3$) coefficients are negligible compared with the A_0, A_1, and A_2 coefficients (on an individual basis) for $k_0 b \leq 1.3$. Near the feed point, however, the contributions of these higher-order terms are cumulative and their neglect leads primarily to an error in the determination of the imaginary component of the current in this region. An error in the determination of the input susceptance will also result.

The current distributions on transmitting loop antennas where $\Omega = 2\ln(2\pi b/b) = 10$ for $k_0 b = 0.2$, 0.4, 0.6, 0.8, 1.0, and 1.2 as determined from (68) and (71) are shown in Figs. 27a and 27b. These distributions are consistent with the more exact results (arrived at by the numerical calculation of the A_n coefficients up to $n = 20$) given in [35], except for the imaginary current component near the feed point.

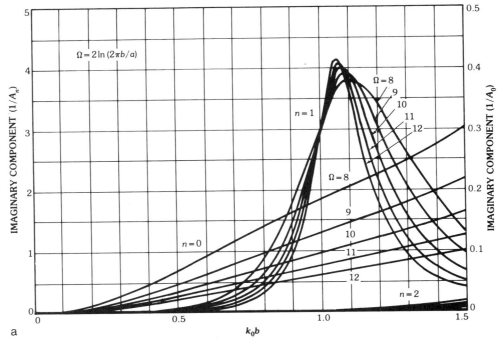

Fig. 26. Approximate real and imaginary components of the inverses of the Fourier coefficients for the current on circular thin-wire loop antennas. (*a*) Approximate imaginary component, with the a_0 components shown ten times actual value. (*b*) Approximate real component.

The input admittance to an electrically small, thin-wire loop can be determined from (68) in the manner:

$$Y_{\text{in}} = I^T(0)/V_0 = J^T(0), \qquad Z_{\text{in}} = 1/Y_{\text{in}} \tag{72}$$

The input resistance and reactance of thin-wire loop antennas where $\Omega = 2\ln(2\pi b/a) = 8$, 10, and 12 are shown in Figs. 28a and 28b, respectively, as functions of the electrical radius $k_0 b = 2\pi b/\lambda_0$.

Fair-Field Radiation from a Circular Loop Antenna

The far-field radiation from a thin-wire loop antenna can be formulated using the general approach outlined in Section 1. In the case of an electrically small loop ($k_0 b \leqq 1.3$) with a voltage source located at $\phi = \phi_0$, the vector far-field pattern is sufficiently well approximated by

$$\mathbf{F}(\theta, \phi; \phi_0) = -j\frac{2b}{\xi_0}\left\{\left[\frac{f_0(\theta)}{A_0} + 2\frac{f_1(\theta)}{A_1}\cos(\phi - \phi_0) + 2\frac{f_2(\theta)}{A_2}\cos 2(\phi - \phi_0)\right]\hat{\mathbf{a}}_\phi \right.$$
$$\left. + \cos\theta\left[2\frac{g_1(\theta)}{A_1}\sin(\phi - \phi_0) + 2\frac{g_2(\theta)}{A_2}\sin 2(\phi - \phi_0)\right]\hat{\mathbf{a}}_\theta\right\} \tag{73}$$

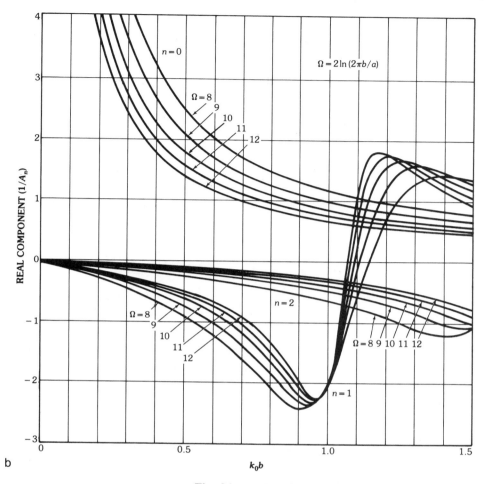

Fig. 26, *continued.*

where

$$f_n(\theta) = (j)^{n-1} J'_n(k_0 b \sin \theta) \tag{74a}$$

$$g_n(\theta) = (j)^{n-1} \frac{n J_n(k_0 b \sin \theta)}{k_0 b \sin \theta} \tag{74b}$$

Here the Bessel function J_n and its derivative J'_n (with respect to its argument) can be easily evaluated using the truncated series

$$J_n(k_0 b \sin \theta) \cong \sum_{\ell=0}^{3} \frac{(-1)^\ell}{\ell!(\ell+n)!} \left(\frac{k_0 b}{2} \sin \theta\right)^{2\ell+n} \tag{75}$$

and

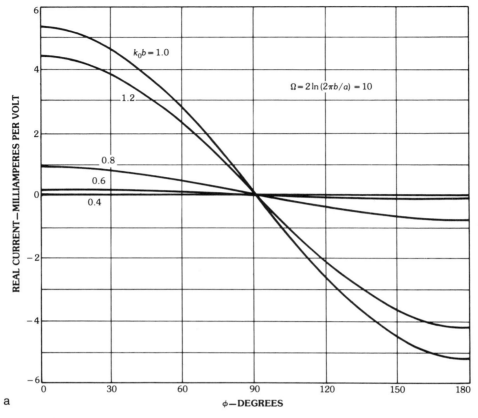

Fig. 27. Current distributions on transmitting loop antennas. (*a*) Real component of the current (approximate) on circular thin-wire antennas. (*b*) Imaginary component of the current (approximate) on circular thin-wire antennas.

$$J'_n(k_0 b \sin\theta) \cong \frac{1}{2}\sum_{\ell=0}^{3}(-1)^\ell \frac{(2\ell+n)}{\ell!(\ell+n)!}\left(k_0\frac{b}{2}\sin\theta\right)^{2\ell+n-1} \quad (76)$$

which are useful for $k_0 b \leqq 1.3$.

The Electrically Small Receiving Loop Antenna

Consider the electrically small ($k_0 b = 2\pi b/\lambda_0 \leqq 1.3$) circular thin-wire loop antenna illuminated by a uniform plane wave as illustrated in Fig. 25b. This plane wave is incident at an angle θ_i with respect to the axis of the loop (z axis) and polarized at an angle ψ_i with respect to the y axis. The electric-field vector of the incident plane-wave field is expressible as

$$\mathbf{E}^i = E^i(-\sin\psi_i \cos\theta_i \hat{\mathbf{a}}_x + \cos\psi_i \hat{\mathbf{a}}_y + \sin\psi_i \sin\theta_i \hat{\mathbf{a}}_z) \quad (77)$$

The incident azimuthal angle is taken here as $\phi_i = 0$ without any loss of generality.

Wire and Loop Antennas

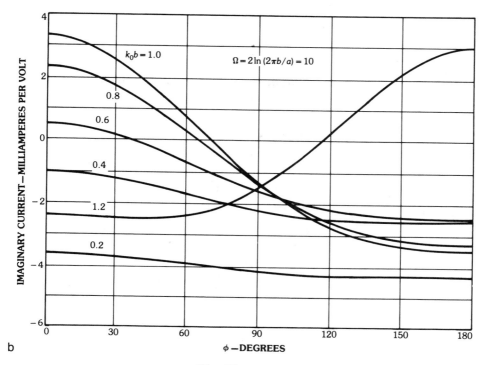

Fig. 27, *continued.*

Based on the discussion in Section 1, using (15), (73), and (77), the receiving current distribution can be simply expressed as

$$I^R(\theta_i, \phi_0) = E^i j \frac{2b}{\xi_0} \left\{ \cos\psi_i \left[\frac{f_0(\theta_i)}{A_0} + 2\frac{f_1(\theta_i)}{A_1}\cos\phi_0 + 2\frac{f_2(\theta_i)}{A_2}\cos 2\phi_0 \right] \right.$$
$$\left. - \sin\psi_i \cos\theta_i \left[2\frac{g_1(\theta_i)}{A_1}\sin\phi_0 + 2\frac{g_2(\theta_i)}{A_2}\sin 2\phi_0 \right] \right\} \quad (78)$$

where ϕ_0 specifies the observation point on the loop.

Loaded Loop Antennas

The introduction of a source or load impedance in a circular transmitting or receiving loop antenna is easily handled in the manner described in Section 1. The current distribution about a thin-wire transmitting loop antenna driven by a voltage source V_0 at $\phi = 0$ having a source admittance $Y_\ell = 1/Z_\ell$ can be written as

$$I^T(\phi) = V_0 \left\{ \frac{Y_\ell}{Y_{in} + Y_\ell} \right\} J^T(\phi) \quad (79)$$

And the current distribution on a receiving loop antenna having a load admittance $Y_\ell = 1/Z_\ell$ located at an azimuth of ϕ_0 on the loop can be written as

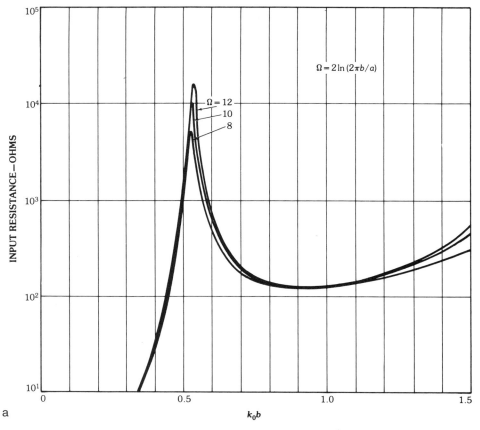

a

Fig. 28. Approximate input impedance of circular thin-wire loop antennas. (*a*) Input resistance. (*b*) Input reactance.

$$I^R(\phi) = E^i \left\{ J^R(\theta_i; \phi) - \frac{J^R(\theta_i; \phi_0)}{Y_{\text{in}} + Y_\ell} J^T(\phi - \phi_0) \right\} \tag{80}$$

where Y_{in} is the input admittance to the loop $Y_{\text{in}} = 1/Z_{\text{in}}$.

6. Concluding Remarks

Various basic thin-wire antenna structures have been discussed and simple mathematical expressions given that permit the calculation of the electrical properties of these antennas with relative ease. A host of composite antenna structures which embody some of these more basic structures can be analyzed through an appropriate combination of the individual analyses. For example, a circular folded loop antenna could be analyzed by combining the elements of the folded dipole and circular loop antenna analyses in Sections 4 and 5, respectively. Partially shielded, coaxially fed loop antennas could be handled by incorporating the "junction effect" in Section 3 into the loop antenna discussion of Section 5.

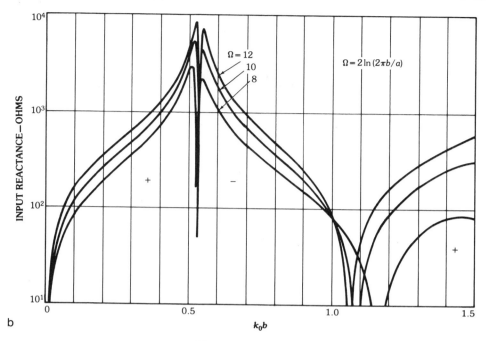

Fig. 28, *continued*.

Other composite structures could be adequately modeled using the basic elements described here.

Based on the general equations for the current on thin-wire antennas given in this chapter, many other related antenna parameters can be obtained. For instance, the charge distribution is easily obtained from the derivative of the current distribution with respect to the axial coordinate. The open-circuit terminal voltage of a receiving antenna is obtained from the product of the short-circuit current and the antenna impedance. Furthermore, the equations given here for the current distributions on thin-wire antennas are, for the most part, analytic with respect to frequency, thereby lending themselves to approximate transient analyses as described in Section 2 for the thin-wire linear dipole.

Acknowledgments

The authors wish to thank Dr. Ahmad Hoorfar for his participation in several helpful discussions related to the transient analysis of linear antennas.

7. References

[1] K. K. Mei, "On the integral equations of thin wire antennas," *IEEE Trans. Antennas Propag.*, vol. 13, no. 3, pp. 374–378, May 1965.
[2] R. W. P. King, *The Theory of Linear Antennas*, Cambridge: Harvard University Press, 1956.

[3] R. W. P. King and C. W. Harrison, *Antennas and Waves: A Modern Approach*, Cambridge: MIT Press, 1969.

[4] R. F. Harrington, *Field Computation by Moment Methods*, New York: Macmillan Co., 1968.

[5] F. M. Teche, "On the analysis of scattering and antenna problems using the singularity expansion technique," *IEEE Trans. Antennas Propag.*, vol. 21, no. 1, pp. 53–62, January 1973.

[6] C. E. Baum, "On the singularity expansion method for solution of electromagnetics problems," *Interaction Notes*, no. 88, Air Force Weapons Lab, Albuquerque, New Mexico, December 1971.

[7] E. A. Wolff, *Antenna Analysis*, chapter 3, New York: John Wiley & Sons, 1966.

[8] S. W. Lee and R. Mittra, "Admittance of a solid cylindrical antenna," *Can. J. Phys.*, no. 47, pp. 1959–1970, 1969.

[9] R. W. P. King, "Current distribution in arbitrarily oriented receiving and scattering antenna," *IEEE Trans. Antennas Propag.*, vol. AP-20, no. 2, pp. 152–159, March 1979.

[10] R. W. P. King and T. T. Wu, "Cylindrical antenna with arbitrary driving point," *IEEE Trans. Antennas Propag.*, vol. AP-13, no. 5, pp. 711–718, September 1965.

[11] R. W. P. King and T. T. Wu, "Currents, charges and near fields of cylindrical antennas," *Radio Sci.*, vol. 69D, no. 3, pp. 429–446, March 1965.

[12] R. Mittra, *Computer Techniques for Electromagnetics*, New York: Pergamon Press, 1973.

[13] W. A. Imbriale and P. G. Ingerson, "On numerical convergence of moment solutions of moderately thick antenna using sinusoidal basis functions," *IEEE Trans. Antennas Propag.*, vol. AP-21, no. 3, pp. 363–366, May 1973.

[14] D. C. Chang, "On the electrically thick monopole, part I," *IEEE Trans. Antennas Propag.*, vol. AP-16, no. 1, pp. 58–64, January 1968.

[15] C. C. Kao, "Electromagnetic scattering from a finite cylinder, numerical solution," *Radio Sci.*, vol. 5, no. 3, pp. 617–624, March 1970.

[16] L. A. Weinstein, *The Theory of Diffraction and the Factorization Method*, Boulder: Golem Press, 1969.

[17] R. Mittra and S. W. Lee, *Analytical Techniques in the Theory of Guided Waves*, New York: Macmillan Co., 1971.

[18] J. B. Anderson, "Admittance of infinite and finite cylindrical metallic antenna," *Radio Sci.*, vol. 3 (New Series), no. 6, June 1968.

[19] C.-L. Chen, "On the scattering of electromagnetic waves from a long wire," *Radio Sci.*, vol. 3 (New Series), no. 7, June 1968.

[20] L. C. Shen, T. T. Wu, and R. W. P. King, "A simple formula for current in dipole antennas," *IEEE Trans. Antennas Propag.*, vol. 16, no. 5, pp. 543–547, September 1968.

[21] O. Einarsson, "Electromagnetic scattering by a thin wire," *Acta Polytech. Scandinavia*, Electrical Engineering Series 23, Stockholm, 1969.

[22] L. C. Shen, "A simple theory of receiving and scattering antennas," *IEEE Trans. Antennas Propag.*, vol. 18, no. 1, pp. 112–114, January 1970.

[23] D. C. Chang, S. W. Lee, and L. W. Rispin, "Simple formula for current on a receiving antenna," *IEEE Trans. Antennas Propag.*, vol. 26, no. 5, pp. 683–690, September 1978.

[24] L. W. Rispin and D. C. Chang, "A unified theory for thin-wire antennas of arbitrary length," *Sci. Rep. No. 38* (N00014-76-C-0318), Department of Electrical Engineering, University of Colorado, Boulder, February 1980.

[25] M. Abramowitz and A. Segun, *Handbook of Mathematical Functions*, New York: Dover Publications, 1972.

[26] E. K. Miller, "Admittance dependence of the infinite cylindrical antenna upon exciting gap thickness," *Radio Sci.*, vol. 2 (New Series), no. 12, pp. 1431–1435, December 1967.

[27] A. Hoorfar and D. C. Chang, "Analytic determination of the transient response of a

thin-wire antenna based upon an SEM representation," *IEEE Trans. Antennas Propag.*, vol. AP-30, no. 6, pp. 1145–1152, November 1982.

[28] L. Marin and T. K. Liu, "A simple way of solving transient thin-wire problems," *Radio Sci.*, vol. 11, no. 2, pp. 149–155, February 1976.

[29] T. K. Liu and K. K. Mei, "A time-domain integral equation for linear antennas and scatterers," *Radio Sci.*, vol. 8, no. 9, pp. 797–804, September 1973.

[30] J. Taylor, *The Sleeve Antenna*, doctoral dissertation, Harvard University, Cambridge, Massachusetts, 1950.

[31] D. C. Chang, "Junction effect of two thin, coaxial cylinders of dissimilar radius," p. 103, *Nat. Radio Sci. Mtg. Dig.*, Seattle, June 18–22, 1979.

[32] A. Hurd, private communication.

[33] R. W. P. King and T. T. Wu, "The thick tubular transmitting antenna," *Radio Sci.*, vol. 2, no. 9, pp. 1061–1066, September 1967.

[34] S. Uda and Y. Mushiake, *Yagi-Uda Antennas*, Tokyo: Maruzen Co., p. 19, 1954.

[35] R. W. P. King, "The loop antenna for transmission and reception," chapter 11 of *Antenna Theory, Part I*, ed. by R. E. Collin and F. J. Zucker, New York: McGraw-Hill Book Co., 1969.

[36] T. T. Wu, "Theory of the thin-circular loop antenna," *J. Math. Phys.*, vol. 3, no. 6, pp. 1301–1304, November–December, 1962.

Chapter 8

Horn Antennas

Constantine A. Balanis
Arizona State University

CONTENTS

1. Introduction — 8-3
2. The *E*-Plane Horn — 8-5
   - *Aperture Fields* 8-5
   - *Radiated Fields* 8-7
   - *Universal Curves* 8-10
   - *Directivity* 8-14
   - *Gain* 8-18
   - *Design Procedure* 8-19
3. The *H*-Plane Horn — 8-20
   - *Aperture Fields* 8-20
   - *Radiated Fields* 8-20
   - *Universal Curves* 8-23
   - *Directivity* 8-29
   - *Design Procedure* 8-33
4. The Pyramidal Horn — 8-34
   - *Aperture and Radiated Fields* 8-34
   - *Directivity* 8-37
5. The Design Procedure for the Pyramidal Horn — 8-43
6. The Conical Horn — 8-46
   - *The Design Procedure* 8-48
7. Special Horns — 8-50
   - *Corrugated Horns* 8-50
   - *Aperture-Matched Horns* 8-64
   - *Multimode Horns* 8-69
   - *Dielectric-Loaded Horns* 8-73
8. Phase Center — 8-73
   - *Procedure to Locate Phase Center* 8-76
9. References — 8-85

**Constantine A. Balanis** was born October 1938 in Trikala, Greece. He received his PhD in electrical engineering from Ohio State University in 1969.

From 1964 to 1970 he was with NASA at Langley Research Center. In 1970 he joined the Department of Electrical Engineering of West Virginia University as a visiting associate professor and held the positions of associate and full professor. Since 1983 he has been a full professor in the Department of Electrical and Computer Engineering at Arizona State University, where he teaches graduate and undergraduate courses in electromagnetic theory, microwave circuits, and antennas.

Formerly he was an associate editor of the *IEEE Transactions on Antennas and Propagation*, and of the *IEEE Transactions on Geoscience and Remote Sensing*, and editor of the *Newsletter* of the IEEE Geoscience and Remote Sensing Society. Dr. Balanis is a Fellow of the IEEE and the author of *Antenna Theory: Analysis and Design* (Harper & Row, 1982). His research interests are in high-frequency asymptotic methods (such as GTD and PTD), radar cross section (RCS), electromagnetic geotomography, wave propagation in microstrip lines, and electromagnetic-wave multipath.

1. Introduction

One of the simplest and probably the most widely used microwave antennas is the horn. Its existence and early use date back to the late 1800s. Although neglected somewhat in the early 1900s its revival began in the late 1930s from the interest in microwaves and waveguide transmission lines during World War II. Since that time a number of articles have been written describing its radiation mechanism, optimization design methods, and applications. Many of the articles published since 1939 which deal with the fundamental theory, operating principles, and designs of a horn as a radiator can be found in a book of reprinted papers [1].

The horn is widely used as a feed element for large radioastronomy, satellite-tracking, and communication dishes found installed throughout the world. In addition to its utility as a feed for reflectors and lenses it is a common element of phased arrays and serves as a universal standard for calibration and gain measurements of other high-gain antennas. Its widespread applicability stems from its simplicity in construction, ease of excitation, versatility, large gain, and preferred overall performance.

An electromagnetic horn can take many different forms, four of which are shown in Fig. 1. The horn is nothing more than a hollow pipe of different cross sections which has been tapered to a larger opening. The type, direction, and amount of taper can have a profound effect on the overall performance of the element as a radiator.

The total field radiated by a conventional horn is a combination of the direct field and the diffractions of it from the edges of the aperture, which can be accounted for using diffraction techniques [2–5]. Techniques found in Chapter 11 of [6] can be used to determine both the direct field and the diffractions from the edges. The edge diffractions, especially those that occur at edges where the electric field is normal to them, influence the antenna pattern structure especially in the back lobe region. The diffractions provide undesirable radiation in the minor lobe structure of the pattern, as well as in that of the main lobe. However, they dominate in low-intensity regions.

A conventional optimum gain horn is usually designed so that its on-axis gain is maximum. This is accomplished by controlling the dimensions of the horn (length and/or flare angle) in such a way that diffractions from the aperture edges add in phase with the direct radiation. In addition, a number of other schemes have been introduced to minimize the effect of diffractions, provide a better pattern symmetry in all planes, and reduce the side lobe intensity. These include the

(a) introduction of corrugations on the inside walls of the horns (pyramidal or conical),
(b) curving of the walls of the horn at its aperture,

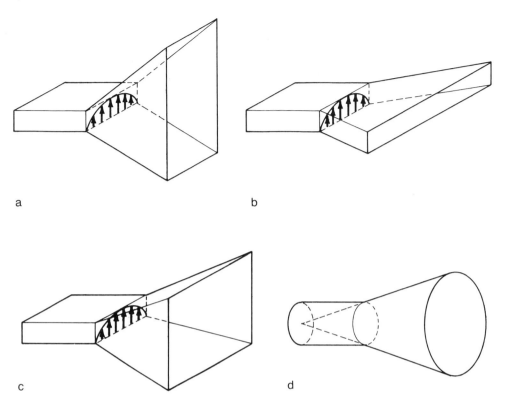

Fig. 1. Typical electromagnetic horn antenna configurations. (*a*) *E*-plane. (*b*) *H*-plane. (*c*) Pyramidal. (*d*) Conical. (*After Balanis [6], © 1982; reprinted by permission of Harper & Row Publishers, Inc.*)

- (*c*) incorporation of corrugations with the curving of the walls at its aperture, and
- (*d*) introduction of higher-order modes within the horn so that the field at the aperture edges is very weak.

In addition to their wide utilization as single-element radiators, horns have also been used frequently as the basic elements of arrays [7–9]. Techniques have been introduced which can be used to suppress grating lobes for limited-scan applications [7, 8]. In addition, a method has been presented, and verified experimentally, which can be used to eliminate blind spots which can possibly exist in rectangular horn arrays with oversized (overmoded) apertures [9]. The blind spots are a result of forced aperture resonances, and they have been verified both analytically and experimentally at 14 GHz in rectangular-grid arrays of tapered rectangular horns. The techniques can possibly be extended to arrays of conical horns and other tapered structures.

Much of the material in this chapter is drawn from an antenna textbook this author has written [6].

Horn Antennas

2. The *E*-Plane Horn

The *E*-plane horn is formed by flaring the walls of a rectangular waveguide in the direction of the **E** field, as shown in Fig. 1a. A more detailed geometry is shown in Figs. 2a and 2b. In this section we want to present, in a summary form, the most pertinent equations related to the aperture fields, radiated fields, and directivity, and to outline a procedure for optimum horn design.

Aperture Fields

The fields at the aperture of an *E*-plane horn will be assumed to be the same as those of the dominant TE_{10} mode of a rectangular waveguide whose dimensions are the same as those of the horn aperture. The only difference is that for the fields of the horn antenna a phase term must also be included to account for the difference in phase that the fields exhibit across the aperture. The phase term is necessary to account for the differences in path that the waves travel from the throat to the different points at the aperture of the horn, as shown in Fig. 2b.

It can be shown [6] that if the (1) fields of the feed waveguide are those of its dominant TE_{10} mode and (2) horn length is large compared with the aperture dimensions, the lowest-order-mode fields at the aperture of the horn are given by

$$E'_z = E'_x = H'_y = 0 \tag{1a}$$

$$E'_y(x', y') \cong E_1 \cos\left(\frac{\pi}{a}x'\right) e^{-j(k/2)(y')^2/\varrho_1} \tag{1b}$$

$$H'_z(x', y') \cong jE_1\left(\frac{\pi}{ka\eta}\right)\sin\left(\frac{\pi}{a}x'\right) e^{-j(k/2)(y')^2/\varrho_1} \tag{1c}$$

$$H'_x(x', y') \cong -\frac{E_1}{\eta} \cos\left(\frac{\pi}{a}x'\right) e^{-j(k/2)(y')^2/\varrho_1} \tag{1d}$$

$$\varrho_1 = \varrho_e \cos\psi_e \tag{1e}$$

where E_1 is a constant. The primes are used to indicate the fields at the aperture of the horn. The expressions are similar to the fields of a TE_{10} mode for a rectangular waveguide with aperture dimensions of a and b_1 (with $b_1 > a$). The only difference is the complex exponential term which is used here to represent the quadratic phase variations of the fields over the aperture of the horn.

The necessity of the quadratic phase term in (1b)–(1d) can be illustrated geometrically. Referring to Fig. 2b, let us assume that at the imaginary apex of the horn (shown dashed) there exists a line source radiating cylindrical waves. As the waves travel in the outward radial direction the constant phase fronts are cylindrical. At any point y' at the mouth of the horn the phase of the field will not be the same as that at the origin ($y' = 0$). The phase is different because the wave has traveled different distances from the apex to the aperture. The difference in travel path, designated as $\delta(y')$, can be obtained by referring to Fig. 2b. For any point y'

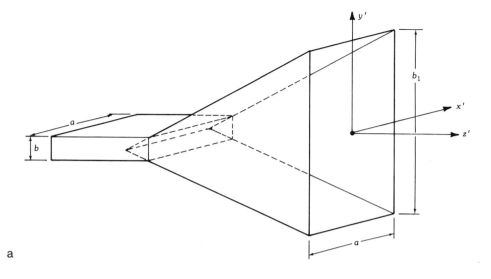

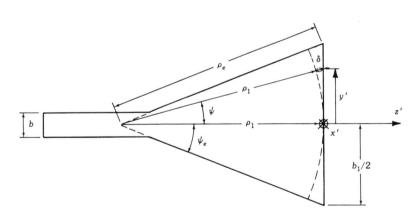

Fig. 2. *E*-plane horn and coordinate system. (*a*) *E*-plane sectoral horn. (*b*) *E*-plane view. (*After Balanis [6]*, © *1982; reprinted by permission of Harper & Row Publishers, Inc.*)

$$[\varrho_1 + \delta(y')]^2 = \varrho_1^2 + (y')^2 \tag{2a}$$

or

$$\delta(y') = -\varrho_1 + [\varrho_1^2 + (y')^2]^{1/2} = -\varrho_1 + \varrho_1[1 + (y'/\varrho_1)^2]^{1/2} \tag{2b}$$

Using the binomial expansion and retaining only the first two terms of (2b) reduces it to

$$\delta(y') \cong -\varrho_1 + \varrho_1\left[1 + \frac{1}{2}\left(\frac{y'}{\varrho_1}\right)^2\right] = \frac{1}{2}\frac{(y')^2}{\varrho_1} \quad (2c)$$

When (2c) is multiplied by the phase factor k, the result is identical with the quadratic phase term in (1b)–(1d).

Radiated Fields

To find the fields radiated by the horn, only the tangential components of the **E** and/or **H** fields over a closed surface need to be known [6]. The closed surface is chosen to coincide with an infinite plane passing through the aperture of the horn. To solve for the fields the approximate equivalent of Section 11.5.2 of [6] is used. That is, we assume that the equivalent current densities $\mathbf{J}_s$ and $\mathbf{M}_s$ exist over the horn aperture, and they are zero elsewhere. Doing this yields

$$\mathbf{J}_s = \hat{\mathbf{n}} \times \mathbf{H}_a = \hat{\mathbf{a}}_y J_y = -\hat{\mathbf{a}}_y \frac{E_1}{\eta}\cos\left(\frac{\pi}{a}x'\right)e^{-jk\delta(y')}, \quad -a/2 \leq x' \leq a/2$$

$$\mathbf{M}_s = -\hat{\mathbf{n}} \times \mathbf{E}_a = \hat{\mathbf{a}}_x M_x = \hat{\mathbf{a}}_x E_1 \cos\left(\frac{\pi}{a}x'\right)e^{-jk\delta(y')}, \quad -b_1/2 \leq y' \leq b_1/2 \quad (3)$$

where $\mathbf{E}_a$ and $\mathbf{H}_a$ represent, respectively, the electric and magnetic fields at the horn aperture, as given by (1a)–(1e), and $\hat{\mathbf{n}} = \hat{\mathbf{a}}_z$ is a unit vector normal to the horn aperture.

Using (3) and the equations of Section 11.3 of Reference 6 it can be shown that the far-zone radiated fields are given by [6]

$$E_r = 0 \quad (4a)$$

$$E_\theta = -j\frac{a\sqrt{\pi k\varrho_1}E_1 e^{-jkr}}{8r}\left\{e^{jk_y^2(\varrho_1/2k)}\sin\phi(1+\cos\theta)\left[\frac{\cos(k_x a/2)}{(k_x a/2)^2 - (\pi/2)^2}\right]F(t_1, t_2)\right\} \quad (4b)$$

$$E_\phi = -j\frac{a\sqrt{\pi k\varrho_1}E_1 e^{-jkr}}{8r}\left\{e^{jk_y^2(\varrho_1/2k)}\cos\phi(\cos\theta+1)\left[\frac{\cos(k_x a/2)}{(k_x a/2)^2 - (\pi/2)^2}\right]F(t_1, t_2)\right\} \quad (4c)$$

where

$$F(t_1, t_2) = [C(t_2) - C(t_1)] - j[S(t_2) - S(t_1)] \quad (4d)$$

$$C(x) = \int_0^x \cos\left(\frac{\pi}{2}t^2\right)dt \quad (4e)$$

$$S(x) = \int_0^x \sin\left(\frac{\pi}{2} t^2\right) dt \tag{4f}$$

$$t_1 = \sqrt{\frac{1}{\pi k \varrho_1}} \left(-\frac{kb_1}{2} - k_y \varrho_1\right) \tag{4g}$$

$$t_2 = \sqrt{\frac{1}{\pi k \varrho_1}} \left(-\frac{kb_1}{2} - k_y \varrho_1\right) \tag{4h}$$

$$k_x = k \sin\theta \cos\phi \tag{4i}$$

$$k_y = k \sin\theta \sin\phi \tag{4j}$$

$C(x)$ and $S(x)$ are known as the cosine and sine Fresnel integrals and are well tabulated [6]. Computer subroutines are also available for efficient numerical evaluation of each [10, 11].

In the principal E- and H-planes the electric field reduces to

$$E\text{-Plane } (\phi = \pi/2)$$

$$E_r = E_\phi = 0 \tag{5a}$$

$$E_\theta = -j\frac{a\sqrt{\pi k \varrho_1} E_1 e^{-jkr}}{8r} \left[-e^{j(k\varrho_1 \sin^2\theta)/2} (2/\pi)^2 (1 + \cos\theta) F(t'_1, t'_2) \right] \tag{5b}$$

$$t'_1 = \sqrt{\frac{k}{\pi \varrho_1}} \left(-\frac{b_1}{2} - \varrho_1 \sin\theta\right) \tag{5c}$$

$$t'_2 = \sqrt{\frac{k}{\pi \varrho_1}} \left(+\frac{b_1}{2} - \varrho_1 \sin\theta\right) \tag{5d}$$

$$H\text{-Plane } (\theta = 0)$$

$$E_r = E_\theta = 0 \tag{6a}$$

$$E_\phi = -j\frac{a\sqrt{\pi k \varrho_1} E_1 e^{-jkr}}{8r} \left\{ (1 + \cos\theta) \left[\frac{\cos[(1/2) ka \sin\theta]}{[(1/2) ka \sin\theta]^2 - (\pi/2)^2} \right] F(t''_1, t''_2) \right\} \tag{6b}$$

$$t''_1 = -\frac{b_1}{2} \sqrt{\frac{k}{\pi \varrho_1}} \tag{6c}$$

$$t''_2 = +\frac{b_1}{2} \sqrt{\frac{k}{\pi \varrho_1}} \tag{6d}$$

To better understand the performance of an E-plane sectoral horn and gain some insight into its performance as an efficient radiator, a three-dimensional normalized field pattern has been plotted in Fig. 3 utilizing (4a)–(4j). As expected, the *E*-plane pattern is much narrower than the *H*-plane because of the flaring and larger dimensions of the horn in that direction. Fig. 3 provides an excellent visual view of the overall radiation performance of the horn. To display additional details the corresponding normalized *E*- and *H*-plane patterns (in decibels) are illustrated in Fig. 4. These patterns also illustrate the narrowness of the *E*-plane and provide information on the relative levels of the pattern in those two planes.

To examine the behavior of the pattern as a function of flaring, the *E*-plane patterns for a horn antenna with $\varrho_1 = 15\lambda$ and with flare angles of $20° \leqq 2\psi_e \leqq 35°$ are plotted in Fig. 5a and for $40° \leqq 2\psi_e \leqq 55°$ in Fig. 5b. In each figure a total of four patterns are illustrated. Since each pattern is symmetrical, only half of each pattern is displayed. For small included angles the pattern becomes narrower as the flare increases. Eventually the pattern begins to widen, to become flatter around the main lobe, and the phase tapering at the aperture is such that even the main maximum does not occur on axis. This is illustrated in Fig. 5a by the pattern with $2\psi_e = 35°$. As the flaring is extended beyond that point the flatness (with certain allowable ripple) increases and eventually the main maximum returns again on axis as shown in Fig. 5b by the pattern with $2\psi_e = 45°$. It is also observed that as the flaring increases, the pattern exhibits much sharper cutoff characteristics. In practice, to compensate for the phase taper at the opening a lens is usually placed at the aperture, making the pattern of the horn always narrower as its flare increases.

Similar pattern variations occur as the length of the horn is varied while the flare angle is held constant. This is illustrated in Figs. 6a and 6b for an included

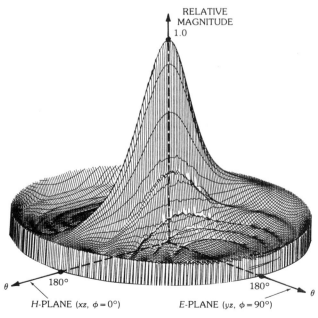

Fig. 3. Three-dimensional field pattern of *E*-plane sectoral horn ($\varrho_1 = 11.773\lambda$, $b_1 = 4.9269\lambda$, $a = 0.7849\lambda$).

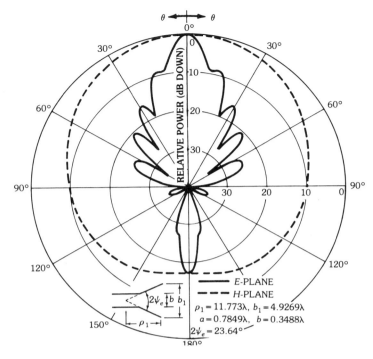

Fig. 4. E- and H-plane patterns of an E-plane sectoral horn.

angle of $2\psi_e = 35°$. As the length increases, the pattern begins to broaden and eventually becomes flatter (with a ripple). Beyond a certain length (see Fig. 6b, $\varrho_1 = 15\lambda$) the main maximum does not even occur on axis, and the pattern continues to broaden and to become flatter (within an allowable ripple) until the maximum return on axis. The process continues indefinitely.

Universal Curves

An observation of the E-plane pattern, as given by (5a)–(5d), indicates that the *magnitude of the normalized pattern, excluding the factor* $(1 + \cos\theta)$, can be written as

$$E_{\theta n} = F(t_1', t_2') = [C(t_2') - C(t_1')] - j[S(t_2') - S(t_1')] \tag{7a}$$

$$\begin{aligned} t_1' &= \sqrt{\frac{k}{\pi\varrho_1}}\left(-\frac{b_1}{2} - \varrho_1\sin\theta\right) = 2\sqrt{\frac{b_1^2}{8\lambda\varrho_1}}\left[-1 - \frac{1}{4}\left(\frac{8\varrho_1\lambda}{b_1^2}\right)\left(\frac{b_1}{\lambda}\sin\theta\right)\right] \\ &= 2\sqrt{s}\left[-1 - \frac{1}{4}\left(\frac{1}{s}\right)\left(\frac{b_1}{\lambda}\sin\theta\right)\right] \end{aligned} \tag{7b}$$

$$\begin{aligned} t_2' &= \sqrt{\frac{k}{\pi\varrho_1}}\left(\frac{b_1}{2} - \varrho_1\sin\theta\right) = 2\sqrt{\frac{b_1^2}{8\lambda\varrho_1}}\left[1 - \frac{1}{4}\left(\frac{8\varrho_1\lambda}{b_1^2}\right)\left(\frac{b_1}{\lambda}\sin\theta\right)\right] \\ &= 2\sqrt{s}\left[1 - \frac{1}{4}\left(\frac{1}{s}\right)\left(\frac{b_1}{\lambda}\sin\theta\right)\right] \end{aligned} \tag{7c}$$

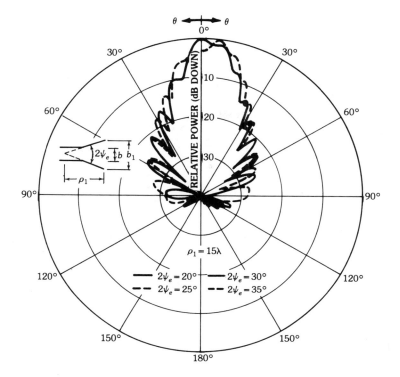

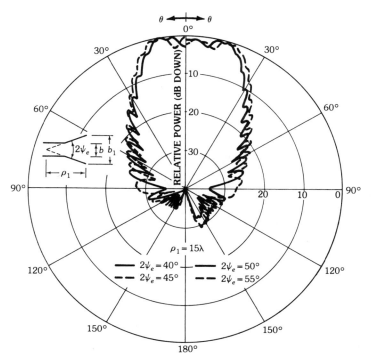

Fig. 5. *E*-plane patterns of *E*-plane sectoral horns with constant length and different included angles. (*a*) $2\psi_e = 20°$ to $35°$. (*b*) $2\psi_e = 40°$ to $55°$. (*After Balanis [6], © 1982; reprinted by permission of Harper & Row Publishers, Inc.*)

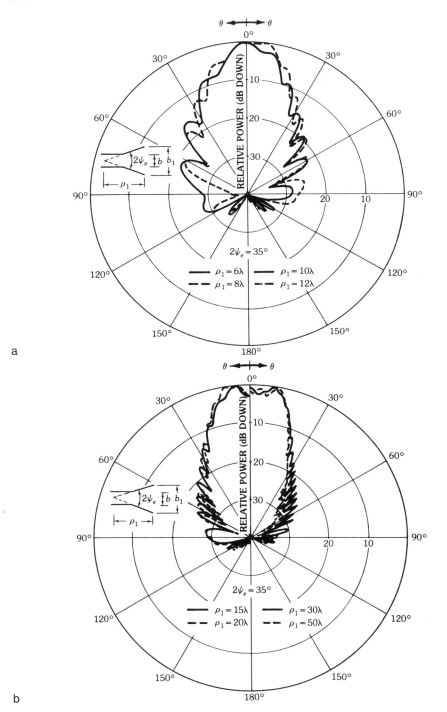

Fig. 6. E-plane patterns of E-plane sectoral horns for constant included angle and different lengths. (a) $\varrho_1 = 6\lambda$ to 12λ. (b) $\varrho_1 = 15\lambda$ to 50λ.

$$s = \frac{b_1^2}{8\lambda\varrho_1} \tag{7d}$$

For a given value of s the field of (7a) can be plotted as a function of $(b_1/\lambda)\sin\theta$, as shown in Fig. 7 for s = 1/64, 1/8, 1/4, 1/2, 3/4, and 1. These plots are usually referred to as *universal curves*, because from them the normalized E-plane pattern of any E-plane sectoral horn can be obtained. This is accomplished by first determining the value of s from a given b_1 and ϱ_1 by using (7d). For that value of s the field strength (in decibels) as a function of $(b_1/\lambda)\sin\theta$ (or as a function of θ for a given b_1) is obtained from Fig. 7. Finally the value of $(1 + \cos\theta)$, normalized to 0 dB and written as $20\log_{10}[(1 + \cos\theta)/2]$, is added to that number to arrive at the required field strength.

Example 1—An E-plane horn has dimensions of $a = 0.7849\lambda$, $b = 0.3488\lambda$, $b_1 = 4.9269\lambda$, and $\varrho_1 = 11.773\lambda$. Find its E-plane normalized field intensity (in decibels *and* as a voltage ratio) at an angle of $\theta = 30°$ using the universal curves of Fig. 7.

Solution—Using (7d)

$$s = \frac{b_1^2}{8\lambda\varrho_1} = \frac{(4.9269)^2}{8(11.773)} = 0.2577 \cong \frac{1}{4}$$

At $\theta = 30°$

$$\frac{b_1}{\lambda}\sin\theta = 4.9269\sin 30° = 4.9269(0.5) = 2.463 \cong 2.5$$

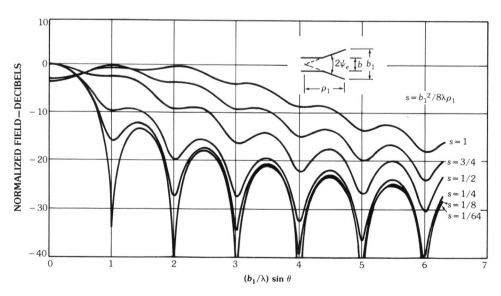

Fig. 7. E-plane universal patterns for E-plane sectoral and pyramidal horns. (*After Balanis* [6], © 1982; reprinted by permission of Harper & Row Publishers, Inc.)

At that point, using the $s = 1/4$ curve of Fig. 7, the field intensity is about -15.6 dB. Therefore the total field intensity at $\theta = 30°$ is equal to

$$E_\theta = -15.6 + 20\log_{10}\left(\frac{1 + \cos 30°}{2}\right) = -15.6 - 0.6 = -16.2 \text{ dB}$$

or as a normalized voltage ratio of

$$E_\theta = 0.155$$

which closely agrees with the results of Fig. 4.

Directivity

The directivity is a parameter which most often is used as the figure of merit to describe the performance of a horn antenna. By definition the directivity can be written as [6]

$$D_0 = \frac{U_{max}}{U_0} = \frac{4\pi U_{max}}{P_r} \tag{8}$$

where

U_{max} = maximum radiation intensity
U_0 = radiation intensity of isotropic source
P_r = total radiated power

Using (4a)–(4c) the directivity of (8) for the E-plane horn can be written as

$$D_E = \frac{4\pi U_{max}}{P_r} = \frac{64 a \varrho_1}{\pi \lambda b_1}\left[C^2\left(\frac{b_1}{\sqrt{2\lambda\varrho_1}}\right) + S^2\left(\frac{b_1}{\sqrt{2\lambda\varrho_1}}\right)\right] \tag{9}$$

The overall performance of an antenna system can often be judged by its beamwidth and/or its directivity. The half-power beamwidth, as a function of flare angle for different horn lengths, is shown in Fig. 8. In addition, the directivity (normalized with respect to the constant aperture dimension a) is displayed in Fig. 9. For a given length the horn exhibits a monotonic decrease in half-power beamwidth and an increase in directivity up to a certain flare. Beyond that point a monotonic increase in beamwidth and decrease in directivity is indicated, followed by rises and falls. The increase in beamwidth and decrease in directivity beyond a certain flare indicate the broadening of the main beam.

If the values of b_1 (in wavelengths), which correspond to the maximum directivities in Fig. 9, are plotted versus their corresponding values of ϱ_1 (in wavelengths), it can be shown that each optimum directivity occurs when

$$b_1 \cong \sqrt{2\lambda\varrho_1} \tag{10a}$$

Horn Antennas

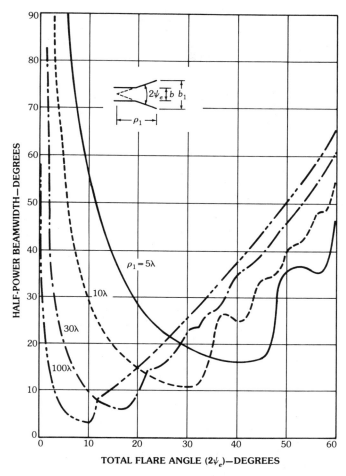

Fig. 8. Half-power beamwidth of E-plane sectoral horns as a function of included angle and for different lengths. (*After Balanis [6], © 1982; reprinted by permission of Harper & Row Publishers, Inc.*)

with a corresponding value of s equal to

$$s\bigg|_{b_1=\sqrt{2\lambda\varrho_1}} = s_{op} = \frac{b_1^2}{8\lambda\varrho_1}\bigg|_{b_1=\sqrt{2\lambda\varrho_1}} = \frac{1}{4} \qquad (10b)$$

The directivity of an E-plane sectoral horn can also be computed by using the following procedure [12].

(*a*) Calculate B by

$$B = \frac{b_1}{\lambda}\sqrt{\frac{50}{\varrho_e/\lambda}} \qquad (11a)$$

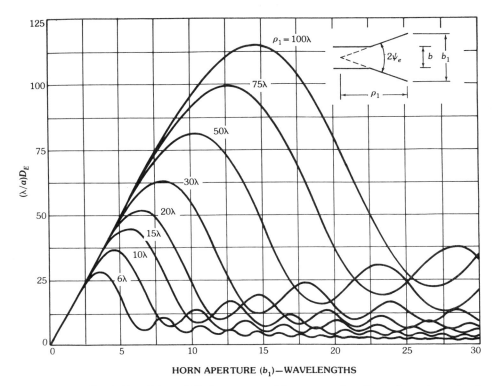

Fig. 9. Normalized directivity of E-plane sectoral horns as a function of aperture size for different lengths. (*After Balanis [6], © 1982; reprinted by permission of Harper & Row Publishers, Inc.*)

(b) Using this value of B, find the corresponding value of G_E from Fig. 10. If, however, the value of B is smaller than 2, compute G_E using

$$G_E = \frac{32}{\pi} B \qquad (11\text{b})$$

(c) Calculate D_E by using the value of G_E from Fig. 10 or from (11b). Thus

$$D_E = \frac{a}{\lambda} \frac{G_E}{\sqrt{50\lambda/\varrho_e}} \qquad (11\text{c})$$

It has been found [13] through comparisons with experimental data that the expression of (9) gives values which are about 25 percent below those measured on fairly large horns. The discrepancy will be larger for smaller horns. A convenient formula, whose values pass through the median of experimental values, has been proposed [13] and is given by

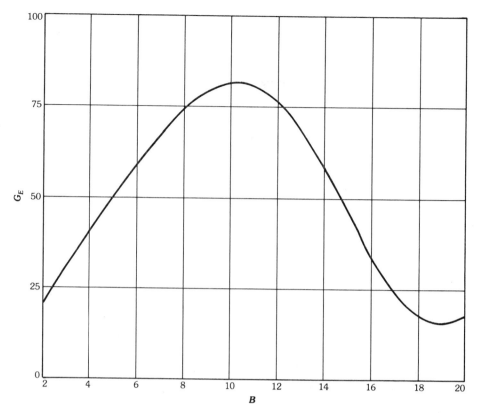

Fig. 10. Term G_E as a function of B. (*Adapted from data by E. H. Braun [12], © 1956 IEEE*)

$$D_E = \frac{16ab_1}{\lambda^2(1 + \lambda_g/\lambda)} \left\{ \frac{2\lambda\varrho_1}{b_1^2} \left[C^2\left(\frac{b_1}{\sqrt{2\lambda\varrho_1}}\right) + S^2\left(\frac{b_1}{\sqrt{2\lambda\varrho_1}}\right) \right] \right\} e^{(\pi a/\lambda)(1 - \lambda/\lambda_g)} \quad (12a)$$

where

$$\lambda_g = \frac{2\pi}{\beta} = \frac{2\pi}{\sqrt{k^2 - (\pi/a)^2}} = \frac{2}{\sqrt{(2/\lambda)^2 - (1/a)^2}} \quad (12b)$$

Two reasons for the discrepancy in the values as given by (9) are provided by the assumptions that the aperture fields are the incident fields with a propagation constant equal to that of free space and the gain formula is based on the Kirchhoff approximation. These are good approximations for horns with large apertures, such as pyramidal horns; however, typical *E*-plane sectoral horns have closely spaced edges which do not satisfy these approximations well.

Example 2—An *X*-band (8.2–12.4 GHz) *E*-plane sectoral horn has dimensions of $a = 0.9$ in (2.286 cm), $b = 0.4$ in (1.016 cm), $b_1 = 5.65$ in (14.35 cm), and

$\varrho_1 = 13.5$ in (34.29 cm). Compute the directivity at $f = 10.3$ GHz using (9), (11c), and (12a). Compare the answers.

Solution—At $f = 10.3$ GHz, $\lambda = 30/10.3 = 2.9126$ cm $= 1.1427$ in.

$$a = 2.286\lambda/2.9126 = 0.7849\lambda$$
$$b = 1.016\lambda/2.9126 = 0.3488\lambda$$
$$b_1 = 14.35\lambda/2.9126 = 4.9269\lambda$$
$$\varrho_1 = 34.29\lambda/2.9126 = 11.7730\lambda$$
$$\varrho_e = \lambda\sqrt{(11.7730)^2 + (2.463\,45)^2} = 12.027\,9\lambda$$
$$\lambda_g = 2\lambda/\sqrt{(2)^2 - (1/0.7849)^2} = 1.2973\lambda$$
$$w = b_1/\sqrt{2\lambda\varrho_1} = 4.9269/\sqrt{2(11.773)} = 1.015$$
$$C(w) = C(1.015) = 0.780$$
$$S(w) = S(1.015) = 0.440$$

Using (9)

$$D_E = \frac{64(0.7849)(11.773)}{\pi(4.9269)}[(0.780)^2 + (0.440)^2] = 30.643 = 14.863 \text{ dB}$$

$$B = 4.9269\sqrt{50/12.0279} = 10.045$$

From Fig. 10, $G_E = 81.55$. Using (11c)

$$D_E = 0.7849(81.55)/\sqrt{50/12.0279} = 31.394 = 14.968 \text{ dB}$$

Using (12a)

$$D_E = \frac{16(0.7849)(4.9269)}{(1 + 1.2973)}\left\{\frac{2(11.773)}{(4.9269)^2}[(0.780)^2 + (0.440)^2]\right\}e^{\pi(0.7849)(1 - 1/1.2973)}$$
$$= 41.168 = 16.146 \text{ dB}$$

From a comparison of the three values of D_E, it is evident that those computed using (12a) is about 32 percent higher than those of (9) and (11c).

Gain

A gain G_0 of a horn, as for any other antenna, is related to the directivity D_0 by the total antenna efficiency e_t [6]. Thus we can write that

$$G_0 = e_t D_0 = e_{cd} e_r D_0 = e_{cd}(1 - |\Gamma|^2)D_0 \qquad (13)$$

where

e_t = total antenna efficiency

e_{cd} = conduction-dielectric (radiation) efficiency

Horn Antennas

e_r = reflection efficiency

Γ = reflection coefficient at transmission line–antenna connection

For most well-matched horns, $e_t \cong 1$.

Design Procedure

In designing an optimum directivity E-plane horn, the usual specifications are the following:

Given: (a) Desired optimum directivity (or gain)
             (b) Center frequency of operation f
              (c) Dimensions a, b of the feed waveguide

Desired: Dimensions ϱ_1, b_1, ϱ_e, and angle ψ_e

Procedure: The design procedure is as follows:
    (a) Substitute (10a) into (9).
    (b) Knowing D_E (as a dimensionless quantity), a, and λ, find ϱ_1 using (9). The term ϱ_1 will have the same dimensions as a and λ.
    (c) Determine b_1 using (10a). The term b_1 will have the same dimensions as ϱ_1 and λ.
    (d) Find ϱ_e and angle ψ_e using the geometry of Fig. 2.

Example 3—Design an optimum gain X-band E-plane sectoral horn such that its directivity at $f = 10$ GHz is 14.437 dB. The dimensions of the feed rectangular waveguide are $a = 0.9$ in (2.286 cm) and $b = 0.4$ in (1.016 cm).

Solution—At $f = 10$ GHz

$$\lambda = \frac{30 \times 10^9}{10 \times 10^9} = 3 \text{ cm}$$

$$a = \frac{2.286}{3}\lambda = 0.762\lambda$$

$$b = \frac{1.016}{3}\lambda = 0.3387\lambda$$

Substituting (10a) into (9) yields

$$D_E = \frac{64a\varrho_1}{\pi\lambda\sqrt{2\lambda\varrho_1}}[C^2(1) + S^2(1)]$$

Since $D_E = 14.437$ dB $= 27.779$,

$$C(1) = 0.77989$$
$$S(1) = 0.43826$$

Then

$$27.779 = \frac{64(0.762)\varrho_1}{\pi\sqrt{2\lambda\varrho_1}}[(0.77989)^2 + (0.43826)^2]$$

$$27.779 = 8.7846\sqrt{\frac{\varrho_1}{\lambda}}$$

so that

$$\varrho_1 = 10\lambda = 30 \text{ cm}$$
$$b_1 = \sqrt{2\lambda(10\lambda)} = 4.472\lambda = 13.416 \text{ cm}$$
$$\psi_e = \tan^{-1}\left(\frac{2.236}{10}\right) = 12.6°$$

The results of this design agree with the data of the $\varrho_1 = 10\lambda$ curve of Fig. 9.

3. The *H*-Plane Horn

The *H*-plane horn is formed by flaring the rectangular waveguide walls in the direction of the **H** field, as shown in Fig. 1b. A more detailed geometry is shown in Figs. 11a and 11b. The aperture fields, radiated fields, and directivity formulas will be summarized in this section.

Aperture Fields

Using a procedure similar to that of an *E*-plane horn and assuming a TE_{10}-mode field distribution, the electric and magnetic field components at the aperture of the *H*-plane sectoral horn of Figs. 11a and 11b can be written as

$$E'_x = H'_y = 0 \tag{14a}$$

$$E'_y(x') = E_2 \cos\left(\frac{\pi}{a_1}x'\right)e^{-jk\delta(x')} \tag{14b}$$

$$H'_x(x') = -\frac{E_2}{\eta}\cos\left(\frac{\pi}{a_1}x'\right)e^{-jk\delta(x')} \tag{14c}$$

$$\delta(x') = \frac{1}{2}\frac{(x')^2}{\varrho_2} \tag{14d}$$

$$\varrho_2 = \varrho_h \cos\psi_h \tag{14e}$$

As with the *E*-plane sectoral horn the $\delta(x')$ function in the exponential term is introduced to account for the phase variation (tapering) across the aperture of the horn.

Radiated Fields

The fields radiated by the *H*-plane sectoral horn are found, as with the *E*-plane sectoral horn, by formulating the equivalent current densities $\mathbf{J}_s$ and $\mathbf{M}_s$ over the

Horn Antennas

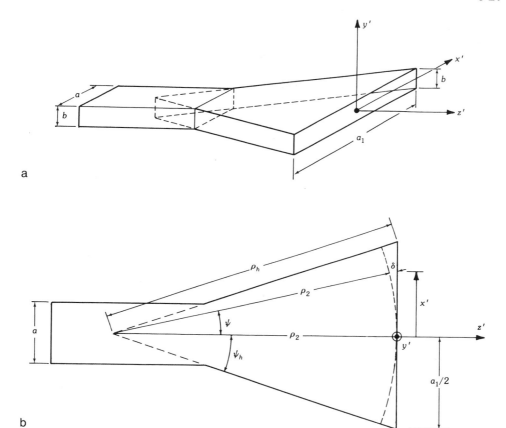

Fig. 11. *H*-plane sectoral horn and coordinate system. (*a*) *H*-plane sectoral horn. (*b*) *H*-plane view. (*After Balanis [6]*, © *1982; reprinted by permission of Harper & Row Publishers, Inc.*)

horn aperture and assuming they are zero outside it. That is,

$$\mathbf{J}_s = \hat{\mathbf{n}} \times \mathbf{H}_a = \hat{\mathbf{a}}_y J_y = -\hat{\mathbf{a}}_y \frac{E_2}{\eta} \cos\left(\frac{\pi}{a_1} x'\right) e^{-jk\delta(x')}, \quad -a_1/2 \leq x' \leq a_1/2$$

$$\mathbf{M}_s = -\hat{\mathbf{n}} \times \mathbf{E}_a = \hat{\mathbf{a}}_x M_x = \hat{\mathbf{a}}_x E_2 \cos\left(\frac{\pi}{a_1} x'\right) e^{-jk\delta(x')}, \quad -b/2 \leq y' \leq b/2$$

(15)

where $\mathbf{E}_a$ and $\mathbf{H}_a$ are the horn aperture fields given by (14a)–(14e) and $\hat{\mathbf{n}} = \hat{\mathbf{a}}_z$ is a unit vector normal to the horn aperture. Using (15) and the equations of Section 11.3 of Reference 6, it can be shown that the far-zone radiated fields can be written as [6]

$$E_r = 0 \qquad (16a)$$

$$E_\theta = jE_2 \frac{b}{8}\sqrt{\frac{k\varrho_2}{\pi}}\frac{e^{-jkr}}{r}\left\{\sin\phi(1+\cos\theta)\frac{\sin Y}{Y}[e^{jf_1}F(t'_1,t'_2)\right.$$
$$\left.+e^{jf_2}F(t''_1,t''_2)]\right\} \tag{16b}$$

$$E_\phi = jE_2 \frac{b}{8}\sqrt{\frac{k\varrho_2}{\pi}}\frac{e^{-jkr}}{r}\left\{\cos\phi(\cos\theta+1)\frac{\sin Y}{Y}[e^{jf_1}F(t'_1,t'_2)\right.$$
$$\left.+e^{jf_2}F(t''_1,t''_2)]\right\} \tag{16c}$$

where

$$Y = \frac{kb}{2}\sin\theta\sin\phi \tag{16d}$$

$$f_1 = \frac{(k'_x)^2 \varrho_2}{2k} \tag{16e}$$

$$F(t_1,t_2) = [C(t_2)-C(t_1)] - j[S(t_2)-S(t_1)] \tag{16f}$$

$$t'_1 = \sqrt{\frac{1}{\pi k\varrho_2}}\left(-\frac{ka_1}{2}-k'_x\varrho_2\right) \tag{16g}$$

$$t'_2 = \sqrt{\frac{1}{\pi k\varrho_2}}\left(+\frac{ka_1}{2}-k'_x\varrho_2\right) \tag{16h}$$

$$k'_x = k\sin\theta\cos\phi + \frac{\pi}{a_1} \tag{16i}$$

$$f_2 = \frac{(k''_x)^2 \varrho_2}{2k} \tag{16j}$$

$$t''_1 = \sqrt{\frac{1}{\pi k\varrho_2}}\left(-\frac{ka_1}{2}-k''_x\varrho_2\right) \tag{16k}$$

$$t''_2 = \sqrt{\frac{1}{\pi k\varrho_2}}\left(+\frac{ka_1}{2}-k''_x\varrho_2\right) \tag{16l}$$

$$k''_x = k\sin\theta\cos\phi - \frac{\pi}{a_1} \tag{16m}$$

and $C(x)$ and $S(x)$ are the cosine and sine Fresnel integrals of (4e) and (4f). The electric field in the principal E- and H-planes reduces to

$$E\text{-Plane } (\phi = \pi/2)$$

$$E_r = E_\phi = 0 \tag{17a}$$

Horn Antennas

$$E_\theta = jE_2 \frac{b}{8}\sqrt{\frac{k\varrho_2}{\pi}}\frac{e^{-jkr}}{r}\left\{(1+\cos\theta)\frac{\sin Y}{Y}[e^{jf_1}F(t'_1,t'_2)+e^{jf_2}F(t''_1,t''_2)]\right\} \quad (17b)$$

$$Y = \frac{kb}{2}\sin\theta \quad (17c)$$

$$k'_x = \frac{\pi}{a_1} \quad (17d)$$

$$k''_x = -\frac{\pi}{a_1} \quad (17e)$$

$$H\text{-Plane } (\phi = 0)$$

$$E_r = E_\theta = 0 \quad (18a)$$

$$E_\phi = jE_2 \frac{b}{8}\sqrt{\frac{k\varrho_2}{\pi}}\frac{e^{-jkr}}{r}\{(\cos\theta+1)[e^{jf_1}F(t'_1,t'_2)+e^{jf_2}F(t''_1,t''_2)]\} \quad (18b)$$

$$k'_x = k\sin\theta + \frac{\pi}{a_1} \quad (18c)$$

$$k''_x = k\sin\theta - \frac{\pi}{a_1} \quad (18d)$$

with f_1, f_2, $F(t'_1,t'_2)$, $F(t''_1,t''_2)$, t'_1, t'_2, t''_1, and t''_2 as defined above.

Computations similar to those for the E-plane sectoral horn were also performed for the H-plane sectoral horn. A three-dimensional field pattern of an H-plane sectoral horn is shown in Fig. 12. Its corresponding E- and H-plane patterns are displayed in Fig. 13. This horn exhibits narrow pattern characteristics in the flared H-plane.

Normalized H-plane patterns for a given length horn ($\varrho_2 = 12\lambda$) and different flare angles are shown in Figs. 14a and 14b. A total of four patterns is illustrated in each figure. Since each pattern is symmetrical, only half of each pattern is displayed. As the included angle is increased, the pattern begins to become narrower up to a given flare. Beyond that point the pattern begins to broaden, attributed primarily to the phase taper (phase error) across the aperture of the horn. To correct this a lens is usually placed at the horn aperture which would yield narrower patterns as the flare angle is increased. Similar pattern variations are evident when the flare angle of the horn is maintained fixed while its length is varied, as shown in Figs. 15a and 15b.

Universal Curves

The *universal curves* for the H-plane sectoral horn are based on (18b), in the absence of the factor $(1+\cos\theta)$. Neglecting the $(1+\cos\theta)$ factor the normalized H-plane electric field of the H-plane sectoral horn can be written as

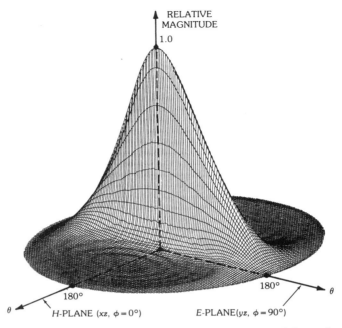

Fig. 12. Three-dimensional field pattern of an H-plane sectoral horn ($\varrho_2 = 12.3841\lambda$, $a_1 = 6.6710\lambda$, $b = 0.3488\lambda$).

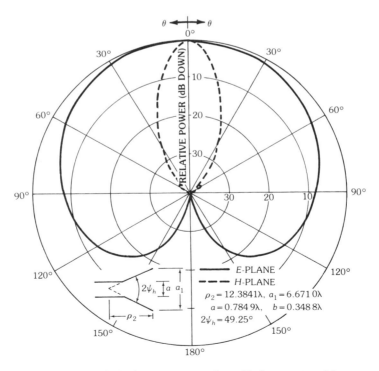

Fig. 13. E- and H-plane patterns of an H-plane sectoral horn.

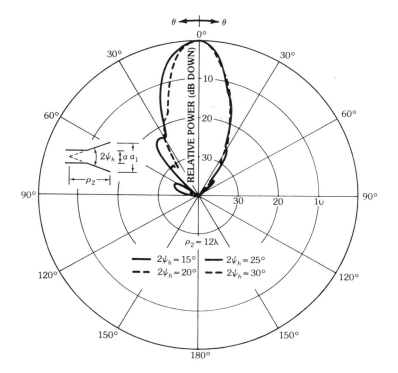

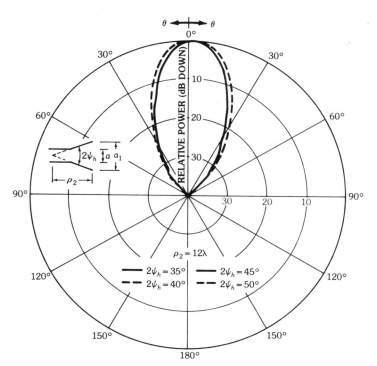

Fig. 14. *H*-plane patterns of *H*-plane sectoral horns for constant length and different included angles. (*a*) $2\psi_h = 15°$ to $30°$. (*After Balanis [6], © 1982; reprinted by permission of Harper & Row Publishers, Inc.*) (*b*) $2\psi_h = 35°$ to $50°$.

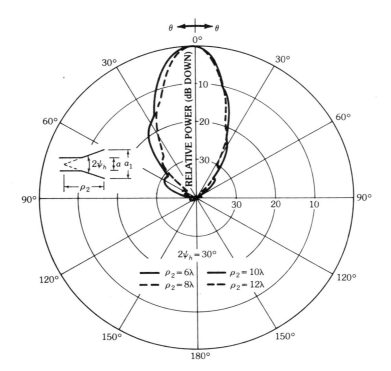

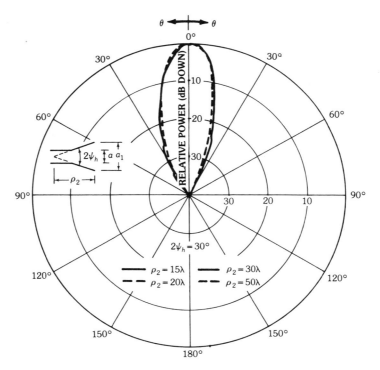

Fig. 15. *H*-plane patterns of *H*-plane sectoral horns of constant included angle and different lengths. (*a*) $\varrho_2 = 6\lambda$ to 12λ. (*b*) $\varrho_2 = 15\lambda$ to 50λ.

Horn Antennas

$$E_{\phi n} = [e^{jf_1}F(t_1', t_2') + e^{jf_2}F(t_1'', t_2'')] \tag{19a}$$

$$F(t_1, t_2) = [C(t_2) - C(t_1)] - j[S(t_2) - S(t_1)] \tag{19b}$$

$$f_1 = \frac{(k_x')^2 \varrho_2}{2k} = \frac{\varrho_2}{2k}\left(k\sin\theta + \frac{\pi}{a_1}\right)^2$$

$$= \frac{\pi}{8}\left(\frac{1}{t}\right)\left(\frac{a_1}{\lambda}\sin\theta\right)^2\left[1 + \frac{1}{2}\left(\frac{\lambda}{a_1\sin\theta}\right)\right]^2 \tag{19c}$$

$$f_2 = \frac{(k_x'')^2 \varrho_2}{2k} = \frac{\varrho_2}{2k}\left(k\sin\theta - \frac{\pi}{a_1}\right)^2$$

$$= \frac{\pi}{8}\left(\frac{1}{t}\right)\left(\frac{a_1}{\lambda}\sin\theta\right)^2\left[1 - \frac{1}{2}\left(\frac{\lambda}{a_1\sin\theta}\right)\right]^2 \tag{19d}$$

$$t_1' = \sqrt{\frac{1}{\pi k \varrho_2}}\left(-\frac{ka_1}{2} - k_x'\varrho_2\right)$$

$$= 2\sqrt{t}\left[-1 - \frac{1}{4}\left(\frac{1}{t}\right)\left(\frac{a_1}{\lambda}\sin\theta\right) - \frac{1}{8}\left(\frac{1}{t}\right)\right] \tag{19e}$$

$$t_2' = \sqrt{\frac{1}{\pi k \varrho_2}}\left(+\frac{ka_1}{2} - k_x'\varrho_2\right)$$

$$= 2\sqrt{t}\left[+1 - \frac{1}{4}\left(\frac{1}{t}\right)\left(\frac{a_1}{\lambda}\sin\theta\right) - \frac{1}{8}\left(\frac{1}{t}\right)\right] \tag{19f}$$

$$t_1'' = \sqrt{\frac{1}{\pi k \varrho_2}}\left(-\frac{ka_1}{2} - k_x''\varrho_2\right)$$

$$= 2\sqrt{t}\left[-1 - \frac{1}{4}\left(\frac{1}{t}\right)\left(\frac{a_1}{\lambda}\sin\theta\right) + \frac{1}{8}\left(\frac{1}{t}\right)\right] \tag{19g}$$

$$t_2'' = \sqrt{\frac{1}{\pi k \varrho_2}}\left(+\frac{ka_1}{2} - k_x''\varrho_2\right)$$

$$= 2\sqrt{t}\left[+1 - \frac{1}{4}\left(\frac{1}{t}\right)\left(\frac{a_1}{\lambda}\sin\theta\right) + \frac{1}{8}\left(\frac{1}{t}\right)\right] \tag{19h}$$

$$t = \frac{a_1^2}{8\lambda\varrho_2} \tag{19i}$$

For a given value of t, as given by (19i), the normalized field of (19a) is plotted in Fig. 16 as a function of $(a_1/\lambda)\sin\theta$ for $t = 1/64, 1/8, 1/4, 1/2, 3/4$, and 1. Following a procedure identical with that for the E-plane sectoral horn, the H-plane pattern of any H-plane sectoral horn can be obtained from these curves. The normalized

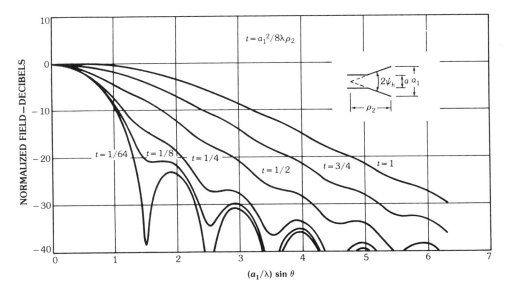

Fig. 16. *H*-plane universal patterns for *H*-plane sectoral and pyramidal horns. (*After Balanis [6], © 1982; reprinted by permission of Harper & Row Publishers, Inc.*)

value of the $(1 + \cos\theta)$ factor in decibels, written as $20\log_{10}[(1 + \cos\theta)/2]$, must also be included.

Example 4—An *H*-plane horn has dimensions of $a = 0.7849\lambda$, $b = 0.3488\lambda$, $a_1 = 6.6710\lambda$, and $\varrho_2 = 12.3841\lambda$. Find the *H*-plane normalized field intensity (in decibels *and* as a voltage ratio) at an angle of $\theta = 30°$ using the universal curves of Fig. 16.

Solution—Using (19i)

$$t = \frac{a_1^2}{8\lambda\varrho_2} = \frac{(6.6710)^2}{8(12.3841)} = 0.449 \cong 0.45$$

None of the curves in Fig. 16 represents $t = 0.45$. However, the curve of $t = 1/2$ is one which should yield very approximate results. More accurate results could be obtained by interpolating between curves.

At $\theta = 30°$

$$\frac{a_1}{\lambda}\sin\theta = 6.6710 \sin 30° = 6.6710(0.5) = 3.3355 \cong 3.34$$

At that point, using the $t = 1/2$ curve of Fig. 16, the field intensity is about -25 dB. Therefore the total field intensity at $\theta = 30°$ is equal to

$$E_\theta = -25 + 20\log_{10}\left(\frac{1 + \cos 30°}{2}\right) = 25 + 0.6 = 25.6 \text{ dB}$$

or as a normalized voltage ratio of

$$E_\theta = 0.0525$$

which closely agrees with the results of Fig. 13.

Directivity

Using the field expressions of (16a)–(16m) we can write the directivity as

$$D_H = \frac{4\pi U_{\max}}{P_r} = \frac{4\pi b \varrho_2}{a_1 \lambda}\{[C(u) - C(v)]^2 + [S(u) - S(v)]^2\} \qquad (20a)$$

where

$$u = \frac{1}{\sqrt{2}}\left(\frac{\sqrt{\lambda \varrho_2}}{a_1} + \frac{a_1}{\sqrt{\lambda \varrho_2}}\right) \qquad (20b)$$

$$v = \frac{1}{\sqrt{2}}\left(\frac{\sqrt{\lambda \varrho_2}}{a_1} - \frac{a_1}{\sqrt{\lambda \varrho_2}}\right) \qquad (20c)$$

The half-power beamwidth as a function of flare angle is plotted in Fig. 17. The normalized directivity (relative to the constant aperture dimension b) for different horn lengths, as a function of the flare angle ψ_h, is displayed in Fig. 18. As for the E-plane sectoral horn the half-power beamwidth exhibits a monotonic decrease and the directivity a monotonic increase up to a given flare; beyond that the trends are reversed.

If the values of a_1 (in wavelengths), which correspond to the maximum directivities in Fig. 18, are plotted versus their corresponding values of ϱ_2 (in wavelengths), it can be shown that each optimum directivity occurs when

$$a_1 \cong \sqrt{3\lambda \varrho_2} \qquad (21a)$$

with a corresponding value of t equal to

$$t\bigg|_{a_1=\sqrt{3\lambda\varrho_2}} = t_{op} = \frac{a_1^2}{8\lambda \varrho_2}\bigg|_{a_1=\sqrt{3\lambda\varrho_2}} = \frac{3}{8} \qquad (21b)$$

The directivity of an H-plane sectoral horn can also be computed by using the following procedure [12].

(a) Calculate A by

$$A = \frac{a_1}{\lambda}\sqrt{\frac{50}{\varrho_h/\lambda}} \qquad (22a)$$

(b) Using this value of A, find the corresponding value of G_H from Fig. 19. If the value of A is smaller than 2, then compute G_H using

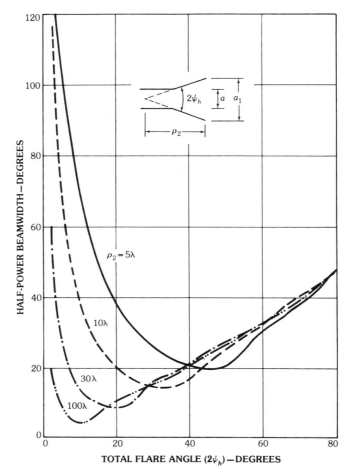

Fig. 17. Half-power beamwidth of *H*-plane sectoral horns as a function of included angle and for different lengths. (*After Balanis [6], © 1982; reprinted by permission of Harper & Row Publishers, Inc.*)

$$G_H = \frac{32}{\pi} A \qquad (22b)$$

(c) Calculate D_H by using the value of G_H from Fig. 19 or from (22b). Thus

$$D_H = \frac{b}{\lambda} \frac{G_H}{\sqrt{50\lambda/\varrho_h}} \qquad (22c)$$

This is the actual directivity of the horn.

Example 5—An *H*-plane sectoral horn has dimensions of $a = 0.9$ in (2.286 cm), $b = 0.4$ in (1.016 cm), $a_1 = 7.65$ in (19.431 cm), and $\varrho_2 = 14.2$ in (36.07 cm).

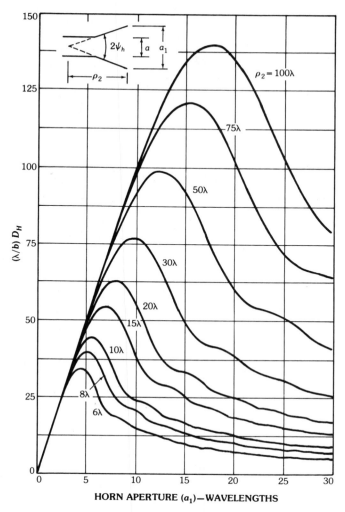

Fig. 18. Normalized directivity of H-plane sectoral horns as a function of aperture size and for different lengths. (*After Balanis [6], © 1982; reprinted by permission of Harper & Row Publishers, Inc.*)

Compute the directivity at $f = 10.3$ GHz using (20a) and (22c). Compare the answers.

Solution—At $f = 10.3$ GHz, $\lambda = 30/10.3 = 2.9126$ cm $= 1.1467$ in.

$$a = 2.286\lambda/2.9126 = 0.7849\lambda$$
$$b = 1.016\lambda/2.9126 = 0.3488\lambda$$
$$a_1 = 19.431\lambda/2.9126 = 6.6710\lambda$$
$$\varrho_2 = 36.07\lambda/2.9126 = 12.3841\lambda$$
$$\varrho_h = \lambda\sqrt{(12.384)^2 + (3.3355)^2} = 12.8254\lambda$$

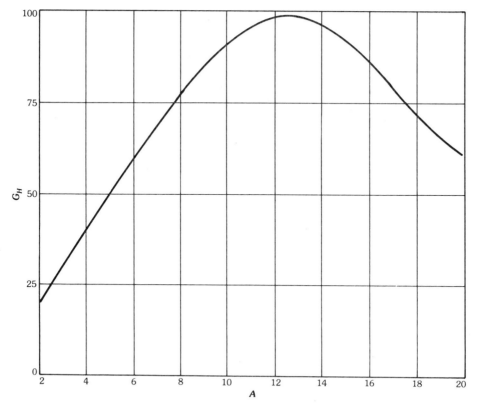

Fig. 19. Term G_H as a function of A. (*Adapted from data by E. H. Braun [12], © 1956 IEEE*)

$$u = \frac{1}{\sqrt{2}}\left(\frac{\sqrt{12.3841}}{6.6710} + \frac{6.6710}{\sqrt{12.3841}}\right) = 1.7134$$

$$v = \frac{1}{\sqrt{2}}\left(\frac{\sqrt{12.3841}}{6.6710} - \frac{6.6710}{\sqrt{12.3841}}\right) = -0.9674$$

$$C(u) = C(1.7134) = 0.3252$$
$$C(v) = C(-0.9674) = -C(0.9674) = -0.7750$$
$$S(u) = S(1.7134) = 0.5359$$
$$S(v) = S(-0.9674) = -S(0.9674) = -0.4063$$

Using (20a)

$$D_H = \frac{4\pi(0.3488)(12.3841)}{6.6710}[(0.3252 + 0.7750)^2 + (0.5359 + 0.4063)^2]$$

$$= 17.073 = 12.323 \text{ dB}$$

$$A = 6.6710\sqrt{50/12.8254} = 13.172$$

Horn Antennas

so that

$$G_H = 98.673$$

From Fig. 19, $G_H = 98.673$. Using (22c)

$$D_H = 0.3488(98.673)/\sqrt{50/12.8254} = 17.431 = 12.413 \text{ dB}$$

A comparison of the values using the two different formulas indicates a very good agreement. For the *H*-plane sectoral horn, there is no need for a formula comparable to (12) because the propagation constant of the dominant mode in the horn is the same as that of free space and because the Kirchhoff methods yield the exact results for the on-axis gain of the TEM mode in an open-ended parallel-plate waveguide. Also the edges of the *H*-plane walls are farther apart than those of the *E*-plane; therefore the interaction between these walls will be weaker.

Design Procedure

In designing an optimum directivity *H*-plane horn the usual specifications are similar to those of an *E*-plane and are the following:

Given: (*a*) Desired optimum directivity (or gain)
         (*b*) Center frequency *f* of operation
         (*c*) Dimensions *a*, *b* of the feed waveguide

Desired: Dimensions ϱ_2, a_1, ϱ_h, and angle ψ_h

Procedure: The design procedure is very similar to that for the *E*-plane, and it is as follows:

(*a*) Substitute (21a) into (20a)–(20c).
(*b*) Knowing D_H (as a dimensionless quantity), *b*, and λ, determine ϱ_2 using (20a). The term ϱ_2 will have the same dimensions as *b* and λ.
(*c*) Find a_1 using (21a). The term a_1 will have the same dimensions as ϱ_2 and λ.
(*d*) Determine ϱ_h and ψ_h using the geometry of Fig. 11.

Example 6—Design an optimum gain *H*-plane sectoral horn such that its directivity at $f = 10$ GHz is 11.884 dB. The dimensions of the feed rectangular waveguide are $a = 0.9$ in (2.286 cm) and $b = 0.4$ in (1.016 cm).

Solution—At $f = 10$ GHz

$$\lambda = \frac{30 \times 10^9}{10 \times 10^9} = 3 \text{ cm}$$

$$a = \frac{2.286}{3}\lambda = 0.762\lambda$$

$$b = \frac{1.016}{3}\lambda = 0.3387\lambda$$

Substituting (21a) into (20b) and (20c), we have

$$u = \frac{1}{\sqrt{2}}\left(\frac{1}{\sqrt{3}} + \sqrt{3}\right) = 1.633$$

$$v = \frac{1}{\sqrt{2}}\left(\frac{1}{\sqrt{3}} - \sqrt{3}\right) = -0.8165$$

Since $D_E = 11.884$ dB $= 15.4312$

$$C(1.633) = 0.35172$$
$$S(1.633) = 0.63889$$
$$C(-0.8165) = -C(0.8165) = -0.72977$$
$$S(-0.8165) = -S(0.8165) = -0.26426$$

(20a) can be written as

$$15.4312 = \frac{4\pi(0.3387)\varrho_2}{\sqrt{3\lambda\varrho_2}}[(0.35172 + 0.72977)^2 + (0.63899 + 0.26426)^2]$$

$$15.4312 = 4.879\sqrt{\frac{\varrho_2}{\lambda}}$$

so that

$$\varrho_2 = 10\lambda = 30 \text{ cm}$$
$$a_1 = \sqrt{3\lambda\varrho_2} = \sqrt{3\lambda(10\lambda)} = 5.477\lambda = 16.432 \text{ cm}$$
$$\psi_h = \tan^{-1}\left(\frac{8.216}{30}\right) = 15.3°$$

The results of this design agree with the data of the $\varrho_2 = 10\lambda$ curve of Fig. 18.

4. The Pyramidal Horn

The most widely used horn is the one which is flared in both directions, as shown in Fig. 1c. It is widely referred to as a *pyramidal horn*, and its radiation characteristics are essentially a combination of the *E*- and *H*-plane sectoral horns. A more detailed geometry of it is shown in Fig. 20.

Aperture and Radiated Fields

To simplify the analysis and to maintain a modeling which leads to computations which have been shown to correlate well with experimental data, the tangential components of the *E*- and *H*-fields over the aperture of the horn are approximated by

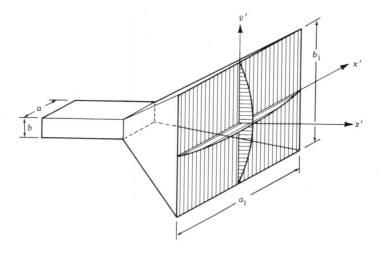

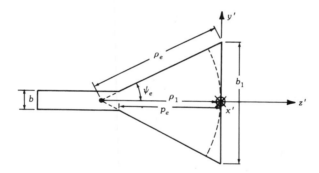

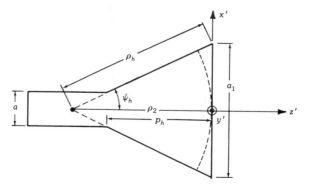

Fig. 20. Pyramidal horn and coordinate system. (*a*) Pyramidal horn. (*b*) *E*-plane view. (*c*) *H*-plane view. (*After Balanis [6], © 1982; reprinted with permission of Harper & Row Publishers, Inc.*)

$$E'_y(x', y') = E_0 \cos\left(\frac{\pi}{a_1}x'\right) e^{-j(k/2)[(x')^2/\varrho_2 + (y')^2/\varrho_1]} \qquad (23a)$$

and

$$H'_x(x', y') = -\frac{E_0}{\eta} \cos\left(\frac{\pi}{a_1}x'\right) e^{-j(k/2)[(x')^2/\varrho_2 + (y')^2/\varrho_1]} \qquad (23b)$$

and the equivalent current densities by

$$\mathbf{J}_s = \hat{\mathbf{n}} \times \mathbf{H}_a = \hat{\mathbf{a}}_y J_y(x', y') = -\hat{\mathbf{a}}_y \frac{E_0}{\eta} \cos\left(\frac{\pi}{a_1}x'\right) e^{-j(k/2)[(x')^2/\varrho_2 + (y')^2/\varrho_1]} \quad (24a)$$

and

$$\mathbf{M}_s = -\hat{\mathbf{n}} \times \mathbf{E}_a = \hat{\mathbf{a}}_x M_x(x', y') = \hat{\mathbf{a}}_x E_0 \cos\left(\frac{\pi}{a_1}x'\right) e^{-j(k/2)[(x')^2/\varrho_2 + (y')^2/\varrho_1]} \quad (24b)$$

The above expressions contain a cosinusoidal amplitude distribution in the x' direction and quadratic phase variations in both the x' and y' directions, similar to those of the sectoral E- and H-plane horns.

Using (24a) and (24b) and the expressions of Section 11.3 of Reference 6, it can be shown that the far-zone electric-field components can be written as follows:

$$E_r = 0 \qquad (25a)$$

$$E_\theta = -j\frac{ke^{-jkr}}{4\pi r}[L_\phi + \eta N_\theta] = j\frac{kE_0 e^{-jkr}}{4\pi r}[\sin\phi(1 + \cos\theta) I_1 I_2] \qquad (25b)$$

$$E_\phi = +j\frac{ke^{-jkr}}{4\pi r}[L_\theta + \eta N_\phi] = j\frac{kE_0 e^{-jkr}}{4\pi r}[\cos\phi(\cos\theta + 1) I_1 I_2] \qquad (25c)$$

where

$$I_1 = \frac{1}{2}\sqrt{\frac{\pi\varrho_2}{k}} \left[e^{j[(k'_x)^2 \varrho_2/(2k)]} \{[C(t'_2) - C(t'_1)] - j[S(t'_2) - S(t'_1)]\} \right. \\
\left. + e^{j[(k''_x)^2 \varrho_2/(2k)]} \{[C(t''_2) - C(t''_1)] - j[S(t''_2) - S(t''_1)]\} \right] \qquad (25d)$$

$$I_2 = \sqrt{\frac{\pi\varrho_1}{k}} e^{j[(k_y)^2 \varrho_1/(2k)]} \{[C(t_2) - C(t_1)] - j[S(t_2) - S(t_1)]\} \qquad (25e)$$

Expressions for t'_1, t'_2, k'_x, t''_1, t''_2, and k''_x are given, respectively, by (16g)–(16i) and (16k)–(16m). Similarly t_1, t_2, and k_y can be evaluated, respectively, using (4g), (4h), and (4j).

The fields radiated by a pyramidal horn, as given by (25a)–(25c), are valid for

Horn Antennas

all angles of observation. An examination of these equations reveals that the principal E-plane pattern ($\phi = \pi/2$) of a pyramidal horn, aside from a normalization factor, is identical with the E-plane pattern of an E-plane sectoral horn. Similarly the H-plane ($\phi = 0$) is identical with the H-plane of an H-plane sectoral horn. Therefore the pattern of a pyramidal horn is very narrow in both principal planes and, in fact, in all planes. This is illustrated in Fig. 21a. The corresponding E-plane pattern is shown in Fig. 4 and the H-plane pattern in Fig. 13.

To demonstrate that the maximum radiation for a pyramidal horn is not necessarily directed along its axis, the three-dimensional field pattern for a horn with $\varrho_1 = \varrho_2 = 6\lambda$, $a_1 = 12\lambda$, $b_1 = 6\lambda$, $a = 0.50\lambda$, and $b = 0.25\lambda$ is displayed in Fig. 21b. The corresponding two-dimensional E- and H-plane patterns are shown in Fig. 22. The maximum does not occur on axis because the phase error taper at the aperture is such that the rays emanating from the different parts of the aperture toward the axis are not in phase.

To physically construct a pyramidal horn the dimension p_e of Fig. 20b given by

$$p_e = (b_1 - b)\left[\left(\frac{\varrho_e}{b_1}\right)^2 - \frac{1}{4}\right]^{1/2} \tag{26a}$$

should be equal to the dimension p_h of Fig. 20c given by

$$p_h = (a_1 - a)\left[\left(\frac{\varrho_h}{a_1}\right)^2 - \frac{1}{4}\right]^{1/2} \tag{26b}$$

The dimensions chosen for Figs. 21a and 21b do satisfy these requirements. For the horn of Fig. 21a, $\varrho_e = 12.0279\lambda$, $\varrho_h = 12.8247\lambda$, and $p_e \cong p_h \cong 10.9\lambda$, while for that of Fig. 21b, $\varrho_e = 6.7082\lambda$, $\varrho_h = 8.4853\lambda$, and $p_e = p_h = 5.75\lambda$.

The fields of (25a)–(25c) provide accurate patterns for angular regions near the main lobe and its closest minor lobes. To accurately predict the field intensity of the pyramidal and other horns, especially in the minor lobes, diffraction techniques can be utilized [2–5]. These methods take into account diffractions that occur near the aperture edges of the horn. The diffraction contributions become more dominant in regions where the radiation of (25a)–(25c) is of very low intensity.

Directivity

As for the E- and H-plane sectoral horns the directivity of the pyramidal configuration is vital to the antenna designer and practicing engineer. The maximum radiation of the pyramidal horn is directed nearly along the z axis ($\theta = 0$). See Fig. 22. Using (25a)–(25c) it can be shown that the directivity for the pyramidal horn defined by (8) can be written as

$$D_p = \frac{4\pi U_{max}}{P_r} = \frac{8\pi\varrho_1\varrho_2}{a_1b_1}\{[C(u) - C(v)]^2 + [S(u) - S(v)]^2\}$$

$$\times \left\{C^2\left(\frac{b_1}{\sqrt{2\lambda\varrho_1}}\right) + S^2\left(\frac{b_1}{\sqrt{2\lambda\varrho_1}}\right)\right\} \tag{27}$$

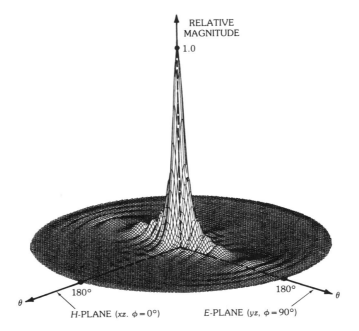

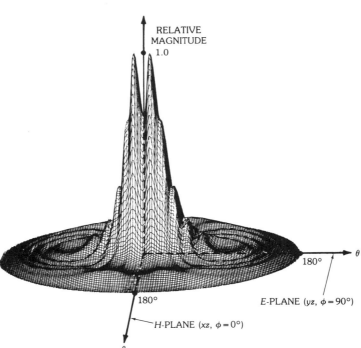

Fig. 21. Three-dimensional patterns of pyramidal horns. (*a*) With maximum on axis: $\varrho_1 = 11.773\lambda$, $\varrho_2 = 12.3841\lambda$, $a_1 = 6.6710\lambda$, $b_1 = 4.9269\lambda$, $a = 0.7849\lambda$, $b = 0.3488\lambda$. (*b*) With maximum not on axis: $\varrho_1 = \varrho_2 = 6\lambda$, $a_1 = 12\lambda$, $b_1 = 6\lambda$, $a = 0.5\lambda$, $b = 0.25\lambda$.

Horn Antennas

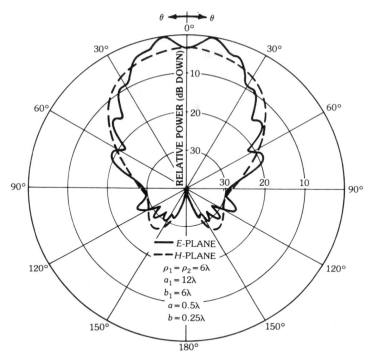

Fig. 22. E- and H-plane amplitude patterns of a pyramidal horn with maximum not on axis. (*After Balanis [6]*, © *1982; reprinted with permission of Harper & Row Publishers, Inc.*)

where u and v are defined, respectively, by (20b) and (20c). Another form of (27) is to write it as

$$D_p = \frac{\pi\lambda^2}{32ab} D_E D_H \tag{28}$$

where D_E and D_H are the directivities of the E- and H-plane sectoral horns as given, respectively, by (9) or (11c) or (12a) and (20a) or (22c). This is a well-known relationship and has been used extensively in the design of pyramidal horns.

The directivity (in decibels) of a pyramidal horn, over isotropic, can also be approximated by [14]

$$D_0(\text{dB}) = 10\left[1.008 + \log_{10}\left(\frac{a_1 b_1}{\lambda^2}\right)\right] - (L_E + L_H) \tag{29}$$

where L_E and L_H represent, respectively, the losses (in decibels) due to phase errors in the E- and H-planes of the horn which are found plotted in Fig. 23.

The directivity of a pyramidal horn can also be calculated by doing the following [12].

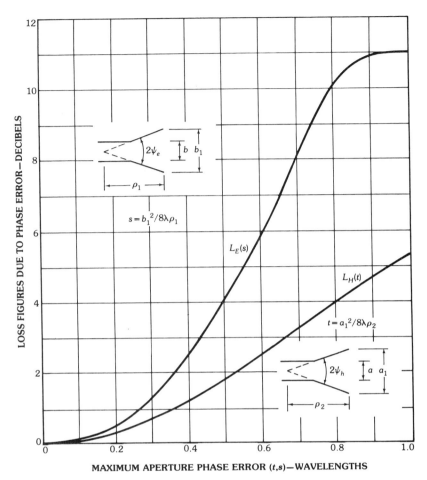

Fig. 23. Loss figures for E- and H-planes of pyramidal horn due to phase errors. (*After W. C. Jakes [14], © 1961; reprinted with permission of McGraw-Hill Book Company*)

(*a*) Calculate

$$A = \frac{a_1}{\lambda}\sqrt{\frac{50}{\varrho_h/\lambda}} \tag{30a}$$

$$B = \frac{b_1}{\lambda}\sqrt{\frac{50}{\varrho_e/\lambda}} \tag{30b}$$

(*b*) Using A and B, find G_H and G_E, respectively, from Figs. 19 and 10. If the values of either A or B or both are smaller than 2, then calculate G_E and/or G_H by

$$G_E = \frac{32}{\pi}B \tag{30c}$$

Horn Antennas

$$G_H = \frac{32}{\pi} A \qquad (30d)$$

(c) Calculate D_p by using the values of G_E and G_H from Figs. 10 and 19 or from (30c) and (30d). Thus

$$D_p = \frac{G_E G_H}{\frac{32}{\pi}\sqrt{50/(\varrho_e/\lambda)}\sqrt{50/\varrho_h/\lambda}} = \frac{G_E G_H}{10.1859\sqrt{50/(\varrho_e/\lambda)}\sqrt{50/\varrho_h/\lambda}} = \frac{\lambda^2 \pi}{32ab} D_E D_H \qquad (30e)$$

where D_E and D_H are, respectively, the directivities of (9) or (11c) or (12a) and (20a) or (22c). This is the actual directivity of the horn. The above procedure has led to results accurate to within 0.01 dB for a horn with $\varrho_e = \varrho_h = 50\lambda$.

A typical X-band (8.2–12.4 GHz) horn is that shown in Fig. 24. It is a precision standard gain horn, which can be used as a

(a) standard for calibrating other antennas,
(b) feed for reflectors and lenses,
(c) pickup horn for sampling power, and
(d) receiving and/or transmitting antenna.

Example 7—A pyramidal horn has dimensions of $\varrho_1 = 11.7730\lambda$, $\varrho_2 = 12.3841\lambda$, $a_1 = 6.6710\lambda$, $b_1 = 4.9269\lambda$, $a = 0.7849\lambda$, and $b = 0.3488\lambda$.

(a) Check to see if such a horn can be constructed physically.
(b) Compute the directivity using (28), (29), and (30e).

Solution—From Examples 2 and 4

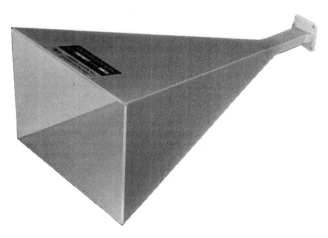

Fig. 24. Commercial X-band (8.2- to 12.4-GHz) pyramidal horn. (*Courtesy Scientific-Atlanta, Atlanta, Georgia*)

$$\varrho_e = 12.0279\lambda$$
$$\varrho_h = 12.8247\lambda$$

Thus

$$p_e = (4.9269 - 0.3488)\lambda \sqrt{\left(\frac{12.0279}{4.9269}\right)^2 - \frac{1}{4}} = 10.939\lambda$$

$$p_h = (6.6710 - 0.7849)\lambda \sqrt{\left(\frac{12.8247}{6.6710}\right)^2 - \frac{1}{4}} = 10.926\lambda$$

Since $p_e \cong p_h \cong 10.9\lambda$, the horn can be constructed physically. The directivity can be computed by utilizing the results of Examples 2 and 4. Using (28) with the values of D_E and D_H computed using (9) and (20a), respectively, gives

$$D_p = \frac{\pi\lambda^2}{32ab} D_E D_H = \frac{\pi}{32(0.7849)(0.3488)} (34.217)(17.073)$$
$$= 209.49 = 23.21 \text{ dB}$$

Utilizing the values of D_E and D_H computed using (11c) and (22c), respectively, the directivity of (30e) is equal to

$$D_p = \frac{\pi\lambda^2}{32ab} D_E D_H = \frac{\pi}{32(0.7849)(0.3488)} (31.394)(17.431)$$
$$= 196.24 = 22.93 \text{ dB}$$

For this horn

$$s = \frac{b_1^2}{8\lambda\varrho_1} = \frac{(4.9269)^2}{8(11.7730)} = 0.2577$$

$$t = \frac{a_1^2}{8\lambda\varrho_2} = \frac{(6.6710)^2}{8(12.3841)} = 0.4492$$

For these values of s and t

$$L_E = 0.625 \text{ dB}$$
$$L_H = 1.50 \text{ dB}$$

from Fig. 23. Using (29)

$$D_0(\text{dB}) = 10\{1.008 + \log_{10}[6.6710(4.9269)]\} - (0.625 + 1.50) = 23.12$$

The agreement is best between the directivities of (28) and (29).

Horn Antennas

5. The Design Procedure for the Pyramidal Horn

The pyramidal horn is widely used as a standard to make gain measurements of other antennas [6], and as such it is often referred to as a *standard-gain horn*. To design a pyramidal horn one usually knows the desired gain G_0 and the dimensions a, b of the rectangular-feed waveguide. The objective of the design is to determine the remaining dimensions (a_1, b_1, ϱ_e, ϱ_h, p_e, and p_h) that will lead to an optimum gain. The procedure that follows can be used to accomplish this [6].

The design equations are derived by first selecting values of b_1 and a_1 that lead, respectively, to optimum directivities for the E- and H-plane sectoral horns using (10a) and (21a). Since the overall efficiency (including both the antenna and aperture efficiencies) of a horn antenna is about 50 percent, the gain of the antenna can be related to its physical area. Thus, it can be written using (10a) and (21a) as [6]

$$G_0 = \frac{4\pi}{\lambda^2} A_{em} = \varepsilon_{ap} \frac{4\pi}{\lambda^2} A_p \cong \frac{1}{2} \frac{4\pi}{\lambda^2} (a_1 b_1)$$

$$= \frac{2\pi}{\lambda^2} \sqrt{3\lambda\varrho_1} \sqrt{2\lambda\varrho_2} = \frac{2\pi}{\lambda^2} \sqrt{3\lambda p_h} \sqrt{2\lambda p_e} \quad (31)$$

since for long horns $\varrho_2 \cong p_h$ and $\varrho_1 \cong p_e$. In (31) ε_{ap} is the aperture efficiency and A_p is the physical area of the horn aperture. In order for a pyramidal horn to be physically realizable, p_e and p_h of (26a) and (26b) must be equal. Using this equality it can be shown that (31) reduces to

$$(\sqrt{2\chi} - b/\lambda)^2 (2\chi - 1) = \left(\frac{G_0}{2\pi}\sqrt{\frac{3}{2\pi}}\frac{1}{\sqrt{\chi}} - \frac{a}{\lambda}\right)^2 \left(\frac{G_0^2}{6\pi^3}\frac{1}{\chi} - 1\right) \quad (32a)$$

where

$$\frac{\varrho_e}{\lambda} = \chi \quad (32b)$$

$$\frac{\varrho_h}{\lambda} = \frac{G_0^2}{8\pi^3}\left(\frac{1}{\chi}\right) \quad (32c)$$

Equation 32a is the horn design equation.

1. As a first step of the design find the value of χ which satisfies (32a) for a desired gain G_0 (dimensionless). Use an iterative technique and begin with a trial value of

$$\chi(\text{trial}) = \chi_1 = \frac{G_0}{2\pi\sqrt{2\pi}} \quad (33)$$

2. Once the correct χ has been found, determine ϱ_e and ϱ_h using (32b) and (32c), respectively.

3. Find the corresponding values of a_1 and b_1 using (10a) and (21a) or

$$a_1 = \sqrt{3\lambda\varrho_2} \cong \sqrt{3\lambda\varrho_h} = \frac{G_0}{2\pi}\sqrt{\frac{3}{2\pi\chi}}\,\lambda \qquad (34a)$$

$$b_1 = \sqrt{2\lambda\varrho_1} \cong \sqrt{2\lambda\varrho_e} = \sqrt{2\chi}\,\lambda \qquad (34b)$$

4. The values of p_e and p_h can be found using (26a) and (26b).

Example 8—Design an optimum gain X-band (8.2–12.4 GHz) pyramidal horn so that its gain (above isotropic) at $f = 11$ GHz is 22.6 dB. The horn is fed by a WR 90 rectangular waveguide with inner dimensions of $a = 0.9$ in (2.286 cm) and $b = 0.4$ in (1.016 cm).

Solution—Convert the gain G_0 from decibels to a dimensionless quantity. Thus

$$G_0(\text{dB}) = 22.6 = 10\,\log_{10}G_0$$

implies that

$$G_0(\text{dimensionless}) = 10^{2.26} = 181.97$$

Since $f = 11$ GHz,

$\lambda = 2.7273$ cm
$a = 0.8382\lambda$
$b = 0.3725\lambda$

1. The initial value of χ is taken, using (33), as

$$\chi_1 = \frac{181.97}{2\pi\,\sqrt{2\pi}} = 11.5539$$

which does not satisfy (32a) for the desired design specifications. After a few iterations a more accurate value is $\chi = 11.1157$.

2. Using (32b) and (32c)

$$\varrho_e = 11.1157\lambda = 30.316 \text{ cm} = 11.935 \text{ in}$$
$$\varrho_h = 12.0094\lambda = 32.753 \text{ cm} = 12.895 \text{ in}$$

3. The corresponding values of a_1 and b_1 are

$$a_1 = 6.002\lambda = 16.370 \text{ cm} = 6.445 \text{ in}$$
$$b_1 = 4.715\lambda = 12.859 \text{ cm} = 5.063 \text{ in}$$

4. The values of p_e and p_h are equal to

$$p_e = p_h = 10.005\lambda = 27.286 \text{ cm} = 10.743 \text{ in}$$

The derived design parameters agree closely with those of a commercial gain horn available in the market.

As a check, the gain of the designed horn was computed using (28) and (29), assuming an antenna efficiency e_t of 100 percent, and (31). The values were

$$G_0 \cong D_0 = 22.4 \text{ dB} \quad \text{for (28)}$$
$$G_0 \cong D_0 = 22.1 \text{ dB} \quad \text{for (29)}$$
$$G_0 = 22.5 \text{ dB} \quad \text{for (31)}$$

All three computed values agree closely with the designed value of 22.6 dB.

Design curves of pyramidal horns with a TE_{10}-mode field distribution, synthesized to maximize the aperture efficiency or to produce maximum power transmission when used as feed elements for reflectors, are shown in Fig. 25 [15]. These curves were derived by expanding the focal plane and feed-horn aperture field distributions into finite-term power series whose coefficients were determined using collocation techniques.

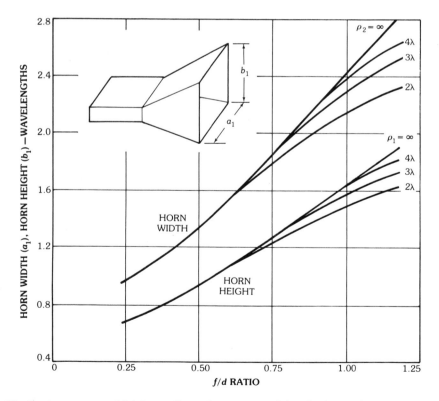

Fig. 25. Optimum pyramidal horn dimensions versus f/d ratio for various horn lengths. (*After Truman and Balanis [15]*, © *1974 IEEE*)

6. The Conical Horn

Another very practical microwave antenna is the conical horn shown in Fig. 26. While the pyramidal, E-, and H-plane sectoral horns are usually fed by a rectangular waveguide, the feed of a conical horn is often a circular waveguide.

The first rigorous treatment of the fields radiated by a conical horn is that of Schorr and Beck [16]. The modes within the horn are found by introducing a spherical coordinate system and are in terms of spherical Bessel functions and Legendre polynomials. The analysis is too involved and will not be attempted here. Data, however, in the form of curves [17] will be presented that give a qualitative description of the performance of a conical horn.

Referring to Fig. 27 it is apparent that the behavior of a conical horn is similar to that of a pyramidal or a sectoral horn. As the flare angle increases, the directivity for a given length horn increases until it reaches a maximum, beyond which it begins to decrease. The decrease is a result of the dominance of the quadratic phase error at the aperture. In the same figure an optimum directivity line is indicated.

The results of Fig. 27 behave as those of Figs. 9 and 18. When the horn aperture (d_m) is held constant and its length (L) is allowed to vary, the maximum directivity is obtained when the flare angle is zero ($\psi_c = 0$ or $L = \infty$). This is equivalent to a circular waveguide of diameter d_m. As for the pyramidal and sectoral horns a lens is usually placed at the aperture of the conical horn to compensate for its quadratic phase error. The result is a narrower pattern as the flare increases.

The directivity (in decibels) of a conical horn, with an aperture efficiency ε_{ap} and aperture circumference C, can be computed using

$$D_c(\text{dB}) = 10\log_{10}[\varepsilon_{ap} \frac{4\pi}{\lambda^2}(\pi a^2)] = 10\log_{10}\left(\frac{C}{\lambda}\right)^2 - L(s) \tag{35a}$$

where a is the radius of the horn at the aperture and

$$L(s) = -10\log_{10}\varepsilon_{ap} \tag{35b}$$

The first term in (35a) represents the directivity of a uniform circular aperture while the second term, represented by (35b), is a correction figure to account for the loss

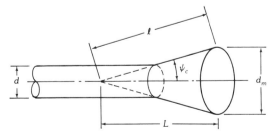

Fig. 26. Geometry of conical horn. (*After Balanis [6]*, © *1982; reprinted with permission of Harper & Row Publishers, Inc.*)

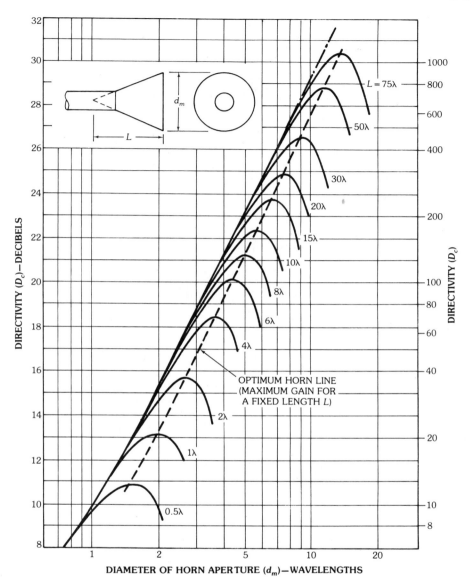

Fig. 27. Directivity of conical horns as a function of aperture diameter and for different horn lengths. (*After King [17]*, © *1950 IEEE*)

in directivity due to the aperture efficiency. Usually the term in (35b) is referred to as the *loss figure*, which can be computed in decibels using [14]

$$L(s) \cong (0.8 - 1.71s + 26.25s^2 - 17.79s^3) \tag{35c}$$

where s is the maximum phase deviation (in number of wavelengths), and it is given by

$$s = \frac{d_m^2}{8\lambda\ell} \tag{36}$$

The gain of a conical horn is optimum when its diameter is equal to

$$d_m \cong \sqrt{3\ell\lambda} \tag{37}$$

which corresponds to a maximum aperture phase deviation of $s = 3/8$ (wavelengths) and a loss figure of about 2.9 dB (or an aperture efficiency of about 51 percent).

The Design Procedure

In designing an optimum directivity (or gain) conical horn, the usual specifications are the following:

Given: (a)   Desired optimum directivity (or gain)
           (b)   Center frequency of operation f
Desired: Dimensions d_m, L, ℓ, d, and angle ψ_c
Procedure: The design procedure is the following:
   (a)   Substitute (37) into (36), which leads to $s = 3/8$ (wavelengths).
   (b)   Find $L(s = 3/8)$ using (35c). For this value of s, the loss figure L is always $L(s = 3/8) = 2.91$ dB.
   (c)   Using the desired directivity value (in decibels) and the value of $L(s) = 2.91$ dB (as obtained in step b), determine the circumference C of the horn aperture utilizing (35a). This will also allow you to determine the radius a and diameter d_m of the horn aperture.
   (d)   For the value of d_m from step c, determine ℓ using (37). In turn find L and ψ_c. The diameter d of the feed circular waveguide can be chosen to satisfy the cutoff characteristics of the desired (usually dominant TE_{11}) mode of the feed waveguides.

Example 9—Design an optimum gain conical horn, using (35)–(37), whose directivity (above isotropic) at $f = 11$ GHz is 22.6 dB. Check your design with the data of Fig. 27. Compare the design dimensions with those of the pyramidal horn design of Example 8.

Solution—Using (37) the maximum phase deviation s of (36) (in number of wavelengths) is equal to

$$s = \frac{d_m^2}{8\lambda\ell} = \frac{3\ell\lambda}{8\ell\lambda} = \frac{3}{8}$$

For $s = 3/8$ the loss figure $L(s = 3/8)$ of (35c) is equal to

$$L(s = 3/8) = 2.91 \text{ dB}$$

Now using (35a) we have

$$22.6 = 10 \log_{10}(C/\lambda)^2 - 2.91$$

so that

$$(C/\lambda)^2 = 10^{2.551} = 355.795$$

and

$$\frac{C}{\lambda} = \frac{\pi d_m}{\lambda} = \sqrt{355.795} = 18.8625$$

means

$$d_m = 6\lambda$$

Since $\lambda = 30 \times 10^9/11 \times 10^9 = 2.7273$ cm, then $d_m = 6\lambda = 16.3636$ cm $= 6.44$ in. Using (37) we have that

$$d_m = \sqrt{3\ell\lambda}$$

and hence

$$\ell = d_m^2/3\lambda = 12\lambda = 32.728 \text{ cm} = 12.88 \text{ in}$$

Thus

$$L = \sqrt{\ell^2 - (d_m/2)^2} = \lambda\sqrt{(12)^2 - (3)^2} = 11.619\lambda = 31.688\lambda = 12.476 \text{ in}$$

and

$$\psi_c = \tan^{-1}\left(\frac{d_m/2}{L}\right) = \tan^{-1}\left(\frac{3}{11.619}\right) = 14.48°$$

which implies that

$$2\psi_c = 28.96°$$

Using Fig. 27 and interpolating between curves, the diameter and length for an optimum directivity of 22.6 dB are equal to

$$d_m \cong 5.8\lambda = 15.818 \text{ cm} = 6.228 \text{ in}$$
$$L \cong 11.5\lambda = 31.364 \text{ cm} = 12.348 \text{ in}$$

The length $\ell = 12\lambda$ and diameter $d_m = 6\lambda$ of the conical horn are contrasted with the lengths $\varrho_e = 11.1157\lambda$, $\varrho_h = 12.0094\lambda$ and the aperture dimensions $a_1 = 6.002\lambda$, $b_1 = 4.715\lambda$ of the pyramidal horn of Example 8.

7. Special Horns

The large emphasis placed on horn antenna research in the 1960s was inspired by the need to reduce spillover efficiency and cross-polarization losses and increase aperture efficiencies of large reflectors used in radioastronomy and satellite communications. In the 1970s high-efficiency and rotationally symmetric antennas were needed in microwave radiometry. Using conventional feeds, aperture efficiencies of 50 to 60 percent were obtained. However, efficiencies on the order of 75 to 80 percent can be obtained with improved feed systems utilizing corrugated horns. In addition, during the same period there was a need for circularly polarized radiators for radar systems, radioastronomy, and ionospheric studies. Most known designs, however, provided circularly polarized waves over a limited angular region, and many efforts were attempted to improve the axial ratio [18–26]. Many of these designs were limited in bandwidth and led to relatively long horns which were delicate in construction.

To overcome some of the above deficiencies and improve the overall radiation characteristics (pattern symmetry, low cross polarization, and low side lobes) of a horn some special horn designs were attempted. Two such designs are the *corrugated* horn [18–25] and the *aperture-matched* horn [26]. The basic concept in each of these designs was to reduce the diffractions at the edges of the aperture. For the corrugated horn aperture diffractions were minimized by reducing the magnitude of the incident field at the edge, while for the aperture-matched horn aperture diffractions were diminished by modifying the horn structure at the aperture so that the magnitude of the diffraction coefficient was reduced.

Corrugated Horns

The unequal *E*- and *H*-plane patterns in a conventional horn result from unfavorable boundary conditions at the top and side walls. The conductive side walls force the tangential components of the electric field to vanish, thus creating for the dominant mode a cosine distribution. The top and bottom walls, however, do not affect the tangential magnetic field, and a uniform distribution is formed between them. If the tangential magnetic field were forced to vanish at the top and bottom walls, the asymmetry of the *E*- and *H*-plane patterns could be corrected.

In 1964 Kay [19] realized that short-circuited quarter-wavelength grooves on the walls of a horn antenna would present the same boundary conditions to all polarizations and would taper the field distribution at the aperture in all the planes. The creation of the same boundary conditions on the walls perpendicular to the **E** field, as those of the *H*-plane, minimizes the spurious diffractions at the edges of the aperture by reducing the incident field at the edge of the aperture. For a square aperture this would lead to an almost rotationally symmetric pattern with equal *E*- and *H*-plane beamwidths over about 60 percent bandwidth and with the 10-dB beamwidth being virtually independent of frequency. In addition, the *E*- and *H*-plane phase centers were found to coincide and be located inside the horn near the throat within a distance of one wavelength from the aperture plane. The bandwidths can be improved by dielectric loading.

For sufficiently long corrugated horns superior pattern performance can be

Horn Antennas

expected over a 2:1 bandwidth. Because the corrugated surface forces the energy away from the horn walls in the E-plane, as the boundary conditions accomplish the same thing in the H-plane, it should be expected that such a horn would have almost identical principal plane patterns. Therefore a corrugated horn can be used as an excellent circularly polarized radiator with axial ratios of 1.05 or smaller over almost the entire beamwidth and with patterns whose maximum radiation is along the axis and with practically no minor lobes.

The Corrugated Pyramidal Horn—A corrugated (grooved) pyramidal horn, with corrugations in the E-plane walls, is shown in Fig. 28a with a side view in Fig. 28b. Since diffractions at the edges of the aperture in the H-plane are minimal, corrugations are usually not placed on the walls of that plane. To form a very effective corrugated surface it usually requires ten or more slots (corrugations) per wave-

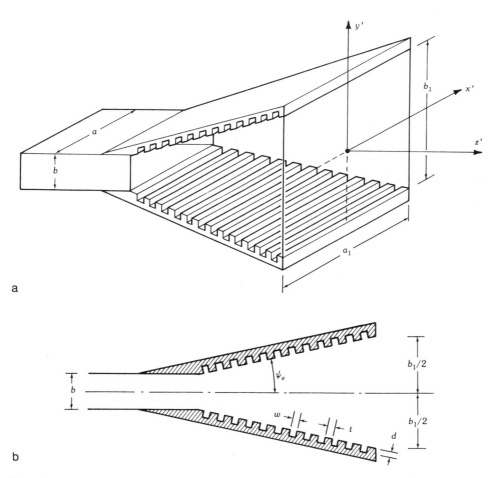

Fig. 28. Pyramidal horn with corrugations in the E-plane. (*a*) Corrugated horn. (*b*) E-plane view. (*After Balanis [6], © 1982; reprinted with permission of Harper & Row Publishers, Inc.*)

length [19]. To simplify the analysis of an infinite corrugated surface, the following assumptions are usually required:

(a) The teeth of the corrugations are vanishingly thin
(b) Reflections from the base of the slot are only those of a TEM mode

The second assumption is satisfied provided the width w of the corrugation is small compared with the free-space wavelength λ_0 and the slot depth d (usually $w < \lambda_0/10$). For a corrugated surface satisfying the above assumptions, its approximate surface reactance is given by [27]

$$X = \frac{w}{w+t} \sqrt{\frac{\mu_0}{\varepsilon_0}} \tan k_0 d \tag{38a}$$

when

$$\frac{w}{w+t} \cong 1 \tag{38b}$$

which can be satisfied provided $t \leqq w/10$.

The surface reactance of a corrugated surface, used on the walls of a horn, must be capacitive in order for the surface to force to zero the tangential magnetic field parallel to the edge at the wall. Thus the surface will not support surface waves, will prevent illumination of the E-plane edges, and will diminish diffractions. This can be accomplished, according to (38a), if $\lambda_0/4 < d < \lambda_0/2$, or more generally when $(2n + 1)\lambda_0/4 < d < (n + 1)\lambda_0/2$, $n = 0, \pm1, \pm2, \ldots$. Even though the cutoff depth is also a function of the slot width w, its influence is negligible if $w < \lambda_0/10$ and $\lambda_0/4 < d < \lambda_0/2$.

To study the performance of a corrugated surface an analytical model was developed and parametric studies were performed [21]. Although the details are numerous, only the results will be presented here. In Fig. 29a, a corrugated surface is sketched and in Fig. 29b its corresponding uncorrugated counterpart is shown.

For a free-space wavelength of $\lambda_0 = 8$ cm the following have been plotted for point B in Fig. 29a relative to point A in Fig. 29b:

(a) In Fig. 30a the surface current decay at B relative to that at $A[J_s(B)/J_s(A)]$ as a function of corrugation number (for 20 total corrugations) due to energy being forced away from the corrugations. As expected, no decay occurs for $d = 0.5\lambda_0$ and the most rapid decay is obtained for $d = 0.25\lambda_0$.
(b) In Fig. 30b the surface current decay at B relative to that in $A[J_s(B)/J_s(A)]$ as a function of the distance z from the onset of the corrugations for four and eight corrugations per wavelength. The results indicate almost an independence of current decay as a function of corrugation density for the cases considered.
(c) In Fig. 30c the surface current decay at B relative to that in $A[J_s(B)/J_s(A)]$ as a function of the distance z from the onset of the corrugations for $w/(w + t)$ ratios ranging from 0.5 to 0.9. For $z < 4$ cm $= \lambda_0/2$, thinner corrugations [larger $w/(w + t)$ ratios] exhibit larger rates of decay. Approximately

Horn Antennas

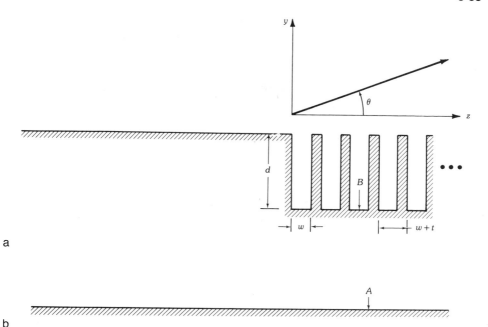

Fig. 29. Geometry of corrugated and plane surfaces. (*a*) Corrugated surface. (*b*) Uncorrugated surface. (*After Mentzer and Peters [21], © 1974 IEEE*)

beyond that point the rate of decay is constant. This would indicate that in a practical design thinner corrugations can be used at the onset followed by thicker ones, which are easier to construct.

The effect of the corrugations on the walls of a horn is to modify the electric field distribution in the *E*-plane from uniform (at the waveguide-horn junction) to cosine (at the aperture). Through measurements it has been shown that the transition from uniform to cosine distribution takes place almost at the onset of the corrugations. For a horn of about 45 corrugations the cosine distribution has been established by the fifth corrugation from the onset and the spherical phase front by the fifteenth [22]. The *E*- and *H*-plane amplitude and phase distributions at the aperture of the horn with 45 corrugations are shown in Figs. 31a and 31b. It is clear that the cosine distribution is well established.

Referring to Fig. 28a, the field distribution at the aperture can be written as

$$E'_y(x', y') = E_0 \cos\left(\frac{\pi}{a_1}x'\right)\cos\left(\frac{\pi}{b_1}y'\right) e^{-j(k/2)[(x')^2/\varrho_2 + (y')^2/\varrho_1]} \quad (39\text{a})$$

$$H'_x(x', y') = -\frac{E_0}{\eta}\cos\left(\frac{\pi}{a_1}x'\right)\cos\left(\frac{\pi}{b_1}y'\right) e^{-j(k/2)[(x')^2/\varrho_2 + (y')^2/\varrho_1]} \quad (39\text{b})$$

corresponding to (23a) and (23b) of the uncorrugated pyramidal horn. Using the above distributions, the fields radiated by the horn can be computed in a manner

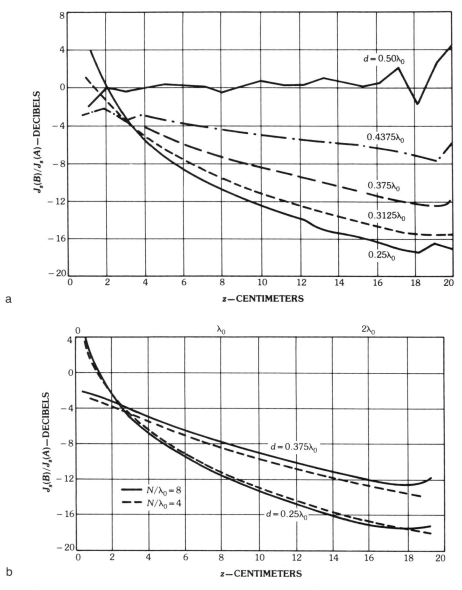

Fig. 30. Surface current decays on corrugated surface. (*a*) Surface current decay due to energy forced away from corrugations. (*b*) Surface current decay as a function of corrugation density. (*c*) Surface current decay on corrugation as a function of corrugation shape. (*After Mentzer and Peters [21], © 1974 IEEE*)

analogous to that of the conventional pyramidal horn of Section 4. Patterns have been computed and compare very well with measurements [22].

In Fig. 32a the measured E-plane patterns of an uncorrugated square pyramidal horn (referred to as the *control horn*) and a corrugated square pyramidal

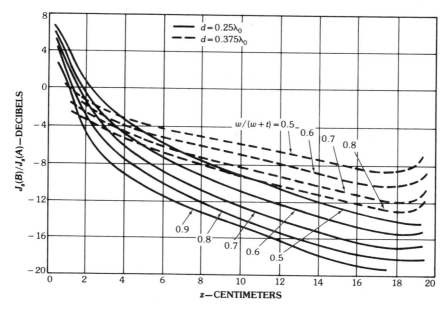

Fig. 30, *continued*

horn are shown. The aperture size on each side was 3.5 in or 8.89 cm ($2.96\lambda_0$ at 10 GHz) and the total flare angle in each plane was 50°. It is evident that the levels of the side lobes and back lobes are much lower for the corrugated horn than those of the control horn. However, the corrugated horn also exhibits wider main beam for small angles, and thus a larger 3-dB beamwidth (half-power beamwidth) but a smaller 10-dB beamwidth. This is attributed to the absence of the diffracted fields from the edges of the corrugated horn which, for nearly on-axis observations, add to the direct wave contribution because of their in-phase relationship. The fact that the on-axis far fields of the direct and diffracted fields are nearly in phase is also evident from the pronounced on-axis maximum of the control horn. The *E*- and *H*-plane patterns of the corrugated horn are almost identical with those of Fig. 32a over the frequency range from 8 to 14 GHz. These suggest that the main beam in the *E*-plane can be obtained from known *H*-plane patterns of horn antennas.

In Fig. 32b the measured *E*-plane patterns of larger control and corrugated square pyramidal horns, having an aperture of 9.7 in (24.64 cm) on each side ($8.2\lambda_0$ at 10 GHz) and included angles of 34° and 31° in the *E*- and *H*-planes, are shown. For this geometry the pattern of the corrugated horn is narrower and its side and back lobes are much lower than those of the corresponding control horn. The saddle formed on the main lobe of the control horn is attributed to the out-of-phase relations between the direct and diffracted rays. The diffracted rays are nearly absent from the corrugated horn and the minimum on-axis field is eliminated. The control horn is a thick-edged horn which has the same interior dimensions as the corrugated horn. The *H*-plane pattern of the corrugated horn is almost identical with the *H*-plane pattern of the corresponding control horn.

In Figs. 32c and 32d the back lobe level and the 3-dB beamwidth for the smaller

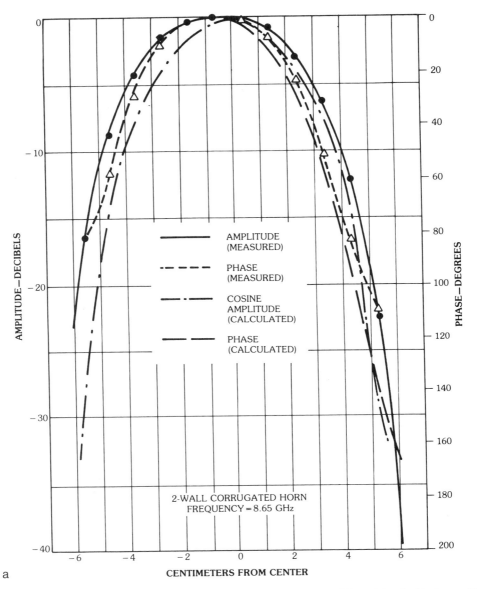

Fig. 31. Amplitude and phase distributions in *H*- and *E*-planes. (*a*) Position in *H*-plane. (*b*) Position in *E*-plane. (*After Mentzer and Peters [22], © 1976 IEEE*)

size control and corrugated horns, whose *E*-plane patterns are shown in Fig. 32a, are plotted as a function of frequency. All the observations made previously for that horn are well evident in these figures.

The presence of the corrugations, especially near the waveguide-horn junction, can affect the impedance and vswr of the antenna. The usual practice is to begin the corrugations at a small distance away from the junction. This leads to low vswr's

Horn Antennas

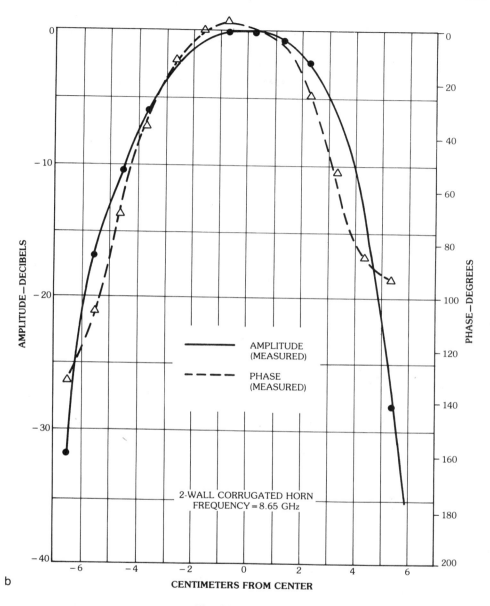

b

Fig. 31, *continued*

over a broadband. Previously it was indicated that the width w of the corrugations must be small (usually $w < \lambda_0/10$) to approximate a corrugated surface. This would cause corona and other breakdown phenomena. However, the large corrugated horn, whose E-plane pattern is shown in Fig. 32b, has been used in a system whose peak power was 20 kW at 10 GHz with no evidence of any breakdown phenomena.

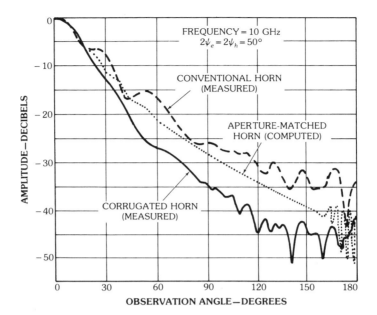

a

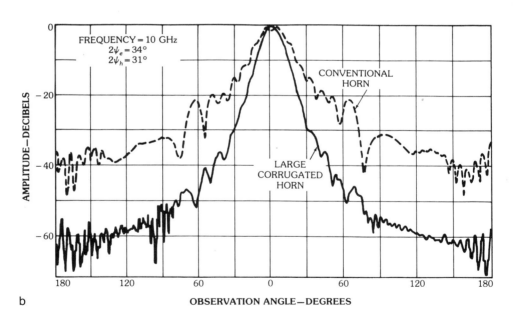

b

Fig. 32. Radiation characteristics of conventional (control), corrugated, and aperture-matched horns. (*a*) *E*-plane patterns of $2.96\lambda_0 \times 2.96\lambda_0$ pyramidal horns. (*After Burnside and Chuang [26], © 1982 IEEE*) (*b*) Measured *E*-plane patterns of $8.2\lambda_0 \times 8.2\lambda_0$ pyramidal horns. (*After Lawrie and Peters [20], © 1974 IEEE*) (*c*) Back lobe to main lobe *E*-plane level of $2.96\lambda_0 \times 2.96\lambda_0$ pyramidal horns. (*After Burnside and Chuang [26], © 1982 IEEE*) (*d*) Half-power beamwidth of $2.96\lambda_0 \times 2.96\lambda_0$ pyramidal horns. (*After Burnside and Chuang [26], © 1982 IEEE*)

Horn Antennas

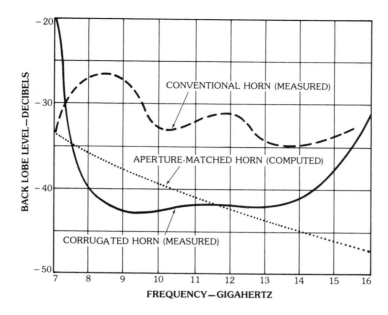

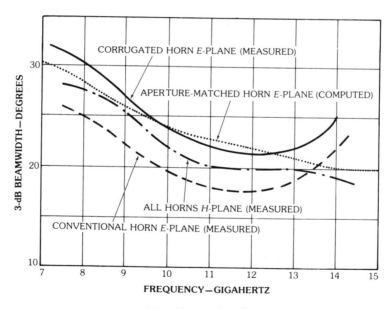

Fig. 32, *continued*

The Corrugated Conical Horn—The design concepts of the pyramidal corrugated horn can be extended to include circumferentially corrugated conical horns. Several designs of conical corrugated horns were investigated [18, 19, 23–25] in terms of pattern symmetry, low cross polarization, low side lobe levels, circular polarization, axial ratio, and phase center. The geometrical configurations of two possible conical corrugated horns are shown in Fig. 33. For small flare angles (ψ_c less than about 20° to 25°) the slots can be machined perpendicular to the axis of the horn, as shown in Fig. 33a, and the grooves can be considered sections of parallel-plate TEM-mode waveguides of depth d. For large flare angles, however, the slots should be constructed perpendicular to the surface of the horn, as shown in Fig. 33b. The groove configuration of Fig. 33a is usually preferred because it is easier to fabricate.

The radiation pattern characteristics of a conical horn can be normalized by the dimensions parameter Δ, which is defined as

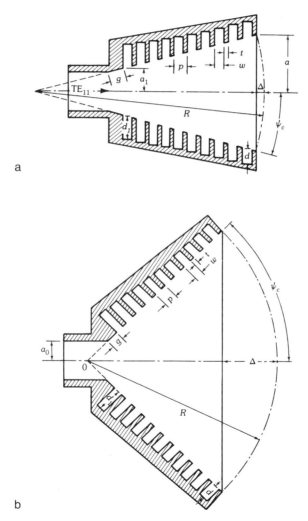

Fig. 33. Geometries of conical corrugated horns of small and large flare angles. (*a*) Small flare angle. (*b*) Large flare angle. (*After Thomas [23], © 1978 IEEE*)

$$\Delta = \frac{a}{\lambda_0}\tan\left(\frac{\psi_c}{2}\right) = \frac{R}{\lambda_0}\sin\psi_c \tan\left(\frac{\psi_c}{2}\right) \qquad (40)$$

As illustrated graphically in Fig. 33, Δ is the difference (in wavelengths) between the spherical wavefront and the plane aperture, and λ_0 is the free-space wavelength.

Depending on the value of Δ, the conical horns can be classified as either "narrowband" or "wideband." Horns with $\Delta < 0.4$ are usually referred to as narrowband, because their characteristics are frequency dependent. For example, it has been found that the beamwidth is determined by the aperture size ka and their phase center moves toward the throat as Δ increases, as shown in Fig. 34. Wideband horns are usually those with $\Delta > 0.75$, because their beamwidth is dependent primarily on the flare angle ψ_c and their phase center is mostly near the throat of the horn. Hence their characteristics are nearly frequency independent, provided $\Delta > 0.75$. In the literature these horns are often referred to as "scalar" horns [19]. Beam efficiencies [23] for narrowband and wideband conical corrugated horns are displayed in Figs. 35a and 35b, respectively.

If the circular waveguide feeding a smooth-surface conical horn operates on its dominant TE_{11} mode and forms a large-diameter (much greater than a wavelength) discontinuity, the first-order forward scattered fields required to match the curved phase front of the TE_{11} mode are those of the TM_{11} and TE_{12} modes, while the second order are those of the TM_{12} and TE_{13} modes [27]. It is assumed that backward scattered modes are negligible. The use of a large diameter (compared to

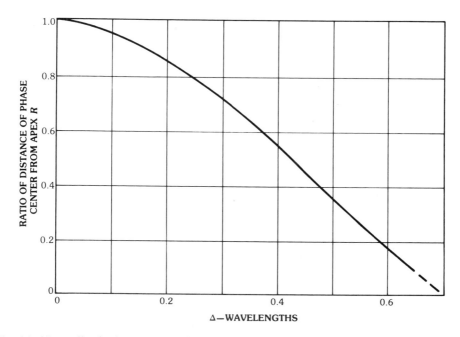

Fig. 34. Normalized phase-center distance (measured from horn apex) of narrowband conical corrugated horn. (*After Thomas [23]*, © *1978 IEEE*)

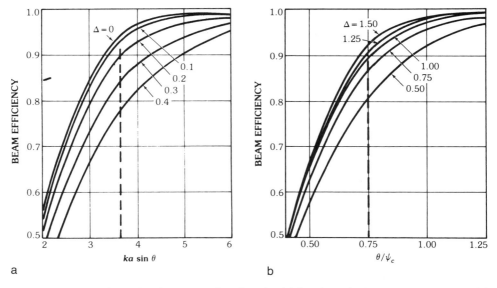

Fig. 35. Beam efficiencies for narrowband and wideband conical corrugated horns. (*a*) Narrowband horn. (*b*) Wideband horn. (*After Thomas [23], © 1978 IEEE*)

the wavelength) ensures the propagation of all of these four higher-order modes. At the center of the junction the electric fields of the TE_{11}, TM_{11}, and TE_{12} modes are oriented in the same direction, while the phases of the electric fields of the TM_{11} and TE_{12} are in quadrature with that of the TE_{11} mode. The phase relations between these modes are required to satisfy the assumed boundary conditions of a distorted phase front and negligible amplitude distortion. To determine the relative amplitudes of the higher-order modes a time-shared computer was employed [27].

However, to obtain a radiation pattern which displays perfect symmetry and zero cross polarization, by the use of a conical corrugated horn, it has been recognized that the field distribution at the horn's aperture should be that of a balanced hybrid HE_{11} mode.* The desired HE_{11}-mode excitation inside a conical horn, when it is connected to a smooth-wall circular waveguide supporting the TE_{11} mode, is achieved in most cases at the discontinuity formed by the feed waveguide and the horn. To facilitate further the TE_{11}-to-HE_{11} mode conversion, corrugations are introduced on the inside surface of the horn spacing from its throat toward its aperture, as shown in Fig. 36a. The corrugations must be such that the longitudinal surface reactance gradually changes from zero inside the smooth-wall circular waveguide to a high value (ideally infinity) near the aperture on the inside of the horn, which is required to support the desired HE_{11} mode. One method which has been suggested to achieve that is to use corrugations whose depth varies gradually from an initial depth of $\lambda/2$ near the throat to $\lambda/4$ near the edge of the TE_{11}-to-HE_{11}

---

*To designate the hybrid mode a mode-content factor γ is defined as the ratio of the longitudinal fields of the TE_{mn} and TM_{mn} components [23]. The term HE_{mn} is used to indicate that the two components are in phase (γ positive) and EH_{mn} when they are out of phase (γ negative). The TM_{mn} modes are represented by $\gamma = 0$ and the TE_{mn}s by $\gamma = \infty$.

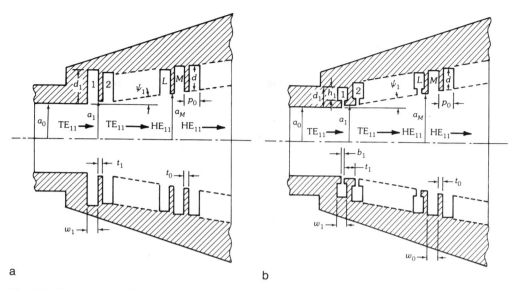

Fig. 36. Geometries of corrugated conical horns with conventional and ring-loaded slots. (*a*) Conventional slots. (*b*) Ring-loaded slots. (*After James [25], © 1982 IEEE*)

mode converter closest to the aperture. The remaining corrugations maintain a constant depth of $\lambda/4$.

The use of $\lambda/2$ deep slots near the throat provides a good match (low vswr) between the TE_{11} and HE_{11} modes over a wide band. To minimize the cross-polarized side lobes, however, the slots should be $\lambda/4$ deep at the center (design) frequency of the bandwidth. In addition, at the discontinuity between the smooth horn and the corrugated surface the EH_{11} surface wave can be introduced [29]. To avoid the generation of it the surface reactance should be negative over the entire desired bandwidth. This can be accomplished by maintaining the depth d_1 of the first slot nearest the throat as $\lambda/4 < d_1 < \lambda/2$ over the desired bandwidth. The design procedure [23] then, for optimum performance, is to choose the depth of the first slot to be slightly less than $\lambda_{max}/2$ at the highest frequency f_{max} of the band. The depth of each successive slot within the TE_{11}-to-HE_{11} converter of Fig. 36a decreases linearly until a depth of $\lambda'/4$ at the center frequency f' is achieved. Even though the surface reactance is negative the undesired EH_{12} mode can still be excited if the diameter of the horn at the first corrugation is too large. For narrow-band horns the radius a_1 of the horn at the first slot should be such that $ka_1 < 4$ at $f = f_{max}$.

The pitch p of the slots for wide-angle "broadband" (scalar) horns should be less than $\lambda_{max}/2$ at $f = f_{max}$, while the wall thickness t between the slots should be very thin so that the width-to-pitch ratio $\delta = w/p$ is near the optimum value of unity. This is a requirement for low cross polarization for wideband operation. For narrow-angle horns the pitch p should be chosen smaller than about $\lambda_{max}/4$ at $f = f_{max}$, which for small ka will avoid increased cross polarization which occurs when wider slots are utilized. For an effective mode conversion ten or more slots should be used within the TE_{11}-to-HE_{11} mode converter. The diameter a_0 of the input smooth-wall circular waveguide is chosen according to the method of excitation of the TE_{11} mode.

The use of varying depth corrugated slots has led to designs [25] with vswr's of less than about 1.065:1 only over a bandwidth ratio of about 1.45. To increase the bandwidth it has been proposed [25, 30, 31] that ring-loaded slots be used, as shown in Fig. 36b, instead of the conventional slots of Fig. 36a. It has been shown that by using ring-loaded slots the bandwidth ratio increases to 1.55 using ten or more slots within the TE_{11}-to-HE_{11} mode converter.

In addition to the excitation of the desired HE_{11} mode an unwanted EH_{12} mode is also usually generated. The radiation pattern of this is entirely cross-polarized in the 45° plane and, if generated, can seriously degrade the cross-polarization efficiency of the horn. The intensity of the cross-polarized side lobe, especially that at the 45° plane, is often used as a sensitive measure of beam circularity. The excitation of the unwanted EH_{12} mode can be minimized if the half-flare angle ψ_c is kept small (typically $\psi_c < 8°$). If the amount of the EH_{12} mode (using ring-loaded slots) is an order of magnitude smaller than its corresponding value using conventional varying depth slots, the bandwidth ratio increases to at least 2.0. The performance of horns having large flare angles toward the aperture can be improved, provided an additional flared section is used between the aperture region and the mode converter [25].

Several other designs of circularly polarized conical corrugated horns were investigated [18] at several frequencies in terms of E- and H-plane patterns, axial ratio, and phase centers. To achieve circular polarization it was recommended that identical dipoles be placed inside the feed guide a $\lambda/4$ guide-wavelength apart and oriented 90° relative to each other. For simplicity in the experiment, however, only one dipole was utilized and the E- and H-plane patterns were compared.

Typical E- and H-plane patterns of one design, at $f = 5.5$ and 6.5 GHz, are shown in Fig. 37. It is evident that the agreement between the E- and H-plane patterns is excellent. To compare better the closeness between the E- and H-plane patterns the axial ratios at several power levels and different frequencies were plotted. The results, measured at -10 and -20 dB for the design whose patterns are displayed in Fig. 37, are shown in Fig. 38. This design exhibits better axial ratios at levels down to -10 dB but deteriorate at lower levels. Additional data for this and other designs can be found in [18].

Aperture-Matched Horns

A horn which provides significantly better performance than an ordinary horn (in terms of pattern, impedance, and frequency characteristics) is that shown in Fig. 39a, which is referred to as an *aperture-matched* horn [26]. The main modification to the ordinary (conventional) horn, which we refer to here as the *control* horn, consists of the attachment of curved surface sections to the outside of the aperture edges, which reduces the diffractions which occur at the sharp edges of the aperture and provides smooth matching sections between the horn modes and the free-space radiation.

In contrast to the corrugated horn, which is complex and costly and reduces the diffractions at the edges of the aperture by minimizing the incident field, the

Horn Antennas

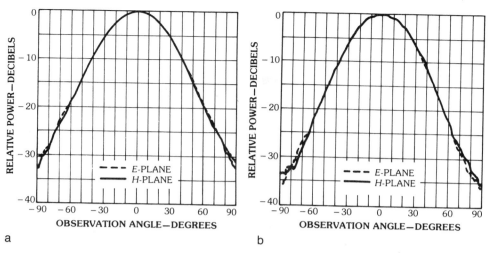

Fig. 37. Measured radiation patterns of a corrugated conical horn. (*a*) Frequency $f = 5.5$ GHz. (*b*) Frequency $f = 6.5$ GHz. (*After Al-Hakkak and Lo [18]*)

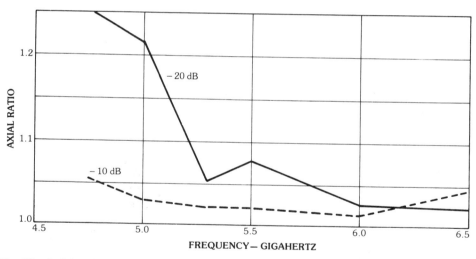

Fig. 38. Axial ratio versus frequency, for a horn whose patterns are shown in Fig. 37, at power levels of -10 dB and -20 dB. (*After Al-Hakkak and Lo [18]*)

aperture-matched horn reduces the diffractions by modifying the structure (without sacrificing size, weight, bandwidth, and cost) so that the diffraction coefficient is minimized. The basic concepts were originally investigated using elliptic cylinder sections, as shown in Fig. 39b; however, other convex curved surfaces, which smoothly blend to the ordinary horn geometry at the attachment point, will lead to similar improvements. This modification in geometry can be used in a wide variety of horns, and includes *E*-plane, *H*-plane, pyramidal, and conical horns. Bandwidths of 2:1 can easily be attained with aperture-matched horns having elliptical,

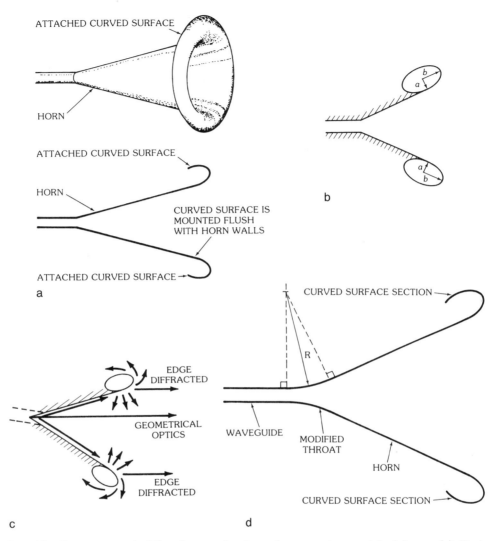

Fig. 39. Geometry and diffraction mechanism of an aperture-matched horn. (*a*) Basic geometry. (*b*) Elliptical cylinder curved surface. (*c*) Diffraction mechanism. (*d*) Modified throat. (*After Burnside and Chuang [26]*, © *1982 IEEE*)

circular, or other curved surfaces. The radii of curvature of the curved surfaces used in experimental models [26] ranged over $1.69\lambda \leq a \leq 8.47\lambda$ with $a = b$ and $b = 2a$. Good results can be obtained by using circular cylindrical surfaces with $2.5\lambda \leq a \leq 5\lambda$.

The basic radiation mechanism of such a horn is shown in Fig. 39c. The introduction of the curved sections at the edges does not eliminate diffractions; instead it substitutes edge diffractions by curved-surface diffractions which have a tendency to provide an essentially undisturbed energy flow across the junction, around the curved surface, and into free space. Compared with conventional horns, this radiation mechanism leads to smoother patterns with greatly reduced back lobes and negligible reflections back into the horn. The size, weight, and construction costs of the aperture-matched horn are usually somewhat larger and can be held to a minimum if half (one-half sections of an ellipse) or quadrant (one-fourth sections of an ellipse) sections are used instead of the complete closed surfaces.

To illustrate the improvements provided by the aperture-matched horns, the E-plane pattern, back lobe level, and half-power beamwidth of a pyramidal $2.96\lambda_0 \times 2.96\lambda_0$ horn were computed and were compared with the measured data of corresponding control and corrugated horns. The data are shown in Figs. 32a, 32c, and 32d. It is evident by examining the patterns of Fig. 32a that the aperture-matched horn provides a smoother pattern and lower back lobe level than conventional horns (referred to here as control horns); however, it does not provide, for the wide minor lobes, the same reduction as the corrugated horn. To achieve nearly the same E-plane pattern for all three horns the overall horn size would have to be increased. If the modifications for the aperture-matched and corrugated horns were only made in the E-plane edges, the H-plane patterns for all three horns would be virtually the same except that the back lobe level of the aperture-matched and corrugated horns would be greatly reduced.

The back lobe level of the same three horns (control, corrugated, and aperture-matched) are shown in Fig. 32c. The corrugated horn has lower back lobe intensity at the lower end of the frequency band, while the aperture-matched horn exhibits superior performance at the high end. However, both the corrugated and aperture-matched horns exhibit superior back lobe level characteristics to the control (conventional) horn throughout almost the entire frequency band. The half-power beamwidth characteristics of the same three horns are displayed in Fig. 32d. Because the control (conventional) horn has uniform distribution across the complete aperture plane, compared with the tapered distributions for the corrugated and aperture-matched horns, it possesses the smallest beamwidth almost over the entire frequency band.

In a conventional horn the vswr and antenna impedance are primarily influenced by the throat and aperture reflections. Using the aperture-matched horn geometry of Fig. 39a the aperture reflection toward the inside of the horn is greatly reduced. Therefore the only remaining dominant factor is the throat reflection. To reduce the throat reflections it has been suggested that a smooth curved surface be used to connect the waveguide and horn walls, as shown in Fig. 39d. Such a transition has been applied in the design and construction of a commercial X-band (8.2–12.4 GHz) pyramidal horn (see Fig. 12.23 of [6]), whose tapering is of exponential nature. The vswr's measured in the 8- to 12-GHz frequency band using the conventional exponential X-band horn (shown in Fig. 12.23 of [6]), with and without curved sections at its aperture, are shown in Fig. 40.

The matched sections used to create the aperture-matched horn were small cylinder sections. The vswr's for the conventional horn are very small (less than 1.1) throughout the frequency band because the throat reflection is negligible compared with the aperture reflection. It is evident, however, that the vswr's of the corresponding aperture-matched horn are much superior to those of the conventional horn because both the throat and aperture reflections are very minimal.

The basic design of the aperture-matched horn can be extended to include

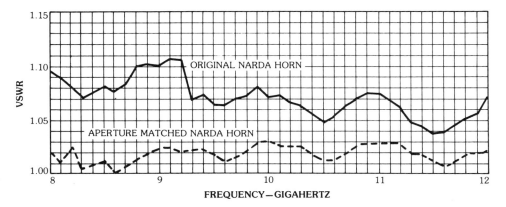

Fig. 40. Measured vswr for exponentially tapered pyramidal horns (conventional and aperture-matched). (*After Burnside and Chuang [26]*, © *1982 IEEE*)

corrugations on its inside surface. A typical configuration [24] of a conical aperture-matched corrugated horn is shown in Fig. 41. This type of design enjoys the advantages presented by both the aperture-matched and corrugated horns with cross-polarized components of less than −45 dB over a significant part of the bandwidth. Because of its excellent cross-polarization characteristics this horn is

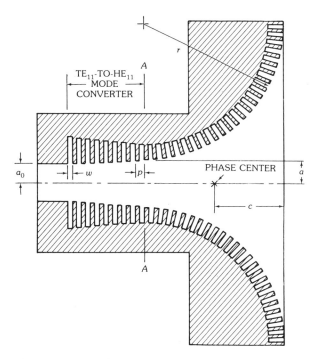

Fig. 41. Cross-section geometry of an aperture-matched corrugated conical horn. (*After Thomas and Greene [24]*, © *1982 IEEE*)

Horn Antennas

recommended for use as a reference and for frequency reuse applications in both satellite and terrestrial applications.

The corrugations are designed in the same manner as for the conventional corrugated conical horns, with the depth of the first being $\lambda/2$ at the highest frequency f_{max} of the bandwidth and that of the remaining decreasing linearly through the TE_{11}-to-HE_{11} mode converter. The last groove in the mode converter should have a depth of $\lambda/4$ at the center frequency f', and the remaining corrugations outside the converter toward the aperture should have a constant depth of $\lambda/4$ at $f = f'$. To achieve very low cross-polarization components the normalized radius of curvature of the curved section $\bar{r}$ (normalized with respect to the radius a at the throat, $\bar{r} = r/a$) should be large. Although horns with $\bar{r} = 1.3$ have been designed and tested, cross-polarized components of -45 dB or less have been obtained over a significant part of the bandwidth. The measured 45° plane patterns (principal and cross-polarized) for a horn with $\bar{r} = 4.3$ operating at $f = 15$ GHz is shown in Fig. 42. The cross-polarization component of this horn is about -50 dB or smaller. The bandwidth is extended as $\bar{r}$ increases until a value of $\bar{r}$ is reached when the undesired EH_{12} mode is generated.

Multimode Horns

Over the years there has been a need in many applications for horn antennas which provide symmetric patterns in all planes, phase center coincidence for the electric and magnetic planes, and side lobe suppression. All of these are attractive features for designs of optimum reflector systems and monopulse radar systems.

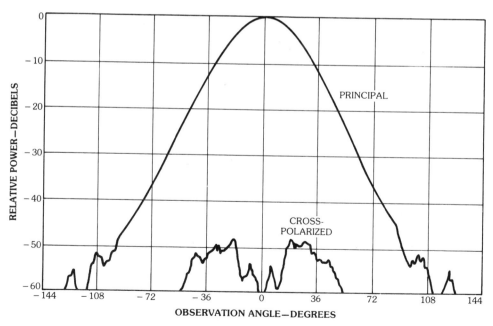

Fig. 42. Principal and cross-polarized 45° plane patterns of a corrugated conical horn with $\bar{r} = 4.3$ at $f = 15$ GHz. (*After Thomas and Greene [24]*, © *1982 IEEE*)

Side lobe reduction is a desired attribute for horn radiators utilized in antenna range, anechoic chamber, and standard gain applications, while pattern plane symmetry is a valuable feature for polarization diversity.

Pyramidal horns have traditionally been used over the years, with good success, in many of these applications. Such radiators, however, possess nonsymmetric beamwidths and undesirable side lobe levels, especially in the E-plane. Conical horns, operating in the dominant TE_{11} mode, have a tapered aperture distribution in the E-plane. Thus they exhibit more symmetric electric- and magnetic-plane beamwidths and reduced side lobes than do the pyramidal horns. One of the main drawbacks of a conical horn is its relative incompatibility with rectangular waveguides.

To remove some of the deficiencies of pyramidal and conical horns and further improve some of their attractive characteristics, corrugations were introduced on the interior walls of the waveguides, which lead to the corrugated horns that were discussed in a previous section of this chapter. In some other cases designs were suggested to improve the beamwidth equalization in all planes and reduce side lobe levels by utilizing horn structures with multiple-mode excitations. These have been designated as *multimode* horns, and some of the designs will be discussed briefly here. For more information the reader should refer to the cited references.

One design of a multimode horn is the "diagonal" horn [32], shown in Fig. 43, all of whose cross sections are square and whose internal fields consist of a superposition of TE_{10} and TE_{01} modes in a square waveguide. For small flare angles the field structure within the horn is such that the **E**-field vector is parallel to one of the diagonals. Although it is not a multimode horn in the true sense of the word because it does not make use of higher-order TE and TM modes, it does possess the desirable attributes of the usual multimode horns, such as equal beamwidths and suppressed beamwidths and side lobes in the E- and H-planes which are nearly equal to those in the principal planes. These attractive features are accomplished, however, at the expense of pairs of cross-polarized lobes in the intercardinal planes which make such a horn unattractive for applications where a high degree of polarization purity is required.

Diagonal horns have been designed, built, and tested [32] such that the 3-, 10-, and 30-dB beamwidths are nearly equal not only in the principal E- and H-planes but also in the 45° and 135° planes. Although the theoretical limit of the side lobe level in the principal planes is 31.5 dB down, side lobes of at least 30 dB down have been observed in those planes. Despite a theoretically predicted level of −19.2 dB in the ±45° planes, side lobes with levels of −23 to −27 dB have been observed. The principal deficiency in the side lobe structure appears in the ±45°-plane cross-polarized lobes whose intensity is only 16 dB down; despite this the overall horn efficiency remains high. Compared with diagonal horns, conventional pyramidal square horns have H-plane beamwidths which are about 35 percent wider than those in the E-plane, and side lobe levels in the E-plane which are only 12 to 13 dB down (although those in the H-plane are usually acceptable).

For applications which require optimum performance with narrow beamwidths, lenses are usually recommended for use in conjunction with diagonal horns. Diagonal horns can also be converted to radiate circular polarization by inserting a differential phase shifter inside the feed guide whose cross section is circular and adjusted so that it produces phase quadrature between the two orthogonal modes.

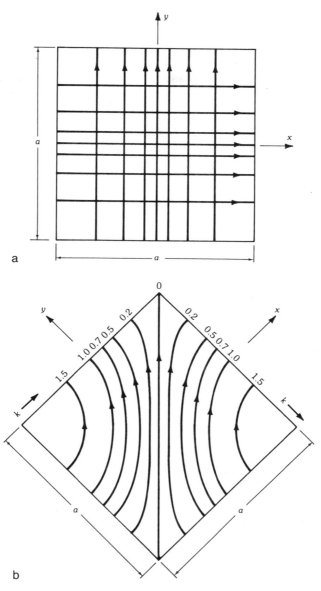

Fig. 43. Electric-field configuration inside square diagonal horn. (*a*) Two coexisting equal orthogonal modes. (*b*) Result of combining the two modes shown in (*a*). (*After Love [32], reprinted with permission of* Microwave Journal, *from March 1962 issue,* © *1962 Horizon House–Microwave, Inc.*)

Another multimode horn which exhibits suppressed side lobes, equal beamwidths, and reduces cross polarization is the *dual-mode* conical horn [33]. Basically this horn is designed so that diffractions at the aperture edges of the horn, especially those in the *E*-plane, are minimized by reducing the fields incident on the aperture edges and consequently the associated diffractions. This is accomplished by utilizing a conical horn which at its throat region is excited in both the dominant TE_{11} and higher-order TM_{11} mode. A discontinuity is introduced at a position within the horn where two modes exist. The horn length is adjusted so that the superposition of the relative amplitudes of the two modes at the edges of the aperture is very small compared with the maximum aperture field magnitude. In addition, the dimensions of the horn are controlled so that the total phase at the aperture is such that, in conjunction with the desired amplitude distribution, it leads to side lobe suppression, beamwidth equalization, and phase center coincidence.

Qualitatively the pattern formation of a dual-mode conical horn operating in the TE_{11} and TM_{11} modes is accomplished by utilizing a pair of modes which have radiation functions with the same argument. However, one of the modes, in this case the TM_{11} mode, contains an additional envelope factor which varies very rapidly in the main beam region and remains relatively constant at large angles. Thus it is possible to control the two modes in such a way that their fields cancel in all directions except within the main beam. The TM_{11} mode exhibits a null in its far-field pattern. Therefore a dual-mode conical horn possesses less axial gain than a conventional dominant-mode conical horn of the same aperture size. Because of that, dual-mode horns render better characteristics and are more attractive for applications where pattern plane symmetry and side lobe reduction are more important than maximum aperture efficiency. A most important application of a dual-mode horn is as a feed of Cassegrainian reflector systems.

Dual-mode conical horns have been designed, built, and tested [33] with relatively good success in their performance. Generally, however, diagonal horns would be good competitors for the dual-mode horns if it were not for the undesirable characteristics (especially the cross-polarized components) that they exhibit in the very important 45° and 135° planes. Improved performance can be obtained from dual-mode horns if additional higher-order modes (such as the TE_{12}, TE_{13}, and TM_{12}) are excited [34] and if their relative amplitudes and phases can be properly controlled. Computed maximum aperture efficiencies of paraboloidal reflectors, using such horns as feeds, have reached 90 percent contrasted with efficiencies of about 76 percent for reflector systems using conventional dominant-mode horn feeds. In practice the actual maximum efficiency achieved will be determined by the number of modes that can be excited and the degree to which their relative amplitudes and phases can be controlled.

The techniques of the dual-mode and multimode conical horns can be extended to the design of horns with rectangular cross sections. In fact a multimode pyramidal horn design has been designed, built, and tested to be used as a feed in a low-noise Cassegrain monopulse system [35]. This rectangular pyramidal horn utilizes additional higher-order modes to provide monopulse capability, side lobe suppression in both the *E*- and *H*-planes, and beamwidth equalization. Specifically the various pattern modes for the monopulse system are formed in a single horn as follows:

    (*a*)   *Sum*: Utilizes $TE_{10} + TE_{30}$ instead of only TE_{10}. When the relative amplitude and phase excitations of the higher-order TE_{30} mode are properly adjusted, they provide side lobe cancellation at the second minor lobe of the TE_{10}-mode pattern

(b) *E-Plane Difference*: Utilizes $TE_{11} + TM_{11}$ modes
(c) *H-Plane Difference*: Utilizes TE_{20} mode

In its input the horn of [35] contained a four-guide monopulse bridge circuitry, a multimode matching section, a difference mode phasing section, and a sum mode excitation and control section. To illustrate the general concept Figs. 44a, 44b, and 44c are plots of three-dimensional patterns of the sum, *E*-plane difference, and *H*-plane difference modes which utilize, respectively, the $TE_{10} + TE_{30}$, $TE_{11} + TM_{11}$, and TE_{20} modes. The relative excitation between the modes has been controlled so that each pattern utilizing multiple modes in its formation displays its most attractive features for its function.

Dielectric-Loaded Horns

Over the years much effort has been devoted in enhancing the antenna and aperture efficiencies of aperture antennas, especially for those that serve as feeds for reflectors (such as the horn). One technique that was proposed and was investigated was to use dielectric guiding structures, referred to as *Dielguides* [36], between the primary feed and the reflector (or subreflector). The technique is simple and inexpensive to implement and provides broadband, highly efficient, and low-noise antenna feeds. The method negates the compromise between taper and spillover efficiencies, and it is based on the principle of internal reflections, which has been utilized frequently in optics. Its role bears a very close resemblance to that of a lens, and it is an extension of the classical parabolic-shaped lens to other geometrical shapes.

Another method which has been used to control the radiation pattern of electromagnetic horns is to insert totally within them various shapes of dielectric material (wedges, slabs, etc.) [37–39] to control in a predictable manner not only the phase distribution over the aperture, as is usually done by using the classical parabolic lenses, but also to change the power (amplitude) distribution over the aperture. The control of the amplitude and phase distributions over the aperture are very essential in the design of very low side lobe antenna patterns.

Symmetrical loading of the *H*-plane walls has also been utilized, by proper parameter selection, to create a dominant longitudinal section electric (LSE) mode and to enhance the aperture efficiency and pattern-shaping capabilities of symmetrically loaded horns [38]. The method is simple and inexpensive, and it can also be utilized to realize high efficiency from small horns which can be used in limited scan arrays. Aperture efficiencies on the order of 92 to 96 percent have been attained, in contrast to values of 81 percent for unloaded horns.

A similar technique has been suggested to symmetrically load the *E*-plane walls of rectangular horns [39], and eventually to line all four of its walls with dielectric slabs. Other similar techniques have been suggested, and a summary of these and other classical papers dealing with dielectric-loaded horns can be found in [1].

8. Phase Center

Each of the far-zone field components radiated by an antenna can be written, in general, as

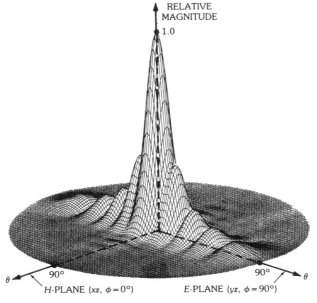

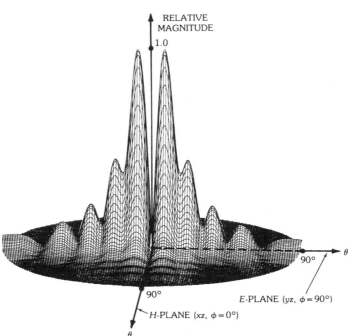

Fig. 44. Three-dimensional patterns of the sum, E-plane difference, and H-plane difference patterns. (*a*) Sum field pattern of a monopulse pyramidal horn operating in the TE_{10} and TE_{30} modes. (*b*) E-plane difference field pattern of a monopulse pyramidal horn operating in the TE_{11} and TM_{11} modes. (*c*) H-plane field pattern of a monopulse pyramidal horn operating in the TE_{20} mode.

Horn Antennas

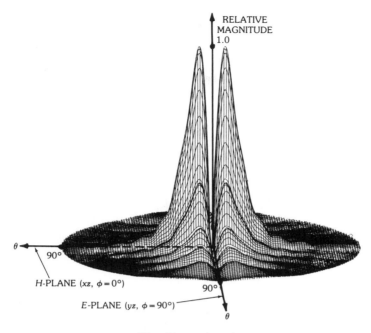

Fig. 44, *continued*

$$\mathbf{E}_u = \hat{\mathbf{u}}\, E(\theta,\phi) e^{j\psi(\theta,\phi)} \frac{e^{-jkr}}{r} \qquad (41)$$

where $\hat{\mathbf{u}}$ is a unit vector. The terms $E(\theta,\phi)$ and $\psi(\theta,\phi)$ represent, respectively, the (θ,ϕ) variations of the amplitude and phase.

In navigation, tracking, homing, landing, and other aircraft and aerospace systems it is usually desirable to assign to the antenna a reference point such that for a given frequency, $\psi(\theta,\phi)$ of (41) is independent of θ and ϕ [i.e., $\psi(\theta,\phi)$ = constant]. The reference point which makes $\psi(\theta,\phi)$ independent of θ and ϕ is known as the *phase center* of the antenna [40–44]. When referenced to the phase center the fields radiated by the antenna are spherical waves with ideal spherical wavefronts or equiphase surfaces. Therefore a phase center is a reference point from which radiation is said to emanate, and radiated fields measured on the surface of a sphere whose center coincides with the phase center have the same phase.

For practical antennas such as arrays, reflectors, and others, a single unique phase center valid for all values of θ and ϕ does not exist; for most, however, their phase center moves along a surface, and its position depends on the observation point. However, in many antenna systems a reference point can be found such that $\psi(\theta,\phi)$ = constant, or nearly so, over most of the angular space, especially over the main lobe. When the phase center position variation is sufficiently small, that point is usually referred to as the *apparent phase center*.

The need for the phase center can best be explained by examining the radiation characteristics of a paraboloidal reflector (parabola of revolution). Plane waves incident on a paraboloidal reflector focus at a single point which is known as the *focal point*. Conversely, spherical waves emanating from the focal point are reflected by the paraboloidal surface and form plane waves. Thus in the receiving mode all the energy is collected at a single point. In the transmitting mode, ideal plane waves are formed if the radiated waves have spherical wavefronts and emanate from a single point.

In practice, no antenna is a point source with ideal spherical equiphases. Many of them, however, contain a point from which their radiation, over most of the angular space, seems to have spherical wavefronts. When such an antenna is used as a feed for a reflector its phase center must be placed at the focal point.

The analytical formulations for locating the phase center of an antenna are usually very laborious and exist only for a limited number of configurations [40–42]. Experimental techniques [43, 44] are available to locate the phase center of an antenna.

The horn is a microwave antenna which is widely used as a feed for reflectors. To perform as an efficient feed for reflectors it is imperative that its phase center is known and it is located at the focal point of the reflector. Instead of presenting analytical formulations for the phase center of a horn, graphical data will be included to illustrate typical phase centers.

Usually the phase center of a horn is not located at its mouth (throat) or at its aperture but mostly between its imaginary apex point and its aperture. The exact location depends on the dimensions of the horn, especially on its flare angle. For large flare angles the phase center is closer to the apex. As the flare angle of the horn becomes smaller, the phase center moves toward the aperture of the horn.

Computed phase centers for an *E*-plane and an *H*-plane sectoral horn are displayed in Figs. 45a and 45b. It is apparent that for small flare angles the *E*- and *H*-plane phase centers are identical. Although each specific design has its own phase center the data of Figs. 45a and 45b are typical. If the *E*- and *H*-plane phase centers of a pyramidal horn are not identical, its phase center can be taken to be the average of the two. Phase centers for narrowband corrugated horns are also displayed in Fig. 34.

Phase center nomographs for conical corrugated and uncorrugated (TE_{11}-mode) horns are available [42], and they are displayed, respectively, in Figs. 46a and 46b. The procedure to use these is documented in [42], and it is repeated here.

Procedure to Locate Phase Center

Given: (a) a/λ = radius of the horn aperture (in wavelengths)
          (b) R_0/λ = distance from the aperture, along the horn axis, to the observation point (in wavelengths)
          (c) ℓ/λ = distance from the aperture to the horn apex (in wavelengths)

Find: The phase center location Z_0 (in wavelengths). This is the phase center location measured from the aperture along the horn axis toward the apex (positive Z_0s indicate the phase center is within the horn).

Solution: The procedure based on the results of Figs. 46a and 46b, follows:
          (a)   Determine $\alpha = a^2/(\lambda\ell)$ and locate the α curve on the appropriate figure.

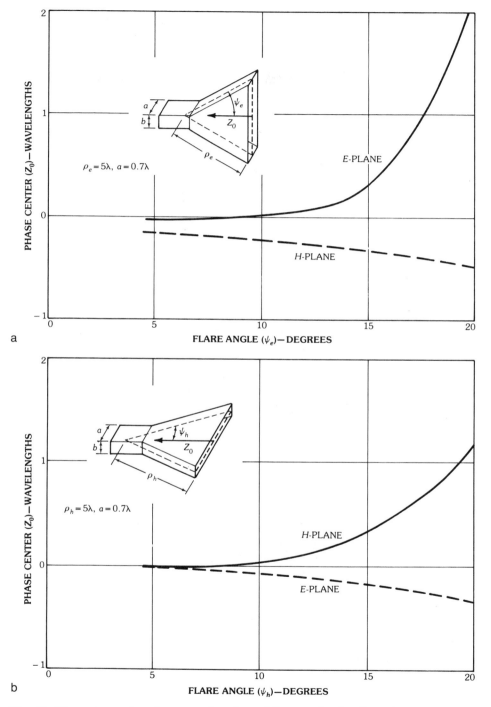

Fig. 45. Phase center location, as a function of flare angle, for E- and H-plane sectoral horns. (*a*) E-plane sectoral horn. (*b*) H-plane sectoral horn. (*Adapted from Hu [40]*)

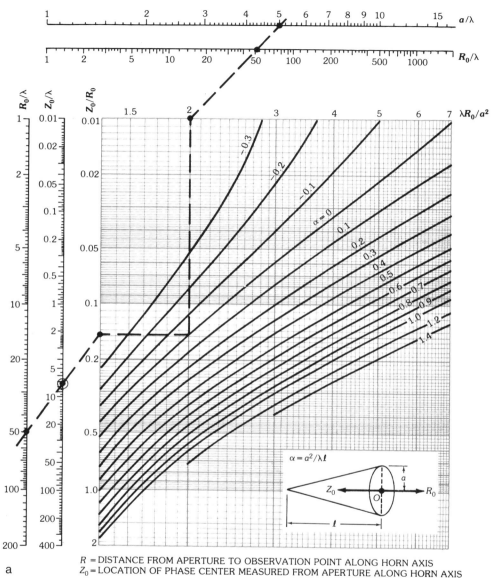

Fig. 46. Phase center nomographs for corrugated and TE_{11}-mode conical horns. (a) Corrugated conical horn. (b) TE_{11} conical horn. (*After Ohtera and Ujiie [42], © 1975 IEEE*)

(b) On the horizontal axes, plot the radius of the aperture a on the a/λ axis and the observation distance R_0 on the R_0/λ axis.

(c) Draw a straight line through the a/λ and R_0/λ points until it intersects the reference $\lambda R_0/a^2$ axis. Read the intersection on the reference axis.

(d) Through the $\lambda R_0/a^2$ axis intersection, draw a vertical line until it

Horn Antennas

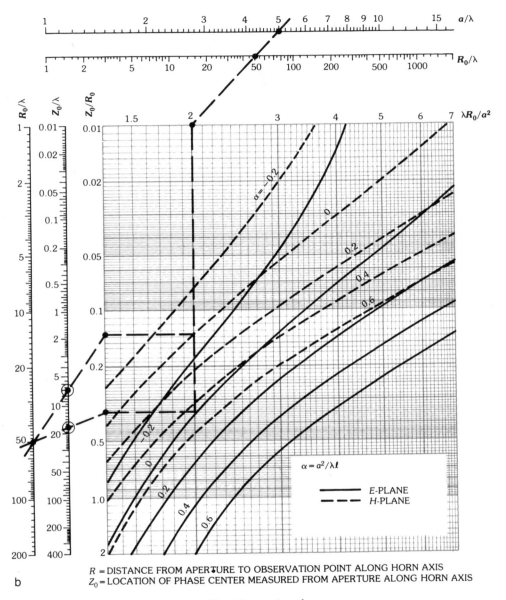

R = DISTANCE FROM APERTURE TO OBSERVATION POINT ALONG HORN AXIS
Z_0 = LOCATION OF PHASE CENTER MEASURED FROM APERTURE ALONG HORN AXIS

Fig. 46, *continued*

        intersects the α curve found in step *a*.
- (*e*) Through the α curve intersection point, draw a horizontal line until it intersects the Z_0/R_0 vertical axis.
- (*f*) On the vertical axes, read the intersection point on the reference Z_0/R_0 axis.
- (*g*) From the intersection point on the Z_0/R_0 axis, draw a straight line to the given point on the vertical R_0/λ axis.

(h) Read Z_0/λ, which is the location of the phase center, from the intersection of the line from step g with the Z_0/λ axis.

Example 10—Find the E- and H-plane phase centers for a corrugated conical horn with $a = 5\lambda$, $R_0 = 50\lambda$, and $\ell = \infty$. For a corrugated horn the E- and H-plane phase centers occur at the same point.
Solution:
(a) For this horn $\alpha = a^2/\lambda\ell = 0$. See Fig. 46a for the $\alpha = 0$ curve.
(b) See Fig. 46a for construction of dashed line.
(c) From Fig. 46a, $\lambda R_0/a^2 = 2$.
(d) See Fig. 46a for the intersection with the $\alpha = 0$ curve.
(e) See Fig. 46a for the intersection with the Z_0/R_0 curve.
(f) From Fig. 46a, $Z_0/R_0 = 0.15\lambda$.
(g) See Fig. 46a for construction of dashed line.
(h) From Fig. 46a $Z_0/\lambda = 7.47$.

Therefore the phase center, for both the E- and H-planes, is located inside the horn at a distance of 7.47λ from the aperture of the horn, along its axis, toward its apex.

Example 11—Find the E- and H-plane phase centers of a conical TE_{11}-mode horn whose dimensions are the same as those of Example 10 (i.e., $a = 5\lambda$, $R_0 = 50\lambda$, and $\ell = \infty$).
Solution:
(a) For this horn $\alpha = a^2/\lambda\ell = 0$. See Fig. 46b for the $\alpha = 0$ curves for the E- and H-planes.
(b) See Fig. 46b for construction of dashed line.
(c) From Fig. 46b, $\lambda R_0/a^2 = 2$.
(d) See Fig. 46b for the intersection with the $\alpha = 0$ curves of the E- and H-planes.
(e) See Fig. 46b for the intersections with the Z_0/R_0 axis.
(f) From Fig. 46b $Z_0/R_0 = 0.35$ (for E-plane) and $Z_0/R_0 = 0.14$ (for H-plane).
(g) See Fig. 46b for constructions of dashed lines (for E- and H-plane).
(h) From Fig. 46b $Z_0/\lambda = 18.5$ (for E-plane) and $Z_0/\lambda = 6.9$ (for H-plane).

Thus the E-plane phase center is located inside the horn at a distance of 18.5λ from the aperture, along the horn axis, toward its apex, while that of the H-plane is also within the horn but a distance of 6.9λ from the aperture. It is evident that the E- and H-plane phase centers do not coincide for this horn, and this is a major deficiency in many applications.

An experimental technique for measuring the phase center of an antenna, based on the work by Dyson [43] and reported also in [18], will be repeated here. The antenna under test is placed in the far field of a transmitting horn, on a positioner that is capable of precise placement of an antenna relative to a rotation axis, as shown in Fig. 47. The relative phase ψ of the unmodulated rf signal e_2 from the test antenna is compared with that of the coherent unmodulated reference signal e_1 in a network analyzer. The output signal e_3, which is proportional to ψ, is then recorded by an X–Y recorder whose X movement is synchronized with the rotation of the antenna under test. A delay line and an attenuator are included in

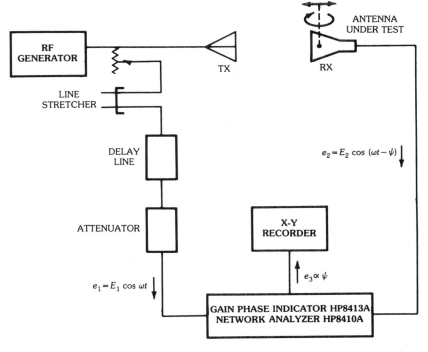

Fig. 47. Circuit diagram for phase center measurement. (*After Al-Hakkak and Lo [18]*)

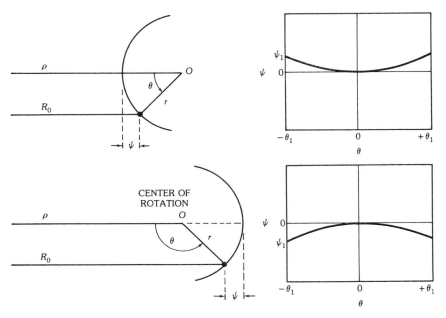

Fig. 48. Geometry and phase change as a displaced antenna is rotated about a given axis. (*After Al-Hakkak and Lo [18]*)

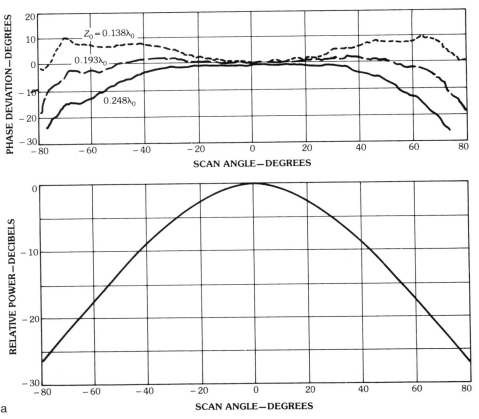

Fig. 49. Relative phase patterns at $f = 5.2$ GHz for the antenna whose patterns and axial ratios are displayed in Figs. 37 and 38. (*a*) *H*-plane. (*b*) *E*-plane. (*After Al-Hakkak and Lo [18]*)

the e_1 branch so that e_1 and e_2 can have equal amplitudes, thus reducing the error in e_3 to a minimum. The procedure to find the phase center is as follows:

(a) The receiving (RX) antenna is first directed so that its axis is pointing in the direction of the transmitting (TX) antenna (i.e., $\theta = 0°$).
(b) The length of the e_1 path is varied, by adjusting the line stretcher, until $\psi = 0$. Thus the phase at $\theta = 0°$ is taken as a reference.
(c) The RX antenna is then rotated from $-\theta_1$ to $+\theta_1$, where θ_1 is a suitable angle, and ψ is recorded. The resulting relative phase pattern obeys the relation

$$\psi \cong k_0 r(1 - \cos\theta) \tag{42}$$

where r is the distance between the phase center and axis of rotation and θ as shown in Fig. 48. If the phase pattern is a straight line, the center of rotation coincides with the phase center of the antenna. If the pattern is not

Horn Antennas

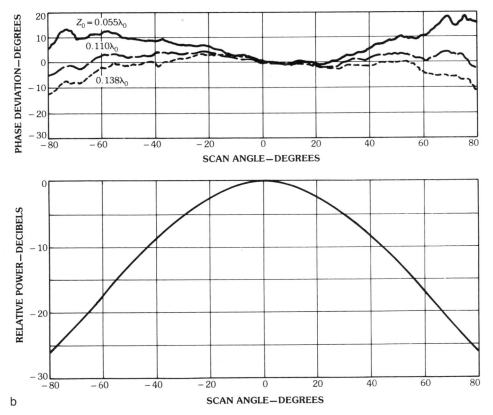

Fig. 49, *continued*

straight, then in theory the position of the phase center can be determined from this pattern by finding r from (42) as

$$r = \frac{\lambda_0}{2\pi} \frac{\psi_1}{1 - \cos\theta_1} \tag{43}$$

where ψ_1 is the relative phase at θ_1. In practice, however, it may be necessary, because of both pattern and experimental anomalies, to record several patterns as the antenna is repositioned along its axis, repeating steps *a–c*. The position that corresponds to the flattest pattern is taken as the phase center.

Using the procedure described above, the phase centers of the conical corrugated horn antenna whose patterns and axial ratios are displayed, respectively, in Figs. 37 and 38 have been determined in both the *E*- and *H*-planes. Typical phase and power patterns measured at 5.2 GHz are shown in Fig. 49. The numbers on the phase patterns refer to the locations of axes of rotation measured in

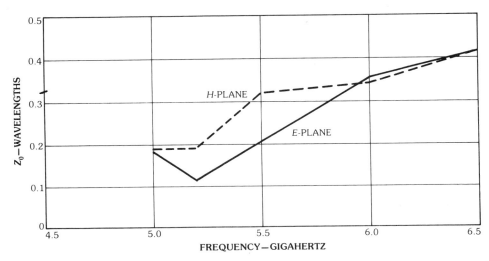

Fig. 50. Phase center versus frequency for the corrugated conical horn whose patterns and axial ratios are shown in Figs. 37 and 38. (*After Al-Hakkak and Lo [18]*)

wavelengths from the aperture's plane inside the horn. The numbers in the middle phase pattern curves refer to the positions Z_0 of the "apparent phase centers." The term Z_0 is the distance of the phase from the horn aperture, along its axis, toward its apex. Fig. 50 shows, for the horn whose patterns and axial ratios were displayed in Figs. 37 and 38, the values of Z_0 measured at several frequencies in the range 4.5 to 6.5 GHz. From the data displayed in Figs. 49 and 50, as well as others found in [18], the following conclusions are drawn:

- (a) The apparent phase centers are located inside the horns at distances Z_0 from the aperture plane, that are within one wavelength, in general. However, the longer the horn, the larger Z_0.
- (b) As the frequency increases, Z_0 (in wavelengths) becomes larger.
- (c) The values of Z_0 for the *E*- and *H*-planes generally differ by less than 0.1 wavelength.
- (d) For shorter horns a constant phase pattern can be found over wider angular regions.

A technique which allows the determination of phase centers of horn antennas at *X*-band, with a very high degree of accuracy, has also been documented [44]. The results indicate that for horn antennas as large as 5 in (12.7 cm), the phase centers can be determined to within ±0.25 electrical degrees. Phase center positions at *X*-band were determined to within 0.002 in (0.005 08 cm).

Acknowledgments

Portions of this chapter were first published in *Antenna Theory: Analysis and Design* by C. A. Balanis [6], copyright 1982, reprinted by permission of Harper & Row Publishers. Inc.

9. References

[1] A. W. Love, *Electromagnetic Horn Antennas*, New York: IEEE Press, 1976.
[2] P. M. Russo, R. C. Rudduck, and L. Peters, Jr., "A method for computing E-plane patterns of horn antennas," *IEEE Trans. Antennas Propag.*, vol. AP-13, no. 2, pp. 219–224, March 1965.
[3] J. S. Yu, R. C. Rudduck, and L. Peters, Jr., "Comprehensive analysis for E-plane of horn antennas by edge diffraction theory," *IEEE Trans. Antennas Propag.*, vol. AP-14, no. 2, pp. 138–149, March 1966.
[4] M. A. K. Hamid, "Diffraction by a conical horn," *IEEE Trans. Antennas Propag.*, vol. AP-16, no. 5, pp. 520–528, September 1966.
[5] M. S. Narasimhan and M. S. Shehadri, "GTD analysis of the radiation patterns of conical horns," *IEEE Trans. Antennas Propag.*, vol. AP-26, no. 6, pp. 774–778, November 1978.
[6] C. A. Balanis, *Antenna Theory: Analysis and Design*, New York: Harper & Row, Publishers, 1982.
[7] R. J. Mailloux and G. R. Forbes, "An array technique with grating-lobe suppression for limited-scan applications," *IEEE Trans. Antennas Propag.*, vol. AP-21, no. 5, pp. 597–602, September 1973.
[8] R. J. Mailloux, L. Zahn, A. Martinez III, and G. R. Forbes, "Grating lobe control in limited scan arrays," *IEEE Trans. Antennas Propag.*, vol. AP-27, no. 1, pp. 79–85, January 1979.
[9] N. Amitay and M. J. Gans, "Design of rectangular horn arrays with oversized aperture elements," *IEEE Trans. Antennas Propag.*, vol. AP-29, no. 6, pp. 871–884, November 1981.
[10] J. Boersma, "Computation of Fresnel integrals," *Math. Comp.*, vol. 14, p. 380, 1960.
[11] Y.-B. Cheng, "Analysis of aircraft antenna radiation for microwave landing system using geometrical theory of diffraction," MSEE thesis, Dept. of Electr. Eng., West Virginia Univ., pp. 208–211.
[12] E. H. Braun, "Some data for the design of electromagnetic horns," *IRE Trans. Antennas Propag.*, vol. AP-4, no. 1, pp. 29–31, January 1956.
[13] E. V. Jull, "Gain of an *E*-plane sectoral horn—a failure of the Kirchhoff theory and a new proposal," *IEEE Trans. Antennas Propag.*, vol. AP-22, no. 2, pp. 221–226, March 1974.
[14] W. C. Jakes, "Horn Antennas," Chapter 10 in *Antenna Engineering Handbook*, ed. by H. Jasik, New York: McGraw-Hill Book Co., 1961.
[15] W. M. Truman and C. A. Balanis, "Optimum design of horn feeds for reflector antennas," *IEEE Trans. Antennas Propag.*, vol. AP-22, no. 4, pp. 585–586, July 1974.
[16] M. G. Schorr and F. J. Beck, Jr., "Electromagnetic field of a conical horn," *J. Appl. Phys.*, vol. 21, pp. 795–801, August 1950.
[17] A. P. King, "The radiation characteristics of conical horn antennas," *Proc. IRE*, vol. 38, pp. 249–251, March 1950.
[18] M. J. Al-Hakkak and Y. T. Lo, "Circular waveguides and horns with anisotropic and corrugated boundaries," *Antenna Laboratory Report No. 73-3*, Depart. Electr. Eng., Univ. of Illinois, Urbana, January 1973.
[19] A. F. Kay, "The scalar feed," *AFCRL Rep. 65-347*, AD601609, March 1964.
[20] R. E. Lawrie and L. Peters, Jr., "Modifications of horn antennas for low side lobe levels," *IEEE Trans. Antennas Propag.*, vol. AP-14, no. 5, pp. 605–610, September 1966.
[21] C. A. Mentzer and L. Peters, Jr., "Properties of cutoff corrugated surfaces for corrugated horn design," *IEEE Trans. Antennas Propag.*, vol. AP-22, no. 2, pp. 191–196, March 1974.
[22] C. A. Mentzer and L. Peters, Jr., "Pattern analysis of corrugated horn antennas," *IEEE Trans. Antennas Propag.*, vol. AP-24, no. 3, pp. 304–309, May 1976.
[23] B. MacA. Thomas, "Design of corrugated conical horns," *IEEE Trans. Antennas Propag.*, vol. AP-26, no. 2, pp. 367–372, March 1978.

[24] B. MacA. Thomas and K. J. Greene, "A curved-aperture corrugated horn having very low cross-polar performance," *IEEE Trans. Antennas Propag.*, vol. AP-30, no. 6, pp. 1068–1072, November 1982.

[25] G. L. James, "TE_{11}-to-HE_{11} mode converters for small-angle corrugated horns," *IEEE Trans. Antennas Propag.*, vol. AP-30, no. 6, pp. 1057–1062, November 1982.

[26] W. D. Burnside and C. W. Chuang, "An aperture-matched horn design," *IEEE Trans. Antennas Propag.*, vol. AP-30, no. 4, pp. 790–796, July 1982.

[27] K. Tomiyasu, "Conversion of TE_{11} mode by a large-diameter conical junction," *IEEE Trans. Microwave Theory Tech.*, vol. MTT-17, pp. 277–279, May 1969.

[28] B. MacA. Thomas, "Mode conversion using circumferentially corrugated cylindrical waveguide," *Electron. Lett.*, vol. 8, pp. 394–396, 1972.

[29] J. K. M. Jansen and M. E. J. Jeuken, "Surface waves in corrugated conical horn," *Electron. Lett.*, vol. 8, pp. 342–344, 1972.

[30] Y. Tacheichi, T. Hashimoto, and F. Takeda, "The ring-loaded corrugated waveguide," *IEEE Trans. Microwave Theory Tech.*, vol. MTT-19, pp. 947–950, December 1971.

[31] F. Takeda and T. Hashimoto, "Broadbanding of corrugated conical horns by means of the ring-loaded corrugated waveguide structure," *IEEE Trans. Antennas Propag.*, vol. AP-24, pp. 786–792, 1976.

[32] A. W. Love, "The diagonal horn antenna," *Microwave J.*, vol. V, pp. 117–122, March 1962.

[33] P. D. Potter, "A new horn antenna with suppressed side lobes and equal beamwidths," *Microwave J.*, pp. 71–78, June 1963.

[34] P. D. Potter and A. C. Ludwig, "Beamshaping by use of higher-order modes in conical horns," *Northeast Electron. Res. and Eng. Mtg*, pp. 92–93, November 1963.

[35] P. A. Jensen, "A low-noise multimode Cassegrain monopulse with polarization diversity," *Northeast Electron. Res. and Eng. Mtg*, pp. 94–95, November 1963.

[36] H. E. Bartlett and R. E. Moseley, "Dielguides—highly efficient low-noise antenna feeds," *Microwave J.*, vol. 9, pp. 53–58, December 1966.

[37] L. L. Oh, S. Y. Peng, and C. D. Lunden, "Effects of dielectrics on the radiation patterns of an electromagnetic horn," *IEEE Trans. Antennas Propag.*, vol. AP-18, no. 4, pp. 553–556, July 1970.

[38] G. N. Tsandoulas and W. D. Fitzgerald, "Aperture efficiency enhancement in dielectrically loaded horns," *IEEE Trans. Antennas Propag.*, vol. AP-20, no. 1, pp. 69–74, January 1972.

[39] R. Baldwin and P. A. McInnes, "Radiation patterns of dielectric loaded rectangular horns," *IEEE Trans. Antennas Propag.*, vol. AP-21, no. 3, pp. 375–376, May 1973.

[40] Y. Y. Hu, "A method of determining phase centers and its applications to electromagnetic horns," *Franklin Inst.*, vol. 271, pp. 31–39, January 1961.

[41] E. R. Nagelberg, "Fresnel region phase centers of circular aperture antennas," *IEEE Trans. Antennas Propag.*, vol. AP-13, no. 3, pp. 479–480, May 1965.

[42] I. Ohtera and H. Ujiie, "Nomographs for phase centers of conical corrugated and TE_{11}-mode horns," *IEEE Trans. Antennas Propag.*, vol. AP-23, no. 6, pp. 858–859, November 1975.

[43] J. D. Dyson, "Determination of the phase center and phase patterns of antennas," in "Radio Antennas for Aircraft and Aerospace Vehicles" (W. T. Blackband, ed.), *AGARD Conf. Proc.*, no. 15, Technivision Services, Slough, England, 1967.

[44] M. Teichman, "Precision phase center measurements of horn antennas," *IEEE Trans. Antennas Propag.*, vol. AP-18, no. 5, pp. 689–690, September 1970.

Chapter 9

Frequency-Independent Antennas

Paul E. Mayes
University of Illinois

CONTENTS

1. Basic Types — 9-3
2. Log-Periodic Dipole Arrays — 9-12
   Design of Log-Periodic Dipole Arrays 9-20
3. Periodic Structure Theory — 9-32
   Design of Log-Periodic Zigzag Antennas 9-37
   Periodically Loaded Lines 9-46
   Log-Periodic Designs Based on Periodic Structure Theory 9-62
4. Log-Spiral Antennas — 9-72
   Conical Log-Spirals 9-79
   Construction Techniques 9-111
5. References — 9-112
6. Bibliography — 9-114

Paul E. Mayes was born in Frederick, Oklahoma, on December 21, 1928. He received his PhD in electrical engineering from Northwestern University in 1955.

From 1950 to 1954 he was employed as a graduate assistant and research associate in the Microwave Laboratory at Northwestern, where his research was on electromagnetic-wave propagation along dielectric-rod waveguides and reflection of electromagnetic waves from curved conducting surfaces. Since 1954 he has been on the faculty of the Department of Electrical Engineering, University of Illinois at Urbana, where he is now a full professor teaching courses in electromagnetic theory and antennas and supervising research in the Electromagnetics Laboratory. His research at Illinois has been concerned with slot antennas, numerical electromagnetic analysis, microwave transmission lines, and frequency-independent antennas.

Dr. Mayes was awarded Certificates of Achievement in 1968 and 1969 for papers published in the *IEEE Transactions on Antennas and Propagation*. He was elected IEEE Fellow in 1975 for "contributions to the theory and development of the log-periodic antennas." He has served as a technical consultant to industry and has eleven patents on antenna inventions.

1. Basic Types

Antennas which theoretically have no limitation on the bandwidth are called *frequency independent*. In practice, the lower frequency limit is determined by the size of the antenna; the upper frequency limit, by the precision of construction. Actually, the electrical performance is not strictly independent of frequency, rather it is periodic with the logarithm of the frequency. Hence these antennas are called *logarithmically periodic*, *log-periodic*, or simply *LP* antennas. Some, which have the shape of equiangular spirals, are called *logarithmic spirals*, or *log-spirals*.

Geometrically, frequency-independent antennas are composed of a multiplicity of adjoining cells, each cell being scaled in dimensions relative to the adjacent cell by a factor which remains fixed throughout the structure. Two examples of planar LP geometries, having only a few cells, are shown in Fig. 1. In Fig. 1a, the nth cell is the annular region between concentric circles having radii R_n and R_{n+1}. In Fig. 1b the nth cell is the region between two concentric squares with sides of length L_n and L_{n+1}. If D_n represents some dimension of the nth cell and D_{n+1} the corresponding dimension of the $(n + 1)$st cell, then the relation

$$D_n/D_{n+1} = \tau$$

holds for all values of the integer n.

While it is true that the first successful, i.e., frequency-independent, LP structures were constructed from planar sheets of thin metal conductor, several of the principles of frequency-independent design are not restricted to structures with planar cells. Fig. 1c shows a conical log-spiral, for example, wherein a cell can be defined as the part of a conical surface bounded by two spheres of radii R_n and R_{n+1}. Fig. 1d shows an LP geometry in which each cell is the surface of a truncated pyramid.

The LP geometry is used to lay out an antenna by first configuring an electrical conductor within any one of the cells. The same configuration of conductor, properly scaled, is then reproduced in the other cells. If we presume this process to be repeated infinitely many times for the smaller cells, the resulting structure will converge to a point. Infinite repetition of the larger cells causes the size of the structure to increase without bound. An important observation can be made about such geometry. If any scale factor, τ^n, where n is an arbitrary integer, is applied to this geometric figure, the result is the identical geometric figure. That is to say, any LP geometry which extends to the limit point and to infinite size scales into itself whenever a scale factor of the form τ^n is applied. If we further presume that a frequency-independent point source of electromagnetic waves is located at the limit point of a perfectly conducting antenna having such geometry, the fields associated with the antenna must be identical at all frequencies that are related by τ^n. This

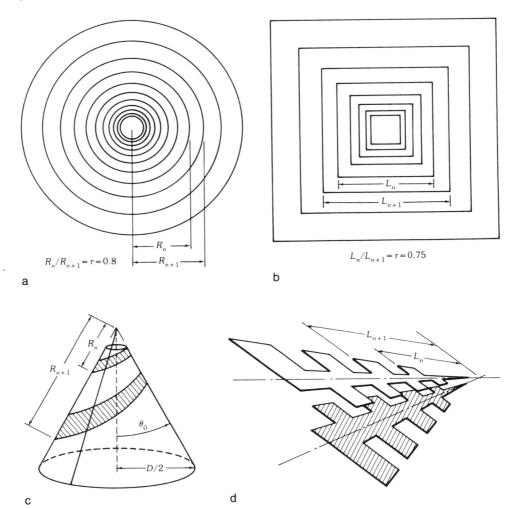

Fig. 1. Some planar and nonplanar log-periodic geometries. (*a*) Circular. (*b*) Square. (*c*) Conical. (*d*) Pyramidal.

demonstrates that the electrical performance of the antenna will be periodic with the logarithm of the frequency, as previously mentioned.

Practically speaking, there are, of course, two impossibilities in the above plan for achieving LP performance: (*a*) cells of near-zero dimension cannot be built, and (*b*) cells of near-infinite size are too large. The trick to achieving log-periodic electrical performance from an antenna with a *finite* LP geometry is associated with the choice of a cell configuration that will minimize the effects of the inevitable truncations of both large and small cells. During the early stages of development of LP antennas, many antennas having finite LP geometry were found not to have LP performance because of large truncation effects. To reduce the effect of the truncation of the large cells, the wave traveling from the excitation point along the

Frequency-Independent Antennas

structure must be attenuated before it encounters the end. Attenuation due to radiation is preferred over reflective attenuation in order to provide a near-constant impedance versus frequency characteristic. This requires that the cell dimensions must increase to a size sufficiently large compared with the wavelength so that appreciable radiation can occur. Generally, at least one dimension of the largest cell on a frequency-independent antenna must be approximately one-half wavelength in order to substantially eliminate the effect of the large-end truncation. It does not follow, however, that any finite LP which contains such a half-wave cell will display LP performance (zero truncation effect).

In order to approximate the condition of excitation at the limit point it is necessary to accurately scale the small cells until their dimensions are a small fraction of the wavelength at the highest frequency of operation. The conductor configuration in each cell must be capable of providing means of propagating the electromagnetic energy from the feed point toward the larger cells at all frequencies in the intended operating band.

Even though all the above conditions for self-scaling and minimal truncation effects are satisfied by a given structure, there is still no guarantee that the resulting antenna will perform with any desired degree of independence of frequency. Some means must be found to control the variations in performance over a period in frequency. For planar antennas, self-complementary geometry can be used to eliminate, at least theoretically, variations in the input impedance. An example of self-complementary geometry is shown in Fig. 2.

In Fig. 2 the planar cells are partitioned by four sinuous radial lines, OA, OB, OC, OD, identical except for rotation by multiples of $90°$. The areas between OA

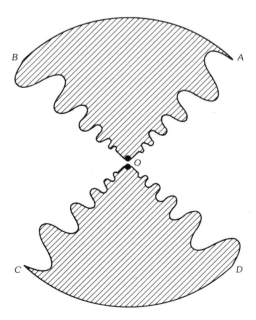

Fig. 2. Example of self-complementary geometry.

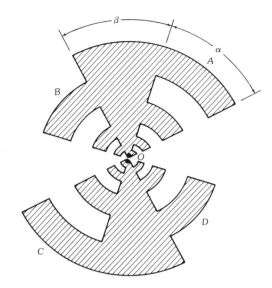

Fig. 3. Self-complementary geometry with flat-top radial lines. (*Adapted from Carrel [8]*)

a

Fig. 4. Multiarm self-complementary antenna example and terminal impedances. (*a*) Self-complementary structure with ninefold symmetry. (*b*) Theoretical values of terminal impedance for several multiarm self-complementary antennas. (*After Deschamps [2]*, © *1959 IEEE*)

Frequency-Independent Antennas

and OB and between OC and OD are conducting sheet. The areas between OD and OA and between OB and OC have the identical shape but are free space. The shape of the radial lines can take many different forms. Fig. 3 shows a case where the four radial lines, OA, OB, OC, and OD, are properly scaled flat-topped pulses.

The self-complementary property of the structures of Figs. 2 and 3 results from the 90° separation between the radial boundary lines of the cells. Using four such lines produces two conducting regions emanating from the origin. Introducing a small gap at the limit point provides two input terminals for the antenna. The conductor is thus divided into two arms that are symmetrically driven by a generator connected to the terminals. The theoretical value for the input impedance of a two-arm, self-complementary antenna with no truncation effect is $60\pi = 189 \, \Omega$ for any frequency [1]. Log-periodic antennas can also be constructed with more than two arms as shown in Fig. 4a. The theoretical values of the terminal impedance for several multiarm self-complementary antennas with various inter-

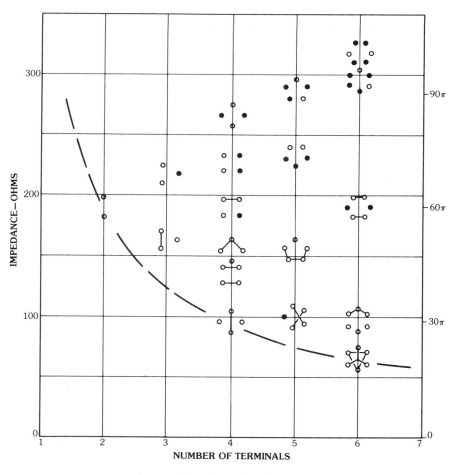

Fig. 4, *continued.*

connections are given in Fig. 4b [2]. In this figure, for each configuration the two groups of terminals connected to the source are represented by small circles. The floating terminals are represented by black dots. Measured values differ somewhat from the theoretical ones due to the finite thickness of the conductors, presence of a feed line, etc. [3].

Fig. 5 shows a different sort of boundary for the cells. In this case the edges of the conductor are defined by logarithmic spirals, each rotated 90° with respect to its neighbor. Usually log-spiral antennas are circularly polarized whereas other log-periodic antennas are linearly polarized. Log-spiral antennas are discussed in detail in Section 4.

Planar LP antennas can be truncated either by eliminating all conductor beyond a certain distance from the limit point or by filling the plane with conductor beyond that distance. In the latter case the antenna can be considered to be a slot antenna.

Symmetry dictates that the radiation occur in the same way on both sides of planar antennas. Patterns typical of planar LP antennas are shown in Fig. 6. These were measured for a slot antenna of the type shown in Fig. 6a with $\alpha = 45°$, $\beta = 45°$, and $\tau = 0.81$. The self-complementary property is preserved for different lengths of the teeth (angle α) as long as the sum of α and β is fixed at 90°. However, the low-frequency limit of frequency-independent operation is increased as α is decreased

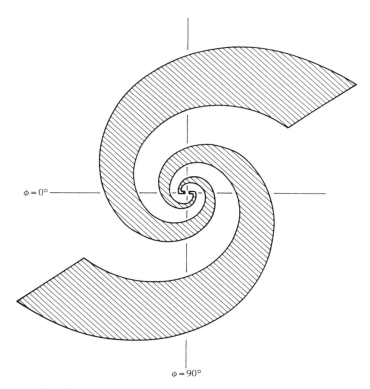

Fig. 5. Logarithmic-spiral geometry. (*After Carrel [8]*)

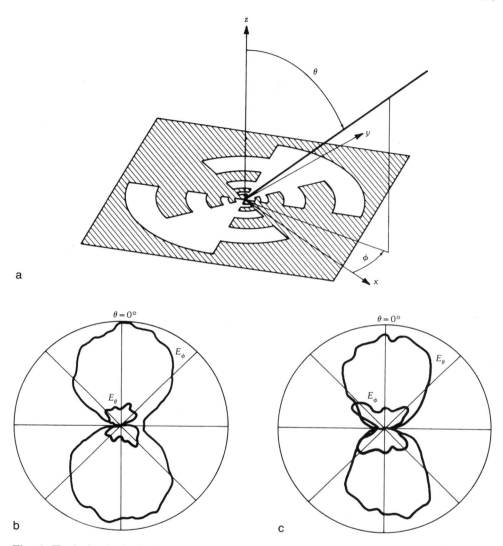

Fig. 6. Typical principal plane radiation patterns. (*a*) Antenna orientation. (*b*) For $\phi = 0°$ and $f = 1530$ MHz. (*c*) For $\phi = 90°$ and $f = 1530$ MHz. (*d*) For $\phi = 0°$ and $f = 1700$ MHz. (*e*) For $\phi = 90°$ and $f = 1700$ MHz. (*After DuHamel and Isbell [3], © 1957 IEEE*)

for a given maximum antenna radius. For example, an antenna with a maximum radius of 10 in had a low-frequency limit of approximately 400 MHz when $\alpha = 45°$, but the low-frequency limit was raised to 800 MHz when α was changed to 20°. It is thus established that the low-frequency limit is determined by the length of the longest tooth.

The principal-plane beamwidths of antennas of the type shown in Fig. 3 can be controlled to some extent by changing the scale factor τ. Table 1 gives data

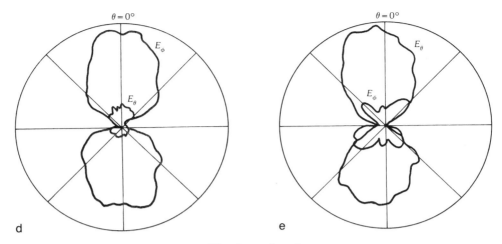

Fig. 6, *continued.*

Table 1. Principal-Plane Beamwidths Versus Scale Factor for Planar Self-Complementary LP Antennas (See Fig. 3) with $\alpha = 45°$

Scale Factor (τ)	Beamwidth
0.81	73°
0.70	70°
0.50	55°
0.25	38°

to illustrate this relationship for self-complementary antennas with $\alpha = 45°$. The bandwidth over which the beamwidth remains constant is reduced as the beamwidth decreases. So while the 10-in (radius) antenna had a frequency-independent beamwidth for frequencies between 400 and 4000 MHz when $\tau = 0.81$, the same size antenna had frequency-independent beamwidth only above 900 MHz when τ was made equal to 0.25. Values of τ below 0.25 produce fields with large cross polarization.

The level of the mean impedance can be reduced by increasing the angle α, e.g., for $\alpha = 75°$ the mean impedance was observed to be approximately 100 Ω. However, the variation of impedance with frequency for such antennas, which are not self-complementary, is greater than it is for antennas which are self-complementary.

For most applications bidirectional patterns, such as those of Fig. 6, are not acceptable. The requirement instead is for a single beam. The symmetry of planar antennas must be eliminated to accomplish this. Planar LP slot antennas can be made unidirectional in a frequency-independent manner by placing an absorbing cavity on one side of the ground plane. For cavities which are sufficiently large the impedance and the patterns in the remaining hemisphere are affected very little by the cavity. Of course, the gain is less by about 3 dB than that of a unidirectional antenna having the same pattern.

Frequency-Independent Antennas

When the two arms of a metal-arm LP antenna are inclined toward each other as shown in Fig. 7, unidirectional patterns can result [4]. Fig. 8 shows E-plane patterns measured for the antenna shown in Fig. 7 with $\tau = 0.81$, $\alpha = \beta = 45°$, and $\phi = 90°$. As the angle between the planar elements, ψ, is reduced, one lobe of the pattern decreases in magnitude. For $\psi < 50°$ this lobe is practically nonexistent. The lobe which remains is in the direction from the larger to the smaller end of the structure. Behavior of the nonplanar LP antenna is thus much different than for wire-type vee antennas or horns that radiate in the direction of increasing size. Fig. 9 shows that the mean impedance drops and the variation in impedance over the band increases as the angle between the two arms decreases.

Log-periodic antennas with arms constructed from planar sheet metal may be practical for frequencies high enough so that the physical structure is satisfactory mechanically. However, there are many applications for frequency-independent antennas in long-distance communications which are conducted in the hf (3–30 MHz) band. In this band the antenna must be so large that construction from planar sheet metal is no longer mechanically feasible. Fortunately, much of the conductor can be removed from the arms without appreciably affecting the radiation patterns. The proper procedure is to retain the conductor on the edges of the teeth in the manner illustrated in Fig. 10 [5]. Not only does this result in a structure that is practical for use at high frequency, it also reduces the capacitance between the arms and lessens the amount that the impedance varies with frequency within the operating band. This makes it possible to bring the two arms of the

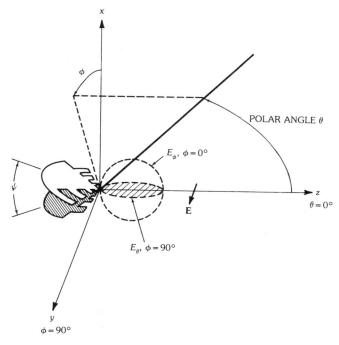

Fig. 7. The first unidirectional LP antenna showing the backfire beam. (*After Isbell [4]*)

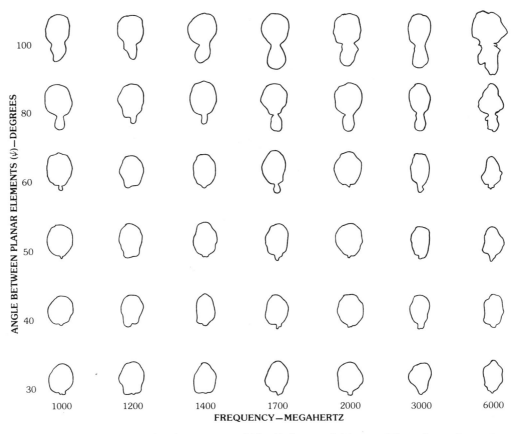

Fig. 8. *E*-plane patterns for the antenna of Fig. 7, for $\tau = 0.81$, $\phi = 90°$, and $\alpha = \beta = 45°$. *(After Isbell [4])*

antenna into a position parallel to each other, with only a small distance between them, without producing wide excursions in the impedance versus frequency behavior.

Another limiting process with the structure of Fig. 10 results in a very important geometry. If the widths of the teeth are made to approach zero, a log-periodic array of dipoles is obtained [6]. The simplicity and excellent performance of the log-periodic dipole array have made it one of the most widely used of frequency-independent antennas. It was the first to be extensively analyzed [7, 8] and, as a result, much design information is available.

2. Log-Periodic Dipole Arrays

Fig. 11a shows the arrangement of elements in an LP dipole (or LPD) array. Fig. 11b shows how the dipoles are fed by a two-wire line. The two basic parameters are the scale factor τ and the angle α between the centerline and the tips of the dipoles. The alternative parameter, σ, is an "aspect ratio" for each cell

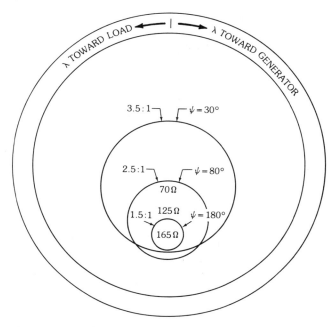

Fig. 9. Boundary circles of the impedance loci as a function of ψ. [*After Isbell [4]*]

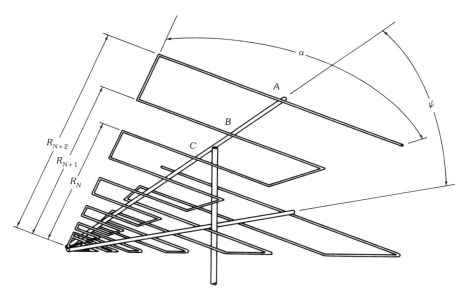

Fig. 10. A wire-outline LP antenna. (*Courtesy Collins Defense Communications, Rockwell International Corporation*)

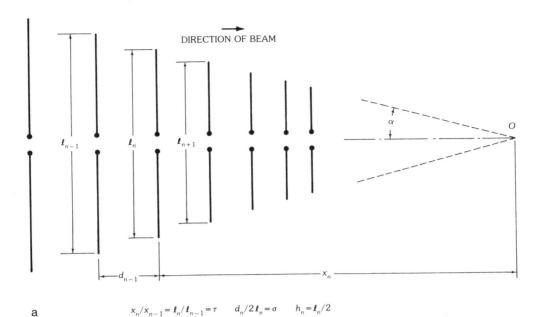

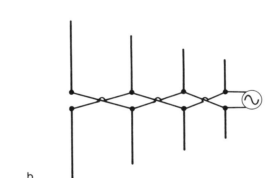

Fig. 11. The LP antenna, including symbols used in its description. (*a*) Schematic of antenna. (*b*) Method of feeding. (*After Carrel [7, 8]*, © *1961 IEEE*)

consisting of a single dipole and the transmission line between that dipole and the next adjacent dipole. Consideration of the geometry discloses that the parameters τ, α, and σ are related by the equation

$$\sigma = \tfrac{1}{4}(1 - \tau)\cot\alpha \qquad (1)$$

The nomograph given in Fig. 12 provides a convenient method for transforming from one set of parameters to another, which may occur frequently during the design process.

The transposition of conductors of the feed line is essential to the frequency-

Frequency-Independent Antennas

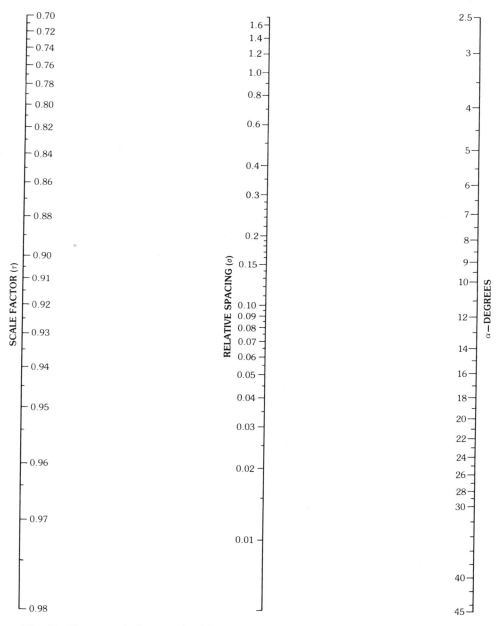

Fig. 12. Nomograph for $\sigma = (1/4)(1 - \tau)\cot\alpha$. *(After Carrel [7, 8], © 1961 IEEE)*

independent performance of LP dipole arrays. One way to achieve the transposition, and provide at the same time mechanical support for the dipoles, is to use twin-boom construction for the feeder, as shown in Fig. 13. It should be noted that, following Carrel, the definition of τ for an LP dipole entails the ratio of the lengths of two adjacent dipoles. However, when the twin-boom construction is used, each

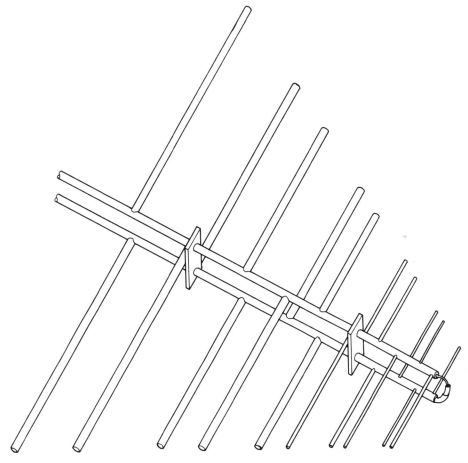

Fig. 13. An LP antenna with twin-boom construction for the feeder. (*After Carrel [8]*)

cell of an LP dipole contains two dipoles. Transition from the balanced drive of the two booms to an unbalanced coaxial cable can be achieved quite simply by passing the cable through the interior of one of the booms as shown in Fig. 14. It is important that symmetry be maintained in this type of "self-balun," i.e., that the actual feed slot occur midway between the two booms. Otherwise, a lumped reactance is introduced at the feed point and this causes rotation of the impedance locus off the real axis of a Smith chart.

Analysis of LP dipole arrays proceeds by separating the dipoles and transmission line as shown in Fig. 15 [7,8]. The terminal properties of the N-element network of dipoles can be represented by an $N \times N$ impedance matrix. Carrel used the formulas of the induced emf method to calculate the impedance matrix. Later work [9–13] used moment methods. The results of the several techniques are not appreciably different [14].

The N-terminal-pair network of the feed line can be readily represented by an admittance matrix. By enforcing continuity of the current at all except the

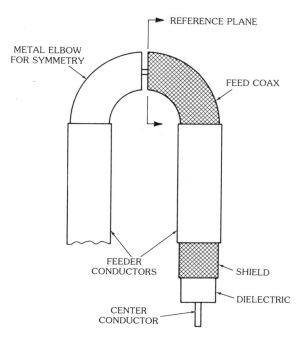

Fig. 14. Details of the symmetrical feed point, showing the reference plane for impedance measurements. (*After Carrel [8]*)

generator terminals when the feeder and antenna networks are paralleled, a set of equations for analysis of the LP dipole results. Fig. 16 shows amplitude and phase of the terminal currents in the several dipoles of an LPD array at the frequency f_3 where the third dipole (counting from the large end) is one-half wavelength. It is to be noted that the input currents of a few dipoles are significantly larger than those of any of the others. These dipoles with highest current are collectively called the *active region*, and they produce most of the radiated field.

When frequency is changed, the active region will move along the axis of the LP dipole in such a way that the dimensions of the active region in wavelengths remain almost constant. For this reason the radiation pattern is insensitive to the frequency changes. Fig. 17 shows the relative amplitude of the input currents at several different frequencies, illustrating the movement of the active region. It is apparent from these results that the pattern will begin to change when the active region encounters either of the two truncation points. The band of frequency-independent patterns is therefore somewhat less than the ratio of longest-to-shortest dipole lengths. This ratio, called the *structure bandwidth*, is given by

$$B_s = h_1/h_N = \tau^{1-N} \qquad (2)$$

The operating bandwidth B and the structure bandwidth B_s are related by

$$B = B_s/B_{ar} \qquad (3)$$

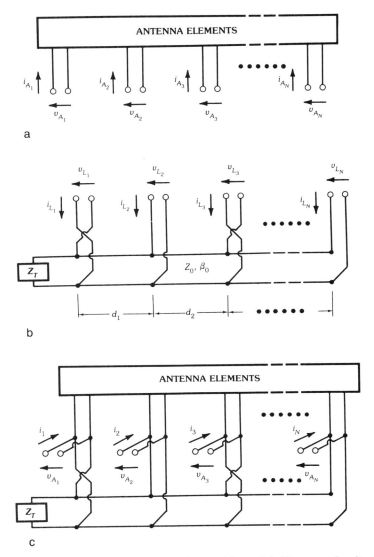

Fig. 15. Schematic circuits for the LP interior problem. (*a*) Element circuit. (*b*) Feeder circuit. (*c*) Complete circuit. (*After Carrel [7, 8], © 1961 IEEE*)

where the term B_{ar} is called the bandwidth of the active region and is dependent on the parameters of the dipole array. Fig. 18 summarizes the results of analyzing many different LP dipole arrays. For $\tau > 0.85$, B_{ar} is almost a linear function of σ. The lines on Fig. 18 show the excellent agreement between the computed data and the empirical formula

$$B_{ar} = 1.1 + 7.7(1 - \tau)^2 \cot \alpha \qquad (4)$$

The nomograph of this relation in Fig. 19 can be used to facilitate design.

Frequency-Independent Antennas

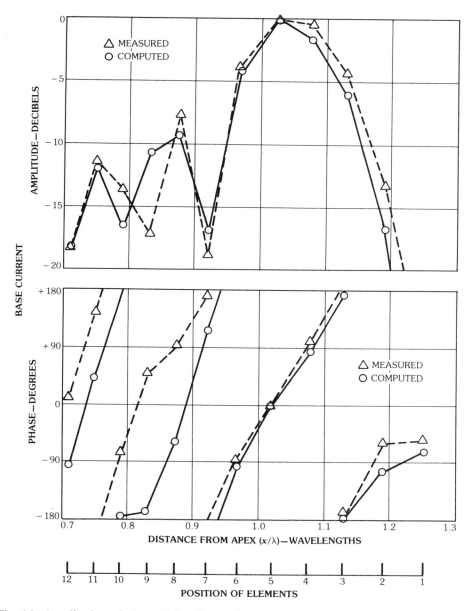

Fig. 16. Amplitude and phase of the element base current versus relative distance from the apex, at frequency f_3, for $\tau = 0.95$, $\sigma = 0.0564$, $Z_0 = 100\ \Omega$, $h/a = 177$, and Z_T a short circuit at $h_1/2$. *(After Carrel [8])*

Once the dipole base currents have been determined, the far fields can be calculated and, from them, the directivity. These data can be represented as plots of constant directivity contours on τ and σ axes as presented in Fig. 20. The contours shown are those first reported by Carrel. The directivity values, however,

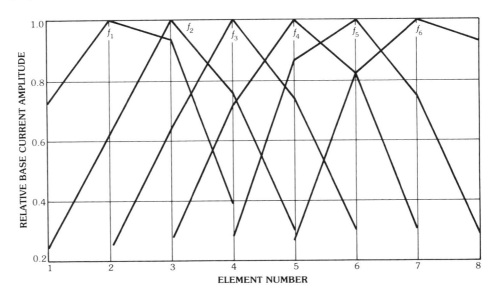

Fig. 17. Relative amplitude of base current in the active region versus element number for frequencies f_1 through f_6, for $\tau = 0.888$, $\sigma = 0.089$, $N = 8$, $Z_0 = 100 \ \Omega$, $h/a = 125$, and Z_T a short at $h_1/2$. (*After Carrel [8]*)

have been reduced in accordance with data published later which are believed to be more accurate [14]. A design procedure can be initiated using the above results.

Design of Log-Periodic Dipole Arrays

To varying degrees all the parameters which specify an LP dipole have an effect on the performance. Table 2 lists the parameters and describes how each affects the performance. The entries in the table denote how the performance changes with an increase in the parameter on the left while all other parameters are held constant.

The directivity of an LP dipole depends primarily on the combination of τ and σ. Since an increase in directivity implies an increased aperture size, high-directivity models are characterized by small α and large L/λ_{max}. For a given τ, σ, and element radius a, the input impedance depends on the characteristic impedance of the feeder. Fortunately the directivity is essentially independent of the feeder impedance. This makes it possible to design an antenna for a given directivity and then, in most cases, the input impedance can be adjusted to the required value. The exceptions occur on models with both low τ and low Z_0. Under these conditions the radiating efficiency of the active region is low, and end effects appear. Input impedances from 50 Ω (for high values of τ) to 200 Ω (for all values of τ) have been obtained.

For most applications the objective is to achieve a given power gain and input impedance over a given frequency band. These specifications do not determine a unique design. The relative importance of minimizing the number of elements or the size of the antenna must also be considered. The number of elements is determined by τ; as τ increases, the number of elements increases. The antenna size is determined by the boom length (the distance between the longest and the

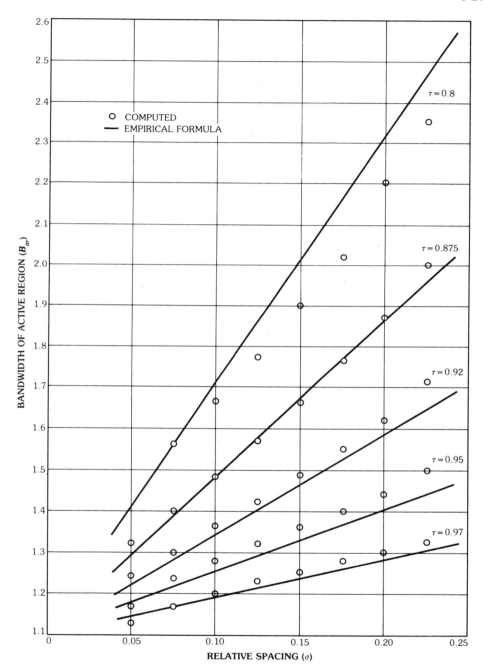

Fig. 18. Bandwidth of the active region versus σ and τ. (*After Carrel [8]*)

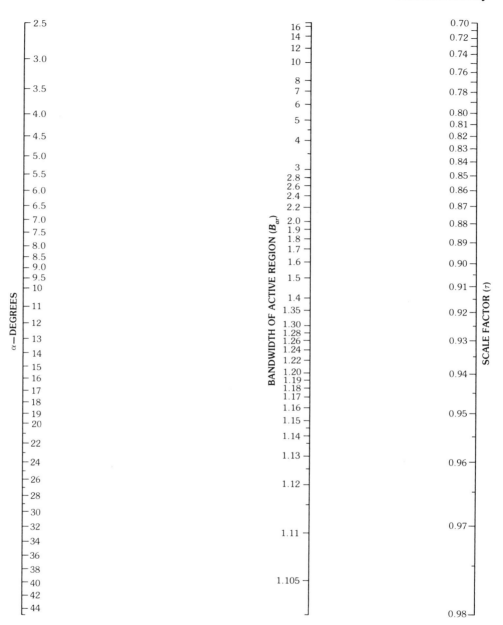

Fig. 19. Nomograph for $B_{ar} = 1.1 + 7.7(1 - \tau)^2 \cot \alpha$. (*After Carrel [8]*)

shortest dipoles), which depends primarily on α. As α decreases, the length increases.

(*a*) Given a value of directivity (desired maximum power gain divided by a reasonable value for efficiency), a set of values for τ and σ can be determined from Fig. 20. Since there are many combinations of τ and σ that will work, it is well to

Frequency-Independent Antennas

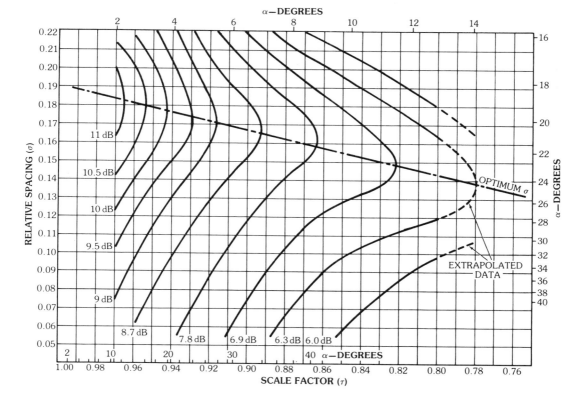

Fig. 20. Computed contours of constant directivity versus τ, σ, and α, for $Z_0 = 100\ \Omega$, $h/a = 177$, and Z_T a short at $h_1/2$. (*Adapted from Carrel [7,8], © 1961 IEEE*)

keep in mind that, for a given bandwidth, a large value σ (small α) will give a very long antenna, and that a large value of τ will give many elements.

(*b*) The value of α for the selected values of τ and σ can be determined from the formula

$$\tan \alpha = (1 - \tau)/4\sigma \tag{5}$$

or read from the nomograph in Fig. 12.

(*c*) Fig. 18 or the nomograph in Fig. 19 can be used to determine the bandwidth of the active region.

(*d*) The required structure bandwidth is next found by using (3).

(*e*) The nomograph of Fig. 21 or the formula

$$L/\lambda_{\max} = \tfrac{1}{4}(1 - 1/B_s)\cot \alpha \tag{6}$$

gives the required length of the array in terms of the wavelength $\lambda_{\max}$ of the low-frequency limit.

(*f*) The formula

Table 2. LP Dipole Parameters and Their Effect on the Observed Performance

LP Dipole Parameter*	Bandwidth of Active Region (B_{ar})	Input Impedance (always less than Z_0)	Directivity	Phase Center Distance to the Apex x_ϱ	Boom Length L/λ_{max} for a Fixed Operating Bandwidth B
τ (σ constant)	Decrease	Small decrease	Increase	Increase (depends on α)	Decrease to a point depending on B, then increase
τ (α constant)	Decrease	Small decrease	Small increase	Independent	Decrease
σ (τ constant)	Increase	Increase	Increase	Increase (depends on α)	Increase
σ (α constant)	Increase	Increase	Small decrease	Independent	Increase
Z_0	Independent but location of AR moves toward apex	Increase	Small decrease	Small decrease	Small decrease
h/a	Independent, but location of AR moves away from apex	Increase	Small decrease	Small increase	Small increase

*The table entries hold true over the following range of parameters for which frequency-independent operation has been verified: $0.875 < \tau < 0.98$, $0.05 < \sigma < \sigma_{optimum}$, $100 < Z_0 < 500$, and $20 < h/a < 10000$. Any one of τ, σ, or Z_0 may take on other values, provided the remaining parameters are suitably restricted as explained in the text.

Frequency-Independent Antennas

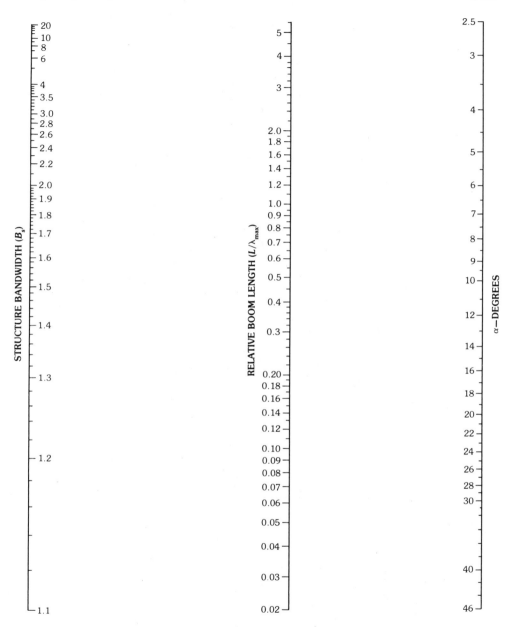

Fig. 21. Nomograph for $L/\lambda_{max} = (1/4)(1 - 1/B_s)\cot\alpha$. (*After Carrel [7, 8], © 1961 IEEE*)

$$N = 1 + (\log B_s)/[\log(1/\tau)] \tag{7}$$

or the nomograph of Fig. 22 can be used to determine the number of elements required to cover the desired band.

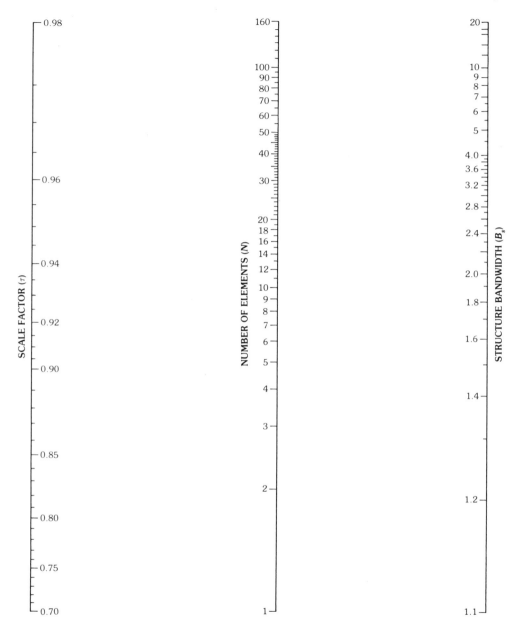

Fig. 22. Nomograph for $N = 1 + (\log B_s)/(\log \tau^{-1})$. (*After Carrel [7, 8]*, © 1961 IEEE)

(g) If either the boom length or the number of elements, but not both, is too large, return to step *a* and select a different τ, σ pair. Repeating the process for different values of τ and σ will establish the trend. For some values of B_s minimum boom length cannot be attained for values of τ and σ within the range of the graph of Fig. 20. In these cases a compromise will have to be made. If a choice of τ and σ exists and there is no apparent basis for a specific selection, it should be noted that the swr increases and the front-to-back decreases as σ departs from the optimum values designated on Fig. 20.

(h) Since the graph of Fig. 20 is based on $Z_0 = 100$ Ω and $h/a = 177$, an adjustment should be made if it is anticipated that Z_0 and h/a will depart from these values by more than a factor of 2. The exact value of Z_0 is yet to be determined. However, it is known that the feeder impedance is always greater than the input impedance, so if R_0 is greater than 100 Ω, the directivity contours of Fig. 20 will read a fraction of a decibel high. If the ratio h/a is much different from 177, another adjustment must be made. According to the curve of Fig. 23 the directivity decreases by about 0.1 dB for each doubling of h/a; for $h/a > 177$ the constant directivity contours of Fig. 20 will read high.

(i) Once a design is achieved that has satisfactory patterns, attention can be given to obtaining a desired value for the input impedance. The achievable minimum swr will be dependent on the scale factor τ as shown in Fig. 24. If the required value of swr is lower than achievable with the value of τ already picked for the desired gain, it will be necessary to start the design procedure again with a new value of τ picked on the basis of impedance performance rather than directivity. It should also be noted from Fig. 24 that low values of τ produce low swr only for a narrow range of σ.

(j) Fortunately, the mean value of the impedance of an LP dipole is adjustable over a rather wide range by merely changing the characteristic impedance of the feeder. Fig. 25 illustrates this point for an LP dipole having $\tau = 0.888$, $\sigma = 0.089$, and $h/a = 125$. As Fig. 25 shows, the mean value of the input impedance, R_0, is less than the characteristic impedance of the feeder, Z_0. This is due to the shunt capacitance that is added to the feeder by the presence of the short (compared to

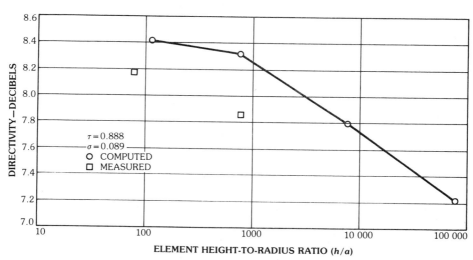

Fig. 23. Example of computed and measured directivity versus height-to-radius ratio. (*After Carrel [8]*)

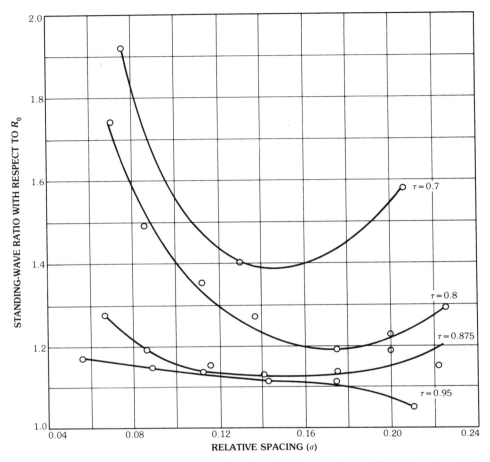

Fig. 24. Computed swr versus σ and τ, for $Z_0 = 100 \ \Omega$ and $h/a = 177$. (*After Carrel [8]*)

wavelength) dipoles in the transmission region, i.e., between the feed point and the active region.

(k) An approximate formula for the mean resistance R_0 is

$$R_0 = Z_0/\sqrt{1 + (Z_0/Z_a)(\tau/4\sigma)} \qquad (8)$$

where Z_0 is the characteristic impedance of the (unloaded) feeder, and Z_a is the average characteristic impedance of a dipole [15]:

$$Z_a = 120[\ln(h/a) - 2.25] \qquad (9)$$

A plot of this equation is presented in Fig. 26. The input impedance thus depends on the density of elements, which is reflected through the mean spacing parameter $\sigma' = \sigma/\sqrt{\tau}$. Using Fig. 26 to determine Z_a for a typical value of h/a, normalizing Z_a to the desired mean resistance R_0, evaluating σ' as determined from the above

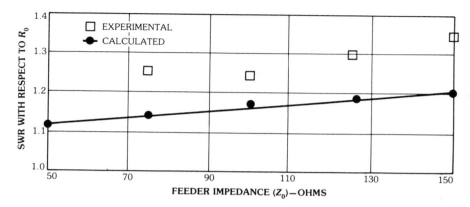

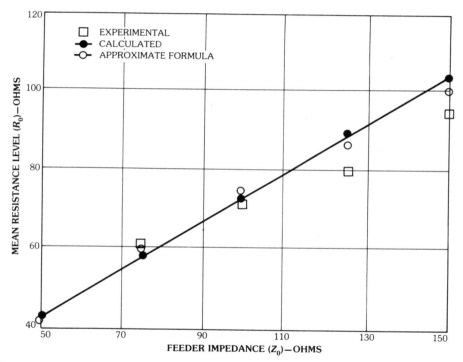

Fig. 25. Input impedance versus feeder impedance, for $\tau = 0.888$, $\sigma = 0.089$, $N = 8$, and $h/a = 125$. (*After Carrel [7, 8], © 1961 IEEE*)

steps related to achieving specified gain, Fig. 27 provides the value of characteristic impedance required for the feeder. For twin-boom or transposed two-wire feeders the spacing-to-diameter ratio of the two conductors determines the characteristic impedance through the relation

$$Z_0 = 120 \cosh^{-1}(b/D) \qquad (10)$$

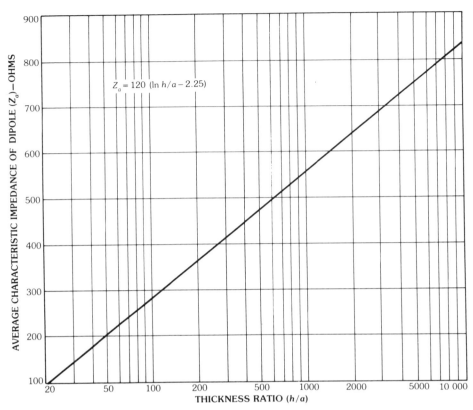

Fig. 26. Average characteristic impedance of a dipole versus height-to-radius ratio h/a. (*After Carrel [7, 8], © 1961 IEEE*)

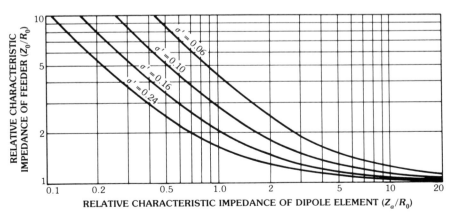

Fig. 27. Relative feeder impedance versus relative dipole impedance from the approximate formula. (*After Carrel [7, 8], © 1961 IEEE*)

Frequency-Independent Antennas

where b is the center-to-center spacing between the parallel conducting cylinders and D is the diameter of each of them.

To determine the physical lengths of the elements the length of one element must be related to a limit on the operating band. The approximation that the longest dipole must be ½ wavelength at the low-frequency limit can be refined to show a slight dependence on the feeder impedance as illustrated in Fig. 28. The length of the longest dipole is then given by

$$\ell_1 = S(\lambda_{max}/2) = 2h_1 \quad (11)$$

and the lengths of all other dipoles are obtained by multiplying the length of each adjacent longer dipole by the scale factor τ. The chosen value of σ provides means

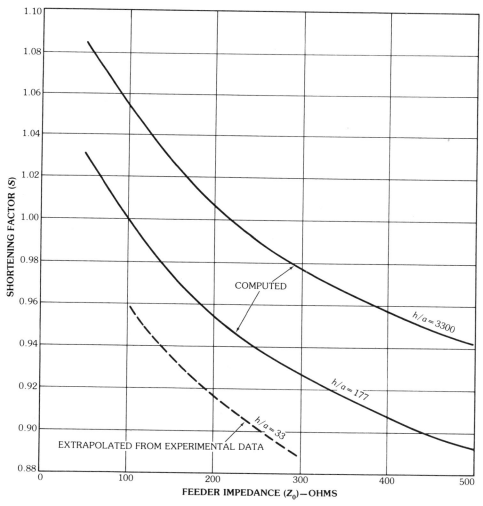

Fig. 28. Shortening factor versus feeder impedance and height-to-radius ratio. (*After Carrel [8]*)

of determining the spacing between each pair of adjacent dipoles. Thus the determination of all dimensions of the antenna has been accomplished and the design procedure is completed.

3. Periodic Structure Theory

The analysis of the log-periodic dipole array, being based on approximations that are valid for thin linear elements, is not readily extendable to other LP structures. Hence it is very useful to have a theory of more general scope. Modifications of the theory of uniform-periodic (UP) structures have provided such a tool [16]. For τ near unity, an LP structure can be considered as a perturbation of a UP structure. The powerful theory of Floquet can be applied to UP structures. This makes it possible either accurately or, at least, approximately to determine the propagation constant for a wave traveling along the UP structure. Although many such waves may be possible, oftentimes the lowest-order one is sufficient to approximate the near fields of an LP structure.

Consider the UP geometry shown in Fig. 29 extended to infinity in both directions along the z axis. Suppose a wave with (complex) propagation constant $\gamma = \alpha + j\beta_0$ is traveling in the positive z direction. Of course, a single term of the form

$$\exp(-\gamma z)$$

would not be sufficient to satisfy whatever boundary conditions may be imposed by the structure in each cell, i.e., the given wave is adequate only for a homogeneous medium. Functions of the form

$$\exp(j2n\pi/d)z, \quad n \text{ an integer}$$

will provide for field variations across each cell without affecting the propagation constant γ. Each vector component of the field is therefore expressible as a superposition of fields having z variation of

$$\exp[-(\gamma - j2n\pi/d)z] = \exp[-\alpha z - j(\beta_0 - 2n\pi/d)z]$$

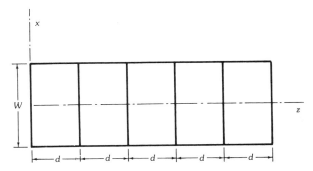

Fig. 29. A periodic structure with five cells.

Frequency-Independent Antennas

In keeping with the terminology of Fourier analysis the wave with

$$\gamma_0 = \alpha + j\beta_0 \tag{12}$$

is called the *fundamental wave* and those with

$$\gamma_n = j\beta_n = j(\beta_0 - 2n\pi/d) \tag{13}$$

are called *space harmonics*. The amplitudes and phases of the space harmonics relative to the fundamental are determined by the structure of a specific cell. The concept that the near fields of a periodic structure can be represented as a superposition of waves having the variations given by (12) and (13) is very useful even when the determination of their relative amplitudes and phases is not pursued. In fact, these basic results of periodic structure theory, when combined with some experimental data, provide considerable insight into the operation of LP antennas. This is particularly true for a class of LP antennas that do not involve the excitation of resonant elements. One such antenna is the zigzag.

Consider the zigzag wire shown in Fig. 30. For thin, highly conducting wires, the results of Pocklington [17] show that the current in the wire propagates with the intrinsic phase velocity of the surrounding (homogeneous) medium. It might be expected that a reflected wave would be generated at each of the corners of the zigzag. However, experimental data taken with the practical realizations of the zigzag, which necessarily involve a nonzero radius of curvature for the wire at each corner, fail to display evidence of such reflections. Assuming that the current along the wire travels with free-space phase velocity, the near fields of the zigzag wire will have a fundamental-wave phase constant given by

$$\beta_0 = k \csc \psi \tag{14}$$

where ψ is the pitch angle defined in Fig. 30. The phase constants of fundamental waves for several zigzags with various pitch angles are shown in a plot of frequency versus phase constant (k-β) diagram in Fig. 31. To the degree that (14) holds, the phase constants on the zigzag wire are directly proportional to frequency, i.e., no dispersion occurs. Normalizing all the space-harmonic phase constants with respect to the free-space wave number k,

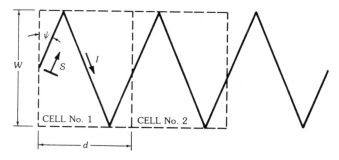

Fig. 30. Basic zigzag wire.

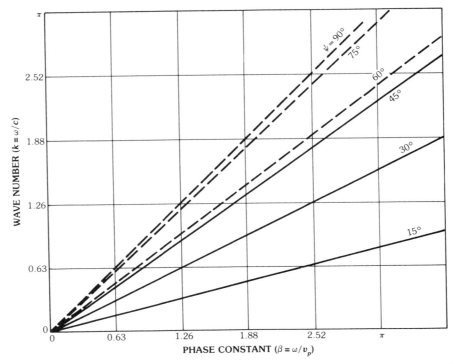

Fig. 31. Approximate phase constants of fundamental waves for several UP zigzag wires.

$$\beta_n/k = \beta_0/k - 2n\pi/kd = \csc\psi - n(\lambda/d) \quad (15)$$

Fig. 32 shows the normalized space spectrum for a nondispersive periodic structure with $\beta_0/k = 2$. As long as $|\beta_n/k| > 1$ the space harmonics are slow waves which produce no radiation. However, $|\beta_n/k| < 1$ corresponds to the visible range of the space spectrum. A wave with normalized phase constant within the visible range is a fast wave which contributes radiation having maximum value in the direction obtained by projecting from the normalized phase constant upward to the unit circle as shown in Fig. 32c. Note that there are no fast waves when $\lambda/d = 4$. The first space harmonic enters the visible range when $\lambda/d = 3$, and the direction of the beam produced is backfire, i.e., opposite to the direction of propagation of the fundamental. When $\lambda/d = 2.5$, the beam produced by the $n = 1$ space harmonic makes an angle of 60° with respect to the axis.

Fig. 33 shows several measured radiation patterns for a monofilar zigzag wire. The zigzag wire was fed against a linear wire in the plane orthogonal to that of the zigzag. By measuring the principal polarization pattern in the E-plane, only the effects of currents on the zigzag are observed. According to the simple approximate theory outlined above, the zigzag wire with $\csc\psi = 4$ and cell length of 4 cm should produce a backfire beam at 1.5 GHz and a broadside beam at 1.875 GHz. The measured patterns in Fig. 33 agree quite well with these predictions. They also indicate that backfire radiation with narrower beamwidths occurs at frequencies

Frequency-Independent Antennas

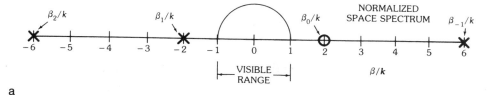

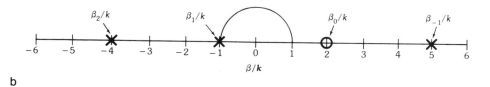

Fig. 32. Variation of space harmonics with frequency, where $\beta_n/k = \beta_0/k - n(\lambda/d)$. (a) For $\lambda/d = 4$. (b) For $\lambda/d = 3$. (c) For $\lambda/d = 2.5$.

below 1.5 GHz. The excess cell-to-cell phase shift at the lower frequencies moves the principal lobe away from the visible region, producing the reduced beamwidth. This implies that the active region on an LP zigzag will begin at cells that are smaller than those phased for backfire. When those cells radiate most of the incident energy, the LP zigzag will have a narrow beam. On the other hand, if energy penetrates to the region where the cells are phased for broadside radiation, the pattern can be expected to be much broader.

The k-β diagrams of Fig. 34 were constructed from data measured for several balanced, bifilar UP zigzags with different pitch angles. In the slow-wave regions, $\beta > k$, the wavelength along the axis can be determined by probing the standing-wave field pattern established by reflection of the incident wave from the open end. As one or more space harmonics enter the visible region, however, the incident wave along the zigzag is attenuated and the wave reflected from the open end soon becomes negligibly small. The $n = 1$ space-harmonic phase constant is then determined from the direction of the peak of the beam using the construction of Fig. 32c.

The variations with frequency observed for the UP zigzag correspond to variations with the cell dimensions for an LP zigzag. The region of small cells in Fig. 35 would correspond to the low-frequency conditions for the UP antenna. The fundamental wave and all space harmonics are slow waves and, therefore, little radiation occurs. The region where the first reverse traveling space harmonic

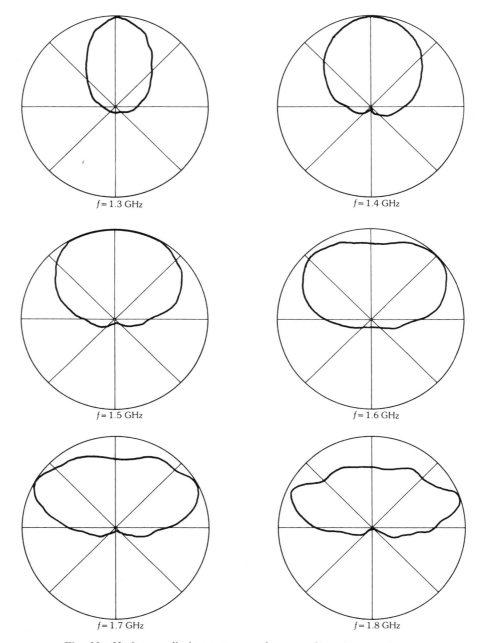

Fig. 33. *H*-plane radiation patterns of a monofilar zigzag antenna.

approaches the backfire condition ($\beta = -k$) produces appreciable radiation and corresponds to the active region. If the radiation is sufficient, the larger cells will be unexcited. This is usually desirable to avoid radiation in directions far from backfire.

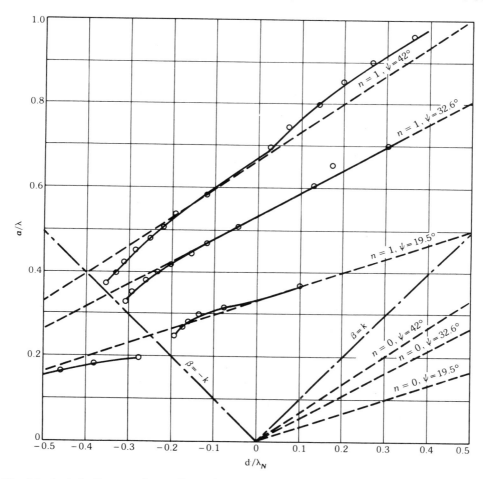

Fig. 34. A k-β diagram for uniform balanced zigzag antennas, showing variation of dispersive curve with pitch angle.

Design of Log-Periodic Zigzag Antennas

The geometric relationships among several types of balanced, two-arm, log-periodic antennas are illustrated in Fig. 36. The wire-outline trapezoidal-tooth LP is shown again in Fig. 36a. A similar antenna with triangular, instead of trapezoidal, teeth is shown in Fig. 36b. The performance of triangular-tooth and trapezoidal-tooth log-periodics of the same parameters (τ, α, and ψ) is very nearly the same [5]. Removing the central boom, the axial conductor of Fig. 36b, from a triangular-tooth, wire-outline LP produces the thin-wire, log-periodic zigzag of Fig. 36c. This antenna does not scale exactly when made using constant-diameter wire. Replacing the wire by a thin metal sheet with tapered width gives the tapered-arm LP zigzag in Fig. 36d. The angle between the center line and the extremities of each arm is called α, as was done for the LPD (see Fig. 11a). However, the angle β is used to define the width of the flat conductor at the tapered-arm zigzag.

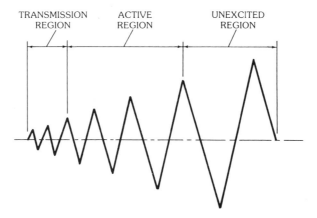

Fig. 35. One arm of a balanced LP thin-wire zigzag antenna, showing various regions for midband operation.

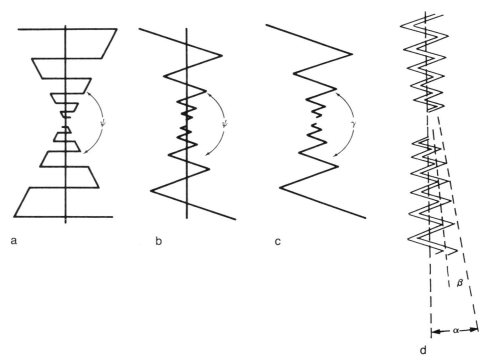

Fig. 36. A comparison of the LP zigzag antenna and the wire, nonplanar trapezoidal-tooth, and triangular-tooth antennas. (*a*) Wire-outline trapezoidal-tooth LP. (*b*) Triangular-tooth LP. (*c*) Thin-wire LP zigzag. (*d*) Tapered-arm LP zigzag. (*After Kuo and Mayes [20]*)

Frequency-Independent Antennas

Analysis of zigzag wire antennas like those of Fig. 36c is feasible [18, 19] and provides guides for design that will be presented later. Using the tapered-width arms as in Fig. 36d gives an additional parameter for control of the antenna performance, but the analysis is much more difficult and has not been reported. However, an extensive program of measurements for tapered-arm antennas, including thin-wire arms as a special case, provides some data on which a design procedure can be based [20]. To varying degrees all the parameters which are necessary to specify an LP zigzag antenna have an effect on the observed performance. Table 3 lists the parameters of tapered-arm LP zigzag antennas and qualitatively describes how each affects performance. (The table entries denote the change in performance for an increase in the parameters of the first column.)

The radiation pattern of an LP zigzag antenna depends primarily on the combination of α and β when τ is held constant. When the angle γ between the planes of the zigzag elements is set equal to 2α, the E-plane and H-plane beamwidths will be almost equal. The H-plane beamwidth can be decreased by increasing γ, but the gain increase that can be obtained in this way is limited by the appearance of side lobes for values of γ that are much larger than 2α. The following design procedure is based on $\tau = 0.9$ and $\gamma = 2\alpha$.

(a) Desired values of directivity can be reduced to required beamwidths by using the nomograph of Fig. 37.

(b) The pattern beamwidths depend primarily on α and β. Values of these two parameters that will give realizable beamwidths are given in the graphs of Fig. 38.

Table 3. Log-Periodic Zigzag Parameters and Their Effect on the Observed Performance*

LP Zigzag Parameters†	Input Impedance	Half-Power Beamwidth	Pitch Angle (ψ)	Boom Length for a Fixed Frequency Band
α (β and τ constant)	Increase	Increase	Decrease	Decrease
β (α and τ constant)	Decrease	Decrease	Increase	Independent
τ (α and β constant)	Increase	Decrease and then increase after τ is greater than 0.9	Decrease	Independent
ψ	Decrease	Depends on α and β		Independent

*Table entries denote the change in performance for an increase in the parameters of the first column.

†The table entries hold true over the following range of parameters for which frequency-independent operation has been verified: $0.85 < \tau < 0.95$, $5° \leq \alpha \leq 15°$, and $0° < \beta \leq \alpha$.

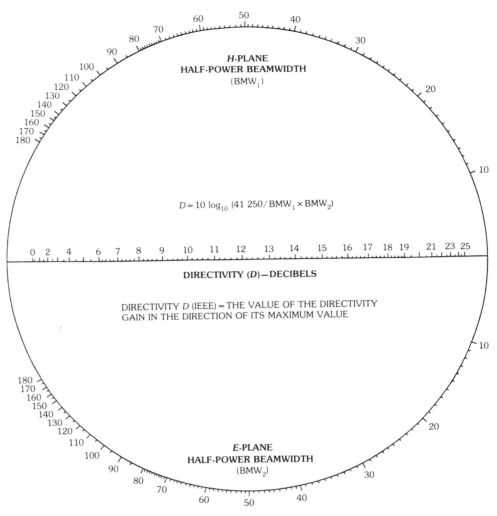

Fig. 37. Nomograph for relating directivity to half-power beamwidths.

The choice of α and β will have an effect on the input impedance which will be given later. Thus if the desired value of impedance cannot be obtained with the chosen values of α and β, a different choice may be required that gives the same beamwidths. The size of antenna that results from certain choices of α and β should also be kept in mind. A small value of α means a long antenna is required to cover a specified frequency band.

(c) The mean input resistance for each combination of α and β can be obtained from Figs. 39 and 40. Combinations of α and β which yield the required beamwidth, even without restrictions on α to meet a limitation on size, will not always produce the desired input resistance. Fig. 41 illustrates the mean input resistance as a function of τ when α and β are held constant. The resistance increases very

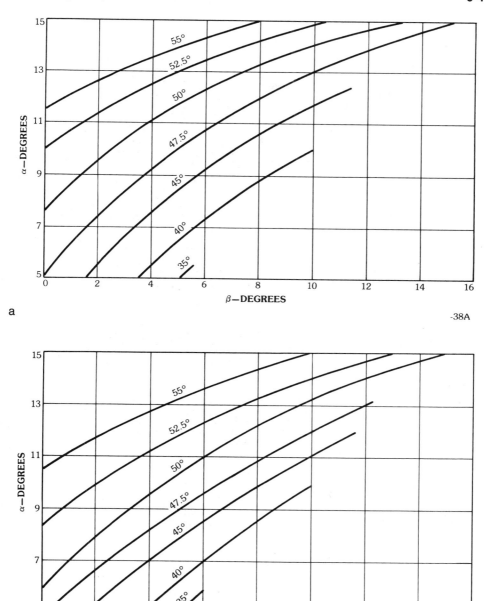

Fig. 38. Experimental contours of constant half-power beamwidth versus α and β, for $\tau = 0.9$. (a) H-plane. (b) E-plane. (*After Kuo and Mayes [20]*)

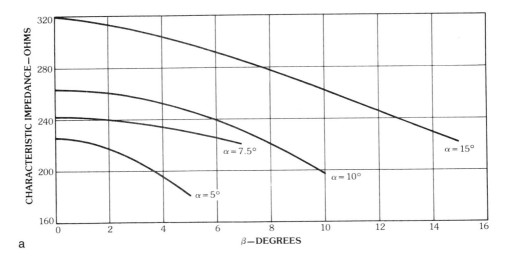

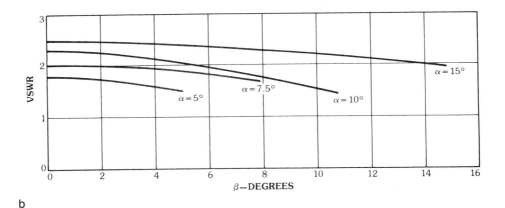

Fig. 39. Characteristic impedance and vswr versus α and β, for $\tau = 0.9$. (*a*) Characteristic impedance. (*b*) Voltage standing-wave ratio. (*After Kuo and Mayes [20]*)

rapidly as τ is increased. Since the radiation pattern does not depend on τ as much, it is possible to modify a design that meets the beamwidth requirements, but does not have the desired input resistance, by simply changing the value of τ while holding α and β fixed.

(*d*) When γ is set equal to $2\alpha = \gamma_0$, the *E*- and *H*-plane beamwidths will be almost the same. The *H*-plane beamwidth can be decreased somewhat to achieve higher directivity by increasing γ beyond γ_0. Fig. 42 provides some information about how the beamwidths vary with increasing γ.

(*e*) Fig. 43 shows how the mean input resistance is affected by changing γ while holding α, β, and τ constant.

Log-periodic zigzag antennas with constant-dimension conductors are attractive for simplicity of construction. The experimental data presented above for the

Frequency-Independent Antennas

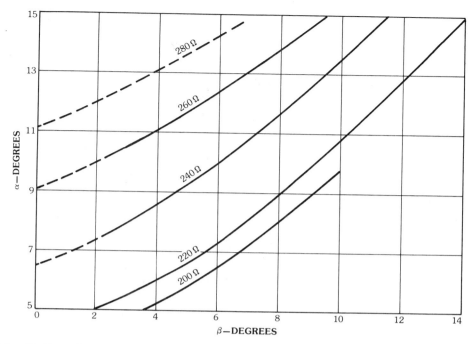

Fig. 40. Experimental contours of constant characteristic impedance versus α and β, for $\tau = 0.9$. (*After Kuo and Mayes [20]*)

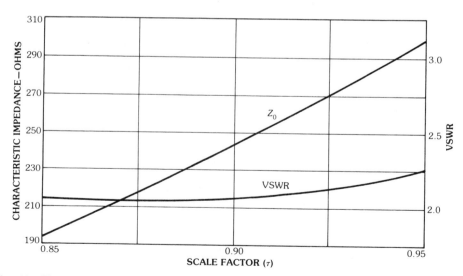

Fig. 41. Characteristic impedance and vswr versus τ, for constants α and β. (*After Kuo and Mayes [20]*)

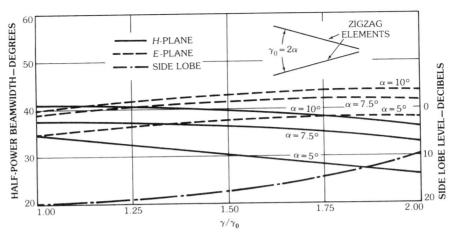

Fig. 42. Half-power beamwidth and side lobe level as a function of γ, for $\tau = 0.9$. (*After Kuo and Mayes [20]*)

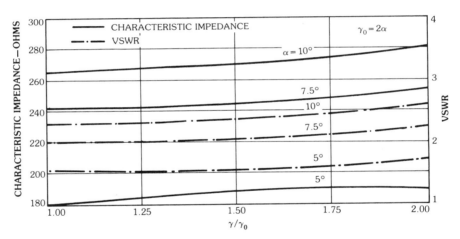

Fig. 43. Characteristic impedance and vswr versus γ, for $\tau = 0.9$. (*After Kuo and Mayes [20]*)

case of small β can be supplemented by theoretical calculations based on solutions of the thin-wire antenna integral equation [18]. Fig. 44 shows the geometry of the thin-wire LP zigzag antenna and the notation used in the literature for its definition. Note that α now refers to the angle from tip to tip rather than tip to centerline. The angle between the planes of the zigzag elements is called ψ.

In Fig. 44 only the zigzag lines designate conductors. The central boom, needed for mechanical support, is nonconducting, and its effects are not included in the analysis. A simple way of fabricating thin-wire LP zigzag antennas is to use small-diameter coaxial cable for the arms. One arm then serves both as a downlead

Frequency-Independent Antennas

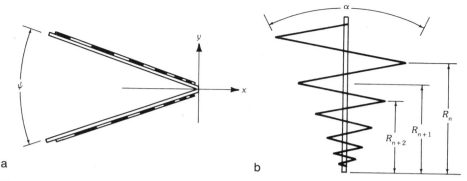

Fig. 44. A balanced LP zigzag antenna. (*a*) Top view. (*b*) Plane view of one half of the antenna. (*After Lee and Mei [19]*, © *1970 IEEE*)

from the actual feedpoint at the apex and as a radiator. Only the outer conductor is used on the second arm, and it serves only as a radiator.

One advantage of the LP zigzag antenna over the LP dipole array is the control of the *H*-plane beamwidth. When $\psi = \alpha$, the *E*- and *H*-plane beamwidths are nearly equal. This is desirable for some applications, such as a feed for a rotationally symmetrical reflector. However, an even narrower *H*-plane beamwidth would produce greater directivity and gain. Values of ψ much greater than α_0 are not usually advantageous because of the rapid increase in *H*-plane side lobes as shown in Fig. 45.

The design of a thin-wire LP zigzag antenna can proceed from *E*- and *H*-plane pattern beamwidths that will produce a desired gain as approximated using the nomograph of Fig. 37. Figs. 46 and 47 indicate the rather narrow range of beamwidths that can be obtained by changing the values of τ, α, and simultaneously changing ψ so that $\psi = \alpha$. As a consequence of this rather limited range of beamwidths the gain of thin-wire LP zigzag antennas falls between 8 and 10 dB as

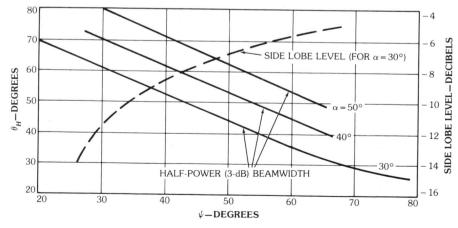

Fig. 45. The *H*-plane beamwidth and side lobe level versus ψ. (*After Lee [18]*)

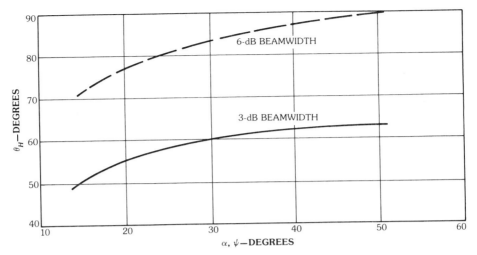

Fig. 46. The *H*-plane beamwidths versus α and ψ, for $\alpha = \psi$. (*After Lee [18]*)

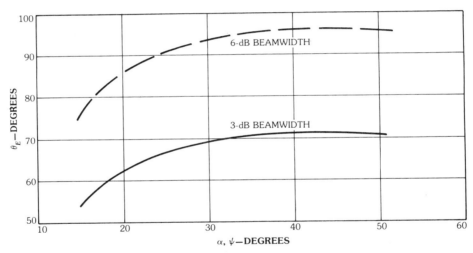

Fig. 47. The *E*-plane beamwidths versus α and ψ, for $\alpha = \psi$. (*After Lee [18]*)

illustrated in Fig. 48. In order to achieve 10 dB gain, large τ (near 0.9) and small α (about 15°) are required.

The mean impedance of thin-wire LP zigzag antennas is controlled by wire diameter and spacing in a manner similar to balanced two-wire transmission lines. The computed values for a range of parameters given in Table 4 can be used as a basis for achieving desired values within the range 250 to 400 Ω.

Periodically Loaded Lines

Some LP antennas are closely modeled by a uniform transmission line loaded with discrete elements. For example, the effect of the dipoles on the waves

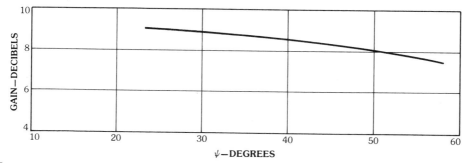

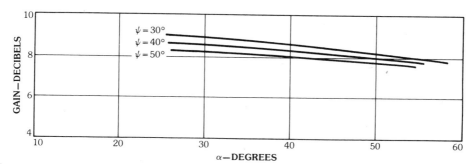

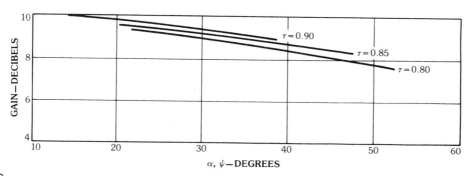

Fig. 48. Thin-wire LP zigzag antenna gain. (*a*) Antenna gain versus ψ, for $\tau = 0.80$ and $\alpha = 30°$. (*b*) Antenna gain versus α, for $\tau = 0.80$. (*c*) Antenna gain versus α and ψ, for $\alpha = \psi$. (*After Lee [18]*)

traveling along the feeder of a uniform dipole array can be represented approximately by a discrete shunt admittance as shown in Fig. 49. Of course, this model fails to account for the field interactions (mutual impedance) among the dipoles. The frequency variation of dipole admittance is similar to that of a lossy, open stub. Although the input conductance of the dipole is a function of frequency, a simplified network composed of a lossless stub and a series resistor can give useful

Table 4. Impedance of Log-Periodic Zigzag Wire Antennas (*After Lee [18]*)

Diameter of Wire	Characteristic Impedance (Ω)
\multicolumn{2}{c}{$\alpha = \psi = 20°$, $\tau = 0.85$}	
0.004λ	$416 + j11$
0.010λ	$350 + j17$
0.016λ	$317 + j30$

Angle ψ (°)	Characteristic Impedance (Ω)
\multicolumn{2}{c}{$\alpha = 30°$, $\tau = 0.80$, Diameter of Wire $= 0.016\lambda$}	
27	$270 + j12$
30	$285 + j16$
40	$305 + j48$
60	$362 + j116$

Scale Factor τ	Characteristic Impedance (Ω)
\multicolumn{2}{c}{Diameter of Wire $= 0.010\lambda$, $\alpha = \psi = 20°$}	
0.80	$340 - j11$
0.85	$350 + j17$
0.90	$370 + j45$

Angle α (°)	Characteristic Impedance (Ω)
\multicolumn{2}{c}{Diameter of Wire $= 0.010\lambda$, $\tau = 0.85$}	
20	$350 + j17$
30	$292 - j3$
40	$236 + j3$

qualitative information. The network model of a periodic array of shunt dipoles would then appear as shown in Fig. 50 and k-β diagrams are readily computed for such a network. Close correspondence between the junction voltages of the periodic and log-periodic networks of shunt lossy stubs has been demonstrated [21, 22] even when τ differs appreciably from unity. The important consideration for this correspondence to hold is that the image impedance of a single symmetric

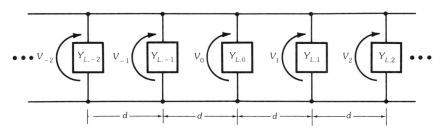

Fig. 49. A transmission line periodically loaded with shunt elements.

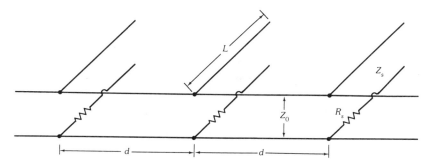

Fig. 50. A transmission line periodically loaded with resistor-stub elements. (*After Ingerson and Mayes [26]*, © *1968 IEEE*)

cell of the periodic structure be essentially independent of frequency below the first resonance.

A distinctive feature appears in the k-β diagrams of networks like Fig. 50 when the elements have resonant behavior. A stopband occurs in the vicinity of each resonance of the elements regardless of the length of transmission line between the elements. Stopbands associated with resonance of the periodic loads are called *two-terminal stopbands* [23]. Stopbands also occur for nonresonant loads that are spaced approximately one-half wavelength apart. The latter are called *structural stopbands*. Fig. 51 shows two typical plots of complex dispersion data for a system like that shown in Fig. 50 with zero loss. Stopbands are identified in lossless systems by nonzero attenuation. Two stopbands are displayed in Fig. 51. The lowest one is due to the resonance of the stubs, which occurs at 275 MHz. The higher stopband is due to the periodic loading, the stubs being one-half wavelength apart at 397 MHz. Note that the width of the stopbands increases with an increase in the characteristic impedance Z_0 of the main line (feeder).

Fig. 52 shows that adding loss to each cell produces a major effect on the k-β diagram. By increasing the value of R in Fig. 50 from zero to 73 Ω the maximum phase shift per cell in the stopband is reduced to about 40° and the maximum attenuation to about 0.6 Np. Although changing the resistance in each cell affects the phase and attenuation of the cell, it is not a convenient way of controlling either in a log-periodic antenna. The resistance in each cell in the antenna case is desirably due to radiation and is affected by mutual coupling. In Fig. 53 it is seen that the maximum phase shift and attenuation are affected very little by changing the characteristic impedance of the stub, i.e., changing the Q of the loads. The width of the stopband, however, is clearly related to the Q of the loads. An effective way of controlling the phase shift per cell when there is no coupling between the loads is to change the length of line in each cell. This is illustrated in Figs. 54a and 54b, where progressive increases in the line length produce increases in the maximum phase shift occurring just below resonance. It is to be noted, however, that increased line length causes the structural stopband to occur at lower frequency. In particular, the value $d = 67.5$ cm in Fig. 54c produces a structural stopband (180 to 220 MHz) below the first resonance. The effect on attenuation per cell at different line lengths is displayed in Fig. 55. The maximum attenuation first increases with increasing

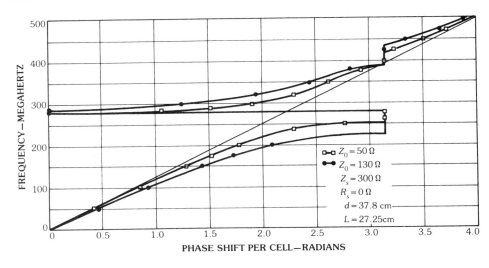

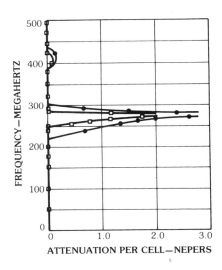

Fig. 51. Dispersion data for a transmission line that is periodically loaded with sections of lossless line, showing dependence on the characteristic impedance.

line length, then decreases as the structural and two-terminal stopbands overlap, increasing again as the structural stopband moves below the two-terminal stopband.

A complete description of symmetrical, reciprocal two-ports requires consideration of image impedances as well as the propagation constant. Fig. 56 shows the image impedance for two examples of cells symmetrically loaded with lossless stubs. The solid line is for the case of smaller spacing ($\sigma = d/4L = 0.25$) and shows how the magnitude of the image impedance changes little until near the first

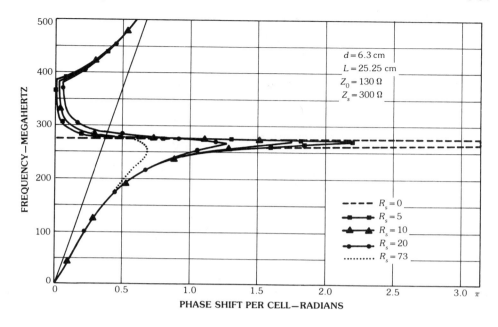
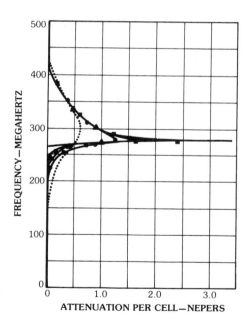

Fig. 52. Dispersion diagram for a transmission line that is periodically loaded with sections of lossy line, showing dependence on loss resistance.

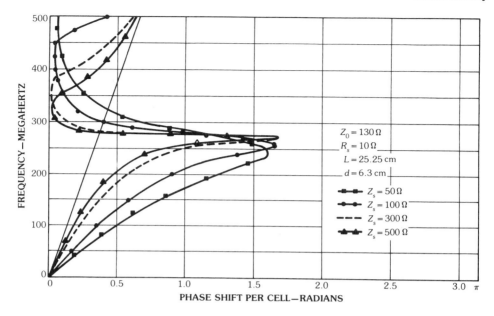
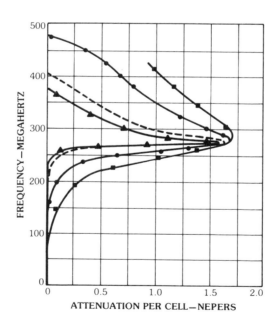

Fig. 53. Computed dispersion curves for stub-loaded transmission line with variable stub characteristic impedance.

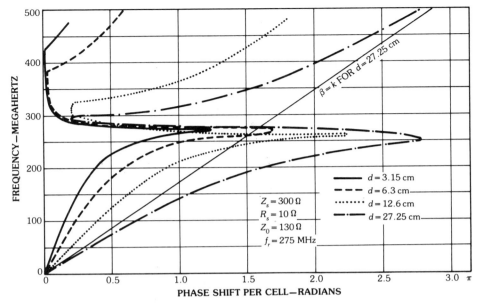

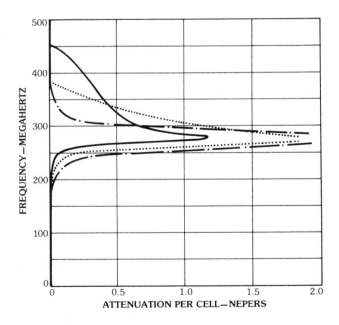

Fig. 54. Dispersion data for transmission line periodically loaded with sections of lossless line, showing dependence on spacing. (*a*) Phase shift, 3.15 cm $\leq d \leq$ 27.25 cm. (*b*) Attenuation, 3.15 cm $\leq d \leq$ 27.25 cm. (*c*) Phase shift, 37.5 cm $\leq d \leq$ 67.5 cm. (*After Ingerson [22]*)

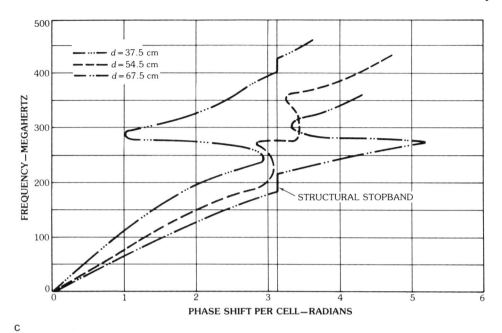

Fig. 54, *continued.*

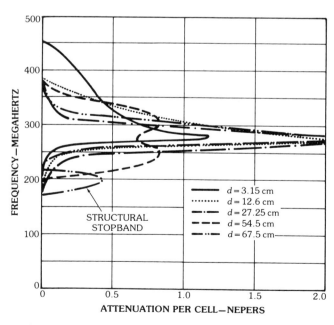

Fig. 55. Computed attenuation curves for stub-loaded transmission line with variable spacing. (*After Ingerson [22]*)

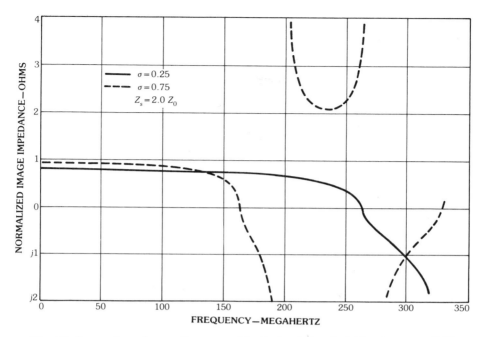

Fig. 56. Image impedance of a symmetrically loaded cell. (*After Ingerson [22]*)

resonance of the loading element. The dashed line ($\sigma = 0.75$) displays the effect of the structural stopband (200 to 275 MHz) wherein the image impedance is purely imaginary. In this case the rapid variation in the magnitude of the image impedance commences just below the stopband (about 160 MHz). When losses are added to the loads the variation of the image impedance with frequency is reduced. A basic principle of achieving low swr in an LP structure is to achieve a minimal variation in the image impedance from one cell to the next, all the way from the feed point to the end of the active region. When this condition is satisfied, the variation in voltage from cell to cell is approximately the same as it would be for individual, image-terminated cells of the corresponding size in terms of wavelengths. Consideration of plots of complex propagation factor and image impedance versus frequency can, therefore, be very helpful in the design of LP antennas. Parameters required to achieve backfire phasing can be determined from the k-β diagrams. Including the attenuation factor in these plots permits one to judge whether appreciable energy will penetrate past the active region and produce truncation effects. Inspection of the variation of image impedance versus frequency for a single cell indicates whether the assumption of minimal reflections at the cell junctions of an LP array of these cells is valid.

Application of the analysis of open-periodic structures to the corresponding log-periodic ones is complicated when appreciable external coupling exists. The base impedance of a radiating element is affected by this coupling. Hence the Q of a resonant radiator in a periodic or log-periodic array may be quite different from its value when the radiator is isolated. Consideration of the k-β diagram is helpful in evaluating this effect. Fig. 57 shows division of the k-β space into visible and

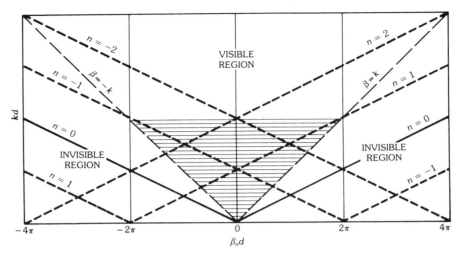

Fig. 57. A k-β diagram showing the fundamental slow wave and some space harmonics. (*After Ingerson [22]*)

invisible regions. In the invisible region neither the fundamental wave nor any of its space harmonics has phase velocity greater than the speed of light. Such waves contribute little to the distant field, and the radiators display high values of Q since the radiative loss is small. On the other hand, a wave in the visible region produces a significant amount of distant field and the radiative loss greatly lowers the Q of elements in an array operating at such points. Hence one important factor when considering the performance of LP antenna arrays with resonant elements is the phasing of the currents on the elements near resonance so that they will be efficient radiators. This requires that the active region phasing correspond to points in or near the visible region of the k-β diagram.

Solving for the modes of propagation on an open structure is also complicated by the mutual coupling. The number of modes on an infinite periodic structure depends on the range of coupling [24]. Interaction limited to adjacent elements produces a single mode; coupling to two adjacent elements produces two modes, and so on. However, experimental measurements on LP structures indicate that only one mode is excited by a source near the tip of such tapered structures.

Fig. 58 shows the k-β diagram of a hypothetical UP dipole array without mutual impedance and the fundamental mode on the same structure when mutual impedances are considered. The main effect of mutual impedance is to change the value of the radiation resistance of the dipoles. To a first approximation the behavior of the dipole in a periodic or log-periodic array can be represented by a lossy stub. When the dipole is short in wavelengths its impedance is approximately that of a stub of characteristic impedance

$$Z_s = 120[\ln(h/a) - 2.25] \qquad (16)$$

where h is the dipole length and a is its radius [7, 8]. For higher frequencies, through the first resonance, the impedance has appreciable real part which can be included approximately by putting a resistor with value R_s in series with the stub.

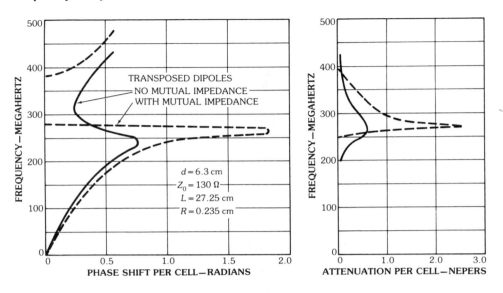

Fig. 58. Dispersion diagram of a periodically loaded transmission line. (*After Ingerson [22]*)

The value of R_s should equal the input resistance of the dipole at resonance. The k-β diagrams for a line symmetrically shunt-loaded by a lossless line in series with the resistor R_s or by a lossy stub with the same characteristic impedance as the lossless one and whose input resistance at resonance is also equal to R_s are found to be essentially identical up to the first antiresonance. Fig. 59 shows the k-β diagrams

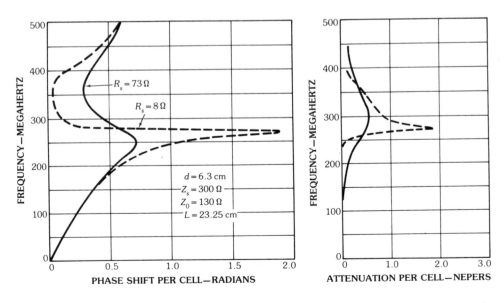

Fig. 59. Dispersion diagram of a periodically stub-loaded transmission line. (*After Ingerson [22]*)

of a stub-resistor approximation to a shunt dipole load on a section of lossless line. The value of Z_s is 300 Ω to correspond to the h/a ratio of the dipole and the values of R_s are 73 Ω (dipole without mutual) and 8 Ω (dipole with mutual). This simulation of the effect of the coupling shows that the presence of coupling increases the phase shift and attenuation in the lower half of the two-terminal stopband. Evidence points to even lower values of R_s when the spacing between dipoles is very small compared with the wavelength. As the spacing parameter σ is increased, the mutual impedance is reduced. The Q is also reduced as the phasing approaches the edge of the visible region. The calculated base input impedances of the dipoles in an LP dipole array are shown in Fig. 60 as a function of σ. The value of the resistance changes as σ changes, but the resonant frequency and the reactive part of the impedance change very little. The swr versus σ curve for an LP dipole array would then be expected to have a minimum for a value of σ other than zero

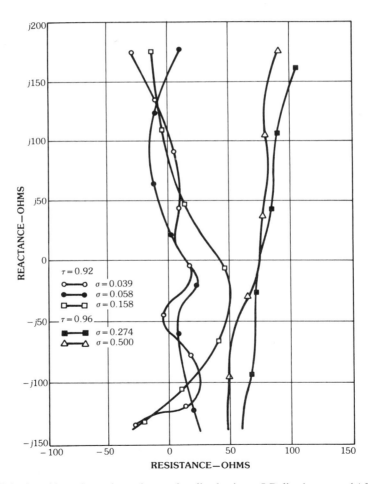

Fig. 60. Calculated base input impedances for dipoles in an LP dipole array. (*After Ingerson [22]*)

(the case without coupling). This is confirmed by the computed swr versus σ curves shown in Fig. 24.

When the dipoles on the feeder of a uniform periodic array are not transposed, a different type of k-β diagram is found for the dominant mode. This is shown in Fig. 61, where two modes are displayed. In the LP structure, however, it again appears as if only the lower wave is excited. The k-β plot for this wave falls inside the invisible region. In the LP structure, then, it would be expected that the dipoles would behave like high-Q stubs and reflect most of the incident energy. A comparison of the transmission-line voltages of the corresponding LP structures with and without transposed dipoles is shown in Fig. 62. It is clearly seen that the untransposed dipoles reflect most of the incident energy producing a high standing wave between the feed point and the resonant dipoles. On the other hand, the transposed dipoles radiate the energy and there is little evidence of a standing wave between the feed point and the active region. The difference between these two cases is attributed to the fact that the currents in the dipoles near resonance are phased close to the visible region when the feeder is transposed.

This view of the dipole excitation being due to waves traveling in opposite directions can be used to relate the swr of LP dipole arrays and the front-to-back ratio of their radiation patterns. It is plausible to expect that the ratio of amplitudes of the energy radiated to the front and rear will be related to the amplitudes of the incident and reflected energy on the feeder. Moreover, except for the scattering due to the larger dipoles, the shape of the patterns would be expected to be similar if τ is not too small. The ratio of incident and reflected power is related to the input vswr by

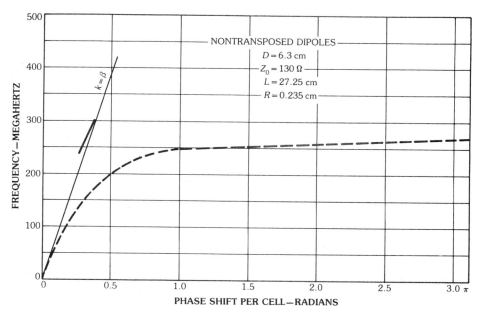

Fig. 61. Dispersion diagram for a periodically loaded transmission line with nontransposed dipoles. (*After Ingerson [22]*)

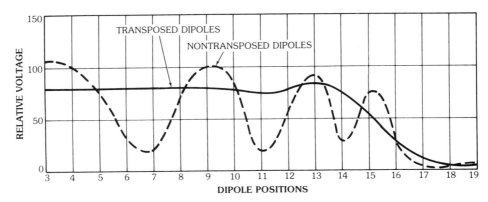

Fig. 62. Relative transmission-line voltage for a transmission line log-periodically loaded with dipoles and with dipoles alternately transposed for additional phase shift. (*After Ingerson [22]*)

$$F/B \cong 10\ln(P_i/P_r) = 20\ln[(\text{vswr} + 1)/(\text{vswr} - 1)] \qquad (17)$$

where P_i and P_r represent the incident and reflected powers. Fig. 63 compares this relationship between front-to-back ratio and the spacing parameter σ with the results previously obtained by computer analysis of many cases [8]. The agreement confirms this as a useful approximation for the front-to-back ratio and the manner in which it will change as a function of antenna parameters. Fig. 64 gives a plot of (17).

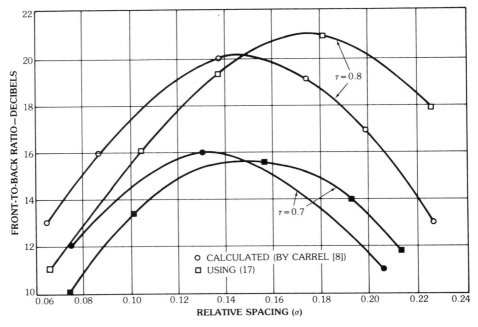

Fig. 63. Calculated front-to-back ratios for LP dipole arrays. (*After Ingerson [22]*)

Frequency-Independent Antennas

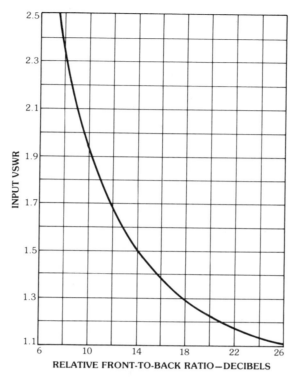

Fig. 64. Simple approximation relating input vswr to front-to-back ratios in LP dipole arrays. (*After Ingerson [22]*)

Several useful design principles can be based on the above arguments about the apparent Q of dipoles in a periodic or log-periodic array environment. For stub-resistor–loaded transmission line sections the swr depends mainly on the Q of the loading elements and not on the characteristic impedance of the feeder. The Q of dipoles in the array depends mainly on σ and only slightly on τ. Hence the vswr of a dipole array is relatively independent of the feeder characteristic impedance. Thus both the vswr and the front-to-back ratio are determined once the minimum Q and τ are known. The directivity, however, is not determined by Q and τ but is a strong function of σ and the characteristic impedance of the feeder.

As mentioned above, the apparent Q of dipoles in a periodic or log-periodic array can be changed by changing σ. For values of σ greater than 0.25 the Q of the dipoles in an array approaches the value for an isolated dipole. The Q of an isolated dipole is primarily dependent on h/a. Fig. 65 shows how the impedance of the equivalent stub depends on h/a. It is clear that to change Q by a factor of 2, h/a must change by a factor of approximately 10. This result substantiates the common practice of using the same radius for several dipole elements having about the same lengths in an LP dipole array. It is also apparent that, unless very thin dipoles have been used, it is not possible to improve the vswr of an LP dipole very much by replacing the dipoles with thicker ones.

In choosing the parameters for an LP antenna it is particularly important that

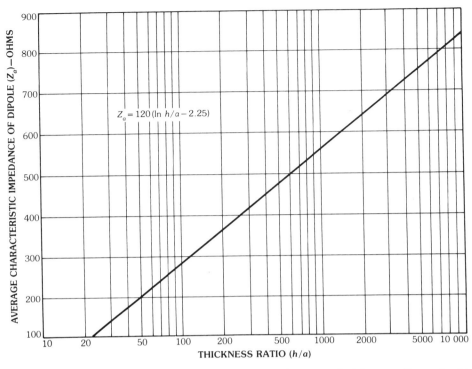

Fig. 65. Average characteristic impedance of a dipole versus height-to-radius ratio h/a. (*After Carrel [8]*)

the stopband which results from a particular choice stop any energy from exciting dipoles which are too long or improperly phased. Since the stopband width is largely determined by the ratio of Z_s/Z_0, the minimum value of τ must be adjusted according to that ratio. To achieve LP dipole designs that have low values of input impedance it is, therefore, necessary to lower Z_s in order to obtain a stopband of sufficient width. Designing LP dipole arrays with low values of input impedance demands that high values of h/a be used. When Z_0 is low compared with Z_s the minimum value of σ must be increased to reduce leakage of energy through the active region.

Log-Periodic Designs Based on Periodic Structure Theory

Parameters for a side-firing LP dipole array can be obtained from study of k-β diagrams. The diagram in Fig. 66 shows that for dipoles with $h/a = 116$, spaced at 50.4 cm, the voltage phasing near resonance approaches 2π radians when a transposed feeder is used between dipoles. Fig. 66 also shows that appreciable attenuation occurs, not only when the dipole is near resonance, but also for shorter dipoles. On an LP structure this would correspond to radiation primarily in the backfire direction. Radiation, however, can be produced in other directions by suppressing radiation in the backfire direction. This is accomplished by using dipoles with very high values of Q in conjunction with a feeder having a low value

Frequency-Independent Antennas

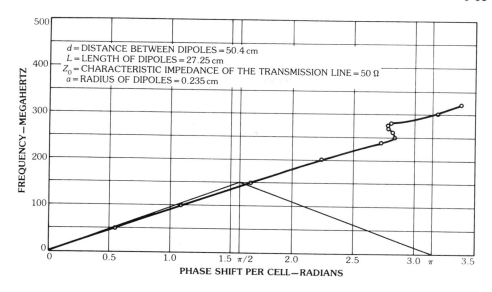

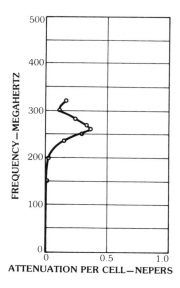

Fig. 66. Dispersion diagram for periodically dipole-loaded transmission line. (*After Ingerson [22]*)

of characteristic impedance. For example, a side-firing LP dipole array has been achieved with $\tau = 0.96$, $\sigma = 0.25$, and $Z_0 = 50$ Ω. In addition, dielectric was inserted between the conductors of the feeder in order to increase the phase shift between the dipoles in the active region. Fig. 67 shows a typical *H*-plane radiation pattern measured for the example side-firing antenna. The split-beam pattern of

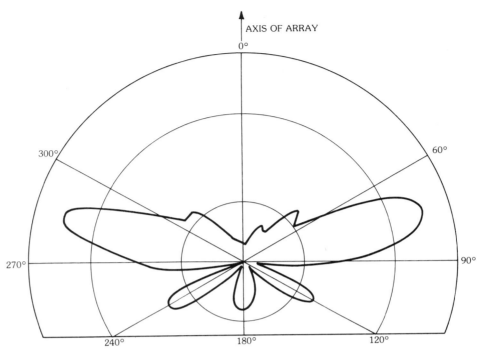

Fig. 67. Typical measured H-plane pattern of a side-firing LP dipole array, for $\tau = 0.96$, $\sigma = 0.25$, and $Z_0 \cong 50\ \Omega$. (*After Ingerson [22]*)

Fig. 67 can be converted to a single-beam pattern by adding an array of parasitic elements along one side of the side-firing LP dipole. The parasitic elements should conform to the LP parameters of the active array. A typical unidirectional H-plane pattern obtained in this way is shown in Fig. 68.

Another use of the k-β diagram is for the design of high-directivity backfire arrays. If the phasing is held nearly constant through an active region having several elements and the attenuation is reduced, it is possible to obtain more elements that radiate in the backfire direction. The solution indicated by study of the trends indicated by the k-β diagrams is to lower the Q of the dipoles by increasing σ and a/h. In addition, the attenuation in the active region should be reduced by lowering the value of the feeder impedance. Fig. 69 shows E- and H-plane patterns that were measured on an LP dipole array having $\tau = 0.95$, $\sigma = 0.175$, and $Z_0 = 50\ \Omega$. Dielectric was also used in the feeder to increase the phase shift. In this way it is possible to obtain gain values that are greater than those indicated for the same parameters in Fig. 20.

The dipole arrays discussed above all have transposed feed lines in order to achieve phasing of the active region that is in or near the visible region. Feeding an LP array of monopole elements over ground with a transposed feeder is not possible. Consideration of the k-β diagrams of Figs. 54a and 54b shows that backfire phasing could also be achieved by using lengths of feeder that are much longer than the separation between the monopoles. These same k-β diagrams, however, show that a structural stopband will occur at frequencies below the first

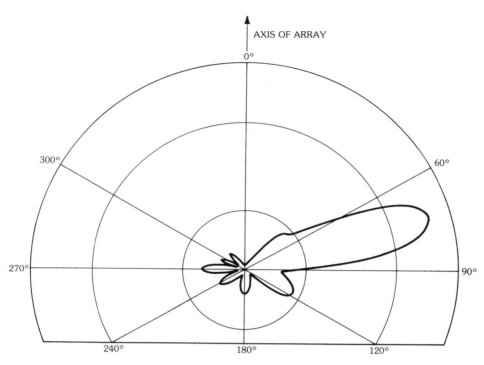

Fig. 68. Measured unidirectional H-plane pattern of a side-firing dipole array with an LP array of parasitic reflectors. (*After Ingerson [22]*)

two-terminal stopband when these longer feeders are used. The ability to get energy from the feed point of an LP monopole array with excess feed-line phasing is then dependent on using some technique to eliminate the structural stopband. One such technique is to modulate the impedance of the feeder as shown in Fig. 70 [26].

Except for the feeder, the design of LP monopole arrays with excess feeder follows the procedures previously given for LP dipole arrays. The nontransposed feeder results in a wider active region for LP monopole arrays than is obtained for LP dipole arrays which use a transposed feeder. Design of the modulated impedance feeder can be facilitated by means of a computer-aided technique. The procedure assumes that τ and σ have been chosen by using the LP dipole data. The feed-design steps are then applied to find the parameters of a single cell, say the largest, and the dimensions of the other cells are determined by applying the appropriate power of τ.

The total length L_n of feeder between any two adjacent elements with heights h_n and h_{n+1} is determined from

$$L_n = 4h_n\lambda_r(m - c\sigma), \qquad m = 0, 1, 2, \ldots \qquad (18)$$

where λ_r is the ratio of the feed-line wavelength to free-space wavelength and c is a coefficient slightly greater than unity. The parameter m is usually chosen to be unity to avoid overlong feeder sections. No attempt is made in this formula to

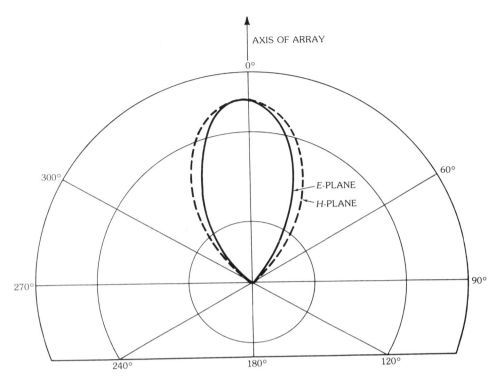

Fig. 69. Measured patterns of a log-periodic dipole array ($\tau = 0.95$, $\sigma = 0.175$), showing increased directivity using low characteristic impedance feeder ($Z_0 \cong 50\ \Omega$). (*After Ingerson [22]*)

consider that the phase velocity along the feeder may be different in sections with different characteristic impedances. Hence, if nonhomogeneous lines, such as microstrip, are used for the feeder, λ_r should be adjusted according to the approximate lengths of line of each different impedance that are expected to be used.

The feeder between any two adjacent monopoles will appear as in Fig. 71. The relative length of the lines with different impedance is

$$r = d_2/d_1 \tag{19}$$

The initial value of r is rather arbitrarily chosen somewhat greater than unity to keep the length of high-impedance line short. For some values of r the image impedance of the subcell may be imaginary and the desired match to R_{01} may not be possible. A different value of r must then be chosen.

Once r has been fixed, d_1 and d_2 can be determined from

$$d_1 = \frac{2L_n}{(1 + r)(1 + \tau)}$$

$$d_2 = rd_1 \tag{20}$$

Frequency-Independent Antennas

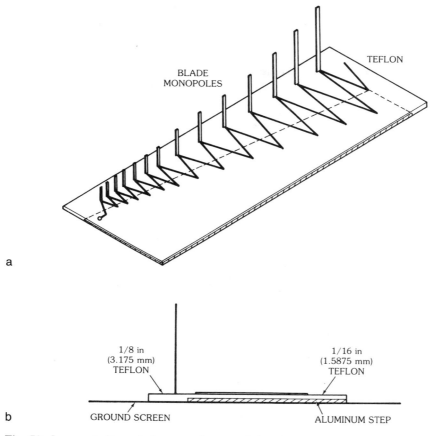

Fig. 70. Log-periodic scaled monopole array fed with a modulated-impedance meandering line. (*a*) Perspective. (*b*) Side view. (*After Ingerson and Mayes [26], © 1968 IEEE*)

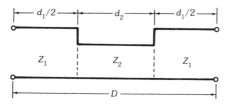

Fig. 71. A section of modulated-impedance feeder for a log-periodic monopole array. (*After Ingerson [22]*)

The dispersion characteristics of the unmodulated cell ($R_{02} = R_{01}$) can be calculated. At the frequency of maximum attenuation in the stopband, set R_{02} equal to the image impedance of the subcell. As a check to ensure that the structural stopband has been minimized, the dispersion characteristics for the modulated cell can be calculated.

A modulated line can also be used to feed an LP array of cavity-backed slots. The first demonstration of such an antenna [26] used an array of 15 cavity-backed slots having $\tau = 0.925$ and $\sigma = 0.15$. Each slot was backed by its own cavity and the depths of the cavities were made equal to one-quarter guide wavelength at the resonant frequency of the slot. The slots were excited by a single-conductor feeder which passed over each slot as shown in Fig. 72. The input vswr was not greatly dependent on whether the feed was at the center of the slot or offset. However, the degree of modulation required to remove the structural stopband was greater with the center feed. Typical H-plane patterns of the LP array of cavity-backed slots are shown in Fig. 73.

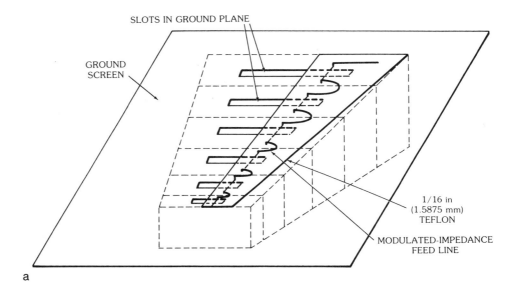

a

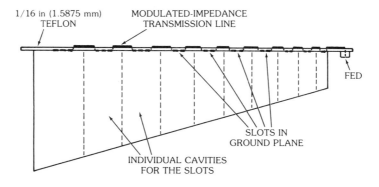

b

Fig. 72. An LP cavity-backed slot array with modulated-impedance feed line. (*a*) Perspective. (*b*) Side view. (*After Ingerson and Mayes [26], © 1968 IEEE*)

Frequency-Independent Antennas

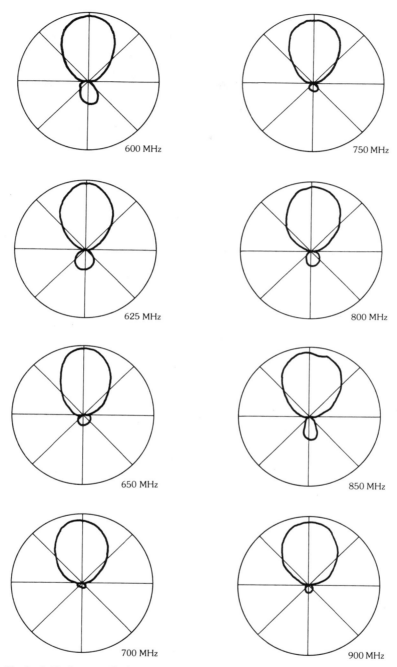

Fig. 73. Typical H-plane radiation patterns of a cavity-backed slot array, for $\tau = 0.925$, $\sigma = 0.15$, and $N = 15$. (*After Ingerson [22]*)

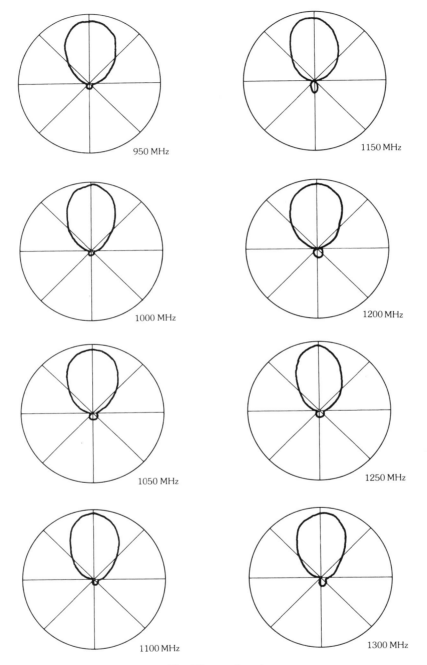

Fig. 73, *continued.*

Frequency-Independent Antennas

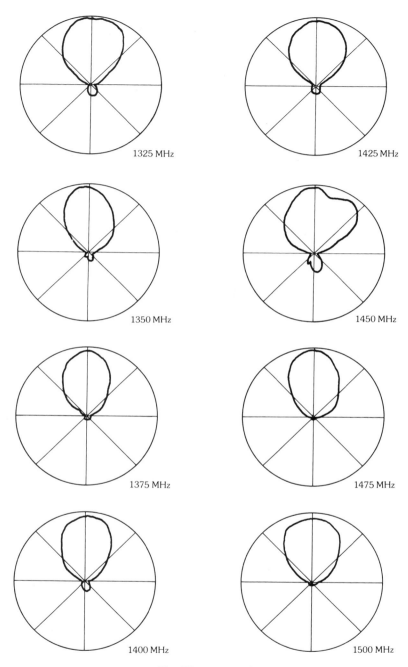

Fig. 73, *continued.*

4. Log-Spiral Antennas*

The shape of planar log-spiral antennas [27] such as that shown in Fig. 5 is based on the logarithmic spiral curve defined by

$$\varrho = \exp(b\phi) \tag{21}$$

where ϱ is the radial distance from the origin in the direction given by the angle ϕ. Note that the spiral-rate constant

$$b = \frac{1}{\varrho}\frac{d\varrho}{d\phi} = \cot\alpha \tag{22}$$

where α is the angle between a tangent to the curve at any point and a line to the origin at that point as shown in Fig. 74. Negative values of b produce spirals that go away from the origin as ϕ decreases. Since α is constant for a given logarithmic spiral, an alternative name is equiangular spiral. A change in scale of log-spirals is equivalent to rotation.

Two log-spiral curves, one rotated through an angle δ, form the edges of one member of an antenna. A balanced structure results when another member, identical with the first, is placed by rotation from the first by 180°. The construction of an antenna solely defined by these four spirals is impractical. Since the spirals would extend from a point at the origin to infinity, this would lead to dimensions that cannot be realized because they are too small near the origin and become too

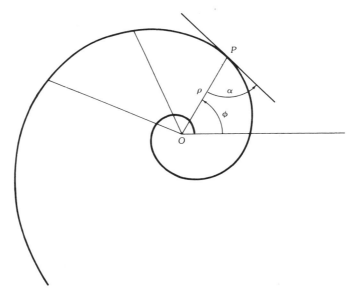

Fig. 74. The equiangular spiral. (*After Dyson [27, 28],* © *1959 IEEE*)

---

*The assistance of John D. Dyson in the preparation of this section is gratefully acknowledged.

Frequency-Independent Antennas

large at some distance far short of infinity. Hence the spiral edges must be limited to finite-length curves. These can be conveniently taken as segments of the infinite spirals that lie between two values of ϕ as shown in Fig. 5, or between two values of ϱ as shown in Fig. 75. Using a circle to define the truncation at the large end apparently minimizes the lower limit of frequency-independent performance for a given antenna diameter. A practical, balanced log-spiral antenna is therefore described by the two circles $\varrho_i = d/2$, $\varrho_0 = D/2$, and the four spirals

$$\varrho_3 = \frac{d}{2} \exp(b\phi)$$

$$\varrho_4 = \frac{d}{2} \exp[b(\phi - \delta)]$$

$$\varrho_1 = \frac{d}{2} \exp[b(\phi - \pi)] \qquad (23)$$

$$\varrho_2 = \frac{d}{2} \exp[b(\phi - \pi - \delta)]$$

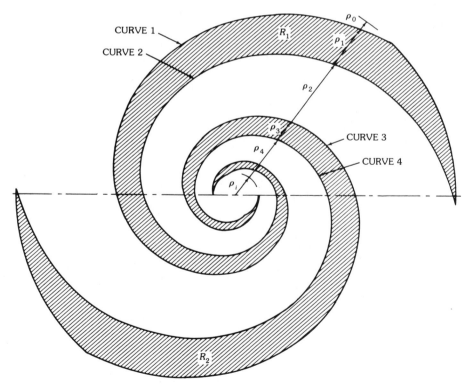

Fig. 75. A balanced, two-arm, log-spiral antenna. (*After Dyson [27]*)

where $d/2 < \varrho_n < D/2$. The regions

R_1: $\varrho_1 < \varrho < \varrho_2$, $\varrho_i < \varrho < \varrho_0$
R_2: $\varrho_3 < \varrho < \varrho_4$, $\varrho_i < \varrho < \varrho_0$

shown cross-hatched in Fig. 75, may be covered with conducting sheet material to make a balanced, two-arm, "dipole" antenna, or these same regions may be cut from a large conducting sheet to make a balanced, two-arm, "slot" antenna.

The sense of rotation of the elliptically polarized field radiated by log-spiral antennas is determined by the direction of winding of the spiral arms. Planar spirals radiate one sense of polarization in one half-space and the opposite sense in the other. The sense can be determined from the hand used when pointing the fingers in the direction the current wave travels and the thumb in the direction of the radiated beam. This procedure gives a result which corresponds to the IEEE standard. For right-hand polarization in the positive-z half-space the currents must travel away from the feed point in the direction of increasing ϕ. The spiral in Fig. 75, having a negative value of b, will produce left-hand polarization in the space above the page and right-hand polarization in the space below.

A symmetrical antenna with more than two arms can be constructed by letting the angle equal to π in (23) be $2\pi/n$ instead, where n is the number of arms. Since the angle δ defines the width of the arms,

$$0° < \delta < 360°/n$$

for a symmetrical n-arm antenna. The case $\delta = 180°/n$ is self-complementary; for the infinite structure the geometry of the arms and the space between the arms is identical except for a rotation of $180°/n$ about the origin.

Like other LP structures, log-spirals must be infinite to fulfill the scaling and self-complementary conditions. For a certain range of parameters, however, log-spiral antennas can be truncated at two distances from the origin, as described above, and still retain over a very wide band the properties of the infinite structures. If the antenna is constructed so that d is small in wavelengths, and is excited at $\varrho = d/2$, a current wave travels along the arms some distance from the point of excitation before producing appreciable radiation. As the energy is radiated, the amplitude of the current decreases. Beyond a certain point the presence or absence of the conductor makes little difference. Since the point of negligible current occurs about one wavelength from the feed point, the result is an antenna having an active region that is constant when measured in wavelengths.

The log-spiral can be expressed in terms of the distance normalized to wavelength:

$$\begin{aligned}\frac{\varrho}{\lambda} &= \frac{\exp(b\phi)}{\lambda} \\ &= \exp\left[b\left(\phi - \frac{\ln \lambda}{b}\right)\right] \\ &= \exp(\phi - \phi_0)\end{aligned} \qquad (24)$$

where $\phi_0 = (\ln\lambda)/b$. The effect of changing the wavelength is equivalent to changing the angle ϕ. Since the active region rotates at a rate dependent on the spiral parameter as the frequency is changed, the radiation pattern also rotates with changing frequency at the same rate. The pattern therefore repeats at frequencies related by integer powers of the scaling factor as for other LP geometries [28, 29]. Except for rotation the radiation characteristics are ideally independent of frequency within the limits imposed by the values of d and D.

A balanced feed is necessary for optimum performance of the balanced antennas. The feed may be brought in perpendicular to the plane of the antenna by a balanced, two-wire line or by an unbalanced (coaxial) line feeding a balanced-to-unbalanced transformer (balun) at the feed point. Of course, the operating bandwidth of the balun must exceed that of the antenna for the latter to be fully realized. The rapid decay of the current along the spiral arms makes possible a convenient frequency-independent method of feeding the balanced antenna with an unbalanced line [27]. The outer conductor (shield) of a coaxial cable may be bonded to the arms of the "dipole" antenna, or to the conductor between the arms of a "slot" antenna. Ideally, this cable would lie along the spiral path midway between the edges of one conductor from $\varrho = D/2$ to $\varrho = d/2$. Only the center conductor crosses the circle $\varrho = d/2$ and is electrically connected to the opposite conductor. A dummy cable, rotated from the feed cable by 180°, can be employed to maintain symmetry. As the frequency is decreased, the diameter of the active region will become greater than D. The current distribution will then be altered by currents on the cable and elsewhere, and the pattern and polarization will differ from that of the frequency-independent mode. The frequency at which the change in performance exceeds some specified value (dependent on the application) becomes the low-frequency limit. The upper frequency limit is determined by the size of the feed region ($\varrho < d/2$) inside which the log-spiral geometry is not continued. Three examples of transitional shapes that have been used in the feed region of log-spiral antennas are shown in Fig. 76. Reducing the dimension of the feed gap in this way will extend somewhat the upper frequency limit. The upper frequency limit is dictated to some degree by the diameter of the feed cable.

The parameters of planar log-spiral antennas are as follows:

b = the rate of expansion (or α, the expansion angle),
$K = \exp(-b\delta)$ = the arm width,
$L = (b^{-2} + 1)^{1/2}(\varrho_2 - \varrho_1)$ = the arm length.

where ϱ_1 and ϱ_2 denote the endpoints. Frequency-independent performance requires that L be approximately equal to one wavelength at the lowest frequency. Hence the maximum diameter, D, is dependent on the rate of spiral, b. The operating characteristics are not strongly dependent on the values of b and K, which typically fall in the ranges 0.2 to 1.2 and 0.375 to 0.97, respectively. The radiation patterns are relatively insensitive to variation in values of b and K, although there are optimum ranges. Good patterns can usually be obtained with only one and one-half turns of the spiral.

Log-spiral antennas radiate two broad lobes in directions perpendicular to the

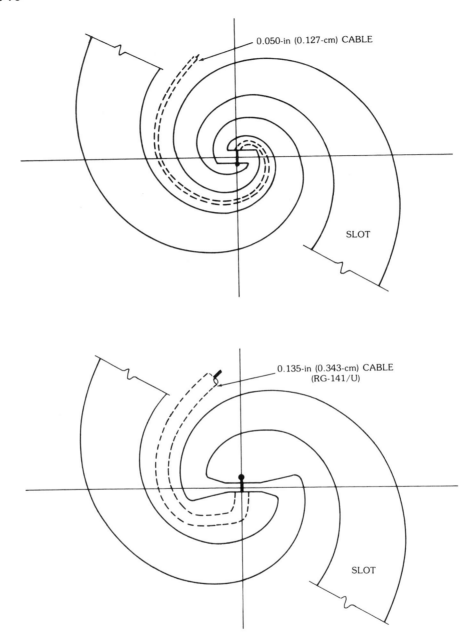

Fig. 76. Examples of feed-gap geometry for balanced, two-arm, planar, log-spiral, slot antennas. (*After Dyson [27]*)

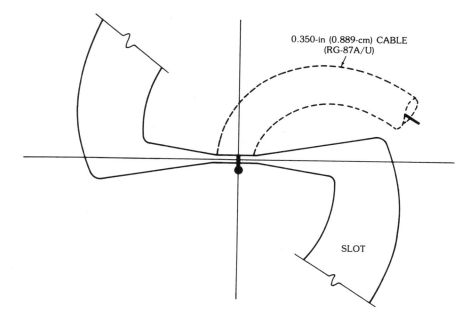

Fig. 76, *continued.*

plane of the antenna. At midband frequencies the beams are circularly polarized on the axis, with axial ratios rising slowly as the off-axis angle increases. Polarization is more sensitive to changes in frequency than is pattern shape. When the antenna arms are very short in wavelengths the polarization is linear, even on the axis. As frequency increases, the axial ratio decreases typically as shown in Fig. 77. The

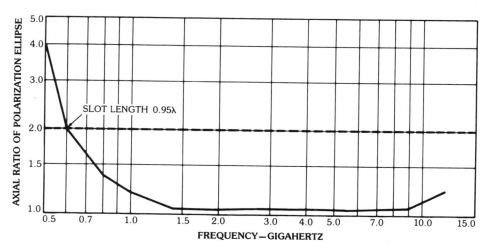

Fig. 77. On-axis polarization of a typical, balanced, two-arm, planar, log-spiral, slot antenna. (*After Dyson [27, 28],* © *1959 IEEE*)

point at which the axial ratio falls below 2 becomes a convenient definition for the lower limit of the operating band.

Although it might be expected that the lower band-limit would be a function of b, δ, and L, it has been found possible to relate the lower limit to just L and K as illustrated in Fig. 78. At a frequency such that ϱ_i approaches one-half wavelength the axial ratio of the field on the axis again increases. An axial ratio of 2 on axis can also be used to define the upper band-limit. Since the upper and lower limits are independent, the only limitations on a design are the required diameter D in which to spiral the necessary length and the allowable size of the feed region.

Although patterns of the balanced spirals are relatively insensitive to variation in parameters, the more tightly spiraled ones with wide arms (b and K small) tend to have smoother and more uniform patterns. These parameters also produce patterns that are more nearly rotationally symmetric, and thus display less variation in beamwidth when observed in a fixed plane.

For frequencies such that the arm lengths of a planar, balanced, log-spiral antenna are greater than one wavelength, or slightly less, the input impedance remains almost constant as frequency is varied. The relationship between the mean impedance and arm width of "slot" antennas is shown in Fig. 79. The data shown pertain directly to antennas made from 1/32-in (0.79-mm) copper, fed with 0.15-in (0.38-mm) diameter coaxial cable bonded to the ground plane midway between the slot arms (no dummy cable). The influence on the standing-wave ratio of using different sizes of feed cables is illustrated in Fig. 80. When $\delta = \pi/2$, the planar log-

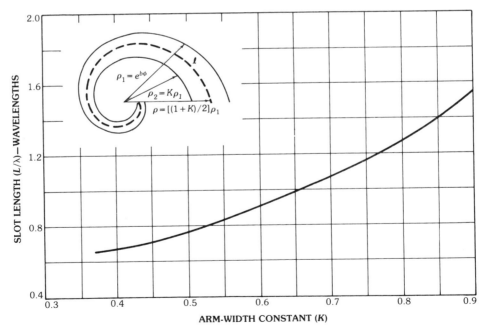

Fig. 78. Minimum slot length necessary to produce a circularly polarized radiated field ($r \leq 2:1$ on axis; $0.2 \leq a \leq 0.45$). (*After Dyson [27, 28]*, © *1959 IEEE*)

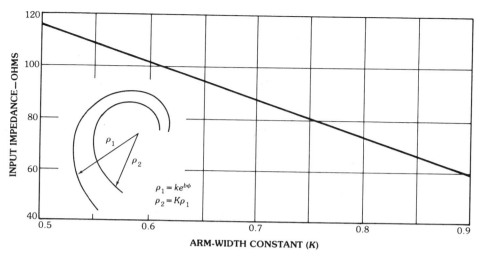

Fig. 79. Input impedance of balanced planar slot antennas. (*After Dyson [27, 28]*, © 1959 IEEE)

spirals are self-complementary, and would be expected to have an input impedance of 189 Ω when made of metal with infinitesimal thickness. Self-complementary antennas of finite thickness invariably display lower values of input impedance.

Conical Log-Spirals

Unidirectional patterns can be achieved, in a manner similar to that for other LP structures, by developing log-spirals on a nonplanar surface. A simple, rotationally symmetrical surface is a cone. A log-spiral arm on the surface of a cone is illustrated in Fig. 81. The parameter θ_0 determines the cone angle. As before, the angle δ determines the angular width of the exponentially expanding arms, and the angle α, the rate of wrap of the arms. These angles are constant for any given antenna. The radius vector to any point on the edge of one arm is given by

$$\varrho = \varrho_0 \exp[b(\phi - \delta)] \tag{25}$$

where

$$\varrho_0 = \frac{d/2}{\sin \theta_0} \tag{26}$$

$$b = \frac{\sin \theta_0}{\tan \alpha} \tag{27}$$

In the conical case the parameters d and D define the diameters of the truncated apex and base of the cone. The orientation of the antenna in a spherical coordinate system used to describe the radiation patterns is also indicated in Fig. 81.

Conical log-spirals with small cone angle θ_0 can be treated using the periodic

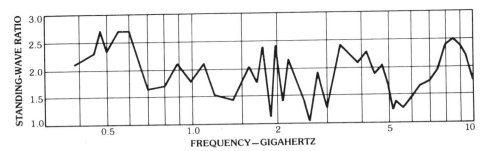

a

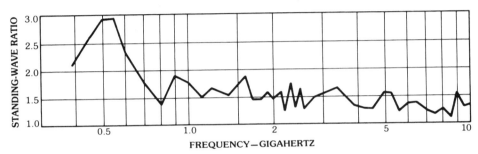

b

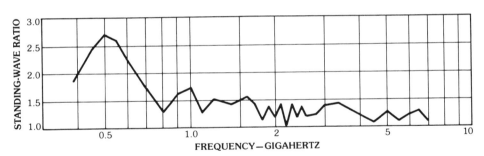

c

Fig. 80. Standing-wave ratios for various feed cable arrangements (50-Ω line). (*a*) Small-diameter cable. (*b*) RG-141/U cable. (*c*) RG-141/U with balancing dummy cable. (*After Dyson [27, 28]*, © *1959 IEEE*)

structure theory of Section 3. The local propagation constant in a small region along the surface of the conical, two-arm, log-spiral antenna is approximately equal to the propagation constant for the corresponding "average" cylindrical bifilar helix. A study of the propagation constant and other characteristics of the near and far fields leads to identification of the active region [30]. Successful design of the antenna depends on knowledge of the relative position and size of the active region as a function of antenna parameters.

The parameters involved in a comparison of the conical spiral and cylindrical

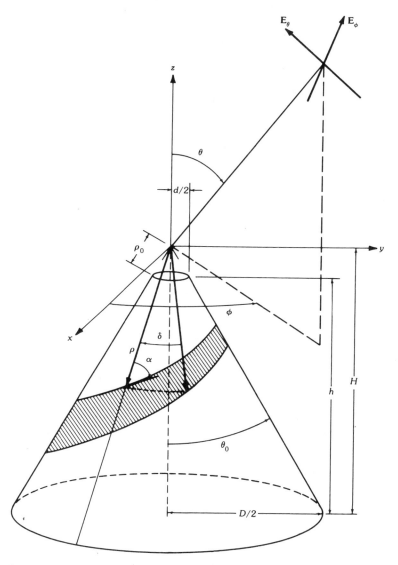

Fig. 81. Conical antenna with associated parameters. (*After Dyson [30]*, © *1965 IEEE*)

helical geometries are indicated in Fig. 82, which shows one turn (or one cell) of a conical antenna with infinitesimally narrow arms and with the following parameters: ℓ, the length of the turn; ξ, the pitch angle; α, the complementary spiral angle; and θ_0, the half-cone angle. Superimposed on this is one cell of a cylindrical helix with the same pitch angle and with a radius a equal to the geometric mean radius of the conical cell.

For observations parallel to the axis the turn-to-turn phasing of the helix is determined to a first approximation by the ratio of the pitch distance P to the turn

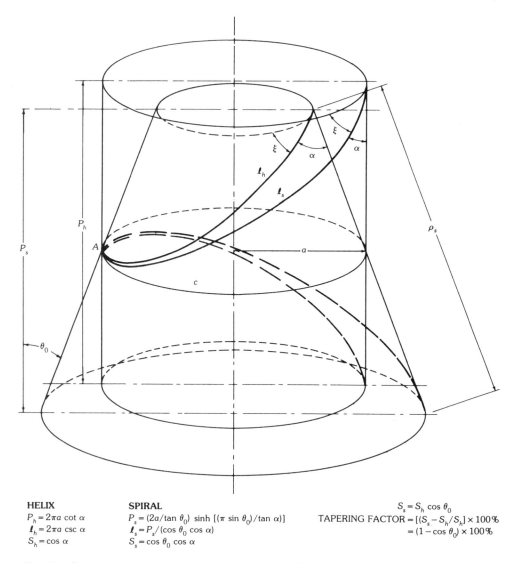

HELIX	SPIRAL	
$P_h = 2\pi a \cot \alpha$	$P_s = (2a/\tan \theta_0) \sinh[(\pi \sin \theta_0)/\tan \alpha]$	$S_s = S_h \cos \theta_0$
$l_h = 2\pi a \csc \alpha$	$l_s = P_s/(\cos \theta_0 \cos \alpha)$	TAPERING FACTOR $= [(S_s - S_h/S_h] \times 100\%$
$S_h = \cos \alpha$	$S_s = \cos \theta_0 \cos \alpha$	$= (1 - \cos \theta_0) \times 100\%$

Fig. 82. Corresponding conical spiral and helical cells. (*After Dyson [30], © 1965 IEEE*)

length ℓ. This assumes a current wave progressing down the arm at the intrinsic velocity of the surrounding medium. On the helix this ratio is the sine of the pitch angle or $\cos \alpha$. The ratio of the pitch distance (parallel to the axis) to the turn length on the conical structure is equal to $\cos \alpha \cos \theta_0$.

The ratio (S) of pitch to turn length is also the ratio of the propagation constant along the arms, k, to that of a wave propagating along the surface of the cylinder, B. If the difference between the k/B ratios for the helix and the conical spiral are expressed as a function of the helix ratio, the result is simply $1 - \cos \theta_0$ (see Fig. 82). For cones with an included angle of 20° this difference is of the order of

Frequency-Independent Antennas

1.5 percent. For all cones which are good unidirectional radiators the difference is only a few percent.

The variation of the propagation constant on periodic antenna structures can be conveniently displayed on the Brillouin (k-β) diagram [16]. One such diagram for the balanced bifilar helix is shown in Fig. 83. The vertical coordinate is given in units $ka/\tan \alpha$ which, since $k = 2\pi/\lambda$, is the pitch distance in free-space wavelengths. The horizontal scale is the pitch distance expressed in the equivalent guide wavelength on the surface of the antenna. For any single helix, since a and α are constant, the only variable involved is the wavelength of operation.

As the frequency of operation is increased, the propagation constant increases. If one considers a single space harmonic to be dominant, and assumes a nondispersive wave along the conductors, there is first a region of slow, closely bound surface waves. As the propagation constant increases still further, there is strong coupling to the first backward space wave, the propagation constant becomes complex, and the structure radiates with a phasing to provide a backfire beam.

If the wavelength is fixed, as a current wave progresses from the point of excitation along a structure with increasing radius, the propagation constant behaves in a manner similar to that observed for the propagation constant for a cylindrical structure as frequency increases. Thus, at a midband frequency the current on the arms near the tip is traveling away from the feed point with constant amplitude. At some distance down the arm, however, where the phasing for backfire radiation is approached, energy is strongly coupled into a radiated wave, and the amplitude of the current diminishes rapidly. The region containing the decaying current of nonnegligible amplitude is the active region. As the frequency is changed, the active region moves on the antenna so that the distance from the apex of the cone to the active region, when expressed in wavelengths, remains constant. The phase center of the radiated field is consistently located in the active region.

Measurements of currents on many conical log-spirals have led to recognition of the region from a point 3 dB below the maximum on the apex side to a point 15 dB below the maximum on the base side as defining the active region. The radii

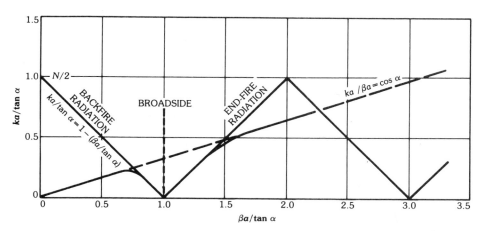

Fig. 83. Brillouin diagram for bifilar helix. (*After Dyson [30], © 1965 IEEE*)

at which these points occur are plotted in Fig. 84 as a function of the included cone angle $2\theta_0$ and the spiral angle α. Since both axes of this graph are normalized to wavelength these curves give the active region bounds. If, however, the vertical axis is considered to be normalized to the shortest wavelength, and the horizontal axis to the longest wavelength, these curves give the required radii of the truncated apex and base of the conical antenna. The circumferences C_3/λ and C_{15}/λ corresponding to these radii are also shown. Fig. 84 also includes a grid indicating the active region bandwidth B_{ar}, previously defined in (3), as a function of the antenna parameters. The data of Fig. 84 are for self-complementary antennas ($\delta = 90°$) and are most accurate for $\theta_0 = 10°$ (20° included cone angle). They tend to become more conservative as the cone angle decreases and slightly more optimistic for larger cone angles.

The rate of decay of the currents decreases as the arm width departs from the self-complementary case, $\delta = 90°$. Fig. 85 gives plots of M, the ratio of maximum radius required for very narrow or very wide arms to that required for arms with $\delta = 90°$. For intermediate values of δ, linear interpolation between the values given in Fig. 85 should be adequate. Since the small-end truncation is affected very little by variations in the arm width, the data for M in Fig. 85 can also be interpreted as the ratio of B_{ar} for other values of δ to that for $\delta = 90°$.

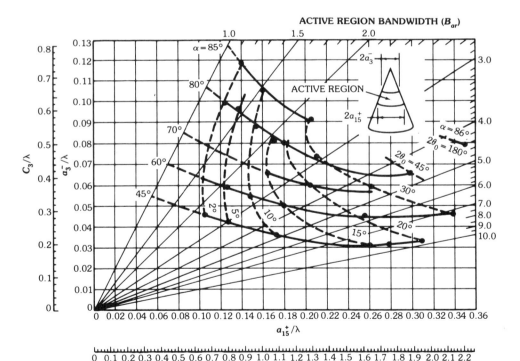

Fig. 84. Bounds of the active region in terms of the radius a or the circumference C of the conical log-spiral antenna. (*After Dyson [30], © 1965 IEEE*)

Frequency-Independent Antennas

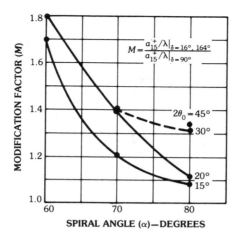

Fig. 85. Modification factor M for very narrow or very wide arm widths. (*After Dyson [30], © 1965 IEEE*)

The conical antenna radiates a single lobe which is directed off the apex along the axis of the cone. In Fig. 86 typical radiation patterns are shown as a function of the included angle of the cone $2\theta_0$ and the spiral angle α. The well-formed, relatively narrow beam for small cone angles and $\alpha = 80°$ is indicative of essentially all turns of the active region being phased for backfire radiation. As the spiral angle (α) decreases and the cone angle ($2\theta_0$) increases, the radiation pattern broadens

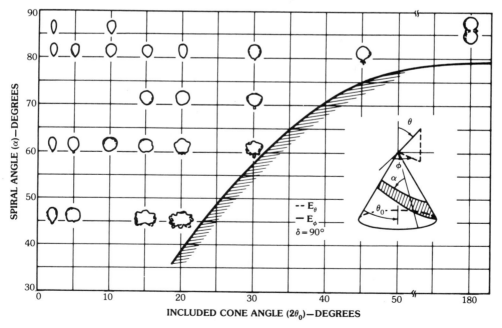

Fig. 86. Electric-field radiation patterns as a function of the spiral angle and cone angle.

and eventually exhibits a tendency to display multiple beams with corresponding irregularities. The shaded area in Fig. 86 represents parameters that are normally unusable.

The approximate half-power beamwidths are plotted in Fig. 87. Although the fields are circularly polarized on the axis the axial ratio increases for off-axis angles. Hence the patterns for the orthogonal components of the electric field differ in beamwidth by 8° to 10°. The values given in Fig. 87 are the averages of these

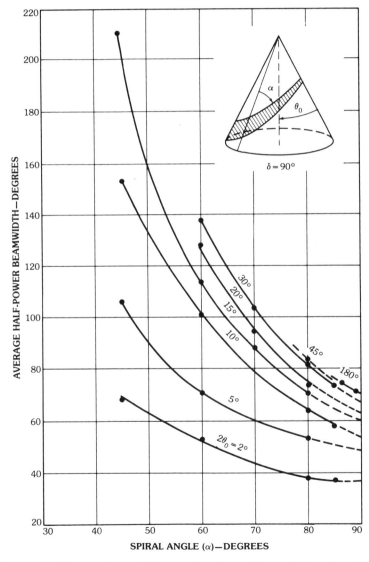

Fig. 87. Average half-power beamwidth as a function of the spiral angle and cone angle. (*After Dyson [30], © 1965 IEEE*)

Frequency-Independent Antennas

beamwidths. In addition, the average beamwidth must be interpreted together with the variation in beamwidth with changing frequency, which is shown in Fig. 88. The beamwidth variation is caused by the rotation with frequency of the slightly asymmetrical pattern.

The approximate directivity with respect to a circularly polarized isotropic source is plotted in Fig. 89. These values were calculated using

$$D \cong \frac{32\,600}{\Delta\phi_1 \times \Delta\phi_2} \quad (\text{dB}) \qquad (28)$$

where the denominator is the product of the average half-power beamwidths in orthogonal planes. The use of the constant 32 600 is considered conservative, resulting in a directivity that is approximately 1 dB less than that obtained with the value of 41 250 that appears frequently in the literature. For ease in converting to

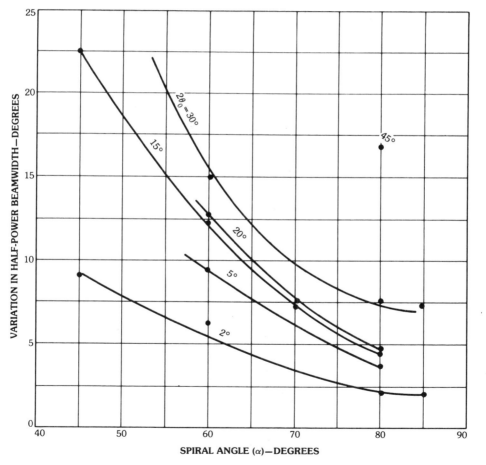

Fig. 88. Variation in average half-power beamwidth ($\delta = 90°$). (*After Dyson [30]*, © *1965 IEEE*)

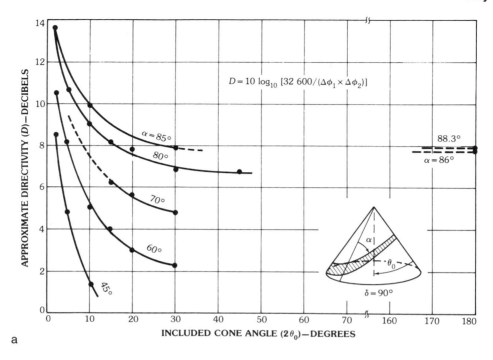
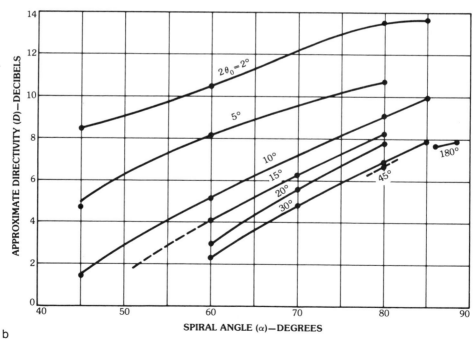

Fig. 89. Directivity as a function of cone angle and spiral angle. (*a*) As a function of cone angle. (*b*) As a function of spiral angle.

the approximate directivity based on the use of other constants in (28), a difference in directivity is plotted in Fig. 90 as a function of the constant. The directivity is directly related to the spiral angle α, rising as α increases. Representative radiation patterns have been plotted in Fig. 91 as a function of the spiral constant b.

In Fig. 92 typical radiation patterns are shown as a function of $2\theta_0$, α, and δ. For each combination of $2\theta_0$ and α, patterns are shown for very narrow arms ($\delta = 16°$), for the self-complementary case ($\delta = 90°$), and for very wide arms ($\delta = 164°$). For each case patterns are shown for two orthogonal polarizations and for two orthogonal planes. The principal change in the radiation patterns with change in arm width from $\delta = 90°$ is an increase in beamwidth.

The patterns shown in Fig. 86 and 92 are typical of those to be expected when the antenna is operated at frequencies such that there is no distortion due to the truncation of the base or tip. As the frequency of operation is decreased and the lower edge of the active region becomes affected by the truncated base, the amplitude of the back lobe will increase rather rapidly. As the leading edge of the active region moves into the truncated tip, the beamwidth may at first decrease and the pattern then becomes rough. Any lack of precision in construction of the apex region will cause pattern tilt and/or distortion. A further increase in frequency may cause the pattern to broaden with a tendency to break into lobes.

Radiation patterns for the very loosely wrapped antennas, $\alpha = 45°$, are shown for self-complementary width arms only. As the arm width deviates from $\delta = 90°$, for antennas with $2\theta_0 \geqq 15°$ and $\alpha \leqq 45°$, the pattern may break into many lobes with a major portion of the energy radiated in the direction of the base. A decrease in the front-to-back ratio may be noted as δ differs from $90°$. For $2\theta_0 \geqq 20°$, the variation with δ is less for $\alpha \geqq 75°$. For these small-cone, tightly wrapped antennas it is possible to use thin arms. The constant-width wire or cable arm versions of the antenna with these parameters perform satisfactorily.

The front-to-back ratios are plotted in Fig. 93. These are typical average values

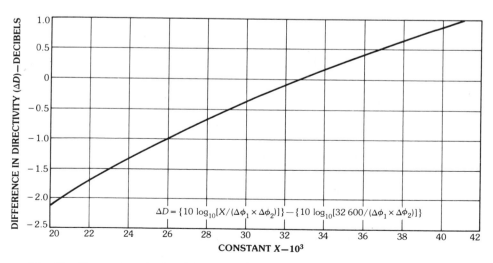

Fig. 90. Change in calculated directivity as the approximate expression is varied.

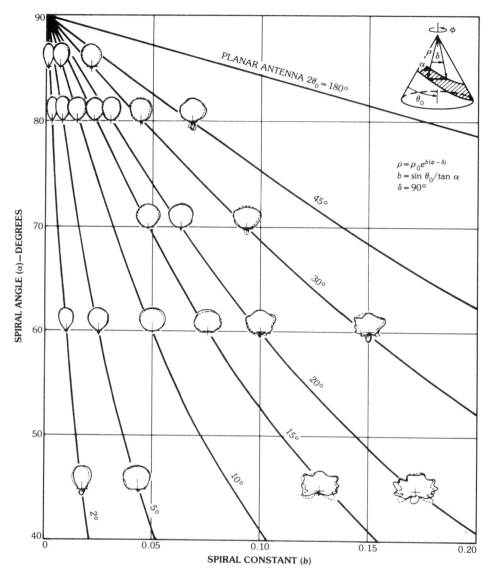

Fig. 91. Electric-field radiation patterns as a function of the spiral constant. (*After Dyson [30], © 1965 IEEE*)

that can be expected from a well-constructed antenna operated at a frequency such that there is no effect due to truncations at the tip or base.

The conical spirals are essentially circularly polarized in any direction where there is substantial radiation. Typical values of axial ratio recorded at angles from the axis of an antenna with $2\theta_0 = 20°$ are shown in Fig. 94. The curves with minimum slope are obtained for the lower values of α. This is indicative of the increased beamwidth and increased energy radiated in directions approaching that

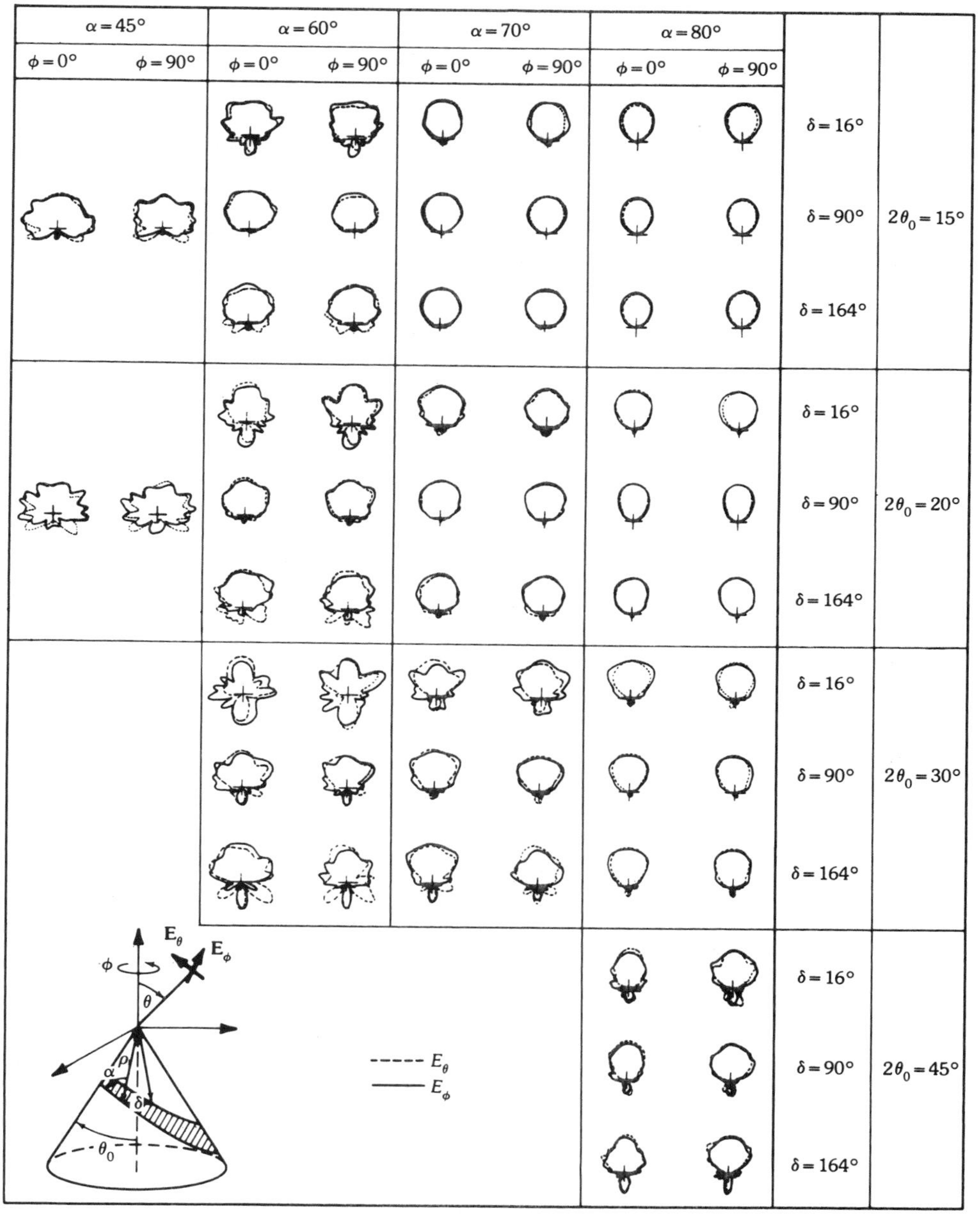

Frequency-Independent Antennas

Fig. 92. Electric-field radiation patterns for three angular arm widths δ, for four cone angles $2\theta_0$, and for four spiral angles α. (*After Dyson [30], © 1965 IEEE*)

perpendicular to the axis as the spiral angle α is decreased. As δ departs from 90° the axial ratio rises slightly as a function of θ_0.

Conical log-spirals do not have a unique phase center. However, a phase center can be found for a major portion of the main beam. Fig. 95 shows the distance from

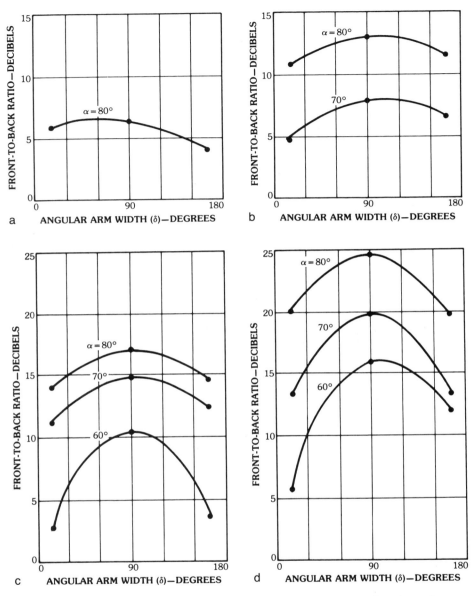

Fig. 93. Minimum front-to-back ratio of radiation patterns as a function of antenna parameters. (*a*) For $2\theta_0 = 45°$. (*b*) for $2\theta_0 = 30°$. (*c*) For $2\theta_0 = 20°$. (*d*) For $2\theta_0 = 15°$. (*After Dyson [30], © 1965 IEEE*)

the virtual apex to the phase centers as measured on several conical log-spirals with $\delta = 90°$. When the conical spirals are used as feeds for parabolic reflectors the antenna should be positioned so that the indicated phase center coincides with the focal point of the reflector. Since the phase center moves as frequency changes, a compromise must be made when operating over a wide frequency band.

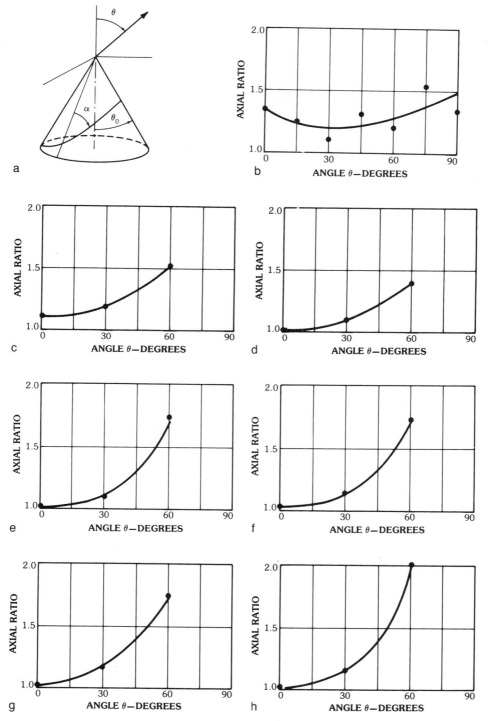

Fig. 94. Typical variation in axial ratio off the axis of the antenna for $2\theta_0 = 20°$. (*a*) Geometry. (*b*) For $\alpha = 45°$, $\delta = 90°$. (*c*) For $\alpha = 70°$, $\delta = 16°$. (*d*) For $\alpha = 70°$, $\delta = 90°$. (*e*) For $\alpha = 70°$, $\delta = 164°$. (*f*) For $\alpha = 80°$, $\delta = 16°$. (*g*) For $\alpha = 80°$, $\delta = 90°$. (*h*) For $\alpha = 80°$, $\delta = 164°$. (*After Dyson [30], © 1965 IEEE*)

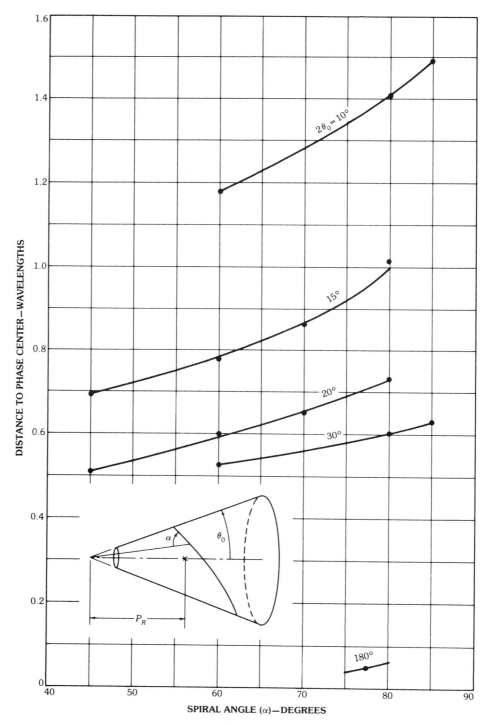

Fig. 95. Location of measured "effective phase center" for balanced, two-arm, conical, log-spiral antennas ($\delta = 90°$).

Frequency-Independent Antennas

The two-arm conical log-spiral, like the planar version, is a symmetrical structure and, when excited in a balanced manner, radiates a beam on the axis of the cone without squint or tilt. A tilt in the radiation pattern can be traced to the physical construction or to unbalance in the feed.

Conical log-spirals can be excited by bringing a balanced transmission line along the axis from the large to the small end and connecting one wire to each arm. It is preferable, although not necessary, that this be a shielded line. The presence of metal *on axis* has a minimum effect on the antenna characteristics if the diameter of the metal is no more than one-third the diameter of the antenna at any point on the axis.

The transition from an unbalanced coaxial line to a balanced line may be made by placing a balun inside the antenna if the balun is nonradiating and does not disturb the fields inside the cone. The taperline balun [32] with its extremely wide bandwidth would seem to be ideally suited for this purpose. However, when it is placed inside conical log-spirals a truly balanced feed is seldom realized. For some purposes, however, the degradation of the pattern can be tolerated.

Coaxial 180° hybrids make satisfactory baluns. These units provide equal-amplitude, out-of-phase signals at the side arms. Fifty-ohm coaxial cables connected to these arms may be carried along the axis of the cone, and the two center conductors of the cables thus become a shielded, balanced, 100-Ω transmission line. For maximum symmetry, and to prevent the two outer sheaths from becoming a line for any unbalanced currents, the outer conductors of these cables should be electrically connected along their lengths.

To overcome the possible limiting bandwidths of baluns the conical spirals can also be fed by carrying a coaxial cable along one arm as described before for the planar spirals. To maintain symmetry a similar "dummy" cable must be placed on the other arm. This remains the most satisfactory method if the presence of the cable does not limit the size of the truncated apex region and if the loss in the relatively long length of cable can be tolerated.

The geometry of the truncated tip, including the possible presence of the feed cable on the antenna arms, has a marked effect on the input impedance. Since there are an unlimited number of cable diameter and truncated-tip dimensions, it is dangerous to expect that published data for input impedance will be realized unless care is taken to ensure that the feed-point geometry is duplicated. The use of coaxial hybrids as baluns makes possible a very convenient method of making balanced impedance measurements, and this is the technique that was used to obtain the data reported here [28, 29]. Identical slotted lines were inserted in the coaxial lines between the antenna and hybrid. If the sections are matched for probe insertion and movement, and for detector characteristics, the phase balance of the lines will be preserved.

Measurements were made over a 5:1 bandwidth. The normalized impedances plotted on a Smith chart were enclosed by a circle. The "center" of this circle, chosen to make the hyperbolic distance to all parts of the circle constant, was considered to be the nominal impedance of the antenna. The maximum vswr referred to the nominal impedance was typically less than 1.5.

Fig. 96 indicates the effect that the feed geometry can have on the measured impedance. The data enclosed by the larger circle were obtained when the center conductors of two RG-141/U cables were extended, bent at right angles at the plane

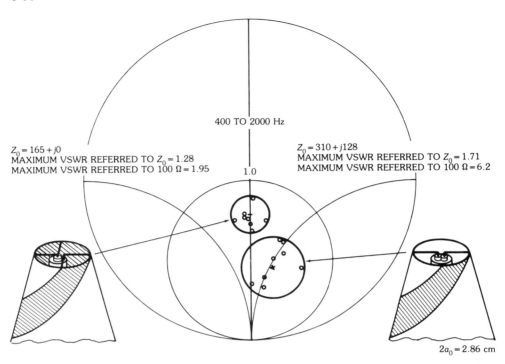

Fig. 96. Effect of feed region geometry on input impedance. (*After Dyson [30]*, © *1965 IEEE*)

of the tip truncation, and carried perpendicular to the axis of the cone out to the antenna arms. The wires were attached at the sharp end of the spiral arm that results when truncated at the circle $\varrho = d/2$. This provides a tapered transition from the small-diameter wire to the much greater width of the spiral arms. The measured values of input impedance referred to a reference plane established at the termination of the shields of the two coaxial cables fell within a reasonably compact area of the Smith chart, but the data display a positive reactive shift off the real axis.

When the transition between the center conductors of the cables and the antenna arms was made by flat, tapered conductors with angular width equal to δ, the scatter of the impedance data as a function of frequency decreased. In addition, the reactive shift disappeared and the resistive level of the impedance was changed significantly. Symmetry, balance, and tapered leads from the axis to the arms on the surface of the cone are all required for best performance.

The variation of the input impedance with arm width is shown in Fig. 97. The impedance is primarily controlled by the arm width, varying from 320 Ω for very narrow arms to 80 Ω for very wide arms. The impedance increases as the cone angle increases. For self-complementary antennas the impedance approaches 189 Ω, the theoretical value for the planar case. With the infinite-balun feed the presence of cable on the arms tends to give the narrow arms near the apex greater cross section and hence shifts the impedance level. Fig. 97 shows values of the measured

Frequency-Independent Antennas

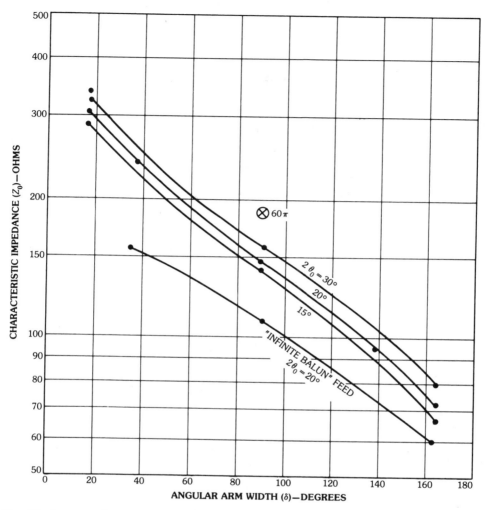

Fig. 97. Average characteristic impedance of conical log-spiral antennas as a function of angular arm width. (*After Dyson [30]*, © *1965 IEEE*)

impedance for an antenna with $2\theta_0 = 20°$, $\alpha = 60°$, and a tip truncated at 0.75 in, fed with RG-141/U cable of diameter 0.141 in. The maximum vswr with respect to the indicated impedances was about 1.75. The impedance variation with change in the spiral angle α is small, although there is a tendency for the impedance to increase with increasing α. The trend in impedance with cone angle shown in Fig. 97 is not obtained for all cone angles. Impedances for cones with angles less than 15° are not lower than those shown for 15°. For $2\theta_0 = 5°$, $\alpha = 80°$, the impedance is about 300 Ω for $\delta = 16°$, 180 Ω for $\delta = 90°$, and 72 Ω for $\delta = 164°$.

Half-power beamwidths of 40° can be obtained with small values of α. When wide bandwidths are to be covered, however, these antennas will become quite long. The structure length in terms of the longest wavelength λ_L is given by

$$\frac{h}{\lambda_L} = \frac{1}{2\tan\theta_0}\frac{D}{\lambda_L} - \frac{d}{B\lambda_H} \qquad (29)$$

Plots of this expression are shown in Fig. 98 for 10:1, 5:1, 2:1, and 1:1 bandwidths as a function of the average half-power beamwidth and the cone angle and spiral angle for $\delta = 90°$. No curve is drawn in the 5:1 case except when $\alpha = 45°$. The use of smaller cone angles requires a cone of greater length but, of course, one of smaller diameter. In Fig. 99 the required maximum base diameter at the lowest frequency has been plotted as a function of the beamwidth and the parameters α and θ_0.

Nomographs are a useful aid to the design of conical log-spirals also. Fig. 100 relates the radius vector ϱ in wavelengths to the half-cone angle θ_0 and the diameter D/λ or the circumference (in wavelengths). Fig. 101 relates the total height of the cone from the virtual apex, H/λ, to the half-cone angle θ_0 and the diameter D/λ. Fig. 102 gives the spiral constant b in terms of the spiral angle α and the cone angle θ_0. Fig. 103 relates the spiral constant b, the angular arm width δ, and the arm-width constant K.

The tightly spiraled, self-complementary log-spiral tends to perform best in the

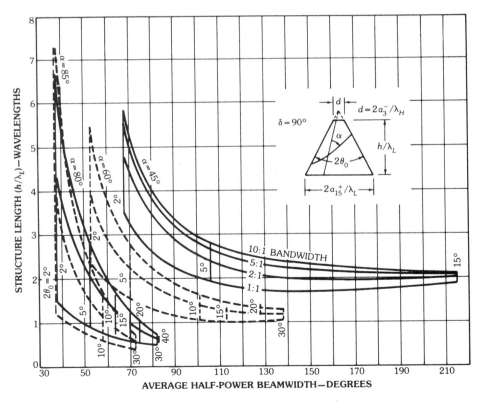

Fig. 98. Approximate cone height normalized to the longest wavelength of operation for bandwidths of 10:1, 5:1, 2:1, and 1:1 when base of cone equals a_{15}^+/λ_L, with $\delta = 90°$.

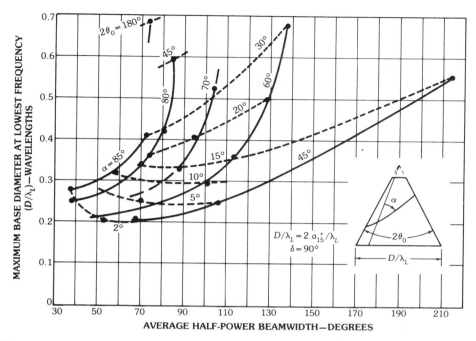

Fig. 99. Maximum diameter of base of cone at longest wavelength of operation when $D/\lambda_L = 2a_{15}^+/\lambda$, with $\delta = 90°$.

sense that it has the narrowest active region. Hence it has the greatest bandwidth for a given physical size. It also tends to have well-formed radiation patterns with the least energy radiated in back lobes. Except for the impedance a change in arm width for cone angles $2\theta_0$ less than 20° and angles of wrap greater than 80° causes only minor changes. Hence the narrow constant-width and cable-arm approximations to the log-spiral antenna can be used effectively. As the cone angle increases and the angle of wrap decreases from these values, a departure from $\delta = 90°$ causes increasingly greater pattern deterioration. The impedance depends primarily on the angular arm width, ranging from 300 Ω for small arm widths to 80 Ω for wide arms. Typical values for the case of $\delta = 90°$ range from 140 to 165 Ω. The impedance bandwidth is consistently greater than the pattern bandwidth.

As shown in Fig. 92 the beamwidth of the two-arm conical spiral can be controlled over a limited range by a suitable choice of α. Typical half-power beamwidths range from 60° to 70° for $\alpha = 82°$, 70° to 80° for $\alpha = 73°$, and 160° to 180° for $\alpha = 60°$. As α is decreased to 45° the beamwidth increases to 180° to 200° and this value of α produces circularly polarized coverage of an entire hemisphere and an omnidirectional pattern in the $\theta = 90°$ plane. To realize this coverage the decrease in α must be accompanied by an increase in δ.

When omnidirectional circular polarization near $\theta = $ constant is desired, it is advantageous to use a four-arm, rather than a two-arm, spiral [31]. When using multiple-arm structures, the number of choices for feeding the antenna increases [2]. There are particular excitations that will produce selected groups from the

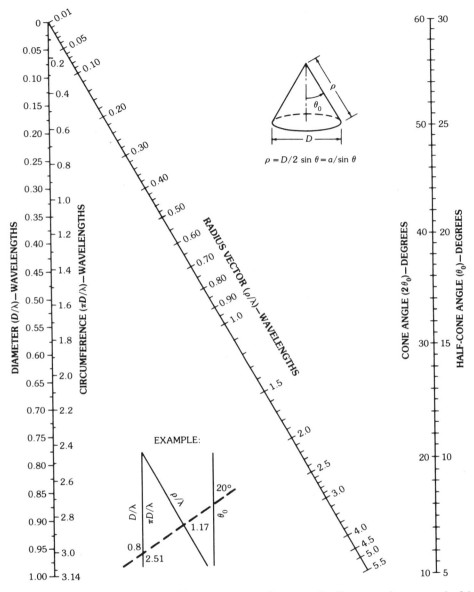

Fig. 100. Nomograph relating radius vector ϱ to diameter D of cone and cone angle $2\theta_0$.

possible azimuthal "modes" having variations $\exp(\pm jm\phi)$, where m is integer. The usual excitation of two-arm spirals shown in Fig. 104a will produce only terms with odd integer values of m since a rotation of 180° is equivalent to 180° phase shift. The remarkable property of the conical log-spirals is that the field is predominantly only the lowest-order term which is consistent with the sense of winding and the excitation. Fig. 105 shows the relative phase of the distant field as a

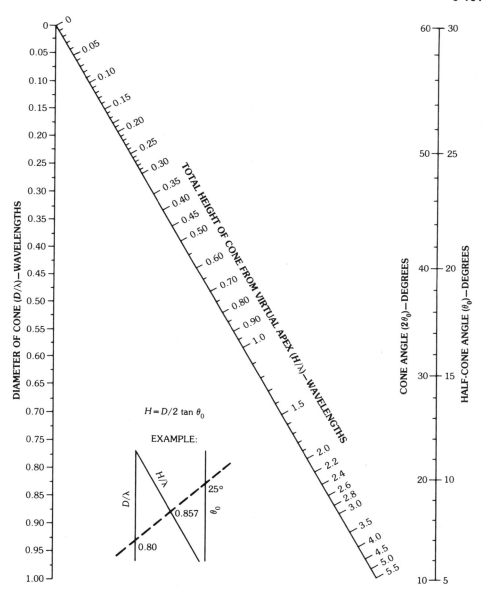

Fig. 101. Nomograph relating full height H of cone to diameter D of cone and cone angle $2\theta_0$.

function of ϕ for two antennas, a two-arm spiral excited with the arms 180° apart corresponding to odd-integer values of m, and a four-arm spiral excited with adjacent arms 180° apart. Note that in each case the phase is almost directly proportional to ϕ, but for the two-arm case the coefficient is 1 while for the four-arm case the coefficient is 2. This indicates that, in each case, there is very little of any of the higher-order terms present.

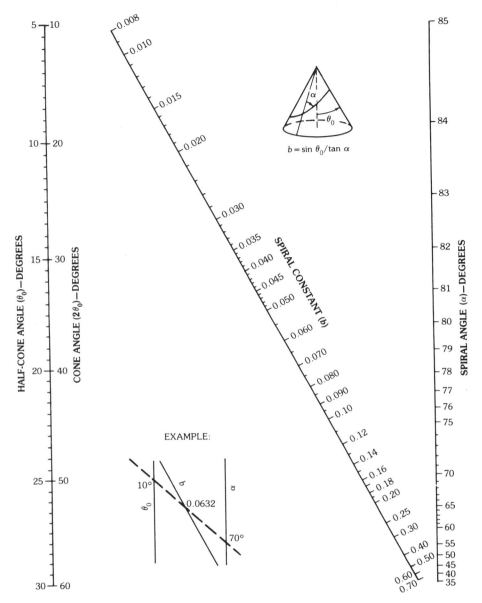

Fig. 102. Nomograph relating spiral constant b to cone angle $2\theta_0$ and spiral angle α.

The Maxwell equations indicate that any field that has m greater than unity will have zero along the polar axis ($\theta = 0$). Any excitation that produces only fields that vary with ϕ according to m greater than one will have a conical pattern with null along the axis and maximum at some nonzero value of θ. The excitation shown in Fig. 104c, producing a minimum value of m of 2, is the simplest case. This excitation is easily achieved by connecting opposite arms together and feeding each

Frequency-Independent Antennas

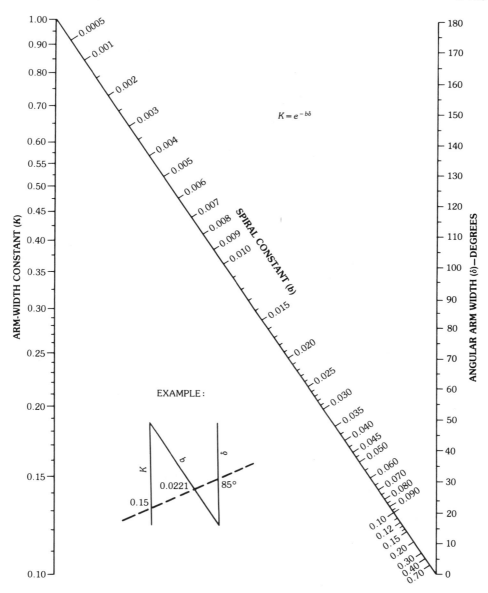

Fig. 103. Nomograph relating arm with constant K to spiral constant b and angular arm width δ.

pair 180° apart. The two terminals created by joining opposite arms can be fed by a balanced line, a coaxial cable connected through a balun, or by carrying a coaxial cable along one of the arms as described previously. Details of the feed-point geometry in the latter case are shown in Fig. 106. Dummy cables on the other arms are required to maintain symmetry.

Fig. 107 shows patterns of symmetrical four-arm spirals fed as shown in

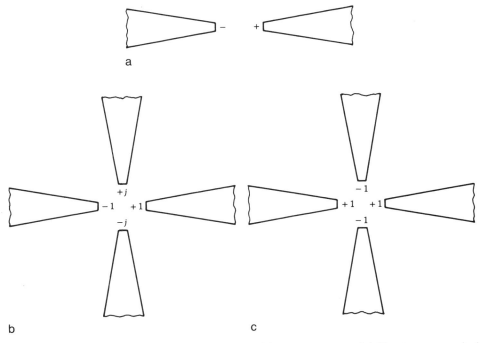

Fig. 104. Possible feeding arrangements for multiarm structures. (*a*) For two-arm spiral, with arms 180° apart. (*b*) For four-arm spiral, with arms 90° apart. (*c*) For four-arm spiral, with arms 180° apart. (*After Dyson and Mayes [31],* © *1961 IEEE*)

Fig. 104c. Just as α could be used to control the beamwidth of two-arm spirals, it can also be used to control the direction of maximum radiation for the conical beams produced by the higher-order excitations. Beam maxima from about 40° to more than 90° off the polar axis can be produced. The case with beam maximum at $\theta = 90°$ is of particular interest since it provides a simple way to achieve a very broadband, circularly polarized, omnidirectional source.

Patterns of a four-arm, balanced, log-spiral constructed on a 15° cone are shown in Fig. 108. This antenna was etched from a flexible, copper-clad substrate and then formed into a cone. The feed and dummy cables were RG-141/U coax. At 550 MHz the base is 0.57 wavelengths in diameter. At 4 GHz the diameter of the truncated apex is approximately 0.2 wavelength. The azimuthal coverage shown in the patterns of Fig. 108 is presented in more detail in Fig. 109, where the deviation from omnidirectional is plotted in decibels for the two orthogonal field components. The axial ratio in the $\theta = 90°$ plane is seen to vary somewhat with ϕ. However, over a considerable bandwidth the amplitude deviation from omnidirectional is less than 3 dB and the axial ratio is also less than 3 dB.

The beamwidth of the conical beam patterns in a $\phi = $ constant plane is relatively insensitive to a change in antenna parameters. Antennas with $2\theta_0$ equal to both 15° and 20°, with α between 45° and 73°, and with cable arms or with arms of expanding width all have half-power beamwidths ranging from 35° to 55°.

Four-arm spirals fed in the manner of Fig. 104c, with $2\theta_0$ of 15° or 20°, typically

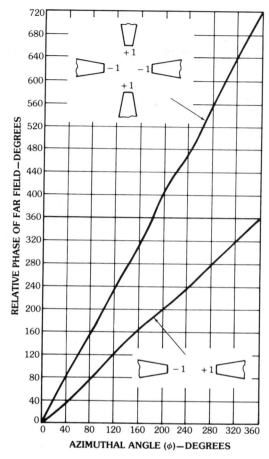

Fig. 105. Measured phase of radiated field as a function of the azimuthal angle ϕ for two antennas, with $\alpha = 45°$ and $\theta = 90°$. (*After Dyson and Mayes [31], © 1961 IEEE*)

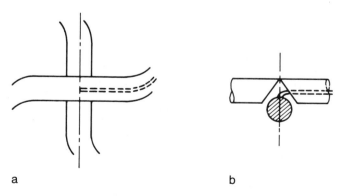

Fig. 106. "Infinite balun" feed used on a four-arm, conical, beam antenna. (*a*) Top view. (*b*) Side view. (*After Dyson and Mayes [31], © 1961 IEEE*)

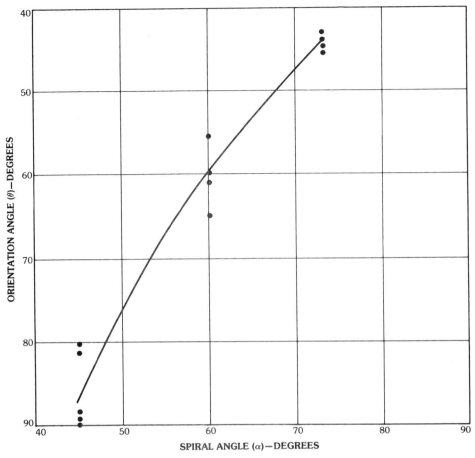

a

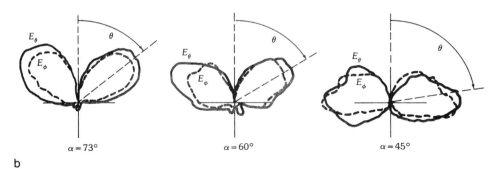

b

Fig. 107. Typical electric-field patterns and orientation of the conical beam as a function of the spiral angle, with $7.5° \leqq \theta_0 \leqq 10°$. (*a*) Graph. (*b*) Some patterns. (*After Dyson and Mayes [31], © 1961 IEEE*)

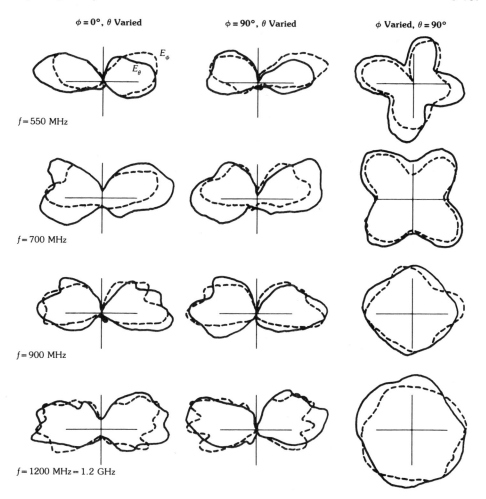

Fig. 108. Electric-field patterns of a balanced, symmetrical, four-arm, conical, equiangular-spiral antenna, with $\theta_0 = 7.5°$, $\alpha = 45°$, $K = 0.925$, $D = 31$ cm, and $d = 1.5$ cm. (*After Dyson and Mayes [31], © 1961 IEEE*)

have an impedance from 45 to 55 Ω for α ranging from 45° to 60°. As α is increased to 73° the impedance rises to the neighborhood of 70 Ω. These values are approximately one-half those noted earlier for similar two-arm antennas. The vswr is typically less than 1.5 (referred to the mean impedance) over most of the pattern bandwidth.

The pattern characteristics of log-spiral antennas are relatively constant over an extended frequency band. The beamwidth of two-arm spirals and the beam maximum of four-arm spirals are directly related to the value of α. Archimedean-spiral curves may also be orthogonally projected onto a conical surface as shown in Fig. 110b. However, the value of α is not constant for the Archimedean spirals, but

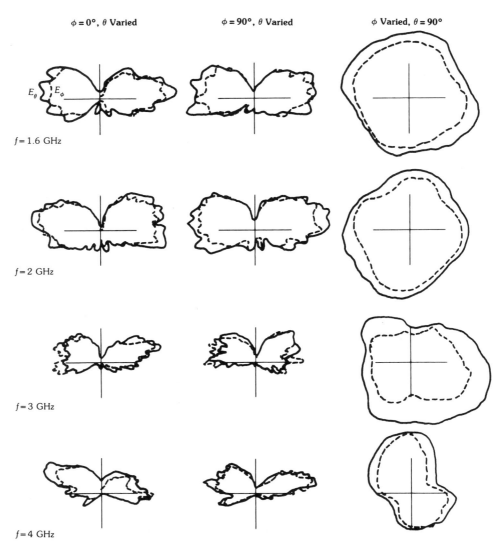

Fig. 108, *continued.*

rather is dependent on the value of ϕ. As the frequency is changed, the active region of a conical Archimedean spiral moves across areas with different values of α. This is manifest in a changing of the beamwidth for a two-arm antenna and a scanning of the beam maximum for the four-arm antennas. Patterns for one particular four-arm Archimedean spiral are shown in Fig. 111a. The value of α for this antenna ranged from about 45° at the apex to 85° at the base. The beam maximum varies from 45° to 90° as the frequency is swept from 1 to 2 GHz. The patterns for a log-spiral with $\alpha = 85°$ are shown in Fig. 111b for comparison.

The various modes of a multiarm spiral could also be selected by carrying an

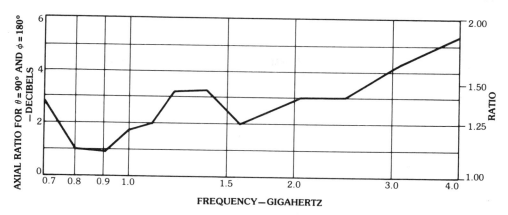

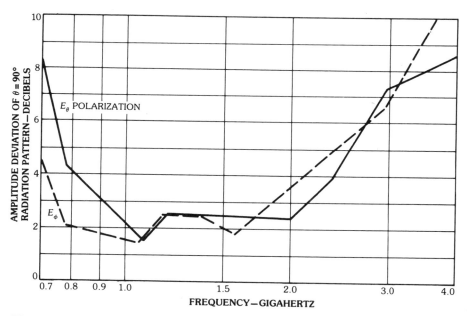

Fig. 109. Azimuthal coverage of the radiation patterns in Fig. 108. (*After Dyson and Mayes [31], © 1961 IEEE*)

active cable for each arm up the axis of the cone and applying the appropriate phase in each line at the base. The phase properties of the far fields for several of the lower-order modes provide a means of determining the direction of arrival of incoming waves. Fields having $\exp(\pm jm\phi)$ far-field behavior have been termed "spiral-phase" fields [33]. In contrast, the fields of a vertical monopole have equiphase contours that are circles, a "circular-phase" field.

Several direction-finding systems can be devised using multiarm log-spirals [34–36]. A system could be based on (*a*) a log-spiral excited in a spiral-phase mode and another antenna operating in the circular-phase mode, (*b*) two log-spiral

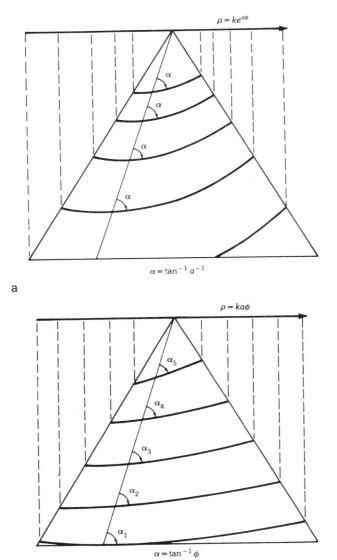

Fig. 110. Projection of equiangular-spiral and Archimedean-spiral curves on a conical surface. (*a*) Equiangular spiral. (*b*) Archimedean spiral. (*After Dyson and Mayes [31]*, © *1961 IEEE*)

antennas excited in $+m$ and $-m$ modes, or (*c*) a single log-spiral antenna operated successively in two or more modes. The various modes of a multiarm antenna can be excited by carrying a feed cable for each arm up the axis of the cone. Switching among various values of phase shift between arms can then be used to sequentially produce any one of the several possible modes. The technique is illustrated for a four-arm spiral in Fig. 112.

Frequency-Independent Antennas

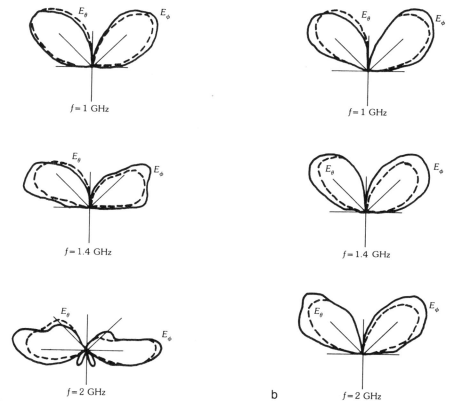

Fig. 111. Electric-field patterns of symmetrical, four-arm, conical antennas, with $\theta_0 = 10°$, $D = 29.5$ cm, and $d = 4.5$ cm ($\phi = 90°$, θ-varied pattern). (*a*) Archimedean spiral, with $\varrho = 1.026\phi$. (*b*) Equiangular spiral, with $\varrho = e^{[(\sin 10°)/(\tan 45°)]\phi}$. (*After Dyson and Mayes [31]*, © *1961 IEEE*)

Construction Techniques

In constructing log-periodic and log-spiral antennas it must be remembered that they are only potentially capable of covering extremely broad bands of frequency. Great care must be taken to use proper techniques if they are to perform with near-constant characteristics over such bandwidths. Generally, all members, both radiating and structural, should scale in size with distance from the apex. The feeders of LP dipole arrays, and other similar transmission means for exciting LP arrays, can be excepted from this rule. Such feeders may require stepped scaling if extremely wide frequency bands are to be covered in a high-performance manner. Element diameters in LP dipole arrays also need not conform exactly to true LP scaling. Symmetry, balance, and precision of construction are all important. An application of log-periodic or log-spiral antennas over extremely wide bandwidths requires all of the precision of construction that can be economically justified. This is particularly true for frequencies above 1 GHz.

Using longitudinal struts for support of the radiating elements has been found

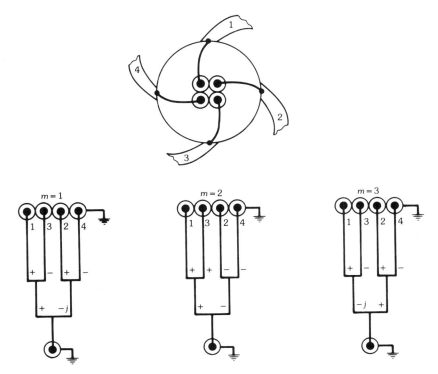

Fig. 112. Multimode feed on axis of cone. (*After Dyson [34], presented at the 1961 National Electronic Conference*)

satisfactory. Ideally, the struts should be tapered. The use of dielectric structural members placed transverse to the axis of the antenna causes some perturbation in performance when the active region passes over these members. Hence such structural members cause greater variation in the performance than members located along a radius vector from the origin (virtual apex).

5. References

[1] Y. Mushiake, *J. Inst. Electron. Eng.*, Japan, vol. 69, pp. 86–88 (in Japanese).
[2] G. A. Deschamps, "Impedance properties of complementary multiterminal planar structures," *IRE Trans. Antennas Propag.*, vol. AP-7, pp. S371–S378, December 1959.
[3] R. H. DuHamel and D. E. Isbell, "Broadband logarithmically periodic antenna structure," *IRE Natl. Conv. Rec.*, pt. I, pp. 119–128, 1957.
[4] D. E. Isbell, "Nonplanar logarithmically periodic antenna structures," *Tech. Rep. 30*, Univ. of Illinois Antenna Lab, Contract AF33(616)-3220, February 1958.
[5] R. H. DuHamel and F. R. Ore, "Logarithmically periodic antenna design," *Tech. Rep. CTR-198*, Collins Radio Co., Cedar Rapids, Iowa, March 1958.
[6] D. E. Isbell, "Log-periodic dipole arrays," *IRE Trans. Antennas Propag.*, vol. AP-8, pp. 260–267, May 1960.
[7] R. L. Carrel, "The design of log-periodic dipole antennas," *IRE Intl. Conv. Rec.*, pt. I, pp. 61–75, 1961.

[8] R. L. Carrel, "Analysis and design of the log-periodic antenna," PhD dissertation, Univ. of Illinois, Urbana, 1961.

[9] W. M. Cheong and R. W. P. King, "Arrays of unequal and unequally spaced dipoles," *Radio Sci.*, vol. 2, no. 11, pp. 1303–1314, November 1967.

[10] W. M. Cheong and R. W. P. King, "Log-periodic dipole antenna," *Radio Sci.*, vol. 2, no. 11, pp. 1315–1325, November 1967.

[11] J. Wolter, "Solution of Maxwell's equations for log-periodic dipole antennas," *IEEE Trans. Antennas Propag.*, vol. AP-18, pp. 734–741, November 1970.

[12] G. DeVito and G. B. Stracca, "Comments on the design of log-periodic dipole antennas," *IEEE Trans. Antennas Propag.*, vol. AP-21, pp. 303–308, May 1973.

[13] G. DeVito and G. B. Stracca, "Further comments on the design of log-periodic dipole antennas," *IEEE Trans. Antennas Propag.*, vol. AP-22, pp. 714–718, September 1974.

[14] P. C. Butson and G. T. Thompson, "A note on the calculation of the gain of log-periodic dipole antennas," *IEEE Trans. Antennas Propag.*, vol. AP-24, pp. 105–106, January 1976.

[15] E. C. Jordan and K. J. Balmain, *Electromagnetic Waves and Radiating Systems*, Englewood Cliffs: Prentice-Hall, 1968, p. 390.

[16] P. E. Mayes, G. A. Deschamps, and W. T. Patton, "Backward-wave radiation from periodic structures and applications to the design of frequency-independent antennas," *Proc. IRE*, vol. 49, pp. 962–963, May 1961.

[17] H. C. Pocklington, "Electrical oscillations in wires," *Camb. Phil. Soc.*, vol. 9, pp. 324–332, October 1897.

[18] S. H. Lee, "Analysis of zigzag antennas," PhD dissertation, Dept. of Electr. Eng., Univ. of California, Berkeley, June 1968.

[19] S. H. Lee and K. K. Mei, "Analysis of zigzag antennas," *IEEE Trans. Antennas Propag.*, vol. AP-18, pp. 760–764, November 1970.

[20] S. C. Kuo and P. E. Mayes, "An experimental study and the design of the log-periodic zigzag antennas," *Tech. Rep. 65-11*, Univ. of Illinois Antenna Lab, Contract AF33(615)-3216, February 1966.

[21] P. G. Ingerson and P. E. Mayes, "Asymmetrical feeders for log-periodic antennas," Seventeenth Annual Symposium, USAF Antenna Res. and Dev. Prog., Allerton Park, Univ. of Illinois, Urbana, October 1967.

[22] P. G. Ingerson, "Analysis of some closed log-periodic antennas with applications to log-periodic antennas," PhD thesis, Univ. of Illinois, Urbana, 1968.

[23] L. Brillouin, *Wave Propagation in Periodic Structures*, New York: Dover Publications, 1953.

[24] E. G. Cristal, "Coupled–transmission-line directional coupler with coupled lines of unequal characteristic impedance," *IEEE Trans. Microwave Theory Tech.*, vol. MTT-14, no. 7, pp. 337–346, July 1966.

[25] R. L. Carrel, "The design of log-periodic dipole antennas," *IRE Intl. Conv. Rec.*, pt. I, pp. 61–75, 1961.

[26] P. G. Ingerson and P. E. Mayes, "Log-periodic antennas with modulated impedance feeders," *IEEE Trans. Antennas Propag.*, vol. AP-16, pp. 633–642, November 1968.

[27] J. D. Dyson, "The equiangular-spiral antenna," Fifth Annual Symposium, USAF Antenna Res. and Dev. Prog., Allerton Park, Univ. of Illinois at Urbana, October 1955.

[28] J. D. Dyson, "The equiangular-spiral antenna," *IRE Trans. Antennas Propag.*, vol. AP-7, pp. 181–187, April 1959.

[29] J. D. Dyson, "Recent developments in spiral antennas," *Proc. IRE Natl. Aeron. Electron. Conf.*, pp. 617–625, 1959.

[30] J. D. Dyson, "The characteristics and design of the conical log-spiral antenna," *IRE Trans. Antennas Propag.*, vol. AP-13, pp. 488–499, July 1965.

[31] J. D. Dyson and P. E. Mayes, "New circularly polarized frequency-independent antennas with conical beam or omnidirectional patterns," *IRE Trans. Antennas Propag.*, vol. AP-19, pp. 334–342, July 1961. Also *Tech. Rep. 60*, Univ. of Illinois

Antenna Lab, Contract AF33(657)-8460, December 1962.
[32] J. W. Duncan and V. P. Minerva, "100:1 bandwidth balun transformer," *Proc. IRE*, vol. 48, pp. 156–164, February 1960.
[33] E. K. Sandeman, "Spiral-phase fields," *Wireless Engineer*, pp. 96–105, March 1949.
[34] J. D. Dyson, "The conical log-spiral antenna in simple arrays," Eleventh Annual Symposium, USAF Res. and Dev. Prog., Allerton Park, Univ. of Illinois, Urbana, October 1961.
[35] J. D. Dyson, "Multimode logarithmic-spiral antennas—possible applications," *Proc. Natl. Electron. Conf.*, pp. 206–213, October 1961.
[36] J. D. Dyson, "The multimode antennas as single-aperture systems," Antenna Forum, Univ. of Illinois, Urbana, January–February 1967.

6. Bibliography

Anders, R., and R. Wohlleben, "Phase velocity on a conical two-armed logarithmic foil-type spiral antenna," *IRE Trans. Antennas Propag.*, vol. AP-17, pp. 233–234, March 1969.

Atia, A. E., and K. K. Mei, "Analysis of multiple-arm conical log-spiral antennas," *IEEE Trans. Antennas Propag.*, vol. AP-19, pp. 320–331, May 1971.

Balmain, K. G., and J. D. Dyson, "The series-fed, log-periodic folded dipole array," *Dig. IEEE PTG-AP Intl. Symp.*, Boulder, Colorado, pp. 143–148, 1963.

Balmain, K. G., and S. W. Mikhail, "Loop coupling to a periodic dipole array," *Electron. Lett.*, vol. 5, no. 11, pp. 228–229, May 29, 1969.

Balmain, K. G., C. C. Bantin, C. R. Oakes, and L. David, "Optimization of log-periodic dipole antennas," *IEEE Trans. Antennas Propag.*, vol. AP-19, pp. 286–288, March 1971.

Balmain, K. G., and J. N. Nkeng, "Asymmetry phenomenon of log-periodic dipole antennas," *IEEE Trans. Antennas Propag.*, vol. AP-24, pp. 402–410, July 1976.

Bantin, C. C., and K. G. Balmain, "Study of compressed log-periodic dipole antennas," *IEEE Trans. Antennas Propag.*, vol. AP-18, pp. 195–203, March 1970.

Barbano, N., "Log-periodic dipole array with parasitic elements," *Microwave J.*, vol. 8, pp. 41–69, October 1965. Also *Tech. Memo EDL-M623*, Electronic Defense Laboratory, Sylvania Electric Products, Inc., Mountain View, Calif., January 1964.

―――. "Log-periodic Yagi-Uda array," *IEEE Trans. Antennas Propag.*, vol. AP-14, pp. 235–238, March 1966.

Bell, R. L., C. T. Elfving, and R. E. Franks, "Near-field measurements on a logarithmically periodic antenna," *IRE Trans. Antennas Propag.*, vol. AP-8, pp. 559–567, November 1960.

Berry, D. G., and F. R. Ore, "Log-periodic monopole array," *IRE Intl. Conv. Rec.*, pt. I, pp. 76–85, 1961. Also technical report, Collins Radio Co., Cedar Rapids, Iowa, October 1960.

Besthorn, J. W., "1.0 to 21.0-GHz log-periodic dipole antenna," Eighteenth Annual Symposium, USAF Antenna Res. and Dev. Prog., Allerton Park, Univ. of Illinois, Urbana, October 1968.

Blume, S., and J. Wolter, "Theoretische und experimentelle untersuchungen an logarithmisch periodischen antennen," *Z. Angew. Phys.*, vol. 23, pp. 386–396, 1967.

Brillouin, L., *Wave Propagation in Periodic Structures*, New York: Dover Publications, 1953.

Brooks, G. P., "A conical log-spiral antenna with resistive terminated arms," MS thesis, Dept. of Electr. Eng., Univ. of Illinois at Urbana, 1965.

Butson, P. C., and G. T. Thompson, "A note on the calculation of the gain of log-periodic dipole antennas," *IEEE Trans. Antennas Propag.*, vol. AP-24, pp. 105–106, January 1976.

Campbell, C. K., I. Traboulay, M. S. Suthers, and H. Kneve, "Design of a stripline log-periodic dipole antenna," *IEEE Trans. Antennas Propag.*, vol. AP-25, pp. 718–721, September 1977.

Carr, J. W., "Some variations in log-periodic antenna structures," *IRE Trans. Antennas*

Propag., vol. AP-9, pp. 229–230, March 1961.

Carrel, R. L., "An analysis of log-periodic dipole antennas," Tenth Annual Symposium, USAF Antenna Res. and Dev. Prog., Allerton Park, Univ. of Illinois, Urbana, October 1960.

———. "The design of log-periodic dipole antennas," *IRE Intl. Conv. Rec.*, pt. I, pp. 61–75, 1961.

———. "Analysis and design of the log-periodic dipole antenna," PhD dissertation, Univ. of Illinois, Urbana, 1961. Also *Tech. Rep. 52*, Univ. of Illinois Antenna Lab, Contract AF33(616)-6079, October 1961.

Chan, K. K., and P. Silvester, "Projective analysis of log-periodic dipole antennas," *IEEE Trans. Antennas Propag.*, vol. AP-21, p. 757, September 1973.

———. "Analysis of the log-periodic V-dipole antenna," *IEEE Trans. Antennas Propag.*, vol. AP-23, pp. 397–401, May 1975.

Chatterjee, J. S., and M. N. Roy, "Helical log-periodic array," *IEEE Trans. Antennas Propag.*, vol. AP-16, pp. 592–593, September 1968.

Cheo, B. R-S., V. H. Rumsey, and W. J. Welch, "A solution to the frequency-independent antenna problem," *IRE Trans. Antennas Propag.*, vol. AP-9, pp. 527–534, November 1961.

Cheo, B. R-S., and V. H. Rumsey, "A solution to the equiangular spiral antenna problem —continued," *IEEE Intl. Symp. Antennas Propag.*, pp. 126–130, December 1966.

———. "Surface waves and mode selection on multielement spiral antennas," *Radio Sci.*, vol. 3, no. 3, pp. 267–271, March 1968.

Cheong, W. M., and R. W. P. King, "Arrays of unequal and unequally spaced dipoles," *Radio Sci.*, vol. 2, no. 11, pp. 1303–1314, November 1967.

———. "Log-periodic dipole antenna," *Radio Sci.*, vol. 2, no. 11, pp. 1315–1325, November 1967.

Cornell, D. D., and B. J. Lamberty, "Multimode planar spiral for DF applications," Antenna Applications Symposium, Allerton Park, Univ. of Illinois, Urbana, September 1981.

Craven, J. H., "Dielectric lens for second-mode spiral," *IRE Trans. Antennas Propag.*, vol. AP-9, p. 499, September 1961.

Cristal, E. G., "Coupled-transmission-line directional coupler with coupled lines of unequal characteristic impedance," *IEEE Trans. Microwave Theory Tech.*, vol. MTT-14, no. 7, pp. 337–346, July 1966.

Curtis, J. W., "Spiral antennas," *IRE Trans. Antennas Propag.*, vol. AP-8, pp. 298–306, May 1960.

Deschamps, G. A., "Impedance properties of complementary multiterminal planar structures," *IRE Trans. Antennas Propag.*, vol. AP-7, pp. S371–S378, December 1959. Also *Tech. Rep. 43*, Univ. of Illinois Antenna Lab, Contract AF33(616)-6079, November 1959.

Deschamps, G. A., and J. Dyson, "The logarithmic spiral in a single-aperture multimode antenna system," *IEEE Trans. Antennas Propag.*, vol. AP-19, pp. 90–96, January 1971.

DeVito, G., and G. B. Stracca, "Comments on the design of log-periodic dipole antennas," *IEEE Trans. Antennas Propag.*, vol. AP-21, pp. 303–308, May 1973.

———. "Further comments on the design of log-periodic dipole antennas," *IEEE Trans. Antennas Propag.*, vol. AP-22, pp. 714–718, September 1974.

DeVito, G., "Influence of the earth in the hf transmission of LPD antennas," *IEEE Trans. Antennas Propag.*, vol. AP-25, pp. 891–896, November 1977.

DiFonzo, D. F., "Reduced size log-periodic antennas," *Microwave J.*, vol. 7, pp. 37–42, December 1964.

Donnellan, J. R., "Second-mode operation of the spiral antenna," *IRE Trans. Antennas Propag.*, vol. AP-8, p. 637, November 1960.

DuHamel, R. H., and D. E. Isbell, "Broadband logarithmically periodic antenna structures," *IRE Natl. Conv. Rec.*, pt. I, pp. 119–128, 1957. Also *Tech. Rep. 19*, Univ. of Illinois Antenna Lab, Contract AF33(616)-3220, May 1957.

DuHamel, R. H., and F. R. Ore, "Logarithmically periodic antenna design," *Tech. Rep.*

CTR-198, Collins Radio Co., Cedar Rapids, Iowa, March 1958.

DuHamel, R. H., and D. G. Berry, "Logarithmically periodic antenna arrays," *IRE Wescon Conv. Rec.*, pt. 1, pp. 161–174, August 1958.

DuHamel, R. H., and F. R. Ore, "Logarithmically periodic antenna design," *IRE Natl. Conv. Rec.*, pt. I, pp. 139–151, 1959.

DuHamel, R. H., and D. G. Berry, "A new concept in high-frequency antenna design," *IRE Natl. Conv. Rec.*, pt. I, pp. 42–50, 1959.

DuHamel, R. H., and F. R. Ore, "Log-periodic feeds for lens and reflectors," *IRE Natl. Conv. Rec.*, pt. I, pp. 128–137, 1959.

Duncan, J. W., and V. P. Minerva, "100:1 bandwidth balun transformer," *Proc. IRE*, vol. 48, pp. 156–164, February 1960.

Dyson, J. D., "The equiangular-spiral antenna," Fifth Annual Symposium, USAF Antenna Res. and Dev. Prog., Allerton Park, Univ. of Illinois, Urbana, October 1955. Also *Tech. Rep. 21*, Univ. of Illinois Antenna Lab, Contract AF33(616)-3220, September 1957. Also PhD dissertation, University of Illinois, Urbana, 1957.

———. "The equiangular-spiral antenna," *IRE Trans. Antennas Propag.*, vol. AP-7, pp. 181–187, April 1959.

———. "Recent developments in spiral antennas," *Proc. IRE Natl. Aeron. Electron. Conf.*, pp. 617–625, 1959.

———. "The unidirectional equiangular-spiral antenna," *IRE Trans. Antennas Propag.*, vol. AP-7, pp. 329–334, October 1959. Also *Tech. Rep. 33*, Univ. of Illinois Antenna Lab, Contract AF33(616)3220, July 1958.

Dyson, J. D., and P. E. Mayes, "New circularly polarized frequency independent antennas with conical beam or omnidirectional patterns," *IRE Trans. Antennas Propag.*, vol. AP-9, pp. 334–342, July 1961. Also *Tech. Rep. 46*, Univ. of Illinois, Urbana, Contract AF33(616)-6079, June 1960.

———. "A note of the difference between equiangular and Archimedes spiral antennas," *IRE Trans.*, vol. MTT-9, pp. 203–205, March 1961.

Dyson, J. D., "The conical log-spiral antenna in simple arrays," Eleventh Annual Symposium, USAF Antenna Res. and Dev. Prog., Allerton Park, Univ. of Illinois, Urbana, October 1961.

———. "Multimode logarithmic-spiral antennas—possible applications," *Proc. Natl. Electron. Conf.*, pp. 206–213, October 1961.

———. "A survey of very wideband and frequency independent antennas—1945 to the present," *J. Res. Natl. Bur. Stand. (US)*, vol. 66D, pp. 1–6, January–February 1962.

———. "Frequency independent antennas: survey of development," *Electronics*, vol. 35, no. 16, pp. 39–44, April 1962.

———. "An antenna to cover the 220 through 2400 Mc telemetry bands," *Proc. Natl. Telemetering Conf.*, vol. 1, sec. 9-4, May 1962.

———. "The coupling and mutual impedance between balanced wire-arm conical log-spiral antennas," *Tech. Rep. 54*, Univ. of Illinois, Urbana, Contract AF33(657)-8460, June 1962.

———. "The coupling and mutual impedance between conical log-spiral antennas in simple arrays," *IRE Intl. Conv. Rec.*, pt. 1, pp. 165–182, 1962.

———. "On a conical quad-spiral array," *Tech. Rept. 63*, Univ. of Illinois Antenna Lab, Contract AF33(657)-8460, September 1962.

Dyson, J. D., and G. L. Duff, "Near field measurements on the conical logarithmic spiral antenna," *G-AP Intl. Symp. Dig.*, pp. 137–142, 1963.

Dyson, J. D., "Research on conical log-spiral antennas," *Proc. Appl. Forum Antenna Res.*, ed. by P. E. Mayes, Univ. of Illinois, Urbana, pp. 215–265, 1964.

———. "The design of conical log-spiral antennas," *IEEE Intl. Conv. Rec.*, pt. 6, pp. 259–273, 1964.

———. "The characteristics and design of the conical log-spiral antenna," *IEEE Trans. Antennas Propag.*, vol. AP-13, pp. 488–499, July 1965. Also *Tech. Rep. 65-4*, Univ. of Illinois Antenna Lab, Contract AF33(657)-10474, May 1965.

———. "The multimode antennas as single-aperture systems," Antenna Forum, Univ. of

Illinois, Urbana, January–February 1967.

Elfving, C. T., "Foreshortened log-periodic dipole array," *1963 Wescon*, San Francisco, August 1963. Also *Tech. Memo. EDL-M401*, Contract DA36-039-AMC-00088(E), Electronic Defense Laboratories, Sylvania Electric Products, Inc., Mountain View, Calif., September 1963.

Elliott, R. S., "A view of frequency-independent antennas," *Microwave J.*, vol. 5, pp. 61–68, December 1962.

Evans, B. G., "The effects of transverse feed displacements on log-periodic dipole arrays," *IEEE Trans. Antennas Propag.*, vol. AP-18, pp. 124–128, January 1970.

Gans, M. J., D. Kajfez, and V. H. Rumsey, "Frequency independent baluns," *Proc. IEEE*, vol. 53, pp. 647–648, June 1965.

Green, P. B., and P. E. Mayes, "A log-periodic monopole array with a modulated impedance microstrip feeder," *Tech. Rep. 73-2*, Univ. of Illinois Antenna Lab, January 1973.

———. "50-ohm log-periodic monopole array with modulated-impedance microstrip feeder," *IEEE Trans. Antennas Propag.*, vol. AP-22, pp. 332–334, March 1974.

Greiser, J. W., and P. E. Mayes, "Vertically polarized log-periodic zig-zag antennas," *Proc. Natl. Electron. Conf.*, vol. 17, pp. 193–204, 1961.

Greiser, J. W., "The bent log-periodic zigzag antenna," technical report, Univ. of Illinois Antenna Lab, Contract NOBSR85243, May 1962.

———. "The bent log-periodic zigzag antenna," MS thesis, Dept. of Elec. Eng., Univ. of Illinois, Urbana, 1962.

———. "A new class of log-periodic antennas," *Proc. IEEE*, vol. 52, pp. 617–618, May 1964.

Greiser, J. W., and P. E. Mayes, "The bent backfire zigzag—a vertically polarized frequency-independent antenna," *IEEE Trans. Antennas Propag.*, vol. AP-12, pp. 281–290, May 1964.

Hahn, G., and R. Honda, "Conical spiral arrays for passive direction finding," Eighteenth Annual Symposium, USAF Antenna Res. and Dev. Prog., Allerton Park, Univ. of Illinois, Urbana, October 1968.

Hessemer, R. A., Jr., "Backward-wave radiation from an equiangular spiral antenna," *IRE Trans. Antennas Propag.*, vol. AP-9, p. 582, November 1961.

Hong, S., and G. Rassweiler, "Size reduction of a conical log-spiral antenna by loading with magneto-dielectric material," *IEEE Trans. Antennas Propag.*, vol. AP-14, pp. 650–651, September 1966.

Hudock, E., and P. E. Mayes, "Near-field investigation of uniform periodic monopole arrays," *IEEE Trans. Antennas Propag.*, vol. AP-13, pp. 840–855, November 1965.

Ingerson, P. G., and P. E. Mayes, "Design of log-periodic structures using complex dispersion data for periodic lines," *Tech. Rep. 66-10*, Univ. of Illinois Antenna Lab, August 1966.

———. "Design of log-periodic structures using dispersion data for periodic lines," Sixteenth Annual Symposium, USAF Antenna Res. and Dev. Prog., Allerton Park, Univ. of Illinois, Urbana, October 1966.

———. "Asymmetrical feeders for log-periodic antennas," Seventeenth Annual Symposium, USAF Antenna Res. and Dev. Prog., Allerton Park, Univ. of Illinois, Urbana, November 1967.

Ingerson, P. G. "Analysis of some closed log-periodic structures with applications to log-periodic antennas," PhD thesis, Univ. of Illinois, Urbana, 1968.

Ingerson, P. G., and P. E. Mayes, "Log-periodic antennas with modulated impedance feeders," *IEEE Trans. Antennas Propag.*, vol. AP-16, pp. 633–642, November 1968. Also *Tech. Rep. AFAL-TR-69-226*, Univ. of Illinois Antenna Lab, Contract F33615-69-C-1122, July 1969.

Ingerson, P. G., "Modulated arm width (MAW) log-spiral antennas," Twentieth Annual Symposium, USAF Antenna Res. and Dev. Prog., Allerton Park, Univ. of Illinois, Urbana, October 1970.

Isbell, D. E., "Nonplanar logarithmically periodic antenna structures," *Tech. Rep. 30*, Univ.

of Illinois Antenna Lab, Contract AF33(616)-3220, February 1958.

———. "Multiple terminal log-periodic antennas," Eighth Annual Symposium, USAF Antenna Res. and Dev. Prog., Allerton Park, Univ. of Illinois, Urbana, October 1958.

———. "A log-periodic reflector feed," *Proc. IRE*, vol. 47, pp. 1152–1153, June 1959.

———. "Log-periodic dipole arrays," *IRE Trans. Antennas Propag.*, vol. AP-8, pp. 260–267, May 1960. Also *Tech. Rep. 39*, Univ. of Illinois Antenna Lab, Contract AF33(616)-6079, 1959.

Jones, K. E., and R. Mittra, "Some interpretations and applications of the k-beta diagram," *IEEE Intl. Conv. Rec.*, pt. 5, pp. 134–139, March 1965. Also *Tech. Rep. 65-1*, Univ. of Illinois Antenna Lab, Contract AF33(657)-10474, January 1965.

Jones, K. E., and P. E. Mayes, "Continuously scaled transmission lines with applications to log-periodic antennas," *IEEE Trans. Antennas Propag.*, vol. AP-17, pp. 2–9, January 1969.

Jordan, E. C., G. A. Deschamps, J. D. Dyson, and P. E. Mayes, "Developments in broadband antennas," *IEEE Spectrum*, vol. 1, pp. 58–71, April 1964.

Jordan, E. C., and K. J. Balmain, *Electromagnetic Waves and Radiating Systems*, Englewood Cliffs: Prentice-Hall, 1968.

Kaiser, J. A., "The Archimedean two-wire spiral antenna," *IRE Trans. Antennas Propag.*, vol. AP-8, pp. 312–323, May 1960.

———. "Dual operation with the two-wire spiral antenna," *IRE Trans. Antennas Propag.*, vol. AP-9, pp. 583–584, November 1961.

Kaiser, P., "On the theoretical relations between periodic and log-periodic structures," PhD dissertation, Dept. of Electr. Eng. Univ. of California, Berkeley, September 1966.

———. "The inclined log-spiral antenna, a new type of unidirectional, frequency independent antenna," *IEEE Trans. Antennas Propag.*, vol. AP-15, pp. 304–305, March 1967.

Keen, K. M., "A planar log-periodic antenna," *IEEE Trans. Antennas Propag.*, vol. AP-22, pp. 489–490, May 1974.

Kieburtz, R. B., "A phase-integral approximation for the current distribution along a log-periodic antenna," *IEEE Trans. Antennas Propag.*, vol. AP-13, pp. 813–814, September 1965.

Kim, O. K., "An experimental investigation of the conical four-arm logarithmic spiral antenna with large cone angle," *Tech. Rep. 66-12*, Univ. of Illinois Antenna Lab, Contract AF33(615)-3216, September 1966. Also MS thesis, Dept. of Electr. Eng., Univ. of Illinois, Urbana, 1966.

Kim, O. K., and J. D. Dyson, "A log-spiral antenna with selectable polarization," *IEEE Trans. Antennas Propag.*, vol. AP-19, pp. 675–677, September 1971.

Klock, P. W., "A study of wave propagation on helices," *Tech. Rep. 68*, Univ. of Illinois Antenna Lab, Contract AF33(657)-10474, March 1963.

Klock, P. W., and R. Mittra, "On the solution of the Brillouin (k-β) diagram of the helix and its application to helical antennas," *G-AP Intl. Symp. Dig.*, pp. 99–103, 1963.

———. "Complex wave analysis of the backfire bifilar helical antenna," *Fall URSI Meeting*, Univ. of Illinois, p. 67, 1964.

Kosta, S. P., M. D. Singh, N. K. Agarwal, and A. Singh, "A note on the theory of log-periodic dipole antenna," *IEEE Trans. Antennas Propag.*, vol. AP-18, p. 701, September 1970.

Kuo, S. C., and P. E. Mayes, "An experimental study and the design of the log-periodic zigzag antenna," *Tech. Rep. 65-11*, Univ. of Illinois Antenna Lab, Contract AF33(615)-3216, February 1966. Also MS thesis, Dept. of Electr. Eng., Univ. of Illinois, Urbana, 1965.

Kuo, S. C., and C. C. Liu, "Dual polarized center-fed multiarm spiral monopulse antenna," *IEEE/G-AP Intl. Symp.*, University of Michigan, Ann Arbor, p. 233, October 1967.

Kuo, S. C., "Size-reduced log-periodic dipole array," *IEEE/G-AP Intl. Symp.*, pp. 151–158, 1970.

Kuo, S. C., "Size-reduced log-periodic dipole array antenna," *Microwave J.*, vol. 15, no. 12, pp. 27–33, December 1972.

Kyle, R. H., "Mutual coupling between log-periodic antennas," *IEEE Trans. Antennas Propag.*, vol. AP-18, pp. 15–22, January 1970.

Lantz, P. A., "A two-channel monopulse reflector antenna system with a multimode logarithmic spiral feed," Sixteenth Annual Symposium, USAF Antenna Res. and Dev. Prog., Allerton Park, Univ. of Illinois, Urbana, 1966.

Laxpati, S. R., and R. Mittra, "Current distribution on a two-arm thin-wire equiangular spiral antenna," *Electron. Lett.*, vol. 1, pp. 213–215, 1965.

———. "A study of the equiangular spiral antenna," *Tech. Rep. 65-20*, Univ. of Illinois Antenna Lab, Contract AF33(615)-3216, February 1966.

———. "Boundary-value problems associated with source-excited planar-equiangular-spiral antennas," *Proc. IEE* (London), vol. 114, pp. 352–359, March 1967.

Lee, S. H., "Analysis of zigzag antennas," PhD dissertation, Dept. of Electr. Eng., Univ. of California, Berkeley, June 1968.

Lee, S. H., and K. K. Mei, "Analysis of zigzag antennas," *IEEE Trans. Antennas Propag.*, vol. AP-18, pp. 760–764, November 1970.

Lee, W. C., "Analysis of nonplanar spiral antennas," *Tech. Rep. 903-15*, Ohio State Univ., Columbus, November 1960.

Liang, C. S., and Y. T. Lo, "A multipole-field study for the multiarm log-spiral antennas," *IEEE Trans. Antennas Propag.*, vol. AP-16, pp. 656–664, November 1968. Also *Tech. Rep. 69-3*, Univ. of Illinois Antenna Lab, July 1969.

Mast, P. E., "A theoretical study of the equiangular-spiral antenna," PhD dissertation, Department of Electrical Engineering, Univ. of Illinois, Urbana, 1958. Also *Tech. Rep. 35*, Univ. of Illinois Antenna Lab, Contract AF33(616)-3220, September 1958.

Mayes, P. E., and R. L. Carrel, "Logarithmically periodic resonant-V arrays," Tenth Annual Symposium, USAF Antenna Res. and Dev. Prog., Allerton Park, Univ. of Illinois, Urbana, October 1960. Also *IRE Wescon Conv. Rec.*, pt. I, 1961, and *Tech. Rep. 47*, Univ. of Illinois Antenna Lab, Contract AF33(616)-6079, July 1962.

Mayes, P. E., G. A. Deschamps, and W. T. Patton, "Backward-wave radiation from periodic structures and application to the design of frequency independent antennas," *Proc. IRE*, vol. 49, pp. 962–963, May 1961. Also *Tech. Rep. 60*, Univ. of Illinois Antenna Lab, Contract AF33(657)-8460, December 1962.

Mayes, P. E., and P. G. Ingerson, "Near-field measurements on backfire periodic dipole arrays," Twelfth Annual Symposium, USAF Antenna Res. and Dev. Prog., Allerton Park, Univ. of Illinois, Urbana, October 1962.

Mayes, P. E., "Broadband backward-wave antennas," *Microwave J.*, vol. 6, pp. 61–71, January 1963.

———. "Balanced backfire zigzag antennas," *IEEE Intl. Conv. Rec.*, pt. 2, pp. 153–165, 1964. Also *Tech. Rep. 82*, Univ. of Illinois Antenna Lab, Contract AF33(657)-10474, October 1964.

———. "Designing an all-channel tv antenna," *Electron. World*, February 1966.

Mayes, P. E. (ed.), "Wave propagation on smooth and periodic structures and applications to antenna design," *Tech. Rep. 66-11*, Univ. of Illinois Antenna Lab, Contract NASA NGR14-005-043, 1966.

McClelland, O. L., "An investigation of the near fields on the conical equiangular spiral antenna," *Tech. Rep. 55*, Univ. of Illinois Antenna Lab, Contract AF33(657)-8460, May 1962.

Mei, K. K., and D. Johnstone, "A broadside log-periodic antenna," *Proc. IEEE*, vol. 54, no. 6, pp. 889–890, June 1966.

Mittra, R., and K. E. Jones, "How to use k-β diagrams in log-periodic antenna design," *Microwaves*, pp. 18–26, June 1965.

Montague, H., M. J. Horrocks, J. W. Margosian, and J. D. Dyson, "The dual-aperture counterwound log-spiral antenna direction-finder system," *IEEE Trans. Antennas Propag.*, vol. AP-21, pp. 224–226, March 1973.

Mosko, J. A., "Reduced size, dual-mode spiral for two-plane monopulse direction finding," *NAVWEPS Rep. 8758*, U. S. Naval Ordinance Test Station, China Lake, California, 1966.

Murphy, L. R., "A shortened log-spiral antenna by the use of resistive termination," MS thesis, Dept. of Electr. Eng., Univ. of Illinois, Urbana, 1967.

Mushiake, Y., *J. Inst. Electron. Eng.*, Japan, vol. 69, pp. 86–88, 1949 (in Japanese).

Ore, F. R., "A coaxial fed unidirectional log-periodic monopole array," technical report, Collins Radio Company, Cedar Rapids, Iowa, August 1961.

———. "Investigation of the log-periodic coaxial fed monopole array," *Tech. Note No. 1*, Univ. of Illinois Radiolocation Research Lab, Contract NOBSR85243, October 1963.

———. "Log-periodic folded monopole array," *Tech. Rep. 4*, Radiolocation Research Lab, Univ. of Illinois, Urbana, Contract NOBSR89229, June 1964.

———. "A wideband vertical incidence radio location array," *Tech. Rep. 7*, Radiolocation Research Lab, Univ. of Illinois, Urbana, Contract NOBSR89229, September 1964.

Ore, F. R., and P. E. Mayes, "A study of periodic and log-periodic series reactance loading with application to a high-efficiency long-wire antenna," *Tech. Rep. 66-5*, Univ. of Illinois Antenna Lab, Contract N123(953)-51806A, June 1966.

Ore, F. R., "A technical report on the investigation of selected frequency independent arrays," *Tech. Rep. 67-273*, Univ. of Illinois Antenna Lab, Contract AF33(615)-3216, November 1967.

Ostertag, E. L., "Experimental study of a log-periodic cavity backed slot array with computer synthesized 50-ohm modulated impedance feeder," MS thesis, Dept. of Electr. Eng., Univ. of Illinois, Urbana, 1972.

Patton, W. T., "The backfire bifilar helical antenna," *Tech. Rep. 61*, Univ. of Illinois Antenna Lab, Contract AF33(657)-8460, September 1962. Also PhD dissertation, Univ. of Illinois, Urbana, 1963.

———. "The backfire bifilar helical antenna," Twelfth Annual Symposium, USAF Antenna Res. and Dev. Prog., Allerton Park, Univ. of Illinois, Urbana, October 1962.

Ransom, P. L., and J. D. Dyson, "Near-field measurements on the planar four-arm log-spiral antenna," Fourteenth Annual Symposium, USAF Antenna Res. and Dev. Prog., Allerton Park, Univ. of Illinois, Urbana, October 1964.

Ransom, P. L., "An experimental investigation of the four-arm planar logarithmic spiral antenna," *Tech. Rep. 65-5*, Univ. of Illinois Antenna Lab, Contract AF33(657)-10474, May 1965. Also MS thesis, Dept. of Electr. Eng., Univ. of Illinois, Urbana, 1965.

Rumsey, V. H., *Frequency Independent Antennas*, New York: Academic Press, 1966.

Sandeman, E. K., "Spiral-phase fields," *Wireless Engineer*, pp. 96–105, March 1949.

Sinnott, D. H., "Multiple-frequency computer analysis of the log-periodic dipole antenna," *IEEE Trans. Antennas Propag.*, vol. AP-22, pp. 592–594, July 1974.

Sivan-Sussman, R., "Various modes of the equiangular spiral antenna," *IEEE Trans. Antennas Propag.*, vol. AP-11, pp. 533–539, September 1963.

Smith, C. E., *Log-Periodic Antenna Design Handbook*, Cleveland: Smith Electronics, Inc., 1966.

Stephenson, D. T., and P. E. Mayes, "Log-periodic helical dipole arrays," IEEE Wescon, 1963. Also Thirteenth Annual Symposium, USAF Antenna Res. and Dev. Prog., Allerton Park, Univ. of Illinois, Urbana, October 1963.

———. "Broadband arrays of helical dipoles," *Tech. Rep. 2*, Univ. of Illinois Antenna Lab, Contract NEL30508A, January 1964.

———. "Investigations of broadband helical dipole arrays," *Tech. Rep. 6*, Univ. of Illinois Antenna Lab, Contract NEL30508A, October 1964.

Stephenson, D. T., "Broadband helical dipole arrays," *Tech. Rep. 65-19*, Univ. of Illinois Antenna Lab, Contract N123(953)-51806A, October 1965.

Stephenson, D. T., and P. E. Mayes, "Variations of broadband helical dipole arrays," *Tech. Rep. 65-3*, Univ. of Illinois Antenna Lab, Contract N123(953)-51806A, April 1966.

Sussman, R., "The equiangular plane spiral antenna," technical report, Electronics Res. Lab, Univ. of California, Berkeley, September 1961.

Tang, C. H., and O. L. McClelland, "Polygonal spiral antennas," *Tech. Rep. 57*, Univ. of Illinois Antenna Lab, Contract AF33(657)-8460, June 1962.

Tang, C. H., "A class of modified log-spiral antennas," *IEEE Trans. Antennas Propag.*, vol. AP-11, pp. 422–427, July 1963.

Turner, E. M., "Spiral slot antenna," *Note WCLR-55-8*, Wright Air Dev. Center, Dayton,

June 1955.

Wheeler, M. S., "On the radiation from several regions in spiral antennas," *IRE Trans. Antennas Propag.*, vol. AP-9, pp. 100–102, January 1961.

Wickersham, A. F., "Recent developments in very broad-band end-fire arrays," *Proc. IRE*, vol. 48, pp. 794–795, April 1960.

Wickersham, A. F., R. E. Franks, and R. L. Bell, "Further developments in tapered ladder antennas," *Proc. IRE*, vol. 49, p. 378, January 1961.

Wohlleben, R., and B. Schumacher, "Randlinien-abwicklung der zweiarmigen, selbstkomplementaren, konischen, logarithmischen spiralantenne," *NTZ*, pp. 585–590, October 1966.

Wolter, J., "Solution of Maxwell's equations for log-periodic dipole antennas," *IEEE Trans. Antennas Propag.*, vol. AP-18, pp. 734–741, November 1970.

Yeh, Y. S., and K. K. Mei, "Theory of conical equiangular-spiral antennas, part I: numerical technique," *IEEE Trans. Antennas Propag.*, vol. AP-15, pp. 634–639, September 1967.

———. "Theory of conical equiangular-spiral antennas, part II: current distributions and input impedances," *IEEE Trans. Antennas Propag.*, vol. AP-16, pp. 14–21, January 1968.

Chapter 10

Microstrip Antennas

William F. Richards
University of Houston

CONTENTS

1. Introduction — 10-5
2. Physical Models — 10-7
    Transmission-Line Model   10-7
    Cavity Model   10-10
3. Pattern — 10-21
    General Properties   10-23
    The (0, 1) Mode   10-24
    The (0, 2) Mode   10-25
    The DC Mode   10-25
    The (1, 1) Mode of a Circular-Disk Element   10-25
    The (0, n) Circular-Disk Modes   10-25
    Other Modes   10-25
4. Impedance and Circuit Models — 10-26
    General Circuit Model   10-26
    Simplified Circuit Model   10-28
    Circuit Model for Near-Degenerate Modes   10-28
    Simple Feed Models   10-28
    Resonant Impedance   10-28
    Feed Reactance   10-31
    Multiport Impedance Parameters   10-34
    Efficient Computation of Impedance Parameters   10-34
5. Resonant Frequency — 10-44
    Rectangular Patch   10-46
    Circular-Disk Patch   10-47
    Circular-Sector Patch   10-47
    Annular Patch   10-47
    Annular-Sector Patch   10-47
6. Efficiency — 10-49
7. Matching — 10-50

William F. Richards was born in Cincinnati, Ohio, in 1950. He received the BS degree in engineering (with a concentration in electrical engineering) in 1970 from Old Dominion University, Norfolk, Virginia, and the MS and PhD degrees in electrical engineering from the University of Illinois in Urbana-Champaign under the direction of Professor Y. T. Lo.

At the University of Illinois he did work on optimization of array designs accounting for the stochastic properties of the excitation network, artificial dielectrics, and did fundamental work with Dr. Lo on the development of the cavity model for microstrip antennas. After obtaining his PhD in 1977, he remained at the University of Illinois until 1980, serving as a visiting assistant professor.

In 1980 he joined the faculty of the Department of Electrical Engineering at the University of Houston, Houston, Texas, where he serves as an associate professor. He has continued his work on microstrip antennas at the University of Houston, developing a theory and a number of applications of loaded microstrip elements. His research interests are microstrip antennas, artificial dielectrics, and numerical techniques in scattering and the on-surface radiation condition. At the time of publication, he was on leave of absence from the University of Houston to attend the University of Miami for its PhD-to-MD program.

Dr. Richards is a member of the IEEE and shared the Antennas and Propagation Society's Best Paper Award in 1979 with his coauthors, Y. T. Lo and D. Solomon, for their paper on the theory and analysis of microstrip antennas.

8. Loaded Microstrip Elements　　10-52
9. Applications　　10-56
　　Circular Polarization　　10-57
　　Dual-Band Elements　　10-63
　　Frequency-Agile Elements　　10-66
　　Polarization-Agile Elements　　10-69
10. References　　10-70

1. Introduction

A class of antennas that has gained considerable popularity in recent years is the *microstrip antenna*. There are many different varieties of microstrip antennas, but their common feature is that they basically consist of four parts:

(a) a very thin flat metallic region often called the *patch*;
(b) a *dielectric substrate*;
(c) a *ground plane*, which is usually much larger than the patch; and
(d) a *feed*, which supplies the element rf power.

A typical microstrip element is illustrated in Fig. 1. Microstrip elements are often made by etching the patch (and sometimes the feeding circuitry) from a single printed-circuit board clad with conductor on both of its sides.

The longest dimension of the patch is typically about a third to a half of a free-space wavelength (λ_0), while the dielectric thickness is usually in the range of $0.003\lambda_0$ to $0.05\lambda_0$. A commonly used dielectric for such antennas is polytetrafluoral ethylene (PTFE), often set in a reinforcing glass fiber matrix. A relative dielectric constant around 2.5 is typical. Sometimes a low-density cellular "honeycomb" material is used to support the patch. This material has a relative dielectric constant much closer to unity and usually results in an element with better efficiency and larger bandwidth [1] though at the expense of an increased element size. Substrate materials with high dielectric constants can also be used. For the radiation modes most used, however, such substrates result in elements which are electrically small in terms of free-space wavelengths and consequently have relatively smaller bandwidths [1] or low efficiencies [2].

The reasons why this class of antennas has become so popular include the following:

1. They are low-profile antennas.
2. They are easily conformable to nonplanar surfaces. Along with their low profile this makes them well suited for use on high-performance airframes.
3. They are easy and inexpensive to manufacture in large quantities using modern printed-circuit techniques.
4. When mounted to a rigid surface they are mechanically robust.
5. They are versatile elements in the sense that they can be designed to produce a wide variety of patterns and polarizations, depending on the mode excited and the particular shape of patch used.
6. Adaptive elements can be made by simply adding appropriately placed pin or varactor diodes between the patch and ground plane. Using such loaded elements one can vary the antenna's resonant frequency [3, 4], polarization [3, 5], impedance [6], and even its pattern by simply changing bias voltages on the diodes [7].

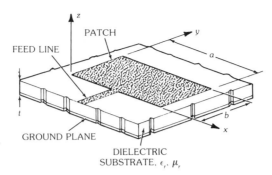

Fig. 1. A typical rectangular microstrip element with dimensional parameters.

These advantages must be weighed against the disadvantages which can be most succinctly stated in terms of the antenna's quality factor, Q. Microstrip antennas are high-Q devices with Qs sometimes exceeding 100 for the thinner elements. High-Q elements have small bandwidths. Also, the higher the Q of an element, the lower is its efficiency. Increasing the thickness of the dielectric substrate will reduce the Q of the microstrip element and thereby increase its bandwidth and its efficiency. There are limits, however. As the thickness is increased, an increasing fraction of the total power delivered by the source goes into a surface wave. This surface-wave contribution can also be counted as an unwanted power loss since it is ultimately scattered at dielectric bends and discontinuities. Such scattered fields are difficult to control and may have a deleterious effect on the pattern of the element [1]. One also needs to be aware that microstrip elements are modal devices. If the band of the element is so large that it encompasses the resonant frequencies of two or more resonant modes, the pattern is likely not to be stable throughout the band even though the vswr at the input could be acceptably low.

Despite the disadvantages the advantages of microstrip antennas have led to their use in many applications in civilian and government systems. In fact, one so-called disadvantage, small bandwidth, is sometimes counted as an advantage instead. For narrow-band applications the antenna itself can act as a filter for unwanted frequency components. Even with the relatively high Q these elements have, a sufficiently thick element with a well-designed external matching circuit can have a bandwidth as large as 35 percent [8]. Finally, if surface-wave loss is not counted as an unwanted power loss (or does not exist because the dielectric is truncated or has a relative dielectric constant near unity), the efficiency of a sufficiently thick element can easily be made larger than 90 percent.

An understanding of the *physical mechanisms* that explain the properties of microstrip antennas is crucial to their creative use in applications. This chapter will introduce the reader to some of the simpler theories and applications of microstrip antennas. No attempt has been made to present a comprehensive review of the literature. Instead, selected additional references are given in the References to assist the reader in obtaining information on theories and applications that exceed the scope of this handbook.

2. Physical Models

In order to design a microstrip element for a given application it is crucial to first have a good understanding of the physical mechanisms that govern these elements. There are several theories for microstrip elements with varying degrees of accuracy and complexity. Among these, two give the best physical insight: (1) the transmission-line model [9, 10] and (2) the cavity model [4, 11]. Of these, the simplest is the transmission-line model. The cavity model, though somewhat more complex, gives a deeper insight into the operation of microstrip antennas. The transmission-line model is considered first.

Transmission-Line Model

The patch element in Fig. 1 can be viewed as a very wide (thus low-impedance) microstrip transmission line of length L. The region between the patch edges and the ground plane at the two ends of the line can be viewed as radiating apertures much like slot antennas. Thus the low-impedance line can be thought of as being loaded at its two ends not by open circuits, but by high-impedance loads. There is also a mutual coupling between the slots. The terminal voltages and currents on the transmission line can be thought of as being coupled through a two-port network representing the propagation of the field in the space exterior to the patch. If the short-circuit parameters of this two-port network are Y_{11} and Y_{12}, then the Π model of the network can be used for the microstrip antenna as illustrated in Fig. 2a.

The self-admittance of the load is $Y_{11} = G_{11} + jB_{11}$. The susceptance B_{11} can be adequately accounted for by extending the length b of the transmission line beyond the physical limits of the patch. One estimate for this extension, ΔL, that has often been used with good success is [12]

$$\Delta L = 0.412 t \frac{(\epsilon_{\text{eff}} + 0.3)(a/b + 0.262)}{(\epsilon_{\text{eff}} - 0.258)(a/b + 0.813)} \quad (1)$$

The ϵ_{eff} is the *effective relative dielectric constant* and is given in terms of the substrate relative dielectric constant ϵ_r by [13]

$$\epsilon_{\text{eff}} = \frac{\epsilon_r + 1}{2} + \frac{\epsilon_r - 1}{2(1 + 10t/w)^{1/2}}$$

An estimate for the self-conductance G_{11} that has been commonly used is [14]

$$G_{11} = \frac{a}{120\lambda_0} \left[1 - \frac{\pi^2}{6} (t/\lambda_0)^2 \right]$$

This is based on the conductance of an infinitely long, uniform slot antenna. Its application to the relatively short "slots" associated with microstrip antennas seems dubious. In fact, another formula yielding quite a different conductance is [1]

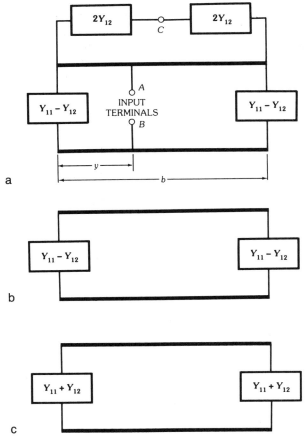

Fig. 2. Transmission-line model. (*a*) Schematic diagram for a rectangular microstrip element including external self and mutual admittances. (*b*) Simplified circuit for symmetric modes. (*c*) Simplified circuit for antisymmetric modes.

$$G_{11} = \begin{cases} (a_{\text{eff}})^2/(90\lambda_0)^2 & \text{for } a_{\text{eff}} < 0.35\lambda_0 \\ w_{\text{eff}}/(120\lambda_0) - 1/(60\pi^2) & \text{for } 0.35\lambda_0 < a_{\text{eff}} < 2\lambda_0 \\ w_{\text{eff}}/(120\lambda_0) & \text{for } 2\lambda_0 < a_{\text{eff}} \end{cases}$$

The effective width of the microstrip line is given by [15]

$$a_{\text{eff}} = \frac{120\pi t}{Z_m \sqrt{\epsilon_{\text{eff}}}}$$

where

$$Z_m = \frac{60\pi}{\sqrt{\epsilon_{\text{eff}}}} \left[\frac{a}{2t} + 0.441 + \frac{1}{\pi} 1.451 + \ln\left(\frac{a}{2t}\right) + 0.94 \right]^{-1}$$

The mutual conductance G_{12} can be found by numerically integrating [10]

$$G_{12} = \frac{1}{120\pi^2} \int_0^\pi \frac{\sin^2[(\pi a_{\text{eff}} \cos\theta)/\lambda_0]}{\cos^2\theta} \sin^3\theta \, J_0([2\pi b \sin\theta]/\lambda_0) \, d\theta$$

where J_0 is the Bessel function. The mutual susceptance has been generally ignored; its effect is to slightly alter the element's resonant frequency.

Symmetric and antisymmetric source-free voltage distributions resonate the circuit in Fig. 2a at complex resonant frequencies which can be determined by finding the complex zeros of the admittance and impedance at terminals A and B. Since symmetric voltage distributions produce a virtual open at point C in the circuit, the network can be simplified to that shown in Fig. 2b in this case. For the antisymmetric case, node C becomes a virtual ground node, reducing the equivalent circuit to that illustrated in Fig. 2c.

For both symmetric (even n) and antisymmetric (odd n) resonant voltage distributions, the input impedance at resonance at a distance y from either end of the patch is given approximately by

$$R \cong \frac{1}{G_{11} - (-1)^n G_{12}} \cos^2(n\pi y/b)$$

The resonant frequency (actually the real part of the complex resonant frequency) is

$$f_{\text{res}} \cong \frac{nc}{2\sqrt{\epsilon_{\text{eff}}}(b + 2\Delta L)}$$

where c is the speed of light in a vacuum. The imaginary part of the resonant frequency is its real part divided by twice Q. The Q is the quality factor of the element; it has been approximated by [1]

$$Q \cong \frac{n\pi}{4Z_m[G_{11} - (-1)^n G_{12}]}$$

Once R and Q have been determined, the input impedance at any frequency f near a resonance f_{res} can be approximated by

$$Z_{\text{in}} \cong \frac{R}{1 + j2Q(f - f_{\text{res}})/f_{\text{res}}}$$

The preceding formula is useful for determining the major features of the variation of input impedance with frequency. Its accuracy depends, of course, on the accurary of slot and mutual conductance parameters used. Its accuracy is also limited by its inability to predict an additional inductive term that is associated with the feed although an *ad hoc* correction term can be introduced; see under "Feed Reactance," in Section 4.

The radiation pattern of the element is computed by the same method used for the cavity model (discussed below) except that in the transmission-line model, only the two ends of the patch are considered as radiating apertures. The expression for the far field is

$$\mathbf{E} \cong (\hat{\boldsymbol{\theta}} \sin \phi - \hat{\boldsymbol{\phi}} \cos \phi \cos \theta)$$
$$\times \frac{\sin[(k_0 a/2) \sin \theta \cos \phi] \sin[n\pi/2 + (k_0 b/2) \sin \theta \sin \phi]}{\sin \theta \cos \phi} \frac{e^{-jk_0 r}}{\pi r} V_0$$

where V_0 is the voltage at either of the two ends of the patch, r is the distance from the antenna, and k_0 is the free-space wave number.

The intuitive appeal of the transmission-line model becomes strained when one attempts to adapt it to other than rectangular microstrip elements with a modal field variation only in the y direction. For example, one *could* think of a mode of an annular-sector microstrip antenna as the voltage distribution on a generalized nonuniform transmission line, but there seems to be little advantage in doing so. Instead, the next model to be considered is both conceptually simpler (in this case) and more accurate and comprehensive than the transmission-line model.

Cavity Model

When an oscillating current is injected into a microstrip element a charge distribution is established on the surface of the ground plane and the *two* surfaces (upper and lower) of the patch as illustrated in Fig. 3a. There are two opposing tendencies which shape this charge distribution. (1) There is an *attractive* tendency between the opposite charges at corresponding points on the lower side of the patch and on the ground plane. This attraction tends to keep the patch charge concentrated on the *bottom* of the patch. (2) However, there is also a *repulsive* tendency between like charges on the bottom of the patch. This tends to push some of the charge around the edge of the patch onto its *top* surface. When the element is very thin the first tendency dominates and almost all of the charge on the patch resides on its bottom side. Correspondingly, most of the current flows on the lower side of the patch, with only a small amount flowing around the edge onto its upper surface. Consequently the component of magnetic field *tangential* to the patch edge is small although not exactly zero. Were it *precisely* zero one could introduce a perfect magnetic conductor in the plane between the patch edge and the ground plane without affecting the fields under the patch. The introduction of such a magnetic wall *will* distort the shape of the magnetic field distribution, but not significantly if the element is thin. Thus, to find the *shape* of the magnetic field distribution under the patch, one can replace the antenna by an ideal cavity as illustrated in Fig. 3b. Of course, from the magnetic field distribution the *shape* of the corresponding electric-field distribution can also be found.

In contrast to its shape the *amplitude* of the field under the patch cannot be found by just analyzing the cavity alone. For example, if the dielectric material and the metal parts within the cavity were assumed to be lossless, then the analysis of the *cavity* would yield a *purely reactive* input impedance. The input impedance of the corresponding microstrip *antenna*, of course, is *not* purely reactive; its resistive part accounts for the power radiated by the antenna.

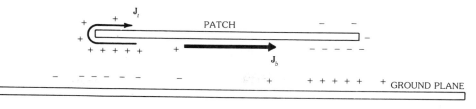

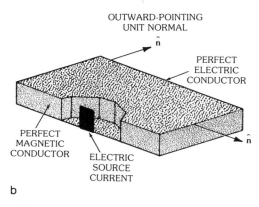

Fig. 3. Development of the cavity model of a microstrip element. (*a*) Charge distributions on the upper and lower sides of the patch and on the ground plane. (*b*) Approximate cavity model of a microstrip antenna.

The impedance function for the ideal *cavity* has only *real* poles. The impedance function for the corresponding *antenna* has *complex* poles. The imaginary parts of these poles account for the power lost by radiation and by dielectric and conduction losses. The real parts of the corresponding cavity and antenna poles are dependent on the *shapes* of their modal field distributions and are consequently almost identical for thin elements. To make the cavity more resemble the antenna it is supposed to model, one can add loss to the cavity dielectric (in one's analysis) by appropriately adjusting the loss tangent of the cavity dielectric. The imaginary parts of the poles of the cavity filled with the lossy dielectric will no longer be zero. The object then becomes to find just how much loss must be added to the cavity so that the imaginary part of its pole near the frequency of interest matches that of the corresponding antenna.

If the dielectric within the cavity (with perfectly conducting electric and magnetic walls) has a dielectric loss tangent δ_{eff}, then at any frequency f near a resonance it has a quality factor of

$$Q = \frac{2\pi f (\text{average total stored energy})}{\text{average power dissipated}} = \frac{1}{2} \frac{\omega_{\text{re}}}{\omega_{\text{im}}} = \frac{1}{\delta_{\text{eff}}}$$

where ω_{re} and ω_{im} are respectively the real and imaginary part of the pole (angular frequency). This result clearly suggests that the appropriate way to choose the cavity dielectric loss tangent δ_{eff} so that the cavity behaves like the antenna is to make it the reciprocal of the *antenna* quality factor.

The most important contribution (one hopes) to this quality factor is from the radiation of the antenna. (In the present context the term "radiation" is used to include power radiated by both the space wave and the surface wave.) This radiation can be attributed, through Huygens' principle, to equivalent magnetic and electric surface currents in the presence of the grounded dielectric slab as illustrated in Fig. 4a. The magnetic surface current density **M** is related to the electric field in the surface between the patch edge and the ground plane by

$$\mathbf{M} = -\hat{\mathbf{n}} \times \mathbf{E}$$

where $\hat{\mathbf{n}}$ is a unit vector pointing outward from the surface of the cavity. The electric surface current densities are $\mathbf{J}_s$ and $\mathbf{J}_t$. The current density $\mathbf{J}_s$ is impressed on the same surface as is **M** and is given by

$$\mathbf{J}_s = \hat{\mathbf{n}} \times \mathbf{H}$$

The current density $\mathbf{J}_t$ is the surface current on the *top* surface of the patch. The preceding discussion argues that both $\mathbf{J}_s$ and $\mathbf{J}_t$ are small, leaving only **M** as the dominant source for the radiated fields. It is emphasized that $\mathbf{J}_t$ is the current *only* on the *top* surface of the patch; it is *not* the *total* patch current. The total patch current, which is the sum of the currents on the top and bottom surfaces of the patch, could be impressed in lieu of all other sources to produce the field radiated by the element [16].

The analysis of the *cavity* yields the *shape* of the electric field **E** under the patch

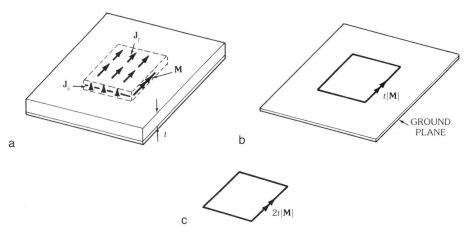

Fig. 4. Source modeling for pattern computations. (*a*) Exact model for the exterior field of a microstrip antenna in terms of equivalent electric and magnetic current. (*b*) Approximate model with the dielectric removed, the electric currents ignored, and the magnetic current condensed to a filamentary magnetic current of strength equal to $t|\mathbf{M}|$. (*c*) Equivalent approximate model with the ground plane removed.

and at its edge. Thus the magnetic current **M** can be determined up to an unknown factor, α; consequently the power P_r it radiates can be determined up to an unknown factor, $|\alpha|^2$. The energy stored in the electric field under the patch is

$$W_E = \frac{1}{2}\epsilon \int_{\text{cavity}} |\mathbf{E}|^2 \, dV$$

which can also be found up to the same unknown factor, $|\alpha|^2$. This unknown amplitude factor cancels in the expression for the quality factor,

$$Q = Q_{\text{ext}} = \frac{4\pi f W_E}{P_r}$$

thus demonstrating that only the shape of the field distribution under the patch is important to find Q.

One can also include the effect of dielectric and conduction losses in the computation of Q. If these are taken into account the expression for antenna quality factor becomes

$$Q = \frac{1}{1/Q_{\text{ext}} + \delta + \Delta/t} \tag{2}$$

where δ is the loss tangent of the *actual* dielectric in the antenna, $\Delta = 0.029(\lambda_0/\sigma)^{1/2}$ is the skin depth in the metal cladding with conductivity σ (with units of siemens per unit length), and t is the thickness of the antenna. Once Q is known, the antenna can be analyzed as if it were a lossy cavity. This is significant since the most commonly used patch shapes correspond to cavities having a separable geometry amenable to simple analytical treatment.

This describes the basic ideas used in the cavity model approximation. A summary of this procedure is given in the flowchart in Fig. 5. A more detailed outline of how to actually carry out a practical analysis is given in the following steps. This analysis has been implemented in FORTRAN and BASIC programs for rectangular, circular-disk, circular-sector, annular, and annular-sector elements. The results show good agreement with experiment for thin elements.

(a) *Find the normalized resonant modes of the associated ideal cavity.* Find the z-independent, source-free solutions $\psi_{mn}(u,v)$ to the homogeneous Helmholz equation,

$$[\nabla^2 + (k_{mn})^2]\psi_{mn}(u,v) = 0$$

subject to the homogeneous boundary condition that

$$\nabla \psi_{mn}(u,v) \cdot \hat{\mathbf{n}} = 0$$

for (u,v) on the edge of the patch. The ψ_{mn}s represent the (purely z-directed and z-independent) modal electric fields that resonate in the ideal cavity. It is convenient to normalize these modes so that the integral of their squares over the

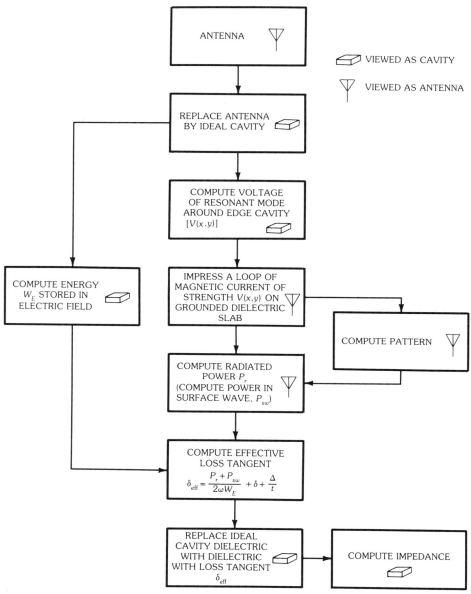

Fig. 5. Flowchart summarizing the steps used in the application of the cavity model.

patch area is 1. The k_{mn}s are the resonant wave numbers of the modes. The point (u, v, z) is a point in the cylindrical coordinate system being used to describe the patch's boundary. The only such systems which are separable are rectangular, circular-cylinder, elliptic-cylinder, and parabolic-cylinder coordinate systems. Table 1 gives the shapes of the patches separable in the rectangular and circular cylinder systems along with their corresponding ψ_{mn}s and characteristic equations for the k_{mn}s. Chart 1 illustrates some of the separable patches in elliptic- and parabolic-cylinder geometries. It also illustrates two triangular patches which can be analyzed in a rectangular system. Little has been done with elliptical disk elements [17] and nothing at all to date is available on other elliptical or parabolic patches. Although limited to only four separable coordinate systems one still has a rich set of geometries available to which the cavity model can be most conveniently applied.

(b) *Choose a resonant mode.* The next step is to choose a resonant mode ($m = M, n = N$) that produces the desired pattern. A more detailed discussion of how to roughly predict the pattern of the element in order to choose the mode is given in Section 3. How to actually compute the pattern for a given mode is given next.

(c) *Compute the far-field pattern.* The far-field pattern is obtained by first setting the surface magnetic current density **M** to be

$$\mathbf{M} = \hat{\mathbf{t}} \psi_{mn}(u, v)$$

at the edge of the patch where $\hat{\mathbf{t}}$ is the unit tangent to the edge: $\hat{\mathbf{t}} = \hat{\mathbf{z}} \times \hat{\mathbf{n}}$. The far-field pattern for the (M, N)th mode can then be found as the response to this magnetic current ribbon in the presence of a grounded dielectric slab. A suitable approximation to the far-field pattern can often be obtained by condensing the magnetic current ribbon shown in Fig. 4a into a magnetic current filament acting in the presence of just the ground plane (Fig. 4b). Then, by doubling the magnetic current, the ground plane can also be eliminated. The approximate source obtained is illustrated in Fig. 4c. This approximation will yield an estimate for the space wave. An estimate for the surface wave can be made by extracting the residue of the surface-wave pole in the spectral integral for the fields when the magnetic current filament is acting in the presence of the grounded dielectric substrate. Often, however, the power carried in the surface wave is small compared with that carried by the space wave and can therefore be ignored in the computation of Q. The far-field computation can be done analytically in the case of rectangular, circular-disk, and annular-disk microstrip elements. One can also sample the magnetic current along the edge of the patch at the points $\{(x_i, y_i)\}, i = 0, 1, 2, \ldots, L + 1, x_{L+1} = x_0, y_{L+1} = y_0$. Then, using linear interpolation to reconstruct the magnetic current distribution from the sample points, one obtains the following approximation for the far field:

$$E_\theta = jk_0 t(F_x \sin \phi - F_y \cos \phi) \frac{1}{4\pi r} e^{-jk_0 r}$$

$$E_\phi = jk_0 t(F_x \cos \phi - F_y \sin \phi) \cos \theta \frac{1}{4\pi r} e^{-jk_0 r}$$

Table 1. Separable Patches, Characteristic Equations, Mode Distributions, and Geometric Parameters and Boundary Conditions*

Patch Shape	Characteristic Equation	$\psi_{mn}(u,v)$	Coordinate Transformation	$g(v)$	u_+	v_-	v_+	Boundary Conditions
Rectangular patch	$k_{mn} = \pi\left[\left(\dfrac{m}{a}\right)^2 + \left(\dfrac{n}{b}\right)^2\right]^{1/2}$	$\dfrac{\cos(m\pi x/a)\cos(n\pi y/b)}{\sqrt{ab/(\epsilon_{0m}\epsilon_{0n})}}$ $\epsilon_{0m} = \begin{cases} 1, & m=0 \\ 2, & m>0 \end{cases}$	$x = u$ $y = v$	1	a	0	b	$U'_m(u_-) = 0$ $U'_m(u_+) = 0$ $V_m^{+'}(v_+) = 0$ $V_m^{-'}(v_-) = 0$
Circular disk patch	$Z'_{mn}(a) = 0$ $Z_{mn}(\varrho) = J_m(k_{mn}\varrho)$	$\dfrac{J_m(k_{mn}\varrho)\cos m\phi}{D_{mn}(a)}$ $D_{mn}(\varrho) = \sqrt{\dfrac{\phi_0}{2\epsilon_{0m}}}\dfrac{\lvert Z_{mn}(\varrho)\rvert\varrho}{\sqrt{1 - \left(\dfrac{ma}{k_{mn}\varrho}\right)^2}}$ $(\alpha = 1;\ \phi_0 = 2\pi)$	$x = ae^v\cos u$ $y = ae^v\sin u$ $\varrho = ae^v$ $\phi = u$	$a^2 e^{2v}$	2π	$-\infty$	0	$U_m(u_-) = U_m(u_+)$ $U'_m(u_-) = U'_m(u_+)$ $\lvert V_m^-(v_-)\rvert < \infty$ $V_m^{+'}(v_+) = 0$

Circular sector patch	$Z'_{mn}(a) = 0$ $Z_{mn}(\alpha) = J_{m\alpha}(k_{mn}\varrho)$	$\dfrac{J_{m\alpha}(k_{mn}\varrho)\cos(m\alpha\phi)}{D_{mn}(a)}$ $\left(\alpha = \dfrac{\pi}{\phi_0}\right)$ (See circular disk for D_{mn})	ϕ_0	$-\infty$	0	$U'_m(u_-) = 0$ $U_m(u_+) = 0$ $\|V_m^{-\prime}(v_-)\| < \infty$ $V_m^+(v_+) = 0$
Annular patch	$Z'_{mn}(a) = 0$ $Z_{mn}(\varrho) = J_m(k_{mn}\varrho)Y'_m(k_{mn}b)$ $\quad - Y_m(k_{mn}\varrho)J'_m(k_{mn}b)$	$\dfrac{Z_{mn}(\varrho)\cos(m\phi)}{D_{mn}(a) - D_{mn}(b)}$ $(\alpha = 1; \phi_0 = 2\pi)$ (See circular disk for D_{mn})	2π	$\ln(b/a)$	0	$U_m(u_-) = U_m(u_+)$ $U'_m(u_-) = U'_m(u_+)$ $V_m^-(v_-) = 0$ $V_m^+(v_+) = 0$
Annular sector patch	$Z'_{mn}(a) = 0$ $Z_{mn}(\varrho) = J_{m\alpha}(k_{mn}\varrho)Y'_{m\alpha}(k_{mn}b)$ $\quad - Y_{m\alpha}(k_{mn}\varrho)J'_{m\alpha}(k_{mn}b)$	$\dfrac{Z_{mn}(\varrho)\cos(m\alpha\phi)}{D_{mn}(a) - D_{mn}(b)}$ $\left(\alpha = \dfrac{\pi}{\phi_0}\right)$ (See circular disk for D_{mn})	ϕ_0	$\ln(b/a)$	0	$U'_m(u_-) = 0$ $U'_m(u_+) = 0$ $V_m^{-\prime}(v_-) = 0$ $V_m^{+\prime}(v_+) = 0$

*Note: 1. $u_- = 0$, and $f(u) \equiv 0$ for the rectangular and all circular geometries.
2. The Js and Ys are Bessel functions of the first and second kinds, respectively.

Chart 1. Other Separable Geometries

Elliptical Geometries

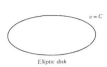
Elliptic disk

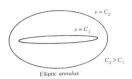
Elliptic annulus

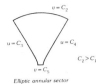
Elliptic annular sector

$$x = d\cos u \cosh v \quad v \geq 0$$
$$y = d\sin u \cosh v \quad 0 \leq u \leq 2\pi$$

Elliptic sector

Constant-u curves are hyperbolas; constant-v curves are ellipses
Solutions expressible in terms of Mathieu functions

Parabolic Geometries

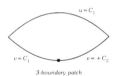
3-boundary patch

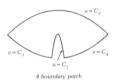
4-boundary patch

$$x = uv$$
$$y = 1/2(u^2 - v^2)$$
$$u \geq 0$$

$$c_2 > c_1$$
$$c_4 > c_3$$

Both constant-u and constant-v curves are parabolas
Solutions expressible in terms of parabolic cylinder functions

Triangular Geometries

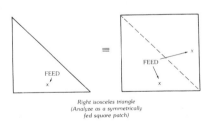
Right isosceles triangle
(Analyze as a symmetrically fed square patch)

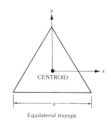
Equilateral triangle

Mode distribution proportional to $\cos[(2\pi x/\sqrt{3}a + 2\pi/3)\ell]\cos[2\pi(m-n)y/3a]$
$\phantom{\text{Mode distribution proportional to }} + \cos[(2\pi x/\sqrt{3}a + 2\pi/3)m]\cos[2\pi(n-\ell)y/3a]$
$\phantom{\text{Mode distribution proportional to }} + \cos[(2\pi x/\sqrt{3}a + 2\pi/3)n]\cos[2\pi(\ell-m)y/3a]$

$$\ell + m + n = 0$$

$$(k_{mn})^2 = (4\pi/3a)^2(m^2 + mn + n^2)$$

where

$$\mathbf{F} = \sum_{\ell=0}^{L} \hat{\tau}_\ell \left[\overline{\psi}_{MN}^{(\ell)} j_0(k_0 \hat{\mathbf{r}} \cdot \Delta \mathbf{r}_\ell/2) + j \frac{\Delta \psi_{MN}^{(\ell)}}{2} j_1(k_0 \hat{\mathbf{r}} \cdot \Delta \mathbf{r}_\ell/2) \right] \Delta s_\ell e^{jk_0 \hat{\mathbf{r}} \cdot \mathbf{r}'_\ell}$$

$$\overline{\psi}_{MN}^{(\ell)} = [\psi_{MN}(u_{\ell+1}, v_{\ell+1}) + \psi_{MN}(u_\ell, v_\ell)]/2$$

$$\Delta \psi_{MN}^{(\ell)} = \psi_{MN}(u_{\ell+1}, v_{\ell+1}) - \psi_{MN}(u_\ell, v_\ell)$$

$$\mathbf{r}_l = x_\ell \hat{x} + y_\ell \hat{y}, \qquad \mathbf{r}'_\ell = (\mathbf{r}_{\ell+1} + \mathbf{r}_\ell)/2, \qquad \Delta \mathbf{r}_\ell = \mathbf{r}_{\ell+1} - \mathbf{r}_\ell$$

$$u_\ell = u(x_\ell, y_\ell), \qquad v_\ell = v(x_\ell, y_\ell), \qquad \Delta s_\ell = |\Delta \mathbf{r}_\ell|, \qquad \hat{\tau}_\ell = \frac{\Delta \mathbf{r}_\ell}{\Delta s_\ell}$$

$$\hat{\mathbf{r}} = \hat{x} \cos\phi \sin\theta + \hat{y} \sin\phi \sin\theta + \hat{z} \cos\theta$$

$$j_0(x) = \frac{\sin x}{x}, \qquad j_1(x) = \frac{\sin x}{x^2} - \frac{\cos x}{x}$$

If the presence of the dielectric slab is included, then the E_θ component should be multiplied by e^{TM}, which is given by

$$e^{TM} = \frac{1 + \Gamma^{TM}}{1 + e^{-j2\zeta(k_0 \sin\theta)t} \Gamma^{TM}}$$

where

$$\Gamma^{TM} = \frac{1 - \zeta(k_0 \sin\theta)/(\epsilon_r k_0 \cos\theta)}{1 + \zeta(k_0 \sin\theta)/(\epsilon_r k_0 \cos\theta)}$$

and the E_ϕ component should be multiplied by e^{TE}. The latter coefficient is the same as e^{TM} except that Γ^{TM} is replaced by

$$\Gamma^{TE} = \frac{1 - \mu_r k_0 \cos\theta/\zeta(k_0 \sin\theta)}{1 + \mu_r k_0 \cos\theta/\zeta(k_0 \sin\theta)}$$

where μ_r is the relative permeability of the dielectric substrate. The function $\zeta(\xi)$ is

$$\zeta(\xi) = \sqrt{\epsilon_r \mu_r k_0^2 - \xi^2}$$

This result is based on condensing the magnetic current into a filament and impressing it on the ground plane in the presence of the dielectric slab. The approximation appears to work reasonably well for soft substrates for dielectric thicknesses up to $0.05\lambda_0$.

(d) *Compute the radiated power*. One must next compute the power carried in the space wave and perhaps also the surface wave. This can be done by integrating the far-field power pattern. This must often be done by using a numerical integration. However, since the microstrip elements are not large-aperture elements their patterns are broad and slowly varying and the numerical effort required to perform the integration is small. The integral for power radiated by the space wave is

$$P_r = \frac{1}{\eta_0} \int_0^{2\pi} \int_0^{\pi/2} (|E_\theta|^2 + |E_\phi|^2) r^2 \sin\theta \, d\theta \, d\phi$$

and the power carried by the surface wave is

$$P_{sw} \cong \frac{1}{4\pi\eta_0} (k_0 t)^3 \frac{\epsilon_r - 1}{\epsilon_r} \int_0^{2\pi} |F_x \sin\phi - F_y \cos\phi|^2 \, d\phi$$

where $\eta_0 = 377\,\Omega$. The approximation given above for the surface wave is valid for soft substrates of thickness up to $0.05\lambda_0$.

(e) *Compute the stored energy.* Because of the normalization of the resonant mode the computation of the energy stored in the electric field under the patch is simple. The stored energy is given by

$$W_E = \frac{1}{2} \epsilon t$$

where ϵ is the substrate permittivity.

(f) *Compute the external quality factor.* The external quality factor is obtained from

$$Q_{ext} = \frac{2c k_{mn} W_E}{(\mu_r \epsilon_r)^{1/2} (P_r + P_{sw})}$$

where c is the speed of light in a vacuum. The radiative quality factor Q_r (i.e., the external quality factor obtained ignoring the surface wave) times the substrate thickness t divided by λ_0 is plotted in Fig. 6 for various modes of a rectangular microstrip element. The radiated power is computed ignoring the presence of the dielectric slab.

(g) *Compute the antenna quality factor.* To get the quality factor of the antenna one must include with the external Q the loss contribution due to the actual dielectric loss and the conduction loss. This quality factor is given in (2).

(h) *Compute the effective loss tangent* δ_{eff}. The effective loss tangent is $1/Q$.

(i) *Analyze the lossy cavity model of the antenna to find the impedance and other antenna parameters obtainable from the interior fields.* The voltage between a point (u, v) on the patch and the point on the ground plane directly below it, due to a unit, z-independent, *filamentary* electric current source impressed between the ground plane and the patch at point (u', v'), is [5]

$$G(u, v | u', v') = -jk\eta t \sum_A \frac{\psi_{mn}(u,v) \, \psi_{mn}(u',v')}{k^2(1 - j/Q) - (k_{mn})^2} \qquad (3)$$

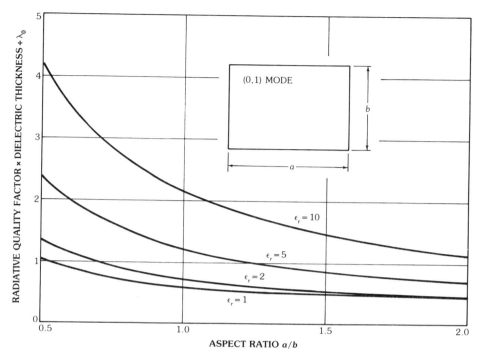

Fig. 6. The dielectric thickness t divided by λ_0 times the radiative quality factor Q for the $(0,1)$ mode versus aspect ratio of a rectangular microstrip element for various relative dielectric constants.

where $\eta = \eta_0(\mu_r/\epsilon_r)^{1/2}$ is the intrinsic wave impedance of the substrate dielectric and $k = 2\pi f(\epsilon_r \mu_r)^{1/2}/c$ is the wave number of the substrate. The summation is over the entire set A of mode indices (m, n). The "$1/Q$" term in the denominator of the summand is the effective loss tangent.

This expression allows one to compute the field everywhere within the cavity due to the filamentary source. If one attempts to set (u', v'), the source point, equal to (u, v), the observation point, to obtain an input impedance, one finds that the series does not converge. This is because there is an infinite inductance associated with a filamentary source. To obtain the correct impedance one must integrate (3) over the perimeter of the actual feed probe or microstrip feed. This integration modifies the summand in (3) by the introduction of an additional factor, $(\xi_{mn})^2$. This factor approaches zero as m and n increase without bound. For the lower-order modes ξ_{mn} is nearly unity. A more detailed discussion of how to actually evaluate the input impedance efficiently is given in Section 4, under "Efficient Computation of Impedance Parameters."

3. Pattern

How to compute the pattern associated with a microstrip element is discussed in Section 2, under (c) "Compute the far-field pattern." The explanation is more qualitative in this section; it guides the designer in making a choice of the shape of the patch and the particular mode he or she should investigate for his or her application.

As noted previously, the far field radiated by the antenna can be attributed, approximately, to a loop of magnetic current in the shape of the patch's outline, flowing in the presence of a grounded dielectric slab, or in free space if the effect of the dielectric is ignored. Fig. 7 illustrates the magnetic current distributions

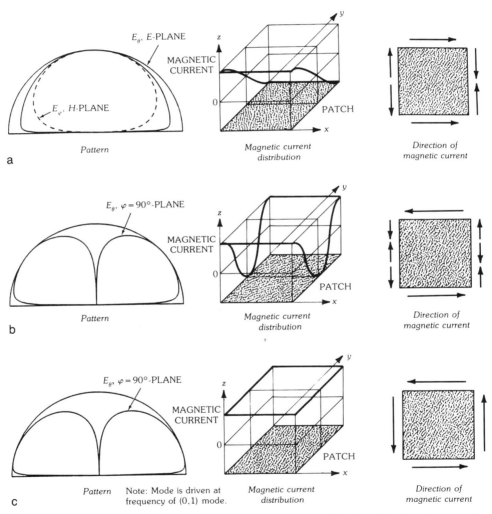

Fig. 7. Principal-plane patterns and magnetic current distributions for the lowest-order modes of rectangular and circular-disk elements. Patterns are plotted on a 40-dB scale and are computed for a $0.01\lambda_0$-thick substrate of relative dielectric constant $\epsilon_r = 2.50$. (*a*) Rectangular element: $(0, 1)$ mode. (*b*) Rectangular element: $(0, 2)$ mode. (*c*) Rectangular element: dc mode (driven at the resonant frequency of the $(0, 1)$ mode). (*d*) Circular-disk element: $(1, 1)$ mode. (*e*) Circular-disk element: $(0, 1)$ mode.

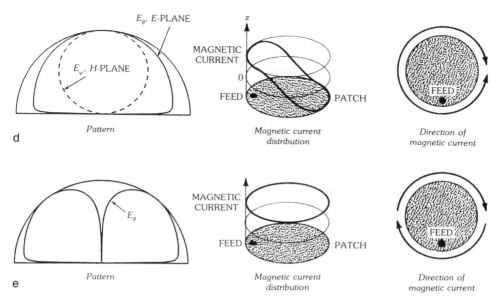

Fig. 7, *continued.*

associated with the lowest few modes of the rectangular and circular disk microstrip antennas. It also includes the corresponding principal-plane patterns. The reader should consult these figures to illuminate the text in the following subsections. These patterns were computed using a magnetic loop current impressed on the ground plane of a $0.01\lambda_0$-thick grounded dielectric slab. The assumed relative dielectric constant of the nonmagnetic substrate was 2.5. Conclusions on the elevation of beam maxima will vary somewhat if different dielectric thicknesses and dielectric constants are assumed. Since the rectangular patch is the most commonly used shape and since its mode functions are the most familiar and the easiest to compute, the discussion in this section centers on this shape of patch. Before considering specific examples, some general discussion is presented on the properties of microstrip-antenna patterns.

General Properties

An important assumption made in the discussion below is that the microstrip elements are mounted on *infinitely large, flat* ground planes. Although some applications could call for microstrip elements mounted on a large, flat, ground plane, it is more likely that the element will be mounted on the curved surface of a missile, or some other curved and more severely abbreviated ground plane. It is clearly impossible to cover all environments in which an element might be used. Thus the designer must be aware of the effects of truncating [18] and bending the ground plane and be ready to compute the pattern radiated by the element in *his or her particular environment*. Fortunately, unless the radius of curvature of the ground plane is small compared with the element's size, one can expect the magnetic current distribution to remain the same when the element is bent [19]. Although some distortion might result from an extreme truncation of the ground plane, the magnetic current distribution at the edges of these high-Q devices remains relatively unaffected.

The effect of the dielectric on the pattern is to cause the pattern to roll off sharply as the observation angle approaches the horizon. This is illustrated in Fig. 8 in which the E-plane pattern produced by a rectangular element is computed when the effect of a $0.05\lambda_0$-thick dielectric is included, and when the effect of the dielectric is ignored. The ratio of the power radiated with the dielectric taken into account and with the dielectric ignored is 0.97. In both cases the magnetic current was condensed into a filament and impressed on the ground plane.

The designer should also recognize that the elevation patterns of thin microstrip antennas are *inherently symmetric* about the broadside direction *regardless of the mode excited or the shape of the patch*. This is because the magnetic current phasor associated with a single excited mode is (or can be considered to be) a purely real-valued function. Only when more than one mode is strongly and simultaneously excited will the magnetic current be complex. The magnetic current is approximately the electric field at the edge of the corresponding ideal closed cavity. The field of a *single mode* within such a cavity can only have phase differences of 0° or 180°. A unidirectional, end-fire pattern cannot be obtained by exciting a *single* mode of any shape of patch. However, by using reactively loaded microstrip elements and exciting an appropriately chosen *pair* of modes, one can, in principle, obtain an end-fire pattern [7].

The (0, 1) Mode

The most commonly used mode of a rectangular patch is the (0, 1) mode. In this mode the magnetic currents on two opposite edges are constant. These edges are called the *radiating edges*. The magnetic currents on the remaining two edges suffer a phase reversal and hence the fields radiated by them largely cancel. The pattern this mode produces is linearly polarized and has a broadside maximum. (In this discussion the term "broadside" is used to indicate a direction normal to the plane of the patch, and "horizon" represents any direction in the plane of the patch.) The "E-plane" is the plane perpendicular to the radiating edges; the electric field for points in this plane is parallel to the plane. The pattern falls in the E-plane by an amount dependent on the dielectric constant of the substrate

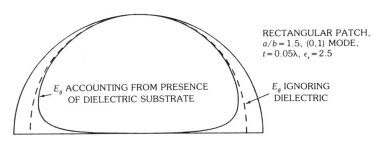

Fig. 8. E-plane far-field pattern of a rectangular element computed when the effect of a $0.05\lambda_0$-thick dielectric slab is taken into account and when it is ignored. Pattern plotted on a 40-dB scale.

Microstrip Antennas

material as the observation direction approaches the horizon. For angles very near the horizon, the pattern depends strongly on the truncation of the dielectric.

The (0, 2) Mode

The $(0, 2)$ mode is similar to the $(0, 1)$ mode except that the magnetic currents on the two radiating edges are antiparallel. This mode is commonly called a "higher-order mode." The physical size of the patch in terms of λ_0 is approximately twice that required to resonate a $(0, 1)$ mode at the same frequency. For a substrate with a relative dielectric constant of 2.5, the patch must be approximately $0.63\lambda_0$ long. The pattern produced by this mode has a broadside null and peaks in the E-plane at an angle of 40° above the horizon. The H-plane pattern is null.

The DC Mode

The rectangular patch (as well as all other unloaded patch geometries) supports a dc mode, that is, a mode which "resonates" at a frequency of $f = 0$ Hz. Of course, one wants to excite the dc mode at an rf frequency, not at dc. This requires an external matching network, or it requires that appropriate reactive lumped loads be placed between the patch and the ground plane [7] as discussed in Section 9, under "Frequency-Agile Elements." Under the cavity model approximation the electric field, $\psi_{0,0}(u, v)$, is a constant. The corresponding magnetic current is constant in magnitude. If one recalls the equivalence between a small loop of constant magnetic current and a short, vertical electric dipole, one can see why this mode produces a pattern resembling that of the vertical monopole.

The (1, 1) Mode of a Circular-Disk Element

This mode of the disk element is very similar to the $(0, 1)$ mode of a rectangular element. Both produce broadside nulls. There are two degenerate $(1, 1)$ disk modes. One has an even field distribution about the diameter on which the feed lies. The other has an odd-symmetric field distribution about this diameter. In the absence of anything that perturbs the symmetry of the disk the odd-symmetric mode cannot be excited by the feed. On the other hand, one can strongly excite this odd mode by making the disk slightly elliptical [20] or by loading the disk appropriately. This has applications for producing circularly polarized patterns as discussed in Section 9, under "Single-Feed CP."

The (0, n) Circular-Disk Modes

These modes are independent of the azimuthal variable ϕ. The magnetic current distribution around the perimeter of the patch is a constant and the pattern has a broadside null. For $n = 1$ and for $\epsilon_r = 2.5$, the beam maximum occurs at 53° above the horizon. For $n >, 1$ grating lobes will appear in the pattern for a relative dielectric constant of 2.5.

Other Modes

Some modes such as the $(1, 1)$ mode of the rectangular patch seem to have little practical application although a similar mode, the $(2, 1)$ mode of a circular-disk element, has been used to produce a conical, circularly polarized pattern [21]. When the sum of the mode indices of a rectangular microstrip element is an even number, a broadside null will always occur. When it is odd, a broadside maximum will always occur. As the mode indices increase, the size of the element in terms of λ_0 increases, thus producing grating lobes in the pattern. This typically renders the pattern useless for most applications. Thus one typically only uses one or more of the lowest-order modes of a microstrip element.

4. Impedance and Circuit Models

The general approach for obtaining the input impedance of a microstrip antenna was discussed in Section 2. In this section practical formulas and techniques are given for computing input and mutual impedances of one or more ports on a microstrip element. A physical explanation of the variation of impedance with feed point is given. Simple circuit models for the microstrip element are presented.

General Circuit Model

If one rewrites the factor

$$\frac{-jk}{k^2(1 - j/Q) - (k_{mn})^2} \quad \text{in (3) as} \quad \frac{-j\omega c/\sqrt{\epsilon_r}}{\omega^2 - (k_{mn})^2 c^2/\epsilon_r - j\omega^2/Q}$$

and compares it with the input impedance of the parallel combination of an inductor, capacitor, and resistor, then one concludes that the impedance of a microstrip line as approximated by (3), after integration over the source current distribution, can be modeled as the input impedance to the network illustrated in Fig. 9a. The element values are

$$C_{mn} = \left[\frac{c}{\sqrt{\epsilon_r}} t\eta \psi_{mn}(u', v')^2 (\xi_{mn})^2\right]^{-1}$$

$$L_{mn} = \frac{\epsilon_r}{C_{mn}(k_{mn})^2 c^2}$$

$$R_{mn} = Q/(C_{mn}\omega)$$

The factor $(\xi_{mn})^2$, it is emphasized, arises from the integration of (3) over the feed current distribution and tends toward zero as m and n increase without bound. The inductive susceptance increases much faster than the capacitive susceptance and the conductance as m and n increase. Thus, for a sufficiently high-order mode, the inductance, L_{mn}, dominates its lossy tank circuit and the corresponding resistance and capacitance can be ignored. The inductances corresponding to all sufficiently high-order modes can be combined into the single series inductance indicated in Fig. 9a. The resistances R_{mn} in this model are frequency dependent. However, over the relatively narrow band of a resonant mode, they can be considered approximately constant with the ω replaced by its value at the band center. The network in Fig. 9a is valid for frequencies in the vicinity of the resonance of a *single*

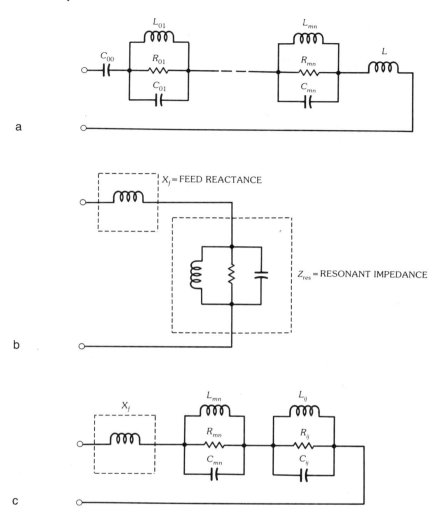

Fig. 9. Circuit models for a microstrip element. (*a*) General circuit model. (*b*) Simplified circuit model valid over the band of a single, isolated mode. (*c*) Circuit model over the overlapping bands of two nearly degenerate modes.

mode of the microstrip element. This is because the effective loss tangent, $1/Q$, was computed for a single [say (M, N)th] mode of the antenna. Since each tank circuit in this network has a relatively high Q, the impedance of any nonresonant tank will be rather insensitive to the exact value of R_{mn}. Thus, one can make the network in Fig. 9a applicable to frequency bands centered about several resonant modes if Q is computed for each of the modes and used to determine the corresponding R_{mn}. This network becomes questionable, however, if one attempts to use it to compute the input impedance of the element at a frequency midway between two well-separated adjacent resonant modes. The problem of analyzing an element far off resonance is considered in Section 8, "Loaded Microstrip Elements."

Simplified Circuit Model

Consider a mode whose resonant band does not significantly overlap the bands of neighboring modes. Then all the tanks can be approximated by an inductor if its resonant frequency is larger than that of the (M, N)th mode, or by a capacitor if its resonant frequency is smaller. If all of the series reactive elements are combined, then over the narrow band of the resonant mode the circuit of Fig. 9a reduces to that of Fig. 9b. The series reactive element is represented schematically as an inductor. Actually, its reactance is slightly nonlinear with frequency, but the reactance is typically inductive. Since it is the nonresonant modes, particularly the very high order modes, which are necessary to represent the rapidly varying currents in the vicinity of the antenna feed, this inductance can be called the *feed inductance* and its associated reactance can be called the *feed reactance*. More will be said about the feed reactance in this section, under "Feed Reactance."

Circuit Model for Near-Degenerate Modes

When two modes have resonant bands that strongly overlap, the modes can be thought of as "near-degenerate" modes. In this case the Smith chart plot of the input impedance in the overlapping bands exhibits a looping or kinking behavior as illustrated in Fig. 10 [11]. Whenever an isolated, unloaded patch exhibits this behavior, expect that muliple modes are present. It has been found experimentally that the model of Fig. 9c yields a resonably good prediction of the input impedance for closely spaced, near-degenerate modes.

Simple Feed Models

The two most common ways of feeding a microstrip antenna are by use of a coaxial-probe feed and a microstripline feed. In the cavity model both of these sources are modeled essentially the same way. They are modeled by z-directed surface currents as illustrated in Fig. 11. For a coaxial feed the current is assumed to flow uniformly over a cylinder of a diameter equal to that of the feed probe. For a microstripline feed the current is assumed to flow uniformly over a strip of width equal to that of the stripline. It can be shown [4] that the impedances obtained by the cavity model for these two source distributions are identical if the cylindrical distribution of diameter d is replaced by a strip distribution of effective feed width $2.24d$.

As the circuit in Fig. 9b suggests, one can view the input impedance of the element as the sum of a resonant component plus a feed-reactive component,

$$Z = Z_{\text{res}} + jX_f$$

This decomposition of the impedance into resonant and nonresonant parts has been found quite useful in explaining the behavior of loaded microstrip elements (see Section 8).

Resonant Impedance

If one will operate the element at a frequency in the vicinity of its (M, N)th resonant mode, then the resonant part of the impedance is

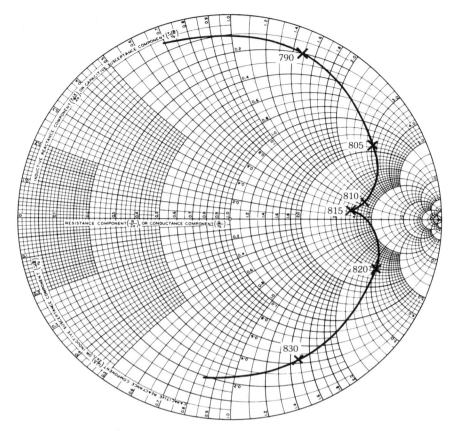

Fig. 10. Typical Smith chart plot of the impedance of an element with two nearly degenerate modes. Measured for an 11.4-cm nearly square patch, with a dielectric 0.16 cm thick and having a relative dielectric constant of 2.62. (*After Lo, Solomon, and Richards [11]*, © *1979 IEEE*)

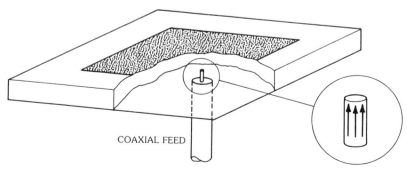

Fig. 11. Coaxially fed element and the simplified feed model obtained by removing the feed probe, closing the aperture of the coaxial cable, and impressing a uniform electric surface current on the surface corresponding the surface of the feed probe.

$$Z_{res} = -jk\eta t \frac{\psi_{MN}(u', v')^2}{k^2 - (k_{MN})^2}$$

where (u', v') is the location of the center of the feed current distribution. A typical plot of resonant resistance and reactance *versus* frequency is shown by the solid and dotted curves in Fig. 12. Near the resonant frequency of the (M, N)th mode the resistance reaches its peak value, denoted R_{res}. This peak resistance does not occur exactly at the resonant frequency of the antenna, as can be seen from Fig. 12, because of the presence of the feed reactance. The value of this resistance is

$$R_{res} = \frac{\eta t}{k} Q\psi_{MN}(u', v')^2 \tag{4}$$

Unless $M = N = 0$ (the dc mode), $\psi_{MN}(u, v)$ will always have a curve $u = h(v)$ along which $\psi_{MN}(u, v)$ is zero. Thus the resonant resistance can be varied from its maximum possible value to zero. This allows one to do some simple matching of the element by feeding the element at a point within the patch at which the R_{res} is equal to (or better yet, slightly larger than) the characteristic impedance of the feed line.

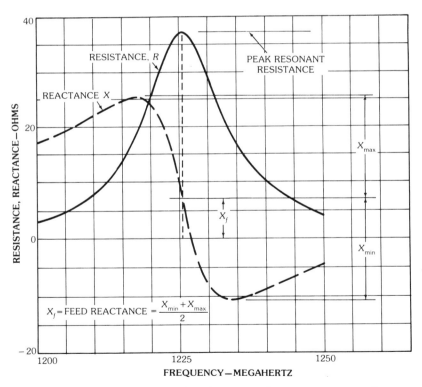

Fig. 12. Typical measured variation of resistance and reactance versus frequency in a 7.62- × 11.43- × 0.16-cm microstrip antenna illustrating one way to empirically determine the feed reactance. (*After Richards et al. [22], reprinted with permission of Hemisphere Publishing Corporation*)

(See Section 7 for more on matching.) Fig. 13 illustrates the theoretical variation of R_{res} with feed location for a rectangular microstrip element in the $(0,1)$ mode.

Feed Reactance

The feed-reactive part [22] X_f of the input impedance represents the contribution to the impedance due to all of the nonresonant modes. The perimeter P of the feed current distribution is an important parameter in determining X_f. If the current were filamentary ($P = 0$), then the feed reactance would be infinite. The smaller the P, the more inductive the feed reactance will be. The feed reactance is given by

$$\frac{1}{P^2} \int_P \int_P G(u, v | u', v') \, ds \, ds' - Z_{res}$$

where the ds and ds' are elements of arc length associated with the source and observation points (u', v') and (u, v) and the integrals are over the perimeter of the current distribution. How to actually carry out these integrations is considered in this section, under "Multiport Impedance Parameters." Typical plots of feed reactance versus frequency can be found in Section 8, "Loaded Microstrip Elements." Fig. 12 illustrates the feed reactance at the resonant frequency.

The feed reactance may or may not be an important parameter in the design of a microstrip element. For thin elements the feed reactance is typically small compared with the resonant resistance of a well-matched element. For thicker elements this may not be true and the feed reactive component may have to be taken into account in matching the element. For loaded-element applications the feed reactance is a critical parameter since it determines, among other parameters, the resonant frequency of the loaded element. (A discussion of loaded elements is given in Section 8.) For loaded elements an understanding of how the feed reactance varies with feed point can be useful in element design.

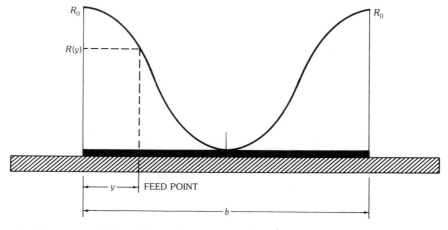

Fig. 13. The theoretical variation of resonant resistance versus feed point on a rectangular microstrip element driven in the $(0,1)$ mode.

To see how feed reactance varies with feed point, one can note that the feed inductance will be roughly proportional to the amount of magnetic energy stored within a small testing volume surrounding the feed (Fig. 14a) [22]. Consider, for example, a rectangular element. The magnetic walls can be removed if the multiple images of the source current are impressed as illustrated in the right-hand side of Fig. 14a. Since these are images through *magnetic* walls, the image currents flow in the same direction as the source current. Suppose the feed current is far from any of the magnetic walls in the cavity model of the element. Then the magnetic fields produced by this current and its multiple images do not strongly overlap (Fig. 14a). However, if the feed current is moved close to one edge, the magnetic fields of the feed current and its nearest image overlap strongly. In the extreme case when the feed current is at the edge of the patch, the feed and image currents coincide and the magnetic field within the testing volume doubles (Fig. 14b). In this case the magnetic energy density in the testing volume quadruples. The volume of the testing volume under the patch, however, is only half as much as it was when the feed was far from a magnetic wall. Thus there is a net doubling of the magnetic energy in the testing volume. One expects from this rough analysis that the feed reactance for a feed near an edge of the patch will be twice that for a feed far from

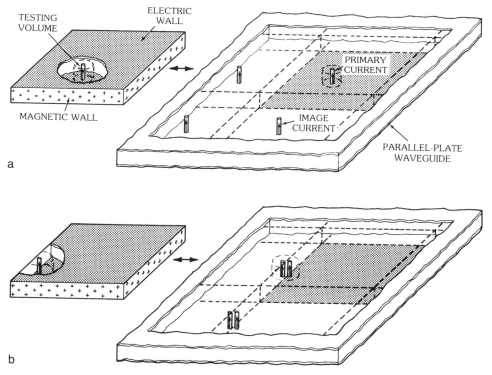

Fig. 14. Multiple-image expansions (right) of cavity model (left) with driving current far from or near to the edge of the patch. (*a*) Feed far from any edge. (*b*) Feed very near one edge. (*After Richards [22], reprinted with permission of Hemisphere Publishing Corporation*)

the edge. If the feed were near a corner of the patch, one would expect a net quadrupling of the inductance because of the presence of three images at the corner.

Although this rough analysis does not prove to be quantitatively accurate, it does precisely predict the trends in feed inductance as a function of feed location. Fig. 15 contains a plot of the feed reactance predicted by the cavity model and the measured feed reactance. The cavity model is able to predict the variation of feed reactance with reasonable accuracy for feeds not too close to the edge. It overestimates the feed reactance for feeds right on the edge. The reason for this is that the cavity model assumes zero tangential component of magnetic field at the patch edge. Actually, for the cavity model to precisely represent the fields under the patch, one would have to introduce an equivalent electric current which is numerically equal to the small tangential magnetic field at the patch edge. When the feed is very close to the edge the tangential field near the probe is not negligible. A strong equivalent electric current *should be* impressed on the magnetic wall. This equivalent current is in a direction opposite to the feed current, which has the effect of reducing the magnetic energy stored in the testing volume. Hence the *actual* feed reactances for feeds very close to the edge are smaller than those predicted by the cavity model.

A formula which is based on the feed reactance within a parallel-plate waveguide (which ignores all image currents) is given by

$$X_f \cong -\frac{\eta k t}{2\pi}[\ln(kd/4) + 0.577]$$

where d is the diameter of the probe. This formula has been used to estimate the feed reactance. It does not predict the variation of feed reactance versus feed

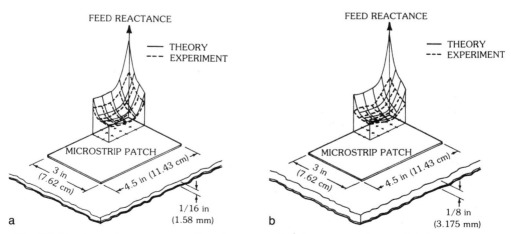

Fig. 15. Feed reactance measured and computed using the cavity model for a 11.43- × 7.62-cm rectangular patch. (*a*) Dielectric thickness 1/16 in (1.6 mm). (*b*) Dielectric thickness 1/8 in (3.2 mm). (*After Richards [22], reprinted with permission of Hemisphere Publishing Corporation*)

position but it is simple and can have some use if only a very rough estimate of the feed reactance is required.

Multiport Impedance Parameters

The discussion above has centered on the computation and the properties of the driving-point or input impedance of a microstrip element. One can also compute the open-circuit parameters of a microstrip element with more than one port. From these open-circuit or z parameters one can compute the corresponding s parameters of the element. The nondiagonal z parameters are essentially independent of the precise feed-current distributions and are given by

$$Z_{ij} = Z_{ji} = G(u_i, v_i | u_j, v_j), \qquad i \neq j \qquad (5)$$

The corresponding s-parameter matrix **S** is given in terms of the z-parameter matrix **Z** by

$$\mathbf{S} = (\mathbf{ZY}_0 + \mathbf{I})^{-1}(\mathbf{ZY}_0 - \mathbf{I})$$

where **I** is the unit matrix and $\mathbf{Y}_0$ is the diagonal matrix containing the characteristic admittances of the feed lines to the element. Typical measured and computed s parameters are illustrated in Fig. 16.

Efficient Computation of Impedance Parameters

Equation 3 for the driving-point impedance and equation 5 for the mutual impedances of a multiple-feed element are not practical to use as is, for two reasons. (1) They both converge very slowly. (2) For all but rectangular elements, computing higher-order modes $\psi_{mn}(u, v)$ becomes a time-consuming process. Fortunately, the equation for $G(u, v | u', v')$ can be accelerated in the way described in this section. As pointed out earlier, under "Feed Reactance," one must average the Green's function G over the feed distribution to obtain a finite driving-point impedance. How to do this averaging on the accelerated G is considered later in this section, under "Multiport Impedance Parameters."

No one will use any but the lowest few modes of any microstrip element. Thus, for a given physical size of an element the maximum frequency of interest is a known, fixed quantity. Let the corresponding maximum wave number of interest in the dielectric substrate be denoted by $k_{\max}$. Divide the set A of all mode indices (m, n) into the dc mode $(0, 0)$, the set A_ξ, and the complement $\bar{A}_\xi$ of A_ξ with respect to A_0. The set A_ξ contains all mode indices (m, n) whose corresponding resonant wave numbers k_{mn} are larger than $\xi k_{\max}$. (The set A_0 is A_ξ with $\xi = 0$.) Then $G(u, v | u', v')$ can be approximated by

$$G(u, v | u', v') \cong -\frac{j\eta t}{k}\frac{1}{S} - jk\eta t \sum_{\bar{A}_\xi} \left\{ \frac{1}{k^2(1 - j/Q_{mn}) - (k_{mn})^2} + a_0\frac{1}{(k_{mn})^2} \right.$$
$$+ k^2 a_1 \frac{1}{(k_{mn})^4} + \ldots + k^{2(L-1)} a_{L-1}\frac{1}{(k_{mn})^{2L}} \right\} \psi_{mn}(u, v)\psi_{mn}(u', v')$$
$$+ \{a_0 G_0(u, v | u', v') + k^2 a_1 G_1(u, v | u', v')$$
$$+ \ldots k^{2(L-1)} a_{L-1} G_{L-1}(u, v | u', v')\} \qquad (6)$$

Microstrip Antennas

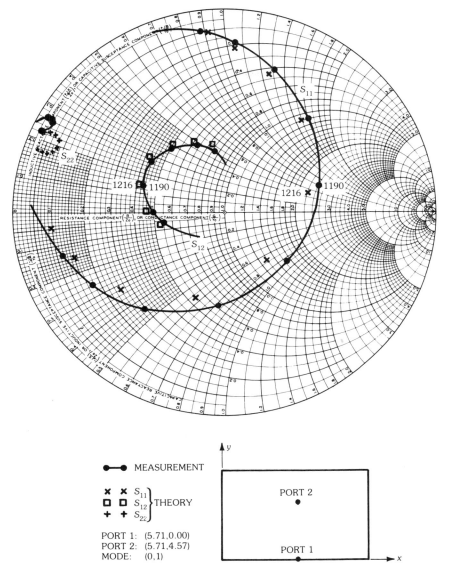

Fig. 16. Typical measured and computed s-parameter Smith chart plots for an 11.43- × 7.62- × 0.16-cm rectangular microstrip element with relative dielectric constant of 2.62. (*After Richards, Lo, and Harrison [5], © 1981 IEEE*)

The parameter S is used to represent both the area of the patch and the set of points comprising the patch. The coefficients $a_0, a_1, \ldots, a_{L-1}$ are the coefficients of the polynomial approximation of

$$\frac{1}{1-x} \cong a_0 + a_1 x + a_2 x^2 + \ldots + a_{L-1} x^{(L-1)}$$

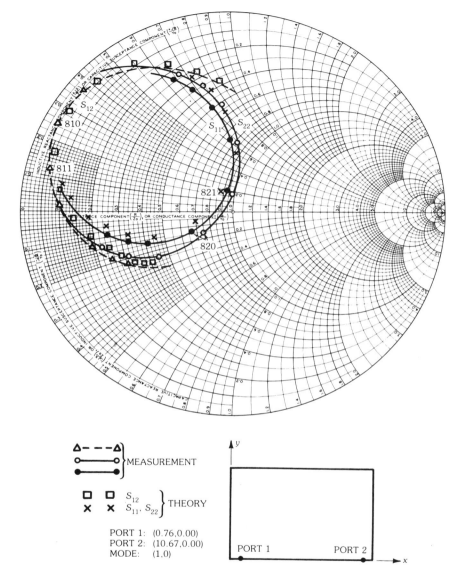

Fig. 16, *continued.*

for $0 \leq x \leq 1/\xi^2$. For $L = 2$, $\xi = 2$, $a_0 = 0.989$, and $a_1 = 1.331$, the maximum error in this approximation is about 1 percent. This has been found to be an adequate choice of parameters for the analysis of the microstrip elements. The G_ℓs are quasi-static terms (independent of frequency except for a linear factor) and are given by

$$G_\ell(u,v\,|\,u',v') = jk\eta t \sum_{A_0} \frac{\psi_{mn}(u,v)\,\psi_{mn}(u',v')}{(k_{mn})^{2\ell}} \qquad (7)$$

Microstrip Antennas

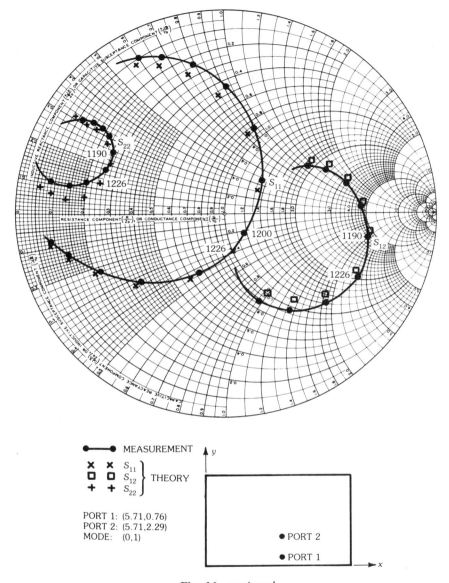

Fig. 16, continued.

Although the series for G_0 is just as slowly converging as the original series for G, G_0 and G_1 need only be computed *once* for a given pair of observation and feed points, (u, v) and (u', v'). Once they are determined, one need only sum over the *finite* set $\overline{A}_\xi$ in (6) to find the G at any frequency.

The G_ℓs can be computed more efficiently than by summing (7). They satisfy the static boundary value problem,

$$\nabla^2 G_0 = -jk\eta t \delta(\mathbf{r} - \mathbf{r}') + jk\eta t/S$$

$$\nabla^2 G_1 = -G_0$$

The vectors in the argument of the Dirac delta function are position vectors to the points (u, v) and (u', v'). The solutions of the differential equations must also satisfy the boundary condition that $\nabla G_\ell(u, v \mid u', v') \cdot \hat{\mathbf{n}} = 0$ for (u, v) on the edge of the patch. For nonseparable geometries the solutions can be determined numerically by a moment-method solution of a line integral equation.

For separable geometries, solutions of these equations can be found which *do not* require the computation of the modal functions ψ_{mn}. Instead, the solutions can be expressed as single-index series of circular, hyperbolic, and exponential functions. Explicit formulas for G_ℓ for rectangular, circular-disk, circular-sector, annular, and annular-sector elements are given in later in this section, under "Rectangular and Circular Geometries." The following "Overview" gives a general overview of how to determine the quasi-static terms.

Overview—One can obtain G_0 and G_1 by evaluating

$$\begin{aligned} G_0 &= k \lim_{k \to 0} \left(\frac{G}{k} + \frac{j\eta t}{k^2 S} \right) \\ G_1 &= k \lim_{k \to 0} \left\{ \frac{\partial}{\partial k^2} \left(\frac{G}{k} + \frac{j\eta t}{k^2 S} \right) \right\} \end{aligned} \quad (8)$$

By expanding G in a Laurent series in k about $k = 0$, one can evaluate these limits. To this end, G is first expressed as

$$G(u, v \mid u', v') = jk\eta t \sum_{m=0}^{\infty} \frac{U_m(u) \, U_m(u') \, V_m^-(v_<) \, V_m^+(v_>)}{N_m W(V_m^-, V_m^+)} \quad (9)$$

where

$$N_m = \int_{u_-}^{u_+} U_m(u)^2 \, du$$

$$W(V_m^-, V_m^+) = W_m = V_m^-(v_+) \, V_m^{+\prime}(v_+) - V_m^{-\prime}(v_+) \, V_m^+(v_+)$$

$$v_< = \begin{cases} v, & \text{for } v < v' \\ v', & \text{for } v > v' \end{cases}$$

$$v_> = \begin{cases} v, & \text{for } v > v' \\ v', & \text{for } v < v' \end{cases}$$

The prime ($'$) on a dependent variable indicates differentiation with respect to argument. The U_ms and V_ms satisfy the separated Helmholtz equations

Microstrip Antennas

$$U_m'' + [(\alpha_m)^2 + k^2 f(u)] U_m = 0$$

$$V_m'' + [-(\alpha_m)^2 + k^2 g(v)] V_m = 0$$

where the + and − superscripts have been suppressed on the V for convenience. The u and v coordinate variables are related to the rectangular x and y coordinate variables through the conformal mapping, $x + jy = w(u + jv)$. This conformal mapping transforms a patch in the xy plane whose edges lie along the curves $u = u_-$, $u = u_+$, $v = v_-$, and $v = v_+$ into a rectangular patch in the uv plane. This is illustrated in Fig. 17.

The square of the common scale factor for the u and v variables is $|w'(u + jv)|^2$ and can be written as $f(u) + g(v)$ for the four cylindrical-coordinate systems in which the Helmholtz equation is separable. These four systems are rectangular, circular-cylinder, elliptic-cylinder, and parabolic-cylinder systems. The $(\alpha_m)^2$ is the separation constant determined by applying the pair of homogeneous boundary conditions that $U_m(u)$ satisfies at $u = u_\pm$. These boundary conditions depend on the geometry of the patch and are listed along with the functions $f(u)$ and $g(v)$ and the $u_\pm$ and $v_\pm$ in Table 1 for the rectangular and circular classes of patch shapes.

These parameters can be expanded in a power series in k^2 as

$$U_m = U_{0m} + k^2 U_{1m} + k^4 U_{2m} + \ldots$$

$$V_m = V_{0m} + k^2 V_{1m} + k^4 V_{2m} + \ldots$$

$$(\alpha_m)^2 = (\alpha_{0m})^2 + k^2 (\alpha_{1m})^2 + k^4 (\alpha_{2m})^2 + \ldots$$

$$N_m = N_{0m} + k^2 N_{1m} + k^4 N_{2m} + \ldots$$

$$W_m = W_{0m} + k^2 W_{1m} + k^4 W_{2m} + \ldots$$

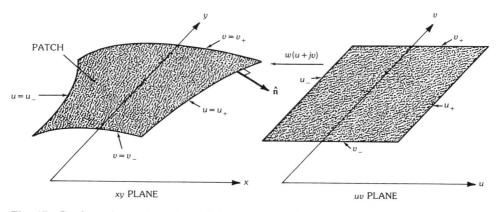

Fig. 17. Conformal mapping of a patch into a rectangular patch for the purpose of obtaining the quasi-static terms.

The U_{nm}s and the $V_{nm}^{\pm}$s satisfy the sequence of differential equations

$$U_{nm}'' + (\alpha_{0m})^2 U_{nm} = -\sum_{\ell=0}^{n-1} [a_{n-\ell,m}]^2 U_{\ell m} - f(u) U_{n-1,m}$$

$$V_{nm}'' - (\alpha_{0m})^2 V_{nm} = \sum_{\ell=0}^{n-1} [a_{n-\ell,m}]^2 V_{\ell m} - g(v) V_{n-1,m}$$

where again the superscript $\pm$ on the V has been dropped for convenience. In addition to the boundary conditions listed in Table 1, the following conditions are applied to obtain a unique solution in the most convenient form:

$$N_{1m} = N_{2m} = \ldots = 0$$

$$V_{nm}^{\pm}(v_{\pm}) = 0, \qquad \text{for } n > 0$$

$$V_{0m}^{\pm}(v_{\pm}) = 1$$

These conditions can be applied since the multiplication of the U and V functions by any arbitrary function of k^2 leaves $G(u,v|u',v')$ unchanged, but *does change* the coefficients of the series in k^2 listed above; i.e., the latter expansions are *not* unique.

Using the solutions of these equations in (8) and (9) yields the expression

$$G_\ell(u,v|u',v') = jk\eta t \left\{ r_{\ell 0} + \sum_{m=0}^{\infty} r_{\ell m} \right\}$$

The r_{ij}s are

$$r_{00} = \frac{1}{S}\left\{ n_{10} - \frac{\Delta u}{S} W_{20} \right\}$$

$$r_{10} = \frac{1}{S}\left\{ n_{20} - n_{10}\frac{\Delta u}{S} W_{20} - \frac{\Delta u}{S} W_{30} + \left(\frac{\Delta u}{S} W_{20}\right)^2 \right\}$$

$$r_{0m} = \frac{2 n_{0m}}{\Delta u W_{0m}}, \qquad m > 0$$

$$r_{1m} = \left\{ n_{1m} - \frac{n_{0m} W_{1m}}{W_{0m}} \right\} \frac{2}{\Delta u W_{0m}}, \qquad m > 0$$

The ns are given by

$$n_{0m} = U_{0m}(u) U_{0m}(u') V_{0m}^{-}(v_<) V_{0m}^{+}(v_>)$$

$$n_{1m} = U_{1m}(u) U_{0m}(u') V_{0m}^{-}(v_<) V_{0m}^{+}(v_>) + U_{0m}(u) U_{1m}(u') V_{0m}^{-}(v_<) V_{0m}^{+}(v_>)$$
$$+ U_{0m}(u) U_{0m}(u') V_{1m}^{-}(v_<) V_{0m}^{+}(v_>) + U_{0m}(u) U_{0m}(u') V_{0m}^{-}(v_<) V_{1m}^{+}(v_>)$$

$$n_{20} = U_{20}(u) + U_{20}(u') + V_{20}^{-}(v_<) + V_{20}^{+}(v_>) + U_{10}(u) U_{10}(u') + V_{10}^{-}(v_<) V_{10}^{+}(v_>)$$
$$+ U_{10}(u) V_{10}^{-}(v_<) + U_{10}(u) V_{10}^{+}(v_>) + U_{10}(u') V_{10}^{-}(v_<) + U_{10}(u') V_{10}^{+}(v_>)$$

Microstrip Antennas

The range of u is $\Delta u = u_+ - u_-$; the range of v is $\Delta v = v_+ - v_-$, if finite. The summand of G_0 always contains the term of the form

$$\frac{jk\eta t}{2\pi} \mathcal{R}\left\{\frac{1}{m}e^{-maz}\right\}$$

where $\alpha_{0m} = m\alpha$ and $z = v_> - v_< + j(u - u')$. This term can be summed in closed form to yield

$$G_0(u,v\,|\,u',v') = \frac{jk\eta t}{2\pi}\left\{\alpha\frac{v_> - v_<}{2} - \ln|\alpha z| - \ln\left|\frac{\sinh(\alpha z/2)}{\alpha z/2}\right|\right\}$$
$$+ \text{remainder terms}$$

For source points (u', v') not on the edge of the patch the remainder terms of G_0 in the preceding expression are nonsingular as $(u,v) \to (u',v')$. The logarithmic singularity of the Green's function is explicitly extracted by this method. The remainder terms of G_0 and the terms of G_1 involve sums over m of terms of the form $e^{-m\alpha z_i}/(m\alpha)^\ell$, where z_i is a complex number whose real and imaginary parts are linear functions of $u, u', v,$ and v'. The series associated with these terms can be evaluated efficiently using the functions listed in Table 2.

This method has been used to obtain explicit formulas for the rectangular patches and the separable patches in the circular-cylinder coordinate system. This analysis can also be applied to the class of patches separable in parabolic- and elliptic-cylinder coordinate systems.

Rectangular and Circular Geometries—For this class of patches the expressions listed above simplify considerably because $(\alpha_m)^2$ can be chosen as a sequence *independent* of k as can the corresponding sequence of eigenfunctions U_m. In this case $U_{nm} \equiv 0$ and $(\alpha_{nm})^2 = 0$ for $n > 0$. For all patches in this class the large-m form of the summands r_{0m} and r_{1m} can be written as

$$r_{0m} \sim \tilde{r}_{0m} = D_1 \frac{1}{m}\left\{\sum_{\ell=1}^{3} e^{-m\alpha\sigma_\ell}\right\} U_m(u, u')$$

$$r_{1m} \sim \tilde{r}_{1m} = \frac{1}{m^2}\left\{\sum_{\ell=1}^{3}\left(D_{2\ell} + \frac{1}{m}D_{3\ell}\right)e^{-m\alpha\sigma_\ell}\right\} U_m(u, u')$$

where

$$D_1 = -\frac{1}{\alpha\Delta u}$$

$$D_{i+1,\ell} = -\frac{4}{\alpha\Delta u}\left[C_{i\ell} + \frac{B_{i-1}}{2\alpha}\right], \quad i = 1, 2; \quad \ell = 1, 2, 3$$

The C coefficients are given by

Table 2. Definition, Properties, and Approximations of $F_n(z)$

n	A	B	$F_n(z)$		
			$0 \leq \mathcal{R}z \leq A$ $0 \leq \mathcal{I}z \leq B$	$0 \leq \mathcal{R}z \leq A$ $B < \mathcal{I}z \leq \pi$	$\mathcal{R}z \geq A$
n			$\sum_{k=1}^{\infty} e^{-kz}/k^n, \quad F_n(z+j2\pi) = F_n(z), \quad F_n^*(z) = F_n(z^*)$		
1			$z/2 - \ln z - \ln\left[\dfrac{\sinh(z/2)}{z/2}\right]$		
2	1.0	2.2	$z\ln z + 1.64493 - z - \dfrac{1}{4}z^2 + \dfrac{1}{72}z^3 - \dfrac{1}{14400}z^5$	$-0.82247 + 0.69315w$ $-\dfrac{1}{4}w^2 + \dfrac{1}{24}w^3 - \dfrac{1}{960}w^5;$ $w = z - j\pi$	$e^{-z} + \dfrac{1}{4}e^{-2z} + \dfrac{1}{9}e^{-3z} + \dfrac{1}{16}e^{-4z}$
3	0.6	2.0	$-0.5z^2\ln z + 1.20206 - 1.64493z + \dfrac{3}{4}z^2 + \dfrac{1}{12}z^3 - \dfrac{1}{288}z^4$	$-0.90154 + 0.82247w - 0.34657w^2 + \dfrac{1}{12}w^3 - \dfrac{1}{96}w^4;$ $w = z - j\pi$	$e^{-z} + \dfrac{1}{8}e^{-2z} + \dfrac{1}{27}e^{-3z} + \dfrac{1}{64}e^{-4z}$

$$C_{11} = \frac{1}{2}[A_1^+ + A_1^-], \qquad C_{21} = \frac{1}{2}[A_2^+ + A_2^-], \qquad C_{12} = \frac{1}{2}[A_1^+ - A_1^-]$$

$$C_{22} = C_{23} = C_{21}, \qquad C_{13} = -C_{12}$$

and the A and B coefficients are listed in Table 3 for rectangular and annular patches. (These coefficients appear in the large-m asymptotic formulas of the Vs and the Ws, respectively.) The function $U_m(u, u')$ is

$$U_m(u, u') = \begin{cases} \cos(m\alpha u)\cos(m\alpha u') & \text{for rectangular, annular-sector,} \\ & \text{and circular-sector patches} \\ \cos[m\alpha(u - u')] & \text{for circular and annular disk patches} \end{cases}$$

The summation of the r_{ij}s can be performed using

$$\sum_{m=1}^{\infty} r_{0m} = \sum_{m=1}^{\infty} [r_{0m} - \tilde{r}_{0m}] + D_1 \sum_{\ell=1}^{3} \mathscr{S}_{1\ell}$$

$$\sum_{m=1}^{\infty} r_{1m} = \sum_{m=1}^{\infty} [r_{1m} - \tilde{r}_{1m}] + \sum_{\ell=1}^{3} [D_{2\ell}\mathscr{S}_{2\ell} + D_{3\ell}\mathscr{S}_{3\ell}]$$

where

$$\mathscr{S}_{i\ell} = \frac{1}{\tau}\mathscr{R}\sum_{k=1}^{\tau} F_i(\alpha z_{k\ell})$$

$$\tau = \begin{cases} 1 & \text{for annular and circular patches} \\ 2 & \text{for rectangular, circular-sector, and annular-sector patches} \end{cases}$$

Table 3. Asymptotic Expansion Coefficients for Rectangular and Annular Patches

	Rectangular	Annular
A_1^+	$-\frac{1}{4\alpha}(b - v)$	$-\frac{a^2}{8\alpha}(1 - e^{2v})$
A_2^+	0	$\frac{1}{\alpha}A_1^+$
A_1^-	$-\frac{1}{4\alpha}v$	$\frac{1}{8\alpha}(b^2 - a^2 e^{2v})$
A_2^-	0	$-\frac{1}{\alpha}A_1^-$
B_0	$\frac{b}{4}$	$\frac{a^2 - b^2}{8}$
B_1	$\frac{1}{4\alpha}$	$\frac{a^2 + b^2}{8\alpha}$

$$z_{k\ell} = (\sigma_\ell + j\omega_k)$$
$$\omega_1 = u - u'$$
$$\omega_2 = u + u'$$
$$\sigma_1 = v_> - v_<$$
$$\sigma_2 = v_> + v_< - 2v_-$$
$$\sigma_3 = 2v_+ - (v_> + v_<)$$
$$F_i(z) = \mathcal{R} \sum_{m=1}^{\infty} \frac{e^{-mz}}{m^i}$$

The expressions listed in Table 2 can be used to efficiently evaluate $F_i(z)$. The remaining infinite sums of the difference between the exact and the asymptotic r_{ij}s converge in just a few (three or four) terms. The parameters needed to construct the exact r_{0m} and r_{1m} can be obtained from Table 4 and Table 5. To obtain the appropriate formulas for circular and circular-sector patches, one simply sets b to zero in the formulas for the annular patch, although this must be done taking appropriate limits in the final expression for the quasi-static terms. One could let b/a be a sufficiently small ratio, such as 10^{-6}, to obtain results for circular and circular-sector patches using the same computer program as for the annular and annular-sector patch. The parameter α used in these tables is

$$\alpha = \begin{cases} 1 & \text{for circular and annular disks} \\ \pi/\phi_0 & \text{for circular- and annular-sector patches (ϕ_0 is the sector angle)} \\ \pi/a & \text{for rectangular patches} \end{cases}$$

Multiport Impedance Parameters—Equation 4 can be used for computing Z_{ij} if the accelerated expression for G is used. To find the driving-point impedance one must average the singular part of the Green's function over the source current distribution. Simply setting $u_2 = u_1$ and $v_2 = v_1$ will yield an infinite inductive part of the impedance. On the other hand, averaging over the nonsingular parts of the Green's function yields nearly the same result as evaluating those parts at the center of the source distribution. Suppose the element is being fed with a coaxial probe of diameter d centered at (u_1, v_1). Then the driving-point impedance is obtained by

$$Z_{11} = \overline{G}(u_1, v_1 | u_1, v_1)$$

where $\overline{G}$ is obtained from G in (6) simply by replacing the z_{11} in $F_1(\alpha z_{11})$ in the expression for G_0 by $d/2$. If the feed distribution is a current strip of width w (such as when one feeds with a microstrip line), replace d by $w/2.24$ in the preceding expression.

5. Resonant Frequency

There are a number of different theories available for computing the resonant frequency of a microstrip element. One of the simplest, though not the most

Microstrip Antennas

Table 4. Solutions to the v-Differential Equations for Rectangular and Annular Patches

	Rectangular	Annular
V_{0m}^+	$\cosh ma(b-v)$	$\cosh mav$
V_{1m}^+	$-\dfrac{1}{2ma}(b-v)\sinh[ma(b-v)]$	$\dfrac{a^2}{8}\left\{\left[\dfrac{ma-2}{ma(ma-1)} - \dfrac{e^{2v}}{ma+1}\right]e^{mav} - \left[\dfrac{ma+2}{ma(ma+1)} - \dfrac{e^{2v}}{ma-1}\right]e^{-mav}\right\}, \quad am \neq 1;$ $-\dfrac{a^2}{16}e^{3v} + \dfrac{a^2}{4}(1-v)e^v - \dfrac{3a^2}{16}e^{-v}, \quad am = 1$
V_{10}^+	$-\dfrac{1}{2}(b-v)^2$	$-\dfrac{a^2}{4}e^{2v} + \dfrac{a^2}{4} + v\dfrac{a^2}{2}$
V_{20}^+	$\dfrac{1}{24}(b-v)^4$	$\dfrac{a^4}{8}\left[\dfrac{e^{4v}}{8} + \dfrac{e^{2v}}{2} - ve^{2v} - \dfrac{v}{2} - \dfrac{5}{8}\right]$
V_{30}^+	$-\dfrac{1}{720}(b-v)^6$	$\dfrac{a^6}{64}\left[-\dfrac{e^{6v}}{36} - \dfrac{e^{4v}}{2} + \dfrac{ve^{4v}}{2} + ve^{2v} + \dfrac{e^{2v}}{4} + \dfrac{1}{6}v + \dfrac{5}{18}\right]$
V_{nm}^-	$V_{nm}^+(v+b)$	$\left(\dfrac{b}{a}\right)^{2n} V_{nm}^+[v - \ln(b/a)]$

Table 5. Wronskians for Rectangular and Annular Patches

	Rectangular	Annular
W_{0m}	$-ma\sinh mab$	$ma\sinh[ma\ln(b/a)]$
W_{1m}	$\dfrac{1}{2ma}\sinh(mab) + \dfrac{b}{2}\cosh(mab)$	$-\dfrac{1}{8}\left\{\left[\dfrac{ma-2}{ma-1}b^2 - \dfrac{ma+2}{ma+1}a^2\right]e^{-ma\ln(b/a)} + \left[\dfrac{ma+2}{ma+1}b^2 - \dfrac{ma-2}{ma-1}a^2\right]e^{ma\ln(b/a)}\right\}, \quad ma \neq 1;$ $-\dfrac{1}{4}ab\ln\left(\dfrac{b}{a}\right) + \dfrac{3}{16}\left[\dfrac{a^3}{b} - \dfrac{b^3}{a}\right], \quad ma = 1$
W_{10}	b	$\dfrac{1}{2}(a^2 - b^2)$
W_{20}	$-\dfrac{1}{6}b^3$	$-\dfrac{a^4 - b^4}{16} - \dfrac{a^2b^2}{4}\ln\left(\dfrac{b}{a}\right)$
W_{30}	$\dfrac{1}{120}b^5$	$\dfrac{a^6 - b^6}{384} + \dfrac{3a^2b^2(a^2 - b^2)}{128} + \dfrac{a^2b^2(a^2 + b^2)\ln(b/a)}{32}$

accurate, is that provided by the cavity model. As discussed in Section 2, under "Cavity Model," the field under the microstrip element is distributed in about the same way as the field in the corresponding magnetic-walled cavity. The resonant wave number k_{mn} of the cavity mode $\psi_{mn}(u, v)$ is approximately the same as that of the corresponding antenna. The characteristic equation for this wave number is obtained from

$$\nabla \psi_{mn}(u, v) \cdot \hat{\mathbf{n}} = 0$$

for (u, v) on the edge of the patch. In terms of the functions used in Section 4, under "Overview," this characteristic equation is

$$\frac{\partial}{\partial v} V_m^-(v_+) = 0$$

where the equation $v = v_+$ describes at least a portion of the edge of the patch in the curvilinear coordinate system (u, v). See Table 1 for a list of characteristic equations for various patch shapes in the rectangular and circular classes.

The resonant frequency predicted by this equation has an accuracy typically in the range of from approximately 0 to 17 percent, depending on the particular shape of patch and mode being used and the thickness of the microstrip element [23]. The thicker the element, the larger the error is. The bandwidth of a thin microstrip antenna is small; designs based solely on the predictions of the cavity model may yield actual operating bands that completely miss the desired resonant frequency. Consequently, the designer must either use a more accurate theory, or actually build a prototype and subsequently adjust the size of the element. Some simple modifications to the cavity-model formulas for rectangular and circular disk elements are listed in the following subsections. More accurate theories for predicting resonant frequency can be found in the References. The option of building a prototype is not difficult. Any complicated feed circuitry can usually be replaced by a very simple feed.

The following five subsections present formulas and results that pertain to rectangular, circular-disk, circular-sector, annular, and annular-sector microstrip elements. The tables, formulas, and graphs are in terms of *normalized resonant frequency* f_n. To obtain a cavity-model estimate of the resonant frequency f_{res} (in megahertz) for your patch, use the formula

$$f_{res} = \frac{4775}{(\epsilon_r \mu_r)^{1/2}} \frac{1}{L} f_n$$

where L is the "characteristic length" (in centimeters) given in the following sections.

Rectangular Patch

See Table 1 for the geometrical parameters for this and the following patches. The characteristic length is $L = a$. The normalized resonant frequency of the (m, n)th mode is

$$f_n = \left[(m\pi)^2 + \left(\frac{n\pi a}{b}\right)^2 \right]^{1/2}$$

Microstrip Antennas

One can usually obtain a much better estimate [23] of the resonant frequency if one replaces a and b by $a + 2\Delta L$ and $b + 2\Delta L$, where ΔL is given in (1) in Section 2.

Circular-Disk Patch

The characteristic dimension is the disk radius, a. The normalized resonant frequencies of the (m, n)th modes are listed in Table 6. A better estimate for resonant frequency can often be obtained if one replaces a by [24]

$$a\left[1 + \frac{2t}{\pi a \epsilon_r}\left(\ln\left(\frac{\pi a}{2t}\right) + 1.7726\right)\right]^{1/2}$$

Circular-Sector Patch

The characteristic dimension is the sector radius, a. The normalized resonant frequencies for the (m, n)th modes as a function of sector angle ϕ_0 can be read off of the graph in Fig. 18. The resonant frequencies of the $(0, n)$ modes are the same as those of the circular disk. Note that the resonant frequency of the (sm, n)th mode of a patch with a sector angle of $s\phi_0$, for $s = 1, 2, \ldots, \ell$ such that $\ell\phi_0 < 2\pi$, has the same resonant frequency as the (m, n)th mode of a patch with a sector angle of ϕ_0.

Annular Patch

The characteristic dimension is the outer radius, a. The normalized resonant frequencies for the $(m, 1)$th modes as a function of b/a can be obtained from the graph in Fig. 19.

Annular-Sector Patch

The characteristic dimension is the outer radius, a. The normalized resonant frequency for the $(1, 1)$th mode as a function of b/a and sector angle ϕ_0 can be obtained from the graphs in Fig. 20. The resonant frequency of the $(0, n)$ mode is the same as that for the annulus. Also, the same comment under "Circular-Sector Patch" above about the equivalence involving the (sm, n)th mode holds here as well.

Table 6. Normalized Resonant Frequencies for a Circular-Disk Microstrip Antenna

Mode		Resonant Frequency			
m \ n	0	1	2	3
0	0	3.832	7.016	10.173
1		1.841	5.331	8.536
2		3.054	6.706	9.969
3		4.201	8.015	11.346

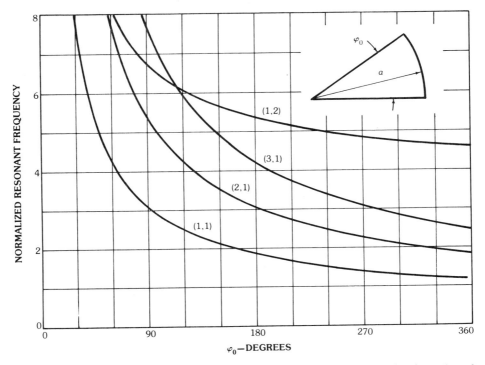

Fig. 18. Normalized resonant frequency of the (1, 1), (2, 1), (3, 1), and (1, 2) modes of a circular-sector microstrip element versus sector angle ϕ_0.

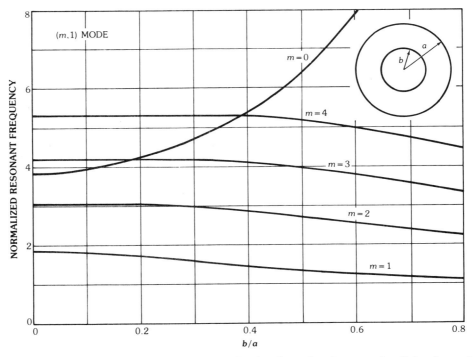

Fig. 19. Normalized resonant frequency of the $(m, 1)$ mode of an annular disk microstrip element versus b/a for $m = 0, 1, 2, 3, 4$.

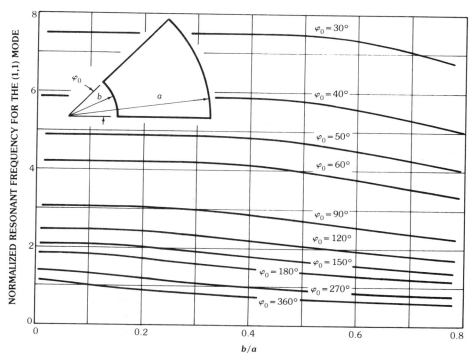

Fig. 20. Normalized resonant frequency of the $(1, 1)$ mode of an annular-sector microstrip element versus b/a for various sector angles ϕ_0.

6. Efficiency

The efficiency of a microstrip element is given by

$$\text{efficiency} = \frac{1/Q_r}{1/Q_r + 1/Q_{sw} + \delta + \Delta/t} \tag{10}$$

where Q_r is the radiative (space-wave) quality factor, Q_{sw} is the surface-wave quality factor, δ is the dielectric loss tangent of the substrate, and Δ is the skin depth in the patch and ground plane. The quality factors can be found in the way described in Section 2, under "Cavity Model." The dielectric loss tangent is usually available from the printed-circuit board manufacturer although the *effective* conductivity of the metal cladding from which Δ can be computed usually is not quoted. An optimistic guess for this conductivity is the conductivity of the pure metal cladding itself. This is usually too high because of the surface roughness of the cladding. A simple experimental method [25] of estimating these parameters from a sample of printed circuit board is available.

The surface-wave quality factor was excluded from the numerator of (10) since the power carried by the surface wave is ultimately scattered at the truncation of the dielectric substrate and combines with the space wave in a way that is not easy to control. A plot based on the cavity model of the efficiency *versus* dielectric thickness is shown in Fig. 21. The fractions of dielectric loss, metal loss, and surface-wave loss to the radiated power are also plotted. One notes that the metal loss is dominant for very thin elements. As the element thickness increases, both metal loss and dielectric loss decrease rapidly. A slowly increasing amount of power is contained in the surface wave. With the efficiency defined by (10), the cavity model indicates that there is a thickness at which maximum efficiency can be obtained. If the dielectric is abruptly truncated at the edge of the patch, or if the substrate is air, then the surface-wave term is zero and the efficiency increases with increasing thickness.

7. Matching

Microstrip elements are typically narrow-band antennas. In most applications the bandwidth limitations are due to an impedance mismatch to the feeding circuitry outside of a narrow band. Usually the pattern itself is stable over a wider band. There are exceptions to this principle; these are pointed out in the applications section, Section 9. This section focuses on matching the antenna impedance.

The simplest form of matching is to choose the feed point of the element

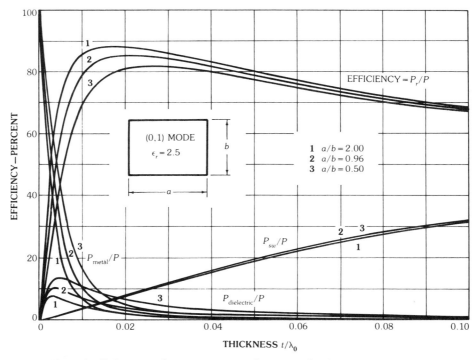

Fig. 21. Plot of efficiency and percent power loss contributions from metal-conduction, dielectric-heating, and surface-wave loss components versus thickness for a rectangular microstrip element with a relative dielectric constant of 2.5 and with three different aspect ratios a/b.

wisely. It was pointed out in Section 4, under "Resonant Impedance," that the resonant resistance of the antenna varies with feed point [see (4)]. By choosing the feed location (u', v') where the resonant resistance is equal to the feed-line impedance (or where it is slightly larger), one can match the element over a limited band. The bandwidth is inversely proportional to the quality factor of the antenna.

This optimum feed location is not unique. For some modes of some elements a suitable feed point can be found on the edge of the patch. For example, one can excite the $(0, 1)$ mode of a rectangular element as illustrated in Fig. 22a and Fig. 22b with the same effect. In other elements, such as a simple, unloaded circular disk, it is impossible to find an optimal feed point on the edge. One must feed at a point interior to the patch. This can be done by using a coaxial feed. If a microstripline feed is desired, then one can feed the element as illustrated in Fig. 22c. Even if it is possible to feed on an edge it still sometimes preferable to feed as in Fig. 22c. Such symmetrical feeding can reduce the excitation of nonresonant modes which can cause a small cross-polarized component of field. The slots cut to inset the microstrip feed should be aligned with the lines of current of the mode being excited to minimize disturbing the modal distribution. The amount that the maximum input resistance is reduced by using this feeding technique will depend on the width of the slots. Other techniques for feeding the element are possible besides coaxial feeds and microstripline feeds. The reader is referred to Section 10, "References," for these.

The matching technique described above does not extract the largest possible bandwidth from the element. It is important for the designer to recognize that there are fundamental theoretical limitations [26] to the bandwidth achievable from

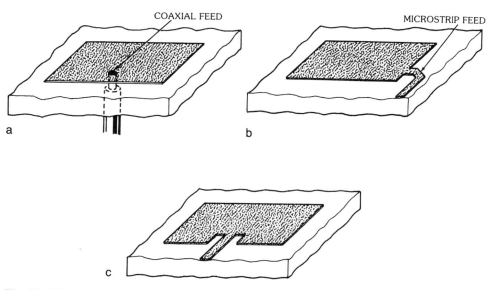

Fig. 22. Three simple matching schemes for a rectangular microstrip element. (*a*) Inset coaxial feed. (*b*) Microstripline feed on nonradiating sides of a patch driven in the $(0, 1)$ mode. (*c*) Inset microstripline feed.

these elements using lossless matching networks. Nevertheless, for a 2:1 vswr a bandwidth improvement by a factor of 3.86 is theoretically possible and an improvement by a factor of 3.18 is practical [8]. This bandwidth improvement can be obtained by feeding through an external matching network. For example, a network as illustrated in Fig. 23, with properly selected element values, will increase the best 2:1 vswr, bandwidth of the patch element alone by a factor of about 2.84. Adding another series LC pair can raise this factor to 3.18. There is little point to using more than three LC pairs of matching elements.

The section of the network in Fig. 23 that is enclosed by the dashed line represents the model of the input impedance of the microstrip element. One notes that part of the inductance required by the first matching section is supplied by the microstrip element itself. This inductive part can be significant, particularly for thicker elements and elements fed at high-inductance points.

The price for this improvement is a reduction in overall efficiency due to unavoidable losses in the matching filter. A greater price is the space that the matching network must occupy. The network can be realized using microstripline or stripline elements. If the latter is used, the network can be placed in a second layer of the circuit board under the patch.

How to actually choose the element values and how to realize them using microstripline or stripline circuits does not fall within the scope of this chapter. The reader is directed to look under the "Matching" category (17) of the References for publications on matching circuit design.

8. Loaded Microstrip Elements

The behavior (pattern, impedance, resonant frequency) of a microstrip element can be adjusted by placing lumped reactive loads between the patch and the ground plane [4]. This has been used for a number of applications discussed in Section 9. A simple theory is presented in this section. It will allow the designer to creatively use reactive loading of elements to control their properties.

A simple example provides the most efficient means of presenting the theory. Suppose that one has a microstrip element shorted at the point (u_1, v_1) as illustrated at the left side of Fig. 24. The resonant frequency of the element and the modal field distribution is approximately the same as that of the corresponding ideal cavity

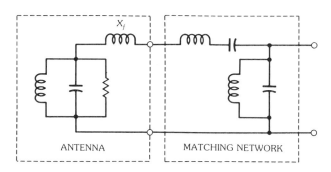

Fig. 23. External matching circuit for a microstrip element using two matching sections.

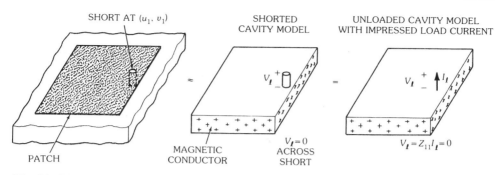

Fig. 24. Modeling a loaded microstrip element. Shorted element (left) is modeled by shorted cavity (middle) and then by an *empty* cavity (right) driven by the load current I_ℓ.

model of the antenna (the middle of Fig. 24). The modal field is the source-free field of the shorted cavity. Associated with this source-free field is an induced electric current I_ℓ on the surface of the shorting probe. The equivalence principle allows the removal of this probe as long as its surface current is impressed in its place (the right side of Fig. 24). The result is an *empty* (i.e., unloaded cavity) element with an impressed electric source current. The voltage induced by this current must be $Z_{11}(f) I_\ell$, where f is the frequency. Since the cavity was shorted at this point, this voltage must vanish. Thus, at the frequency of the source-free mode, $Z_{11}(f)$ must be zero. That is, the resonant frequencies of the shorted cavity are the zeros of the driving-point impedance.

As discussed in Section 4, under "Simple Feed Models," the driving-point impedance Z_{11} can be written as the sum of a resonant impedance, jX_{res}, and a feed-reactive term, jX_f. (In this analysis the effective loss tangent is not introduced since the modal distribution and resonant frequency are not strongly dependent on this parameter. Thus Z_{11} in this context is purely reactive.) The characteristic equation for the resonant frequency is

$$X_{\text{res}} = -X_f$$

Fig. 25 is a plot of X_{res} and $-X_f$ versus frequency. The resonant frequency of the shorted element is at the intersection of these two curves. Since X_f is typically inductive the effect of shorting the cavity is to raise its resonant frequency.

The modal field distribution Φ_{MN} of the loaded element is

$$\Phi_{MN}(u,v) = \left\{ \sum_A \frac{\psi_{mn}(u,v)\,\psi_{mn}(u_1,v_1)}{(k_{\ell MN})^2 - (k_{mn})^2} \right\} \bigg/ \left\{ \sum_A \frac{[\psi_{mn}(u_1,v_1)]^2}{[(k_{\ell MN})^2 - (k_{mn})^2]^2} \right\}^{1/2} \quad (11)$$

where $k_{\ell MN}$ is the resonant wave number of the loaded element. This distribution is normalized so that the integral of the square of $\Phi_{MN}(u,v)$ over the patch is unity. The series in the numerator of (11) is best computed using the accelerated expressions for $G(u,v\,|\,u',v')$ found in Section 4, under "Efficient Computation of Impedance Parameters." The Q of the loaded antenna can be found from Φ_{MN} in the way described in Section 2, under "Cavity Model." From this Q, an effective

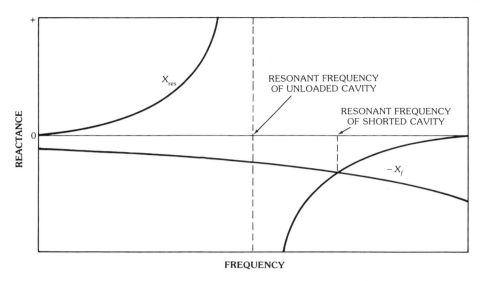

Fig. 25. Plot of X_{res} and $-X_f$ versus frequency. Intersection of the curves shows the resonant frequency of a shorted element.

loss tangent of the loaded element can be found and the resonant part of the impedance of the loaded element can be determined. The expression for this is

$$Z_{res} = -jk\eta t \frac{[\Phi_{MN}(u_2, v_2)]^2}{k^2(1 - j/Q) - (k_{\ell MN})^2} \qquad (12)$$

where (u_2, v_2) is the location of the feed point of the loaded element. The feed-reactive part of the input impedance at the resonant frequency of the loaded element is given by

$$X_f = \text{Im}\left\{-\frac{(Z_{12})^2}{Z_{11}} + Z_{22}\right\}$$

where Z_{11}, Z_{12}, and Z_{22} are the z parameters of the unloaded cavity. A typical input impedance and plot of the edge magnetic current distribution for a shorted element is shown in Fig. 26.

If the load is not a short but a reactive load, X_ℓ, then the characteristic equation for the resonant frequency is just

$$X_{res} = -X_f - X_\ell \qquad (13)$$

The normalized loaded-element resonant mode is still given by (11) and the corresponding resonant part of the input impedance is still given by (12). The feed-reactive part of the input impedance to the loaded element, $X_{\ell f}$, is given by

$$X_{\ell f} = \text{Im}\left\{\frac{-(Z_{12})^2}{Z_{11} + jX_\ell} + Z_{22}\right\}$$

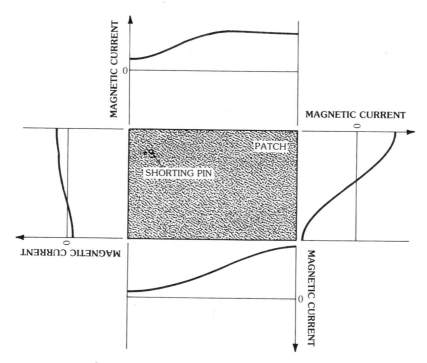

Fig. 26. The edge magnetic current distribution and input impedance at two feed locations of a shorted rectangular element of dimensions 11.43 × 7.62 × 0.152 cm with relative dielectric constant of 2.43 (*a*) Computed magnetic current distribution for a short located at coordinates (1.52, 6.10) cm. (*b*) Computed and measured input impedance with a feed located at coordinates (2.54, 1.27) cm. (*c*) Computed and measured input impedance with a feed located at coordinates (8.89, 6.35) cm. (*After Richards and Lo [4], reprinted with permission of Hemisphere Publishing Corporation*)

If there are L loads with reactances $X_1, X_2, \ldots, X_L$, located at (u_1, v_1), $(u_2, v_2), \ldots, (u_L, v_L)$, then the characteristic equation for the resonant frequency is that the determinant of the matrix

$$\mathbf{C} = \begin{bmatrix} jX_1 + Z_{11} & Z_{12} & \ldots & Z_{1L} \\ Z_{12} & jX_2 + Z_{22} & \ldots & Z_{2L} \\ \vdots & & & \\ Z_{1L} & Z_{2L} & \ldots & jX_L + Z_{LL} \end{bmatrix}$$

must vanish. To find the normalized resonant mode of this general loaded element, first compute

$$\mathbf{I} = -\mathbf{A}^{-1}\mathbf{B}$$

where $\mathbf{I}$ is the $(L-1)$-tuple $(I_2, I_3, \ldots, I_L)$, $\mathbf{A}$ is the $(L-1) \times (L-1)$ matrix obtained from $\mathbf{C}$ by removing its first row and column, and $\mathbf{B}$ is the $(L-1)$-tuple $(Z_{12}, Z_{13}, \ldots, Z_{1L})$. Then one must form the series

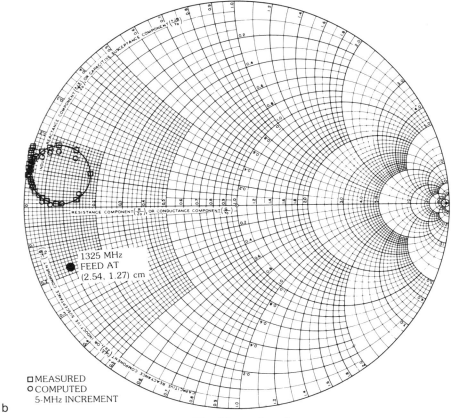

Fig. 26, *continued.*

$$\Psi_{mn} = \psi_{mn}(u_1, v_1) + \sum_{i=2}^{L} I_i \psi_{mn}(u_i, v_i)$$

The normalized resonant mode of the loaded cavity is then given by (11) with the $\psi_{mn}(u_1, v_1)$ replaced by Ψ_{mn}. The resonant part of the input impedance of this multiply loaded element is again given by (12). Its feed-reactive part at resonance is

$$X_f = Z_{L+1,L+1} - \mathbf{Z}^t \mathbf{C}^{-1} \mathbf{Z}$$

where Z is the L-tuple $(Z_{1,L+1}, Z_{2,L+1}, \ldots, Z_{L,L+1})$. The $Z_{i,L+1}$s are the z parameters of the unloaded element with ports at $(u_1, v_1), (u_2, v_2), \ldots, (u_L, v_L)$ and with the feed port at (u_{L+1}, v_{L+1}).

9. Applications

In this section a number of selected applications of microstrip elements are presented. To the extent possible, simple design formulas for the applications are given. More importantly, the physical concepts which support the applications

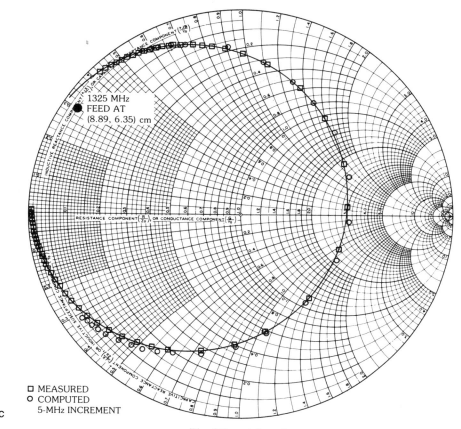

Fig. 26, *continued.*

are presented to aid the reader in developing his or her own designs. Because of the physical insight afforded by it, and the relative simplicity of the theory, the design concepts are derived from the cavity model of the microstrip element.

Circular Polarization

There are two ways of achieving circular polarization (CP) from microstrip elements. The most conventional way, and the way that has the widest CP bandwidth, is to feed the element with two or more feeds in phase quadrature. It is somewhat surprising that under the conditions described in the following subsection one can feed the microstrip element with just a *single* feed and also obtain CP. While no quadrature hybrid circuits are required by this second method, the CP bandwidth is much smaller than even the impedance bandwidth of the element. Common to both methods of producing CP is the requirement that the element be of such a shape that it possesses *two* degenerate modes resonant at the CP operating frequency.

Single-Feed CP—To see how a single-feed, CP patch works, consider the example of a nearly square patch illustrated in Fig. 27. Assume that the dimensions *a* and *b*

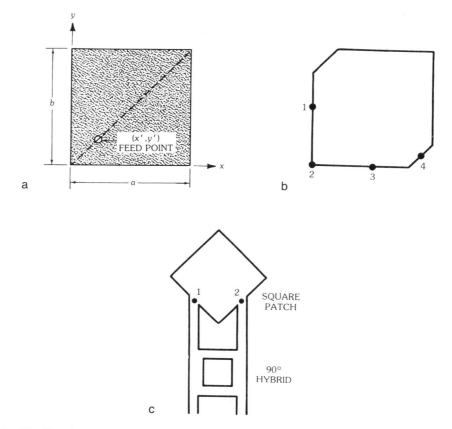

Fig. 27. Circular-polarization elements. (*a*) Nearly square element. (*b*) Square element perturbed at opposite corners. (*After Richards, Lo, and Harrison [5]*, © *1981 IEEE*) (*c*) Square element with quadrature feed.

of the patch are so nearly the same that the bands of the resonant $(0,1)$ and $(1,0)$ modes overlap significantly. In the broadside direction, the $(0,1)$ mode produces an electric far field E_y polarized in the y direction. The $(1,0)$ mode produces an x-polarized far field E_x. These fields satisfy the approximate proportionalities

$$E_x = a\frac{\sin(\pi y'/b)}{k^2(1 - j/Q) - (k_{01})^2}, \qquad E_y = a\frac{\sin(\pi x'/a)}{k^2(1 - j/Q) - (k_{10})^2}$$

The proportionality constant a is identical in both cases if the fields are measured at the same distance in the broadside direction. The effective loss tangent, $1/Q$, is the same in both cases since one of the two nearly degenerate modes is just the other one rotated by 90°. The terms $k_{01} = \pi/b$ and $k_{10} = \pi/a$ are the resonant wave numbers of the $(0,1)$ and $(1,0)$ modes. Suppose for the time being that the feed point (x', y') is along the diagonal of the element so that $y'/b = x'/a$. Then the ratio of the x and y components of the field is approximately

Microstrip Antennas

$$\frac{E_y}{E_x} \cong \frac{k(1 - j/2Q) - k_{10}}{k(1 - j/2Q) - k_{01}} \tag{14}$$

In order to achieve CP, the magnitude of this ratio should be 1 and its phase $\pm 90°$. Since Q is typically large for microstrip elements (at least 10 or larger in most cases), the complex phasor $1 - j/2Q$ is almost parallel to the real axis. A plot of the locus of the point $k(1 - j/2Q)$ in the complex plane as k varies over the very narrow band between k_{10} and k_{01} is illustrated in Fig. 28. One can see from this figure that the condition for CP will be met when the phasors in the numerator and the denominator of (14) have equal length and intersect at right angles. This picture immediately gives in geometrical terms the conditions under which this can occur: (1) The difference $k_{01} - k_{10}$ must be $\bar{k}/Q$. (2) The operating frequency must be midway between the resonant frequencies of the $(0, 1)$ and $(1, 0)$ modes. The first condition yields the simple formula

$$a = b(1 + 1/Q) \tag{15}$$

Feeding the antenna along the lower-left-corner-to-upper-right-corner diagonal will produce left-hand CP. Fed along the opposite diagonal, the element will produce right-hand CP.

This geometrical picture also quickly demonstrates why exciting CP in this way is extremely narrowband. For an axial ratio less than or equal to AR (specified in dB), the CP bandwidth in percent is approximately equal to

$$\text{percent bandwidth} \cong 12 \times AR/Q \tag{16}$$

Equations 15 and 16 have been found to work remarkably well even for elements with as low a value of Q as 10.

The results discussed above pertain to CP along the broadside direction. However, the CP actually remains quite good over a fairly large angular region, as the experimental results shown in Fig. 29 illustrate. These patterns were measured over a very large ground plane. If the ground plane is small, even better CP performance can be obtained.

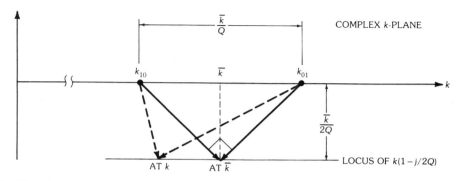

Fig. 28. Phasor diagram illustrating the conditions necessary for single-feed CP for a nearly square patch fed along a diagonal.

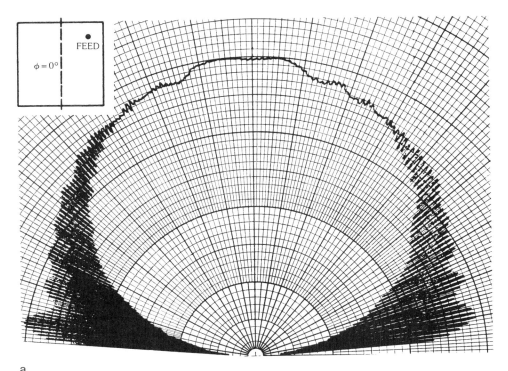

a

Fig. 29. Spinning-dipole CP patterns measured over a very large ground plane for a nearly square microstrip element. (*a*) Elevation pattern in the $\phi = 0°$ plane. (*b*) Elevation pattern in the $\phi = 90°$ plane. (*After Richards, Lo, and Harrison [5], © 1981 IEEE*)

One can also feed the element off the diagonal. In this case the ratio of the dimensions of the element are related by

$$a = b\left(1 + \frac{A + 1/A}{2Q}\right),$$

where

$$A = \frac{\cos(\pi y'/b)}{\cos(\pi x'/a)}$$

If one understands the physical mechanism for producing CP, one can see how there are many different ways to produce CP using this mechanism. For example, one could begin with a square patch and then disturb the symmetry by trimming a pair of opposite corners as illustrated in Fig. 27b. In this case the two near-degenerate modes can be thought of as hybrids formed by taking the sum and difference of the $(0,1)$ and $(1,0)$ modes. The sum-hybrid mode would be unaffected by the trimming while the difference-hybrid mode's resonant frequency would be raised slightly. By feeding at either point 1 or point 3 as indicated in

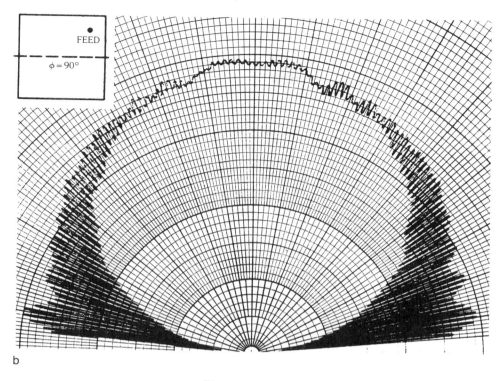

Fig. 29, *continued.*

Fig. 27b, one can achieve CP if the resonant wave numbers of the hybrid modes satisfy the conditions stated above. Similarly, one can achieve CP by disturbing the symmetry of a circular-disk element by adding tabs or by making it slightly elliptical. Adding loads to an otherwise symmetrical element can also be used to disturb the symmetry of otherwise precisely degenerate modes to produce CP. This latter technique is considered in more detail in this section, under "Polarization-Agile Elements."

Multiple-Feed CP Elements—One can generate the required phase quadrature between orthogonal far-field components by using external quadrature hybrids. In this case the element should support exactly degenerate modes instead of the nearly degenerate modes called for by single-feed CP elements. Each mode is excited independently by placing its respective feed at a node of the other mode. For example, Fig. 27c illustrates a dual-feed, square, CP microstrip element. Port 1 is located at a point where mode $(0, 1)$ is zero. Thus port 1 does not load or excite any part of mode $(0, 1)$. Similarly, the second feed port 2 is located where mode $(1, 0)$ is zero.

Because the bandwidth of the external phase shifter is typically larger than even the impedance bandwidth of the microstrip element itself, the CP bandwidth of this type of element will exceed the impedance bandwidth. The price, of course, is the extra volume required by the hybrid and the reduction of efficiency due to its losses.

An interesting application is the conical-beam CP element [21]. This element uses the even and odd (2, 1) modes of the circular-disk microstrip element. The two degenerate modal fields are

$$J_2(k_{21}\varrho)\cos 2\phi \quad \text{and} \quad J_2(k_{21}\varrho)\sin 2\phi, \qquad k_{21} = 3.054/a$$

Each of these modes has nodal planes every 45°. The nodal planes of the odd (sine) mode are 22½° from those of the even mode. By placing feeds in the configuration illustrated in Fig. 30a, one can achieve the theoretical CP pattern

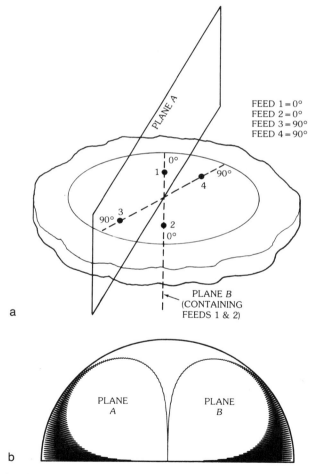

Fig. 30. Circular-disk antenna using degenerate (2, 1) modes to produce a conical CP beam. (*a*) Element and feed phasing to suppress unwanted modes. (*b*) Computed spinning dipole pattern for an element mounted on an infinite grounded $0.01\lambda_0$-thick dielectric slab. The assumed relative dielectric constant is 2.5. Pattern is plotted on a 40-dB scale.

illustrated in Fig. 30b. Four feeds were used in this case to help suppress the excitation of other lower- and higher-order modes.

In principle, one could incorporate these techniques for obtaining CP with the dual-band techniques described next to produce dual-band CP elements.

Dual-Band Elements

This section describes the physical and design principles of dual-band microstrip elements using reactive loads. Other loading techniques using stacked elements have been reported [27] for obtaining dual-band operation. For these the reader is directed to Section 10, "References." The discussion in this section uses the theory of loaded microstrip elements presented in Section 8.

Suppose one loads an element with a lossless resonator of reactance X_ℓ. The characteristic equation for the resonant frequency of the loaded element is (13). This equation is graphically represented in Fig. 31a. The intersection of the two solid curves occurs at the resonant frequency of the loaded element. It is clear from the figure that by adjusting the resonant frequency of the load to be somewhat less than that of the unloaded element, *two* intersections will occur representing *two* resonant frequencies of the loaded element.

By adjusting the resonant frequency of the load, one can shift the locations of the two resonances of the loaded patch. It has been found that there is an optimal pair of frequencies of all that could be achieved by adjusting the resonant frequency of the load. For this optimal pair the impedance characteristics over the dual bands are almost identical as the measured results in Fig. 32 illustrate.

The magnetic current distributions at the two resonant frequencies can be computed using (11) and have been plotted in Fig. 33 for the element corresponding to Fig. 32. In both cases the magnetic current distribution can be viewed as the sum of the magnetic current distribution of the parent (0, 1) unloaded cavity mode and a magnetic current distribution which varies rapidly at points nearest the load. The latter component does not produce a very strong radiated field. Thus the radiation patterns of the two modes of the dual-band elements are expected to be nearly the same, and nearly the same as the pattern of the parent (0, 1) mode of the unloaded cavity. In fact, the measured results verify that this is the case. A typical pattern is given in Fig. 33c.

The resonator load could be an opened or shorted transmission line of an appropriately chosen length. If the transmission line in the resonant load has its characteristic impedance raised, then the reactance plot of Fig. 31a changes to that of Fig. 31b. This illustrates one way to decrease the distance between the dual bands. A simpler way is to simply move the location of the load nearer to a nodal plane of the unloaded cavity mode. When this is done the reactance plot changes to that illustrated in Fig. 31c. By using these two techniques one can produce dual bands spaced as close to each other as one wishes.

The transmission-line resonator is best chosen to be a shorted line since it can be made less than one quarter-wavelength long at the lowest of the two operating bands. This line can be microstripline. It is best located symmetrically so as not to produce a large cross-polarized component of the radiated field. If it is necessary to achieve the desired band separation, the line can be inset as illustrated in Fig. 34.

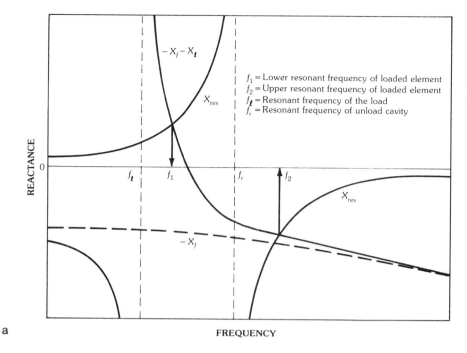

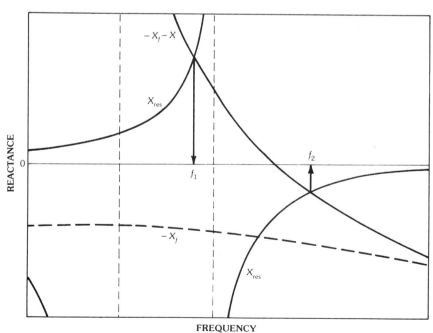

Fig. 31. Reactance plots for determining the resonant frequency of dual-band elements. (*a*) The X_{res} and $-X_\ell - X_f$ curves for a dual-band element using a shorted stub as its reactive load. (*b*) Same curve but for a higher characteristic impedance shorted stub load. (*c*) Same curve but for the load placed nearer to a nodal curve of the resonant mode of the unloaded cavity.

Microstrip Antennas

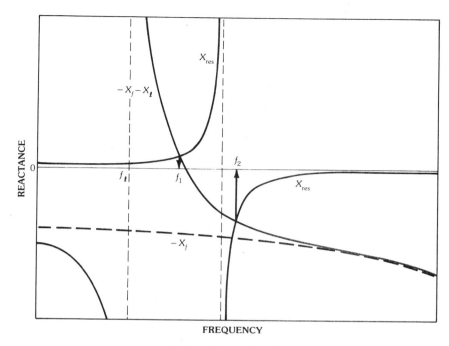

c

Fig. 31, *continued.*

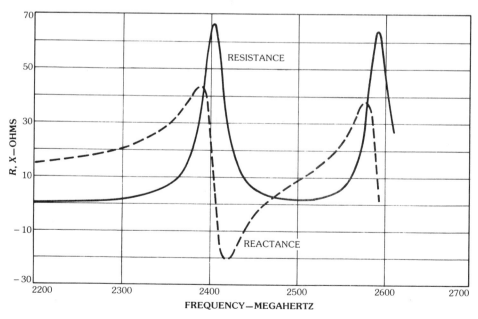

Fig. 32. The measured input resistance and reactance plotted versus frequency for an optimally-loaded rectangular dual-band element. The patch dimensions are 6.0 × 4.0 cm mounted on a 0.079-cm dielectric with relative dielectric constant of 2.17 and loaded by a 5.1-cm, 24.3-Ω shorted stub inset 1.5 cm from the patch edge. (*After Richards, Davidson, and Long [28],* © *1985 IEEE*)

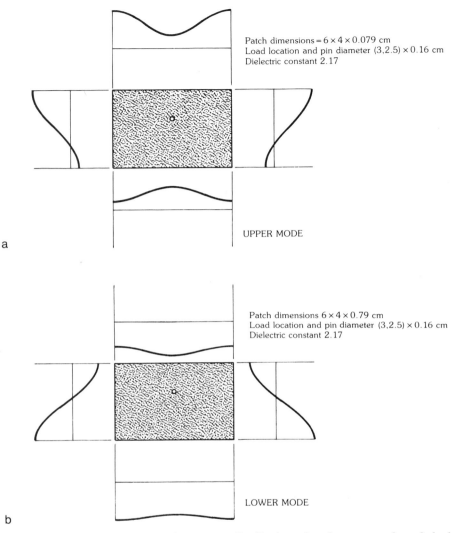

Fig. 33. The computed magnetic current distributions for the two modes of dual-band element corresponding to Fig. 32 and the measured *E*- and *H*-plane patterns. (*a*) Magnetic current distribution for the upper mode. (*b*) Magnetic current distribution for the lower mode. (*c*) Measured pattern for lower mode. (*After Richards, Davidson, and Long [28], © 1985 IEEE*) (Patterns for upper mode are essentially identical.)

Frequency-Agile Elements

As was suggested in Section 8, shorting posts placed between the patch and the ground plane can be used to alter the resonant frequency of a microstrip element [3]. This can be made dynamic by implementing the shorts as pin diodes which can be biased to represent an rf open or short.

It was seen from Fig. 25 in Section 8 that, because the feed reactance is

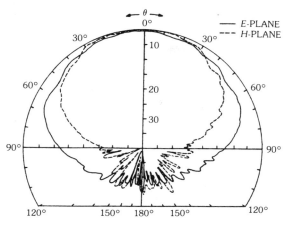

c

Fig. 33, *continued.*

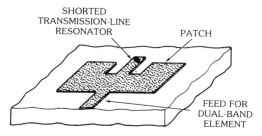

Fig. 34. A monolithic (except for shorting pin) dual-band element with an inset load location to produce narrow-band separation. (Completely monolithic version omits the short and adjusts the line length.)

inductive, shorting the element will raise its resonant frequency. If another short is added, the resonant frequency will be increased even further. In fact, it has been found that one can adjust the resonant frequency of a rectangular element over a 1.7:1 range [3] while still maintaining good vswr and a stable pattern. An implementation of the loaded element theory for multiple loads should give the designer a good quantitative guide for producing a tunable or frequency-agile element.

One problem with this approach is that the diodes should be biased independently to step in small increments through the tuning range. This will require that the diodes be capacitively coupled to the patch, or capacitively coupled to the ground plane. This may require an additional layer of dielectric to support the capacitors and dc biasing lines.

An alternative is to use varactor diodes as the loads and adjust their capacity by adjusting their reverse bias. If all diodes are biased identically, no dc blocking capacitors on the patch are necessary. One need only bias the patch itself through a

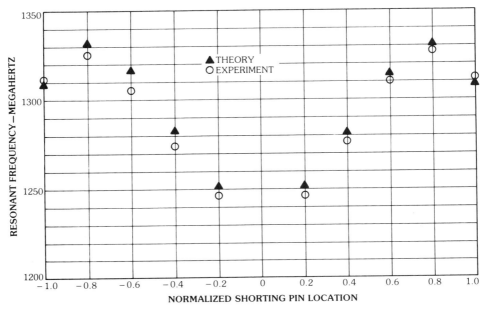

a

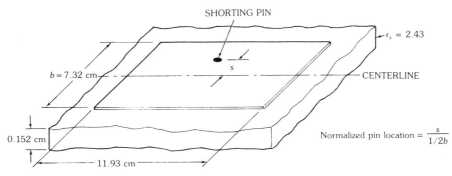

b

Fig. 35. Measured and computed resonant frequencies of a rectangular element with a single shorting pin versus normalized pin location. (*a*) Frequency variation. (*After Richards and Lo [4], reprinted with permission of Hemisphere Publishing Corporation*) (*b*) Element.

microwave inductive choke. This choke could be on the same level as the patch itself or be placed further back in the microwave feed network.

A good quantitative guide to the design of such frequency-agile elements can be obtained by writing a computer program implementing the theory presented in Section 8. The following qualitative guidelines are also available. The closer to a nodal curve (curve along which the modal field is zero) of the unloaded resonant cavity mode the short is placed, the thinner the plot of the X_{res} versus frequency becomes, as illustrated by the transformation of X_{res} in Fig. 31a to Fig. 31c. Thus the loaded element's resonant frequency will be closer to the unloaded cavity's resonant frequency. This is obvious since a load placed right on a nodal curve has no effect on the unloaded cavity mode. Consequently, the farther away from a nodal curve the load is placed, the higher the resonant frequency of the loaded element will be. However, because the feed reactance increases very rapidly close to the edge of a patch for reasons discussed in Section 4, under "Feed Reactance," the loading effect of a short is diminished somewhat when it is placed very near or right on the patch edge. These trends are illustrated in Fig. 35. These results pertain to a single short; similar qualitative results will also hold for multiple shorts.

In applications where a broadside maximum is not desired, but a monopolelike pattern is sought, one can use the dc mode of a patch element. Of course, the dc mode must somehow be excited (resonated) not at dc but at the desired operating frequency. This can be done by adding shorting pins using exactly the same principles as discussed above.

Polarization-Agile Elements

This application is similar in spirit to the frequency-agile element except the shift in frequency is very small. It was seen in the discussion of single-feed CP that a single-feed CP element could be produced if the ratio of the difference of the two near-degenerate modes to their average value could be made to be $1/Q$. This can easily be done by placing pin diodes as illustrated in Fig. 36a. Based on the cavity model the distance s of the two diodes from the center of the patch should be

$$s = a \left(\frac{a}{t} \frac{X_f}{\eta} \frac{1}{2\pi Q} \right)^{1/2}$$

By turning diode 1 on (rf short) and diode 2 off (rf open), the element will produce right-hand CP. By switching the roles of the two diodes, one produces left-hand CP. By biasing both diodes either on or off, linear polarization is obtained.

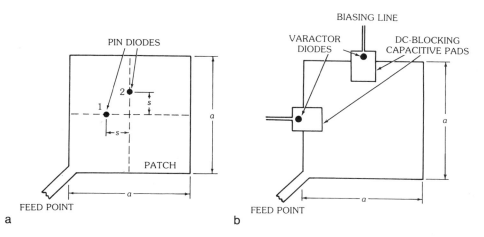

Fig. 36. Polarization-agile elements. (*a*) Square element using pin diodes. (*b*) Square element using varactor diodes.

Another version of the same idea can be accomplished by placing varactor diodes on the edge of the patch as in Fig. 36b. In this case one can sweep the element continuously through whatever axial-ratio elliptical polarization (including circular and linear, of course) one wants. Two diodes are used and biased such that the mean frequency of the two near-degenerate modes is constant. This leaves the operating frequency of the elliptical polarization independent of the selected axial ratio. Clearly, these ideas can be extended to differently shaped patches.

10. References

The following is an index of publications by category. The numbers in square brackets given in each category refer to the numbered references which follow category (19).

(1) Reviews of literature and books on microstrip antennas: [1, 31, 32, 33, 34].
(2) Approximate analytical treatments of microstrip antennas: [1, 3, 4, 5, 10, 11, 17, 22, 24, 28, 29, 31, 34, 47, 48, 49, 50, 56, 62, 63, 64, 69, 73, 74].
(3) Numerical analysis of microstrip elements: [35, 36, 37, 38, 39, 40, 41, 42, 43, 44, 45, 46, 58, 67, 71].
(4) Rectangular microstrip elements: [1, 3, 4, 5, 9, 10, 11, 19, 22, 28, 30, 31, 33, 34, 35, 36, 37, 41, 42, 43, 46, 47, 48, 50, 53, 54, 62, 64, 68, 69, 77, 78, 79, 83].
(5) Circular-disk microstrip patches: [1, 2, 5, 11, 16, 21, 24, 29, 31, 33, 34, 38, 39, 56, 57, 63, 67, 70, 75].
(6) Circular-sector microstrip patches: [11, 74].
(7) Annular-ring microstrip patches: [11, 40, 55, 74].
(8) Annular-sector microstrip patches: [11, 74].
(9) Elliptical microstrip antennas: [11, 17, 20, 54, 59].
(10) Circularly polarized microstrip antennas: [1, 3, 5, 17, 20, 21, 30, 31, 33, 34, 54, 78].
(11) Adaptable microstrip elements: [3, 4, 5, 6, 7].
(12) Dual-band microstrip elements: [27, 28, 31, 33].
(13) Parametric studies (e.g., thickness, dielectric, loss): [1, 23, 31, 34, 51, 52, 57, 60, 61, 63, 65, 68].
(14) Covered microstrip elements: [65, 72].
(15) Feeding microstrip elements: [22, 31, 66, 79, 80].
(16) Mutual coupling between microstrip elements: [30, 43, 67, 68, 73, 76].
(17) External matching of microstrip elements: [8, 81].
(18) Effects of dielectric truncation, ground-plane truncation, and bending of element: [13, 19, 69, 70, 77].
(19) Ferromagnetic substrates and superstrates: [2, 4, 65].

[1] J. R. James, P. S. Hall, and C. Wood, *Microstrip Antenna Theory and Design*, IEE, London: Peter Peregrinus, 1981.
[2] N. D. S. K. Chowdhury, and J. S. Chatterjee, "Circular microstrip antenna on ferrimagnetic substrate," *IEEE Trans. Antennas Propag.*, vol. AP-31, no. 1, pp. 188–190, January 1983.
[3] D. H. Schaubert, F. G. Farrar, A. Sindoris, and S. T. Hayes, "Microstrip antennas with frequency agility and polarization diversity," *IEEE Trans. Antennas Propag.*, vol. AP-29, no. 1, pp. 118–123, January 1981.
[4] W. F. Richards and Y. T. Lo, "Theoretical and experimental investigation of a microstrip radiator with multiple lumped linear loads," *Electromagnetics*, vol. 3, no. 3–4, pp. 371–385, July–December 1983.
[5] W. F. Richards, Y. T. Lo, and D. D. Harrison, "An improved theory of microstrip antennas with applications," *IEEE Trans. Antennas Propag.*, vol. AP-29, no. 1, pp. 38–46, January 1981.
[6] W. F. Richards and S. A. Long, "Impedance control of microstrip antennas utilizing reactive loading," *Proc. Intl. Telemetering Conf.*, pp. 285–290, Las Vegas, 1986.

[7] W. F. Richards and S. A. Long, "Adaptive pattern control of a reactively loaded, dual-mode microstrip antenna," *Proc. Intl. Telemetering Conf.*, pp. 291–296, Las Vegas, 1986.

[8] D. A. Pascen, "Broadband microstrip matching techniques," *Proc. 1983 Antenna Appl. Symp.*, EM Laboratory, Univ. of Illinois, Urbana (no page numbers in digest).

[9] R. E. Munson, "Conformal microstrip antennas and microstrip phased arrays," *IEEE Trans. Antennas Propag.*, vol. AP-22, no. 1, pp. 74–78, January 1974.

[10] A. G. Derneryd, "A theoretical investigation of the rectangular microstrip antenna element," *IEEE Trans. Antennas Propag.*, vol. AP-26, no. 4, pp. 532–535, July 1978.

[11] Y. T. Lo, D. Solomon, and W. F. Richards, "Theory and experiment on microstrip antennas," *IEEE Trans. Antennas Propag.*, vol. AP-27, no. 2, pp. 137–145, March 1979.

[12] E. O. Hammerstad, "Equations for microstrip circuit design," *Proc. Fifth European Microwave Conf.*, pp. 268–272, September 1975.

[13] M. V. Scheider, "Microstrip dispersion," *Proc. IEEE*, vol. 60, no. 1, pp. 144–146, January 1972.

[14] R. F. Harrington, *Time-Harmonic Electromagnetic Fields*, New York: McGraw-Hill Book Co., 1961, p. 183.

[15] H. A. Wheeler, "Transmission-line properties of a strip on a dielectric plane," *IEEE Trans. Microwave Theory Tech.*, vol. MTT-13, pp. 172–185.

[16] S. L. Chuang, L. Tsang, J. A. Kong, and W. C. Chew, "The equivalence of the electric and magnetic surface current approaches in microstrip antenna studies," *IEEE Trans. Antennas Propag.*, vol. AP-28, no. 4, pp. 569–571, July 1980.

[17] L. C. Shen, "The elliptical microstrip antenna with circular polarization," *IEEE Trans. Antennas Propag.*, vol. AP-29, no. 1, pp. 91–94, January 1981.

[18] J. Huang, "The finite ground-plane effect on the microstrip antenna radiation patterns," *IEEE Trans. Antennas Propag.*, vol. AP-31, no. 4, pp. 649–653, July 1983.

[19] C. M. Krowne, "Cylindrical and rectangular microstrip antenna," *IEEE Trans. Antennas Propag.*, vol. AP-31, no. 1, pp. 194–199, July 1983.

[20] Y. T. Lo and W. F. Richards, "Perturbation approach to design of circularly polarized microstrip antennas," *Electron. Lett.*, vol. 17, no. 11, pp. 383–385, May 28, 1981.

[21] J. Huang, "Circularly polarized conical patterns from circular microstrip antennas," *1984 IEEE AP-S Symp. Dig.*, pp. 271–274.

[22] W. F. Richards, J. R. Zinecker, R. D. Clark, and S. A. Long, "Experimental and theoretical investigation of the inductance associated with a microstrip antenna feed," *Electromagnetics*, vol. 3, no. 3–4, pp. 327–346, July–December 1983.

[23] E. Chang, S. A. Long, and W. F. Richards, "An experimental investigation of electrically thick rectangular microstrip antennas," *IEEE Trans. Antennas Propag.*, vol. AP-34, no. 6, pp. 767–772, June 1986.

[24] L. C. Shen, S. A. Long, M. R. Allerding, and M. D. Walton, "Resonant frequency of a circular-disc, printed-circuit antenna," *IEEE Trans. Antennas Propag.*, vol. AP-25, no. 4, pp. 595–596, July 1977.

[25] W. F. Richards, Y. T. Lo, and J. Brewer, "A simple experimental method for separating loss parameters of a microstrip antenna," *IEEE Trans. Antennas Propag.*, vol. AP-29, no. 1, pp. 150–151, January 1981.

[26] R. M. Fano, "Theoretical limitations on the broadband matching of arbitrary impedances," *J. Franklin Inst.*, vol. 249, pp. 57–83, January 1950; pp. 139–154, February 1950.

[27] S. A. Long and M. D. Walton, "A dual-frequency stacked circular-disc antenna," *IEEE Trans. Antennas Propag.*, vol. AP-27, no. 2, pp. 270–273, March 1979.

[28] W. F. Richards, S. Davidson, and S. A. Long, "Dual-band, reactively loaded microstrip antennas," *IEEE Trans. Antennas Propag.*, vol. AP-33 (accepted for publication, 1985).

[29] W. C. Chew, J. A. Kong, and L. C. Shen, "Radiation characteristics of a circular microstrip antenna," *J. Appl. Phys.*, vol. 51, no. 7, pp. 3907–3915, July 1980.

[30] R. E. Munson, "Conformal microstrip antennas and microstrip phased arrays," *IEEE*

Trans. Antennas Propag., vol. AP-22, no. 1, pp. 74–78, January 1974.

[31] K. R. Carver and J. W. Mink, "Microstrip antenna technology," *IEEE Trans. Antennas Propag.*, vol. AP-29, no. 1, pp. 2–24, January 1981.

[32] R. J. Malloux, J. F. McIlvenna, and N. P. Kernweis, "Microstrip array technology," *IEEE Trans. Antennas Propag.*, vol. AP-29, no. 1, pp. 25–37, January 1981.

[33] *Proceedings of the Workshop on Printed-Circuit Antenna Technology* (Keith Carver, Technical Program Chairman), October 17–19, 1979, New Mexico State University, N Mex.

[34] I. J. Bahl and P. Bhartia, *Microstrip Antennas*, Dedham, Mass.: Artech House, 1980.

[35] P. K. Agrawal and M. C. Bailey, "An analysis technique for microstrip antennas," *IEEE Trans. Antennas Propag.*, vol. AP-25, no. 6, pp. 756–759, November 1977.

[36] E. H. Newman and P. Tulyathan, "Analysis of microstrip antennas using moment methods," *IEEE Trans. Antennas Propag.*, vol. AP-29, no. 1, pp. 47–53, January 1981.

[37] T. Itoh and W. Menzel, "A full-wave analysis method for open microstrip structures," *IEEE Trans. Antennas Propag.*, vol. AP-29, no. 1, pp. 63–68, January 1981.

[38] W. C. Chew and J. A. Kong, "Analysis of a circular microstrip disk antenna with a thick dielectric substrate," *IEEE Trans. Antennas Propag.*, vol. AP-29, no. 1, pp. 68–76, January 1981.

[39] K. Araki and T. Itoh, "Hankel transform domain analysis of open circular microstrip radiating structures," *IEEE Trans. Antennas Propag.*, vol. AP-29, no. 1, pp. 84–89, January 1981.

[40] S. M. Ali, W. C. Chew, and J. A. Kong, "Vector Hankel transform analysis of annular-ring microstrip antenna," *IEEE Trans. Antennas Propag.*, vol. AP-30, no. 4, pp. 637–644, July 1982.

[41] M. D. Deshpande and M. C. Baily, "Input impedance of microstrip antennas," *IEEE Trans. Antennas Propag.*, vol. AP-30, no. 4, pp. 645–650, July 1982.

[42] M. C. Baily and M. D. Deshpande, "Integral equation formulation of microstrip antennas," *IEEE Trans. Antennas Propag.*, vol. AP-30, no. 4, pp. 651–656, July 1982.

[43] D. M. Pozar, "Input impedance and mutual coupling of rectangular microstrip antennas," *IEEE Trans. Antennas Propag.*, vol. AP-30, no. 6, pp. 1191–1196, November 1982.

[44] R. Chandra and K. C. Gupta, "Segmentation method using impedance matrices for analysis of planar microwave circuits," *IEEE Trans. Microwave Theory Tech.*, vol. MTT-29, no. 1, pp. 71–74, January 1981.

[45] P. C. Sharma and K. C. Gupta, "Desegmentation method for analysis of two-dimensional microwave circuits," *IEEE Trans. Microwave Theory and Techniques*, vol. MTT-29, no. 10, pp. 1094–1097, October 1981.

[46] D. M. Pozar, "Improved computational efficiency for the moment method solution of printed dipoles and patches," *Electromagnetics*, vol. 3, no. 3–4, pp. 299–307, July–December 1983.

[47] D. C. Chang, "Analytical theory of an unloaded rectangular microstrip patch," *IEEE Trans. Antennas Propag.*, vol. AP-29, no. 1, pp. 54–62, January 1981.

[48] E. F. Kuester, R. T. Johnk, and D. C. Chang, "The thin-substrate approximation for reflection from the end of a slab-loaded parallel-plate waveguide with application to microstrip patch antennas," *IEEE Trans. Antennas Propag.*, vol. AP-30, no. 5, pp. 910–917, September 1980.

[49] E. F. Kuester and D. C. Chang, "A geometrical theory for the resonant frequencies and Q-factors of some triangular microstrip patch antennas," *IEEE Trans. Antennas Propag.*, vol. AP-31, no. 1, pp. 27–34, January 1983.

[50] J. Venkataraman and D. C. Chang, "Imput impedance to a probe-fed rectangular microstrip patch antenna," *Electromagnetics*, vol. 3, no. 3–4, pp. 387–399, July–December 1983.

[51] D. M. Pozar, "Considerations for millimeter-wave printed antennas," *IEEE Trans. Antennas Propag.*, vol. AP-31, no. 5, pp. 740–747, September 1983.

[52] I. J. Bahl, P. Bhartia, and S. S. Stuchly, "Design of microstrip antennas covered with

a dielectric layer," *IEEE Trans. Antennas Propag.*, vol. AP-30, no. 2, pp. 314–318, March 1982.

[53] N. Das and S. K. Chowdhury, "Rectangular microstrip antenna on a ferrite substrate," *IEEE Trans. Antennas Propag.*, vol. AP-30, no. 3, pp. 499–502, May 1982 (plus correction, vol. AP-30, no. 6, p. 1268, November 1982).

[54] S. A. Long, L. C. Shen, D. H. Schaubert, and F. G. Farrar, "An experimental study of the circular-polarized elliptical printed-circuit antenna," *IEEE Trans. Antennas Propag.*, vol. AP-29, no. 1, pp. 95–99, January 1981.

[55] W. C. Chew, "A broadband annular-ring microstrip antenna," *IEEE Trans. Antennas Propag.*, vol. AP-30, no. 5, pp. 918–922, September 1982.

[56] S. Yano and A. Ishimaru, "A theoretical study of the input impedance of a circular microstrip disk antenna," *IEEE Trans. Antennas Propag.*, vol. AP-29, no. 1, pp. 77–83, January 1981.

[57] J. S. Dahele and K. Lee, "Effect of substrate thickness on the performance of a circular-disk microstrip antenna," *IEEE Trans. Antennas Propag.*, vol. AP-31, no. 2, pp. 358–360, March 1983.

[58] J. R. Mosig and F. E. Gardiol, "Analytical and numerical techniques in the Green's function treatment of microstrip antennas and scatterers," *IEEE Proc.*, vol. 130, pt. H, no. 2, pp. 175–182, March 1983.

[59] S. A. Long and M. W. McAllister, "The impedance of an elliptical printed-circuit antenna," *IEEE Trans. Antennas Propag.*, vol. AP-30, no. 6, pp. 1197–1200, November 1982.

[60] A. G. Derneryd and I. Karlsson, "Broadband microstrip antenna element and array," *IEEE Trans. Antennas Propag.*, vol. AP-29, no. 1, pp. 140–141, January 1981.

[61] J. W. Mink, "Sensitivity of microstrip antennas to admittance boundary variations," *IEEE Trans. Antennas Propag.*, vol. AP-29, no. 1, pp. 142–144, January 1981.

[62] P. Hammer, D. Van Bouchaute, D. Verschraeven, and A. Van de Capelle, "A model for calculating the radiation field of microstrip antennas," *IEEE Trans. Antennas Propag.*, vol. AP-27, no. 2, pp. 267–270, March 1979.

[63] A. G. Derneryd, "Analysis of the microstrip disk antenna element," *IEEE Trans. Antennas Propag.*, vol. AP-27, no. 5, pp. 660–664, September 1979.

[64] A. G. Derneryd, "Extended analysis of rectangular microstrip resonator antennas," *IEEE Trans. Antennas Propag.*, vol. AP-27, no. 6, pp. 846–849, November 1979.

[65] N. G. Alexopoulos and D. R. Jackson, "Radiation efficiency optimization for printed-circuit antennas using magnetic superstrates," *Electromagnetics*, vol. 3, no. 3–4, pp. 255–269, July–December 1983.

[66] J. Rivera and T. Itoh, "Analysis of a suspended patch antenna excited by an inverted microstrip line," *Electromagnetics*, vol. 3, no. 3–4, pp. 289–298, July–December 1983.

[67] T. M. Habashy and J. A. Kong, "Coupling between two circular microstrip disk resonators," *Electromagnetics*, vol. 3, no. 3–4, pp. 347–370, July–December 1983.

[68] C. M. Krowne, "Dielectric and width effect on H-plane and E-plane coupling between rectangular microstrip antennas," *IEEE Trans. Antennas Propag.*, vol. AP-31, no. 1, pp. 39–47, January 1983.

[69] E. Lier and K. R. Jakobsen, "Rectangular microstrip patch antennas with infinite and finite ground-plane dimensions," *IEEE Trans. Antennas Propag.*, vol. AP-31, no. 6, pp. 978–984, November 1983.

[70] S. B. De Assis Fonseca and A. J. Giarola, "Microstrip disk antennas, part I: efficiency of space wave launching," and "Microstrip disk antennas, part II: the problem of surface wave radiation by dielectric truncation," *IEEE Trans. Antennas Propag.*, vol. AP-32, no. 6, pp. 561–573, June 1984.

[71] Y. Suzuki and T. Chiba, "Computer analysis method for arbitrarily shaped microstrip antenna with multiterminals," *IEEE Trans. Antennas Propag.*, vol. AP-32, no. 6, pp. 585–590, June 1984.

[72] N. G. Alexopoulos and D. R. Jackson, "Fundamental superstrate (cover) effects on printed-circuit antennas," *IEEE Trans. Antennas Propag.*, vol. AP-32, no. 8,

pp. 807–815, August 1984.

[73] E. H. Van Lil and A. R. Van de Capelle, "Transmission-line model for mutual coupling between microstrip antennas," *IEEE Trans. Antennas Propag.*, vol. AP-32, no. 8, pp. 816–821, August 1984.

[74] W. F. Richards, J. D. Ou, and S. A. Long, "A theoretical and experimental investigation of annular, annular-sector, and circular-sector microstrip antennas," *IEEE Trans. Antennas Propag.*, vol. AP-32, no. 8, pp. 864–867, August 1984.

[75] K. F. Lee, K. Y. Ho, and J. S. Dahele, "Circular-disk microstrip antenna with an air gap," *IEEE Trans. Antennas Propag.*, vol. AP-32, no. 8, pp. 880–884, August 1984.

[76] P. B. Katehi and N. G. Alexopoulos, "On the modeling of electromagnetically coupled microstrip antennas—the printed-strip dipole," *IEEE Trans. Antennas Propag.*, vol. AP-32, no. 11, pp. 1179–1186, November 1984.

[77] K. R. Jakobsen, "The radiation from rectangular microstrip antennas mounted on two-dimensional objects," *IEEE Trans. Antennas Propag.*, vol. AP-32, no. 11, pp. 1255–1259, November 1984.

[78] Y. T. Lo, B. Engst, and R. Q. H. Lee, "Circularly polarized microstrip antennas," Antenna Applications Symposium, Allerton Park, Univ. of Illinois, September 22–24, 1984.

[79] P. S. Hall, C. Wood, and C. Garrett, "Wide-bandwidth microstrip antennas for circuit integration," *Electron. Lett.*, vol. 15, no. 15, pp. 458–460, July 19, 1979.

[80] J. R. James, P. S. Hall, C. Wood, and A. Henderson, "Some recent developments in microstrip antenna design," *IEEE Trans. Antennas Propag.*, vol. AP-29, no. 1, pp. 124–128, January 1981.

[81] G. L. Matthaei, L. Young, and E. M. T. Jones, *Microwave Filters, Impedance-Matching Networks, and Coupling Structures*, New York: McGraw-Hill Book Co., 1965.

[82] S. S. Zhong and Y. T. Lo, "Single-element rectangular microstrip antenna for dual-frequency operation," *Electron. Lett.*, vol. 19, no. 8, pp. 298–300, April 14, 1983.

[83] S. M. Wright and Y. T. Lo, "Efficient analysis for infinite microstrip dipole array," *Electron. Lett.*, vol. 19, no. 24, pp. 1043–1045, November 24, 1983.

Chapter 11

Array Theory

Y. T. Lo
University of Illinois

CONTENTS

1. Introduction — 11-5
2. General Formulation — 11-5
3. Linear Arrays — 11-8
    - Arrays with Prescribed Nulls   11-9
    - Binomial Arrays   11-10
    - Uniform Arrays   11-11
    - Dolph-Chebyshev Arrays   11-13
4. Linear Transformations in Antenna Arrays — 11-23
    - Linear Transformations in Array Geometry   11-23
    - Application to Planar Periodic Arrays   11-26
    - Nonuniform Excitation and Relation between Aperture Antenna and Discrete Array   11-29
    - Hexagonal Arrays   11-30
    - Periodic Arrays with Minimum Number of Elements   11-31
    - Transformation between Circular and Elliptical Arrays   11-33
    - Beam and Pattern Distortion Due to Scanning   11-36
    - Linear Transformations on Excitations   11-39
    - Circular Arrays   11-41
    - Cophasal Uniform Circular Arrays   11-46
    - Nonuniformly Excited Circular Arrays   11-47
    - Elliptical Arrays with Nonuniform Excitations   11-48
5. Planar Arrays — 11-48
    - Two-Dimensional Dolph-Chebyshev Arrays   11-49
    - A Few Major Results   11-52
    - General Discussion of the Transformation   11-57
6. Optimization of Directivity (D) and Signal-to-Noise Ratio (SNR) — 11-58
    - Formulation and Solution   11-58
    - Planar Array with Isotropic Elements or Vertical Dipoles in the (x, y) Plane   11-63

**Yuen T. Lo** is a professor and the director of the Electromagnetics Laboratory (formerly the Antenna Laboratory) in the Electrical and Computer Engineering Department, University of Illinois at Urbana-Champaign. He is a member of the National Academy of Engineering, a Fellow of IEEE, and a member of the International Union of Radio Science. He received the 1964 IEEE AP-S Bolljahn Memorial Award, and the 1964 IEEE AP-S Best Paper Award, the 1979 IEEE AP-S Best Paper Award, the IEEE Centennial Medal, and the Halliburton Education Leadership award. He served as an AP-S AdCom member, the Chairman of the AP-S Education and Tutorial Papers Committee, and twice (1979–1982 and 1984–1987) as the IEEE AP-S National Distinguished Lecturer. Dr. Lo is an honorary professor of the Northwest Telecommunication Engineering Institute and also the Northwestern Polytechnical University, both at Xian, China. He has published over 100 technical articles in refereed journals covering a wide spectrum, from theoretical to experimental works. His works include large-antenna arrays, radiotelescopes, multiple-beam antennas, multiple scattering, antenna synthesis, antijamming antennas, antenna in plasmas, corrugated guides and horns, artificial dielectrics, and microstrip antennas. He designed the University of Illinois Radiotelescope, considered to be the world's largest antenna, in the early 1960s.

    A Typical Example for Maximum Directivity   11-65
    An Example for Maximum SNR   11-70
    Extensions   11-74
7. Pattern Synthesis in the Probabilistic Sense   11-76
8. References   11-86

1. Introduction

Single-element antennas are discussed in other chapters in the book but their performance is somewhat limited. To obtain high directivity, narrow beams, low side lobes, steerable beams, particular pattern characteristics, etc., commonly a group of antenna elements, called an *antenna array*, or simply *array*, is used. The design of an array involves mainly first the selection of elements and array geometry, and then the determination of the element excitations required for achieving a particular performance, sometimes under a given constraint. The realization of the desired excitation requires a detailed knowledge of element input impedance characteristics as well as the mutual impedance between any two elements in the given array environment. In general, this is a difficult problem which may be solved approximately for large arrays with an infinite array model and for small arrays in a two-element environment model. While discussions on these models for some simple elements can be found elsewhere in the book, in this chapter we confine our discussion to the problems stated above.

There is no reason why all elements in an array must be of the same type other than simplicity in fabrication and analysis. In fact, radioastronomers have made use of two or more existing radiotelescopes of different types for interferometry measurement. Furthermore, even with the same type of elements, the shape of current or aperture field distribution of the element near the edge of the array can be different from that of the element in the central portion of the array, depending on the array geometry, element spacing and orientation, and, of course, the element type. Generally such a difference may not cause a serious deterioration in the array performance in some applications.

There is also no particular reason why the array structure must be periodic other than the two reasons just stated above. In fact, a periodic structure can result in grating lobes, frequency sensitivity, blind angles, etc. Even a periodic structure of scatterers, such as those used in artificial dielectrics, possesses some interesting but undesirable properties, such as birefringence, anisotropy, and dispersion. For analytic simplicity, however, uniformly spaced arrays, particularly the linear ones, have been studied in a great detail [1–120]. At least for these arrays the performance can be predicted accurately. Therefore we shall consider uniformly spaced linear arrays first. With some understanding of linear arrays, for convenience in later discussions, we then consider a general transformation theory and its application to planar, circular, and elliptical arrays. Finally, a few array synthesis problems are discussed.

2. General Formulation

For radiation characteristics, only the far field is of interest. In all the following discussions the use of the far-field approximation will be understood. Let $\mathbf{A}(\mathbf{r})$ be the magnetic vector potential due to a typical element. Then

$$\mathbf{A}(\mathbf{r}) \cong \frac{\mu}{\pi} \frac{e^{-jkr}}{r} \int_V \mathbf{J}(\mathbf{r}') e^{jkr' \cos \xi} dV' \qquad (1)$$

where

$\mathbf{J}(\mathbf{r}')$ = the electrical current distribution in the element
$\mathbf{r}$ = the position vector of an observation point
$\mathbf{r}'$ = the position vector of a typical source point
V = the volume of the source element
dV' = a differential volume element of the source
$\cos \xi = \hat{\mathbf{r}} \cdot \hat{\mathbf{r}}'$
$k = \omega \sqrt{\mu \epsilon} = 2\pi/\lambda$ = free-space wave number

The important consequence of the far-field approximation is that the dependence of $\mathbf{A}(\mathbf{r})$ on $\mathbf{r}$ is separated into two parts: one depending on r only and the other on (θ, ϕ) through $\hat{\mathbf{r}}$ (or $\cos \xi$ in the integral). Thus, for a sphere with fixed r, we can write the factor $(jkr' \cos \xi)$ as $(j\mathbf{k} \cdot \mathbf{r}')$ and $\mathbf{A}(\mathbf{r})$ as $\mathbf{A}(\mathbf{k})$ with $\mathbf{k} = k\hat{\mathbf{r}}$, the direction in which $\mathbf{A}$ is to be evaluated. In practice, only the directional characteristic is of main interest; therefore, we shall focus our attention on the integral and rewrite (1) in the following form:

$$\mathbf{A}(\mathbf{r}) = \hat{\mathbf{r}} A_r + \hat{\boldsymbol{\theta}} A_\theta + \hat{\boldsymbol{\phi}} A_\phi \cong \frac{\mu e^{-jkr}}{4\pi r} [\hat{\mathbf{r}} f_r(\hat{\mathbf{k}}) + \hat{\boldsymbol{\theta}} f_\theta(\hat{\mathbf{k}}) + \hat{\boldsymbol{\phi}} f_\phi(\hat{\mathbf{k}})] \qquad (2a)$$

where

$$\begin{Bmatrix} f_\theta(\theta, \phi) \\ f_\phi(\theta, \phi) \end{Bmatrix} = \begin{Bmatrix} f_\theta(\hat{\mathbf{k}}) \\ f_\phi(\hat{\mathbf{k}}) \end{Bmatrix} = \begin{Bmatrix} \hat{\boldsymbol{\theta}} \\ \hat{\boldsymbol{\phi}} \end{Bmatrix} \cdot \int \mathbf{J}(\mathbf{r}') e^{j\mathbf{k} \cdot \mathbf{r}'} dV' \qquad (2b)$$

Then the far fields are

$$\mathbf{E}(\mathbf{r}) = -j\omega \mathbf{A} + \frac{1}{j\omega\epsilon} \nabla \nabla \cdot \mathbf{A} \cong -j\omega[\hat{\boldsymbol{\theta}}\hat{\boldsymbol{\theta}} + \hat{\boldsymbol{\phi}}\hat{\boldsymbol{\phi}}] \cdot \mathbf{A}$$

$$\mathbf{H}(\mathbf{r}) \cong \frac{1}{\eta} \hat{\mathbf{r}} \times \mathbf{E} \qquad (3)$$

where $\eta = \sqrt{\mu/\epsilon}$.

The angular-dependent parts $f_\theta(\theta, \phi)$ and $f_\phi(\theta, \phi)$ give the directional characteristics of E_θ and E_ϕ. It should be noted that f_θ and f_ϕ are complex functions. Their absolute values, or magnitudes, $|f_\theta(\theta, \phi)|$ and $|f_\phi(\theta, \phi)|$, are

Array Theory

generally referred to as the *pattern functions*; in fact, often one of them is termed the major component (or copolarization) pattern function and the other the cross-polarization pattern function. Their phases, namely $\angle f_\theta(\theta,\phi)$ and $\angle f_\phi(\theta,\phi)$, are referred to as the *phase pattern functions*. Obviously the significance of a phase pattern function should be weighted by its associated pattern function $|f(\theta,\phi)|$. The phase pattern is of interest in some applications, such as the determination of the phase center. Sometimes one is also interested in evaluating the patterns in terms of circularly, instead of linearly, polarized components. In that case

$$\mathbf{E}(\hat{r}) \propto \hat{\mathbf{L}} f_L(\theta,\phi) + \hat{\mathbf{R}} f_R(\theta,\phi) \tag{4}$$

where

$$\begin{aligned} f_L(\theta,\phi) &= f_\theta(\theta,\phi) - jf_\phi(\theta,\phi) = \text{LCP pattern} \\ f_R(\theta,\phi) &= f_\theta(\theta,\phi) + jf_\phi(\theta,\phi) = \text{RCP pattern} \end{aligned} \tag{5}$$

Finally, it may also be noted that, in practice, for simplicity, $f(\theta,\phi)$ is often called the pattern function, rather than the precise term *complex* pattern function.

For an array of N arbitrary elements, as shown in Fig. 1, the far field is

$$\mathbf{E}(\mathbf{r}) \cong \frac{-j\omega\mu e^{-jkr}}{4\pi r} \mathbf{f}(\theta,\phi) \tag{6}$$

where

$$\mathbf{f}(\theta,\phi) = \sum_{n=1}^{N} \mathbf{f}_n(\theta,\phi)$$

$$\mathbf{f}_n(\theta,\phi) = (\hat{\theta}\hat{\theta} + \hat{\phi}\hat{\phi}) \int_{\substack{n\text{th} \\ \text{element}}} \mathbf{J}_n(\mathbf{r}'_n) e^{jk\hat{\mathbf{r}}\cdot(\mathbf{r}'-\mathbf{r}_n)} dV'_n \tag{7}$$

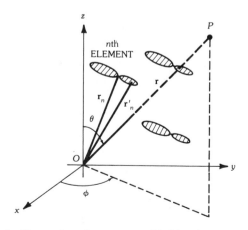

Fig. 1. Geometry of an array with identical elements.

$\mathbf{r}_n$ = a convenient reference point, such as the phase center, of the nth element

$\mathbf{r}'_n$ = a typical point on the nth element

$\mathbf{f}_n(\theta,\phi)$ = the nth-element pattern function

If all elements are identical and also identically oriented, with the assumption that the current distribution of all elements in the array have identical shape except for a constant multiplier, one can write

$$\mathbf{f}_n(\theta,\phi) = I_n \mathbf{f}_0(\theta,\phi) \tag{8}$$

where

$\mathbf{f}_0(\theta,\phi)$ = complex pattern function of a single element

I_n = relative complex excitation to the nth element

It should be noted again that element patterns in different array environments could be significantly different from each other (see Chapters 13 and 14). For large arrays or arrays with simple elements the neglect of this difference may still lead to a useful approximation. Then, using (8), one can rewrite (6) as follows:

$$\mathbf{E}(\mathbf{r}) \cong -\frac{j\omega\mu e^{-jkr}}{4\pi r}\mathbf{f}_0(\theta,\phi)F(\theta,\phi) \tag{9}$$

where

$$F(\theta,\phi) = \sum_n I_n e^{jk\hat{\mathbf{r}}\cdot\mathbf{r}_n} \tag{10}$$

is usually called the (complex) array factor, or the array pattern function for N isotropic point sources at $\{\mathbf{r}_n\}$. Equation 9 also states the *pattern multiplication principle*, namely, that the array pattern for N identical elements, similarly oriented, is equal to the product of the element pattern and the array factor (or the array pattern for isotropic sources). The study of an array usually implies the study of the array factor, which is often called simply the *array pattern function*. In the following we shall devote most of our discussions to this function.

3. Linear Arrays

Linear arrays occupy a unique position in array theory and have received great attention. This is probably because for uniformly spaced elements the pattern function can simply be expressed in terms of a polynomial for which the analytic tool is well-known. Let the elements be placed along the z axis with interelement spacing d; then for the nth element $\mathbf{r}_n = \hat{\mathbf{z}}nd$ and for an array of $(N+1)$ elements

Array Theory

$$F(\theta) = \sum_{n=0}^{N} I_n e^{jknd\cos\theta} = \sum_{n=0}^{N} I_n Z^n \qquad (11)$$

where

$$Z = e^{j\psi} \quad \text{and} \quad \psi = kd\cos\theta \qquad (12)$$

For the physically observable region $0 \leq \theta \leq \pi$, the function $F(\theta)$ is given by the value of the polynomial in (11) with Z only on a unit circle and its phase angle ψ bounded between $-kd$ and $+kd$, which is called the *visible region*. Polynomials have been thoroughly studied and it is therefore not surprising to find a large number of contributions in the literature on linear arrays. A few important results are summarized below.

Arrays with Prescribed Nulls

These are useful for antijamming and interference elimination. Let $\{\theta_n\}$, $n = 1, 2, \ldots, N$, be the set of null angles. Then the desired array pattern function is

$$F(\theta) = c \prod_{n=1}^{N} (Z - Z_n) = c[I_0 + I_1 Z + \cdots + I_N Z^N] \qquad (13)$$

where c = a constant, commonly a normalization factor such that

$\max|F(\theta)| = 1,$
$Z_n = e^{j\psi_n},$
$\psi_n = kd\cos\theta_n,$
I_n = the required complex excitation for the nth element

Comments:

(a) The magnitude pattern is

$$|F(\theta)| = 20\left(\log|c| + \sum_{n=1}^{N} \log|Z - Z_n|\right)$$

in decibels, where $|Z - Z_n|$ is the length between a typical point $Z(\theta)$ on the unit circle and Z_n, which is also on the circle as shown in Fig. 2.

(b) Excitations are given by the coefficients of the polynomial expansion as in (13).

(c) Because of the axial symmetry the 3D pattern is the generation of the above pattern function rotated about the array axis.

(d) To obtain a deep null at, say, θ_n, multiplicity of the root Z_n can be imposed. In fact, $(Z - Z_n)^p$, with p an integer, implies that all derivatives up to the $(p-1)$th order vanish at $Z = Z_n$.

(e) These excitations are determined only by the desired null directions without

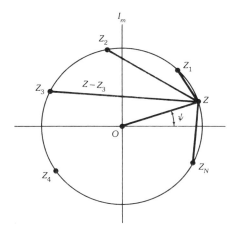

Fig. 2. Zeros of an array polynomial on a unit circle.

considering the desired signal. For a more meaningful solution see Section 6, where signal-to-interference ratio is to be maximized.

(*f*) Since each element can be a subarray, the element could be designed against the jamming signals while the entire array is designed for receiving the desired signal, or vice versa, provided the desired signal is in a direction different from that of any of the jamming signals.

(*g*) Null steering can be achieved by varying Z_n on a unit circle, but in general this change will alter excitations of many elements.

(*h*) Of course, some roots $\{Z_n\}$ need not be on the unit circle nor in the visible region. In that case the array will not have deep nulls at $\{\theta_n\}$.

Binomial Arrays

If all roots $\{Z_n\}$ coincide and are equal to Z_1, then

$$F(\theta) = c(Z - Z_1)^N$$
$$= c[Z^N + N(-Z_1)Z^{N-1} + \frac{N(N-1)}{2}(-Z_1)^2 Z^{N-2}$$
$$+ \cdots + C_k^N(-Z_1)^k Z^{N-k} + \cdots + (-Z_1)^N] \tag{14}$$

where

$Z_1 = e^{jkd \cos \theta_1}$

$\theta_1 =$ null angle

$C_k^N =$ binomial coefficient, which can be found more easily by using the Pascal triangle

Thus $I_1 = 1$, $I_2 = NZ_1$, $\ldots$, $I_k = C_k^N(-Z_1)^k$, $\ldots$, $I_N = (-Z_1)^N$.

Comments:

(a) The magnitude pattern in decibels is

$$|F(\theta)| = 20(\log|c| + N\log|Z - Z_1|)$$

(b) This array has no "side lobes" if $d \leq \lambda/2$. For example, if $d = \lambda/2$ and $\theta_1 = 0$, then $Z_1 = -1$ and all element excitations are in phase with weight $\{C_k^N\}$. The pattern has only a broadside beam, irrespective of the array length. If $\theta_1 = \pi/2$, $Z_1 = 1$, all element excitations will be alternatively opposite in phase. The pattern will consist of two separate beams along $\theta = 0$ and $\theta = \pi$, respectively, and a null at $\theta = \pi/2$. Again there are no other lobes, no matter how long the array is. If d is near or larger than λ, grating lobes appear partially or totally.

Uniform Arrays

When all excitations are equal, say 1,

$$F(\theta) = \sum_{n=0}^{N-1} Z^n = \frac{Z^N - 1}{Z - 1} \tag{15}$$

If the array center is chosen as the origin,

$$F(\theta) = \frac{\sin(N\psi/2)}{\sin(\psi/2)} \tag{16}$$

where

$$\psi = kd\cos\theta \tag{17}$$

The normalized pattern function of (16) as a function of ψ for a few values of N is shown in Fig. 3. The graphical method for determining $F(\theta)$ through (17) is demonstrated in Fig. 4 for two cases: broadside and scan angle θ_0.

Comments:

(a) As will be seen in Section 6 the uniform array for d nearly equal to or greater than $\lambda/2$ has a directivity close to maximum. For $d < \lambda/2$, a substantial increase in directivity over that of the uniform excitation is only possible in theory but not in practice. Thus the directivity of a uniformly excited array is about the maximum achievable in practice and therefore is often used as a reference for comparing directivities of various designs.

(b) The beam maximum of (16) appears at $\theta = \pi/2$ and, therefore, the array is called a *broadside array*. The beam can be steered to any direction θ_0 if the excitation I_n contains a progress phase factor $nkd\cos\theta_0$, for all ns, as will be seen in a general discussion in Section 4, under "Application to Planar Periodic Arrays." This phase shift results in a translation of ψ by $kd\cos\theta_0$. Thus, instead of (17),

$$\psi = kd(\cos\theta - \cos\theta_0) \tag{18}$$

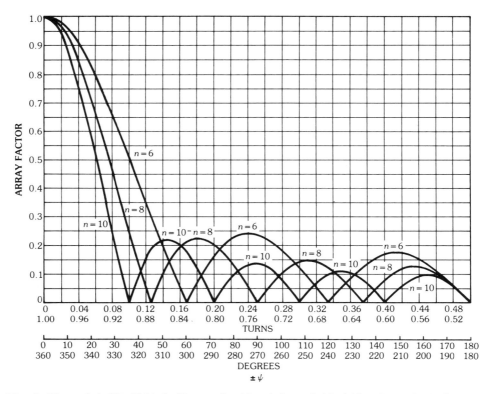

Fig. 3. Plots of $\sin(N\psi/2)/\sin(\psi/2)$ vs ψ for $N = 6$ through 11. (*After Kraus [60]*, © *1950; reprinted with permission of McGraw-Hill Book Co.*)

In other words, what happened at $\theta = \pi/2$ before introducing the phase shift will happen afterwards at $\theta = \theta_0$. The pattern function does not change with respect to ψ except for a simple translation. But this is not true when plotted against θ due to the nonlinear functional dependence of ψ on θ. Therefore it is often more convenient to study the pattern as a function of ψ, or the so-called u space in Section 4. When $\theta_0 = 0°$, the main beam will be along the array axis and the array will be called *end-fire*.

(c) A few important formulas, such as directivity D, half-power beamwidth, beamwidth between first nulls, null angular position, and side lobe maximum position, for broadside and end-fire uniform arrays are listed in Table 1.

(d) Also listed in Table 1 are the formulas for Hansen-Woodyard end-fire arrays. Hansen and Woodyard [50] found that for a long uniform end-fire array when element spacing is small the directivity in the $\theta = 0°$ direction can be increased from that of the ordinary end-fire if the phase shift per element is [4]

$$\beta = -\left(kd + \frac{2.94}{N}\right) \cong -\left(kd + \frac{\pi}{N}\right) \tag{19}$$

and if

Array Theory

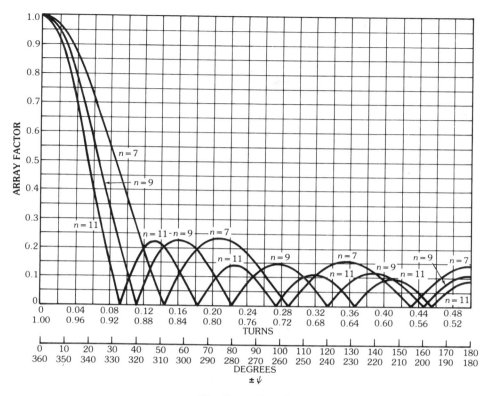

Fig. 3, *continued.*

$$|\psi| = |kd\cos\theta + \beta|_{\theta=0°} \cong \pi/N$$
$$|\psi| = |kd\cos\theta + \beta|_{\theta=180°} \cong \pi \tag{20}$$

It will be seen in Section 6 that this excitation does not give the maximum directivity and, in fact, for $d \cong \lambda/2$ or larger, its directivity is even smaller than that of the ordinary end-fire.

Dolph-Chebyshev Arrays

It is well known that the antenna aperture distribution and pattern function (in the wave vector **k** space, or the u space) are a Fourier transform pair. Since for any practical antenna the aperture function must vanish outside a finite region, its Fourier transform, namely the pattern function, is analytic. An analytic function which is zero (or constant) over a finite region must vanish (or be constant) everywhere. This result, if translated into antenna language, implies that any physically realizable antenna pattern must either have a broad beam covering the entire visible region, or have side lobes. From the Parseval's theorem the L_2 norm of the pattern function must be finite. Thus if one side lobe is pushed down, somewhere else the pattern function must go up. Therefore a meaningful optimum

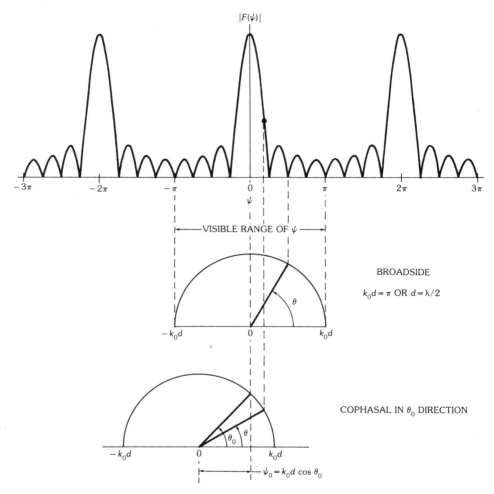

Fig. 4. Graphical method of constructing the radiation pattern from a universal pattern for a uniformly excited broadside array and a uniformly excited cophasal array with the beam in the θ_0 direction.

design would be one in which no one side lobe has a level higher than any others, i.e., equal side lobe levels. For an aperture antenna this is impossible because of the finite L_2 norm of the pattern function for $|u| < \infty$, and the pattern function must approach zero as $|u|$ approaches infinity. Thus the best design one can hope for will be one in which only a *finite* number of side lobes are of equal level. Taylor [100] has provided a solution to this problem which is discussed in Chapter 4.

Since a discrete array can be considered as an aperture antenna sampled at a discrete set of points, all previously stated results are generally applicable. However, when all elements are uniformly spaced, the pattern function becomes periodic (see Chapters 13 and 14). In this case the L_2 norm can be defined over a period, and the Parseval's theorem becomes the well-known power relation. It is thus possible to realize a design of a pattern function with a finite L_2 norm and all

Table 1. Approximate Formulas for a Few Linear Arrays

	Uniform Broadside	Uniform End-fire	Hansen-Woodyard End-fire
Directivity $\cong$ ($Nd \gg \lambda$)	$2N(d/\lambda)$	$4N(d/\lambda)$	$1.79 \times [4N(d/\lambda)]$
Half-power beamwidth $\cong$ ($\pi d/\lambda \ll 1$)	$2\left[\dfrac{\pi}{2} - \cos^{-1}\left(\dfrac{1.391\lambda}{\pi Nd}\right)\right]$	$2\cos^{-1}\left(1 - \dfrac{1.391\lambda}{\pi Nd}\right)$	$2\cos^{-1}\left(1 - 0.1398\dfrac{\lambda}{Nd}\right)$
Beamwidth between nulls	$2\left[\dfrac{\pi}{2} - \cos^{-1}\left(\dfrac{\lambda}{Nd}\right)\right]$	$2\cos^{-1}\left(1 - \dfrac{\lambda}{Nd}\right)$	$2\cos^{-1}\left(1 - \dfrac{\lambda}{2Nd}\right)$
Null angular position $n = 1, 2, \ldots$ $n \neq N, 2N, \ldots$	$\cos^{-1}\left(\pm\dfrac{n}{N}\dfrac{\lambda}{d}\right)$	$\cos^{-1}\left(1 - \dfrac{n\lambda}{Nd}\right)$	$\cos^{-1}\left[1 + (1 - 2n)\dfrac{\lambda}{2Nd}\right]$
Side lobe maximum position $\cong$ ($\pi d/\lambda \ll 1$) ($s = 1, 2, \ldots$)	$\cos^{-1}\left[\pm\dfrac{(2s+1)\lambda}{2Nd}\right]$	$\cos^{-1}\left[1 - \dfrac{(2s+1)\lambda}{2Nd}\right]$	$\cos^{-1}\left(1 - \dfrac{s\lambda}{Nd}\right)$

(After Balanis [4], © 1982 Harper & Row, Publishers, Inc.; reprinted by permission of the publisher)

side lobes in one period of equal level. This is achieved by Dolph [22] by making use of the Chebyshev polynomials.

(a) For the Dolph-Chebyshev array first consider the Chebyshev polynomials. They can be expressed in either of two equivalent forms, the use of which depends on a particular consideration. For the mth degree, they are

$$T_m(x) = \begin{cases} \cos(m\alpha) = \cos(m\cos^{-1}x), & |x| \leq 1 \\ \cosh(m\alpha) = \cosh(m\cosh^{-1}x), & |x| \geq 1 \end{cases} \quad (21)$$

or

$$\begin{aligned} T_m(x) &= \text{Re}\{e^{jm\alpha}\} = \text{Re}\{(\cos\alpha + j\sin\alpha)^m\} \\ &= \cos^m\alpha - \binom{m}{2}(\cos^{m-2}\alpha)(\sin^2\alpha) + \cdots \\ &\quad + (-1)^n\binom{m}{2n}(\cos^{m-2n}\alpha)(\sin^{2n}\alpha) + \cdots + \text{Re}\{(j)^m \sin^m\alpha\} \\ &= x^m - \binom{m}{2}x^{m-2}(1-x^2) + \cdots + \text{Re}\{(j)^m(1-x^2)^{m/2}\} \\ &= A_m \prod_{p=1}^{m}(x - x_p) \end{aligned} \quad (22)$$

where

$$x = \cos\alpha \qquad (23)$$

$$x_p = \cos\alpha_p = \cos\left(\frac{2p-1}{2m}\pi\right) = p\text{th root, with } p = 1, 2, \ldots, m \qquad (24)$$

$$\binom{m}{2n} = \frac{m!}{(2n)!(m-2n)!} = \text{binomial coefficient} \qquad (25)$$

$$A_0 = A_1 = 1$$
$$A_m = 2A_{m-1}, \qquad m \geqq 2 \qquad (26)$$

A few of these polynomials are sketched in Fig. 5. Note that $T_m(x)$ is a

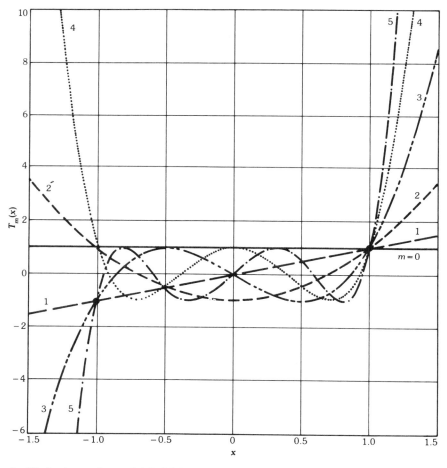

Fig. 5. Chebyshev polynomial $T_m(x)$ for $m = 0, 1, \ldots, 5$. (*After Balanis [4], © 1982, Harper & Row, Publishers, Inc.; reprinted by permission of the publisher*)

polynomial of mth degree, having only even-power terms of x if m is even and only odd-power terms of x if m is odd. The recursion formula, useful for generation, is

$$T_{m+1}(x) + T_{m-1}(x) = 2xT_m(x) \tag{27}$$

with $T_0(x) = 1$ and $T_1(x) = x$. Thus $T_m(x)$ is $1/x$ times the arithmetic mean of its two adjacent neighbors, $T_{m-1}(x)$ and $T_{m+1}(x)$.

(b) Pattern function of an array with N equally spaced and symmetrically excited elements:

If N is odd,

$$\begin{aligned} F(\theta) &= I_0 + (I_1 e^{j\psi} + I_{-1} e^{-j\psi}) + (I_2 e^{j2\psi} + I_{-2} e^{-j2\psi}) \\ &\quad + \cdots + [I_{N-1/2} e^{j(N-1)\psi/2} + I_{-N-1/2} e^{-j(N-1)\psi/2}] \\ &= I_0 + \sum_{n=1}^{(N-1)/2} I_n (Z^n + Z^{-n}) \end{aligned} \tag{28}$$

If N is even,

$$\begin{aligned} F(\theta) &= I_1 e^{j\psi/2} + (I_{-1} e^{-j\psi/2}) + (I_2 e^{j3\psi/2} + I_{-2} e^{-j3\psi/2}) \\ &\quad + \cdots + [I_{N/2} e^{j(N-1)\psi/2} + I_{-N/2} e^{-j(N-1)\psi/2}] \\ &= \sum_{n=1}^{N/2} I_n [Z^{(2n-1)/2} + Z^{-(2n-1)/2}] \end{aligned} \tag{29}$$

where $\psi = kd \cos \theta$, $Z = e^{j\psi}$, and d is the interelement spacing.

(c) To relate (a) and (b), let

$$x = a \cos \psi/2 = a(Z^{1/2} + Z^{-1/2})/2 \tag{30}$$

Then

$$\begin{aligned} T_{N-1}(x) &= A_{N-1} \prod_{p=1}^{N-1} (x - x_p) \\ &= A_{N-1} \prod_{p=1}^{N-1} a(Z^{1/2} + Z^{-1/2} - Z_p^{1/2} - Z_p^{-1/2})/2 \end{aligned} \tag{31}$$

$$x_p = a \cos(\psi_p/2) = a \cos\left[\frac{2p-1}{2(N-1)}\pi\right] \tag{32}$$

$$\begin{aligned} Z_p &= e^{j\psi_p} \\ \psi_p &= kd \cos \theta_p \end{aligned} \tag{33}$$

θ_p = angular position of the pth pattern null (34)

Note that the pattern function as given by (28) or (29) is a polynomial in $Z^{1/2}$ of power from $-(N-1)$ to $(N-1)$, and so is $T_{N-1}(x)$ as given by (31). They can be equated to determine the excitation $\{I_n\}$ once the spacing d/λ and the parameter a are chosen. To find a, one needs to specify the desired side lobe level. Let the main beam to side lobe level ratio be b. Then

$$b = T_{N-1}(a) = \cosh[(N-1)\cosh^{-1}a]$$

or

$$a = \cosh\left(\frac{1}{N-1}\cosh^{-1}b\right) \tag{35}$$

as θ varies from $\quad 0 \to \pi/2 \to \pi$
ψ varies from $\quad 2\pi d/\lambda \to 0 \to -2\pi d/\lambda$
x varies from $a\cos(\pi d/\lambda) \to a \to a\cos(\pi d/\lambda)$ $\tag{36}$

From Fig. 6 it is seen that to avoid grating lobes

$$a\cos(\pi d/\lambda) \leqq -1 \tag{37}$$

Although tables for $\{I_n\}$ are available [93], today, with the availability of computers, it is simple to compute them according to the formula given by Elliott [28]:

$$I_n = I_{-n} = \sum_{p=n}^{(N-1)/2} (-1)^{(N-1)/2-p} \frac{N-1}{N-1+2p} \begin{bmatrix} (N-1)/2 + p \\ 2p \end{bmatrix} \begin{bmatrix} 2p \\ p-n \end{bmatrix} a^{2p},$$
if N is odd $\tag{38}$

$$I_n = I_{-n} = \sum_{p=n}^{N/2} (-1)^{(N/2)-p} \frac{N-1}{N+2(p-1)} \begin{bmatrix} N/2 + p - 1 \\ 2p - 1 \end{bmatrix} \begin{bmatrix} 2p-1 \\ p-n \end{bmatrix} a^{2p-1},$$
if N is even $\tag{39}$

(d) Design procedure: Choose number of elements N and main beam-to-sidelobe ratio b; then successively determine the following quantities, using the equations shown in the boxes, and finally the required excitation $\{I_n\}$

$$\downarrow \boxed{(35)}$$
$$a$$
$$\downarrow \boxed{(37)}$$

the largest permissible value of d

Array Theory

$$\downarrow \boxed{(32)}$$

roots $\{x_k\}$ of $T_{N-1}(x)$

$$\downarrow \boxed{(33) \text{ and } (34)}$$

ψ_k and Z_k

$$\downarrow \boxed{(31)}$$

expand $T_{N-1}(x)$ in terms of $Z^{\pm 1/2}$ (ignoring the factor A_{N-1} and a)

$$\downarrow$$

compare with (28) or (29) to obtain $\{I_n\}$, or use (38) and (39).

(e) To find the pattern, use (30) and (31) [or (28) and (29)]. Or, graphically, as shown in Fig. 6, first construct $T_{N-1}(x)$ and a circle with radius a and center at $x = 0$, and then use (36) to find a point on the circle corresponding to a given value θ and project on the x axis to find $T_{N-1}(x)$, which is $F_{N-1}(\theta)$ except for a constant multiplier.

(f) Cophasal Dolph-Chebyshev array. Let θ_0 be the main beam angle and replace ψ in the above discussion by $\psi - \psi_0 = (2\pi d/\lambda)(\cos\theta - \cos\theta_0)$; then what happened at $\theta = \pi/2$ (i.e., the main beam) will occur at $\theta = \theta_0$. However, for the visible region,

$$\theta = 0 \quad\quad\quad \to \theta_0 \to \pi$$

$$\psi - \psi_0 = \frac{2\pi d}{\lambda}(1 - \cos\theta_0) \quad \to 0 \to -\frac{2\pi d}{\lambda}(1 + \cos\theta_0)$$

$$x = a\cos[(\psi - \psi_0)/2] = a\cos\left[\frac{\pi d}{\lambda}(1 - \cos\theta_0)\right] \to a \to a\cos\left(\frac{\pi d}{\lambda}\right)(1 + \cos\theta_0) \quad (40)$$

To avoid the presence of grating lobes, neither of the end points of x in the above range should be less than -1. This will determine the largest allowable element spacing d.

Comments:

(1) For large arrays and for side lobe levels in the range of -20 to -60 dB, the excitation transformation method of Elliott [28] gives an approximation solution for the half-power beamwidth in terms of the beam-broadening factor f, which is defined as the ratio of the half-power beamwidth of the array with the Dolph-Chebyshev excitation and that with a uniform excitation. Fig. 7a shows the plot of f versus side lobe level in decibels.

(2) A Dolph-Chebyshev array is optimum only in the sense of narrowest beamwidth for a given side lobe level, or lowest side lobe level for a given beamwidth, but not for maximum directivity. Elliott [28] gives an approximate expression for the directivity of a large Dolph-Chebyshev array:

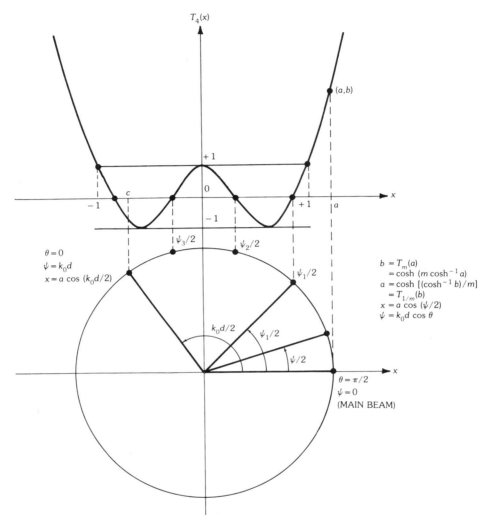

Fig. 6. Graphical method for constructing the radiation pattern of a five-element Dolph-Chebyshev array with side lobe level $20\log b$ and element spacing d.

$$D \cong \frac{2b^2}{1 + (b^2 - 1)f\lambda/(L + d)}$$

where

L = the total array length

f = the beam-broadening factor given in Fig. 7a

b = determined by the desired side lobe level [see (35)]

Fig. 7b shows the plots of D versus $(L + d)/\lambda$ for several side lobe levels. The directivity is generally lower than that of a uniform array, particularly for large L and high side lcbe levels.

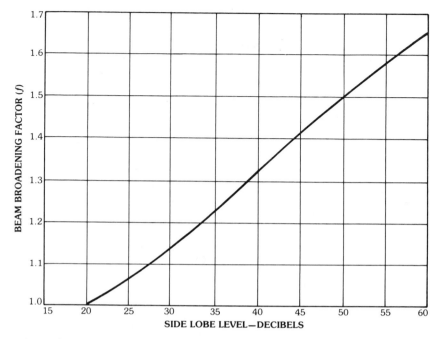

a

Fig. 7. Beam broadening factor and directivity of a Chebyshev array as functions of side lobe level. (*a*) Beam broadening factor *f* versus side lobe level. (*b*) Directivity versus ($L + d$)/λ for various side lobe levels from 15 to 40 dB. (*After Elliott [121], reprinted with permission of* Microwave Journal, *from the December 1963 issue,* © *1963 Horizon House–Microwave, Inc.*)

(*g*) *An alternative Dolph-Chebyshev array design.* So far the Chebyshev polynomial in (*a*) and the pattern function in (*b*) are related by (30) as shown graphically in Fig. 6, where the center of the circle is at $x = 0$. The visible region as given by (36) may not necessarily reach the point $x = -1$, depending on the value of $a\cos\pi d/\lambda$. To make full use of the interval $(-1, a)$ for x, one may let

$$x = c\cos\psi + h = 2c(Z + Z^{-1}) + h \tag{41}$$

where

$$\psi = kd\cos\theta$$
$$a = \cosh\left(\frac{2\cosh^{-1}b}{N-1}\right)$$
$$c = \frac{a+1}{1-\cos kd} \tag{42}$$
$$h = -\frac{a\cos kd + 1}{1-\cos kd}$$

In this case,

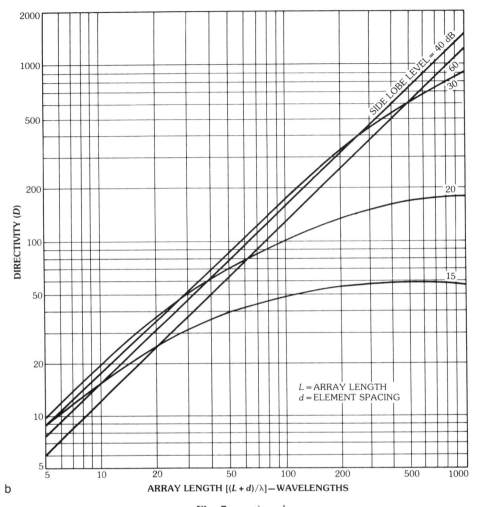

Fig. 7, *continued.*

$$\theta = 0 \to \pi/2 \to \pi$$

$$\psi = kd \to 0 \to -kd$$

$$x = -1 \to a \to -1$$

To find $\{I_n\}$, the term x defined by (41), instead of (30), must be substituted in (31) and the polynomial is compared with (28) for odd N, since now the polynomials in (28) and (31) are both in powers of Z, not $Z^{1/2}$, and also there is a zero-degree term due to h in (41). Furthermore, the power of Z in (28) ranges from $(N-1)/2$ to $-(N-1)/2$ [or similarly from $(N-1)$ to 0 by multiplying all terms by $Z^{-(N-1)/2}$], whereas that in (31) with x substituted by (41) ranges from $(N-1)$ to $-(N-1)$.

Since the number of terms in the polynomial must equal the number of elements, N, only the Chebyshev polynomial of $(N-1)/2$ degree, i.e., $T_{(N-1)/2}(x)$, is needed for (31). Drane [23] has given the following formula for the excitation of an array with N (odd) number of elements:

$$I_n = \frac{\varepsilon_n}{4M} \sum_{m=0}^{M_1} \varepsilon_m \varepsilon_{m_{2-m}} T_n(Y_m) S_M^n(c, h, y_m) \qquad (43)$$

where

$\varepsilon_0 = 1$ and $\varepsilon_n = 2$ for $n \neq 0$,
$y_m = \cos(m\pi/M)$
$M = (N-1)/2$
$M_1, M_2 =$ the integer parts of $M/2$ and $(M+1)/2$, respectively,

$$S_M^n(c, h, y_m) = T_M(cy_m + h) + (-1)^n T_M(h - cy_m) \qquad (44)$$

If M is even,

$$S_M^n(c, h, y_{M/2}) = T_M(cy_{M/2} + h) \qquad (45)$$

4. Linear Transformations in Antenna Arrays

A linear transformation can be applied to array pattern functions in two different ways: (1) When applied to element positions it relates the pattern function of an array of one geometry to that of another [64]. In other words, the pattern of one member of a family of arrays determines the patterns of the entire family, if their geometries are linearly transformable. (2) When applied to excitations it can express the pattern function of an array with one type of excitation in terms of the patterns of the same array with different types of excitations, in particular, the patterns for some simple canonical excitations, such as uniform cophasal excitations. Of course, the application of some of these transformations is valid only if the mutual coupling effect among elements can be ignored, since this effect, in general, is not linearly dependent on element spacings. But the transformation can also be used if the feeding network is redesigned to compensate for the coupling effect after transformation. We shall discuss some fundamental results in this section.

Linear Transformations in Array Geometry

Let us consider a Cartesian coordinate system with bases $\{\hat{\mathbf{e}}_1, \hat{\mathbf{e}}_2, \hat{\mathbf{e}}_3\}$ in which all lengths are expressed in terms of free-space wavelength λ. First we define the following vectors in matrix notations:

$$\mathbf{E} = (\hat{\mathbf{e}}_1, \hat{\mathbf{e}}_2, \hat{\mathbf{e}}_3) \qquad (46)$$

$$\mathbf{X} = (x_1, x_2, x_3)^t \qquad (47)$$

$$\mathbf{U} = (u_1, u_2, u_3)^t \tag{48}$$

where the superscript t, as usual, designates the transpose, and the us are Cartesian components of a unit vector $\hat{\mathbf{u}}$ in spherical coordinates, namely

$$u_1 = \sin\theta \cos\phi, \qquad u_2 = \sin\theta \sin\phi, \qquad u_3 = \cos\theta \tag{49}$$

In the following discussion we shall use either vector notations or matrix notations, depending on which one is more convenient in a particular discussion. Thus

$$\mathbf{x} = x_1 \hat{\mathbf{e}}_1 + x_2 \hat{\mathbf{e}}_2 + x_3 \hat{\mathbf{e}}_3 = \mathbf{EX} \tag{50}$$

$$\hat{\mathbf{u}} = u_1 \hat{\mathbf{e}}_1 + u_2 \hat{\mathbf{e}}_2 + u_3 \hat{\mathbf{e}}_3 = \mathbf{EU} \tag{51}$$

Then from (2a) for each component, the pattern, due to a source called x, with distribution $J(\mathbf{x})$, as a function of (θ, ϕ), or $\hat{\mathbf{u}}$, is

$$F_x(\hat{\mathbf{u}}) = \int_{\text{source } x} J(\mathbf{x}) \, e^{j2\pi \hat{\mathbf{u}} \cdot \mathbf{x}} \, d\mathbf{x} \tag{52}$$

where $d\mathbf{x} = dx_1 \, dx_2 \, dx_3$, the subscript x of $F_x(\cdot)$ denotes the pattern function for source x, and $J(\mathbf{x})$ can be any source distribution function, with continuous, sectionally continuous, or discrete point sources. For the latter case,

$$J(\mathbf{x}) = \sum_n I(\mathbf{x}_n) \, \delta(\mathbf{x} - \mathbf{x}_n) \tag{53}$$

and

$$F_x(\hat{\mathbf{u}}) = \sum_n I(\mathbf{x}_n) \, e^{j2\pi \hat{\mathbf{u}} \cdot \mathbf{x}_n} \tag{54}$$

which is simply the array factor for point elements at $\{\mathbf{x}_n\}$ with excitations $\{I_n\}$. Thus the so-called array factor can be regarded as a special case of (52).

Equation 52 indicates that $F_x(\hat{\mathbf{u}})$ and $J(\mathbf{x})$ are a Fourier transform pair. However, in physical space (θ, ϕ) are real, and, because of (49), the observable pattern function is described by only a part of the Fourier transform of $J(x)$ which lies on a unit sphere in the transform space u.

Now let $\mathbf{X}$ and $\mathbf{U}$ undergo linear transformations described by matrices $\mathbf{A}$ and $\mathbf{B}$, respectively, which takes $\mathbf{X}$ to $\mathbf{Y} = \mathbf{AX}$ and $\mathbf{U}$ to $\mathbf{V} = \mathbf{BU}$. If

$$\mathbf{B}^{-1} = \mathbf{A}^t \tag{55}$$

then the scalar product

Array Theory

$$\mathbf{v} \cdot \mathbf{y} = \mathbf{V}'\mathbf{Y} = (\mathbf{BU})'(\mathbf{AX}) = \mathbf{U}'\mathbf{X} = \mathbf{u} \cdot \mathbf{x} \tag{56}$$

remains invariant. Hence (52) can be written as

$$F_x(\hat{\mathbf{u}}) = \int_{\text{source } x} J(\mathbf{x}) e^{j2\pi \hat{\mathbf{u}} \cdot \mathbf{x}} d\mathbf{x}$$

$$= \int_{\text{source } y} J(\mathbf{A}^{-1}\mathbf{Y}) e^{j2\pi \mathbf{v} \cdot \mathbf{y}} |\mathbf{A}|^{-1} d\mathbf{Y} = F_y(\mathbf{v}) \tag{57}$$

where $|\mathbf{A}|$ is the determinant of $\mathbf{A}$. This relation states in effect that the field $F_x(\hat{\mathbf{u}})$ observed in the direction $\hat{\mathbf{u}}$ due to the source x is the same as that in the direction $\mathbf{v}$ due to the source y, that is, $J(\mathbf{y}) = |\mathbf{A}|^{-1} J(\mathbf{A}^{-1}\mathbf{Y})$. Thus the pattern function of a member of a family of arrays which are related to each other by a linear transformation in geometry determines the patterns of all members of the family. However, since in general $\mathbf{u} \neq \mathbf{v}$, part of the invisible region in the u space may become visible in the v space after the transformation, or vice versa. To see this transformation in detail, let us consider first the scalar product invariant transformation, from (55):

$$\mathbf{B}'\mathbf{A} = \mathbf{A}'\mathbf{B} = \mathbf{E}$$

Let $\mathbf{a}_1, \mathbf{a}_2, \mathbf{a}_3$ be the column vector of $\mathbf{A}$ and $\mathbf{b}_1, \mathbf{b}_2, \mathbf{b}_3$ be those of $\mathbf{B}$. Then $\mathbf{Y} = \mathbf{AX}$ implies that

$$\mathbf{y} = y_1 \hat{\mathbf{e}}_1 + y_2 \hat{\mathbf{e}}_2 + y_3 \hat{\mathbf{e}}_3 = x_1 \mathbf{a}_1 + x_2 \mathbf{a}_2 + x_3 \mathbf{a}_3 \tag{58}$$

This transformation can be regarded as a *mapping* (rotation and linear stretching) of $\mathbf{x}$ into $\mathbf{y}$, or, by comparing the right-hand side of (58) with (50), simply a *relabeling* of the base vectors of $\mathbf{x}$ from $\mathbf{e}_n$ into $\mathbf{a}_n$. Similar interpretations can be given to $\mathbf{V} = \mathbf{BU}$,

$$\mathbf{v} = v_1 \hat{\mathbf{e}}_1 + v_2 \hat{\mathbf{e}}_2 + v_3 \hat{\mathbf{e}}_3 = u_1 \mathbf{b}_1 + u_2 \mathbf{b}_2 + u_3 \mathbf{b}_3 \tag{59}$$

For invariant scalar product, (55) states that

$$\mathbf{b}_i \cdot \mathbf{a}_j = \delta_{ij}, \quad i, j = 1, 2, 3 \tag{60}$$

where the Kronecker delta $\delta_{ij} = 0$ if $i \neq j$ and 1 if $i = j$. Thus $\mathbf{b}_i$ has a projection on $\mathbf{a}_i$ equal to $1/a_i$ and a direction perpendicular to all other $\mathbf{a}_j$s. Specifically,

$$\mathbf{b}_1 = |\mathbf{A}|^{-1}(\mathbf{a}_2 \times \mathbf{a}_3), \quad \mathbf{b}_2 = |\mathbf{A}|^{-1}(\mathbf{a}_3 \times \mathbf{a}_1), \quad \mathbf{b}_3 = |\mathbf{A}|^{-1}(\mathbf{a}_1 \times \mathbf{a}_2) \tag{61}$$

Here $\{\mathbf{a}_i\}$ and $\{\mathbf{b}_i\}$ are called *reciprocal bases* to each other. These are very useful in the study of periodic structures and Floquet space harmonics.

Application to Planar Periodic Arrays

Fig. 8a shows a broadside planar array with elements at the square grid intersection points in the $x_1 x_2$ plane and spacing d (in wavelengths). Its pattern function in the $u_1 u_2$ space is also periodic as shown in Fig. 8b, where the small circles indicate the locations of the grating lobes (including the main beam). If the array is of infinite extent, each circle corresponds to a space harmonic. The spacing between two circles along the u_1 or u_2 axis is $1/d$. The visible region is bounded by a unit circle and a typical point in this $u_1 u_2$ plane has the polar coordinates $(\sin\theta, \phi)$. Thus a given point (u_1, u_2) determines uniquely (θ, ϕ) in the physical region $0 \leq \theta \leq \pi$, $0 \leq \phi < 2\pi$, and vice versa.

For a cophasal array with main beam at (θ_0, ϕ_0)

$$J(\mathbf{x}) = |J(\mathbf{x})| \exp(-j2\pi \hat{\mathbf{u}}_0 \cdot \mathbf{x}) \tag{62}$$

where

$$u_{01} = \sin\theta_0 \cos\phi_0, \quad u_{02} = \sin\theta_0 \sin\phi_0 \tag{63}$$

This additional phase results simply in a translation of the origin of the $u_1 u_2$ space by $(-u_{01}, -u_{02})$ as shown in Fig. 9 with all grating beam circles fixed, or with the origin remains fixed and all grating beam circles translated by (u_{01}, u_{02}). The visible region in this case is given by a unit disc with (u_{01}, u_{02}) as center as shown in Fig. 9a. For a phased array with full scan range, $0 \leq \theta_0 \leq \pi$ and $0 \leq \phi_0 < 2\pi$, the *overall* visible region is bounded by a disc with radius 2 and center at the original origin as shown in Fig. 9b. For a particular scan range, say $0 \leq \theta_0 \leq \theta_0''$ and $\phi_0' \leq \phi_0 \leq \phi_0''$, the total visible region is bounded by that part of the $u_1 u_2$ plane which is covered by all unit discs with their centers (u_{01}, u_{02}) satisfying the inequalities

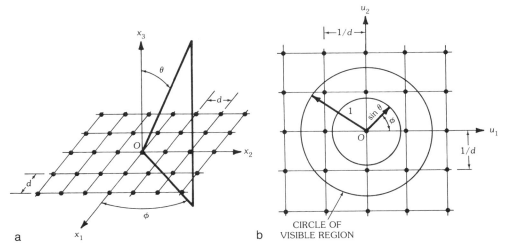

Fig. 8. Broadside planar array and pattern function. (*a*) A periodic planar array with square cells in the $x_1 x_2$ plane. (*b*) Periodic structure of the pattern function of the array (*a*) in the $u_1 u_2$ plane. (*After Lo and Lee [64], © 1965 IEEE*)

Array Theory

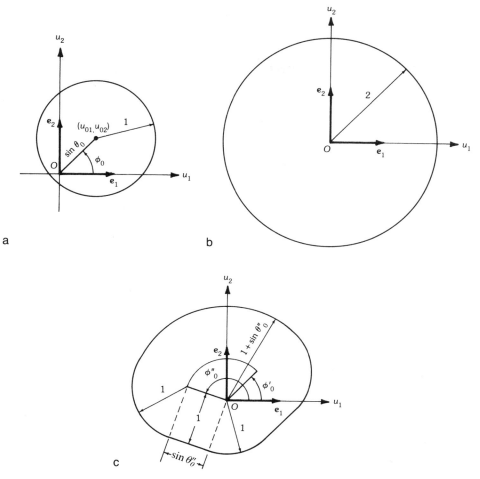

Fig. 9. Visible region in the u_1u_2 plane. (*a*) As the beam is scanned to (θ_0, ϕ_0). (*b*) For a full scan: $0 \leq \theta_0 \leq \pi$ and $0 \leq \phi_0 \leq 2\pi$. (*c*) For a limited scan: $\phi'_0 \leq \phi_0 \leq \phi''_0$ and $0 \leq \theta_0 \leq \theta''_0$. (*After Lo and Lee [64], © 1965 IEEE*)

$$0 \leq (u_{01}^2 + u_{02}^2)^{1/2} \leq \sin\theta''_0 \quad \text{and} \quad \phi'_0 \leq \tan^{-1}\left(\frac{u_{02}}{u_{01}}\right) \leq \phi''_0 \tag{64}$$

as illustrated in Fig. 9c.

An Example—Consider a uniform array (called the *x* array) with square grid size $d = 1$ as shown in Fig. 8a. Then the *mn*th element position is

$$\mathbf{x}_{mn} = m\hat{\mathbf{e}}_1 + n\hat{\mathbf{e}}_2, \quad 1 \leq m \leq M, \quad 1 \leq n \leq N \tag{65}$$

Its pattern is

$$F_x(\mathbf{u}) = \frac{\sin M\pi u_1}{\sin \pi u_1} \frac{\sin N\pi u_2}{\sin \pi u_2} \qquad (66)$$

Next we consider a more general uniform planar array, called the *y* array, as shown in Fig. 10a, whose *mn*th element is at

$$\mathbf{y}_{mn} = m\mathbf{a}_1 + n\mathbf{a}_2, \qquad 1 \leq m \leq M, \qquad 1 \leq n \leq N \qquad (67)$$

where $\mathbf{a}_1$ and $\mathbf{a}_2$ need not be orthogonal, nor of equal length. From (58) we see that the *y* array can be obtained from the *x* array with the transformation

$$\mathbf{A} = (\mathbf{a}_1, \mathbf{a}_2, \hat{\mathbf{e}}_3) \qquad (68)$$

since for planar arrays there is no need for the transformation for the axis normal to the array, i.e., $\mathbf{a}_3 = \hat{\mathbf{e}}_3$. Now, from (55),

$$\mathbf{U} = \mathbf{B}^{-1}\mathbf{V} = \mathbf{A}'\mathbf{V}$$

or

$$\mathbf{u}_1 = \mathbf{a}_1 \cdot \mathbf{v}, \qquad \mathbf{u}_2 = \mathbf{a}_2 \cdot \mathbf{v}, \qquad \mathbf{u}_3 = \hat{\mathbf{e}}_3 \cdot \mathbf{v} \qquad (69)$$

where

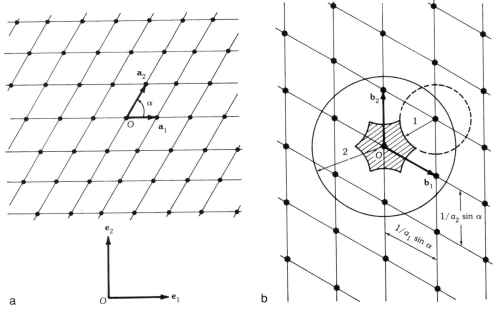

Fig. 10. Periodic planar array and pattern function. (*a*) Periodic planar array with parallelogram cells defined by vectors $\mathbf{a}_1$ and $\mathbf{a}_2$ in *y* space. (*b*) Periodic pattern function structure in the v_1v_2 plane. (*After Lo and Lee [64], © 1965 IEEE*)

$$\mathbf{v} = \sin\theta\cos\phi\hat{\mathbf{e}}_1 + \sin\theta\sin\phi\hat{\mathbf{e}}_2 + \cos\theta\hat{\mathbf{e}}_3 \tag{70}$$

Therefore from (66) and (69) the pattern of the y array is simply

$$F_y(\mathbf{v}) = \frac{\sin(M\pi\mathbf{a}_1\cdot\mathbf{v})}{\sin(\pi\mathbf{a}_1\cdot\mathbf{v})}\frac{\sin(N\pi\mathbf{a}_2\cdot\mathbf{v})}{\sin(\pi\mathbf{a}_2\cdot\mathbf{v})} \tag{71}$$

Since $F_x(\mathbf{u})$ has grating lobes at u_1 and u_2 equal to $0, \pm 1, \pm 2, \ldots$, or

$$\mathbf{u} = m\hat{\mathbf{e}}_1 + n\hat{\mathbf{e}}_2, \quad m, n = 0, \pm 1, \pm 2, \ldots, \tag{72}$$

$F_y(\mathbf{v})$ will have grating lobes at $(\mathbf{a}_1\cdot\mathbf{v})$ and $(\mathbf{a}_2\cdot\mathbf{v})$ equal to $0, \pm 1, \pm 2, \ldots$, or

$$\mathbf{v} = m\mathbf{b}_1 + n\mathbf{b}_2 \tag{73}$$

where $\mathbf{b}_1$ and $\mathbf{b}_2$ are determined from $\mathbf{a}_1$ and $\mathbf{a}_2$ from (61) as shown in Fig. 10b. Note that $\mathbf{b}_1$ and $\mathbf{b}_2$ are perpendicular to $\mathbf{a}_2$ and $\mathbf{a}_1$, respectively, and that the lengths of $\mathbf{b}_1$ and $\mathbf{b}_2$ are equal to the reciprocals of $a_1\sin\alpha$ and $a_2\sin\alpha$, respectively. If $a_1 = a_2$ and the angle between $\mathbf{a}_1$ and $\mathbf{a}_2$, namely α, is 60°, the y array becomes an equilateral-triangular grid. It is very simple to determine the range of scan for which no grating lobe will appear. This region is bounded by the arcs of unit circles with centers at all adjacent grating lobe locations shown as the shaded parts in Fig. 10b. To translate the boundary of this region into one in the $\theta\phi$ plane, one needs to recall that any point in the v_1v_2 plane has the polar coordinates $(\sin\theta, \phi)$. Thus by placing a transparent overlay with ordinary polar coordinates (except that the concentric circles are marked with θ according to the value of $\sin\theta$ as their radii) on the top of Fig. 10b, one can determine immediately the scanning region in θ and ϕ (see Fig. 14). This method is not only simpler but also more illuminating than others [88].

Nonuniform Excitation and Relation between Aperture Antenna and Discrete Array

Often an array with discrete elements can be considered as a sampled set of an aperture antenna with a distribution function $f_0(\mathbf{y})$. A question of interest will be how close the pattern of the sampling array is to that of the aperture antenna. To see this, let $f_0(\mathbf{y})$ be the given excitation for $\mathbf{y}$ inside the aperture and zero outside the aperture. Then for the array $\{y\}$ one may have

$$f(\mathbf{y}) = f_0(\mathbf{y})\sum_{m,n}\delta(\mathbf{y} - \mathbf{y}_{mn}) \tag{74}$$

where $\delta(\mathbf{y} - \mathbf{y}_{mn})$ is the two-dimensional delta function and the summation is taken over all the integers for both m and n. Now applying the transformation $\mathbf{A}$, one obtains a corresponding array $\{x\}$ with excitation

$$f(\mathbf{AX}) = f_0(\mathbf{AX})\sum_{m,n}\delta(\mathbf{AX} - \mathbf{AX}_{mn}) \tag{75}$$

By substituting (75) into (52) and making use of the convolution integral theorem, one has

$$F_x(\mathbf{u}) = F_{0x}(\mathbf{u}) * \sum_{m,n} \delta(\mathbf{U} - \mathbf{X}_{mn})|\mathbf{A}|^{-1} \tag{76}$$

where the asterisk denotes the convolution integration and $F_{0x}(\mathbf{u})$ is the Fourier transform or the pattern function of an aperture antenna with excitation $f_0(\mathbf{AX})$. As in (67), if $\mathbf{y}_{mn} = m\mathbf{a}_1 + n\mathbf{a}_2$, then $\mathbf{x}_{mn} = m\hat{\mathbf{e}}_1 + n\hat{\mathbf{e}}_2$ and $F_x(\mathbf{u})$ is periodic in $\mathbf{u}_1$ and $\mathbf{u}_2$ of unity period. Equation 76 states that $F_x(\mathbf{u})$ is the sum of infinitely many $F_{0x}(\mathbf{u})$s, each displaced from the other by a unit along the u_1 and u_2 axes. Since $f_0(\mathbf{AX})$ is an aperture limited function, $F_{0x}(\mathbf{u})$ becomes negligible for large u. Therefore, for (u_1, u_2) in the neighborhood of (m, n), $F_x(\mathbf{u})$ is essentially equal to $F_{0x}(\mathbf{u})$. From (76) the pattern function of array $\{y\}$ is given immediately by*

$$F_y(\mathbf{v}) = F_x(\mathbf{B}^{-1}\mathbf{V}) = |\mathbf{B}| F_{0x}(\mathbf{B}^{-1}\mathbf{V}) * \sum_{m,n} \delta(\mathbf{V} - \mathbf{B}\mathbf{X}_{mn})$$
$$= |\mathbf{B}| F_{0x}(\mathbf{B}^{-1}\mathbf{V}) * \sum_{m,n} \delta(\mathbf{v} - m\mathbf{b}_1 - n\mathbf{b}_2) \tag{77}$$

As in Fig. 10b the pattern function of array $\{y\}$ is also a periodic function with periods inversely proportional to $a_1 \sin \alpha$ and $a_2 \sin \alpha$ along $\mathbf{b}_1$ and $\mathbf{b}_2$, respectively, and it is thus a linearly distorted pattern of $F_x(\mathbf{u})$. Obviously (76) and (77) also show the exact difference between the pattern function for continuous (or sectionally continuous) excitation and that for discrete excitation with periodically spaced elements.

Hexagonal Arrays

Periodic arrays with unit cells other than parallelograms may also be analyzed in a similar manner. As an example, consider a hexagonal array of uniform excitation, as shown in Fig. 11a, where the element positions are marked by crosses. The array may be regarded simply as a superposition of two parallelogram arrays with equal excitations but opposite in sign. The sides of the two parallelograms are defined by

$$\begin{aligned} \mathbf{a}_1 &= \hat{\mathbf{e}}_1, & \mathbf{a}_2 &= \cos(60°)\hat{\mathbf{e}}_1 + \sin(60°)\hat{\mathbf{e}}_2 \\ \mathbf{a}_1' &= 3\hat{\mathbf{e}}_1, & \mathbf{a}_2' &= \sqrt{3}\cos(30°)\hat{\mathbf{e}}_1 + \sqrt{3}\sin(30°)\hat{\mathbf{e}}_2 \end{aligned} \tag{78}$$

From the relation $\mathbf{B} = (\mathbf{A}')^{-1}$, one immediately obtains

---

*It is understood that the convolution integral in (77) refers to the variable $\mathbf{v}$ whereas in (76) it refers to the variable $\mathbf{u}$.

Array Theory

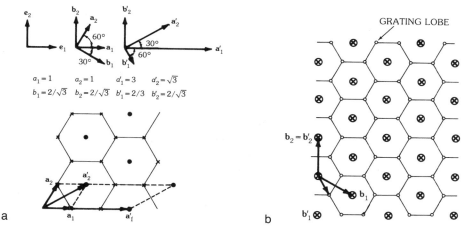

Fig. 11. Periodic array with hexagonal cells and hexagonal pattern function structure (a) Array with cells in y space. (b) Pattern function structure in v space. *(After Lo and Lee [64], © 1965 IEEE)*

$$\mathbf{b}_1 = \frac{2}{\sqrt{3}}\cos(-30°)\,\hat{\mathbf{e}}_1 + \frac{2}{\sqrt{3}}\sin(-30°)\,\hat{\mathbf{e}}_2, \qquad \mathbf{b}_2 = \frac{2}{\sqrt{3}}\hat{\mathbf{e}}_2$$

$$\mathbf{b}'_1 = \frac{2}{3}\cos(-60°)\,\hat{\mathbf{e}}_1 + \frac{2}{3}\sin(-60°)\,\hat{\mathbf{e}}_2, \qquad \mathbf{b}'_2 = \frac{2}{\sqrt{3}}\hat{\mathbf{e}}_2 \tag{79}$$

as shown in Fig. 11b. It is seen that the grating lobes, denoted by the small circles without crosses, also form a hexagonal periodic structure.

Periodic Arrays with Minimum Number of Elements

In practice it is often desirable to have only a single main beam in the visible region. This condition sets an upper limit on the element spacing of a period array. For the array with elements at the vertices of square cells, as shown in Fig. 8a, it is well known that for the case of full scan the element spacing must be smaller than half wavelength, as seen from Fig. 8b. As a result this sets a lower limit on the required number of elements. In the following we shall show that, for some other periodical arrays, this lower limit can be reduced.

First, it should be understood that the previous statement about a single main beam in the visible region, in fact, has different meanings, depending on the types of arrays under discussion. For example, for a linear array, due to symmetry, a single main beam may actually be a disklike beam when it is broadside, a cone-shaped beam when it is fired at an angle, and a unidirectional beam when it is end-fired. On the other hand, for a planar array, whenever there is beam firing above the plane of the array, there is always one below the plane, due to symmetry, except for the limiting case of end-fire. Customarily, one may regard all of these as being degenerated cases of a three-dimensional array which, in general, can possess a truly single beam in the visible range.

Now consider a periodic array with parallelograms as unit cells, as shown in Fig. 10a. Then, as previously shown, the grating lobes will appear at

$$\mathbf{u} = m\mathbf{b}_1 + n\mathbf{b}_2, \quad m, n = \text{integers} \tag{80}$$

Let

$$\mathbf{b}_1 = b_{11}\hat{\mathbf{e}}_1 + b_{12}\hat{\mathbf{e}}_2$$
$$\mathbf{b}_2 = b_{21}\hat{\mathbf{e}}_1 + b_{22}\hat{\mathbf{e}}_2 \tag{81}$$

Then, in matrix notation, (80) becomes

$$\begin{bmatrix} u_1 \\ u_2 \end{bmatrix} = \begin{bmatrix} b_{11} & b_{21} \\ b_{12} & b_{22} \end{bmatrix} \begin{bmatrix} m \\ n \end{bmatrix}$$

The objective is to determine all bs under the condition that no grating lobes appear in the visible region* and then determine all as which fix the array arrangement. In order to make the number of elements minimum for a given aperture one must add a condition that the area of each unit cell, namely $|\mathbf{A}|$, should be made as large as possible, or $|\mathbf{B}|$ as small as possible.

To give an example, first consider the case of *full* scan. Then it is required that:

(1) for all grating lobes outside the *full* scan circle in u space: $(b_{11}m + b_{21}n)^2 + (b_{12}m + b_{22}n)^2 \geqq 4$, for m, n, equal to all integers except $m = n = 0$;
(2) for minimum number of elements:
$|\mathbf{B}| = b_{11}b_{22} - b_{12}b_{21}$ to be minimum

The solution is

$$\mathbf{a}_1 = \frac{1}{\sqrt{3}}\hat{\mathbf{e}}_1 \quad \text{and} \quad \mathbf{a}_2 = \frac{1}{\sqrt{3}}[\cos(60°)\hat{\mathbf{e}}_1 + \sin(60°)\hat{\mathbf{e}}_2] \tag{82}$$

This is the well-known equilateral-triangular grid structure.

If the scan range is limited to

$$0 \leqq \theta_0 \leqq \theta_0'' \quad \text{and} \quad 0 \leqq \phi_0 \leqq 2\pi$$

the solution can be determined similarly as

$$\mathbf{a}_1 = \frac{2}{\sqrt{3}(1 + \sin\theta_0'')}\hat{\mathbf{e}}_1$$

$$\mathbf{a}_2 = \frac{2}{\sqrt{3}(1 + \sin\theta_0'')}(\hat{\mathbf{e}}_1 \cos 60° + \hat{\mathbf{e}}_2 \sin 60°) \tag{83}$$

---

*This condition is regarded as one where the grating lobe begins to show up in the visible region at the extreme scanning angle. This determines the upper limit of the element spacings. Actually, to avoid the grating lobe *completely*, the spacings should be somewhat *smaller*, depending on the beam shape, width, or the aperture and the excitation function.

Array Theory

Now comparing this array with a conventional one with square cells, as shown in Fig. 8a, the number of elements is reduced approximately by

$$\frac{|\mathbf{B}_{sq}| - |\mathbf{B}|}{|\mathbf{B}_{sq}|} = \frac{4 - 2\sqrt{3}}{4} = 13.4 \text{ percent}$$

where $|\mathbf{B}_{sq}|^{-1}$, $|\mathbf{B}|^{-1}$ are, respectively, the areas of a square cell and a parallelogram given by (83). It can be easily shown that this percentage is dependent on ϕ_0'' only, and the previous value applies when ϕ_0 covers the whole range $(0, 2\pi)$. If ϕ_0 covers only part of this range, condition 1 should be revised. For the special case, $0 \leq \theta_0 < 45°$, $0 \leq \phi_0 < 2\pi$, $a_1 = a_2 = 0.676$.

It is easy to extend the previous results to a three-dimensional periodic array with elements over vertices of parallelopiped cells. In this case, for *full* scan, $\mathbf{b}_1$, $\mathbf{b}_2$, and $\mathbf{b}_3$ are found to form an oblique coordinate system with 60° between any two vectors, and each with length 2. With respect to the orthonormal basis $\{\hat{\mathbf{e}}_1, \hat{\mathbf{e}}_2, \hat{\mathbf{e}}_3\}$, they are

$$\mathbf{b}_1 = 2\hat{\mathbf{e}}_1, \quad \mathbf{b}_2 = \hat{\mathbf{e}}_1 + \sqrt{3}\hat{\mathbf{e}}_2, \quad \mathbf{b}_3 = \hat{\mathbf{e}}_1 + \frac{1}{\sqrt{3}}\hat{\mathbf{e}}_2 + \frac{2\sqrt{2}}{\sqrt{3}}\hat{\mathbf{e}}_2 \quad (84)$$

From the relation between as and bs, one obtains

$$\mathbf{a}_1 = \frac{1}{2}\hat{\mathbf{e}}_1 - \frac{1}{2\sqrt{3}}\hat{\mathbf{e}}_2 - \frac{1}{2\sqrt{6}}\hat{\mathbf{e}}_3, \quad \mathbf{a}_2 = \frac{1}{\sqrt{3}}\hat{\mathbf{e}}_2 - \frac{1}{2\sqrt{6}}\hat{\mathbf{e}}_3, \quad \mathbf{a}_3 = \frac{\sqrt{3}}{2\sqrt{2}}\hat{\mathbf{e}}_3 \quad (85)$$

It is readily verified that all as have a length equal to $\sqrt{3}/2\sqrt{2}$ and make an angle of $\cos^{-1}(-1/3) = 109.5°$ between any two of them. In comparison with a conventional array with cubic cells, the number of elements is reduced approximately by 29.3 percent.

Transformation between Circular and Elliptical Arrays

The pattern of a circular array or aperture antenna can be expressed in terms of Bessel functions (see later in this section, under "Circular Arrays"). In a similar manner the pattern of an elliptical array or aperture antenna can be expressed in terms of Mathieu functions. But the numerical computation of the latter is much more difficult than that of the former. By using linear transformation, this difficulty can be alleviated.

Let the Cartesian coordinates of the nth element with excitation I_n of an elliptical array, called the x array, be $\mathbf{x}_n = (x_{n_1}, x_{n_2})$ in the $\hat{\mathbf{e}}_1\hat{\mathbf{e}}_2$ plane; then

$$(x_{n_1}/a)^2 + (x_{n_2}/b)^2 = 1 \quad (86)$$

where $2a$ and $2b$ are the major and minor axes, respectively. The pattern function of the x array is

$$F_x(\mathbf{u}) = \sum_n I_n e^{j2\pi\mathbf{u}\cdot\mathbf{x}_n} \quad (87)$$

Now, applying the transformation

$$\mathbf{A} = \begin{bmatrix} 1 & 0 \\ 0 & \tau \end{bmatrix} \quad \text{and} \quad \mathbf{B} = (\mathbf{A}^{-1})^t = \begin{bmatrix} 1 & 0 \\ 0 & t \end{bmatrix} \tag{88}$$

where $\tau = 1/t = a/b$ = axial ratio, one obtains

$$F_x(\mathbf{u}) = \sum_n I_n e^{j2\pi \mathbf{u} \cdot \mathbf{x}_n} = \sum_n I_n e^{j2\pi \mathbf{v} \cdot \mathbf{y}_n} = F_y(\mathbf{v}) \tag{89}$$

where

$$\mathbf{y}_n = \begin{bmatrix} y_{n_1} \\ y_{n_2} \end{bmatrix} = \mathbf{AX} = \begin{bmatrix} 1 & 0 \\ 0 & \tau \end{bmatrix} \begin{bmatrix} x_{n_1} \\ x_{n_2} \end{bmatrix} = \begin{bmatrix} x_{n_1} \\ \tau x_{n_2} \end{bmatrix} \tag{90}$$

and

$$\mathbf{V} = \begin{bmatrix} v_1 \\ v_2 \end{bmatrix} = \mathbf{BU} = \begin{bmatrix} 1 & 0 \\ 0 & t \end{bmatrix} \begin{bmatrix} u_1 \\ u_2 \end{bmatrix} = \begin{bmatrix} u_1 \\ t u_2 \end{bmatrix} \tag{91}$$

From (86) and (90)

$$y_{n_1}^2 + y_{n_2}^2 = a^2 \tag{92}$$

Thus the y array is a circular array whose nth element excitation is still I_n. Similarly, since

$$u_1^2 + u_2^2 = \sin^2\theta^{(u)} \cos^2\phi^{(u)} + \sin^2\theta^{(u)} \sin^2\phi^{(u)} = \sin^2\theta^{(u)} \tag{93}$$

one obtains from (91)

$$v_1^2 + (\tau v_2)^2 = \sin^2\theta^{(v)} \cos^2\phi^{(v)} + \tau^2 \sin^2\theta^{(v)} \sin^2\phi^{(v)} = \sin^2\theta^{(u)} \tag{94}$$

where the superscript u is used to denote the values of θ and ϕ associated with a certain point in the u space and the superscript v is used to denote the values of θ and ϕ for its transformed point in the v space. All the above results can be summarized as in the following:

"The pattern of an elliptical array, with excitation $\{I_n\}$, major axis $2a$, and axial ratio τ when computed over a circle of radius $\sin\theta^{(u)}$ and $0 \leq \phi^{(u)} < 2\pi$ in the u plane is the same as that of a circular array, with excitations $\{I_n\}$ and radius a when computed over an ellipse with major axis $2\sin\theta^{(u)}$ and axial ratio τ in the v plane, as shown in Fig. 12."

Sometimes it is more convenient to specify element position and observation point by azimuthal angles as shown in Fig. 12. One finds that the element located at angle

Array Theory

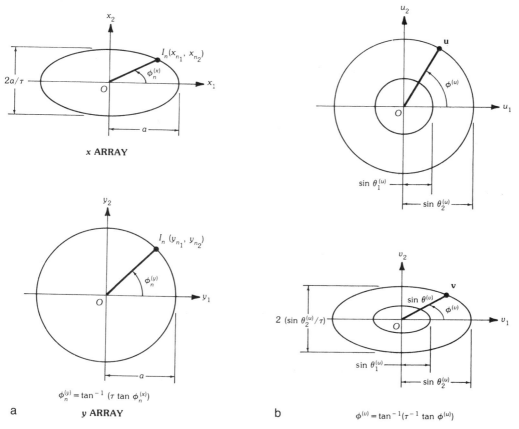

Fig. 12. Far field at u with polar coordinates $(\sin\theta^{(u)}, \phi^{(u)})$ in the $u_1 u_2$ plane due to an elliptical array x is the same as that at v with polar coordinates $(\sin\theta^{(v)}, \phi^{(v)})$ in the $v_1 v_2$ plane due to a circular array y. (*a*) Array plane. (*b*) Field plane. (*After Lo and Hsuan [66]*, © *1965 IEEE*)

$\phi_n^{(x)}$ of the x array will be located for the y array at

$$\phi_n^{(y)} = \tan^{-1}(\tau \tan \phi_n^{(x)}) \tag{95}$$

Similarly, an observation at $\phi^{(u)}$ on the circle of radius $\sin\theta^{(u)}$ in the u plane will be on the ellipse with major axis $2\sin\theta^{(u)}$ and axial ratio τ in the v plane at angle

$$\phi^{(v)} = \tan^{-1}(\tau^{-1}\tan\phi^{(u)}) \tag{96}$$

All of these are shown in Fig. 12, and also Fig. 13 for $\tau = 3, 7,$ and 10. A few examples will be given later in this section, under "Circular Arrays." This technique can also be applied to aperture antennas. In particular the Taylor distribution for a circular aperture [99, 43, 45] can be extended to that for an elliptical aperture [66].

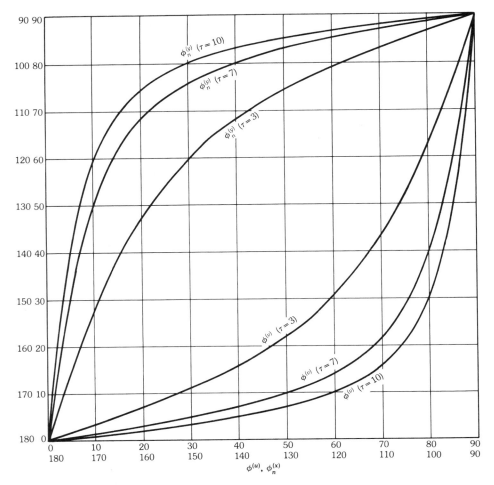

Fig. 13. Transformation of the angular position $\phi_n^{(x)}$ of the nth element of the x array to that of the y array, $\phi_n^{(y)}$, for various axial ratios τ; also, transformations of the observation angle $\phi^{(u)}$ in the u plane to that in the v plane, $\phi^{(v)}$, for various values of τ. (*After Lo and Hsuan [66], © 1965 IEEE*)

Beam and Pattern Distortion Due to Scanning

It is well known that the beam broadens and the pattern distorts in the $\theta\phi$ space as the beam scans away from the broadside direction [28]. However, if the pattern is presented in the u space, the beam scanning due to progressive phase shift as in (62) results only in a translation of the origin of the u space, leaving the beam and the pattern unchanged. The so-called distortion is really a result of the nonlinear-functional relationship between (θ, ϕ) and $\mathbf{u} - \mathbf{u}_0$. An understanding of this relationship is therefore important for determining the distortion, no matter which array is considered.

Consider a hemisphere of unit radius as shown in Fig. 14. A point on the sphere

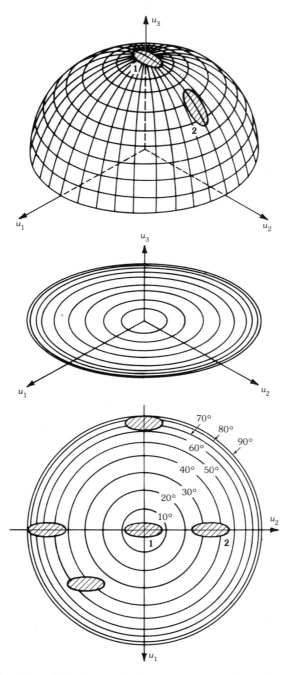

Fig. 14. Beam and pattern distortion as the beam scans away from the polar axis, a result from the nonlinear projection of the u_1u_2 plane on the sphere (the beam shape in the u_1u_2 plane remains unchanged as the beam scans).

with coordinates $(1, \theta, \phi)$ when projected on the $u_1 u_2$ plane will have the Cartesian coordinates $(u_1, u_2) = (\sin\theta\cos\phi, \sin\theta\sin\phi)$. To find how a beam (or pattern) appears on the sphere, it need only project the beam contour lines in the $u_1 u_2$ plane onto the spherical surface. One can even construct a three-dimensional pattern model by making the radial length in the $u_1 u_2$ plane proportional to the level of the contour in that direction.

Suppose in the broadside direction the beam cross section at a certain level looks like the cross-hatched region 1 at the bottom of Fig. 14. If this beam is scanned to $\theta = 33°$ and $\phi = 90°$ by a progressive phase shift, the origin of the $u_1 u_2$ plane will be moved by $\sin 33°$ in the $-u_2$ direction, or the beam is moved by $\sin 33°$ along the $+u_2$ direction as shown by region 2 in the figure (with *no* distortion). However, when beam 2 is projected on the sphere, the width along the ϕ direction remains unchanged but that along the θ direction is widened. Since the azimuthal angle of the sphere is the same as that of the $u_1 u_2$ coordinate system, there is no distortion in the ϕ direction. The distortion is therefore due to $\sin\theta$ only.

It is somewhat awkward to make projections from a plane onto a sphere. Instead, one can project latitude circles of the sphere on the $u_1 u_2$ plane, or simply draw a series of concentric circles with radius equal to $\sin\theta$ as shown at the bottom of Fig. 14. If the circles are marked with their corresponding values of θ, one obtains a set of nonlinear scales which can measure any beam distortion due to scan simply by moving the same beam contour structure to the desired scan angle (θ_0, ϕ_0) in this set of scales.

One can also compute the beam broadening from the following formula:

$$\sin\Delta\theta_p = \sin(\theta_0 + \Delta\theta_p') - \sin\theta_0 = 2\cos(\theta_0 + \Delta\theta_p'/2)\sin(\Delta\theta_p'/2) \qquad (97)$$

where

$\Delta\theta_p$ = the polar-angular position of a point at the p-dB level of the beam when the beam is in the broadside direction

$\Delta\theta_p' + \theta_0$ = the polar-angular position of the p-dB level point of the beam when it is scanned to θ_0

Once $\Delta\theta_p'$ is found for $\Delta\theta_p$ and θ_0 the amount of beam broadening is determined. For *narrow* beams the above formula can be approximated as

$$\Delta\theta_p'/\Delta\theta_p \cong \sec\theta_0$$

in particular, for half-power

$$\Delta\theta_{HP}' = \Delta\theta_{HP}/\cos\theta_0 \qquad (98)$$

which is the well-known result; namely, the half-power beamwidth broadens according to the cosine of the scan angle from the broadside direction.

Array Theory

Linear Transformations on Excitations

For simplicity, consider a linear array of N elements at $z = z_1, z_2, \ldots, z_N$ (in wavelengths) and with excitations $I_1, I_2, \ldots, I_N$, respectively. Consider further a nonsingular matrix $\mathbf{A}$ and let

$$\mathbf{I}' = \mathbf{A}^{-1}\mathbf{I} \quad \text{or} \quad \mathbf{I} = \mathbf{A}\mathbf{I}' \tag{99}$$

where $\mathbf{I}$ is the column vector with elements $I_1, I_2, \ldots, I_N$ and $\mathbf{I}'$ is another column vector, with elements $I'_1, I'_2, \ldots, I'_N$. Then the pattern of the array can be written as

$$\begin{aligned}
f(\theta) &= \sum_n^N I_n e^{j\psi_n} = \mathbf{V}^t \mathbf{I} = \mathbf{V}^t \mathbf{A} \mathbf{I}' \\
&= \mathbf{V}^t (\mathbf{A}_1, \mathbf{A}_2, \ldots, \mathbf{A}_N) \mathbf{I}' \\
&= \mathbf{V}^t \mathbf{A}_1 I'_1 + \mathbf{V}^t \mathbf{A}_2 I'_2 + \cdots + \mathbf{V}^t \mathbf{A}_N I'_N \\
&= F_1(\theta) I'_1 + F_2(\theta) I'_2 + \cdots + F_N(\theta) I'_N
\end{aligned} \tag{100}$$

where the superscript t denotes the transpose, and

$$\mathbf{V}^t = (e^{j\psi_1}, e^{j\psi_2}, \ldots, e^{j\psi_N}) \tag{101}$$

$$\psi_n = 2\pi z_n \cos\theta \tag{102}$$

$\mathbf{A}_1, \mathbf{A}_2, \ldots, \mathbf{A}_N$ are the first, second, $\ldots$, and Nth *columns* of matrix $\mathbf{A}$

$$F_n(\theta) = \mathbf{V}^t \mathbf{A}_n = \sum_m^N A_{mn} e^{j\psi_m} \tag{103}$$

Equation 103 states that $F_n(\theta)$ is the pattern of an array with excitations $A_{1n}, A_{2n}, \ldots, A_{Nn}$, and (100) states that the pattern of the array with excitations $I_1, I_2, \ldots, I_N$ is the sum of $F_1(\theta), F_2(\theta), \ldots, F_N(\theta)$ weighed by $I'_1, I'_2, \ldots, I'_N$, respectively. In other words, the pattern of an array with excitations $\{I_n\}$ can be expressed in terms of the patterns associated with some other, say canonical, excitations. An interesting application is given below.

Consider a linear uniformly spaced array of N elements with interelement spacing d (in wavelengths) and excitations $I_1, I_2, \ldots, I_N$. Let

$$\mathbf{A} = \begin{bmatrix}
1 & 1 & 1 & 1 & \cdots & 1 \\
1 & e^{-j2\pi/N} & e^{-j4\pi/N} & e^{-j6\pi/N} & \cdots & e^{-j2\pi(N-1)/N} \\
1 & e^{-j4\pi/N} & e^{-j8\pi/N} & e^{-j12\pi/N} & \cdots & e^{-j4\pi(N-1)/N} \\
\vdots & \vdots & \vdots & \vdots & & \vdots \\
1 & e^{-j2\pi(N-1)/N} & e^{-j4\pi(N-1)/N} & e^{-j6\pi(N-1)/N} & \cdots & e^{-j2\pi(N-1)^2/N}
\end{bmatrix} \tag{104}$$

Then

$$F(\theta) = \frac{1}{N}\sum_{n}^{N} I_n e^{j\psi_n} = I_1' F_1(\theta) + I_2' F_2(\theta) + \cdots + I_N' F_N(\theta) \quad (105)$$

where the factor $1/N$ has been added for convenience, and

$$F_1(\theta) = \mathbf{V}^t \mathbf{A}_1 = \frac{1}{N}\sum_{n=0}^{N-1} e^{j2\pi nd\cos\theta} = \frac{1}{N}\frac{\sin(N\psi/2)}{\sin(\psi/2)} \quad (106)$$

$$F_2(\theta) = \mathbf{V}^t \mathbf{A}_2 = \frac{1}{N}\sum_{n=0}^{N-1} e^{j2\pi nd(\cos\theta - 1/Nd)} = \frac{1}{N}\frac{\sin[N(\psi - \psi^{(1)})/2]}{\sin[(\psi - \psi^{(1)})/2]} \quad (107)$$

$$F_3(\theta) = \mathbf{V}^t \mathbf{A}_3 = \frac{1}{N}\sum_{n=0}^{N-1} e^{j2\pi nd(\cos\theta - 2/Nd)} = \frac{1}{N}\frac{\sin[N(\psi - \psi^{(2)})/2]}{\sin[(\psi - \psi^{(2)})/2]} \quad (108)$$

$$F_N(\theta) = \mathbf{V}^t \mathbf{A}_N = \frac{1}{N}\sum_{n=0}^{N-1} e^{j2\pi nd[\cos\theta - (N-1)/Nd]} = \frac{1}{N}\frac{\sin[N(\psi - \psi^{(N-1)})/2]}{\sin[(\psi - \psi^{(N-1)})/2]} \quad (109)$$

$$\psi = 2\pi d\cos\theta$$
$$\psi^{(m)} = 2\pi m/N$$

Equations 106–109 are simply uniform array patterns with a progressive phase shift such that the main beam appears at

$$\psi = \psi^{(m)} \quad \text{or} \quad \theta = \cos^{-1}(m/Nd) \quad (110)$$

The set of functions $\{F_n(\theta)\}$ has the following interesting properties:

(a) All their nulls coincide.

(b) Let the beam maximum of $F_n(\theta)$ be at $\theta_{\max}^{(n)}$. Then at this angle all other members of $\{F_n(\theta)\}$ vanish, i.e., $F_p(\theta_{\max}^{(n)}) = 1$ as $p = n$, and 0 as $p \neq n$. This result has been used for pattern synthesis [117, 118]. From (105)

$$F(\theta_{\max}^{(n)}) = I_n', \quad n = 1, 2, \ldots, N \quad (111)$$

Thus let the desired pattern be sampled at equispaced points in $\psi_{\max}^{(n)} = 2\pi d\cos\theta_{\max}^{(n)}$, for $n = 1, 2, \ldots, N$; then (111) states that these samples are simply $\{I_n'\}$ and from (99) one determines the required excitations:

$$\mathbf{I} = \mathbf{A}\mathbf{I}' = \mathbf{A}F(\theta_{\max}^{(n)}) \quad (112)$$

Of course, the synthesizing pattern agrees with the desired pattern exactly *only* at this set of points. Since in general there is some arbitrariness in selecting this set (such as the starting point, sampling rate, and sampling numbers—see "An Example," below), the solution is not unique. One should in general try a few possibilities, bearing in mind that the sampling rate depends on d, sampling number is the number of elements N, and array size is Nd (in wavelengths). For close agreement with the desired pattern, a high sampling rate and a large sampling

Array Theory

number are required, resulting in closely spaced elements. In that case impractical unstable solutions, as for the so-called superdirectivity problem, may occur (see Section 7 for more discussion).

To recapitulate, the philosophy of this method can be viewed as a linear transformation from the N degrees of freedom of $\{I_n\}$ to another N degrees of freedom of $\{I'_n\}$ as shown in (99). Therefore its application can be very general. For example, $F(\theta)$ can also be expressed in terms of a *set* of patterns, each associated with a *uniform aperture* excitation function $f_n(z)$ along z:

$$I'_n f_n(z) = I'_n e^{j2\pi z \cos\theta_n}, \qquad n = 1, 2, \ldots, N \tag{113}$$

The array so synthesized will be an aperture antenna. Mathematically speaking, the difference between this and the discrete array is rather trivial, mainly in the choice of "basis" functions only.

An Example—Fig. 15 shows an example considered by Balanis [4] in which the desired pattern is

$$F_d(\theta) = \begin{cases} 1 & \text{if } \pi/4 \leq \theta \leq 3\pi/4 \\ 0 & \text{elsewhere} \end{cases}$$

It is synthesized first by a linear array with 11 elements and $d = 1/2$, then by a continuous aperture distribution of length 5λ, both sampled at 11 points as shown in Table 2. Fig. 15b shows seven nonzero composing functions $\{F_n(\theta)\}$ and their sum.

Another Example—Fig. 16 shows two among many possible solutions for synthesizing a $\csc\theta$ pattern with a linear array of 21 elements and spacing $d = 1/2$. The first solution is obtained by sampling $\csc\theta$ such that the synthesizing array pattern is zero at $\theta = 0°$. The second solution is obtained by choosing the sampling points such that the synthesizing pattern is 0.45 at $\theta = 0°$. It is seen that the latter gives a closer solution to the desired pattern $\csc\theta$ over the useful region.

Circular Arrays

First consider a circular array of N elements uniformly excited and uniformly distributed over a circle of radius a (in wavelengths). Then the nth element will be located at

$$\mathbf{r}_n = \hat{\mathbf{x}} a \cos\phi_n + \hat{\mathbf{y}} a \sin\phi_n \tag{114}$$

where

$$\phi_n = 2\pi n/N, \qquad n = 1, 2, \ldots, N \tag{115}$$

Then the pattern function is

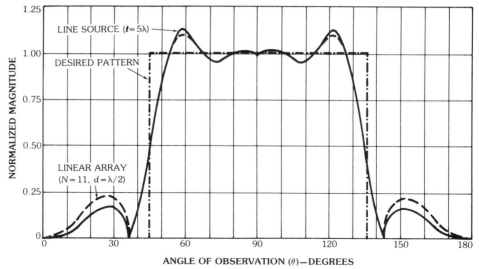

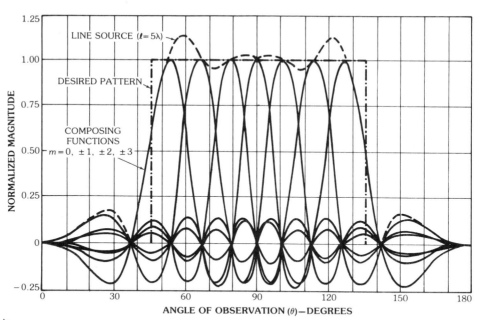

Fig. 15. Desired and synthesized patterns and composing functions, using Woodward's method [117, 4] for a linear array with 11 elements at $\lambda/2$ spacing and a line aperture 5 wavelengths long. (*a*) Normalized amplitude patterns. (*b*) Composing functions for line source ($l = 5\lambda$). (*After Balanis [4], © 1982, Harper & Row, Publishers, Inc.; reprinted by permission of the publisher*)

Table 2. Pattern Synthesized at $\{\theta_m\}$

m	θ_m	$F(\theta_m)$	m	θ_m	$F(\theta_m)$
0	90°	1			
1	78.46°	1	−1	101.54°	1
2	66.42°	1	−2	113.58°	1
3	53.13°	1	−3	126.87°	1
4	36.87°	0	−4	143.13°	0
5	0°	0	−5	180°	0

(After Balanis [4], © 1982 Harper & Row, Publishers, Inc., reprinted by permission of the publisher)

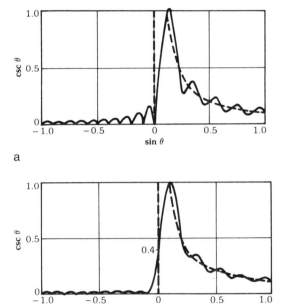

Fig. 16. Two possible synthesized patterns for the csc θ in dashed line. (*a*) With a value of 0 at $\theta = 0°$. (*b*) With a value of 0.45 at $\theta = 0°$. (*After Woodward and Lawson [119], © British Crown Copyright 1948*)

$$F(\theta, \phi) = \frac{1}{N} \sum_{n=1}^{N} e^{j2\pi \mathbf{r}_n \cdot \mathbf{u}} = \frac{1}{N} \sum_{n=1}^{N} e^{j2\pi a \sin\theta \cos(\phi - \phi_n)}$$

$$= \sum_{m=-\infty}^{\infty} J_{mN}(2\pi a \sin\theta) e^{jmN(\pi/2 - \phi)} \quad (116)$$

$$\cong J_0(2\pi a \sin\theta) \quad \text{if } 2\pi a/N \ll 1 \quad (117)$$

where use has been made of the expansion

$$e^{jz\cos\phi} = \sum_{m=-\infty}^{\infty} (j)^m J_m(z) e^{jm\phi} \tag{118}$$

and

$$\frac{1}{N}\sum_{n=1}^{N} e^{j2\pi mn/N} = \begin{cases} 1 & \text{if } m = Np, p = 0, \pm 1, \pm 2, \ldots \\ 0 & \text{otherwise} \end{cases} \tag{119}$$

The last approximation in (117) applies if $N \gg 2\pi a$, i.e., if the circumferential distance between any two adjacent elements, $2\pi a/N$, is sufficiently less than λ. In that case the pattern varies with $\sin\theta$ according to the $J_0(\cdot)$ function and independent of ϕ, as shown in Fig. 17a, where $|J_0(\cdot)|$ is plotted. When the above

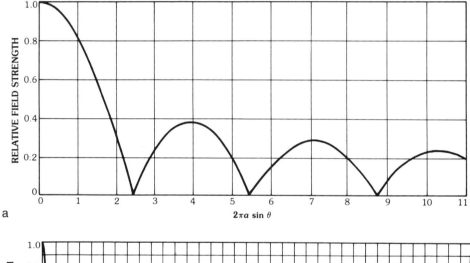

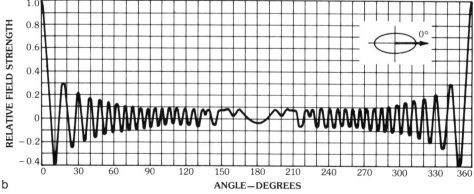

Fig. 17. Patterns of uniformly excited circular and elliptical arrays. (*a*) Pattern of a circular array with radius *a* and small element spacing. (*b*) Pattern of a cophasal elliptical array with major axis $2a = 20\lambda$, axial ratio $\tau = 3$, and main beam at $\phi_0 = 0°$. (*c*) Same as (*b*) but with $\phi_0 = 30°$. (*d*) Same as (*b*) but with $\phi_0 = 60°$. (*e*) Same as (*b*) but with $\phi_0 = 90°$. (*After Lo and Hsuan [66], © 1965 IEEE*)

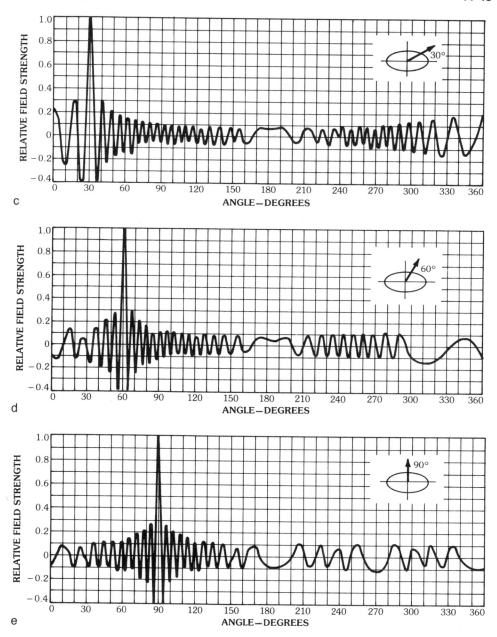

Fig. 17, *continued.*

condition is not satisfied, higher-order terms become significant and the pattern varies with ϕ. Using the transform discussed earlier in this section, under "Transformation between Circular and Elliptical Arrays," one can readily find the patterns of a uniformly excited cophasal elliptical array with major axis $2a = 20\lambda$, axial ratio = $\tau = 3$, and main beam in $\phi_0 = 0°$, $30°$, $60°$, and $90°$ directions as shown in Figs. 17b to 17e.

Cophasal Uniform Circular Arrays

For an array cophasal in the (θ_0, ϕ_0) direction the nth-element excitation is

$$I_n = e^{j2\pi \mathbf{r}_n \cdot \mathbf{u}_0} \tag{120}$$

where $\mathbf{u}_0 = \hat{\mathbf{x}} \sin\theta_0 \cos\phi_0 + \hat{\mathbf{y}} \sin\theta_0 \sin\phi_0$. Then

$$F(\theta, \phi) = \frac{1}{N} \sum_{n=1}^{N} e^{j2\pi \mathbf{r}_n \cdot (\hat{\mathbf{u}} - \mathbf{u}_0)} \tag{121}$$

Using (114) and the geometry in Fig. 18, one finds that

$$\mathbf{r}_n \cdot (\hat{\mathbf{u}} - \mathbf{u}_0) = a|\mathbf{u}_{xy} - \mathbf{u}_0| \cos(\xi - \phi_n) \tag{122}$$

where

$$\mathbf{u}_{xy} = \hat{\mathbf{u}} \cdot (\hat{\mathbf{x}}\hat{\mathbf{x}} + \hat{\mathbf{y}}\hat{\mathbf{y}}) = \text{projection of } \hat{\mathbf{u}} \text{ on the } xy \text{ plane}$$

$$\xi = \text{the azimuthal angle of } (\mathbf{u}_{xy} - \mathbf{u}_0) = \cos^{-1} \frac{\hat{\mathbf{x}} \cdot (\hat{\mathbf{u}} - \mathbf{u}_0)}{|\mathbf{u}_{xy} - \mathbf{u}_0|}$$

$$= \cos^{-1} \left\{ \frac{\sin\theta\cos\phi - \sin\theta_0\cos\phi_0}{[(\sin\theta\cos\phi - \sin\theta_0\cos\phi_0)^2 + (\sin\theta\sin\phi - \sin\theta_0\sin\phi_0)^2]^{1/2}} \right\} \tag{123}$$

Inserting (122) in (121) and comparing it with (116), nothing is really changed except that $|\mathbf{u}_{xy} - \mathbf{u}_0|$ replaces $\sin\theta$, and ξ replaces ϕ. Therefore

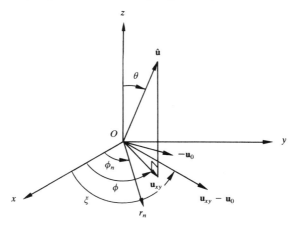

Fig. 18. Geometrical relation for cophasal circular array formulas (123)–(125).

Array Theory

$$F(\theta, \phi) = \sum_{m=-\infty}^{\infty} J_{mN}(2\pi a |\mathbf{u}_{xy} - \mathbf{u}_0|) e^{jmN(\pi/2 - \xi)}$$
$$\cong J_0(2\pi a |\mathbf{u}_{xy} - \mathbf{u}_0|), \quad \text{if } 2\pi a/N \ll 1 \qquad (124)$$

where ξ is given by (123) and

$$|\mathbf{u}_{xy} - \mathbf{u}_0| = [(\sin\theta\cos\phi - \sin\theta_0\cos\phi_0)^2 + (\sin\theta\sin\phi - \sin\theta_0\sin\phi_0)^2]^{1/2} \qquad (125)$$

Nonuniformly Excited Circular Arrays

A nonuniformly excited circular array can be analyzed in terms of uniformly excited circular arrays by applying a linear transformation on excitation. For example let $\{I_n\}$ be any given excitation. Now applying the transformation

$$\mathbf{I}' = \mathbf{A}^{-1}\mathbf{I} \quad \text{or} \quad \mathbf{I} = \mathbf{A}\mathbf{I}' \qquad (126)$$

where

$$\mathbf{A} = \begin{bmatrix} 1 & e^{j2\pi/N} & e^{j4\pi/N} & \cdots & e^{j2\pi(N-1)/N} \\ 1 & e^{j4\pi/N} & e^{j8\pi/N} & \cdots & e^{j4\pi(N-1)/N} \\ \cdot & \cdot & \cdot & \cdots & \cdot \\ 1 & e^{j2\pi N/N} & e^{j4\pi N/N} & \cdots & e^{j2\pi(N-1)N/N} \end{bmatrix}$$

Then, similar to (105) through (109),

$$F(\theta, \phi) = \frac{1}{N} \sum_{n=1}^{N} I_n e^{j2\pi \mathbf{r}_n \cdot \hat{\mathbf{u}}}$$
$$= I'_1 F_1(\theta, \phi) + I'_2 F_2(\theta, \phi) + \cdots + I'_N F_N(\theta, \phi) \qquad (127)$$

where

$$F_1(\theta, \phi) = \mathbf{V}^t \mathbf{A}_1 = \frac{1}{N} \sum_{n=1}^{N} e^{j2\pi a \sin\theta \cos(\phi - \phi_n)}$$
$$= \sum_{m=-\infty}^{\infty} J_{mN}(2\pi a \sin\theta) e^{jmN(\pi/2 - \phi)} \cong J_0(2\pi a \sin\theta) \qquad (128)$$

$$F_2(\theta, \phi) = \mathbf{V}^t \mathbf{A}_2 = \frac{1}{N} \sum_{n=1}^{N} e^{j2\pi[a \sin\theta \cos(\phi - \phi_n) + (n/N)]}$$
$$= \sum_{m=-\infty}^{\infty} J_{mN+1}(2\pi a \sin\theta) e^{j(mN+1)(\pi/2 + \phi)}$$
$$\cong J_1(2\pi a \sin\theta) e^{j(\pi/2 + \phi)} \qquad (129)$$

$$F_3(\theta,\phi) = \mathbf{V}^t \mathbf{A}_3 = \frac{1}{N}\sum_{n=1}^{N} e^{j2\pi[a\sin\theta\cos(\phi-\phi_m)+(2n/N)]}$$

$$= \sum_{m=-\infty}^{\infty} J_{mN+2}(2\pi a \sin\theta)\, e^{j(mN+2)(\pi/2+\phi)}\, e^{j4\pi/N}$$

$$\cong J_2(2\pi a \sin\theta)\, e^{j2(\pi/2+\phi)}$$

$$F_N(\theta,\phi) = \mathbf{V}^t \mathbf{A}_N = \sum_{m=-\infty}^{\infty} J_{mN+N-1}(2\pi a \sin\theta)\, e^{j(mN+N-1)(\pi/2+\phi)}$$

$$\cong J_{N-1}(2\pi a \sin\theta)\, e^{j(N-1)(\pi/2+\phi)} \tag{130}$$

As before, all the approximate solutions are valid only if $2\pi a/N \ll 1$. Similar to the linear array in (105), the nonuniform excitation is now expressed in terms of the uniform excitation given by (126). This transformation is known as the method of symmetrical components in power engineering for analyzing an unbalanced polyphase system in terms of the balanced ones and also multiple-arm spiral antennas [122].

Elliptical Arrays With Nonuniform Excitations

By applying the linear transformations (95) and (96) to each of the $\{F_n(\theta,\phi)\}$ of (127) one can then analyze an elliptical array with nonuniform excitation in terms of that with *uniformly* excited *circular* arrays. This case serves as an interesting example for applying simultaneously both transformations, one for the geometry and the other for the excitations. If the minor axis of the ellipse approaches zero, the transformation will give a closed-form solution for an *unequally* spaced *nonuniformly* excited linear array.

5. Planar Arrays

When a planar array has a separable excitation, say in x and y, an analysis can be made simply by regarding each row (or column) subarray as a single element and then considering all rows (or columns) to form a "linear" column (or row) array. In so doing, the theory for linear arrays applies. For example, the pattern function of a uniform rectangular array cophasal in the (θ_0,ϕ_0) direction is given by

$$F(\theta,\phi) = \frac{1}{M}\frac{\sin(M\psi_x/2)}{\sin(\psi_x/2)}\frac{1}{N}\frac{\sin(N\psi_y/2)}{\sin(\psi_y/2)} \tag{131}$$

where M and N are the number of elements along the x and y axes, respectively, and

$$\psi_x = 2\pi d_x(\sin\theta\cos\phi - \sin\theta_0\cos\phi_0)$$
$$\psi_y = 2\pi d_y(\sin\theta\sin\phi - \sin\theta_0\sin\phi_0)$$

and d_x and d_y are interelement spacings (in wavelengths) along the x and y axes, respectively.

Array Theory

A three-dimensional pattern for a 5 × 5 element array with $d_x = d_y = \frac{1}{2}$ and $\theta_0 = 0°$ is shown in Fig. 19. The first side lobe level in the $\phi = 0°$ and $90°$ planes is approximately -12 dB (as compared with -13.2 dB for a uniformly excited linear aperture) while that in the $\phi = 45°$ and $135°$ planes is -24 dB. Fig. 20 shows the pattern for the same array but with $d_x = d_y = 1$. In this case four grating lobes appear at $\phi = 0°, 90°, 180°,$ and $270°$ in the $\theta = 90°$ plane.

For nonseparable excitations the pattern function in general cannot be expressed in terms of a polynomial, and the pattern may have to be evaluated numerically.

Two-Dimensional Dolph-Chebyshev Arrays

One of the interesting problems is how to design a planar array which will have side lobes of equal level in the three-dimensional pattern. Evidently this cannot be achieved with two separable Dolph-Chebyshev excitations along the x and y directions as in the method discussed above, because in any plane other than $\phi = 0°$ and $90°$, the side lobes will not have the same level. Baklanov [123] and Tseng and Cheng [107] have given a simple but interesting solution to this problem. In essence, they introduced a transformation which generates from a linear Dolph-Chebyshev array a planar array with a pattern having equal side lobe levels

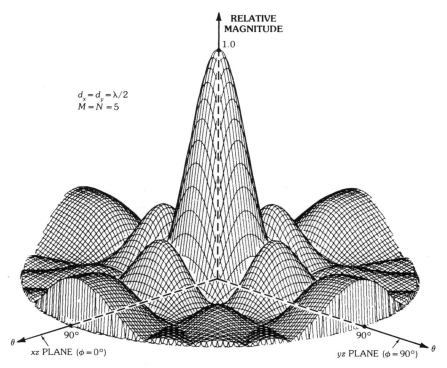

Fig. 19. Three-dimensional antenna pattern of a uniform planar array of isotropic elements with spacing $d_x = d_y = \lambda/2$. (*After Balanis [4]*, © 1982, Harper & Row, Publishers, Inc.; *reprinted by permission of the publisher*)

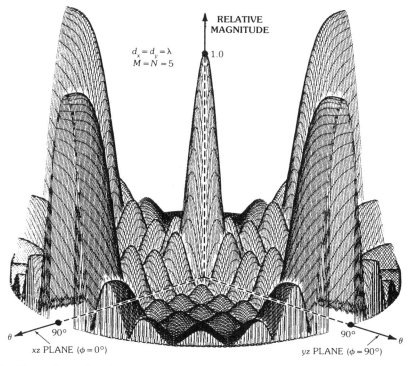

Fig. 20. Three-dimensional antenna pattern of a uniform planar array of isotropic elements with spacing $d_x = d_y = \lambda$. (*After Balanis [4], © 1982, Harper & Row, Publishers, Inc.; reprinted by permission of the publisher*)

in *every* cut of the ϕ = constant plane. Because of the nature of the transformation, however, the side lobe peaks do not occur at the same polar angle θ in all cuts, thus forming the so-called ringlike side lobes (i.e., not in concentric circles) in the three-dimensional patterns [28]. This, as will be seen later, is simply a result of the distortion introduced by the transformation in the u plane.

First, for simplicity, consider a square array in the xy plane with $2N \times 2N$ elements and interelement spacing d (in wavelengths) along both x and y directions. (If the unit cell is not a square, i.e., $d_x \neq d_y$, the pattern can be deduced from that of the square one simply by changing the scale of one of the u axes, not the magnitude of the pattern function as discussed in Section 4.) Let the center of the array be the origin and assume the excitation to be symmetrical with respect to both the x and y axes. Further let the excitation of the element at $x = md$ and $y = nd$ be I_{mn}, for $m, n = -N, \ldots, N$. Then

$$I_{mn} = I_{-m,n} = I_{m,-n} = I_{-m,-n} \tag{132}$$

and the pattern function

Array Theory

$$F(\theta, \phi) = \sum_{m,n=1}^{N} I_{mn} \cos[(2m-1)\cos^{-1}\alpha]\cos[(2n-1)\cos^{-1}\beta]$$

$$= \sum_{m,n=1}^{N} I_{mn} T_{2m-1}(\alpha) T_{2n-1}(\beta) \qquad (133)$$

where

$$\cos^{-1}\alpha = u_1 = \pi d \sin\theta \cos\phi \qquad (134)$$

$$\cos^{-1}\beta = u_2 = \pi d \sin\theta \sin\phi \qquad (135)$$

$T_m(\cdot)$ = the Chebyshev polynomial of degree m

Equation 133 states that $F(\theta, \phi)$ is a polynomial in α and β; thus one can write

$$F(\theta, \phi) = \sum_{m,n=1}^{N} A_{mn} \alpha^{2m-1} \beta^{2n-1} \qquad (136)$$

If I_{mn} can be chosen such that

$$A_{mn} = \delta_{mn} A_{mm} \qquad (137)$$

i.e., all $A_{mn} = 0$ if $m \neq n$, then

$$F(\theta, \phi) = \sum_{m=1}^{N} A_{mm}(\alpha\beta)^{2m-1} \qquad (138)$$

which is a polynomial of degree $2N-1$ in $\alpha\beta$. Let us equate this polynomial to the Chebyshev polynomial

$$F(\theta, \phi) = \sum_{m=1}^{N} A_{mm}(\alpha\beta)^{2m-1} = T_{2N-1}(w_0 \alpha\beta) \qquad (139)$$

with

$$T_{2N-1}(w_0) = R \qquad (140)$$

Then, for every plane cut ϕ = constant, $\alpha\beta$ is a function of θ only; thus $F(\theta, \phi)$ follows a Chebyshev pattern in θ. As $\theta = 0$, $\alpha\beta = 1$, and, therefore,

$$F(0, \phi) = T_{2N-1}(w_0) = R \qquad (141)$$

which is the main-beam magnitude with the side lobe level being 1. To recapitulate, the key point in this method is to reduce (136) to (139) by imposing (137). The former is a general planar array pattern while the latter is that of a linear array. One may thus consider this process as one which generates a two-dimensional

Dolph-Chebyshev pattern from a one-dimensional one, $T_{2n-1}(z)$, by using the transformation

$$z = w_0 \alpha \beta = w_0 \cos u_1 \cos u_2 \tag{142}$$

Beam Scanning—From the discussion for the one-dimensional case, if the main beam is scanned to (θ_0, ϕ_0), it is sufficient only to replace the definitions of u_1 and u_2 of (134) and (135) in the above analysis by

$$\cos^{-1}\alpha = u_1 - u_{01} = \pi d(\sin\theta\cos\phi - \sin\theta_0\cos\phi_0)$$
$$\cos^{-1}\beta = u_2 - u_{01} = \pi d(\sin\theta\sin\phi - \sin\theta_0\cos\phi_0)$$

Odd Number of Elements—From the basic principle given above, it is also evident that the method can also be applied to an array with $(2N + 1) \times (2N + 1)$ elements.

Symmetry—As a result of (137) one will find an additional symmetry of the excitation to those shown in (132):

$$I_{mn} = I_{nm} \tag{143}$$

Thus for such an array one needs to determine only $N(N + 1)/2$ excitation currents for an array with $2N \times 2N$ elements and $(N + 1)(N + 2)/2$ excitation currents for an array with $(2N + 1) \times (2N + 1)$ elements.

A Few Major Results*

(a) For an array with $L \times L$ elements, the excitation for the mnth element as determined from (133), (138), and (139) can be reduced to the following form:

$$I_{mn} = \left(\frac{4}{L}\right)^2 \sum_{p=1}^{N} \sum_{q=1}^{N} T_{L-1}\left(w_0 \cos p - \frac{1}{2}\frac{\pi}{L}\cos q - \frac{1}{2}\frac{\pi}{L}\right)$$
$$\times \cos\left(\frac{2\pi}{L}m - \frac{1}{2}p - \frac{1}{2}\right)\cos\left(\frac{2\pi}{L}n - \frac{1}{2}q - \frac{1}{2}\right), \quad \text{for } L = 2N \tag{144}$$

and

$$I_{mn} = \left(\frac{2}{L}\right)^2 \sum_{p=1}^{N+1} \sum_{q=1}^{N+1} \varepsilon_p \varepsilon_q T_{L-1}\{w_0 \cos[(p-1)\pi/L]\cos[(q-1)\pi/L]\}$$
$$\times \cos\left[\frac{2\pi}{L}(m-1)(p-1)\right]\cos\left[\frac{2\pi}{L}(n-1)(q-1)\right], \quad \text{for } L = 2N + 1 \tag{145}$$

---

*These are taken from Tseng and Cheng [107].

Array Theory

where $w_0 = \cosh\left(\dfrac{1}{L-1}\cosh^{-1}R\right)$.

(b) Beamwidth $\Delta\theta_c$ at level c/R, or $20\log_{10}(R/c)$ dB below the main-beam maximum at (θ_0, ϕ_0). Let θ_c be the solution to the equation

$$\cosh\left(\frac{1}{L-1}\cosh^{-1}c\right) = w_c = w_0 \cos u_{1c} \cos u_{2c}$$

where

$$u_{1c} = \pi d(\sin\theta_c - \sin\theta_0)\cos\phi_0$$
$$u_{2c} = \pi d(\sin\theta_c - \sin\theta_0)\sin\phi_0$$

Because of the nonlinearity in transformation (i.e., nonlinear distortion from the u plane to the $\theta\phi$ plane) the beamwidth at level c/R, as determined by θ_c in the above equations, varies with ϕ_0. In particular, if $\phi_0 = 0$,

$$\Delta\theta_c = \sin^{-1}\left[\sin\theta_0 + \frac{1}{\pi d}\cos^{-1}\left(\frac{w_c}{w_0}\right)\right]$$
$$- \sin^{-1}\left[\sin\theta_0 - \frac{1}{\pi d \cos\theta_0}\cos^{-1}\left(\frac{w_c}{w_0}\right)\right]$$
$$\cong 2\sin^{-1}\left[\frac{1}{\pi d \cos\theta_0}\cos^{-1}\left(\frac{w_c}{w_0}\right)\right], \quad \text{when } L \text{ is large} \quad (146)$$

If $\phi_0 = \pi/4$,

$$\Delta\theta_c = \sin^{-1}\left[\sin\theta_0 + \frac{\sqrt{2}}{\pi d}\cos^{-1}\left(\frac{w_c}{w_0}\right)\right] - \sin^{-1}\left[\sin\theta_0 - \frac{\sqrt{2}}{\pi d}\cos^{-1}\left(\frac{w_c}{w_0}\right)\right]$$
$$\cong 2\sin^{-1}\left(\frac{\sqrt{2}}{\pi d \cos\theta_0}\right)\cos^{-1}\left(\frac{w_c}{w_0}\right), \quad \text{when } L \text{ is large} \quad (147)$$

(c) Minimum number of elements. This is determined by the maximum permissible spacing d for a given array area so that no grating lobe will appear in a given scanning range. As discussed in Section 4, under "Beam and Pattern Distortion Due to Scanning," since there is no distortion in azimuthal angles in the transformation from the (θ, ϕ) plane to the u plane, one may expect that the maximum value of d depends only on the maximum scan angle of θ_0, designated by θ_M. Then

$$d \leqq \frac{1 - (1/\pi)\cos^{-1}(1/w_0)}{1 + \sin\theta_M} \quad (148)$$

(d) Some numerical results for (c). Figs. 21 and 22 show the largest beamwidth $(\Delta\theta_c)_M$ at the 30-dB level below the main beam for maximum scanning angles $\theta_M =$

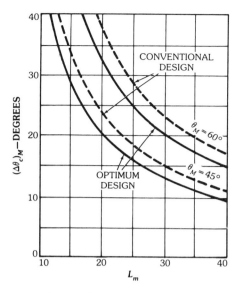

Fig. 21. Beamwidth versus minimum element number L_m on each side of a square array ($c/R = 0.03$, 30-dB side lobe level). (*After Tseng and Cheng [107]*, © *1968 IEEE*)

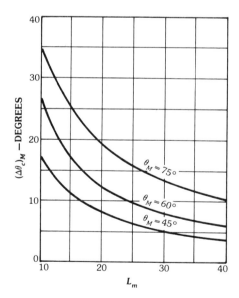

Fig. 22. Beamwidth versus minimum element number L_m on each side of a square array (optimum design: $c/R = 0.707$, 30-dB side lobe level). (*After Tseng and Cheng [107]*, © *1968 IEEE*)

Array Theory

60° and 45° and an array with minimum number of elements $L_m \times L_m$, or largest permissible spacing as given by (148). The design is for an equal side lobe level of -30 dB in all planes cutting through the main-beam maximum. Also shown for comparison is the beamwidth (in dashed curves) under the same condition but based on the conventional separable product of two Dolph-Chebyshev linear array designs. The beamwidth of the Baklanov-Tseng-Cheng design is always narrower than that of the conventional design, but the side lobes of the latter in planes other than the principal ones are lower. Fig. 22 shows the half-power beamwidth versus L_m for the same array condition as in Fig. 21. Fig. 23 shows the largest element spacing d_M permissible versus L for various scanning ranges $\theta_M = 45°$, $60°$, and $75°$ and for a -30-dB side lobe level. Directivity of this design may be higher or lower than that of the conventional design.

A Numerical Example—An array with 10×10 elements, spacing $d_x = 1/2$, $d_y = 3/4$, -20-dB side lobe level, and $\theta_0 = 0°$ has been considered by Elliott [28]. Table 3 shows the relative currents in all elements, and Figs. 24a through 24d show the patterns in four different planes $\phi = 0°$, $30°$, $60°$, and $90°$. It is seen that all side lobes have equal levels of -20 dB but are displaced along the θ axis. Since for this example $d_x \neq d_y$, the displacements are different in these planes. In fact the beamwidth is narrower in the $\phi = 90°$ plane than that in the $\phi = 0°$ plane because $d_y > d_x$. If $d_x = d_y$, the patterns in these two planes would be identical and the beamwidth would become the widest in the $\phi = 45°$ plane.

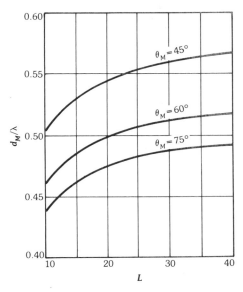

Fig. 23. Largest element spacing d_M (in wavelengths) versus L, the element number on each side of a square array, for nonappearance of grating lobes for a 30-dB side lobe level. (*After Tseng and Cheng [107], © 1968 IEEE*)

Table 3. Current Distributions for 10 × 10 Array to Give 20-dB Tseng-Cheng Pattern [28]

I_{mn}	20-dB Tseng-Cheng
I_{11}	0.773
$I_{21} = I_{12}$	0.569
$I_{31} = I_{13}$	0.796
$I_{41} = I_{14}$	0.029
$I_{51} = I_{15}$	1.000
I_{22}	0.946
$I_{32} = I_{23}$	0.119
$I_{42} = I_{24}$	0.618
$I_{52} = I_{25}$	0.667
I_{33}	0.486
$I_{43} = I_{34}$	0.777
$I_{53} = I_{35}$	0.286
I_{44}	0.387
$I_{54} = I_{45}$	0.071
I_{55}	0.008

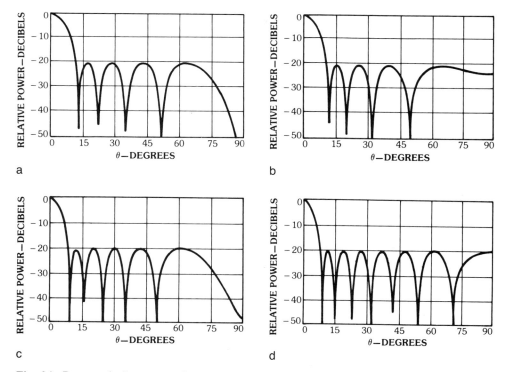

Fig. 24. Patterns in four ϕ cuts for a 10 × 10 rectangular Baklanov-Tseng-Cheng array with $d_x = 0.5\lambda$, $d_y = 0.75\lambda$, and side lobe level of 20 dB. (*a*) For $\phi = 0°$. (*b*) For $\phi = 30°$. (*c*) For $\phi = 60°$. (*d*) For $\phi = 90°$. (*After Tseng and Cheng,* Radio Science, *vol. 12, pp. 653–57,* © *American Geophysical Union*)

General Discussion of the Transformation

To see the distortion introduced by the transformation (142), a set of universal contour lines defined by

$$\cos u_1 \cos u_2 = c \quad \text{(constant)}$$

for various values of c from 0.01 to 1.00 are plotted in Fig. 25, with u_1 and u_2 defined by (134) and (135). Along the u_1 axis (i.e., $\phi = 0$) the pattern, as given by (139), is exactly that of a linear Dolph-Chebyshev array $T_{2N-1}(w_0 \alpha)$ with $\alpha = \cos(\pi d_x \sin \theta)$. Along the u_2 axis the pattern is $T_{2N-1}(w_0 \beta)$ with $\beta = \cos(\pi d_y \sin \theta)$. They are identical if $d_x = d_y$. (If $d_x \neq d_y$, only a linear change of the scale in u is needed.) For any $\phi = $ constant plane, the pattern is given by $T_{2N-1}(w_0 \cos u_1 \cos u_2)$, which is again the same as that of a linear Dolph-Chebyshev array except that the scale in $\sin \theta$ is changed, depending on

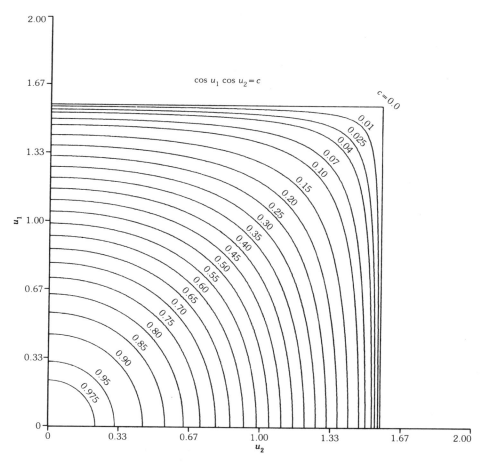

Fig. 25. Plots of $\cos u_1 \cos u_2 = c$ in the $u_1 u_2$ plane for various values of c. *(Prepared by Michael L. Oberhart)*

the value of ϕ. Fig. 25 shows exactly the dependence of this scale change on ϕ. (Note that the polar angle of any point (u_1, u_2) in the plane is simply ϕ.) As shown, the scale stretches to the largest extent at $\phi = 45°$. To find how this change is reflected in the $\theta\phi$ scale, one needs only to place an overlay on this figure as shown in Fig. 14.

As discussed earlier, a polynomial $P(z)$ can always be factorized in terms of its roots z_k:

$$P(z) = A \prod_k (z - z_k)$$

The pattern magnitude is thus proportional to the product of all the phasor magnitudes $|z - z_k|$. When z passes through a certain z_k to z_{k+1} one obtains either a main beam or a side lobe. In the latter case the locations of z_k and z_{k+1} will generally have a major influence on the height of that side lobe level, since all other roots are presumably farther away from them. Using this fact Elliott [28] is able to perturb certain roots of a Chebyshev polynomial for achieving a certain side lobe topology. For an array with an even number of elements, if the excitations of any two symmetrically located elements are equal in magnitude and opposite in phase, one obtains a sharp null, instead of a main beam, in the broadside direction. Optimum designs for nearly equal side lobes, except for the first two next to the broadside null of the so-called difference pattern, have been considered in great detail by Elliott [28] and Bayliss [6].

6. Optimization of Directivity (*D*) and Signal-to-Noise Ratio (*SNR*)

When an array is used for transmitting, the maximization of D is of interest, and when an array is used for reception, the maximization of SNR is of interest. These two problems can be solved in the same manner. In fact, the problem of jamming or interference can be treated similarly.

Formulation and Solution

Consider an arbitrary array with N elements. Let its nth element be located at $\mathbf{r}_n = (x_n, y_n, z_n)$ all measured in wavelengths and be excited with current

$$I_n = J_n e^{-j\mathbf{r}_n \cdot \mathbf{u}_0} \tag{149}$$

where

$$\mathbf{r}_n \cdot \mathbf{u}_0 = 2\pi(x_n \sin\theta_0 \cos\phi_0 + y_n \sin\theta_0 \sin\phi_0 + z_n \cos\theta_0) \tag{150}$$

(θ_0, ϕ_0) = angular direction in which D is to be maximized, or from which the signal is to be received

In (149), $\{J_n\}$ can be complex and will be determined. If $\{J_n\}$ is real, $\{I_n\}$ will be cophasal excitation.

Array Theory

For convenience, we shall use *bra-ket* notation as defined below:

$$\langle J = a\ row\ \text{vector} = (J_1, J_2, \ldots, J_N) \tag{151}$$

$$\langle V = a\ row\ \text{vector} = (\exp[j(\psi_1^0 - \psi_1)], \exp[j(\psi_2^0 - \psi_2)], \ldots, \exp[j(\psi_N^0 - \psi_N)]) \tag{152}$$

$J\rangle$ and $V\rangle$ are transposes of $\langle J$ and $\langle V$, respectively, and are therefore column vectors

where

$$\psi_n = 2\pi(x_n \sin\theta \cos\phi + y_n \sin\theta \sin\phi + z_n \cos\theta) \tag{153}$$

$$\psi_n^0 = \psi_n\big|_{\theta=\theta_0, \phi=\phi_0} \tag{154}$$

Then the pattern function is given by

$$F(\theta, \phi) = \sum_n^N J_n \exp[j(\psi_n - \psi_n^0)] = \langle JV^*\rangle \tag{155}$$

The power density, or the received signal in the direction (θ_0, ϕ_0), is proportional to

$$|F(\theta, \phi)|^2 = \langle JV_1^*\rangle\langle V_1 J^*\rangle = \langle JCJ^*\rangle \tag{156}$$

where

$V_1\rangle = V\rangle\big|_{\theta=\theta_0, \phi=\phi_0}$

$\mathbf{C}$ = a dyad, or a special $N \times N$ matrix:

$$\mathbf{C} = V_1^*\rangle\langle V_1 = V^*\rangle\langle V\big|_{\phi=\theta_0, \phi=\phi_0} = \begin{bmatrix} 1 & 1 & \cdots & 1 \\ 1 & 1 & \cdots & 1 \\ \cdot & \cdot & \cdots & \cdot \\ 1 & 1 & \cdots & 1 \end{bmatrix} \tag{157}$$

If the array is in a noisy environment with temperature distribution $T(\theta, \phi)$, the total received noise power is proportional to

$$P_n = \langle JAJ^*\rangle \tag{158}$$

where $\mathbf{A}$ is an $N \times N$ matrix given by

$$\mathbf{A} = \frac{1}{4\pi}\int_{4\pi} V^*\rangle\langle V\, T(\theta, \phi)\, d\Omega \tag{159}$$

Therefore the signal-to-noise ratio is proportional to

$$SNR = \frac{\langle JCJ^* \rangle}{\langle JAJ^* \rangle} \tag{160}$$

Similarly, the average of total radiated power is proportional to

$$P_{av} = \frac{1}{4\pi} \int_{4\pi} \langle JV^* \rangle \langle VJ^* \rangle \, d\Omega = \langle JBJ^* \rangle \tag{161}$$

where **B** is an $N \times N$ matrix given by

$$\mathbf{B} = \frac{1}{4\pi} \int_{4\pi} V^* \rangle \langle V \, d\Omega = \mathbf{A}\big|_{T=1} \tag{162}$$

Therefore the directivity is given by

$$D = \frac{\langle JCJ^* \rangle}{\langle JBJ^* \rangle} \tag{163}$$

The objective is to maximize D for directivity, or to maximize the SNR for reception in the given noise environment $T(\theta, \phi)$, with the N degrees of freedom $J_1, J_2, \ldots, J_N$. Before proceeding, it may be noted that

$$SNR = \frac{\langle JCJ^* \rangle}{\langle JBJ^* \rangle} \bigg/ \frac{\langle JAJ^* \rangle}{\langle JBJ^* \rangle} = D/T_{av} \tag{164}$$

where

$$T_{av} = \langle J(1/4\pi) \int_{4\pi} V^* \rangle \langle VT(\theta,\phi) \, d\Omega J^* \rangle \big/ \langle J(1/4\pi) \int_{4\pi} V^* \rangle \langle V \, d\Omega J^* \rangle$$

= temperature of the received noise power if it were distributed uniformly in space

= equivalent uniform temperature distribution for the same noise power received by the array (165)

It may also be noted that for interference, or jamming signal, with power density distribution, $I(\theta, \phi)$, uncorrelated in space, the signal-to-interference (S/I) ratio is also given by (160) with $T(\theta, \phi)$ replaced by $I(\theta, \phi)$.

Therefore (a) the problem of maximizing D/T_{av} is exactly the same as that of maximizing the SNR, (b) the problem of maximizing D is the same as that of maximizing the SNR with $T = 1$, i.e., uniform temperature distribution in space, and (c) the problem of maximizing S/I is the same as that of maximizing the SNR. Hence, for all these problems, one needs only to consider the maximization of the SNR; but the optimum solution for $\{I_n\}$ depends on the matrix **A**, or **B**, or both.

Solutions to all these problems may lead to unrealistic results which are generally referred to as the "supergain," "superdirectivity," "ill-conditioned," or "improperly posed" problems [9, 65, 120]. To make the solutions more physically meaningful, a constraint may be imposed on the so-called array Q factor

$$Q = \frac{\langle JJ^* \rangle}{\langle JBJ^* \rangle} \quad (166)$$

which follows from Taylor's definition of *supergain ratio* for an aperture antenna [100]. Except for a constant and the quality factor of a single element, the above definition of the array Q factor has the usual meaning of 2π times the total stored energy divided by the energy radiated per cycle if the mutual impedance effect can be neglected.

The most general optimization problem can thus be stated as follows:

Given: Array geometry, direction of desired signal (θ_0, ϕ_0), $T(\theta, \phi)$, and constraint on the Q factor.

To find: $\langle J$ (or $\langle I$) such that the *SNR* is maximized.

Solution: Using the fact that matrices **A**, **B**, and **C** are Hermitian, the solutions can be found for various cases as summarized in Table 4 [65]. The optimum $\langle J$ as listed in the last row in Table 4 is for maximum *SNR* with a prescribed Q factor. When there is a constraint on the Q factor, one needs first to solve for p in the following eigenvalue equation:

$$\det(V_1\rangle, W_2\rangle, \ldots, W_N\rangle) = 0 \quad (167)$$

where

$\mathbf{I}$ = identity matrix
$V_1\rangle = (1, 1, \ldots, 1)^t$
$W_n\rangle = p^2(Q\mathbf{B} - \mathbf{I})V_n^*\rangle + 2p\mathbf{A}V_n^*\rangle + \mathbf{A}(Q\mathbf{B} - \mathbf{I})^{-1}\mathbf{A}V_n^*\rangle$,
    for $n = 2, 3, \ldots, N$
$V_2\rangle = (-1, 1, 0, 0, \ldots, 0)^t$
$\cdots \quad \cdots$
$V_N\rangle = (-1, 0, 0, \ldots, 1)^t$

In words, $V_n\rangle$, for $n = 2, \ldots, N$, is a column vector with -1 as its first element, $+1$ as its nth element, and 0 for all other elements. Once p is found, one can compute the matrix

$$\mathbf{K} = \mathbf{A} + p(Q\mathbf{B} - \mathbf{I}) \quad (168)$$

and the optimum $J\rangle$ as shown in the last row in Table 4:

$$J\rangle = \mathbf{K}^{-1*} V_1\rangle \quad (169)$$

Table 4. Formulas for Optimum Gain and SNR of an Arbitrary Array

	Current	D	SNR	Q Factor
Definition	$J\rangle$	$\dfrac{\|\langle JV_1^*\rangle\|^2}{\langle JBJ^*\rangle}$	$\dfrac{\|\langle JV_1^*\rangle\|^2}{\langle JAJ^*\rangle}$	$\dfrac{\langle JJ^*\rangle}{\langle JBJ^*\rangle}$
Uniform current excitation	$V_1^*\rangle$	$\dfrac{N^2}{\langle V_1 BV_1^*\rangle}$	$\dfrac{N^2}{\langle V_1 AV_1^*\rangle}$	$\dfrac{N}{\langle V_1 BV_1^*\rangle}$
Optimum D without constraint on Q	$\mathbf{B}^{*-1}V_1^*\rangle$	$\langle V_1 \mathbf{B}^{-1} V_1^*\rangle$	$\dfrac{\|\langle V_1 \mathbf{B}^{-1} V_1^*\rangle\|^2}{\langle V_1 \mathbf{B}^{-1}\mathbf{A}\mathbf{B}^{-1} V_1^*\rangle}$	$\dfrac{\langle V_1 \mathbf{B}^{-2} V_1^*\rangle}{\langle V_1 \mathbf{B}^{-1} V_1^*\rangle}$
Optimum D with a prescribed Q	$\mathbf{F}^{*-1}V_1^*\rangle$	$\dfrac{\|\langle V_1 \mathbf{F}^{-1} V_1^*\rangle\|^2}{\langle V_1 \mathbf{F}^{-1}\mathbf{B}\mathbf{F}^{-1} V_1^*\rangle}$	$\dfrac{\|\langle V_1 \mathbf{F}^{-1} V_1^*\rangle\|^2}{\langle V_1 \mathbf{F}^{-1}\mathbf{A}\mathbf{F}^{-1} V_1^*\rangle}$	A given constant
Optimum SNR without constraint on Q	$\mathbf{A}^{*-1}V_1^*\rangle$	$\dfrac{\|\langle V_1 \mathbf{A}^{-1} V_1^*\rangle\|^2}{\langle V_1 \mathbf{A}^{-1}\mathbf{B}\mathbf{A}^{-1} V_1^*\rangle}$	$\langle V_1 \mathbf{A}^{-1} V_1^*\rangle$	$\dfrac{\langle V_1 \mathbf{A}^{-2} V_1^*\rangle}{\langle V_1 \mathbf{A}^{-1}\mathbf{B}\mathbf{A}^{-1} V_1^*\rangle}$
Optimum SNR with a prescribed Q	$\mathbf{K}^{*-1}V_1^*\rangle$	$\dfrac{\|\langle V_1 \mathbf{K}^{-1} V_1^*\rangle\|^2}{\langle V_1 \mathbf{K}^{-1}\mathbf{B}\mathbf{K}^{-1} V_1^*\rangle}$	$\dfrac{\|\langle V_1 \mathbf{K}^{-1} V_1^*\rangle\|^2}{\langle V_1 \mathbf{K}^{-1}\mathbf{A}\mathbf{K}^{-1} V_1^*\rangle}$	A given constant

Symbolism

Actual current in the nth element $= J_n e^{-j\psi_n^0}$
$\langle J = (J_1, J_2, \cdots, J_N)$, $\quad \langle V_1 = (1, 1, \cdots, 1)$
$\psi_n = 2\pi(x_n \sin\theta\cos\phi + y_n \sin\theta\sin\phi + z_n\cos\theta)$
$\psi_n^0 = \psi_n\big|_{\theta=\theta_0,\,\phi=\phi_0}$
$\mathbf{K} = \mathbf{A} + p(Q\mathbf{B} - \mathbf{I})$, $\qquad \mathbf{F} = \mathbf{K}\big|_{\mathbf{A}=\mathbf{B}}$
$\mathbf{A} = (1/4\pi)\int_{4\pi} V^*\rangle\langle VT(\theta,\phi)\,d\Omega$, $\qquad \mathbf{B} = \mathbf{A}\big|_{T=1}$

(After Lo, Lee, and Lee [65], © 1966 IEEE)

From (168) one can compute its associated directivity

$$D = \frac{|\langle V_1 \mathbf{K}^{-1} V_1^*\rangle|^2}{\langle V_1 \mathbf{K}^{-1}\mathbf{B}\mathbf{K}^{-1} V_1^*\rangle} \tag{170}$$

and its associated SNR

$$SNR = \frac{|\langle V_1 \mathbf{K}^{-1} V_1^*\rangle|^2}{\langle V_1 \mathbf{K}^{-1}\mathbf{A}\mathbf{K}^{-1} V_1^*\rangle} \tag{171}$$

One of the eigenvalues p of (167) which gives the largest value of SNR is the optimum solution. Since the solution $J\rangle$ is determined for maximum SNR, D as computed from (170) using that $J\rangle$ is not necessarily maximum.

- For maximum *SNR* without constraint on Q the solution is given in row 5 of Table 4. For this case, $p = 0$ and no eigenvalue needs to be computed. Therefore from (168) $\mathbf{K} = \mathbf{A}$ and

$$J\rangle = \mathbf{A}^{-1*} V_1\rangle \tag{172}$$

- For maximum D with and without constraint on the Q factor, the solutions are listed in rows 4 and 3 of Table 4, respectively. The solutions are exactly the same as those for maximum *SNR* except that $T(\theta, \phi) = 1$ and $\mathbf{A} = \mathbf{B}$. The corresponding $\mathbf{K}$ matrix is denoted by

$$\mathbf{F} = \mathbf{B} + p(Q\mathbf{B} - \mathbf{I}), \quad \text{for a constraint on the } Q \text{ factor} \tag{173}$$

$$\mathbf{F} = \mathbf{B}, \quad \text{for no constraint} \tag{174}$$

- Since $V_1\rangle = (1, 1, \ldots, 1)'$, the optimum solution for J_n is simply the *sum* of all the elements in the nth row of $\mathbf{B}^{*-1}$, $\mathbf{F}^{-1*}$, $\mathbf{A}^{-1*}$, and $\mathbf{K}^{-1*}$, respectively, for each case. The actual current for the nth element, by definition, is given by (149).

Planar Array with Isotropic Elements or Vertical Dipoles in the (x, y) *Plane*

Referring to Fig. 26, the elements of matrix $\mathbf{B}$ for this array can be integrated in closed form:

$$b_{nm} = b^*_{mn} = e^{-j\psi^0_{nm}} \left[\frac{\sin 2\pi \varrho_{nm}}{2\pi \varrho_{mn}} + q \frac{\cos 2\pi \varrho_{nm}}{(2\pi \varrho_{nm})^2} - q \frac{\sin 2\pi \varrho_{nm}}{(2\pi \varrho_{nm})^2} \right],$$
$$\text{for } n < m \tag{175}$$

$$b_{nn} = 1 - q/3 \tag{176}$$

where

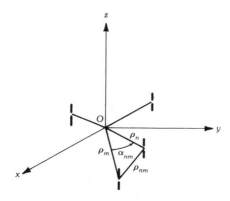

Fig. 26. A planar array with elements (vertical dipoles or isotropic) in the *xy* plane.

$$\psi_{mn}^0 = 2\pi\varrho_{nm} \sin\theta_0 \cos(\phi_0 - \alpha_{nm})$$
$$\varrho_{nm} = [(x_n - x_m)^2 + (y_n - y_m)^2]^{1/2}$$
$$= \text{distance between the } n\text{th and } m\text{th elements}$$
$$(x_n, y_n) = \text{coordinates of the } n\text{th element}$$
$$\alpha_{nm} = \tan^{-1}[(y_n - y_m)|(x_n - x_m)], \text{ where } 0 \leq \alpha_{nm} < \pi$$
$$q = \begin{cases} 0 & \text{for isotropic elements} \\ 1 & \text{for vertical dipoles} \end{cases}$$

From the above a few interesting results can be obtained:

(a) In particular for a linear array, say along the x axis, with N isotropic elements,

$$b_{nm} = b_{mn}^* = e^{-j\psi_{nm}^0} \frac{\sin 2\pi\varrho_{nm}}{2\pi\varrho_{mn}}, \quad n < m \tag{177}$$

$$b_{nn} = 1 \tag{178}$$

where

$$\varrho_{nm} = x_n - x_m$$
$$\psi_{nm}^0 = 2\pi\varrho_{mn} \sin\theta_0 \cos\phi_0$$

For uniform interelement spacing equal to $\lambda/2$ or its multiples,

$$b_{nm} = \begin{cases} 0 & \text{for } n \neq m \\ 1 & \text{for } n = m \end{cases}$$

Thus

$$\mathbf{B} = \mathbf{I} = \text{identity matrix}$$
$$Q = 1, \text{ independent of } J\rangle$$
$$\max D = N \tag{179}$$

and

$$\text{optimum } \langle J = (1, 1, \ldots, 1) \tag{180}$$

In other words, if $d = m/2$ (a multiple of $\lambda/2$), the *cophasal uniform array is optimum*. This conclusion is approximately true for dipole arrays since, for $d = m/2$, the diagonal elements of $\mathbf{B}$ dominate over all others, and $\mathbf{B}$ is approximately $\mathbf{I}$.

(b) For broadside planar array $\psi_{nm}^0 = 0$. Thus the $\mathbf{B}$ matrix is real and the optimum excitation is real. In other words, for broadside arrays *optimum*

Array Theory

excitations are either in phase or antiphase. A few examples are shown in Figs. 27 through 31. In Figs. 30 and 31 the term "cophasal" implies that only the excitation amplitudes are optimized while the phases are kept cophasal in the end-fire direction.

(c) For planar arrays with large interelement spacings, say much greater than a wavelength, $\mathbf{B} \cong \mathbf{I}$. *The uniform cophasal excitation is nearly optimum, max* $D \cong N$ *and* $Q \cong 1$. Some workers have sought optimum spacings and excitations for a thinned array; this result shows that the uniform cophasal excitation is the solution no matter what element spacings are as long as they are sufficiently large as in a thinned array.

(d) From the definition of the array Q factor, the value of $1/Q$ is bounded between the smallest and the largest eigenvalue of the $\mathbf{B}$ matrix. For equally spaced linear arrays with isotropic elements, $Q = 1$ as $d = m/2$, $m = 1, 2, \ldots$, and $Q \cong 1$ as $d > 1$, no matter what $\langle J$ is.

(e) From the example to be discussed next, it will be seen that the optimum excitation for a linear end-fire array with uniform spacings is nearly *antiphase for* $d \leqq 1/2$ *and nearly cophasal for* $d \geqq 1/2$.

(f) Gilbert and Morgan [124] showed that the *average* of *maximum* directivity over all directions is equal to N, the number of isotropic elements, i.e.,

$$\frac{1}{4\pi} \int_{4\pi} \max D(\theta_0, \phi_0) \, d\Omega = N \tag{181}$$

Thus if max D is greater than N in some direction, it must be smaller than N in some other direction.

(g) In general, ohmic losses in most antennas with low Q are of little importance in antenna efficiency. However, for antennas with strong local fields and large circulating currents, such as in superdirective arrays, the efficiency for radiation in (θ_0, ϕ_0) may be defined as

$$\eta = \frac{\text{radiated power density in } (\theta_0, \phi_0)}{\text{radiated power density in } (\theta_0, \phi_0) + \text{ohmic power loss}} = \frac{1}{1 + rS} \tag{182}$$

where

r = ohmic resistance of each element,

$$S = \frac{Q}{D} = \frac{\langle JJ^* \rangle}{\langle JCJ^* \rangle} \tag{183}$$

The S parameter, called the *sensitivity factor* [109], is a measure of the mean-square variation of the maximum field with respect to the mean-square deviation of the excitation. Thus large Q results in not only low efficiency but also high sensitivity, as will be seen in the following example, Section 7, and [8, 9, 53, 120].

A Typical Example for Maximum Directivity

Consider an equally spaced linear end-fire array with 10 isotropic elements along the x axis. Using the results in row 3 of Table 4 the maximum directivity can be

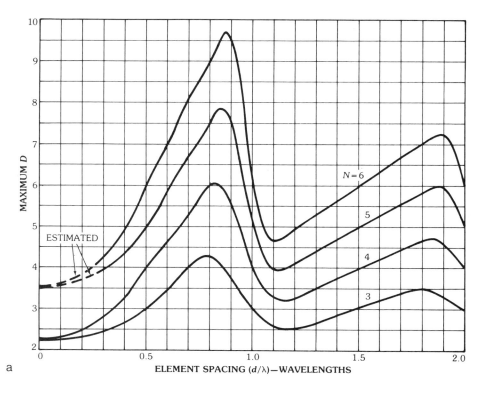

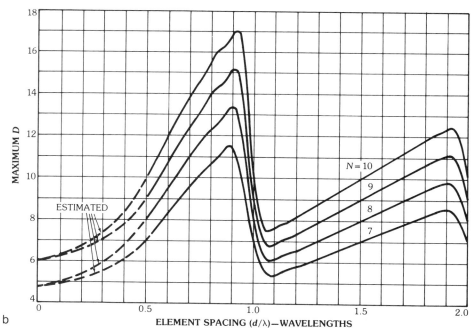

Fig. 27. Maximum directivity D of a linear broadside array of N isotropic elements versus element spacing d in wavelengths. (*a*) For $N = 3$ through 6. (*b*) For $N = 7$ through 11. (*Courtesy C. T. Tai*)

Array Theory

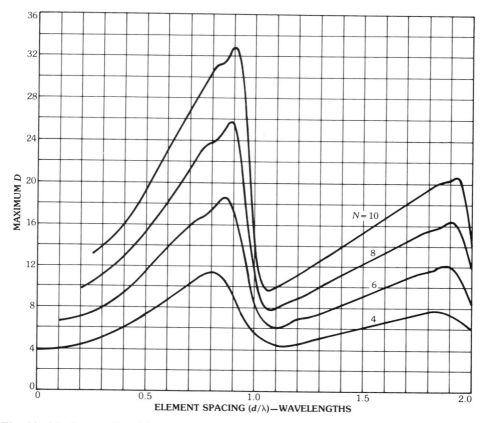

Fig. 28. Maximum directivity of a linear broadside array of N parallel dipoles versus element spacing d (in wavelengths) for $N = 4, 6, 8,$ and 10. (*Courtesy C. T. Tai*)

computed for various values of spacing d as shown by the curve $D(O)$ in Fig. 32. As expected, max $D = N = 10$ at $d = m/2$, $m = 1, 2, \ldots$. As d decreases below ½, D increases rapidly and approaches $N^2 = 100$ as d approaches zero. For comparison a few other cases are also shown in the figure:

$D(OC)$ = maximum directivity with only the excitation *magnitudes* subject to optimization and with the phases confined to the cophasal condition in the end-fire direction

$D(U)$ = directivity for a uniform cophasal excitation

$D(HW)$ = directivity for the Hansen-Woodyard (HW) excitation (see Section 3, under "Uniform Arrays With N Elements")

Fig. 33 shows the Q factor versus d for all the cases just stated. From these two figures the following remarks can be made:

(*a*) $D(O)$ is considerably higher than $D(U)$ only when $d \leqq 0.4$. A moderate

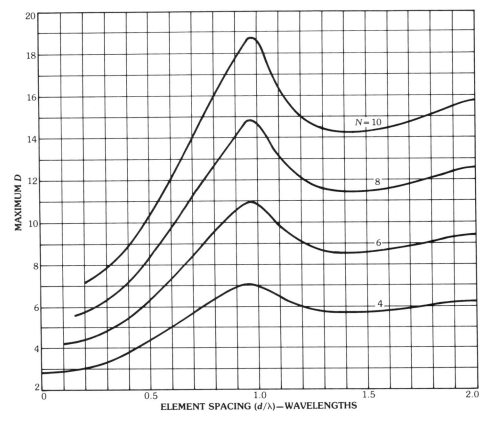

Fig. 29. Maximum directivity of a linear broadside array of N collinear dipoles versus element spacing d (in wavelengths) for $N = 4, 6, 8,$ and 10. (*Courtesy C. T. Tai*)

improvement of $D(O)$ over $D(U)$ can be obtained only at the price of astronomically large Q, which implies extremely high local field, dissipation loss, sensitivity, and extremely low efficiency and narrow bandwidth. Therefore a superdirective array is more a fantasy than reality.

(*b*) Hansen-Woodyard excitation is not truly optimum; in fact its directivity is much lower than $D(U)$ for $d > 1/2$.

(*c*) $D(U) \to 1$ as $d \to 0$, and $D(U) = N$ as $d = m/2, m = 1, 2, \ldots$.

(*d*) $D(O) \cong D(U)$ as $d \geqq 1/2$. In view of the simplicity of uniform excitation the uniform array is indeed an excellent practical antenna so far as the directivity is concerned.

(*e*) A superdirective antenna, according to the IEEE Standard, is one whose directivity is "significantly" higher than that of a uniform excitation. A top-loaded monopole, for example, is not a superdirective antenna since the top loading is in effect to make the excitation more uniform. The radiation leak from an open-circuit, balanced, two-wire transmission line with small spacing can be considered as a superdirective antenna, but obviously it is a very inefficient poor

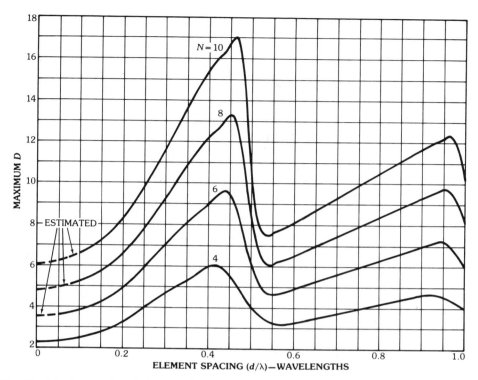

Fig. 30. Maximum directivity of a linear cophasal end-fire array of N isotropic elements versus element spacing d (in wavelengths) for $N = 4$, 6, 8, and 10. (*Courtesy C. T. Tai*)

antenna. So are some other small antennas whose largest dimensions are smaller than a few thousandths of a wavelength.

For academic interest the optimum excitation magnitude and phase for the above example are shown in Figs. 34 and 35. In both figures the curves for $d = 0.2\lambda$ should not be taken seriously, and in Fig. 35 the phases of the excitations of elements 9 and 10 are omitted for brevity, but they can be obtained by extrapolation from the curves shown. From these two figures the following remarks can be made:

(*a*) The optimum excitation for small spacing is highly tapered toward the ends of the array and approximately uniform for spacing equal to or larger than $\lambda/2$. This is consistent with the directivity characteristics stated above, namely, for $d \geqq 1/2$ the maximum directivity is nearly that of a uniform array.

(*b*) The optimum phase is nearly cophasal for $d \geqq 1/2$, again in agreement with the directivity characteristics stated above. However, for $d < 1/2$ the optimum phase distribution is nearly antiphase, about 170°, which results in large local field, low efficiency, etc.

(*c*) For $d = 0.4$ wavelength, where the strong superdirectivity begins to show $[D(O) - D(U) \cong 3.5 \text{ dB}]$, the optimum pattern is shown in Fig. 36a. For com-

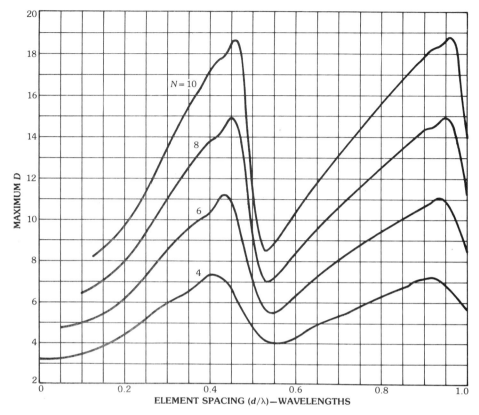

Fig. 31. Maximum directivity of a linear cophasal end-fire array of N parallel dipoles versus element spacing d (in wavelengths) for N = 4, 6, 8, and 10. (*Courtesy C. T. Tai*)

parison, the pattern for the same array but with Hansen-Woodyard excitation is shown in Fig. 36b. It is seen that the latter has a much broader beam.

(*d*) Fig. 37 shows how the maximum D varies with the angle for which D is maximized for two different cases: $D(O)$ with both excitation magnitudes and phases subject to optimization and $D(OC)$ with only the magnitudes subject to optimization and cophasal phase distribution. It is seen that for both cases D is larger than N, namely 10 here, for small angles (i.e., near end-fire) and smaller than N, namely 10, in some other directions, as expected from (181).

An Example for Maximum SNR

Superdirective arrays are impractical as stated above. However, for reception, it is the *SNR*, not D, that is of concern. In particular, one is interested in finding out whether a significant improvement of *SNR* over that of a uniform array is possible without paying a high price on the Q factor. To show that this is possible a simple semicircular array, consisting of nine uniformly distributed isotropic elements in the xz plane is considered. Let the signal come from the z axis and $T(\theta, \phi)$ = constant for $z < 0$ and 0 for $z > 0$ as shown in Fig. 38. Using the formulas in Table 4

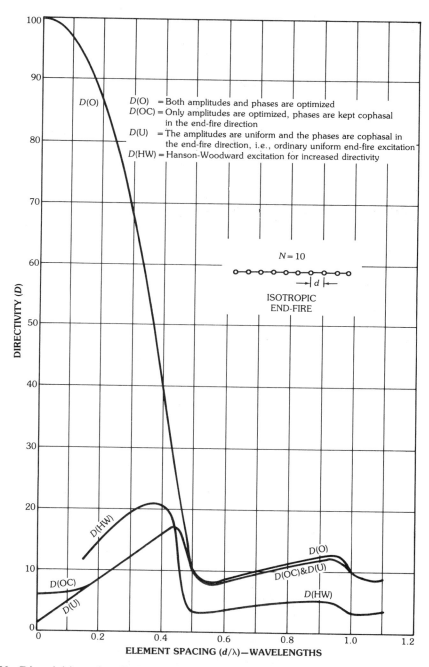

Fig. 32. Directivities of a linear end-fire array of 10 isotropic elements with various excitations. (*After Lo, Lee, and Lee [65],* © *1966 IEEE*)

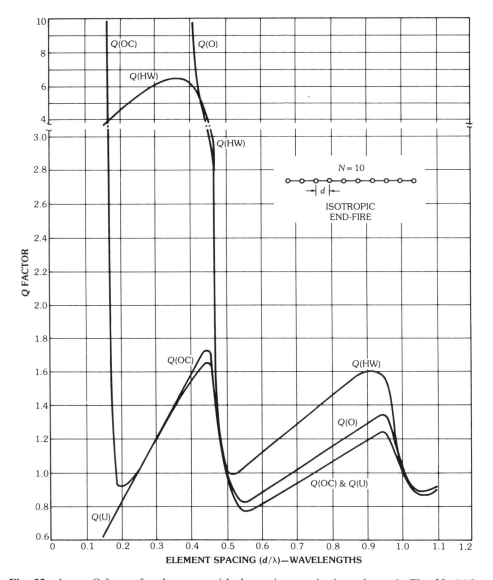

Fig. 33. Array Q factor for the array with the various excitations shown in Fig. 32. (*After Lo, Lee, and Lee [65], © 1966 IEEE*)

various results are computed as shown in Table 5 for the circle radius equal to λ and Table 6 for the circle radius equal to $\lambda/4$. From these the following remarks can be made:

(*a*) From Table 5 the highest *SNR* of 81.6 with $Q = 1.14$ is obtained, as compared with 35.5 with $Q = 0.916$ for a uniform excitation. Thus an improvement of 3.6 dB is obtained without paying a high price on the Q factor.

(*b*) From Table 6 where the elements are closely spaced, the highest *SNR* of

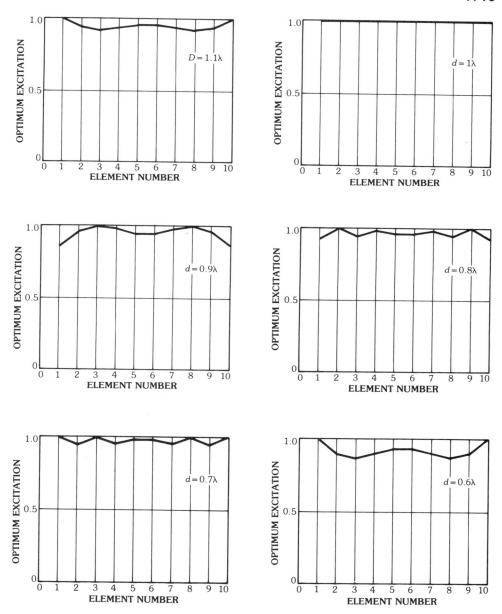

Fig. 34. Relative amplitudes of the optimum excitations for $D(O)$ in Fig. 32 for various element spacings d. (*After Lo, Lee, and Lee [65]*, © *1966 IEEE*)

47.1 is obtained at the price of $Q = 3.26 \times 10^3$. But with a prescribed value of 20 for the Q factor, $SNR = 21.8$, which is about 5.2 dB higher than that of a uniform excitation.

(c) If D, rather than the SNR, is optimized for the array to receive a signal from

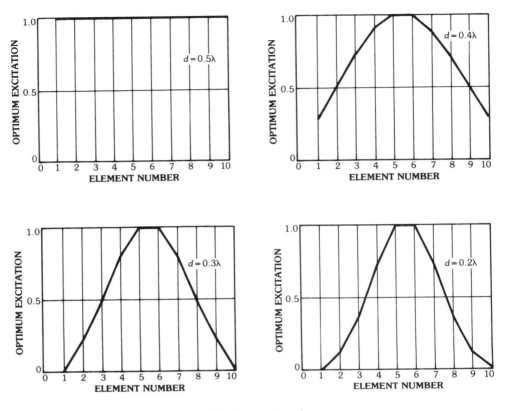

Fig. 34, *continued.*

the z axis in a noise environment as stated above, one would obtain a significantly lower *SNR*. Therefore, the criterion of maximum *SNR* may not be substituted with that of maximum D.

(*d*) In general, large Q is always associated with antiphase distribution.

Extensions

The theory and technique developed above can also be applied to the problem of maximizing the *beam efficiency* where the power radiated in a solid angle Ω_0 about the direction (θ_0, ϕ_0) for a given total radiated power, rather than D, is maximized. For this case, only the **C** matrix need be redefined as

$$\mathbf{C} = \int_{\Omega_0} V^* \rangle \langle V d\Omega \tag{184}$$

This solution is given by solving the eigenvalue problem $(\mathbf{C} - \lambda \mathbf{B})(J) = 0$. The optimization of an aperture antenna can also be solved in a similar manner except that a set of basis functions, or modal functions, over the aperture, instead of elements, should be considered [65].

The technique can also be applied for pattern synthesis. In that case the

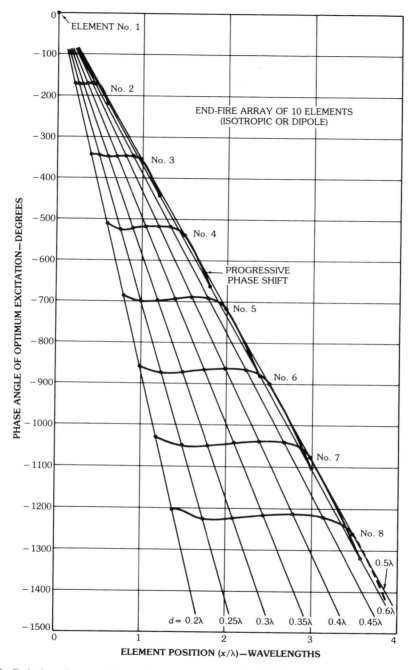

Fig. 35. Relative phases of the optimum excitations for $D(O)$ in Fig. 32 for various element spacings d, where only the phases for the first eight elements are shown.

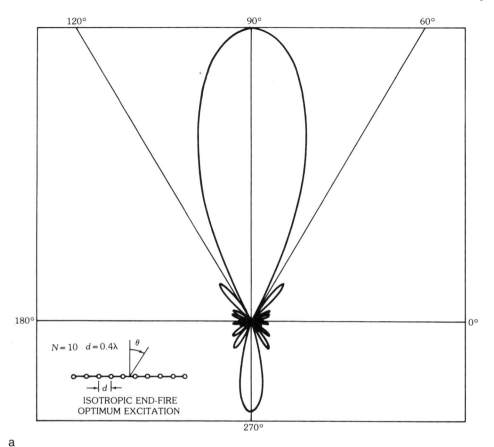

a

Fig. 36. Radiation patterns for end-fire isotropic array. (*a*) For the array in Fig. 32 with optimum excitations shown in Figs. 34 and 35 for $d = 0.4\lambda$. (*b*) For the array in (*a*) but with Hansen-Woodyard's increased directivity excitation. (*After Lo, Lee, and Lee [65], © 1966 IEEE*)

mean-square difference between the array pattern and the desired pattern is minimized, as discussed in the next section.

7. Pattern Synthesis in the Probabilistic Sense

In the last section an array Q factor was defined to indicate the degree of superdirectivity. It was then shown that a moderate improvement in directivity over the uniform excitation can be obtained only at the price of an astronomically large Q factor. To make the design practical a constraint on the Q factor should be imposed. However, the question of how large a value of the Q factor can be considered practical is left unanswered. One possible approach to this problem is to take the parametric uncertainty into consideration. To illustrate this, a pattern synthesis problem is discussed. Let

Array Theory

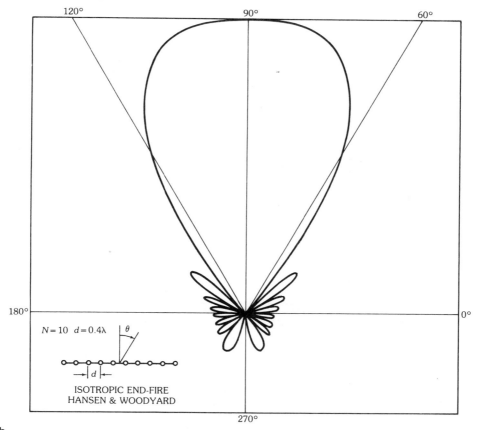

Fig. 36, *continued.*

$$f_d(\theta, \phi) = \text{desired pattern function}$$
$$f(\theta, \phi) = \text{actually realizable pattern of the array} = \langle JV^* \rangle \qquad (185)$$

where

$\langle J = $ a row vector $= (J_1, J_2, \ldots, J_N)$

$J_n = $ excitation current for the nth element

$V \rangle = $ a column vector $= (V_1, V_2, \ldots, V_N)^t$

$V_n^* = $ pattern function of the nth element

Then, for least-square optimization, the following norm in the L_2 space is to be minimized:

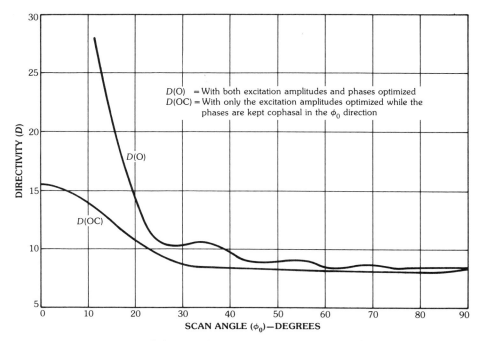

Fig. 37. The directivities $D(O)$ and $D(OC)$ versus scan angle ϕ_0 in which the directivity is maximized for a linear array with $N = 10$ isotropic elements along the x axis and element spacing $d = 0.4\lambda$. $D(O)$: with both excitation magnitudes and phases optimized. $D(OC)$: with only the excitation magnitudes optimized while the phases are kept cophasal in the ϕ_0 direction. (*After Lo, Lee, and Lee [65], © 1966 IEEE*)

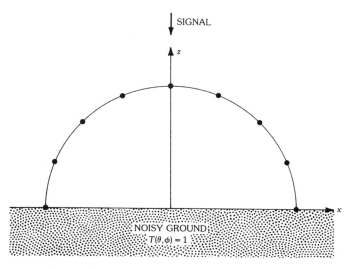

Fig. 38. Geometry of a semicircular array above a noisy ground to receive a signal from the z axis. (*After Lo, Lee, and Lee [65], © 1966 IEEE*)

Array Theory

Table 5. Optimum Semicircular Array ($r = \lambda$) of Nine Cophasally Excited Isotropic Elements

	Current	D	SNR	Q Factor
Uniform excitation	$J_1 = J_2 = J_3 = J_4 = J_5 =$ $J_6 = J_7 = J_8 = J_9 = 1$	8.24	35.5	0.916
Optimum D without constraint on Q factor	$J_1 = J_9 = 1.123$ $J_2 = J_8 = 1.29$ $J_3 = J_7 = 0.881$ $J_4 = J_6 = 0.757$ $J_5 = 0.600$	8.71	55.0	1.03
Optimum D with a prescribed Q factor	$J_1 = J_9 = 1.082$ $J_2 = J_8 = 1.218$ $J_8 = J_7 = 0.898$ $J_4 = J_6 = 0.816$ $J_5 = 0.659$	8.67	50.5	1.0 (prescribed)
Optimum SNR without constraint on Q factor	$J_1 = J_9 = 11.436$ $J_2 = J_8 = 15.396$ $J_3 = J_7 = 10.446$ $J_4 = J_6 = 3.746$ $J_5 = -0.421$	7.76	81.6	1.14
Optimum SNR with a prescribed Q factor	$J_1 = J_9 = 5.835$ $J_2 = J_8 = 7.719$ $J_3 = J_7 = 7.451$ $J_4 = J_6 = 5.223$ $J_5 = 2.664$	8.44	55.1	1.0 (prescribed)

Element positions (in wavelengths):
$x_1 = -x_9 = 1.000 \quad z_1 = z_9 = 0.0$
$x_2 = -x_8 = 0.924 \quad z_2 = z_8 = 0.383$
$x_3 = -x_7 = 0.707 \quad z_3 = z_7 = 0.707$
$x_4 = -x_6 = 0.383 \quad z_4 = z_6 = 0.924$
$x_5 = 0.0 \quad z_5 = 1.0$

Main beam: $\theta_0 = 0$

Thermal noise distribution: $T(\theta, \phi) = \begin{cases} 1 & \text{for } \pi/2 < \theta \leq \pi \\ 0 & \text{otherwise} \end{cases}$

(After Lo, Lee, and Lee [65], © 1966 IEEE)

$$\begin{aligned}
\varepsilon &= \|f_d - f\|^2 = \int_{4\pi} (f_d - f)(f_d - f)^* w \, d\Omega \\
&= \|f_d\|^2 + \int_{4\pi} |f|^2 w \, d\Omega - 2\,\text{Re}\left\{\int ff_d^* w \, d\Omega\right\} \\
&= \|f_d\|^2 + \langle J \int_{4\pi} V^* \rangle \langle V_d w \, d\Omega J^* \rangle - 2\,\text{Re}\left\{\langle J \int_{4\pi} V^* \rangle f_d^* w \, d\Omega\right\} \\
&= \|f_d\|^2 + \langle JGJ^* \rangle - 2\,\text{Re}\{\langle JC \rangle\}
\end{aligned} \quad (186)$$

where

Table 6. Optimum Semicircular Array ($r = 0.25\lambda$) of Nine Cophasally Excited Isotropic Elements

	Current	D	SNR	Q Factor
Uniform excitation	$J_1 = J_2 = J_3 = J_4 = J_5 = J_6 = J_7 = J_8 = J_9 = 1$	2.19	6.63	0.244
Optimum D without constraint on Q factor	$J_1 = J_9 = 5.23$ $J_2 = J_8 = -15.74$ $J_3 = J_7 = 34.81$ $J_4 = J_6 = -55.83$ $J_5 = 66.69$	3.63	37.8	3.76×10^3
Optimum D with a prescribed Q factor	$J_1 = J_9 = 2.24$ $J_2 = J_8 = -2.92$ $J_3 = J_7 = 3.35$ $J_4 = J_6 = -2.23$ $J_5 = 2.37$	3.25	20.2	20.0 (prescribed)
Optimum SNR without constraint on Q factor	$J_1 = J_9 = 58.86$ $J_2 = J_8 = -179.6$ $J_3 = J_7 = 412.72$ $J_4 = J_6 = -686.83$ $J_5 = 836.80$	3.52	47.1	3.26×10^3
Optimum SNR with a prescribed Q factor	$J_1 = J_9 = 12.80$ $J_2 = J_8 = -15.58$ $J_3 = J_7 = 19.70$ $J_4 = J_6 = -18.96$ $J_5 = 25.87$	3.19	21.8	20.0 (prescribed)

Element positions (in wavelengths):
$x_1 = -x_9 = r$, $z_1 = z_9 = 0$
$x_2 = -x_8 = 0.9239r$, $z_2 = z_8 = 0.3827r$
$x_3 = -x_7 = 0.7071r$, $z_3 = z_7 = 0.7071r$
$x_4 = -x_6 = 0.3827r$, $z_4 = z_6 = 0.9239r$
$x_5 = 0$, $z_5 = r$

Main beam: $\theta_0 = 0$

Thermal noise distribution: $T(\theta, \phi) = \begin{cases} 1 & \text{for } \pi/2 < \theta \leq \pi \\ 0 & \text{otherwise} \end{cases}$

(After Lo, Lee, and Lee [65], © 1966 IEEE)

w = a possible weighting function that stresses a closer approximation of f to f_d for some angular regions

$$\mathbf{G} = \int_{4\pi} V^* \rangle \langle V \, w \, d\Omega \text{ (a positive-definite Hermitian matrix)} \quad (187)$$

$$\mathbf{C} = \int_{4\pi} V^* \rangle f_d^* w \, d\Omega \quad (188)$$

Using the variational method, the solution of $J\rangle$ which minimizes ε is found to be

Array Theory

$$|J_0\rangle = (\mathbf{G}^{-1}\mathbf{C}\rangle)^* = U \operatorname{diag}[(\lambda_1)^{-1}, (\lambda_2)^{-1}, \ldots, (\lambda_N)^{-1}] U^\dagger \mathbf{C}^*\rangle \qquad (189)$$

where

$\lambda_1, \lambda_2, \ldots, \lambda_N$ = the eigenvalues, in descent order, of $\mathbf{G}^*$

U = a unitary matrix which diagonalizes $\mathbf{G}^*$

$U^\dagger$ = the complex conjugate transpose of U

Since $\mathbf{G}$ depends on the array geometry, its largest to smallest eigenvalue ratio, λ_1/λ_N, can increase very rapidly to an extremely large value as the array element spacing decreases. This finally leads to an ill-conditioned $\mathbf{G}$ and a very unstable solution $|J_0\rangle$. Mathematically, this phenomenon is identical with the superdirectivity. Such a solution is not only impractical, but is also difficult to compute accurately. Therefore, the so-called optimum solution (189) in such a case should not be taken seriously.

Because of the physical limitation, instrumental error, environmental variation, etc., the excitation $|J\rangle$ can only be adjusted in practice within an uncertain random error, say $|\delta \tilde{J}\rangle$. Therefore for a more practically meaningful design, this error should be taken into consideration. Assume that the error $|\delta \tilde{J}\rangle$ is, as in most practical cases, only *relative* to $|J\rangle$:

$$|\delta \tilde{J}\rangle = \tilde{\mathbf{A}} |J\rangle \qquad (190)$$

where $\tilde{\mathbf{A}}$ is a matrix with stochastic elements, which may result from, for example, some uncertainties in the feeding network. Thus the pattern function due to the actual current $|J\rangle + |\delta \tilde{J}\rangle$ will be a random function,

$$\tilde{f}(\theta, \phi) = \langle (J + \delta \tilde{J}), V^*(\theta, \phi) \rangle \qquad (191)$$

and so must be the difference between $\tilde{f}(\theta, \phi)$ and $f_d(\theta, \phi)$. Following (186), let

$$\tilde{\varepsilon} = \| f_d - \tilde{f} \|^2 \qquad (192)$$

Then for a given $|J\rangle$ and a given realization of the random error $|\delta \tilde{J}\rangle$, there is a realization of $\tilde{f}$ and also $\tilde{\varepsilon}$. Hence, for the totality of all these realizations there is a probability distribution $F(\varepsilon; |J\rangle)$ which defines the probability for $\tilde{\varepsilon} < \varepsilon$ for the given $|J\rangle$. With this distribution function one can define the probability mean of a quantity in the following formula:

$$E\{\cdot\} = \int_{\text{entire prob. space}} (\cdot) \, dF(\varepsilon; |J\rangle) \qquad (193)$$

With this preparation one can define the optimization problem in many different ways. Among them we may state three theoretically possible situations as shown in Fig. 39, each giving an optimum solution different from the others,

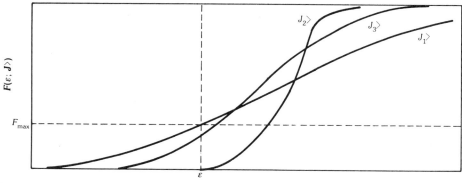

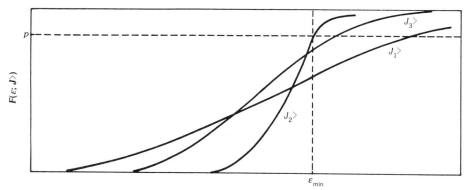

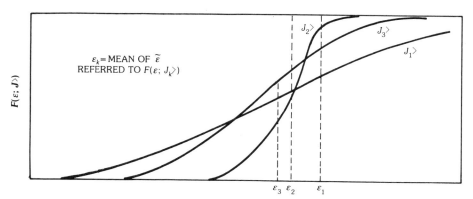

Fig. 39. Three possible philosophies of optimization. (*a*) Vertical optimization: $J_1\rangle$ is the optimum. (*b*) Horizontal optimization: $J_2\rangle$ is the optimum. (*c*) Optimization of $E\{\tilde{\varepsilon}\}$: $J_3\rangle$ is the optimum. (*After Richards and Lo [80], © 1975 IEEE*)

Array Theory

depending on the sense of optimization. The first scenario shown in Fig. 39a, which may be called the *vertical optimization*, illustrates that for a *given* tolerable error ε, $J\rangle$ is to be determined under the condition $F(\varepsilon;J\rangle)$ is the maximum, i.e., the largest probability for having error less than the given tolerance ε. Comparing the three possible distribution curves, $J_1\rangle$ is apparently the best solution since it gives the largest value of $F(\varepsilon;J\rangle)$ for the given ε.

The second possible optimization, which may be called the *horizontal optimization*, is defined by seeking $J\rangle$ such that for a given probability, say p, the error ε is the minimum as shown in Fig. 39b. It is seen that for this scenario $J_2\rangle$ is the best solution. The third possible optimization is to seek $J\rangle$ which minimizes the mean value of ε, namely $\bar{\varepsilon} = E\{\bar{\varepsilon}\}$. As can be seen from Fig. 39c, $J_3\rangle$ is the best solution. As stated, these three distribution curves are hypothetical and whether they would occur in a certain problem is not clear. For simplicity, in what follows the last sense of optimization is considered and it is found later that the solution so obtained also leads to approximately the optimum solutions in the other two senses.

Without loss of generality one may assume that

$$E\{\delta \tilde{J}\rangle\} = 0 \tag{194}$$

Since now the excitation is $J\rangle + \delta \tilde{J}\rangle$, similar to (186), the random error between the realized pattern f and desired pattern f_d is simply

$$\tilde{\varepsilon} = \|f_d\|^2 + \langle (J + \delta \tilde{J}) \mathbf{G} (J + \delta \tilde{J})^* \rangle - 2 \operatorname{Re}\{\langle (J + \delta \tilde{J}) C \rangle\} \tag{195}$$

Since $\delta \tilde{J}\rangle = \tilde{A} J\rangle$,

$$\bar{\varepsilon} = E\{\tilde{\varepsilon}\} = \|f_d\|^2 + \langle JGJ^* \rangle - 2\operatorname{Re}\{\langle JC \rangle\} + E\{\langle \delta \tilde{J} \mathbf{G} \delta \tilde{J}^* \rangle\} \tag{196}$$

where

$$E\{\langle \delta \tilde{J} \mathbf{G} \delta \tilde{J}^* \rangle\} = \langle J \mathbf{K}_2 J^* \rangle \quad \text{and} \quad \mathbf{K}_2 = E\{\tilde{A}^t \mathbf{G} \tilde{A}^*\} \tag{197}$$

The *mn*th element of matrix $\mathbf{K}_2$ is given by

$$(\mathbf{K}_2)_{mn} = \sum_{i,j} G_{ij} E\{\tilde{A}_{jm} \tilde{A}_{in}^*\} \tag{198}$$

where G_{ij} is the *ij*th element of $\mathbf{G}$, and $E\{\tilde{A}_{jm}\tilde{A}_{in}^*\}$ is the covariance (or joint moment) of elements (jm) and (in) of $\tilde{\mathbf{A}}$. Using (196)–(198) the optimum solution of $J\rangle$ which minimizes $\bar{\varepsilon}$ for a given probabilistic property of $\delta \tilde{\mathbf{J}}$ (or $\tilde{\mathbf{A}}$) is given by

$$J\rangle = [(\mathbf{G} + \mathbf{K}_2)^{-1} C\rangle]^* \tag{199}$$

For a typical simple case where all array elements are identical, $\delta \tilde{J}\rangle$ has uncorrelated elements and identical variance σ^2, the matrix $\mathbf{K}_2$ becomes diagonal:

$$\mathbf{K}_2 = \sigma^2 G_{11}\mathbf{I} = \alpha\mathbf{I} \qquad (200)$$

where $\mathbf{I}$ is an identity matrix and α is a scalar equal to $\sigma^2 G_{11}$. Therefore, following (189), the optimum solution for this case is simply

$$|J\rangle = \mathbf{U}\,\mathrm{Diag}[(\lambda_1 + \alpha)^{-1}, (\lambda_2 + \alpha)^{-1}, \ldots, (\lambda_N + \alpha)^{-1}]\mathbf{U}^\dagger|C^*\rangle \qquad (201)$$

where $\lambda_1, \ldots, \lambda_N$ are, as before, the eigenvalues of $\mathbf{G}$. It is interesting to compare (201) with (189) and note that the two solutions are identical except for the constant α. Now even if $\lambda_1, \ldots, \lambda_N$ march off to zero as the element spacing decreases, the solution $|J\rangle$ is no longer unstable and dominated by those small λ_ns. Tihonov [125] introduced the so-called regularization method for ill-conditioned problems such as the one in (189). Cabayan et al [126], in making use of that method for the pattern synthesis problem, proposed to minimize $\|f_d - f\|^2 + \alpha\langle JJ^*\rangle$, instead of $\|f_d - f\|^2$. It is clear, however, that α must assume a proper value because if too large the minimization will be applied mainly for $\langle JJ^*\rangle$ and, if too small, the solution may be again unstable. For a given probability distribution of $\delta\bar{J}\rangle$, Cabayan used the Monte Carlo method to determine the proper value of α numerically. The above analysis shows that the proper value is simply $\sigma^2 G_{11}$.

An Example—Richards and Lo [80] considered a planar array with 10 concentric rings as shown in Fig. 40. The radius of the kth ring is

$$\varrho_k = 2\pi\left[\frac{1}{4}\left(k - \frac{2}{\pi}\right) - \frac{1}{2\pi}\left(\frac{1}{2}\right)^{10-k}\right], \qquad k = 1, 2, \ldots, 10$$

A total of 200 vertical dipoles are placed over the intersecting points between the rings and 20 rays from the origin uniformly spaced in angles over 360° as illustrated in the figure. The objective is to determine the optimum excitation which will produce a pattern "closest" to a secant pattern:

$$f_d(\theta) = \begin{cases} \sec\theta & \text{for } 0 < \theta \leq 70° \\ \sec 70° & \text{for } 70° \leq \theta \leq 90° \end{cases}$$

The solid curve in Fig. 41a is the nominal elevation pattern for $\sigma = 0$ using the solution given by (189) for no tolerance. The solid curve in Fig. 41b is the nominal pattern for $\sigma = 5$ percent using the solution given by (201). It is seen that the former is closer than the latter to the desired pattern f_d which is also shown in both figures. However, if the actual error $\delta\bar{J}\rangle$ is added to $J\rangle$, the results are completely different. Using the Monte Carlo method, five sets of sample errors of $\delta\bar{J}\rangle$ with zero mean and 5 percent standard deviation are generated. From these five sets five sample patterns are computed for both cases, shown as the dotted curves in Fig. 41a and dots only in Fig. 41b. It is clearly seen that in the former case, where the unavoidable error is ignored in the optimization, none of the five realizations even remotely resembles the desired, or the nominal, pattern. As for the latter case, although the nominal pattern is less close to f_d than that of the first case, with all five sets of real errors included, the five realizations of the pattern are so close to

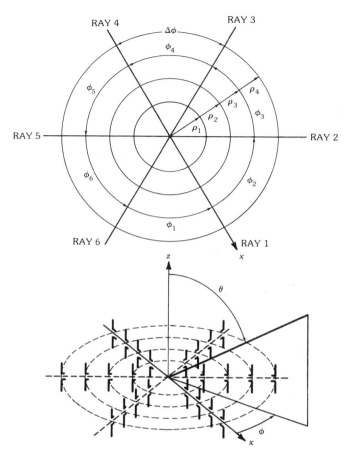

Fig. 40. Geometry of a planar array of vertical dipoles (or monopoles above the ground). (*After Richards and Lo [80]*, © *1975 IEEE*)

the nominal pattern that, to avoid confusion, no attempt was made to connect the dots for each realization. This result vividly confirms that in some design problems, where the stability is questionable, to demand less may *actually* obtain a better optimum solution.

For the example discussed above, Richards and Lo [80] has derived an Edgeworth series representation for the distribution function $F(\varepsilon; J\rangle)$ and also showed the validity of this solution with 2500 samples using the Monte Carlo simulation with a computer. They also verified the theoretical solution for one value of σ while using Monte Carlo simulations for a wide range of σ values. For the example studied, the sample distribution curves for $\sigma = 3, 5,$ and 10 percent are very close to each other, indicating that a precise information of the statistics of $\delta \bar{J}$ may not be very important for determining the optimum solution $J\rangle$. Another interesting result shows that $\|J\|$ for $\sigma = 0$ is several orders of magnitude larger than that for $\sigma = 5$ percent, implying very high Q, low efficiency, and unstable solution for the former, in agreement with previous discussion for maximum D.

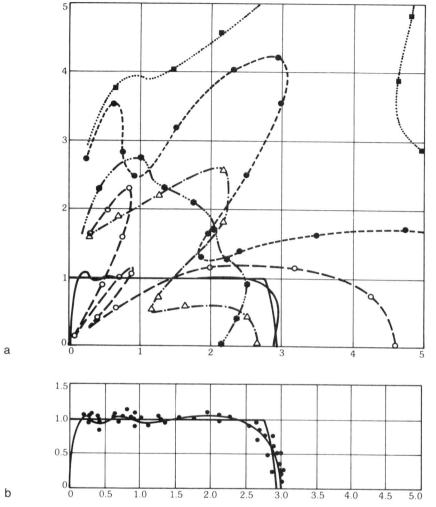

Fig. 41. Synthesis for a desired pattern function: $f_d = \sec\theta$ for $0° < \theta \leq 70°$ and $\sec 70°$ for $70° \leq \theta \leq 90°$. (*a*) Nominal pattern (solid curve), in polar plot, corresponding to the deterministic optimum solution with zero tolerance and five actual sample patterns (dotted) for 5% standard deviation of the tolerance, generated by the Monte Carlo method. (*b*) Nominal pattern (solid curve), in polar plot, corresponding to the probabilistic optimum with the 5% tolerance taken into consideration and five actual sample patterns (dots only), generated as in (*a*) by the Monte Carlo method. (*After Richards and Lo [80]*, © *1975 IEEE*)

8. References

[1] J. L. Allen et al., "Phased array radar studies," *Tech. Rep. 238*, MIT Lincoln Labs, August 1960.

[2] N. Amitay et al., *Theory and Analysis of Phased Array Antennas*, New York: John Wiley & Sons, 1972.

[3] H. Bach and J. E. Hansen, "Uniformly spaced arrays," chapter 5 in *Antenna Theory*,

Part 1, ed. by R. E. Collin and F. J. Zucker, New York: McGraw-Hill Book Co., 1969.
[4] C. A. Balanis, *Antenna Theory: Analysis and Design*, New York: Harper & Row, 1982.
[5] D. Barbiere, "A method for calculating the current distribution of Tschebyscheff arrays," *Proc. IRE*, vol. 40, pp. 78–82, 1952.
[6] E. T. Bayliss, "Design of monopulse antenna difference patterns with low side lobes," *Bell Syst. Tech. J.*, vol. 47, pp. 623–640, 1968.
[7] A. Bloch, R. G. Medhurst, and S. D. Pool, "A new approach to the design of super-directive aerial arrays," *IEE Proc.*, vol. 100, pt. III, pp. 303–314, 1953.
[8] A. Bloch, R. G. Medhurst, and S. D. Pool, "Superdirectivity," *Proc. IRE*, vol. 48, p. 1164, 1960.
[9] C. J. Bouwkamp and N. G. De Bruijn, "The problem of optimum antenna current distribution," *Philips Res. Rep.*, vol. 1, p. 135, 1946.
[10] P. A. Bricout, "Pattern synthesis using weighted functions," *IRE Trans. Antennas Propag.*, vol. AP-8, pp. 441–444, 1960.
[11] J. L. Brown, "A simplified derivation of the Fourier coefficients for Chebyshev patterns," *Proc. IEE*, vol. 105C, pp. 167–168, 1957.
[12] J. L. Brown, "On the determination of excitation coefficients for a Tchebycheff pattern," *IRE Trans. Antennas Propag.*, vol. AP-10, pp. 215–216, 1962.
[13] L. B. Brown and G. A. Scharp, "Tschebyscheff antenna distribution, beam-width, and gain tables," *NOLC Rep. 383*, February 1958.
[14] D. K. Cheng, "Optimization techniques for antenna arrays," *Proc. IEEE*, vol. 59, pp. 1664–1674, 1971.
[15] D. K. Cheng and M. T. Ma, "A new mathematical approach for linear array analysis," *IRE Trans. Antennas Propag.*, vol. AP-8, pp. 255–259, 1960.
[16] L. J. Chu, "Physical limitations of omnidirectional antennas," *J. Appl. Phys.*, vol. 19, pp. 1163–1175, 1948.
[17] H. P. Coleman, "An iterative technique for reducing side lobes of circular arrays," *IEEE Trans. Antennas Propag.*, vol. AP-18, pp. 566–567, July 1970.
[18] R. E. Collin and S. Rothschild, "Evaluation of antenna Q," *IEEE Trans. Antennas Propag.*, vol. AP-12, pp. 23–27, January 1964.
[19] A. G. Dermerud and J. J. Gustincic, "The interpolation of general active array impedance from multielement simulators," *IEEE Trans. Antennas Propag.*, vol. AP-27, pp. 68–71, 1979.
[20] B. L. Diamond, "A generalized approach to the analysis of infinite planar array antennas," *Proc. IEEE*, vol. 56, pp. 1837–1851, 1968.
[21] B. L. Diamond and G. H. Knittel, "A new procedure for the design of a waveguide element for a phased-array antenna," *Phased Array Antennas*, ed. by A. A. Oliner and G. H. Knittel, Dedham, Mass.: Artech House, 1972, pp. 149–156.
[22] C. L. Dolph, "A current distribution for broadside arrays which optimizes the relationship between beam width and side lobe level," *Proc. IRE*, vol. 34, pp. 335–348, 1946.
[23] C. J. Drane, "Derivation of excitation coefficients for Chebyshev arrays," *Proc. IEE*, vol. 110, pp. 1755–1758, 1963.
[24] C. J. Drane, "Dolph-Chebyshev excitation coefficient approximation," *IEEE Trans. Antennas Propag.*, vol. AP-12, pp. 781–782, 1964.
[25] C. J. Drane, "Useful approximations for the directivity and beamwidth of large scanning Dolph-Chebyshev arrays," *Proc. IEEE*, vol. 56, pp. 1779–1787, 1968.
[26] R. H. DuHamel, "Optimum patterns for end-fire arrays," *Proc. IRE*, vol. 41, pp. 652–659, 1953.
[27] R. H. DuHamel, "Pattern synthesis for antenna arrays on circular, elliptical, and spherical surfaces," *Tech. Rep. no. 16*, E. E. Research Lab, Univ. of Illinois, Urbana, 1952.
[28] R. S. Elliott, *Antenna Theory and Design*, Englewood Cliffs: Prentice-Hall, 1981.
[29] R. S. Elliott, "Mechanical and electrical tolerances for two-dimensional scanning

antenna arrays," *IRE Trans. Antennas Propag.*, vol. AP-6, pp. 114–120, 1958.

[30] R. S. Elliott, "The theory of antenna arrays," in chapter 1 of *Microwave Scanning Antennas, Volume II*, ed. by R. C. Hansen, New York: Academic Press, 1966.

[31] R. S. Elliott, "Design of line-source antennas for sum patterns with side lobes of individually arbitrary heights," *IEEE Trans. Antennas Propag.*, vol. AP-24, pp. 76–83, 1976.

[32] R. S. Elliott, "On discretizing continuous aperture distributions," *IEEE Trans. Antennas Propag.*, vol. AP-25, pp. 617–621, 1977.

[33] R. S. Elliott and R. M. Johnson, "Experimental results on a linear array designed for asymmetric side lobes," *IEEE Trans. Antennas Propag.*, vol. AP-26, pp. 351–352, 1978.

[34] J. E. Evans, "Synthesis of equiripple sector antenna patterns," *IEEE Trans. Antennas Propag.*, vol. AP-24, pp. 347–353, 1976.

[35] R. J. Evans and T. E. Fortmann, "Design of optimal line-source antennas," *IEEE Trans. Antennas Propag.*, vol. AP-23, pp. 342–347, 1975.

[36] R. L. Fante, "Optimum distribution over a circular aperture for best mean-square approximation to a given radiation pattern," *IEEE Trans. Antennas Propag.*, vol. AP-18, pp. 177–181, 1970.

[37] G. R. Forbes, "An end-fire array continuously proximity-coupled to a two-wire line," *IEEE Trans. Antennas Propag.*, vol. AP-8, pp. 518–519, 1960.

[38] A. C. Gately, Jr., et al., "A network description for antenna problems," *Proc. IEEE*, vol. 56, pp. 1181–1193, 1968.

[39] W. S. Gregorwich et al., "A waveguide simulator for the determination of a phased-array resonance," *IEEE G-AP Intl. Symp. Dig.*, pp. 134–141, 1968.

[40] J. J. Gustincic, "The determination of active array impedance with multielement waveguide simulators," *IEEE Trans. Antennas Propag.*, vol. AP-20, pp. 589–595, 1972.

[41] P. W. Hannan and M. A. Balfour, "Simulation of a phased-array antenna in waveguide," *IEEE Trans. Antennas Propag.*, vol. AP-13, pp. 342–353, 1965.

[42] P. W. Hannan, "Discovery of an array surface wave in a simulator," *IEEE Trans. Antennas Propag.*, vol. AP-15, pp. 574–576, 1967.

[43] R. C. Hansen, "Tables of Taylor distributions for circular aperture antennas," *IRE Trans. Antennas Propag.*, vol. AP-8, pp. 23–26, 1960.

[44] R. C. Hansen, "Comparison of square array directivity formulas," *IEEE Trans. Antennas Propag.*, vol. AP-20, pp. 100–102, 1972.

[45] R. C. Hansen, "A one-parameter circular aperture distribution with narrow beamwidth and low sidelobes," *IEEE Trans. Antennas Propag.*, vol. AP-24, pp. 477–480, 1976.

[46] R. C. Hansen, "Gain limitations of large antennas," *IRE Trans. Antennas Propag.*, vol. AP-8, pp. 490–495, 1970; see also correction in vol. AP-13, p. 997, 1965.

[47] R. C. Hansen, *Microwave Scanning Antennas, Volume 1*, chapter 1, New York: Academic Press, 1964.

[48] R. C. Hansen (ed.), Special issue on "Electronic scanning," *Proc. IEEE*, vol. 56, pp. 1761–2038, 1968.

[49] R. C. Hansen, *Significant Phased Array Papers*, Dedham: Artech House, 1973.

[50] W. W. Hansen and J. R. Woodyard, "A new principle in directional antenna design," *Proc. IRE*, vol. 26, pp. 333–345, 1938.

[51] R. F. Harrington, "Effect of antenna size on gain, bandwidth, and efficiency," *J. Res. Natl. Bur. Stand.* (US), vol. 64D, pp. 1–12, 1960.

[52] W. G. Jaeckle, "Antenna synthesis by weighted Fourier coefficients," *IEEE Trans. Antennas Propag.*, vol. AP-12, pp. 369–370, 1964.

[53] E. C. Jordan and K. G. Balmain, "Electromagnetic waves and radiating systems," section 14.08, Englewood Cliffs: Prentice-Hall, 1968.

[54] W. K. Kahn, "Ideal efficiency of a radiating element in an infinite array," *IEEE Trans. Antennas Propag.*, vol. AP-15, pp. 534–538, 1967.

[55] W. K. Kahn, "Impedance-match and element-pattern constraints for finite arrays," *IEEE Trans. Antennas Propag.*, vol. AP-25, pp. 747–755, 1977.

[56] H. E. King, "Directivity of a broadside array of isotropic radiator," *IRE Trans. Antennas Propag.*, vol. AP-7, pp. 197–198, 1959.

[57] M. J. King and R. K. Thomas, "Gain of large scanned arrays," *IRE Trans. Antennas Propag.*, vol. AP-8, pp. 635–636, 1960.

[58] R. W. P. King, *The Theory of Linear Antennas*, Cambridge: Harvard Univ. Press, 1956.

[59] H. L. Knudsen, "Radiation from ring quasi-arrays," *IRE Trans. Antennas Propag.*, vol. AP-4, pp. 452–472, 1956.

[60] J. D. Kraus, *Antennas*, New York: McGraw-Hill Book Co., 1950.

[61] H. N. Kritikos, "Optimal signal-to-noise ratio for linear arrays by the Schwartz inequality," *J. Franklin Inst.*, vol. 276, pp. 195–304, 1963.

[62] A. Ksienski, "Maximally flat and quasi-smooth sector beams," *IRE Trans. Antennas Propag.*, vol. AP-8, pp. 476–484, 1960.

[63] L. Lapaz and G. A. Miller, "Optimum current distributions on vertical antennas," *Proc. IRE*, vol. 31, pp. 214–232, 1943.

[64] Y. T. Lo and S. W. Lee, "Affine transformation and its application to antenna arrays," *IEEE Trans. Antennas Propag.*, vol. AP-13, pp. 890–896, 1965.

[65] Y. T. Lo, S. W. Lee, and Q. H. Lee, "Optimization of directivity and signal-to-noise ratio of an arbitrary antenna array," *Proc. IEEE*, vol. 54, pp. 1033–1045, 1966.

[66] Y. T. Lo and H. C. Hsuan, "An equivalence theory between elliptical and circular arrays," *IEEE Trans. Antennas Propag.*, vol. AP-13, pp. 247–256, 1965.

[67] M. T. Ma, *Theory and Application of Antenna Arrays*, New York: John Wiley & Sons, 1974.

[68] M. T. Ma, "A new mathematical approach for linear array analysis and synthesis," PhD dissertation, Syracuse University, 1961.

[69] M. T. Ma and L. C. Walters, "Synthesis of concentric ring antenna arrays yielding approximately equal sidelobes," *Radio Sci.*, vol. 3, pp. 465–470, May 1968.

[70] R. B. Mack, "A study of circular arrays," technical report, Cruft Lab, Harvard University, May 1963.

[71] B. A. Munk et al., "Scan independent phased arrays," *Radio Sci.*, vol. 14, pp. 979–990, 1979.

[72] E. A. Nelson, "Quantization sidelobes of a phased array with a triangular element arrangement," *IEEE Trans. Antennas Propag.*, vol. AP-17, pp. 363–365, 1969.

[73] E. H. Newman et al., "Superdirective receiving arrays," *IEEE Trans. Antennas Propag.*, vol. AP-26, pp. 629–635, 1978.

[74] R. L. Pritchard, "Optimum directivity patterns for linear point arrays," *J. Acoust. Soc. Am.*, vol. 25, pp. 879–891, 1953.

[75] D. R. Rhodes, "The optimum line source of the best mean-square approximation to a given radiation pattern," *IEEE Trans. Antennas Propag.*, vol. AP-11, pp. 440–446, 1963.

[76] D. R. Rhodes, "On an optimum line source for maximum directivity," *IEEE Trans. Antennas Propag.*, vol. AP-17, pp. 485–492, 1969.

[77] D. R. Rhodes, "On a fundamental principle in the theory of planar antennas," *Proc. IEEE*, vol. 52, pp. 1013–1021, 1965.

[78] D. R. Rhodes, *Synthesis of Planar Antenna Sources*, London: Clarendon Press, 1974.

[79] H. J. Riblet, "Discussion on 'A current distribution for broadside arrays which optimizes the relationship between beam width and side-lobe level,'" *Proc. IRE*, vol. 35, pp. 489–492, 1947.

[80] W. F. Richards and Y. T. Lo, "Antenna pattern synthesis based on optimization in a probabilistic sense," *IEEE Trans. Antennas Propag.*, vol. AP-23, pp. 165–172, 1975.

[81] G. M. Royer, "Directive gain and impedance of a ring array of antennas," *IEEE Trans. Antennas Propag.*, vol. AP-14, pp. 566–573, 1966.

[82] L. A. Rondinelli, "Effects of random errors on the performance of antenna arrays of many elements," *IRE Natl. Conv. Rec.*, pt. 1, pp. 174–189, 1959.

[83] J. Ruze, "Antenna tolerance theory—a review," *Proc. IEEE*, vol. 54, pp. 633–640, 1966.

[84] H. E. Salzer, "Calculating Fourier coefficients for Chebyshev patterns," *Proc.

IEEE, vol. 63, pp. 195–197, 1975.

[85] S. A. Schelkunoff, "A mathematical theory of linear arrays," *Bell Syst. Tech. J.*, vol. 22, pp. 80–107, 1943.

[86] S. A. Schelkunoff and H. T. Friis, *Antenna Theory and Practice*, New York: John Wiley & Sons, 1942, pp. 368 and 401.

[87] A. C. Schell and A. Ishimaru, "Antenna pattern synthesis," chapter 7. in *Antenna Theory*, ed. by R. E. Collin and F. J. Zucker, New York: McGraw-Hill Book Co., 1969.

[88] E. D. Sharp, "A triangular arrangement of planar-array elements that reduces the number needed," *IRE Trans. Antennas Propag.*, vol. AP-9, pp. 126–129, 1961.

[89] J. C. Simon, "Application of periodic functions approximation to antenna pattern synthesis and circuit theory," *IRE Trans. Antennas Propag.*, vol. AP-4, pp. 429–440, 1956.

[90] G. C. Southworth, "Arrays of linear elements," in *Antenna Engineering Handbook*, ed. by H. Jasik, New York: McGraw-Hill Book Co., 1961.

[91] R. J. Spellmire, "Tables of Taylor aperture distributions," *Rep. TM 581*, Hughes Aircraft Co., Culver City, CA, 1958.

[92] L. Stark, "Radiation impedance of a dipole in an infinite planar phased array," *Radio Sci.*, vol. 1, pp. 361–377, 1966.

[93] R. J. Stegen, "Excitation coefficients and beamwidths of Tschebyscheff arrays," *Proc. IRE*, vol. 41, pp. 1671–1674, 1953.

[94] R. J. Stegen, "Gain of Tchebyscheff arrays," *IRE Trans. Antennas Propag.*, vol. AP-8, pp. 629–631, 1960.

[95] W. L. Stutzman, "Synthesis of shaped-beam radiation patterns using the iterative sampling method," *IEEE Trans. Antennas Propag.*, vol. AP-19, pp. 36–41, 1971.

[96] W. L. Stutzman, "Side lobe control of antenna patterns," *IEEE Trans. Antennas Propag.*, vol. AP-20, pp. 102–104, 1972.

[97] W. L. Stutzman and G. A. Thiele, *Antenna Theory and Design*, New York: John Wiley & Sons, 1981.

[98] C. T. Tai, "The optimum directivity of uniformly spaced broadside arrays of dipoles," *IEEE Trans. Antennas Propag.*, vol. AP-12, pp. 447–454, 1964.

[99] T. T. Taylor, "Design of circular apertures of narrow beamwidth and low side lobes," *IRE Trans. Antennas Propag.*, vol. AP-8, pp. 17–22, 1960.

[100] T. T. Taylor, "Design of line-source antennas for narrow beamwidth and low side lobes," *IRE Trans. Antennas Propag.*, vol. AP-3, pp. 16–28, 1955.

[101] T. T. Taylor, "A synthesis method for circular and cylindrical antennas composed of discrete elements," *IRE Trans. Antennas Propag.*, vol. AP-1, pp. 251–261, 1952.

[102] J. D. Tillman, Jr., "The theory and design of circular antenna arrays," Univ. of Tennessee, Eng. Exp. Station, Knoxville, 1966.

[103] G. Toraldo Di Francia, "Directivity, super-gain, and information," *IRE Trans.*, 1956, pp. 473–478; also "Theory of antenna arrays," special issue, *Radio Sci.*, vols. 3–5, 1968.

[104] G. N. Tsandoulas, "Tolerance control in an array antenna," *Microwave J.*, vol. 20, pp. 24–30, 1977.

[105] F. I. Tseng, "Design of array and line-source antennas for Taylor patterns with a null," *IEEE Trans. Antennas Propag.*, vol. AP-27, pp. 474–479, 1979.

[106] F. I. Tseng and D. K. Cheng, "Antenna pattern response to arbitrary time signals," *Can. J. Phys.*, vol. 42, pp. 1358–1368, 1964.

[107] F. I. Tseng and D. K. Cheng, "Optimum scannable planar arrays with an invariant side lobe level," *Proc. IEEE*, vol. 56, pp. 1771–1778, 1968.

[108] A. I. Uzkov, "An approach to the problem of optimum directive antenna design," *C. R. Acad. Sci. URSS*, vol. 53, p. 35, 1946.

[109] M. Uzsoky and L. Solymar, "Theory of super-directive linear arrays," *Acta Phys. Hung.*, vol. 6, pp. 185–205, 1956.

[110] G. J. van der Maas, "A simplified calculation for Dolph-Tchebycheff arrays," *J. Appl. Phys.*, vol. 25, pp. 121–124, 1954.

[111] R. C. Voges and J. K. Butler, "Phase optimization of antenna array gain with constrained amplitude excitation," *IEEE Trans. Antennas Propag.*, vol. AP-20, pp. 432–436, 1972.

[112] W. H. von Aulock, "Properties of phased arrays," *Proc. IRE*, vol. 48, pp. 1715–1727, 1960.

[113] W. Wasylkiwskyj and W. J. Kahn, "Element pattern bounds in uniform phased arrays," *IEEE Trans. Antennas Propag.*, vol. AP-25, pp. 597–604, 1977.

[114] W. Wasylkiwskyj and W. K. Kahn, "Element patterns and active reflection coefficient in uniform phased arrays," *IEEE Trans. Antennas Propag.*, vol. AP-22, pp. 207–212, 1974.

[115] H. A. Wheeler, "A survey of the simulator technique for designing a radiating element," in *Phased Array Antennas*, ed. by A. A. Oliner and G. H. Knittel, pp. 132–148, Dedham: Artech House, 1972.

[116] L. P. Winkler and M. Schwartz, "A fast numerical method for determining the optimum snr of an array subject to a Q factor constraint," *IEEE Trans. Antennas Propag.*, vol. AP-20, pp. 503–505, 1972.

[117] P. M. Woodward, "A method for calculating the field over a plane aperture required to produce a given polar diagram," *Proc. IEE*, vol. 93, pt. IIIA, pp. 1554–1558, 1946.

[118] P. M. Woodward, "A method of calculating the field over a plane aperture required to produce a given polar diagram," *Proc. IEE*, vol. 93, pt. III, pp. 1554–1558, 1947.

[119] P. M. Woodward and J. D. Lawson, "The theoretical precision with which an arbitrary radiation pattern may be obtained from a source of finite size," *Proc. IEE*, vol. 95, pt. III, pp. 363–370, 1948.

[120] N. Yaru, "A note on supergain antenna arrays," *Proc. IRE*, vol. 39, pp. 1081–1085, 1951.

[121] R. S. Elliott, "Beamwidth and directivity of large scanning arrays," first of two parts, *Microwave J.*, December 1963.

[122] C. S. Liang and Y. T. Lo, "A multipole-field study for the multiarm log-spiral antennas," *IEEE Trans. Antennas Propag.*, vol. AP-16, pp. 656–664, 1968.

[123] Ye. V. Baklanov, "Chebyshev distribution of currents for a plane array of radiators," *Radio Eng. Electron. Phys.*, vol. 11, pp. 640–642, April 1966 (English translation from Russian).

[124] E. N. Gilbert and S. P. Morgan, "Optimum design of directive antenna arrays subject to random variations," *Bell Syst. Tech. J.*, vol. 34, pp. 637–661, 1955.

[125] A. H. Tihonov, "Solution of incorrectly formulated problems and the regularization method," *Sov. Math.*, vol. 4, pp. 1035–1038, 1964.

[126] H. S. Cabayan, P. E. Mayes, and G. A. Deschamps, "Techniques for computation and realization of stable solutions for synthesis of antenna patterns," *Antenna Lab Rep. 70-13*, Univ. of Illinois at Urbana, October 1970.

Chapter 12

The Design of Waveguide-Fed Slot Arrays

Robert S. Elliott
University of California at Los Angeles

CONTENTS

1. Introduction — 12-3
2. The E-Field Distribution in a Longitudinal Slot — 12-6
3. The Three Design Equations for Arrays of Longitudinal Slots — 12-10
4. The Design of a Linear Array of Resonantly Spaced Longitudinal Slots (Standing-Wave Feed) — 12-13
5. The Design of a Linear Array of Nonreasonantly Spaced Longitudinal Slots (Traveling-Wave Feed) — 12-16
   - Case 1. All Slots on the Same Side of the Center Line — 12-17
   - Case 2. Slots Alternately Displaced, $\beta_{10}d < \pi$ — 12-18
   - Case 3. Slots Alternately Displaced, $\beta_{10}d > \pi$ — 12-18
6. The Design of a Planar Array of Longitudinal Slots — 12-20
7. The Achieved Aperture Distribution of Arrays of Longitudinal Slots — 12-24
8. The Design of Arrays of Centered-Inclined Broad Wall Slots — 12-24
9. The Design of Arrays of Inclined Narrow Wall Slots — 12-25
10. Difficulties in the Design of Large Arrays — 12-28
11. Second-Order Effects — 12-34
    - *Infinite Ground Plane* — 12-34
    - *Wall Thickness* — 12-35
    - *E-Field Distribution in the Slot Aperture* — 12-35
    - *Internal Higher-Order-Mode Mutual Coupling between Radiating Slots* — 12-35
    - *Higher-Order-Mode Coupling in Junctions* — 12-36
12. Far-Field and Near-Field Diagnostics as Design Tools — 12-36
13. References — 12-37

Robert S. Elliott has been a full professor of electrical engineering at UCLA since 1957. In addition to his BS, MS, and PhD degrees in electrical engineering, with the latter two being obtained from the University of Illinois, he holds an AB in English and an MA in economics.

His prior experience includes periods at the Applied Physics Laboratory of Johns Hopkins University and at the Hughes Research Laboratories, where he headed antenna research. Dr. Elliott has also been on the faculty of the University of Illinois and was a founder of Rantec Corporation, serving as its first vice president and technical director. He has recently completed a second two-year stint as a distinguished lecturer for the IEEE and was chairman of the coordinating committee for the 1981 IEEE Symposium, held in Los Angeles. In addition to being a Fellow of the IEEE he is also a member of Sigma Xi, Tau Beta Pi, and the National Academy of Engineering. Dr. Elliott is the author of 60 journal papers and two textbooks. He is the recipient of two Best Paper Awards from the IEEE.

1. Introduction

Many different types of wave-guiding structures have found practical application in a variety of microwave devices. These guiding structures include the coaxial line, rectangular and circular waveguides, stripline, and microstrip. A characteristic of each of these structures is that it will support a set of modes, usually described as consisting of two subsets: TE (transverse electric) and TM (transverse magnetic). For the multiple-conductor structures (coaxial line, stripline, and microstrip) a TEM mode is also possible—indeed, is the dominant mode.

For simple geometries the space-time description of the electric and magnetic fields which comprise a particular mode is obtained readily from Maxwell's equations. When the guiding structure is composed of good conductor, as is usually the case, the lineal current density **K** in the conductor surface can be found from **K** = **n** × **H**, where **n** is a unit vector normal to the conductor surface and **H** is the magnetic field for the particular mode, evaluated at the surface of the conductor.

If a hole is cut in the conductor surface such as to interrupt **K**, and if this hole is open to outer space, radiation can occur. A family of such holes constitutes an antenna array. Judicious choice of the shape, orientation, and relative placement of these holes can produce a variety of useful antenna patterns. Efficient extraction of power from the mode and its transformation to radiation must also be considered. These design questions are the subject of the present chapter.

Most applications of the foregoing idea involve the use of rectangular waveguide as the guiding structure, with transverse dimensions chosen so that only the TE_{10} mode can propagate. The wall currents associated with this mode are sketched in Fig. 1a. The most commonly used slots are shown in Figs. 1b through 1d. Slot A, which interrupts negligible current, is essentially nonradiative and is thereby ideal for use with an inserted movable probe, thus effecting a measurement of the fields within the guide (vswr indicator).

Slot B, which is offset, primarily interrupts transverse current; the induced **E** field in slot B is increased by increasing the offset; the polarity of the induced **E** field is reversed by reversing the direction of offset.

Slot C, which is tilted, primarily interrupts longitudinal current; the induced **E** field in slot C is increased by increasing the tilt; the polarity of the induced **E** field is reversed by reversing the direction of the tilt.

The inclination of slot D causes interruption of the transverse current in the narrow wall; the induced **E** field in slot D is increased by increasing the inclination and is reversed by reversing the direction of inclination.

The three slot types B, C, and D can all radiate, and one can see from the foregoing that the intensity of radiation as well as its polarity is controllable by the amount and direction of offset or tilt. Thus, if λ_g is the guide wavelength for the TE_{10} mode, and if slots are placed $\lambda_g/2$ apart and alternately offset or tilted,

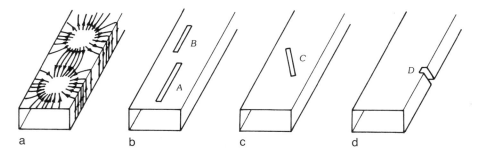

Fig. 1. Rectangular waveguide, TE_{10} mode. (*a*) Wall current distribution. (*b*) Longitudinal broad wall slots. (*c*) Centered, inclined broad wall slot. (*d*) Inclined narrow wall slot.

as shown in Fig. 2, the result is a linear array for which the induced **E** fields in the various slots can have a common phase and an arbitrary amplitude distribution. Linear arrays of this type produce a useful antenna pattern consisting of a main beam and side lobes of governed heights.

If linear arrays of the types shown in Fig. 2a or 2b or 2c are placed side by side, the result is a planar slot array. An example is shown in Fig. 3. By controlling the relative excitation of the individual linear arrays, one is able to achieve a governable two-dimensional aperture distribution and thereby produce a desired radiation pattern.

When the slots are spaced $\lambda_g/2$ apart in a common waveguide the array is said to be *standing-wave fed*. But they need not be $\lambda_g/2$ apart, and many applications

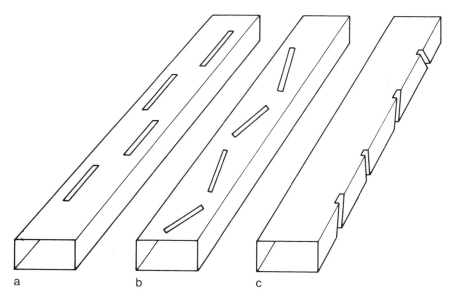

Fig. 2. Linear slot arrays. (*a*) Longitudinal broad wall slots. (*b*) Centered, inclined broad wall slots. (*c*) Inclined narrow wall slots.

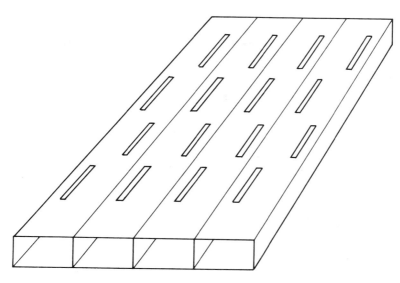

Fig. 3. Planar array of longitudinal slots.

require that they have a spacing other than this. Such arrays are said to be *traveling-wave fed*. Both types will be discussed in this chapter.

Pattern requirements, such as side lobe level in a sum or difference pattern, or ripple in a shaped-beam pattern, are usually sufficiently stringent that external mutual coupling between slots in an array cannot be ignored. There are basically three design procedures currently in vogue which account for mutual coupling:

(1) The first procedure employs an infinite array assumption and posits that the mutual coupling experienced by every slot in the array is the same, being what that slot would experience if it were imbedded in an infinite array, with the array excited in a uniform-amplitude/uniform-progressive-phase distribution. Mutual coupling effects are usually accounted for by an experimental measurement of active impedance or active admittance, using large sample arrays. This approach is acceptable when the array is large, when the actual amplitude distribution is gently tapered, and when the side lobe level requirement is modest (say -20 dB to 30 dB). It is not applicable when the side lobe level is low and/or asymmetric, nor is it applicable when the needed excitation exhibits considerable variability, such as for shaped-beam patterns.

(2) The second procedure relies on a theoretical formulation and calculation of the external mutual coupling between any two slots in the array, with these effects summed for each slot. This approach is, at the present time, fully developed only for cases in which the slots lie in a ground plane sufficiently large to be modeled by an infinite ground plane, or when they lie in a conducting surface of large curvature. Thus the method focuses on broad wall slots. It uses a computer program which is moderately difficult to construct and render error-free. But once that is done, the design costs are modest when compared with the "infinite array" procedure, even for arrays containing several thousand slots, and the inherent accuracy is much higher. Because of this the "infinite array" approach is usually

undertaken only for edge slot arrays, where the second method is not applicable.

(3) The third procedure in reality improves on the design achieved by either the first or second procedure. It takes the best slot array produced by one of those methods and subjects it to a near-field probing, the result of which reveals errors in excitation that can be corrected, at least partially, in the construction of a second array. Iteration of this procedure can result in impressive array performance, notably the attainment of very low side lobe levels.

All three design methods will be described herein, with primary emphasis on the second procedure. But regardless of which approach, or combination of approaches, is adopted in a particular case, the central problem in designing waveguide-fed slot arrays can be stated very simply. How does one choose the dimensions of the individual slots so as to produce a desired radiation pattern and a prescribed input impedance (usually a match)?

Solutions to this problem for various types of slot arrays will be presented in the sections which follow.

2. The E-Field Distribution in a Longitudinal Slot

Longitudinal broad wall slot arrays of the types shown in Figs. 2a and 3 are attractive because of their flush-mounting capability. Additionally, the sturdy box-beam construction evidenced by Fig. 3 is a desirable feature. For these reasons they are widely used in fuze antenna, seeker antenna, and various other radar applications. The design of such arrays begins with an understanding of the E-field distribution in the aperture of a single longitudinal slot when it is excited by a TE_{10} mode.

The slot shown in Fig. 4 is assumed to be followed by a matched load and to be fed by a matched generator which causes a TE_{10} mode, traveling in the $+z$ direction, to be incident on the slot. The slot has a length 2ℓ, a width w, and is offset a distance x_o from the side wall. It is cut in the upper broad wall of the rectangular waveguide; this wall is assumed to be imbedded in a large ground plane of good conductivity. Under these circumstances, what E-field distribution is induced in the slot and how does it depend on the slot's offset and length?

When the waveguide walls are very thin (a modern trend) this question can be answered by modeling the actual situation with a "zero" wall thickness replica and matching internal and external expressions for tangential H at all points P in the slot aperture. In mathematical terms, one imposes the boundary condition

$$\mathbf{1}_y \times [\mathbf{H}_{int}^{inc}(P) + \mathbf{H}_{int}^{scat}(P)] = \mathbf{1}_y \times \mathbf{H}^{ext}(P) \tag{1}$$

where $\mathbf{1}_y$ is a unit vector in the y direction, $\mathbf{H}_{int}^{inc}$ is the field of the incident TE_{10} mode, $\mathbf{H}_{int}^{scat}$ is the field scattered by the slot into the waveguide, and $\mathbf{H}^{ext}$ is the field scattered by the slot into the outer half-space.

Equation 1 can be written in the component form

$$H_x^{ext}(P) - H_x^{scat}(P) = H_x^{inc}(P) \tag{2}$$

$$H_z^{ext}(P) - H_z^{scat}(P) = H_z^{inc}(P)$$

The Design of Waveguide-Fed Slot Arrays

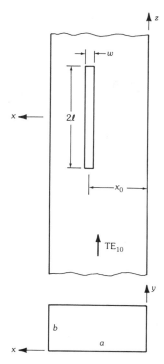

Fig. 4. A single longitudinal slot excited by a TE_{10} mode.

The incident field serves as the driving function in (2) and (3) and is given by

$$H_x^{inc} = -\frac{\beta_{10}}{\pi/a} A_{10} \sin\left(\frac{\pi x}{a}\right) e^{-j\beta_{10}z} \quad (4)$$

$$H_z^{inc} = jA_{10}\cos\left(\frac{\pi x}{a}\right) e^{-j\beta_{10}z} \quad (5)$$

with $\beta_{10} = [k^2 - (\pi/a)^2]^{1/2}$ the propagation constant of the TE_{10} mode, where $k = \omega(\mu_0\epsilon)^{1/2}$ is the wave number and ϵ is the permittivity of the dielectric filling the waveguide.

The internal scattered field which appears in (1)–(3) can be expressed in terms of Stevenson's Green's functions [1]. Similarly, the external field can be expressed in terms of the Green's function for a half-space [2]. These introductions permit (2) and (3) to be rewritten in the forms

$$\int_{slot} [G_{xx}(P,P')E_x(P') + G_{xz}(P,P')E_z(P')]dS' = -\frac{\beta_{10}}{\pi/a} A_{10}\sin\left(\frac{\pi x}{a}\right) e^{-j\beta_{10}z} \quad (6)$$

$$\int_{slot} [G_{zx}(P,P')E_x(P') + G_{zz}(P,P')E_z(P')]dS' = jA_{10}\cos\left(\frac{\pi x}{a}\right) e^{-j\beta_{10}z} \quad (7)$$

in which $P(x,z)$ is any field point in the slot aperture and $P'(x',z')$ is any source point in the slot aperture. The Green's functions $G_{ij}(P,P')$ which appear in (6) and (7) are known functions [1,2].

These integral equations in the unknown aperture field $E_x(P')$, $E_z(P')$ can be solved using the method of moments [3]. When $w \ll 2\ell$, which is the usual case, one finds that $E_z(P')$ is negligible compared with $E_x(P')$.* The problem is then considerably simplified and attention can be confined to (7) in the reduced form

$$\int_{slot} G_{zx}(P,P')E_x(P')dx'dz' = jA_{10}\cos\left(\frac{\pi x}{a}\right)e^{-jB_{10}z} \quad (8)$$

where

$$G_{zx}(P,P') = \frac{1}{2\pi j\omega\mu_0}\left(\frac{\partial^2}{\partial z'^2} + k_o^2\right)\frac{e^{-jk_o R}}{R}$$

$$+ \frac{2}{j\omega\mu_0 ab}\sum_{m=0}^{\infty}\sum_{n=0}^{\infty}\frac{\varepsilon_{mn}^2}{\gamma_{mn}}\cos\left(\frac{m\pi x}{a}\right)\cos\left(\frac{m\pi x'}{a}\right)$$

$$\times \left(\frac{\partial^2}{\partial z'^2} + k^2\right)e^{-\gamma_{mn}|z-z'|} \quad (9)$$

with $R = \overline{PP'}$ and k_o the free-space wave number. The propagation constant γ_{mn} of the mnth mode is given by $\gamma_{mn} = [(m\pi/a)^2 + (n\pi/b)^2 - k^2]^{1/2}$ and the Neumann numbers ε_{mn}^2 are such that $\varepsilon_{00}^2 = 1/4$, $\varepsilon_{0n}^2 = \varepsilon_{m0}^2 = 1/2$, $\varepsilon_{mn}^2 = 1$ otherwise.

Various method of moments types of solution of (8) give results typified by the curves shown in Fig. 5, where E_x along the center line of a longitudinal slot is plotted as a function of z'/ℓ. It can be observed that the magnitude of E_x is an almost-symmetric function, peaking near the center of the slot, and that the phase of E_x is nearly constant over the length of the slot, these observations applying for slot lengths around the first resonance ($2\ell_{res} \cong \lambda_o/2$ for air-filled guide, less for dielectric-filled guide). The peak amplitude is greatest at resonance and the phase distribution is most nearly constant at resonance.

The three amplitude distributions shown in Fig. 5 (for $\ell = 0.95\ell_{res}$, $1.00\ell_{res}$, $1.05\ell_{res}$) are redrawn in Fig. 6 to a common normalized scale. It can be seen that the curves differ slightly over the half of the slot closer to the generator but are indistinguishable over the other half. Also shown in Fig. 6 is a half-cosinusoid distribution. The normalized results from a method of moments calculation can be characterized as slightly bulged out when compared to a half-cosinusoid, with the bulge slightly asymmetric. Investigation has disclosed that this bulge is inconsequential[†] in determining the dimensions of a typical slot array [4]. Thus the important conclusion is reached that the electric-field distribution along the center line of a narrow longitudinal slot, when the slot is excited by a TE_{10} mode, can be represented by the function

---
*This is not true for tilted slots.
[†]It is *not* inconsequential in determining the equivalent circuit of an isolated slot.

The Design of Waveguide-Fed Slot Arrays

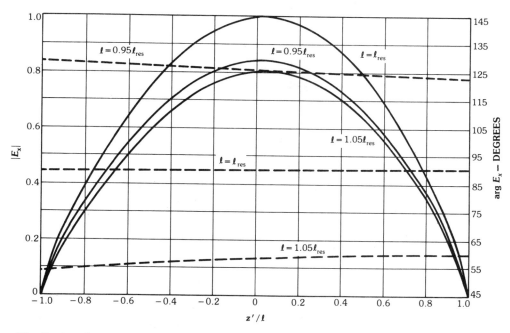

Fig. 5. Amplitude and phase distribution of **E** field in longitudinal slot versus its length. (*After Elliott [5]*)

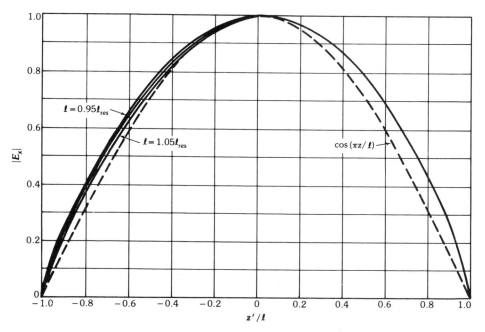

Fig. 6. Normalized **E**-field amplitude distribution in longitudinal slot compared with half-cosinusoid.

$$E_x = \frac{V^s}{w} \cos\left(\frac{\pi z'}{2\ell}\right) \qquad (10)$$

in which V^s is called the slot voltage and is given by

$$V^s = \int_{x_o - w/2}^{x_o + w/2} E_x(x', 0) dx' \qquad (11)$$

The term V^s is a complex quantity whose magnitude and phase are dependent on slot length and offset.

3. The Three Design Equations for Arrays of Longitudinal Slots

When the assumption is made that the electric-field distribution in a longitudinal slot is given by (10), the internal scattered field in the dominant mode can be expressed [5] in the form

$$B_{10} = C_{10} = -K f(x, \ell) V^s \qquad (12)$$

in which B_{10} and C_{10} are, respectively, the complex amplitudes of the back and forward scattered TE_{10} modes. The constant K and the function $f(x, \ell)$ are given by

$$K = \frac{2(\pi/a)^2}{j\omega\mu_0(\beta_{10}/k)(ka)(kb)} \qquad (13)$$

$$f(x, \ell) = \frac{(\pi/2k\ell)\cos\beta_{10}\ell}{(\pi/2k\ell)^2 - (\beta_{10}/k)^2} \sin\left(\frac{\pi x}{a}\right) \qquad (14)$$

with $x = x_o - a/2$ the slot offset measured from the center line of the broad wall of the waveguide.

All the foregoing has been deduced as the response of a single slot to a TE_{10} mode of amplitude A_{10} incident from $z < -\ell$, with the slot terminated in a matched load beyond $z = \ell$. These results can be applied to the design of linear and planar arrays of longitudinal slots. Consider the nth slot in an array of N slots. We shall assume that its aperture field is in the form (10) with its slot voltage V_n^s consisting of three parts, that is,

$$V_n^s = V_{n,1}^s + V_{n,2}^s + V_{n,3}^s \qquad (15)$$

where $V_{n,1}^s$ is due to a TE_{10} mode of amplitude A_{10}^n incident from $z < -\ell n$, $V_{n,2}^s$ is due to a TE_{10} mode of amplitude D_{10}^n incident from $z > \ell_n$, and $V_{n,3}^s$ is due to external mutual coupling to the other $N - 1$ slots in the array.

The total dominant mode scattering off the nth slot is given by

$$B_{10}^n = C_{10}^n = -K f(x_n, \ell_n) V_n^s \qquad (16)$$

This symmetrical scattering can be modeled by a shunt admittance on an equivalent lossless transmission line [6] for which

$$B_n = C_n = -\frac{1}{2}\frac{Y_n^a}{G_o}V_n \qquad (17)$$

with Y_n^a the *active* admittance of the nth slot (including external mutual coupling). The term V_n is the mode voltage on the equivalent transmission line at the site of Y_n^a; G_o is the characteristic conductance of the equivalent transmission line.

Since B_{10}^n and B_n should be proportional, the combination of (16) and (17) yields

$$\frac{Y_n^a}{G_o} = K_1 f(x_n, \ell_n)\frac{V_n^s}{V_n} \qquad (18)$$

where K_1 is a proportionality constant. Equation 18 is the first of three design equations needed to determine the lengths and offsets of all slots in the array.

For an isolated slot

$$\frac{Y}{G_o}(x, \ell) = -\frac{2B_{10}/A_{10}}{1 + (B_{10}/A_{10})} \qquad (19)$$

which means that

$$B_{10}^{n,1} = -\frac{Y(x_n, \ell_n)/G_o}{2 + Y(x_n, \ell_n)/G_o} A_{10}^n \qquad (20)$$

with $B_{10}^{n,1}$ that part of B_{10}^n due to $V_{n,1}^s$. Since $B_{10}^{n,1} = -Kf(x_n, \ell_n)V_{n,1}^s$, it follows that

$$V_{n,1}^s = \frac{1}{Kf(x_n, \ell_n)}\frac{Y(x_n, \ell_n)/G_o}{2 + Y(x_n, \ell_n)/G_o} A_{10}^n \qquad (21)$$

Similarly,

$$V_{n,2}^s = \frac{1}{Kf(x_n, \ell_n)}\frac{Y(x_n, \ell_n)/G_o}{2 + Y(x_n, \ell_n)/G_o} D_{10}^n \qquad (22)$$

A development [5] which entails use of the reciprocity theorem reveals that the third component of slot voltage is given by

$$V_{n,3}^s = -j(\beta_{10}/k)(k_o b)(a/\lambda)^3 \frac{1}{f^2(x_n, \ell_n)}\frac{Y(x_n, \ell_n)/G_o}{2 + Y(x_n, \ell_n)/G_o}$$

$$\times \sum_{m=1}^{N}{}' V_m^s g_{mn}(x_m, \ell_m, x_n, \ell_n) \qquad (23)$$

where

$$g_{mn} = \int_{-k_o\ell_m}^{k_o\ell_m} \cos\left(\frac{u'_m}{4\ell_m/\lambda_o}\right)\left\{\frac{1}{(4\ell_n/\lambda_o)}\left[\frac{e^{-jk_oR_1}}{k_oR_1} + \frac{e^{-jk_oR_2}}{k_oR_2}\right]\right.$$
$$\left. + \left[1 - \frac{1}{(4\ell_n/\lambda_o)^2}\right]\int_{-k_o\ell_n}^{k_o\ell_n}\cos\left(\frac{u'_n}{4\ell_n/\lambda_o}\right)\frac{e^{-jk_oR}}{k_oR}du'_n\right\}du'_m \quad (24)$$

In (23) the prime on the summation sign means that the term $m = n$ is to be excluded. In (24) the surrogate variable $u' = k_o z'$ has been introduced. The variable distances R, R_1, and R_2 are defined in Fig. 7. Since generally

$$\frac{Y_n^a}{G_o} = -\frac{2B_{10}^n}{A_{10}^n + D_{10}^n + B_{10}^n} \quad (25)$$

where $B_{10}^n = B_{10}^{n,1} + B_{10}^{n,2} + B_{10}^{n,3}$, one can relate all quantities which appear on the right side of (25) to $V_{n,1}^s$, $V_{n,2}^s$, and $V_{n,3}^s$. When this is done, the second design equation results, viz.,

$$\frac{Y_n^a}{G_o} = \frac{2f^2(x_n,\ell_n)}{\frac{2f^2(x_n,\ell_n)}{Y(x_n,\ell_n)/G_o} + j\frac{\beta_{10}}{k}(k_ob)\left(\frac{a}{\lambda}\right)^3\sum_{m=1}^{N}{}'\frac{V_m^s}{V_n^s}g_{mn}} \quad (26)$$

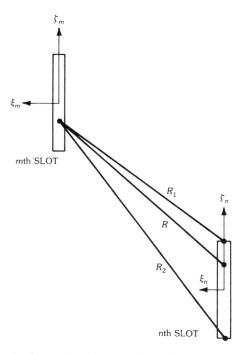

Fig. 7. Geometry for mutual coupling calculation.

The Design of Waveguide-Fed Slot Arrays

If there are P slots in the waveguide containing the pth slot, the third design equation is simply

$$\sum_{p=1}^{P} \frac{Y_p^a}{G_o} = \text{specified constant} \tag{27}$$

For a linear array (single waveguide) the specified constant will usually be unity, indicating that an input match is desired. But in a planar array the individual branch line waveguides need not be matched since the sum of the active admittances is transformed down to an equivalent load on the main line waveguide; it is the sum of these equivalent loads that is often specified to be a match.

4. The Design of a Linear Array of Resonantly Spaced Longitudinal Slots (Standing-Wave Feed)

When all of the slots are in a common waveguide and resonantly spaced ($\lambda_g/2$) on centers, the mode voltage V_n is common to all slots except for an alternation in sign. This alternation is compensated by an alternation in the direction of slot offset, which causes $\sin(\pi x_n/a)$ to change sign at successive slots. In this case the first design equation (18) can be written in ratio form

$$\frac{Y_n^a}{G_o} \bigg/ \frac{Y_m^a}{G_o} = \frac{|f(x_n, \ell_n)|V_n^s}{|f(x_m, \ell_m)|V_m^s} \tag{28}$$

with m and n any two slots in the linear array and $f(x, \ell)$ given by (14).

To use the second design equation one needs first to express in functional form the input data that has been accumulated on the isolated self-admittance Y/G_o versus slot length and offset. That data can be obtained in either of two ways:

(1) Experimentally, by constructing a family of waveguides, each with a single slot but with different offsets and shorter-than-resonant lengths. After Y/G_o has been measured for each member of the family, the slots can be lengthened and the measurements repeated, etc. For most applications it is sufficient to obtain data in the range $0.95\,\ell_{res} \leq \ell \leq 1.05\,\ell_{res}$ and seven well-spaced data points per family member are usually sufficient.

(2) Computationally, by solving (8) via the method of moments to find $E_x(P')$ as a function of slot offset and length. The knowledge thus gained can be used to calculate B_{10}/A_{10} and then, using (19), to deduce $Y(x, \ell)/G_o$. If need be, a correction for wall thickness can be applied to these calculations [7].

Whichever of these two approaches is used, one should find that the input data can be normalized in the following way. Let

$$g(x) = \frac{G_{res}}{G_o}(x, \ell_{res}) \tag{29}$$

$$v(x) = k\,\ell_{res}(x) \tag{30}$$

$$h(y) = h_1(y) + jh_2(y) \tag{31}$$

$$h_1(y) = \frac{G(x,y)/G_o}{g(x)} \tag{32}$$

$$h_2(y) = \frac{B(x,y)/G_o}{g(x)} \tag{33}$$

$$y = \ell/\ell_{\text{res}} \tag{34}$$

The function $g(x)$ gives the conductance of a resonant-length slot versus offset and can be fitted with a few terms of a polynomial (polyfitted) or often better fitted by a transcendental function. The function $v(x)$ is seen to be the resonant length versus offset multiplied by the wave number. The function $h(y)$ is a complex function such that

$$Y(x,y)/G_o = g(x)\, h(y) \tag{35}$$

The two components of $h(y)$ are real functions, with $h_1(y)$ representing $G(x,y)/G_o$ in ratio to $g(x)$ and $h_2(y)$ similarly representing $B(x,y)/G_o$ in ratio to $g(x)$. The virtue of assembling the data in this form, with Y/G_o normalized to $g(x)$, is that when the data are plotted versus $y = \ell/\ell_{\text{res}}$, the result is found to be essentially independent of offset. Since $h_1(y)$ and $h_2(y)$ are easily polyfitted, the representation of Y/G_o given in the form (35) is helpful in the ensuing calculations.

With Y/G_o now a known input function, the next step is to find *starting* values for the lengths and offsets of all the slots. Since the design procedure converges so rapidly, it would be sufficient to start with all the slots on the center line and resonant-length. However, it is prudent, and a good benchmark, to begin instead by finding the slot dimensions which would prevail were there no mutual coupling. To do this, one need work only with (28), which takes on the special form*

$$\frac{g(x_n)}{g(x_m)} = \frac{|f(x_n, \ell_n^{\text{res}})| V_n^s}{|f(x_m, \ell_m^{\text{res}})| V_m^s} \tag{36}$$

The desired slot voltage ratio V_n^s/V_m^s is known from pattern specifications. If we let $m = 1$, and choose x_1 arbitrarily, the only unknown in (36) is x_n, which can be obtained by a search routine. In this fashion a family of slot offsets $(x_1, x_2, \ldots, x_n)$ is determined. The corresponding resonant conductances can be found from $g(x)$ and the input conductance from

$$\frac{G_{\text{in}}}{G_o} = \sum_{n=1}^{N} g(x_n) \tag{37}$$

---

*It is being assumed here, because this is a resonantly spaced array, that the aperture distribution is equiphase.

The Design of Waveguide-Fed Slot Arrays

If this sum, which is a use of the third design equation, does not satisfy the input admittance specification (usually a match), x_1 can be adjusted upward or downward as appropriate, and the process repeated. A relatively few iterations will usually produce a set of offsets $(\mathring{x}_1, \mathring{x}_2, \ldots, \mathring{x}_N)$ which do cause agreement with the input admittance specification. Then the corresponding resonant lengths $(\mathring{\ell}_1^{\text{res}}, \mathring{\ell}_2^{\text{res}}, \ldots, \mathring{\ell}_N^{\text{res}})$ can be found by using the known function $v(x)$.

The next step is to compute g_{mn} from (24) for $m = 1, 2, \ldots, N$ and $n = 1, 2, \ldots, N$ but $m \neq n$, with the starting values $(\mathring{x}_n, \mathring{\ell}_n^{\text{res}})$ and $(\mathring{x}_m, \mathring{\ell}_m^{\text{res}})$ used in the determination of R, R_1, and R_2. The values of g_{mn} should be stored and retrieved to compute the mutual coupling terms

$$MC_n = j(\beta_{10}/k)(k_o b)(a/\lambda)^3 \sum_{m=1}^{N}{}' \frac{V_m^s}{V_n^s} g_{mn} \tag{38}$$

Equation 37 can be recognized as the second term in the denominator of the second design equation (26). Absent this term, (26) reduces to the simple identity that the active admittance of the nth slot equals its self-admittance. Thus MC_n, accounts for external mutual coupling and is seen to contain the sum of the g_{mn} values weighted by the aperture distribution. In calculating MC_n, one should use the desired ratio V_m^s/V_n^s.

After these starting values of MC_n are known for each slot in the array, a search needs to be undertaken to find a couplet (x_n, ℓ_n) which will cause

$$\mathscr{I}m\left\{\frac{2f^2(x_n, \ell_n)}{Y(x_n, \ell_n)/G_o}\right\} = -\mathscr{I}m\{MC_n\} \tag{39}$$

Satisfaction of (39) will make Y_n^a/G_o in (26) be pure real, resulting in what one might call active resonance.*

It will be found that a continuum of couplets (x_n, ℓ_n) satisfies (39). Similarly, a continuum of couplets (x_m, ℓ_m) will satisfy (39) when the mth slot is considered in place of the nth slot. But for a given choice (x_n', ℓ_n'), there is only one couplet (x_m', ℓ_m') which also satisfies the first design equation in the form (28). Using this criterion, one can assemble a compatible family of slot offsets and lengths $(x_1', \ell_1'), (x_2', \ell_2'), \ldots, (x_N', \ell_N')$.

But was the right family selected? This question can be answered by seeing if the sum of the active admittances satisfies the specification implicit in the third design equation (27). If not, a new family needs to be chosen by starting with a different couplet (x_n', ℓ_n').

After the right family of couplets has been found, the process starting with (37) must be iterated, because now it is possible to improve on the calculation of g_{mn}, using $(x_n', \ell_n', x_m', \ell_m')$ in place of $(\mathring{x}_n, \mathring{\ell}_n, \mathring{x}_m, \mathring{\ell}_m)$. The iterations should be continued until the latest set of slot dimensions differs from the penultimate set by

---

*It is somewhat arbitrary to require that Y_n^a/G_o be pure real for all n. Doing so ensures that the aperture distribution V_n^s/V_m^s will be equiphase. But that result could be accomplished by requiring that all the active admittances have any common phase. However, demanding that their common phase be zero degrees helps to optimize the bandwidth of the array.

less than the achievable construction tolerances.* At this point the design of the linear array of longitudinal slots is completed.

A simple example which illustrates how the input data on the admittance of a single slot can be functionally fitted to give $g(x)$, $v(x)$, and $h_1(y) + jh_2(y)$, and how the three design equations can be used to determine the slot dimensions for a two-by-four array, can be found in [4].

5. The Design of a Linear Array of Nonresonantly Spaced Longitudinal Slots (Traveling-Wave Feed)

When all of the slots are in a common waveguide and equispaced a distance d, but with $d \neq \lambda_g/2$, and when the last slot is terminated by a matched load, the array is said to be traveling-wave fed. Here, in contrast to the situation considered in the previous section, it is not correct to say that the mode voltage V_n is common to all slots except for an alternation in sign. This introduces an additional complexity into the design.

The equivalent circuit of a linear traveling-wave array is shown in Fig. 8. With reference to that figure, let Y_n be the total admittance seen looking into the nth junction toward the matched load. Then

$$\frac{Y_n}{G_o} = \frac{Y_n^a}{G_o} + \frac{(Y_{n-1}/G_o)\cos\beta_{10}d + j\sin\beta_{10}d}{\cos\beta_{10}d + j(Y_{n-1}/G_o)\sin\beta_{10}d} \qquad (40)$$

The mode voltages at successive junctions are connected by the relation

$$V_n = V_{n-1}[\cos\beta_{10}d + j(Y_{n-1}/G_o)\sin\beta_{10}d] \qquad (41)$$

When the first design equation (18) is written in ratio form for two successive slots and (41) is employed, one finds that

$$\frac{Y_n^a/G_o}{f(x_n, \ell_n)} = \frac{Y_{n-1}^a/G_o}{f(x_{n-1}, \ell_{n-1})} \cdot \frac{V_n^s/V_{n-1}^s}{\cos\beta_{10}d + j(Y_{n-1}/G_o)\sin\beta_{10}d} \qquad (42)$$

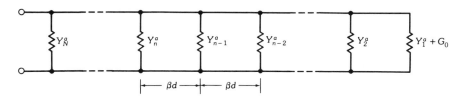

Fig. 8. Equivalent circuit of a linear array of longitudinal slots, traveling-wave fed.

---
*Usually only a few iterations are needed to achieve this.

The Design of Waveguide-Fed Slot Arrays

Equation 42 is a recurrence relation which, in conjunction with the second design equation (26), permits determination of the length and offset of the nth slot, once the length and offset of the $(n-1)$st slot are known.*

As in the case of the standing-wave array discussed in the previous section, the first step in the design is to find starting values for the slot lengths and offsets. Once again it is advisable to do this by ignoring (temporarily) all external mutual coupling. Then, if self-resonant slots are assumed initially, (42) becomes

$$\frac{g(\mathring{x}_n)}{f(\mathring{x}_n, \mathring{l}_n^{\text{res}})} = \frac{g(\mathring{x}_{n-1})}{f(\mathring{x}_{n-1}, \mathring{l}_{n-1}^{\text{res}})} \cdot \frac{V_n^s/V_{n-1}^s}{\cos\beta_{10}d + j(Y_{n-1}/G_o)\sin\beta_{10}d}$$

It is necessary to digress at this point in order to discuss an important consideration which impacts on the design process. Most practical applications involving traveling-wave linear slot arrays are such that the number of slots N is large (say 30 or more) and the fraction of the power absorbed in the terminating load is modest (say 5 to 10 percent). With 90 to 95 percent of the power being radiated by N slots, and with a typical aperture distribution designed to give a sum pattern with low side lobes, the slot closest to the matched load will be called on to radiate less than 1 percent of the power. Thus $g(x_1)$ will typically be an order of magnitude less than unity and Y_1/G_o will be close to unity. As one considers the total admittance at junctions farther and farther removed from the matched load, since $g(x_n)$ is small compared to unity for all n, it follows that $Y_{n-1}/G_o \cong 1 + j0$, for all n. Thus (42) is given to good approximation by

$$\frac{g(\mathring{x}_n)}{f(\mathring{x}_n, \mathring{l}_n^{\text{res}})} = \frac{g(\mathring{x}_{n-1})}{f(\mathring{x}_{n-1}, \mathring{l}_{n-1}^{\text{res}})} \cdot \frac{V_n^s/V_{n-1}^s}{e^{j\beta_{10}d}} \tag{43}$$

This equation puts a limitation on the *phase progression* one can specify for V_n^s/V_{n-1}^s. Three cases need to be examined:

Case 1. All Slots on the Same Side of the Center Line

When this is so, $f(\mathring{x}_n, \mathring{l}_n^{\text{res}})$ and $f(\mathring{x}_{n-1}, \mathring{l}_{n-1}^{\text{res}})$ have the same sign and a solution of (43) is only possible if V_n^s/V_{n-1}^s has a phase progression of $-\beta_{10}d$ radians. Since the array factor for a sum pattern is given by

$$S(\theta) = \sum_{n=1}^{N} (V_n^s/V_1^s) e^{-jnk_o d \cos\theta}$$

with $\theta = 0°$ directed along the array toward the matched load, we get for this case

$$S(\theta) = \sum_{n=1}^{N} |V_n^s/V_1^s| e^{-jnd(k_o \cos\theta - \beta_{10})}$$

---

*It is assumed here that an input match is desired; thus by implication the third design equation is also being taken into account.

and the main beam points at an angle θ_o given by

$$\theta_o = \cos^{-1}(\beta_{10}/k_o), \qquad (44)$$

a result which is seen to be independent of the spacing d. For economy reasons, d should be chosen as large as possible without permitting an extra main beam to occur [8].

Case 2. Slots Alternately Displaced, $\beta_{10}d < \pi$

When this is so, (43) becomes

$$\frac{g(\mathring{x}_n)}{|f(\mathring{x}_n, \mathring{\ell}_n^{res})|} = \frac{g(\mathring{x}_{n-1})}{|f(\mathring{x}_{n-1}, \mathring{\ell}_{n-1}^{res})|} \cdot \frac{V_n^s/V_{n-1}^s}{e^{j(\beta_{10}d - \pi)}}$$

The phase progression of V_n^s/V_{n-1}^s must be $\beta_{10}d - \pi$ radians and the main beam points at an angle θ_o, which can be determined from

$$\theta_o = \cos^{-1}\left(\frac{\beta_{10}d - \pi}{k_o d}\right) \qquad (45)$$

With $\beta_{10}d < \pi$, the angle θ_o lies somewhere between broadside and reverse end-fire. In this case, θ_o is not independent of d. However, there is a practical lower limit on d because the slots should not overlap, otherwise internal higher-order-mode coupling becomes nonnegligible. Similarly, there is an upper limit on d because, as $\beta_{10}d - \pi \to 0$, scattering off the different slots tends to add up in phase at the input, causing an unsatisfactory mismatch.

Case 3. Slots Alternately Displaced, $\beta_{10}d > \pi$

Here the analysis is the same as in case 2, and equation (46) still applies but now, with $\beta_{10}d > \pi$, θ_o lies somewhere between broadside and forward end-fire. Once again, θ_o is not independent of d. However, d has a lower bound because $\beta_{10}d - \pi$ should not be so close to zero as to cause the scattering from the various slots to add up at the input. The term d also has an upper bound since multiple main beams usually should be avoided.

The discussion of these three cases has been based on (42), which ignores mutual coupling. However, it applies equally well to (41). As will be seen shortly, when mutual coupling is taken into account the individual slots are detuned to make the appropriate adjustments so that the desired aperture distribution V_n^s/V_{n-1}^s is achieved. But if what is desired does not fit one of the three natural cases just described, the range of adjustment possible in the slot lengths and offsets is not adequate to accomplish the task.

This digression can be summarized by saying that the desired aperture distribution must be restricted to be in the form

$$\frac{V_n^s}{V_{n-1}^s} = \left|\frac{V_n^s}{V_{n-1}^s}\right| e^{j(\beta_{10}d - p\pi)} \tag{46}$$

where, if $p = 0$, the main beam position θ_o points somewhere between broadside and forward end-fire and is given by (44). If $p = 1$, θ_o points somewhere between broadside and reverse (forward) end-fire if $\beta_{10}d$ is less (greater) than π. No other values of p are allowed. If $p = 0$, θ_o is independent of d; if $p = 1$, θ_o is dependent on d. More will need to be said about the choice of the value of d shortly.

With these restrictions on the allowable desired aperture distributions, (43) reduces to

$$\frac{g(\mathring{x}_n)}{f(\mathring{x}_n, \mathring{\ell}_n^{\text{res}})} = \frac{g(\mathring{x}_{n-1})}{f(\mathring{x}_{n-1}, \mathring{\ell}_{n-1}^{\text{res}})} \cdot \left|\frac{V_n^s}{V_{n-1}^s}\right| \tag{47}$$

with $(\mathring{x}_n, \mathring{\ell}_n^{\text{res}})$ the sought-for starting couplet for the nth slot. These starting values can be found by the following procedure:

First, *assume* a value for $\mathring{x}_1$, that is, for the offset of the slot closest to the matched load. Then find $v(\mathring{x}_1)$ from (30). This yields $\mathring{\ell}_1^{\text{res}}$. Next find $f(\mathring{x}_1, \mathring{\ell}_1^{\text{res}})$ from (14). At this point the right side of (47) can be computed for the case $n = 2$, using the desired slot voltage ratio $|V_2^s/V_1^s|$. One now searches for the couplet $(\mathring{x}_2, \mathring{\ell}_2^{\text{res}})$ which brings the left side of (47) into harmony with the right side. This process is then repeated for $n = 3$, etc., until values have been obtained for all slot offsets and lengths.

This set of couplets needs to be examined for appropriateness in the following sense: If $\mathring{x}_1$ has been chosen too low, the maximum offset, say $\mathring{x}_m$, is below the upper limit for which the input data on $g(x)$ and $v(x)$ are considered reliable. In this case the fraction of the input power that is dissipated in the load is unnecessarily high. Conversely, if $\mathring{x}_1$ has been chosen too high, the maximum offset, say $\mathring{x}_m$, is beyond the upper limit for which the input data on $g(x)$ and $v(x)$ are trusted. In this case the power dissipated in the load is too small. In either event the direction to be taken for the next guess for the value of $\mathring{x}_1$ is indicated. Usually a few such trials are sufficient to yield the proper set of starting slot lengths and offsets.

With the starting couplets ascertained, attention can be focused on the second design equation. The quantities g_{mn} should be calculated for $m = 1, 2, \ldots, N$ and $n = 1, 2, \ldots, N$, but $m \neq n$. (Before performing these computations one should be sure to input whether or not the slots are alternately offset. See the earlier discussion in this section of the three cases.) In order to find g_{mn}, one must first decide on the value of d. For case 1, d should be as large as possible, consistent with the avoidance of extra main beams. For cases 2 and 3, d should be chosen so as to satisfy (45), with beam placement at the desired angle θ_o.

Knowledge of the g_{mn} values permits calculation of the mutual coupling terms MC_n, given by (38). With the offset of the first slot left unchanged, one should search for a new slot length ℓ_1 such that

$$\mathscr{I}m\left\{\frac{2f^2(\mathring{x}_1, \ell_1)}{Y(\mathring{x}_1, \ell_1)/G_o}\right\} = -\mathscr{I}m\{MC_1\} \tag{48}$$

With this length known, Y_1^a/G_o can be found from (26) and Y_1/G_o from (40). The next step is to find a new couplet (x_2, ℓ_2) which simultaneously satisfies (39) and (42) with $n = 2$. This process is then repeated to find a new couplet (x_3, ℓ_3), etc., until finally new values exist for all the couplets and the input admittance Y_N^a/G_o is known.

At this point it is wise to pause and take stock of several factors. Is Y_N^a/G_o close to $1 + j0$? If not, there is probably an error in the computer program. Is the largest slot offset comfortably close to the upper limit of reliable self-admittance input data? If not, there is probably too much (or too little) power going into the matched load. This can be checked easily by using the formula

$$\frac{P_{\text{load}}}{P_{\text{in}}} = \frac{V_1 V_1^*}{V_N V_N^* G_N^a/G_o} \tag{49}$$

with the ratio V_1/V_N found through recursive use of (41).

One is now ready to iterate the design since the new couplet values (x_n, ℓ_n) allow an improved calculation of g_{mn}. This time it may be desirable to adjust x_1 up or down, depending on what was found in taking stock. The entire process should be iterated as many times as necessary to reach the point at which the ultimate set of slot offsets and lengths differs from the penultimate set by less than construction tolerances. Experience has shown that usually this requires only a few iterations.

It may prove desirable, as a final polish on the design, to make a minor adjustment in the value of d in order to optimize the input admittance Y_N^a/G_o. It can be shown [9] that an input match is achieved when

$$\sum_{n=1}^{N} \frac{Y_n^a}{G_o} e^{j2n\beta_{10}d} = 0 \tag{50}$$

If the ultimate values of Y_n^a/G_o are used in (50), a value of d can be found very close to the original value, which should improve on the input admittance. Indeed, one could choose to change d at every iteration, using (50), thus converging on a final value of d in concert with converging on final values of the couplets (x_n, ℓ_n).

An actual example of the use of this design technique can be found in [9].

6. The Design of a Planar Array of Longitudinal Slots

Two-dimensional arrays of slots of the type shown in Fig. 3 can be designed by a simple extension of the procedures detailed in the previous two sections. The analysis to be presented here will be limited to situations in which all waveguides containing radiating slots have the same a and b dimensions and a common vanishingly thin wall thickness. Adjacent waveguides share a common narrow wall, giving a box-beam construction. All slots have a rectangular periphery and a common width w and all are longitudinal, with individual offsets and lengths (x_{mn}, ℓ_{mn}). There is a common longitudinal spacing d_z and a common waveguide-to-waveguide spacing d_x.

The new ingredient, not present in the earlier designs of linear arrays, is the

The Design of Waveguide-Fed Slot Arrays

need to couple to the different waveguides in the correct ratios. A common coupling mechanism is shown in Fig. 9. It can be seen from a study of this figure that a single waveguide (hereafter called the "main line guide") has been placed behind the array and transverse to it. It shares a broad wall sequentially with all the radiating waveguides (hereafter called the "branch line guides"). In that shared wall a sequence of centered-inclined slots has been cut. These serve to couple power from the main line guide to the branch line guides. The relative amounts of coupled powers are governed by the tilts θ_n and lengths ℓ_n of the coupling slots. It is customary (and desirable for bandwidth purposes) to make the coupling slots resonant-length. The scattering matrix of the nth coupling junction is simply [10]

$$\mathbf{S} = \begin{bmatrix} s_{11} & (1-s_{11}) & s_{13} & -s_{13} \\ (1-s_{11}) & s_{11} & -s_{13} & s_{13} \\ s_{13} & -s_{13} & (1-s_{11}) & s_{11} \\ -s_{13} & s_{13} & s_{11} & (1-s_{11}) \end{bmatrix} \quad (51)$$

in which $s_{11}(\theta_n)$ can be deduced by measurement or calculation, and in which $s_{13} = [s_{11}(1 - s_{11})]^{1/2}$. The back-scattering coefficient $s_{11}(\theta_n)$ is obtained by placing matched loads in a branch line devoid of radiating slots, as well as placing a matched load in the main line guide just beyond the nth coupling slot, and measuring or calculating B_1/A_1, with A_1 the voltage wave leaving the matched generator and B_1 the voltage wave returning to it, both referenced at the slot center. For a given tilt θ_n, as the coupling slot length is varied, resonance is said to occur when s_{11} is pure real.

Since

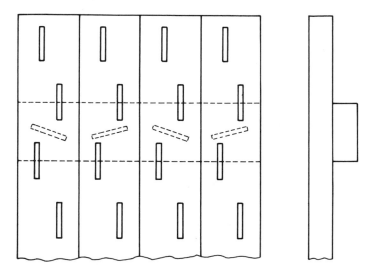

Fig. 9. Planar array of longitudinal broad wall slots fed from below by a transverse waveguide containing centered, inclined coupling slots.

$$[B] = [S][A] \qquad (52)$$

with $[A]$ an arbitrary set of voltage waves incident on the four ports and $[B]$ the corresponding set of back-scattered voltage waves, if a total active admittance

$$\frac{Y_n^a}{G_o} = \sum_{m=1}^{M(n)} \frac{Y_{mn}^a}{G_o} \qquad (53)$$

occurs at a cross section in the nth radiating branch line a distance $\lambda_g/4$ from the coupling slot center, it is a simple exercise to show that this total active admittance is transformed to appear as a series load in the main line, the relation being

$$\frac{Z_n^a}{R_o} = \varkappa_n^2 \frac{Y_n^a}{G_o} \qquad (54)$$

where $\varkappa_n^2 = s_{11}(\theta_n)/[1 - s_{11}(\theta_n)]$. Because equivalent expressions for complex power flow must equate, it follows that $I_n I_n^* Z_n^a/R_o = V_n V_n^* Y_n^a/G_o$ and thus

$$V_n = \varkappa_n I_n \qquad (55)$$

with I_n the mode current in the main line at the nth junction and V_n the mode voltage in the nth branch line at a cross section $\lambda_g/4$ from the junction center. Equation 55 is a key result which must be included in the design of a planar array. Obtained here for the case of a centered-inclined slot, it is equally valid for other coupling mechanisms, such as iris-coupled slots on the broad wall center line, and corporate feeds.

The new ingredient (the set of coupling junctions) does not fundamentally alter the design process described in the previous two sections. If the branch lines and main line are both standing-wave fed, which implies that $|I_n|$ is common to all junctions, and that $|V_n|$ is common to all M radiating slots, then the starting couplets $(\mathring{x}_{mn}, \mathring{\ell}_{mn}^{res})$ can be found from

$$\frac{g(\mathring{x}_{mn})}{g(\mathring{x}_{pq})} = \frac{|f(\mathring{x}_{mn}, \mathring{\ell}_{mn}^{res})|}{|f(\mathring{x}_{pq}, \mathring{\ell}_{pq}^{res})|} \frac{V_{mn}^s}{V_{pq}^s} \frac{\varkappa_n}{\varkappa_q} \qquad (56)$$

which is seen to be a form of the first design equation. For the nth branch line, the appropriate starting form for the third design equation is

$$\frac{G_{in}^n}{G_o} = \sum_{m=1}^{M(n)} g_{mn}(\mathring{x}_{mn}) \qquad (57)$$

whereas for the main line it is

$$\frac{R_{in}}{R_o} = \sum_{n=1}^{N} \frac{R_n}{R_o} = \sum_{n=1}^{N} \varkappa_n^2 \sum_{m=1}^{M(n)} g(\mathring{x}_{mn}) \qquad (58)$$

The Design of Waveguide-Fed Slot Arrays

One can find the starting offsets by considering (56)–(58) in concert, but it is important to note that there is not a unique solution to this set of equations. It is recommended that the slot which is to have the greatest slot voltage be singled out and assigned initially an offset near the upper limit of reliable input data. Equation 56 can then be used to obtain initial offsets for all the other slots in the same branch line.

At this point it is necessary to guess a set of values for the coupling coefficients $\varkappa_n$. Having done this, one can then use (56) to obtain initial offsets for all the other slots in all the other branch lines.

Equation 58 can be examined for an initial estimate of the input resistance to the main line. If this is too high (low), the coupling coefficients $\varkappa_n$ can be collectively adjusted down (up). Further, if the slots in a particular branch line are too little (much) offset, that particular coupling coefficient can be decreased (increased). A certain amount of adjusting of the various coupling coefficients is usually required before all slots have offsets in the most reliable range* and the desired main line input resistance is achieved, but this is an important step in the design process. On the other hand, too much accuracy is not called for since these are only starting values and they will need to be modified to include the effects of external mutual coupling.

With $\mathring{x}_{mn}$ selected for every slot, the corresponding resonant lengths can be determined from $v(\mathring{x}_{mn})$. The couplets $(\mathring{x}_{mn}, \mathring{\ell}_{mn}^{res})$ and $(\mathring{x}_{pq}, \mathring{\ell}_{pq}^{res})$ can then be used in the calculation of g_{mnpq}, using (24) but with double subscript notation. A starting value for the mutual coupling term MC_{mn} can next be calculated from

$$MC_{mn} = j(\beta_{10}/k)(k_o b)(a/\lambda)^3 \sum_{p=1}^{M(q)} \sum_{q=1}^{N} \frac{V_{pq}^s}{V_{mn}^s} g_{mnpq} \tag{59}$$

for every slot in the planar array. Equation 59 is seen to be a restatement of (38) in double subscript notation, and is the second term in the denominator of the second design equation.

Equation 39 is used to find suitable couplets (x_{mn}, ℓ_{mn}) and then acceptable families of couplets are assembled by satisfying

$$\frac{G_{mn}^a(x_{mn}, \ell_{mn})/G_o}{G_{pq}^a(x_{pq}, \ell_{pq})/G_o} = \frac{|f(x_{mn}, \ell_{mn})|}{|f(x_{pq}, \ell_{pq})|} \cdot \frac{V_{mn}^s}{V_{pq}^s} \cdot \frac{\varkappa_n}{\varkappa_q} \tag{60}$$

The proper family is the one which satisfies (58) with G_{mn}^a/G_o replacing $g(\mathring{x}_{mn})$. The procedure will need to be iterated since improved values of g_{mnpq} are now possible. Some adjustment of the coupling coefficients $\varkappa_n$ may prove necessary as the iterations progress.

If the branch lines are traveling-wave fed, the procedure outlined in Section 5 needs to be folded into the design process. If the main line is traveling-wave fed, that procedure can be applied to the main line as well.

---

*An alternate criterion for determining the coupling coefficients involves choosing the admittance levels of the branch lines so as to optimize input vswr and pattern side lobe level over a specified frequency range.

In many planar array applications the aperture is divided into four quadrants for the purpose of providing sum and difference patterns. In such cases, if the sum pattern is to have quadrant symmetry, the slot offsets and lengths display quadrant I/quadrant III and quadrant II/quadrant IV symmetry, and one needs to find the offsets and lengths of only half the slots. However, external mutual coupling with *all* the other slots must still be included.

7. The Achieved Aperture Distribution of Arrays of Longitudinal Slots

Given the lengths and offsets of all slots in a linear or planar array it is possible to find the resulting aperture distribution. This is accomplished by equating the right sides of the first and second design equations which, after rearrangement, gives

$$\frac{2f^2(x_{mn}, \ell_{mn})}{Y(x_{mn}, \ell_{mn})/G_o} V^s_{mn} + j(\beta_{10}/k)(k_o b)(a/\lambda)^3 \sum_{p=1}^{M(q)} \sum_{q=1}^{N}{}' g_{mnpq} V^s_{pq}$$
$$= V_{mn} f(x_{mn}, \ell_{mn}) \tag{61}$$

(The immaterial constant $2/K_1$ has been suppressed on the right side of this equation.)

Equation 61 is given in double subscript notation but it applies equally well for linear arrays. With the slot lengths and offsets known, (61) is seen to be a set of simultaneous linear equations in the unknown slot voltages V^s_{mn}. Matrix inversion gives the aperture distribution.

This equation is useful as a design check, but in the case of planar arrays with quadrantal symmetry it has further utility. Suppose such an array has been designed to produce a specified sum pattern and that (61) has provided a gratifying check on the aperture distribution. If half of the mode voltages are reversed in sign, the *E*-plane or *H*-plane difference pattern can be computed by first using (61) to deduce the relevant aperture distribution. In a similar vein, if the array has been designed to produce a specified *E*-plane (*H*-plane) difference pattern, reversing half of the mode voltages permits calculation of the aperture distributions, and subsequently the patterns, for the sum mode and *H*-plane (*E*-plane) difference mode.

8. The Design of Arrays of Centered-Inclined Broad Wall Slots

All of the analysis presented in Sections 4 through 7 for longitudinal slots can be duplicated for arrays which use centered-inclined slots in the broad wall. This has been done by Orefice in an earlier version of the theory [11]. The use of centered-inclined slots is less common because, unless the array is large, the tilt angles are big enough to cause a sizable cross-polarized radiation pattern, a side effect which is often not desirable.

Cross polarization can be eliminated by leaving the slots on the center line but untilted. (Such slots can also be viewed as longitudinal but not offset.) In this case the slots are excited by irises [12]. The theory described in earlier sections is then applicable with Y/G_o represented in terms of slot length and iris dimensions.

9. The Design of Arrays of Inclined Narrow Wall Slots

Linear arrays of edge slots of the type shown in Fig. 2c find application as the line source feeds for cylindrical reflector antennas. *Planar* arrays of edge slots are attractive when beam scanning is desired in the *H*-plane. The reason for this is that adjacent waveguides can be placed close together ($\lambda_o/2$ spacing), permitting wide-angle scanning without the intrusion of extra main beams.

Unfortunately, unlike broad wall slots, inclined narrow wall slots cannot be assumed to lie in an infinite ground plane, because they actually "wrap around" and invade both broad walls in order to achieve resonant length. For this reason adjacent waveguides in planar arrays of edge slots need to be spaced apart, with metallic barriers often placed in the interstices, as suggested in Fig. 10. Although this results in a conducting surface which, viewed from the outer half-space, is closed except for the slots, that surface is far from a plane.* Thus the theoretical computation of external mutual coupling embodied in formulas such as (24) is not valid for planar arrays of narrow wall slots. Even in the case of a linear array, embedment in a ground plane is not feasible.

One is forced to seek an *experimental* estimation of mutual coupling in arrays of this type. Depending on the accuracy desired, the complexity of the measurements can vary. Consider first a standing-wave-fed linear array of inclined narrow wall slots. Because these slots are almost parallel, mutual coupling is significant. If the array is long, and the desired aperture distribution is equiphase and not extreme in its amplitude variation, one can argue that each slot has immediate neighbors whose excitations are essentially the same as its own. If one further assumes that most of the mutual coupling is due to nearest neighbors, then it follows that the active admittance of any edge slot in the linear array is practically the same as if it were part of an infinitely long array of identically excited slots.

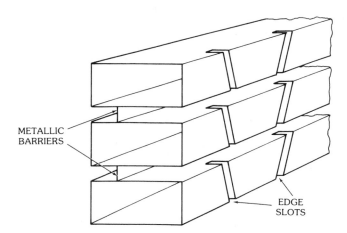

Fig. 10. Planar array of edge slots with baffles placed in the interstices.

---

*To complicate matters still further, in some applications baffles or wire grids are placed in front of the array to cut down further on cross polarization.

This argument paves the way for an experimental procedure to determine $Y^a(\theta, \ell)/G_o$ for an edge slot. One constructs a family of linear arrays of edge slots. Each member of the family is characterized by having N slots of common length and common but alternating tilt angle θ. The number N should be large ($N \geqq 10$). The slots are spaced $\lambda_g/2$ apart and the last one is followed by a short-circuit $\lambda_g/4$ beyond. With the slot closest to the generator covered over with conducting tape, the normalized input admittance is measured for each family member. Then with the first slot uncovered, the measurement is repeated. Arguably, the difference in the two readings is $Y^a(\theta, \ell)G_o$ for a central slot in the array. Alternatively, with N slots in the array, one can assume that $Y^a(\theta, \ell)/G_o = N^{-1} Y_{in}/G_o$.

After this, ℓ is lengthened and the measurements repeated on all members of the family. In this manner a mosaic of active admittance values can be assembled with θ and ℓ the independent variables.

Usually one is interested in extracting from this mosaic the resonant-length information,* that is, the function $G^a_{res}(\theta, \ell_{res})/G_o$. Since $VV^* G^a_{res}(\theta_n, \ell^{res}_n)/G_o$ measures the power radiated by the nth slot, with V the common mode voltage, and since this power is approximately proportional to $(V^s_n)^2$, it follows that

$$\frac{G^a_{res}(\theta_n, \ell^{res}_n)/G_o}{G^a_{res}(\theta_m, \ell^{res}_m)/G_o} = \left(\frac{V^s_n}{V^s_m}\right)^2 \tag{62}$$

With the desired aperture distribution known, (62) can be used to determine all the slot tilt angles and the corresponding slot lengths once these quantities are known for a reference slot. A unique solution results when the input admittance is specified. This is usually a match; in that case

$$\sum_{n=1}^{N} G^a_{res}(\theta_n, \ell^{res}_n)/G_o = 1 \tag{63}$$

and (62) and (63) combine to give a complete design.

The limitations of this procedure are obvious. The array must be long and the variations in the aperture distribution modest. The approximations inherent in this method become less acceptable as the demands on the desired pattern (such as very low side lobes) increase.

With only a minor change the experimental procedure just described can be extended to the determination of active admittance in a traveling-wave-fed linear array of edge slots. This time the family members contain N slots that are nonresonantly spaced and followed by a matched load. Under the argument that every slot sees the same environment, one can say that the fractional loss in power is the same at every slot, that is,

$$\frac{P_{1N} - P_N}{P_{1N}} = \frac{P_N - P_{N-1}}{P_N} = \cdots = \frac{P_n - P_{n-1}}{P_n} = \cdots = \frac{P_1 - P_{LOAD}}{P_1} \tag{64}$$

---

*For this reason the procedure just described is generally referred to as the incremental conductance technique.

The Design of Waveguide-Fed Slot Arrays

Solution of this train of equations gives

$$\frac{P_{1N}}{P_{\text{LOAD}}} = \left(\frac{P_1}{P_{\text{LOAD}}}\right)^{N+1} \tag{65}$$

But the power radiated by the slot closest to the load is $1/2\,\text{Re}\{V_1 V_1^* Y_1^a\} = P_1 - P_{\text{LOAD}}$, whereas the power absorbed in the load is $1/2 V_1 V_1^* G_o = P_{\text{LOAD}}$. Therefore

$$\text{Re}\left\{\frac{Y_1^a}{G_o}\right\} = \frac{P_1 - P_{\text{LOAD}}}{P_{\text{LOAD}}} \tag{66}$$

The combination of (65) and (66) yields

$$\text{Re}\left\{\frac{Y_1^a}{G_o}\right\} = \left(\frac{P_{1N}}{P_{\text{LOAD}}}\right)^{1/(N+1)} - 1 \tag{67}$$

If one measures the ratio P_{LOAD}/P_{1N}, (67) can be used to measure the normalized active conductance of a slot in the test array. This can be repeated as the lengths of all the slots in the test array are changed to a common longer value. Resonance can be defined as occurring for that length which results in minimum power entering the load (maximum power being radiated). This is the condition usually desired, for which

$$\frac{G_{\text{res}}^a(\theta, \ell^{\text{res}})}{G_o} = \left(\frac{P_{1N}}{P_{\text{LOAD}}^{\text{MIN}}}\right)^{1/(N+1)} - 1 \tag{68}$$

With data assembled on G_{res}^a/G_o versus tilt angle one can proceed to the design of a traveling-wave-fed linear array. Fig. 8 and (40) and (41) are applicable, but now (62) is replaced by

$$\frac{V_n V_n^* G_{\text{res}}^a(\theta_n, \ell_n^{\text{res}})/G_o}{V_m V_m^* G_{\text{res}}^a(\theta_m, \ell_m^{\text{res}})/G_o} = \left|\frac{V_n^s}{V_m^s}\right|^2 \tag{69}$$

Implicit in (69) is the assumption that a traveling-wave aperture distribution is desired (arbitrary amplitude variation, but uniform progressive phase).

One begins by assuming a tilt angle for the slot closest to the load. With $m = n - 1$, (41) is used to determine V_n/V_{n-1} and then (69) is used to determine the tilt angle of the second slot from the load. This process is repeated until the inclinations of all the slots are known. At this point the fraction of the input power going into the load can be calculated and the maximum tilt angle noted. A change in the guess for θ_1 might be indicated, in which case the process can be repeated.

For planar arrays of edge slots, if one assumes that external mutual coupling between waveguides can be ignored, all of the discussion just concluded about linear arrays can be carried over intact.* If it is felt that a more exact accounting for

---

*An additional ingredient is that the optimum set of coupling coefficients will also need to be determined. See Section 6.

mutual coupling is needed, small planar arrays (say 10 by 10) can be constructed, each containing slots with common but alternating tilt. The active resonant conductance of a central slot can be measured, with all waveguides equally excited, using the incremental conductance technique described earlier in this section. The resulting information can be used in (62) and/or (69) to design the array, given the desired aperture distribution. Here again, the applicability is limited to situations in which the array is large and variations in the aperture distribution are not extreme.

What does one do when these conditions are not met? One solution is offered in Section 12.

10. Difficulties in the Design of Large Arrays

A study of the second design equation for longitudinal slot arrays (26), plus an appreciation of the design procedure described in Sections 4 through 6, reveals that the calculation of mutual coupling involves a computation of the double integral g_{mn} for all values* of m and n. This can produce a sizable computer cost for large arrays.[†]

A possible way around this difficulty exists if the mutual coupling experienced by a slot is mostly due to its nearest neighbors, for then the calculation of g_{mn} could be truncated for some maximum value of m, given n. For a central element in a large planar array one could consider square rings of surrounding slots, as suggested in Fig. 11, and ask how many rings would need to be included in the calculation of MC_n before the value of MC_n stabilized. It can be anticipated that convergence would be slow because the magnitude of g_{mn} is roughly inversely proportional to R, whereas the number of slots in a ring is proportional to R.

If the special case of uniform excitation is considered, the question of convergence reduces to a study of the summation

$$S = \frac{MC_n}{j(\beta_{10}/k)(k_o b)(a/\lambda)^3} = \sum_m{}' g_{mn} \qquad (70)$$

The term S has been evaluated for the case of longitudinal slots in air-filled waveguide with a common length $\lambda_o/2$ and a common spacing $0.7\lambda_o$. (Offsets were ignored.) The results are shown in Fig. 12 for a central element, an end element, and a penultimate element in a linear array. Fig. 13 shows equivalent results for a central element, a corner element, and middle-of-a-side elements in a planar array.

One can observe from Figs. 12a and 12b that convergence for a linear array is fairly rapid in the H-plane but less so in the E-plane. A frequently posed question is: Does the second element in from the end of a linear array see a mutual coupling environment similar to that seen by a central element? A study of Figs. 12c and 12d indicates that the answer is clearly no for short arrays but is approximately yes for

---

*Symmetries can reduce the number of calculations, but not significantly.
[†]However, this cost is not prohibitive. As an example, for a slot array of one thousand elements the entire design procedure only entails a computer cost of about one thousand US dollars.

The Design of Waveguide-Fed Slot Arrays

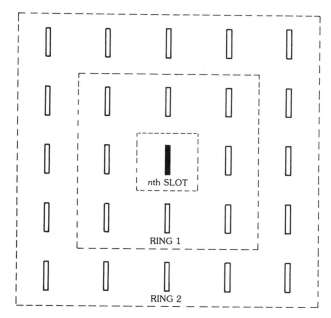

Fig. 11. A longitudinal slot surrounded by rings of similar slots.

longitudinal arrays of eight or more elements and for transverse arrays of fourteen or more elements.

Fig. 13a shows that the variability in mutual coupling for a central element in a planar array is still significant even after 24 rings are included. This is also true for a corner element, as Fig. 13b attests. Also it is interesting to note that a corner element has mutual coupling that is about half that of a central element, not one-quarter. Surprisingly, Fig. 13c reveals that a middle-of-a-side element, with the slots perpendicular to that side, has mutual coupling comparable to that of a central element. For a middle-of-a-side element in a side parallel to the slots, the mutual coupling is at about the level of a corner element, that is, half that of a central element, as indicated by Fig. 13d.

The results given in Figs. 12 and 13 are for a uniform distribution. When a tapered distribution is assumed, convergence is affected slightly [13]. It is clear from this study that truncation must be used with caution if one wishes to obtain accurate values for mutual coupling.

If the procedure outlined in Sections 4 through 6 is to be used for the design of large arrays, it is recommended that a routine be set up for the calculation of MC_n that involves the use of rings of neighboring elements, but that the rings be rectangular,* rather than square as in Fig. 11. The desired slot voltage distribution should be used as a weighting function on g_{mn}. More specifically, the full expression (38) should be used, with truncation occurring when the variation in MC_n has become less than a prescribed value.

---

*This will accommodate the fact that E-plane mutual coupling is stronger than H-plane mutual coupling, resulting in slower convergence in the E-plane direction.

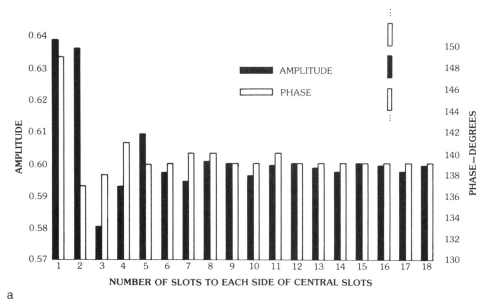

a

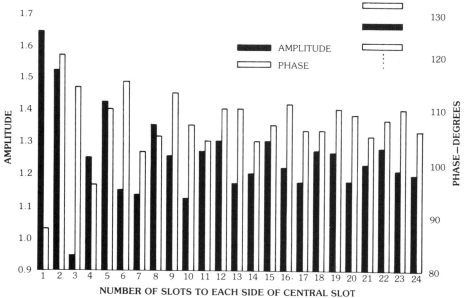

b

Fig. 12. Mutual coupling in a linear array, with $2\ell = \lambda_o/2$, $d = 0.7\lambda_o$. (*a*) Longitudinal slots, central element (divide by 2 for end element). (*b*) Transverse slots, central element (divide by 2 for end element). (*c*) Longitudinal slots, penultimate element. (*d*) Transverse slots, penultimate element.

The Design of Waveguide-Fed Slot Arrays

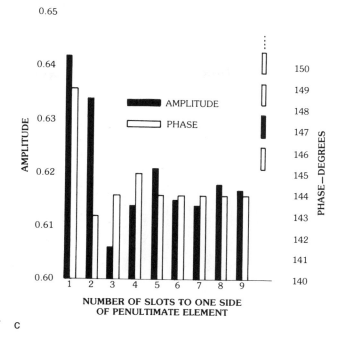

c

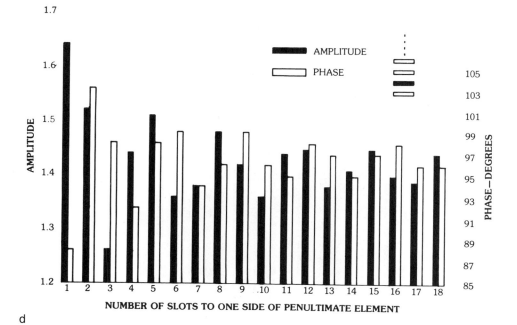

d

Fig. 12, *continued.*

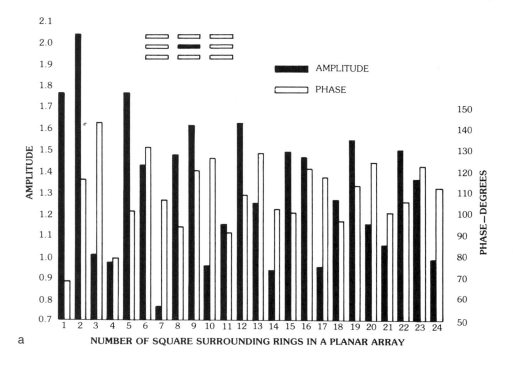

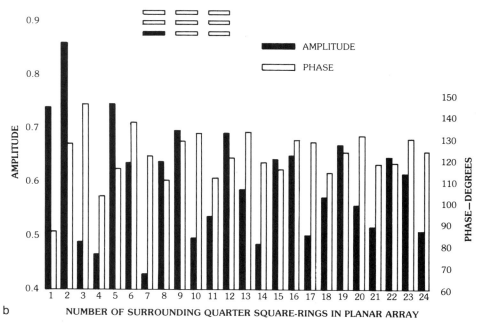

Fig. 13. Mutual coupling in a planar array, with $2\ell = \lambda_o/2$, $d = 0.7\lambda_o$. (*a*) Central element. (*b*) Corner element. (*c*) Element in middle of longitudinal side. (*d*) Element in middle of transverse side.

The Design of Waveguide-Fed Slot Arrays

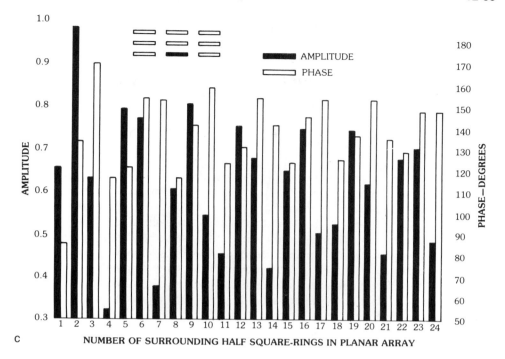

c

NUMBER OF SURROUNDING HALF SQUARE-RINGS IN PLANAR ARRAY

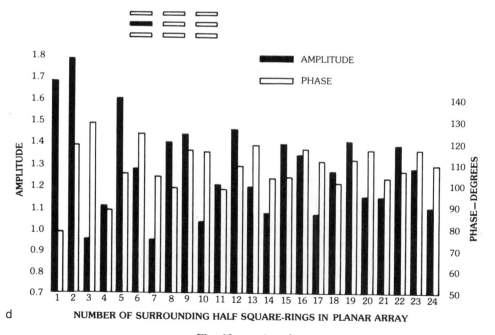

d

NUMBER OF SURROUNDING HALF SQUARE-RINGS IN PLANAR ARRAY

Fig. 13, *continued.*

A different type of truncation can be considered when using (61) to find the slot voltage distribution, given the offsets and lengths of the slots. The value g_{mnpq} diminishes as the distance between the mnth and pqth slots increases. This means that the matrix in (61) has terms which decrease with offset from the main diagonal. One can set a fraction of the far-out matrix elements equal to zero and note the effect on the solution $[V_{mn}^s]$ as the fraction is varied.

An alternative approach to the design of arrays of longitudinal slots that has been widely used is the incremental conductance technique, described for edge slots in Section 9. For a long linear array of longitudinal slots which is to have an aperture distribution with uniform progressive phase (perhaps equiphase) and modest variation in amplitude, $G_{\text{res}}^a(x, \ell_{\text{res}})/G_o$ can be found from families of linear arrays, in each of which all slots have a common offset. For a large planar array of longitudinal slots, with similar restrictions on the aperture distribution, $G_{\text{res}}^a(x, \ell_{\text{res}})/G_o$ can be found by putting baffles on each side of linear arrays comprising the aforementioned family; the baffles serve to create an equivalent infinite environment in the E-plane direction. This knowledge of resonant active conductance can be used to design the arrays in the same manner as described in Section 9. However, the information contained in Figs. 12 and 13 cautions that the incremental conductance technique should not be used unless the arrays are very large and even then is not accurate for slots near the end of a linear array nor near the periphery of a planar array, nor when the variability in amplitude and/or phase distribution is significant.

From this discussion it can be seen that errors in slot lengths and offsets should be expected if truncation is used in calculating MC_n or if the incremental conductance technique is assumed to apply for all slots in the array. A method which will reduce these errors substantially is described in Section 12.

This discussion applies equally well to centered, inclined slots in the broad wall, and limitations on the use of the incremental conductance technique apply also to edge slot arrays. In situations where a choice is possible, one should weigh the costs and errors of a truncated computer program against the costs and errors of an experimental program which gathers data on incremental active conductance. The choice will usually tip in favor of the computational approach.

11. Second-Order Effects

Some of the assumptions made in the design procedures described in earlier sections of this chapter are not strictly met in practice. The result of this is to introduce errors which can be identified but not easily overcome. Fortunately these errors are small and are thus called second-order effects. Even second-order effects, however, can become significant if very high performance (such as an ultralow side lobe level) is demanded of the array. A listing of the principal second-order effects follows.

Infinite Ground Plane

This assumption is made in deriving a formula for g_{mn} for broad wall slots of either the longitudinal or centered, inclined type. For linear arrays not embedded in ground planes it is most questionable in the transverse direction. For planar arrays not embedded in a ground plane (a frequently encountered situation) it has the least validity for peripheral slots.

In some applications, e.g., fuze antennas, a linear broad wall slot array is embedded in a conducting cylinder of small enough radius that the infinite ground plane assumption is inadequate. Lee and Safari-Naini [14] have derived an appropriate equivalent expression for g_{mn} for such cases.

Wall Thickness

The theory presented in Section 2 assumes that the upper broad wall, in which the slots are cut, is vanishingly thin. Finite wall thickness can be taken into account [15, 7] but the additional complexity adds to the computer cost of finding $Y(x, \ell)/G_o$ computationally for a single longitudinal slot, or $Z(\theta, \ell)/R_o$ computationally for a single centered, inclined slot, and the result is approximate. Of course, if these quantities are measured, the effect of wall thickness is automatically included. However, measurement of the input data is inherently less accurate and more costly than a theoretical determination. Either way a small error in $Y(x, \ell)/G_o$ or $Z(\theta, \ell)/R_o$ creeps into the design procedure. Since the modern trend is toward thinner and thinner walls, the computational approach should gain in favor.

E-Field Distribution in the Slot Aperture

Figs. 5 and 6 show the transverse **E** field along the center line of a longitudinal slot in amplitude and phase as a function of the longitudinal coordinate. This **E** field is caused by an incident TE_{10} mode and is seen to be almost equiphase and almost symmetrical in amplitude. The asymmetries are small and are ignored in deriving a design procedure based on (10), as is the symmetrical bulging of the amplitude distribution relative to a half-cosinusoid, as is the transverse variation of transverse **E**, as is the small longitudinal component of the **E** field in the slot aperture.

These small ignored quantities, if included, would combine to cause $B_{10} \neq C_{10}$ for a moderately longitudinal broad wall slot. Similarly, they have the effect of causing $B_{10} \neq -C_{10}$ for a centered, inclined broad wall slot. This means that the simple equivalent circuits which represent the longitudinal slot as a shunt element and the centered-inclined slot as a series element are not strictly correct.

For small-offset longitudinal slots, the longitudinal **E** field component in the slot aperture cannot be ignored, further weakening the shunt element assumption.

A more rigorous formulation requires T or π networks. That representation is so cumbersome that it is then preferable to design the arrays in terms of scattered waves rather than equivalent admittances and impedances. As pattern and input match requirements become more demanding, it will become increasingly important to take these scattering asymmetries into account.

Internal Higher-Order-Mode Mutual Coupling between Radiating Slots

All of the design procedures presented in this chapter contain the underlying premise that only TE_{10} modes matter—that higher-order-mode scattering off any radiating slot has died out to a negligible level before reaching a neighboring slot.*

---

*This premise implies that the *a* dimension is chosen so that all modes but TE_{10} are cut off.

This is not strictly true; particularly, neglect of TE_{20} scattering introduces an error which is more serious as the *b* dimension lessens. This is because the resonant length of a longitudinal slot increases as *b* decreases, thus putting the tips of adjacent slots closer together. The TE_{20} mode effects can be included at negligible cost in complexity of the design procedures [16]. They affect array performance in two ways: (1) TE_{20} mode scattering off one slot induces an additional electric field in its two neighboring slots in the same guide. That affects the pattern. (2) This additional electric field in turn causes a change in the TE_{10} scattering off the neighboring slots. That affects the overall slot voltage and the input impedance. When high performance is required, these effects cannot be ignored.

Higher-Order-Mode Coupling in Junctions

This only occurs in planar arrays, but it can be the most serious of all the second-order effects. A classic example occurs in slot arrays of the type shown in Fig. 9. The tilted coupling slot is seen to be only $\lambda_g/4$ from the two radiating slots which straddle it. Compared to the problem of higher-order-mode internal coupling between neighboring radiating slots, this situation is more severe because the cutoff modes have only half the distance in which to die out. Further complicating the situation, the coupling slot and the two radiating slots are essentially transverse, so that more than the TE_{20} mode is involved in the supplemental coupling mechanism. If one ignores this effect, the results are that the straddling slots are improperly excited, thereby degrading the pattern, and the main line match is not achieved.

12. Far-Field and Near-Field Diagnostics as Design Tools

Sections 10 and 11 chronicled some of the difficulties in the design of linear and planar slot arrays—difficulties due to imperfect assumptions in theory and experiment, and difficulties due to prohibitive costs in computation or in the amassing of extensive experimental input data. The result, particularly for high-performance arrays, is that actual pattern and input impedance fall somewhat short of the design goals. When this is the case, far- and near-field measurements can be used to improve on the design procedures.

As an example, a seeker antenna is often fed by four main lines, one per quadrant, so that sum and difference patterns can be produced. Each main line contains a sequence of coupling junctions, one per branch line. The radiating slots which straddle a junction have their slot voltages perturbed by higher-order-mode scattering (see also Section 11). These perturbations occur at regularly spaced positions in the aperture and are the cause of grating lobes. In some applications they can be seen unambiguously in the far-field pattern. Their levels and positions can be used to deduce the errors in slot voltages of the straddling slots and these errors can be corrected by small changes in the offsets and lengths of the affected slots.

More potent is the use of a near-field probe facility to determine the amplitude and phase of each slot in the array. An illustration of the way in which near-field data can improve on the design is the case of a traveling-wave linear array of edge slots. Section 9 contains a description of how the real part of the normalized active admittance, $\text{Re}\{Y^a(\theta,\ell)/G_o\}$, is obtained using a family of test arrays, each with N slots of alternating tilt, but with a different θ value for each family member. If one takes near-field data on each test array as ℓ is altered, one finds that the uniform progressive phase changes with ℓ. If the uniform progressive phase corresponding to resonant length (minimum power into the load) is taken as reference, the differential phase can be related to the phase of $Y^a(\theta,\ell)/G_o$. Thus one possesses more information than just $\text{Re}\{Y^a(\theta,\ell)/G_o\}$. If the actual array is now designed according to the procedure described in Section 9, and if its performance falls short of expectation, one can then probe the near field of the actual array and determine the errors in the various slot voltages. With the foreknowledge of $Y^a(\theta,\ell)/G_o$, the error in slot voltage for the slot closest to the load can be compensated. One then moves back to the next slot and institutes another compensation, proceeding in this way to the slot farthest from the load. Experience has shown that a second array, built to incorporate these compensations, has an improved performance over the first array.

Acknowledgments

The author wishes to thank M. Armstrong, L. Josefsson, L. Kurtz, R. Mailloux, and R. Shavit for criticizing all or portions of the manuscript. Discussions with W. Lange and J. Thomas were helpful and appreciated. Phyllis Parris typed the manuscript with her customary skill and her cheerfulness is warmly acknowledged.

13. References

[1] A. F. Stevenson, "Theory of slots in rectangular waveguides," *J. Appl. Phys.*, vol. 19, pp. 24–38, 1948.

[2] T. V. Khac, "A study of some slot discontinuities in rectangular waveguides," PhD dissertation, Monash University, Australia, November 1974.

[3] R. F. Harrington, *Field Computation by Moment Methods*, New York: Macmillan Co., 1968.

[4] Y. U. Kim, "Electric field distribution in a longitudinal slot and its effect on the design of slot array antennas," MS thesis, University of California, Los Angeles, June 1983.

[5] R. S. Elliott, "An improved design procedure for small arrays of shunt slots," *IEEE Trans. Antennas Propag.*, vol. AP-31, pp. 48–54, January 1983.

[6] R. S. Elliott, *Antenna Theory and Design*, Englewood Cliffs: Prentice-Hall, 1981, pp. 95–96.

[7] H. Y. Yee, "Impedance of a narrow longitudinal shunt slot in a slotted waveguide array," *IEEE Trans. Antennas Propag.*, vol. AP-22, pp. 589–592, July 1974. See also G. J. Stern and R. S. Elliott, "Resonant length of longitudinal slots and validity of circuit representation: theory and experiment," *IEEE Trans. Antennas Propag.*, vol. AP-33, pp. 1264–1271, November 1985.

[8] See, for example, Reference 6, pp. 134–136.

[9] See Reference 6, p. 471.

[10] R. S. Elliott, "Dominant mode analysis of coupling junctions for flat-plate slot array antennas," *Report No. IDC 56 B1. 30/68*, Hughes Missile Systems Group, Canoga Park, California, July 1983.

[11] M. Orefice and R. S. Elliott, "Design of waveguide-fed series slot arrays," *IEE Proc.*, pt. H, vol. 129, pp. 165–169, August 1982.

[12] R. Tang, "A slot with variable coupling and its application to a linear array," *IRE Trans. Antennas Propag.*, vol. AP-8, pp. 97–101, January 1960.

[13] Y. U. Kim and R. S. Elliott, "External mutual coupling in large arrays of longitudinal slots," *Report No. AP-201*, Dept. of Electr. Eng., University of California, Los Angeles, July 15, 1983.

[14] S. W. Lee and S. Safari-Naini, "Approximate asymptotic solution of surface field due to magnetic dipole on a cylinder," *IEEE Trans. Antennas Propag.*, vol. AP-26, pp. 593–598, July 1978.

[15] A. A. Oliner, "The impedance properties of narrow radiating slots in the broad face of rectangular waveguide," *IRE Trans. Antennas Propag.*, vol. AP-5, pp. 1–20, January 1957.

[16] R. S. Elliott and W. R. O'Loughlin, "The design of slot arrays including internal mutual coupling," *IEEE Trans. Antennas Propagat.*, vol. AP-34, pp. 1149–1154, September 1986.

Chapter 13

Periodic Arrays

R. J. Mailloux
Rome Air Development Center, Electromagnetic Sciences Division

CONTENTS

1. Introduction   13-5
   - *Pattern and Excitation*   13-5
   - *Time Delay and Phase Steering*   13-7
   - *Examples of Array Collimation*   13-8
   - *Quality of the Array Beam*   13-10
2. Patterns of Periodic Arrays   13-12
   - *Characteristics of an Array Scanned in One Plane*   13-12
   - *Scanning in Two Planes*   13-15
   - *Pattern Shape and Beam Broadening*   13-17
   - *Phased Array Bandwidth*   13-19
   - *Antenna Pattern Synthesis*   13-20
   - *General Procedures*   13-21
   - *Synthesis With Orthogonal Beams*   13-22
   - *Low Side Lobe Solutions: Basic Formulas and Engineering Data*   13-23
   - *Dolph-Chebyshev Synthesis*   13-24
   - *Taylor Line Source Synthesis*   13-25
   - *Bayliss Line Source Synthesis*   13-27
3. Array Organization: Subarrays and Broadband Feeds   13-30
   - *Aperture Illumination Control at Subarray Input Ports*   13-30
   - *Wideband Characteristics of Time-Delayed Subarrays*   13-32
   - *Contiguous Subarrays of Discrete Time-Delay Devices*   13-32
   - *Overlapped Subarrays*   13-35
   - *Broadband Array Feeds with Time-Delayed Offset Beams*   13-39
4. Practical Arrays   13-41
   - *Mutual Coupling and Element Patterns*   13-41
   - *Array Blindness*   13-45
   - *Conformal Arrays*   13-49
   - *Array Errors and Phase Quantization*   13-52

 **Robert J. Mailloux** was born in Lynn, Massachusetts. He received the BS degree in electrical engineering from Northeastern University, Boston, Massachusetts, in 1961, and the SM and PhD degrees from Harvard University, Cambridge, Massachusetts, in 1962 and 1965, respectively.

He was with the NASA Electronics Research Center, in Cambridge, from 1965 to 1970, and with the Air Force Cambridge Research Laboratories from 1970 to 1976. He is presently acting chief of the Antennas and Components Division, Rome Air Development Center, Electromagnetics Directorate. His research interests are in the area of periodic structures and antenna arrays. He has published numerous papers on antennas and arrays, and book chapters on antenna research topics, on hybrid systems of arrays and reflectors or lenses, and on conformal arrays. He was elected to the grade of Fellow of the IEEE in 1978.

Dr. Mailloux is a member of Tau Beta Pi, Eta Kappa Nu, Sigma Xi, and Commission B of URSI. He was Technical Activities Chairman for Commission B of URSI from 1979 to 1982, and President of the Hanscom Chapter, Sigma Xi, in 1980. He was President of the Antenna and Propagation Society in 1983 and has previously been AP-S Distinguished Lecturer, AdCom Member, and Meetings Chairman.

    Array Elements   13-57
    Passive Components for Arrays: Polarizers and Power Dividers   13-60
    Array Phase Control   13-62
5. References   13-64

1. Introduction

Most scanning array antennas are composed of rows and columns of periodically spaced antenna elements. Periodic arrays can be designed to provide extremely low side lobes and high gain when element spacing is kept relatively small, and they are chosen in preference to arrays with aperiodic lattices for many radar and communications applications. In addition, when hundreds or thousands of closely spaced elements are required, it is also simpler and cheaper to construct periodic than aperiodic arrays. Aperiodic arrays, treated in the next chapter, have advantages in high-resolution thinned configurations, and when used to achieve equivalent amplitude taper without the use of a complex feed network.

Chapter 11 deals with a number of theoretical aspects of phased arrays and establishes a basis in mathematics by which one can compute or synthesize detailed radiation patterns. This chapter will provide some background in theory to increase comprehension, but is intended mainly to present fundamental engineering data to aid in antenna design and evaluation.

Although phase-scanned arrays are the main topic of the chapter, some consideration is given to time-delay scanned arrays. There are discussions of pattern distortions resulting from grating lobes and array errors and a brief treatment of mutual coupling effects. Pattern synthesis is covered in more detail in Chapter 11, so the treatment in this chapter is restricted to giving engineering data and some examples of the effects of discreting line source data. The results of organizing an array into subarrays are presented and several approaches described that achieve wide instantaneous bandwidth.

The discussion of specific devices for arrays is restricted to passive components (phase shifters, power dividers, array elements, etc.) and does not include amplifiers or any of a wide variety of array control systems, mixer steering, or signal processing means of pattern control.

Pattern and Excitation

The most general form of radiation pattern used in this section is given below. Each element at position (x_i, y_i, z_i) of Fig. 1 is excited by a complex weighting a_i and radiates with a vector element pattern $\mathbf{f}_i(\theta, \phi)$ so that the total far-field radiated field is given by

$$\mathbf{E}(\mathbf{r}) = \frac{e^{-jkR_0}}{R_0} \sum_i a_i \mathbf{f}_i(\theta, \phi) e^{+jk\mathbf{r}_i \cdot \hat{\boldsymbol{\varrho}}} \tag{1}$$

where $k = 2\pi/\lambda$, the vectors $\mathbf{r}_i$ define the locations of array elements relative to the element with index zero, and the unit vector $\hat{\boldsymbol{\varrho}}$ is the position vector locating the observation point P a distance R_0 from the origin at the zeroth element:

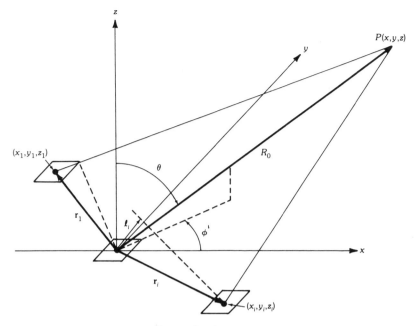

Fig. 1. Generalized array geometry.

$$\hat{\varrho} = \hat{x}u + \hat{y}v + \hat{z}\cos\theta$$
$$\mathbf{r}_i = \hat{x}x_i + \hat{y}y_i + \hat{z}z_i \qquad (2)$$
$$\mathbf{R}_0 = |r - r_0| = \sqrt{x^2 + y^2 + z^2}$$

The direction cosines u and v are

$$u = \sin\theta\cos\phi$$
$$v = \sin\theta\sin\phi$$

with $u^2 + v^2 \leq 1$.

In this generalized form the weighting a_i is the *applied* excitation, an incident mode amplitude for a transmission-line or waveguide-fed array, or an *applied* voltage for an array of wire elements. The feeding sources are assumed matched to their characteristic impedance. The variables $\mathbf{f}_i(\theta,\phi)$ are the radiation patterns of each ith element, in the presence of all other elements, and so are called *element patterns* (sometimes termed "active element patterns"). The vector element patterns $\mathbf{f}_i$ are in general different, even in an array of like elements. The difference results from the interaction between individual elements and proximity to the array edge. In the formal treatment of antenna radiation the element patterns are unknowns, obtained by solving an electromagnetic boundary value problem.

Before leaving the topic of interpreting (1), it should be noted that an alternative but fully as general formalism considers the a_i to be the actual (unknown) currents for wire elements, or electric fields for slots or apertures. From this perspective, when mutual coupling is considered, the $a_i f_i(\theta, \phi)$ are replaced by the radiation pattern of the unknown distribution of current or field on the ith element: a distribution that is found from the solution of a boundary value problem, and will not, in general, be directly proportional to the applied excitations. In the absence of mutual coupling, however, the two formalisms are identical.

Throughout this section the a_i will be the applied excitation, as in the element pattern formalism described earlier. Furthermore, throughout Sections 2, 3, and 5, the element patterns will be assumed equal and isotropic over the hemisphere $z \geq 0$, but element pattern distortion will be described in Section 4, in connection with array mutual coupling. In addition, it is assumed that all expressions relate to the axis of principal polarization, and so the patterns are written as scalar functions.

The usual purpose of an array is to form a beam at some specific angle in space (θ_0, ϕ_0). In the absence of mutual coupling this can be done at all frequencies by choosing the excitation

$$a_i = |a_i| e^{-jk\mathbf{r}_i \cdot \hat{\varrho}_0} = |a_i| e^{-jk\ell_i} \tag{3}$$

with ϱ_0 given by (2) using θ_0 and ϕ_0 in the direction cosine expressions. At such a point the fully collimated beam field strength is a simple vector summation of the element patterns weighted by the amplitudes $|a_i|$

$$\mathbf{E}_0(\mathbf{R}_0) = \frac{e^{-jkR_0}}{R_0} \sum_i \mathbf{f}_i(\theta_0, \phi_0)|a_i| \tag{4}$$

and if the element patterns $\mathbf{f}_i$ are equal and isotropic, ($|\mathbf{f}_i| = 1$); this is the largest possible value of the field $\mathbf{E}(\mathbf{r})$ for any given R_0 in the far field. Selection of the excitations of (3) is understood intuitively by considering that the projected distance to the observer at (R_0, θ_0, ϕ_0) is different for each array element by the length ℓ_i in Fig. 1. Removal of this path length difference will cause the contributions from each element to add in phase in the far field. The envelope of coefficients $|a_i|$ is the array illumination and is the primary determinant of the radiation side lobe levels, just as it is for aperture antennas.

Time Delay and Phase Steering

Applying signals of the form of (3) is called *time-delay steering* because the phase of the excitation signals exactly compensates for the time delay of a signal traveling in the projected distances ℓ_i. Time-delay steering results in a fully collimated beam at all frequencies, but is extremely expensive, bulky, and lossy since it depends on switching relatively long delay lines. For this reason true time delay is not often used at the array element level but more commonly incorporated into the feed circuits of arrays divided into subarrays. Examples of subarray excitation are described in Section 3.

Alternatively, at some fixed frequency f_0, with wavelength λ_0 and wave number $k_0 = 2\pi/\lambda_0$, phase weighting can be substituted for the time-delay steering. In such a case the weighting factors a_i are

$$a_i = |a_i| e^{-jk_0 \mathbf{r}_i \cdot \hat{\varrho}_0} \tag{5}$$

and

$$\mathbf{E}_0(\mathbf{r}) = \frac{e^{-jkR_0}}{R_0} \sum_i \mathbf{f}_i(\theta,\phi)|a_i| e^{+j\mathbf{r}_i \cdot (k\hat{\varrho} - k_0\hat{\varrho}_0)}$$

which represents exact collimation only at fixed frequencies $\lambda = \lambda_0$ and is called *phase steering*. Most arrays are phase steered, but when wide operating bandwidths are required it may be necessary to investigate options for time delay. Section 3 describes several broadbanding approaches.

Examples of Array Collimation

Several examples of array collimation are given below for the arrays of Fig. 2. Note m and n are half-integers, $\pm\frac{1}{2}, \ldots,$ to $\pm(N_y - 1)/2$ or $\pm(N_x - 1)/2$ for arrays with even numbers of elements, or integers, $0, \pm 1, \ldots, (N_y - 1)/2$, etc.

Periodic Column Array in One Plane

$$\mathbf{r}_m = \hat{\mathbf{x}} x_m = \hat{\mathbf{x}} m d_x, \qquad u_0 = \cos\theta_0 \qquad (6)$$

Steering excitation:

$$a_m = |a_m| e^{-jk_0 m d_x u_0}$$

Radiation pattern:

$$\mathbf{E}(\mathbf{r}) = \frac{e^{-jkR_0}}{R_0} \sum_m \mathbf{f}_m(\theta,\phi)|a_m| e^{jmd_x(ku - k_0 u_0)}$$

Periodic Two-Dimensional Array (Rectangular Lattice)

$$\mathbf{r}_{mn} = \hat{\mathbf{x}} m d_x + \hat{\mathbf{y}} n d_y, \qquad u_0 = \sin\theta_0 \cos\phi_0, \qquad v_0 = \sin\theta_0 \sin\phi_0$$

Steering excitation:

$$a_{mn} = |a_{mn}| e^{-jk_0(m d_x u_0 + n d_y v_0)} \qquad (7)$$

Radiation pattern:

$$\mathbf{E}(\mathbf{r}) = \frac{e^{-jkR_0}}{R_0} \sum_{m,n} \mathbf{f}_{mn}(\theta,\phi)|a_{mn}| e^{j[md_x(ku - k_0 u_0) + nd_y(kv - k_0 v_0)]}$$

Circular Array Section

The circular array section of Fig. 2c is another characteristic array shape that requires a simple regular excitation vector to form a beam in the principal plane $(\theta, 0)$:

Periodic Arrays

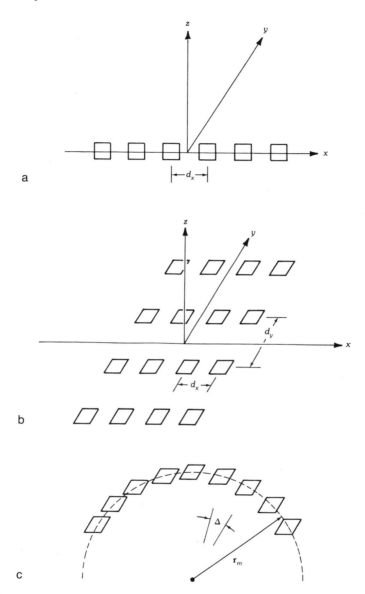

Fig. 2. Common array orientations. (*a*) One-dimensional array. (*b*) Two-dimensional rectangular grid array. (*c*) Sector of circular array.

$$\mathbf{r}_m = \hat{\mathbf{x}} a \cos \theta_m + \hat{\mathbf{y}} a \sin \theta_m, \qquad \hat{\boldsymbol{\varrho}} = \hat{\mathbf{x}} \cos \theta + \hat{\mathbf{y}} \sin \theta \qquad (8)$$

Steering excitation:

$$a_m = |a_m| e^{-jk_0 a \cos(\theta_0 - \theta_m)}, \qquad \theta_m = m\Delta$$

Radiation pattern:

$$\mathbf{E}(\mathbf{r}) = \frac{e^{-jkR_0}}{R_0} \sum_m \mathbf{f}_m(\theta) |a_m| e^{ja[k\cos(\theta-\theta_m) - k_0\cos(\theta_0-\theta_m)]}$$

In each of the above cases the phase steering beam collimation is changed to time-delayed collimation by substituting $k = 2\pi/\lambda$ for the phase steered term $k_0 = 2\pi/\lambda_0$.

Quality of the Array Beam

The quality of the beam formed by an array is measured by a number of factors. Chief among these are the directivity, beamwidth, and side lobe level of the array pattern, and the bandwidth over which satisfactory radiation characteristics can be obtained.

The *directivity* is the ratio of power density at the peak of the main beam ($\mathbf{r} = \mathbf{r}_0$) to the average power density, or in terms of (1):

$$D = E(\mathbf{r}_0) E^*(\mathbf{r}_0) \bigg/ (4\pi)^{-2} \int_0^{2\pi} \int_0^{\pi} E(\mathbf{r}) E^*(\mathbf{r}) \sin\theta \, d\theta \, d\phi \qquad (9)$$

The integral over θ is often carried only to $\pi/2$ for most planar arrays with a ground screen, as it is assumed that radiation is negligible for $\theta > \pi/2$, which is consistent with the assumption of hemispherical element patterns.

Equation 9 can be reduced to much simpler forms for linear and planar arrays.

Elliott [1] gives convenient formulas for the directivity of linear dipole arrays and derives an especially simple form for arrays of isotropic elements with half-wave spacing and currents a_m as in (6).

$$D = \left(\sum a_m\right)^2 \bigg/ \sum a_m^2 \qquad (10)$$

The directivity is increased by the factor 2 for hemispherical element patterns. This expression shows the directivity of a linear array to be independent of scan angle. As pointed out by Elliott, this behavior is peculiar to the linear array and results from the broad pattern perpendicular to the array axis. As the array is scanned toward end-fire the area of this conical shape is reduced and the effect offsets the beam broadening in the plane of scan that tends to reduce gain.

Fig. 3 shows the broadside directivity (dashed curve) of a linear array of eight isotropic elements with uniform excitation for various array element spacings [2]. It indicates that directivity is a severe function of element spacing. The solid curve shown in the figure is the maximum directivity for the array, as derived by Tai, and serves to indicate that the directivity of a broadside array is very nearly optimum except in the supergain region ($d/\lambda < 0.5$). The reduced directivity occurring near $d/\lambda = 1$ results directly from the radiation of an additional set of primary lobes, called *grating lobes*, which will be discussed in detail later. Since the presence of grating lobes is unacceptable for most applications, the directivity of a uniformly

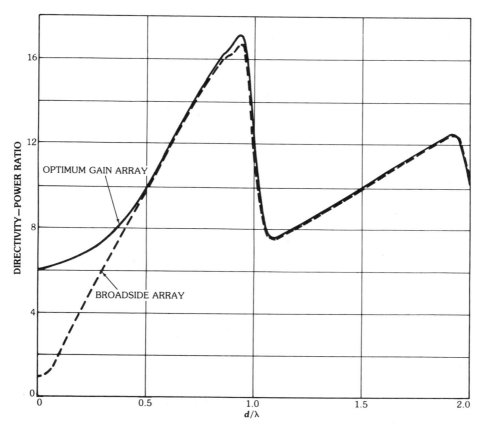

Fig. 3. Array gain for eight isotropic elements versus element spacing. (*After Tai [2], © 1964 IEEE*)

illuminated roadside array is well approximated by the following expression throughout the linear (useful) portion of the curve shown in Fig. 3:

$$D \cong 2Nd/\lambda \qquad (11)$$

The directivity of an ideal planar array (one that is perfectly matched and has $\cos\theta$ hemispherical element patterns) is given approximately by

$$D = 32\,400/B \qquad (12)$$

where B is the 3-dB beamwidth product for scan angle θ

$$B = \theta_{x_3}\theta_{y_3}\sec\theta \qquad (13)$$

with beamwidths given in degrees.

2. Patterns of Periodic Arrays

Array antennas, with their numerous closely spaced elements, provide a degree of pattern control not achievable with reflector or lens apertures. Most early developments in this area emphasized electronic scanning, but more recently there has been substantial effort devoted to producing low side lobe and adaptively controlled patterns.

Characteristics of an Array Scanned in One Plane

The array illumination determines the pattern beamwidth, side lobes, and directivity. In general, the beamwidth is proportional to the inverse of the normalized array length, or

$$\Delta\theta = K\lambda/L \tag{14}$$

with $\Delta\theta$ in radians, K a constant, and L defined to be Nd, for N elements in the θ plane. This equation gives the half-power beamwidth of this radiation pattern for a linear array or in the principal planes of a rectangular array at broadside. Uniform illumination ($|a_i| = 1$), with $K = 0.886$, produces the highest directivity and narrowest beam of any illumination (except for certain special "superdirective" illuminations associated with rapid phase fluctuations and closely spaced elements), but this illumination produces relatively high side lobe patterns (about -13 dB). Selection of various tapered illuminations can result in much lower side lobes, accompanied by wider beamwidths and lower directivity. As the array is scanned from broadside the beamwidth widens, again like $\sec\theta_0$ except near end-fire. A more general formula and a plot of beamwidth versus scan angle are given later, but it is important to note that the array beamwidth does increase with scan when observed in the (θ,ϕ) space. The solid-line curves of Fig. 4 show the radiation pattern of an eight-element, one-dimensional array at broadside and scanned to 45° ($u = 0.7071$). The element spacing is 0.5λ. The comparison shows that the array pattern is invariant in the parameter $(d/\lambda)(u - u_0)$, so that no beam broadening or other pattern change is evident when plotted in direction cosine space. The observed $\sec\theta$ beam broadening factor for large arrays is thus the result of mapping the uniform beamwidth in u space onto the θ plane. The advantage of plotting antenna patterns in u ($\sin\theta$) space is that as far as the pattern function is concerned, there is no need to recompute patterns for any other scan angle. Later we shall see that mutual coupling introduces angle-dependent effects that can drastically alter the radiation characteristics.

Similarly, the form of the pattern is not dependent on the spacing d/λ, except that its scale in $u - u_0$ space expands or contracts with the choice of spacing d/λ. This subject is addressed in the next sections.

Array Lattice Spacings and Comparison With Continuous Line Source—The 0.5λ spacing used in the array of Fig. 4 was chosen to minimize the effects of periodicity and makes that array pattern very little different from that of a continuous aperture. Normalized broadside radiation patterns for an N-element array [of actual length $(N - 1)d_x$] and a line source of length L are given as follows:

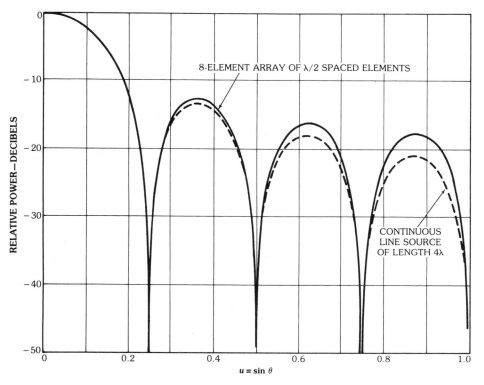

Fig. 4. Patterns of uniformly illuminated array and line source.

Linear array:

$$f(\theta) = \frac{\sin(N\pi d_x u/\lambda)}{N \sin \pi d_x u/\lambda} \tag{15}$$

Line source:

$$f(\theta) = \frac{\sin(L\pi u/\lambda)}{L\pi u/\lambda} \tag{16}$$

The radiation patterns for a continuous line source of length 4λ and an eight-element array of $\lambda/2$ spaced elements with uniform excitation are shown in Fig. 4. The line source pattern differs very little from the array pattern up to the second side lobe, and the null positions are unchanged, since these are determined from the numerators of the two expressions in (15) and (16). The similarity of these expressions makes it convenient to define the length parameter $L = Nd$ for arrays.

Fig. 5 shows the pattern (solid) of an eight-element array with 4λ spacings, as compared with the (dashed) pattern of a continuous line source with $L = 32\lambda$. Here the dramatic difference is the occurrence of lobes in the periodic array pattern at the zeroes of the denominator of (15). These lobes, which are called *grating*

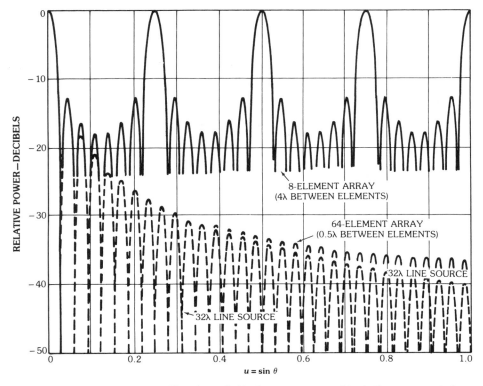

Fig. 5. Patterns of uniformly illuminated 64-element array with 0.5λ spacing, 32λ line source, and 8-element array with 0.5λ spacing.

lobes, have the same peak value as the main beam and are located at distances $p\lambda/d_x$ from the main beam in u space, for all integers p that define angles in real space ($|u| \leq 1$). Grating lobes are a direct consequence of the periodicity and occur independently of any chosen amplitude distribution across the array. When the array is scanned to u_0 they occur at angles

$$u_p = u_0 + p\frac{\lambda}{d_x}, \quad \text{for } p = \pm 1, \pm 2, \cdots \qquad (17)$$

subject to $-1 \leq u_p \leq +1$.

The array factor is thus completely periodic in u space (for $|u| \leq 1$), with period equal to grating lobe separation (λ/d_x). The region $|u| \leq 1$ is called *real space* because $\sin\theta$ has a geometric interpretation in this regime. Before leaving the comparison of continuous apertures and discrete arrays it should be noted that the similarity is maintained almost halfway between the main beam and the nearest grating lobe, because it is on this scale that the pattern of the periodic array is repeated. Fig. 5 also shows (solid) the side lobe peaks of the 64-element array with 0.5λ spacing.

Fig. 6 further emphasizes the pattern invariance in the scale parameters

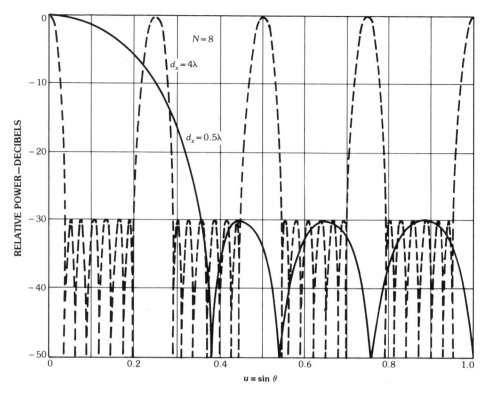

Fig. 6. Patterns of eight-element arrays with −30-dB Chebyshev illuminations: 0.5λ spacing, and 4λ spacing.

$d_x(u - u_0)/\lambda$ by showing the patterns of two eight-element arrays with one of the low side lobe excitations that will be described in a later section, a −30-dB Chebyshev taper. The two patterns are for arrays with $d_x = 0.5\lambda$ and 4λ. Increasing d_x shrinks the scale of the pattern and so brings additional grating lobes into real space but produces no other change in the pattern.

Scanning in Two Planes

Equation 7 gives the pattern of a rectangular planar array like that of Fig. 7a scanned in two planes. The occurrence of grating lobes in rectangular-grid two-dimensional arrays is directly apparent from (7), where the substitutions

$$u_p = u_0 + p\frac{\lambda}{d_x}, \qquad v_q = v_0 + q\frac{\lambda}{d_y} \tag{18}$$

leave these expressions unchanged. Not all values of q and p correspond to allowed angles of radiation, however, for the direction θ of radiation measured from the array normal is given by

$$\cos\theta_{pq} = \sqrt{1 - u_p^2 - v_q^2} \tag{19}$$

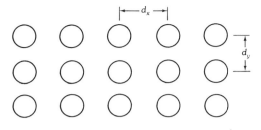

Rectangular Grid Array

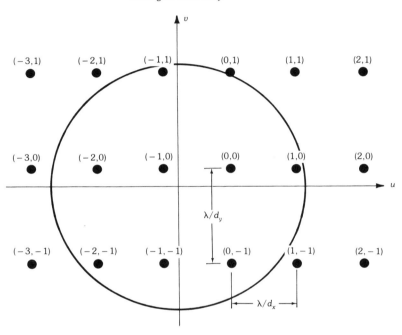

a

Fig. 7. Rectangular and triangular grid planar arrays and their grating lobe lattices. (*a*) Grating lobe lattice for rectangular grid. (*b*) Grating lobe lattice for triangular grid.

and so real angles of radiation θ_{pq} require that the allowed values of u_p and v_q are constrained by the condition

$$u_p^2 + v_q^2 < 1 \qquad (20)$$

These points are shown in (u, v) space as a regularly spaced grating lobe lattice about the main beam location (u_0, v_0) in Fig. 7a. The circle with unity radius represents the bounds of the above inequality; all grating lobes within the circle represent those radiating into real space, and those outside do not radiate.

Fig. 7b shows a triangular array lattice and pertinent grating lobe locations for that lattice. In this case, (5) is still valid and there are still grating lobes, but the nearest lobes in the azimuth scan plane are removed. The grating lobe locations shown in Fig. 7b are given by

Periodic Arrays

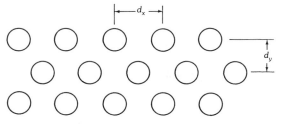

Triangular Grid Array

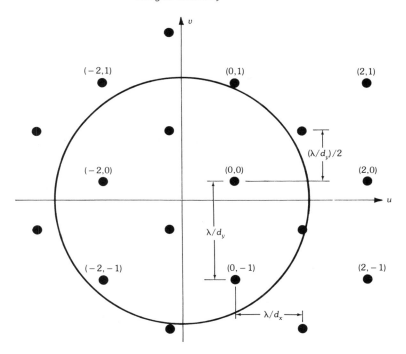

b

Fig. 7, *continued.*

$$u_p = u_0 + p\frac{\lambda}{d_x}$$

$$v_q = \begin{cases} v_0 + q\dfrac{\lambda}{d_y} & \text{for } p = 0, \pm 2, \cdots \\ v_0 + \left(q - \dfrac{1}{2}\right)\dfrac{\lambda}{d_y} & \text{for } p = \pm 1, \pm 3, \cdots \end{cases} \quad (21)$$

Pattern Shape and Beam Broadening

In the previous section it was pointed out that for a periodic array the pattern shape does not change with scan angle if it is plotted in u space, and changes in array spacing alter only the scale (width) of the pattern. The actual shape of the pattern, its relative width, side lobes, and slope values depend only on the applied amplitude and phase illumination. Fig. 8 indicates that one of the major results of reducing the side lobes for a given size array is to broaden the beamwidth. The figure shows three different Chebyshev patterns, with -20-, -30-, and -40-dB side lobe levels, and indicates beam broadening factors $B = K/0.886$ of 1.12, 1.29, and 1.43 relative to the pattern of the array with uniform illumination. In general, then, side lobes can be lowered by employing tapered array illuminations, but this is achieved at the expense of broadening the beamwidth and reducing the array gain. Fig. 9 gives the beamwidth in degrees for a scanned array with arbitrary taper as a function of the array length and beam broadening factor. By using the relationship

$$\Delta\theta = \sin^{-1}(u_0 + 0.443B\lambda_0/L) - \sin^{-1}(u_0 - 0.443B\lambda_0/L) \qquad (22)$$

which is valid to within a beamwidth of end-fire. The curves are not plotted beyond that point, because the array factor is symmetric about $\theta = 90°$. Here $F(\theta) = F(180° - \theta)$ and the pattern for $\theta < 90°$ coalesces with the pattern for $\theta > 90°$ to form a broader beamwidth. Elliott [1] gives the expression below for the beamwidth at end-fire. Further narrowing of the beam can be obtained using slow-wave excitation [3]:

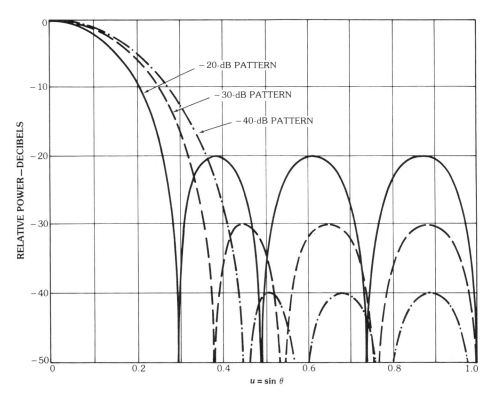

Fig. 8. Chebyshev patterns for an 8-element array ($\lambda/2$ spacing) with illuminations for -20 dB, -30 dB, and -40 dB.

Periodic Arrays

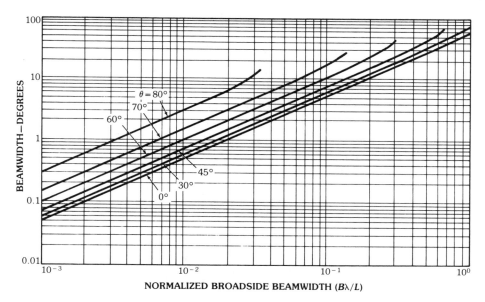

Fig. 9. Beamwidth versus normalized broadside beamwidth for a scanned array.

$$\Delta\theta = 2\cos^{-1}(1 - 0.443 B\lambda_0/L) \tag{23}$$

Phased Array Bandwidth

As indicated in Section 1, most arrays are designed using phase shifters, not time-delay units. A phase-steered array establishes a progressive phase front to match a wave at a single frequency. At a different frequency the progressive phase front corresponds to a wave at a different angle. This effect is shown schematically in Fig. 10.

Equations 3 and 5 give array excitation coefficients for time-delay and phase-

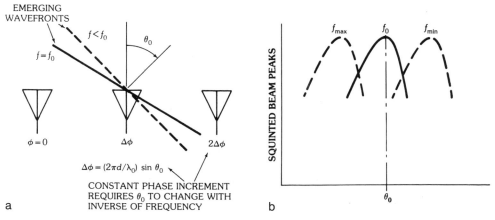

Fig. 10. Beam squint for a phase-scanned array. (*a*) Wavefronts. (*b*) Beam-pointing errors.

steered arrays. Scanning with phase shifters causes the peak gain angle of a phase-steered beam to not occur at the established (θ_0, ϕ_0) position except at $\lambda = \lambda_0$. For example, at center frequency f_0 the interelement phase shift ϕ_0 is given by

$$\phi_0 = \frac{2\pi}{\lambda_0} d_x \sin \theta_0 \quad (24)$$

but the interelement phase required for this scan angle at some other frequency is

$$\phi = \frac{2\pi}{\lambda} d_x \sin \theta_0 \quad (25)$$

The net effect, a beam-pointing error referred to as *beam squint*, is that the beam scans away from the desired θ_0 to an angle with $\sin \theta = (\lambda/\lambda_0) \sin \theta_0$; this is the most significant bandwidth limiting effect in array antennas.

Assuming an approximate half-power beamwidth $K\lambda/L$ (note $K = 0.886$ for uniform illumination) one can solve for the array bandwidth under the assumption that the gain at each frequency limit is reduced to half-power (that the squint is equal to a half-beamwidth at each limit). The resulting fractional bandwidth is given by

$$\frac{\Delta f}{f_0} = 0.886B \left(\frac{\lambda_0}{L}\right) \frac{1}{\sin \theta} \quad (26)$$

and for small scan angles

$$\frac{\Delta f}{f_0} \cong \frac{1}{n_B} \quad (27)$$

where n_B is the number of beamwidths scanned.

Fig. 11 shows how the 3-dB beamwidth varies with array length and scan angle for arrays with arbitrary side lobe levels.

For very large arrays it is necessary to divide the array into subarrays in order that some time-delay correction can be applied for each subarray. This is discussed in more detail in Section 3.

Antenna Pattern Synthesis

One of the primary advantages of the use of a phased array is the flexibility to form desired antenna patterns and so to match the radiation pattern to the technical requirement. The theory of antenna pattern synthesis is addressed in more detail in Chapter 2, hence the treatment that follows is restricted to remarks about several generalized procedures and a listing of relevant formulas and engineering data for the special case of low side lobe methods.

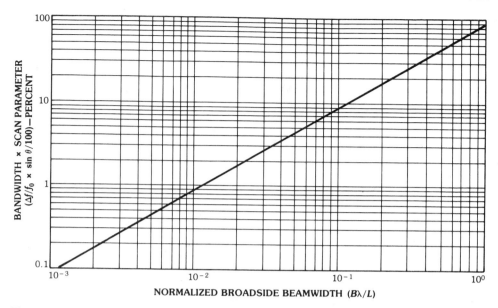

Fig. 11. Bandwidth scan product versus normalized broadside beamwidth for a scanned array.

General Procedures

Schelkunoff's method [4], introduced in Chapter 3, is based on the manipulation of zeros of the pattern function in the complex plane. The procedure is very useful of itself, and in addition is the foundation of computer-based methods with great power and flexibility. Elliott [5, 6] has described a generalized method based on a null perturbation approach to modify the side lobe structure of Taylor and Bayliss line source patterns. The procedure develops simultaneous linear equations solved by matrix inversion for a perturbed pattern and is repeated iteratively to converge to the desired result.

Fourier series methods can be readily applied to array synthesis problems by recognizing that the pattern function

$$E(u) = \sum_{m=-(M-1)/2}^{(M-1)/2} a_m e^{jkmud_x} \qquad m = \begin{cases} \pm\tfrac{1}{2}, \pm\tfrac{3}{2}, \ldots, & \text{for } M \text{ even} \\ 0, \pm 1, \ldots, & \text{for } M \text{ odd} \end{cases} \qquad (28)$$

is a finite Fourier series and is periodic in u space with the interval of the grating lobe distance λ/d_x. Thus, given a desired field distribution $E(u)$, one can obtain an expression for the currents from orthogonality, with

$$a_m = \frac{d_x}{\lambda} \int_{-\lambda/2d_x}^{\lambda/2d_x} e^{-j(2\pi/\lambda)ud_x m} E(u)\, du \qquad (29)$$

Stutzman and Thiele [7] show the application of this procedure to synthesis of a sector pattern. The method provides the least mean squared error approximation to the desired pattern as long as the element spacing $d_x \geq 0.5\lambda$. For closer spacing the domain of integration exceeds the visible region, and definition of a pattern function is ambiguous. The method provides a convenient test of required array size for synthesizing a given pattern, because one can vary the number of terms in the Fourier series and observe pattern convergence.

Synthesis with Orthogonal Beams

One of the most insightful methods, this approach is particularly useful for the synthesis of generalized sector coverage patterns, patterns without nulls, and certain low side lobe patterns. The method is due to Woodward [8] and Woodward and Lawson [9] and is referenced in several texts, so it will not be described in detail here. Although usually carried out for line source excitations, the method is very appropriate for array synthesis. It consists of using the multiple orthogonal beams of the periodic array as a basis for expanding the desired radiation field.

If the aperture length of an N-element array is defined to be $Nd_x = L$, N beams will fill a sector of width $(N-1)\lambda/L$ in u space, as shown in Fig. 12. The desired shaped pattern can then be matched at N points by selecting the amplitude of each beam. The specific patterns shown in Fig. 12 are for $N = 8$.

To excite the ith beam the elements are excited by progressive phase distribution

$$a_n = e^{-j(2\pi/\lambda)d_x u_i n} \tag{30}$$

where

$$u_i = (\lambda/L)i = (1/N)(\lambda/d_x)i \quad \text{and} \quad i, n = \pm\tfrac{1}{2}, \cdots, \pm(N-1)/2$$

The ith beam is given by

$$\begin{aligned}
g_i(u) &= \sum_{n=-(N-1)/2}^{(N-1)/2} e^{j(2\pi/\lambda)x_n(u-u_i)} \\
&= N\left\{\frac{\sin[(N\pi d_x/\lambda)(u-u_i)]}{N\sin[(\pi d_x/\lambda)(u-u_i)]}\right\}
\end{aligned} \tag{31}$$

The set of beams $g_i(u)$ is also orthogonal and occupies the beam positions shown in Fig. 12 (the example was done for $N = 8$). Fig. 12b shows two of the normalized orthogonal beams ($i = -7/2$ and $i = +1/2$) and clearly indicates that the domain of pattern synthesis must be restricted to $|Nd_x \sin\theta| \leq 0.5(N-1)$, for beyond that the grating lobes of the outermost beams present an ambiguity that leads to significant pattern distortion. One of the most significant results of synthesis with orthogonal beams is that the resultant feed networks can be lossless, since the progressive phase sequences can be formed by lossless Butler matrices or orthogonal beam lenses.

Section 3 shows an example of the use of orthogonal beams in a subarraying feed configuration to synthesize a pulse-type subarray pattern.

Periodic Arrays

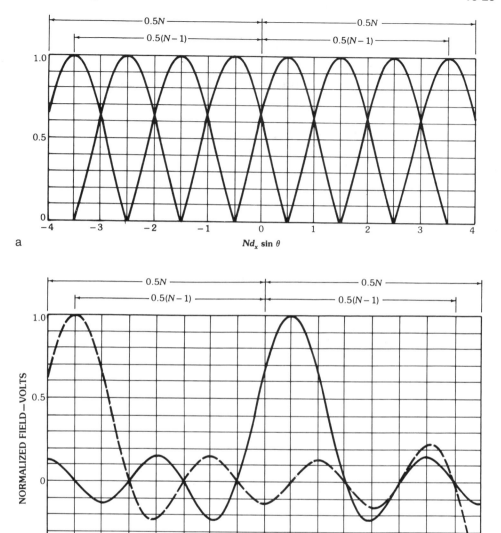

Fig. 12. Orthogonal beams for an N-element array. (*a*) Complete set of beams (plotted to first zeros). (*b*) Two orthogonal beams (plotted over domain of orthogonality) with $i = 1/2$ (solid) and $i = -7/2$ (dashed).

Low Side Lobe Solutions: Basic Formulas and Engineering Data

Among the most important and useful synthesis procedures are those that result in minimum beamwidth consistent with very low side lobes. From this class of methods three of the most important are Chebyshev and Taylor line source methods for symmetric patterns and Bayliss line source synthesis for antisymmetric difference patterns. The basic formulas for these three methods follow.

Dolph-Chebyshev Synthesis

Based on equating the array polynomial expression to a given Chebyshev polynomial, the method [10] synthesizes patterns for an N-element array (N any integer) with the narrowest beamwidth subject to a given side lobe level. A characteristic of the patterns is that all side lobes are of equal amplitude.

The synthesized pattern is given by the expression below for an array of $M + 1$ elements, spaced d apart:

$$F(z) = T_M(z) \tag{32}$$

where $T_M(z)$ is the Chebyshev polynomial of order M and is defined by

$$\begin{aligned} T_M(z) &= \cos(M \cos^{-1} z) \quad \text{for } |z| \leq 1 \\ &= \cosh(M \cosh^{-1} z) \quad \text{for } |z| \geq 1 \end{aligned} \tag{33}$$

The parameter z is given by

$$z = z_0 \cos[(\pi d/\lambda) \sin \theta] \tag{34}$$

where

$$z_0 = \cosh(M^{-1} \cosh^{-1} R)$$

for voltage side lobe level R ($SL_{dB} = 20 \log_{10} R$).

The original formulation of this problem is attributable to Dolph, who derived results for $\lambda/2$-spaced elements. Later Riblett [11] extended the analysis to elements greater than $\lambda/2$. Barbiere [12] derived the expression for z_0 above, and derived convenient relations for the currents. Stegen [13] derived the most widely used exact expressions for the required current distribution, and extensive tabulations of his results are given by Brown and Scharp [14], who also include gain and beamwidth values for arrays of up to 40 elements.

Stegen also gives a formula for Chebyshev array beamwidth, valid for large arrays. The following equation, derived by Drane [15], gives the beamwidth (here converted to degrees) for an array of length L with side lobe level $|SL|$(dB)

$$\text{BW} = \left(\frac{\lambda}{L}\right)(10.314)(|SL| + 4.52)^{1/2} \tag{35}$$

Fig. 13 shows the normalized beamwidth parameter (L/λ)BW (in degrees) as a function of side lobe level R as computed from the above equation. Drane also gives an equation for directivity for array spacings greater or less than a half-wavelength. For spacings greater than $\lambda/2$ the directivity D is [15]

Periodic Arrays

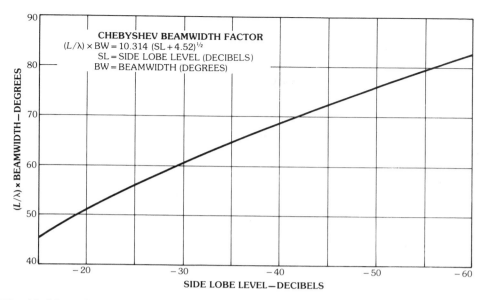

Fig. 13. Normalized beamwidth versus side lobe level for Chebyshev arrays. (*After Drane [15]*, © *1968 IEEE*)

$$D \cong \frac{2R^2}{1 + (\lambda/L)R^2\sqrt{(\ln 2R)/\pi}} \tag{36}$$

This result is in close agreement with the equation of Elliott [1]. For large arrays this directivity reaches the maximum value $2R^2$, or 3 dB greater than the specific side lobe level.

Chebyshev patterns have been used for purposes of illustration throughout this chapter, and so no additional patterns will be shown here.

Taylor Line Source Synthesis

The classic paper by Taylor [16] investigated the synthesis of equal side lobe patterns with continuous line source excitations and satisfying the same criteria as do the Chebyshev linear array patterns. He showed that the idealized pattern is the solution to this problem is not physically realizable but could be approximated arbitrarily closely by a function of two parameters, A and $\bar{n}$. The family of patterns derived by Taylor has the first $\bar{n}$ side lobes at the desired level, and all side lobes beyond the $\bar{n}$th fall off as $(\sin \pi z)/\pi z$ (for $z = Lu/\lambda$) for a line source of length L.

The synthesized pattern, normalized to unity, is given by

$$F(z, A, \bar{n}) = \frac{\sin \pi z}{\pi z} \prod_{n=1}^{\bar{n}-1} \frac{1 - z^2/z_n^2}{1 - z^2/n^2} \tag{37}$$

for $z = (L/\lambda)u$.

The zeros of the function are at

$$z_n = \begin{cases} \pm\sigma\sqrt{A^2 + (n - \tfrac{1}{2})^2} & \text{for } 1 \leq n \leq \bar{n} \\ \pm n & \text{for } \bar{n} \leq n < \infty \end{cases} \qquad (38)$$

where

$$\beta = \frac{\bar{n}}{\sqrt{A^2 + (\bar{n} - \tfrac{1}{2})^2}}$$

The parameter A is defined so that $\cosh(\pi A)$ is the voltage side lobe ratio, or

$$A = \frac{1}{\pi}\cosh^{-1} R \qquad (39)$$

An approximation for the beamwidth is given by

$$\Delta\theta \cong \sigma\beta_0(\lambda/L) = \frac{\beta_0 \bar{n}(\lambda/L)}{\sqrt{A^2 + (\bar{n} - \tfrac{1}{2})^2}} \qquad (40)$$

where

$$\beta_0 = (2/\pi)\sqrt{(\cosh^{-1} R)^2 - (\cosh^{-1} R/\sqrt{2})^2}$$

The current or aperture field distribution necessary to produce this pattern family is

$$g(x) = F(0, A, \bar{n}) + 2 \sum_{m=1}^{\bar{n}-1} F(m, A, \bar{n}) \cos(2m\pi x/L) \qquad \text{for } -L/2 \leq x \leq L/2 \quad (41)$$

where the coefficients $F(z, A, \bar{n})$ are the pattern values of (37) at the integers ($z = m \leq \bar{n}$). In abbreviated form these are

$$F(m, A, \bar{n}) = \frac{[(\bar{n} - 1)!]^2}{(\bar{n} - 1 + m)!(\bar{n} - 1 - m)!} \prod_{n=1}^{\bar{n}-1} (1 - m^2/z_n^2) \qquad (42)$$

These coefficients $F(m, A, \bar{n})$ are tabulated in Hansen [17] for various n for side lobe levels between −30 and −40 dB in 5-dB increments.

Fig. 14 shows two Taylor patterns. The solid one is the line source pattern computed from (37), while the dashed curve is the radiation pattern of the $g(x)$ distribution sampled at the points: $(L/N)i$ for $\pm i = \tfrac{1}{2}, \tfrac{3}{2}, \ldots, (N-1)/2$ for an array of N elements using the tabulated values of F. The patterns are for a −30-dB Taylor distribution with $\bar{n} = 4$ and $N = 16$. The figure indicates that the sampled line source distribution is a very good approximation of the line source pattern, even for an array with only 16 elements.

The practical selection of the $\bar{n}$ parameter is discussed by Taylor. This parameter must be large enough or the beam will be broader than necessary; $\bar{n}$

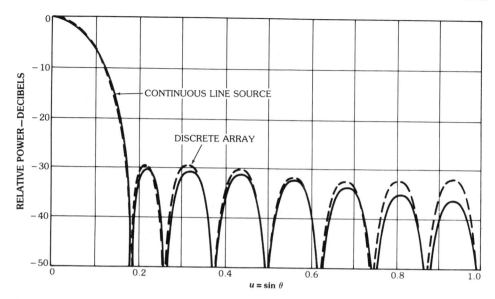

Fig. 14. Taylor $\bar{n} = 4$ line source patterns: continuous line source and discrete array pattern ($N = 16$) using sampled Taylor illumination.

should be at least 3 for -25-dB side lobes, and at least 6 for -40 dB. Increasing $\bar{n}$ further narrows the main beam, but supergaining results if it is made too large.

Bayliss Line Source Synthesis

E. T. Bayliss [18] has developed a method of synthesizing line source difference patterns that parallel the essential features of Taylor line source patterns. The difference pattern is fully described by two parameters, A and $\bar{n}$, which completely control the side lobe level and decay behavior. Bayliss' method results in a pattern of the following form:

$$F(z) = \pi z \cos \pi z \prod_{n=1}^{\bar{n}-1} [1 - (z/\sigma z_n)^2] \bigg/ \prod_{n=0}^{\bar{n}-1} \{1 - [z/(n + \tfrac{1}{2})]\} \qquad (43)$$

where

$$z = (L/\lambda)u$$
$$\sigma = \frac{\bar{n} + \tfrac{1}{2}}{z_{\bar{n}}}$$
$$z_{\bar{n}} = (A^2 + \bar{n}^2)^{1/2}$$

Here, as in the Taylor method, the first few side lobes are at the design level, and beyond about $z = \bar{n}$ the side lobes decay as $z^{-3/2}$. Unlike the Taylor line source method, the z_n terms are not available in closed form but are given in terms of the coefficients of fourth-order polynomials relating the null positions and side lobe levels. The null positions are given by:

$$\sigma z_n = \begin{cases} 0 & \text{for } n = 0 \\ \pm \sigma \xi_n & \text{for } n = 1, 2, 3, 4 \\ \pm \sigma (A^2 + n^2)^{1/2} & \text{for } n = 5, 6, \ldots \end{cases} \quad (44)$$

Bayliss computed ξ_n and A by iterative methods but presented coefficients of fourth-order polynomials to evaluate these constants and the location of the difference peak P_0 using Table 1 and the expression

$$\text{polynomial name} = \sum_{n=0}^{4} c_n (\text{SL})^n \quad (45)$$

where SL is the side lobe level in decibels.

Elliott [19] gives a convenient table of A and ξ_n for patterns with SL = -15 through -40 dB in increments of 5 dB. The table is given here as Table 2.

The line source excitation required to produce this pattern is given by

$$g(x) = \sum_{n=0}^{\bar{n}-1} B_n \sin[(2\pi x/L)(n + \tfrac{1}{2})], \quad -L/2 \leq x \leq L/2 \quad (46)$$

with Fourier coefficients

Table 1. Polynomial Coefficients

Polynomial Name	c_0	c_1	c_2	c_3	c_4
A	0.303 875 30	−0.050 429 22	−0.000 279 89	−0.000 003 43	−0.000 000 02
ξ_1	0.985 830 20	−0.033 388 50	0.000 140 64	0.000 001 90	0.000 000 01
ξ_2	2.003 374 87	−0.011 415 48	0.000 415 90	0.000 003 73	0.000 000 01
ξ_3	3.006 363 21	−0.006 833 94	0.000 292 81	0.000 001 61	0.000 000 00
ξ_4	4.005 184 23	−0.005 017 95	0.000 217 35	0.000 000 88	0.000 000 00
p_0	0.479 721 20	−0.014 566 92	−0.000 187 39	−0.000 002 18	−0.000 000 01

Table 2. Values of A and ξ_n

Polynomial Name	Side Lobe Level (dB)					
	15	20	25	30	35	40
A	1.0079	1.2247	1.4355	1.6413	1.8431	2.0415
ξ_1	1.5124	1.6962	1.8826	2.0708	2.2602	2.4504
ξ_2	2.2561	2.3698	2.4943	2.6275	2.7675	2.9123
ξ_3	3.1693	3.2473	3.3351	3.4314	3.5352	3.6452
ξ_4	4.1264	4.1854	4.2527	4.3276	4.4093	4.4973

Periodic Arrays

$$B_m = \frac{1}{2j}(-1)^m(m+\tfrac{1}{2})^2 \prod_{n=1}^{\bar{n}-1}\{1 - [(m+\tfrac{1}{2})/\sigma z_n]^2\} \Big/$$
$$\prod_{\substack{n=0 \\ n \neq m}}^{\bar{n}-1}\{1 - [(m+\tfrac{1}{2})/(n+\tfrac{1}{2})]^2\} \qquad (47)$$

for $m = 0, 1, \cdots, \bar{n}-1$. Also

$$B_m = 0, \quad \text{for } m \geq \bar{n}$$

Fig. 15 shows two Bayliss patterns. The solid one is computed directly from (43) and is the line source pattern, while the dashed one is the radiation pattern of the $g(x)$ distribution sampled at the points (L/Ni) for $\pm i = \tfrac{1}{2}, \tfrac{3}{2}, \ldots, (N-1)/2$. The patterns are for a -30-dB illumination with $\bar{n} = 4$ and $N = 16$.

Although the discretized pattern is a good approximation to the line source pattern, even for as few as 16 elements, in some cases it may be important to improve the discretized patterns. Elliott has applied perturbation methods for this purpose and derived a set of linear equations from the perturbations of peak side lobe values. The procedure is iterated until convergence is adequate.

The above methods have been extended to treat circular apertures by Taylor [20] and Bayliss [18], and in addition Tseng and Cheng [21] have synthesized a class of circularly rectangular arrays with rectangular lattices and circularly symmetric patterns.

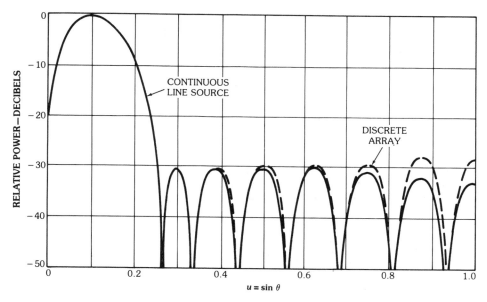

Fig. 15. Bayliss $\bar{n} = 10$ line source patterns: continuous line source and discrete array pattern ($N = 16$) using sampled Bayliss illumination.

3. Array Organization: Subarrays and Broadband Feeds

Often it is convenient to treat a large array as an array of smaller arrays. This is done to simplify power distribution networks, to incorporate low-power, lightweight, or compact circuitry that may have high losses, to integrate amplifier stages into the feed network, or to introduce time-delay networks to improve the broadband properties of the array.

Grouping the array into subarrays may be quite advantageous, but for several reasons it tends to increase the array side lobe level. In Section 4 it is shown that random errors in subarray excitation result in larger side lobes than errors at the array elements because there are so few subarrays. Furthermore, the periodic phase errors that occur in arrays of equal-size time-delayed subarrays produce grating lobes in the array factor. Techniques for producing wide-band behavior with an array of phase-steered elements are described in this section.

Aperture Illumination Control at Subarray Input Ports

One example of an application of subarrays is shown in Fig. 16. An array of $m \times M$ elements is divided into m arrays of M elements each. For simplicity the identical M-way power dividers provide in-phase, equal-amplitude output signals. The m-way beamformer provides feed coefficients a_i applied at the subarray ports. Since the subarrays are formed by equal-amplitude power dividers the array illumination has a staircase appearance shown in the figure, with steps at the a_i level. The normalized radiation pattern at center frequency, unscanned, is

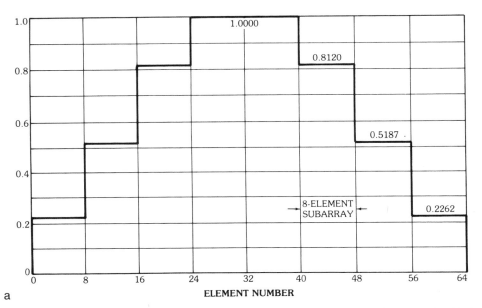

Fig. 16. An array of contiguous uniformly illuminated subarrays with −30-dB and −40-dB Chebyshev array illuminations. (*a*) Array illumination (−30-dB case) for eight subarrays with eight elements each. (*b*) Subarray pattern and array factor (for −30-dB case). (*c*) Total array pattern for −30-dB case (−40-dB case shown partially).

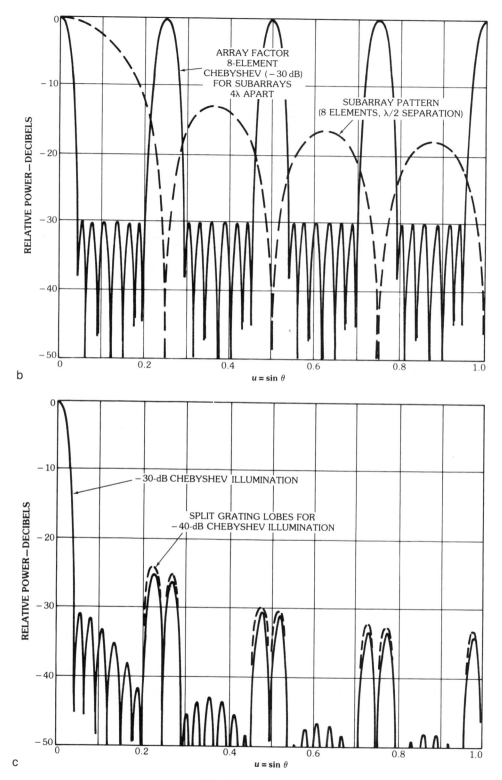

Fig. 16, *continued.*

$$E(r) = \frac{e(u,v)}{m} \frac{\sin[Mu\pi(d_x/\lambda)]}{M\sin[\pi u(d_x/\lambda)]} \sum_{i=-(m-1)/2}^{(m-1)/2} a_i e^{ji(2\pi/\lambda)uD_x} \qquad (48)$$

Fig. 16 shows the pattern for a 64-element array with $M = m = 8$ and with the coefficients a_i shown in Fig. 16a selected to produce a -30-dB Chebyshev array factor. The final pattern, shown in Fig. 16c, is the product of the array factor 16b (solid) with low side lobes but large grating lobes and the subarray pattern 16b (dashed), and exhibits characteristic split grating lobe peaks that occur because of the broadened beamwidth of the low side lobe array factor. If the a_i coefficients were chosen for still lower near side lobes, the main beam and grating lobes in the array factor would broaden further and the subarray pattern nulls would be much narrower than the width of each grating lobe, thus leading to increased values of the split-peak grating lobes. Sample values corresponding to the -40-dB Chebyshev results are indicated dashed on the figure.

The above grating lobes could be lowered by using more and therefore smaller subarrays. The end result would be to broaden the subarray pattern nulls and so reduce the product of subarray pattern times array factor in the vicinity of the grating lobes.

In all of the cases treated in this section the grating lobe peaks can be suppressed by using unequally spaced subarrays. This modification leaves higher average side lobes than the array would have without subarraying, but is often the method selected for introducing time delay in a large array.

Wideband Characteristics of Time-Delayed Subarrays

The bandwidth limitations implied by (26) are often reasonable for small arrays but are usually too restrictive for large arrays. For this reason it is common practice to combine phase and time-delay steering by organizing the array into a relatively small number of subarrays and to use time-delay devices at the subarray input ports and phase steering at all the array elements. The resulting array bandwidth is a compromise between the cost of providing time-delay devices for a large number of subarrays and the pattern deterioration and bandwidth limitations of dividing the array into too few subarrays.

Contiguous Subarrays of Discrete Time-Delay Devices

The array of contiguous subarrays (Fig. 17) is conceptually simpler than other subarray approaches, and uses separate distribution networks to feed adjacent sections of the array. Phase shifters control the subarray pattern to produce a beam tilt, and the time-delay devices produce true time delay between the subarray centers. Fully equivalent is an array with time-delay devices behind each element, but with only a fixed number of discrete time-delay bits. The situations are mathematically equivalent if the number of available time-delay steps is made equal to the number of subarrays. In this case, for equal-size subarrays, the following results give peak grating lobe levels.

To consider an example, assume a one-dimensional array of elements spaced d_x apart, with element pattern $e(u,v)$. The elements are grouped into subarrays of M elements. The entire array has m equally spaced subarrays. Each of the subarrays has a subarray pattern that is the same as the middle term in (48), and

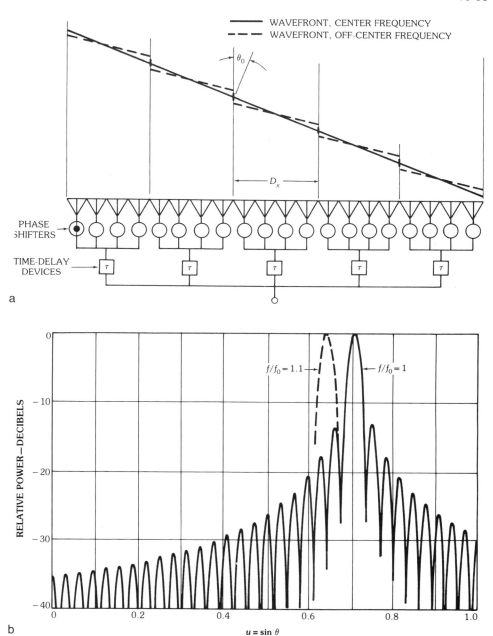

Fig. 17. Broadband characteristics of an array with time delay at subarray level, and contiguous phase-steered subarrays. (*a*) Array geometry, showing wavefront at center and off-center frequencies. (*b*) Patterns of 64-element array with phase shift steering. (*c*) Patterns of array organized with eight time-delayed subarrays.

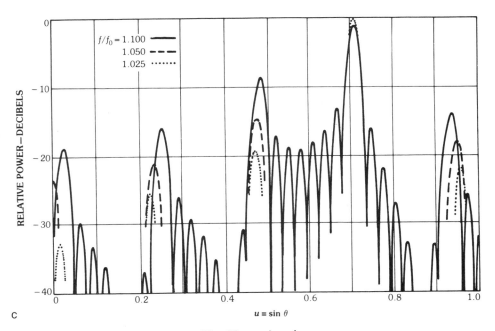

Fig. 17, *continued.*

when these subarrays are arrayed with time delay appropriate for beam collimation the complete field pattern is given by the expression

$$E(r) = e(u,v) \left\{ \frac{\sin[M\pi d_x(u/\lambda - u_0/\lambda_0)]}{M \sin[\pi d_x(u/\lambda - u_0/\lambda_0)]} \right\} \left\{ \frac{\sin[(m\pi D_x/\lambda)(u - u_0)]}{m \sin[(\pi D_x/\lambda)(u - u_0)]} \right\} \quad (49)$$

where $D_x = Md_x$ is the subarray size.

This expression shows the total field as the product of element pattern, phase-steered subarray pattern, and time-delayed array factor. If the array were purely phase controlled, with $m \times M$ elements each spaced d_x apart, its bandwidth given by (26) would be

$$\frac{\Delta f}{f_0} = \frac{K\lambda_0}{Mm d_x \sin\theta_0} \quad (50)$$

In its present subarrayed form, however, the time-delayed array factor exactly collimates the subarray contributions at all frequencies, and the system bandwidth is essentially the same as the subarray bandwidth:

$$\frac{\Delta f}{f_0} = \frac{K\lambda_0}{M d_x \sin\theta_0} \quad (51)$$

For example, an array of ten subarrays of ten elements each has approximately ten times the bandwidth of the phase-steered array of one hundred elements.

Periodic Arrays

The above description emphasizes bandwidth based on gain, but in fact subarraying can introduce severe pattern degradation in the form of grating lobes that arise as frequency is changed. Grating lobes exist in this case, even though the subarray phase centers are appropriately delayed to form a beam at θ_0. This is because each subarray has a phase squint that causes the peak of the subarray pattern to move off the position θ_0 and the subarray pattern nulls to move so that they do not suppress the array pattern grating lobes. Fig. 17 shows the pattern of a uniformally illuminated array of 64 elements, arranged in subarrays of 8 elements, with each element 0.5λ apart. The array is scanned to $45°$. In Fig. 17b, the array is steered by phase controls alone, and its main beam squints from the desired $45°$ to $40°$ for $f/f_0 = 1.10$. Fig. 17c shows that the same array with time delay at the subarray input ports exhibits no beam squint, but that large grating lobes (about 8 dB below the main beam) seriously distort the pattern and cause a loss of gain at $f/f_0 = 1.1$. Grating lobes at $f/f_0 = 1.05$ and 1.025 are shown dashed and dotted, respectively. Clearly, the use of contiguous time-delayed subarrays leads to intolerable pattern deterioration for all but extremely small fractional bandwidths.

Overlapped Subarrays

A technique for implementing time-delay steering at the subarray level without the occurrence of large grating lobes involves the synthesis of subarray illuminations that are not merely contiguous but actually overlap. By using an aperture illumination wider than the intersubarray period it is possible to produce subarray patterns that have flat tops and are narrow enough to suppress the array pattern grating lobes [22]. This synthesis is achieved using two back-to-back transform networks in order to form a number of flat-topped subarray patterns, using the orthogonal beams as in a Woodward-type synthesis. The transform networks could be Butler matrices, as described here, or confocal lenses (Fig. 18), or reflectors, or some combination of these.

Fig. 18a shows the basic configuration of two Butler matrixes back to back used to excite an array that has phase shifters at each array element. The phase shifters are controlled in accordance with (24). A signal applied to the ith input port of the matrix at right (the $M \times M$ matrix) produces a progressive set of phases at the N array elements and radiates with the pattern

$$g_i(u) = N f^e(u) \frac{\sin[(N\pi d_x/\lambda_0)(fu/f_0 - u_i)]}{N \sin[(\pi d_x/\lambda_0)(fu/f_0 - u_i)]} \qquad (52)$$

where $f^e(u)$ is the array element pattern (assumed equal for all elements) and

$$u_i = i \frac{\lambda_0}{N d_x} + u_0 \qquad (53)$$

Each of the orthogonal beams is displaced from the angle of its peak radiation with all phase shifters set to zero by the amount u_0. When the matrix at left is used to provide the signals at the input to the $M \times N$ matrix at right, each input J_m excites a whole set of signals I_{im}, given by

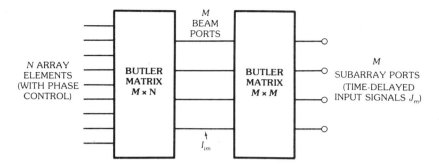

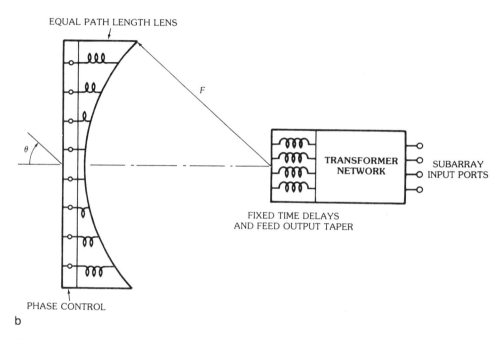

Fig. 18. Constrained and space-fed systems for overlapped subarray formation. (*a*) Constrained network. (*b*) Space-fed network.

$$I_{im} = \frac{J_m}{\sqrt{M}} e^{j2\pi(m/M)i}, \qquad -\left(\frac{M-1}{2}\right) \leq i \leq \frac{M-1}{2} \tag{54}$$

The aperture illumination (phase scanned to u_0) corresponding to the mth subarray is, for each nth element of the N-element array,

$$i_m^{(n)} = e^{-j(2\pi/\lambda_0)\mu_0 n d_x} \frac{1}{\sqrt{N}} \sum_{i=-(M-1)/2}^{(M-1)/2} I_{im} e^{-j2\pi(n/N)i}$$

$$= \frac{NJ_m e^{-j(2\pi/\lambda_0)\mu_0 n d_x}}{\sqrt{NM}} \frac{\sin M\pi[(mN-nM)/MN]}{M \sin \pi[(mN-nM)/MN]} \tag{55}$$

Periodic Arrays

This illumination has a maximum at the element with index $n = m(N/M)$, and overlaps all the elements of the array. An example of one such subarray illumination is the dashed curve of Fig. 19a for the subarray ($m = 4$) of an array of 64 elements ($N = 64$) with $\lambda/2$ separation. The array has eight subarrays ($M = 8$).

The radiated subarray patterns are given by

$$f_m(u) = \sum_i I_{im} g_i(u)$$

$$= \frac{NJ_m f^e(u)}{\sqrt{MN}} \sum_{i=-(M-1)/2}^{(M-1)/2} e^{j2\pi(m/M)i} \left(\frac{\sin[(N\pi d_x/\lambda_0)(fu/f_0 - u_i)]}{N \sin[(\pi d_x/\lambda_0)(fu/f_0 - u_i)]} \right) \quad (56)$$

This expression is a sum of M orthogonal pencil beams arranged to fill the sector, and taken together to form a flat-topped pattern for the mth subarray, which is shifted in angle so that its center is at u_0 at center frequency.

Fig. 19b shows two subarray patterns at center frequency for the same 64-element array. The selected subarrays are an edge ($m = -7/2$) and one of the two central subarrays ($m = 1/2$). The edge subarray has higher side lobes and a highly rippled pass region because its illumination is truncated at the array edge.

The array excitation with all subarrays excited is

$$\mathcal{I}_n = \sum_{m=-(M-1)/2}^{(M-1)/2} J_m i_n^{(m)} \quad (57)$$

where the $i_n^{(m)}$ are given in (55). An example of such a composite excitation is shown in the solid curve of Fig. 19a, for all subarrays weighted with a -30-dB Chebyshev illumination. This excitation is much smoother than the illumination

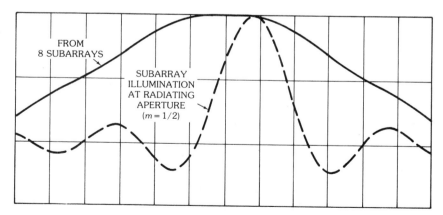

Fig. 19. Broadband characteristics of an array with time delay at the subarray level and completely overlapped phase-steered subarrays. (*a*) Typical subarray illumination at radiating aperture and total illumination from eight subarrays with -30-dB Chebyshev weighting. (*b*) Radiated subarray patterns near array edge and array center. (*c*) Array radiation pattern at broadside and 45° scan ($f = f_0$) and 45° scan ($f = 1.1 f_0$).

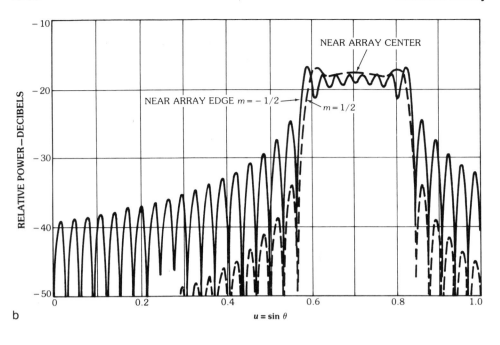

b

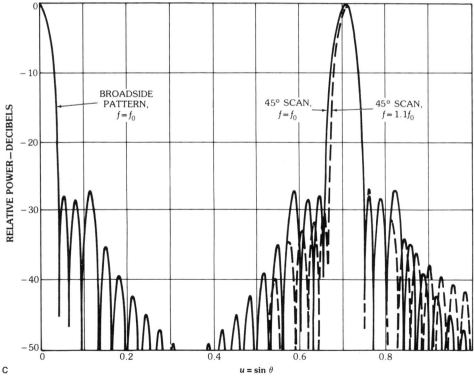

c

Fig. 19, *continued.*

Periodic Arrays

shown in Fig. 16a, which shows the same weightings applied to an array of contiguous subarrays. So it can be expected that the side lobes would be lower for the overlapped subarray case.

The radiated array pattern is given by

$$F(u) = \sum_{M=-(M-1)/2}^{(M-1)/2} J_m f_m(u) \tag{58}$$

To scan the array beam to u_0 with time delay, one applies the signals

$$J_m = |J_m| e^{-j(2\pi/\lambda_0) D_x (f/f_0 - 1) u_0} \tag{59}$$

where $D_x = (N/M) d_x$ is the distance between subarray centers at the array face.

The absolute values of the input signals $|J_m|$ are weighted directly to provide the appropriate near side lobe distribution.

Figs. 19c and 19d show several array patterns for the array with the -30-dB illumination. As shown in the figures the side lobe levels exceed -30 dB because of the rippled subarray patterns. At center frequency ($f/f_0 = 1$) the pattern scanned to $u = 0.707$ ($45°$) has the same form as the broadside pattern, and even at 10 percent above center frequency ($f/f_0 = 1.1$) the main beam is not altered in location or gain, though the side lobe structure is.

The bandwidth of such a subarraying is on the order of

$$\frac{\Delta f}{f_0} \cong \frac{(M-1)}{N} \frac{\lambda}{d_x} \frac{1}{\sin \theta_0} \tag{60}$$

Overlapped subarray systems have been implemented using multiple-beam lens systems [22, 23, 24] and Butler matrix [25] networks. With the emergence of digital beam-forming technology it is likely that it will be convenient to form subarrayed patterns digitally for future system applications.

The networks described above produce completely overlapped subarrays; each subarray extends over the whole array. However, convenient networks have also been developed to overlap small groups of elements. Such techniques form approximations of the ideal flat-topped pattern and are useful for limited scan applications [26, 27].

Broadband Array Feeds with Time-Delayed Offset Beams

Equation 26 gives the fractional bandwidth of a phase-steered array as a function of its beamwidth and the maximum scan angle θ_0. The bandwidth can be relatively large if the array scan remains small, so that the product $(L/\lambda_0) \sin \theta_0$ does not become a large number. Similarly, if an array is fed by a system that produces a time-delayed beam at some angle θ_T, and the beam is phase steered to an angle θ_0 by phase shifters, then the bandwidth is given by

$$\frac{\Delta f}{f_0} = \frac{0.866 B \lambda_0}{L |\sin \theta_T - \sin \theta_0|} \tag{61}$$

This equation shows that by using a feed system that provides a number of fixed-offset time-delayed beams it is possible to scan those beams over the limited angular regions between the beams, and so operate over substantially increased instantaneous bandwidth. One such implementation, suggested by Rotman and Franchi [28], is indicated in Fig. 20a, which shows an active lens with four feed horns equally spaced along the focal arc of a two-dimensional microwave-cons-

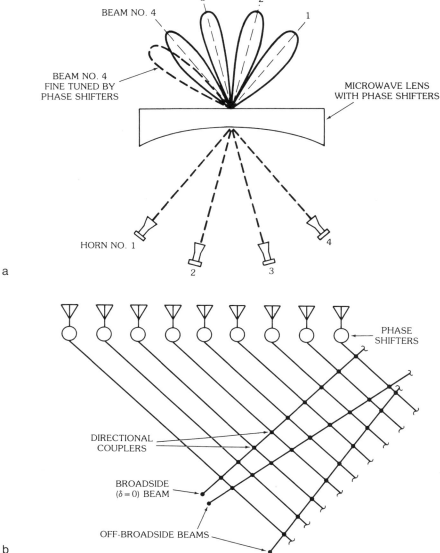

Fig. 20. Wideband scanning using phase shifter and fixed time-delayed beams. (*a*) Microwave lens configuration. (*b*) True time-delay matrix.

trained lens. Energy from a transmitter can be directed to any one of the horns by means of the switching tree. Each horn, in turn, will form a beam in a different azimuth direction for the zero phase shifter setting. A typical beam for the Mth horn is sketched as the solid curve (Mth beam). As the phase of the lens illumination is changed by the phase shifters the beam scans to either side of its no-phase-shift position. This effect is illustrated by the dotted curve (phase-scanned Mth beam). The bandwidth limitation imposed by this phase scanning is given by the following equation, with θ_0 the maximum scan angle and N the number of beam positions (4 for the example in the figure). The system bandwidth, given below and readily derived from (26), is wide because of the limited scan angle:

$$\frac{\Delta f}{f_0} \cong \frac{0.886 BN}{(L/\lambda_0)\sin\theta_0} \qquad (62)$$

Other means for achieving fixed offset beams include the use of constrained multiple-beam systems, such as the true time-delay matrix of Fig. 20b, or precut switched time delays at each array element. In addition to providing wideband gain, systems that use offset beams have no phase discontinuities and therefore no grating lobes. In principle they can have very low side lobes.

4. Practical Arrays

The previous sections on periodic arrays treat the array in an idealized case, with perfectly regular lattices, prescribed exact phase controls, half-space isotropic element patterns, and, most importantly, with all element patterns equal.

This section deals with a number of problems that confront array designers. Array mutual coupling leads to unequal element patterns and to a need to solve coupled integral equations before applying any of the synthesis methods mentioned earlier. Conformal nonplanar arrays have lattices that are, at most, periodic in one plane, and so present special problems in synthesis and pattern control. Finally, the section addresses array component errors of several types, and it reviews components used to distribute power and scan the beam of a phased array.

Mutual Coupling and Element Patterns

One of the most important and complex aspects of modern array design is that element excitation coefficients are not proportional to applied sources (voltages or currents) and that the element patterns are nonisotropic and not equal to the pattern of an isolated element. These phenomena occur because each of the elements couples through radiation to all of the others, and hence the relationship between applied sources and element excitation must be expressed in terms of a complex matrix. This phenomenon is called *mutual coupling*.

A detailed treatment of mutual coupling is beyond the scope or intent of this chapter. The solution of wire antenna problems, such as the dipole array of Fig. 21a, is carried out by satisfying a boundary condition (usually that the tangential electric field is zero) at the surface of the wire and equal but opposite to the applied field at the source point. For the dipole array of Fig. 21a, with dipole axes along the $\hat{y}$ direction and their centers located at $(x_m, y_m, c/2)$ with and without a ground

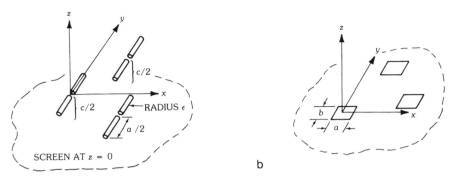

Fig. 21. Elements for scanning arrays. (*a*) Dipoles over ground screen. (*b*) Waveguide apertures in ground screen.

plane, it is convenient to introduce a single component of the vector potential defined at $\mathbf{r} = \hat{x}x + \hat{y}y + \hat{z}z$ as

$$A_y(x,y,z) = \frac{\mu_0}{4\pi} \sum_m \int_{V_m} I_m(y')G(r,r')dy' \qquad (63)$$

where

$$G(r,r') = \frac{e^{-jk_0|\mathbf{r}-\mathbf{r}_m|}}{|\mathbf{r}-\mathbf{r}_m|}$$

$$|\mathbf{r}-\mathbf{r}_m| = \sqrt{(x-x_m)^2 + (y-y')^2 + (z-c/2)^2}$$

for the array without a ground plane, and

$$G(r,r') = \frac{e^{-jk_0|\mathbf{r}-\mathbf{r}_m|}}{|\mathbf{r}-\mathbf{r}_m|} - \frac{e^{-jk_0|\mathbf{r}-\mathbf{r}_m^{(i)}|}}{|\mathbf{r}-\mathbf{r}_m^{(i)}|}$$

$$|\mathbf{r}-\mathbf{r}_m^{(i)}| = \sqrt{(x-x_m)^2 + (y-y')^2 + (z-c/2)^2}$$

for the array with ground plane at $z = 0$.

The second expression accounts for the image dipole at $z = -c/2$.

The set of integral equations equating the tangential **E** field to zero at each dipole radius is written at the nth dipole

$$E_y(x_n, y, c/2) = -V^{(n)}\delta(y-y_n) = -j\frac{\omega}{k_0^2}\left[\frac{\partial^2 A_y}{\partial y^2} + k^2 A_y\right],$$

$$\text{for } -a/2 \leqq y-y_n \leqq a/2 \quad (64)$$

where $V^{(n)}$ is the potential across the source at the nth antenna. The source is assumed to be a delta function of the electric field located at the center of each dipole, and the equation sets the induced field equal to the negative of the source field at this point.

In this form the integral equation is often called *Pocklington's equation* and is frequently chosen for digital computer solution of dipole and wire antenna problems. In this procedure each dipole is considered as made up of a number of connected segments, with each segment radiating and coupling to all other segments in all of the radiators. For detailed descriptions of this procedure the reader is referred to [29] and [30].

Another traditional form of equation for dipole arrays is obtained by constructing a solution to the above differential form to obtain a generalized form of Hallen's equation [31]:

$$\left.\begin{array}{l} A_y(x_n, y, z_n) = c_1^{(n)}\cos ky + c_2^{(n)}\sin ky - j\dfrac{k}{2}\dfrac{V^{(n)}}{\omega}\sin ky, \quad 0 \leq y - y_n \leq a/2 \\ = c_3^{(n)}\cos ky + c_4^{(n)}\sin ky + j\dfrac{k}{2}\dfrac{V^{(n)}}{\omega}\sin ky, \quad -a/2 \leq y - y_n \leq 0 \end{array}\right\} \quad (65)$$

The potential A_y on the left side of the equation is evaluated at the center line of each dipole (using $|\mathbf{r} - \mathbf{r}_m| = [\varepsilon^2 + (y - y')^2]^{1/2}$ for $n = m$). The evaluation of the potential function in this manner is valid for thin dipoles (with ε much less all other dimensions, including wavelength) and assumes the potential outside of and even at the surface of the dipole is the same as if there were a filament of current at the dipole axis.

The constant c_4 is equal to c_2.

The above is a set of M integral equations, written along each dipole, and consisting of M unknown current functions and $3M$ constants ($c_1^{(n)}, c_2^{(n)}, c_3^{(n)}$) that arose from the integration of the integrodifferential equations. The solution of the above equations is necessarily approximate and there have been numerous forms of published solutions, each with different degrees of validity and complexity. The simplest approximation is one that assumes that the form of the current distribution is the same on all dipoles,

$$I_m(y) = I(m)f(y) \quad (66)$$

This approach is most appropriate for resonant dipoles with their source points all in the same plane ($y_n = 0$). Examples of this sort of solution exist in the literature and have been carried out by assuming a form $f(y)$ and solving the resulting simultaneous equations.

The end result after elimination of constants is an impedance matrix

$$\mathbf{V} = \mathbf{ZI} \quad (67)$$

with column vectors $\mathbf{I}$ and $\mathbf{V}$ and square impedance matrix $\mathbf{Z}$ that one can use for computation of currents, given applied voltages, or, in the case of synthesizing required antenna patterns, can be used to compute the required voltages to provide desired current terms. Commonly used single-mode impedance formulas are given by Brown and King [32], Carter [33], and Tai [34].

King and his colleagues have employed several higher-order solutions that are more realistic for computing near-field effects or coupling between column arrays of dipoles with parallel axes. The most comprehensive of these is the five-term theory [35, 36] that includes asymmetric current terms for evaluation of the radiation properties of arrays scanned in two dimensions.

Of these two basic methods, the expansion of the current by a finite number of functions that span the entire dipole or by a piecewise approximation in sections across the dipole, solutions based on higher-order current expansions have to date seen far wider application to large arrays than the multisegment solutions, because the former involve the inversion of much smaller matrices.

Finite waveguide arrays can be treated in a similar manner. In this case there are no electric current sources in the half-space $z \geqq 0$, and the only sources are the magnetic current sources as represented by the tangential aperture fields. In this case there is no single vector component that serves to completely represent the fields except for special two-dimensional cases. In general, however, for a finite waveguide aperture the solution is vector and is formulated by expanding the aperture field in a set of functions and matching fields in the waveguides and free space. For open-ended waveguides it is convenient to choose as basis functions the waveguide normal-mode fields, and for unloaded rectangular waveguides one can choose the orthogonally polarized transverse electric fields. The transverse electric field for the waveguide at the origin of the coordinate system of Fig. 21 is

$$\mathbf{E}^T = \mathbf{e}_0(x,y)e^{-jk_z(0)z} + \sum_n V_n \mathbf{e}_n(x,y)e^{+jk_z(n)z} \tag{68}$$

where the $\mathbf{e}_n(x,y)$ are the transverse-mode functions for the two possible polarizations, the $k_z(n)$ are the modal propagation constants, and V_n are undetermined modal amplitude coefficients. This expression represents a single incident mode in the waveguide and an infinite series of reflected modes. Typically all but the $k_z(0)$ propagation constants are imaginary, indicating that those are beyond waveguide cutoff, but they enter into the solution to match boundary conditions. The solution proceeds by expanding the transverse magnetic waveguide fields in terms of these and writing the half-space fields ($z > 0$) as the aperture field. Construction of the free-space Green's function ensures that the tangential $\mathbf{E}$ field is continuous, and imposed continuity of the magnetic field components results in a vector integrodifferential equation

$$\hat{\mathbf{z}} \times \mathbf{B}(z=0^-) = \hat{\mathbf{z}} \times \mathbf{B}(z=0^+)$$
$$= j2\omega\varepsilon \hat{\mathbf{z}} \times \sum_s \bar{\bar{\Gamma}}^0(\mathbf{r},\mathbf{r}') \cdot (\hat{\mathbf{z}} \times \mathbf{E}) ds' \tag{69}$$

where the free-space dyadic Green's function is given by

$$\bar{\bar{\Gamma}}^0(\mathbf{r},\mathbf{r}') = \left(\bar{\bar{\mathbf{U}}} + \frac{1}{k^2}\nabla\nabla\right)G(\mathbf{r},\mathbf{r}')$$

where

$$G(\mathbf{r},\mathbf{r}') = \frac{e^{-jk|\mathbf{r}-\mathbf{r}'|}}{4\pi|\mathbf{r}-\mathbf{r}'|}$$

$$\mathbf{r} - \mathbf{r}' = \sqrt{(x-x')^2 + (y-y')^2 + z^2}$$

is the scalar Green's function. The unit dyad $\bar{\bar{\mathbf{U}}}$ is defined in rectangular coordinates as

$$\bar{\bar{\mathbf{U}}} = \hat{\mathbf{x}}\hat{\mathbf{x}} + \hat{\mathbf{y}}\hat{\mathbf{y}} + \hat{\mathbf{z}}\hat{\mathbf{z}}$$

The magnetic field **B** for $z = 0$ can be obtained from the expansion (68) using waveguide modal admittances. By following Galerkin's procedure the above can then be reduced to a matrix equation and solved for aperture fields. The details of this procedure will not be carried further, but are described in a number of available references [37, 38].

Waveguide array solutions using single-mode approximations in each aperture have much more limited applicability than the simple theories for dipole arrays. Single-mode waveguide coupling solutions fail to predict blindness effects (see next section) but can be used successfully for small arrays and for relatively closely spaced elements at scan angles far from the grating lobe onset. Single-mode solutions have also been used by Golden [39], Steyskal [40], and others for elements conformal to curved surfaces.

Fig. 22 [41] shows several of the most significant effects due to mutual coupling. Fig. 22a shows the element pattern of the central element in an array of N parallel plane elements with the incident waveguide fields in the plane of scan. The presence of multiple ripples in the element pattern is due to reflections from the array edge, as indicated by the higher angular frequency for increased N. The infinite array results, also shown in the figure, demonstrate element pattern narrowing due to mutual coupling that forces the pattern to be zero at the horizon. Fig. 22b shows the associated reflection coefficient for the central element and again evidences the rippling effect for finite arrays and the unity reflection coefficient for the infinite array model at end-fire (about $\psi = 140°$) and throughout the slow-wave region $\psi > 140°$.

Array Blindness

In certain circumstances it is possible to have mutual coupling effects that actually create a null in the array element patterns so that the array cannot transmit energy in given directions.

Fig. 23 shows the basic phenomena as described by Farrell and Kuhn [42, 43] in the first published analytical work on the subject. The figure shows a measured deep null in an element pattern of a waveguide array and compares the data with results computed using a single-mode grating lobe series for an infinite array and a full modal array solution. The null is due to the cumulative effects of mutual coupling and can be related to surface wave type behavior at the array face. In many cases the existence of the null is understood as a cancellation process involving waveguide higher-order modes, and this is why the single-mode grating lobe solution bears little correlation to the data in Fig. 23. In the years since this initial discovery these blindnesses have been found in most waveguide array

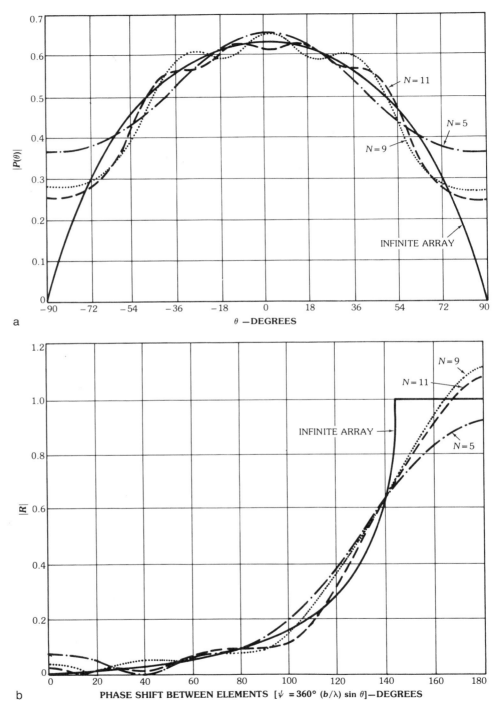

Fig. 22. Element patterns $P(\theta)$ and reflection coefficients R of center elements in unloaded waveguide array ($b/\lambda = a/\lambda = 0.4$). (*a*) Radiation patterns. (*b*) Reflection coefficients. (*After Wu [41], © 1970 IEEE*)

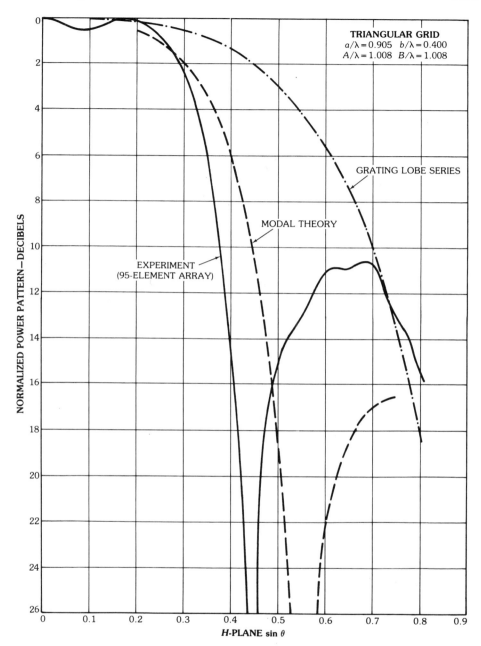

Fig. 23. Array element power pattern showing array blindness. (*After Farrell and Kuhn [42, 43], © 1968 IEEE*)

configurations and in some dipole and stripline arrays [44]. The problem can usually be reduced or eliminated by keeping the element lattice dimensions d_x and d_y small enough so that grating lobes are well beyond the maximum scan angle throughout the operating frequency range [45].

Reported blindnesses in dipole arrays seem to be related to the presence of dipole supports. Analytical studies of infinite dipole arrays [46] without supports do not exhibit array blindness. Mayer and Hessel [47] analyze a stripline dipole structure and show that for practical spacings the balanced stripline dipole feed structure supports a propagating TM mode in addition to two TEM modes. The TEM mode propagation constant is scan dependent, and for certain parameter selections it occurs before the onset of the grating lobe. It is conjectured that this mode might be the cause of blindness in dipole arrays.

Experience with array blindness has led to the practice of performing infinite-array studies, measuring the array element in simulator, or fabricating a small array for element pattern tests before embarking on the construction of a large array. Multimodal infinite array solutions are far simpler to obtain than multimode solutions for large arrays and such solutions have been obtained for many array types. Fig. 24 shows a few of the basic array configurations for which infinite array

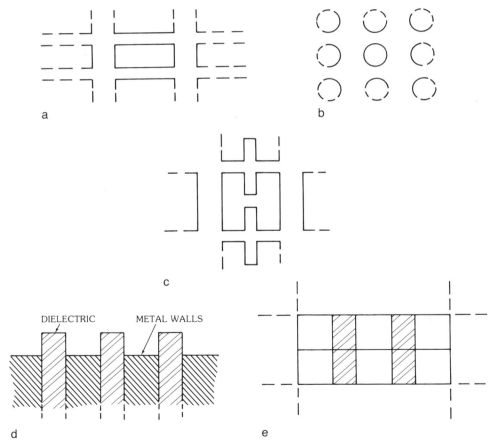

Fig. 24. Several configurations with existing infinite array solutions. (*a*) Rectangular waveguide array. (*b*) Circular waveguide array. (*c*) Ridge loaded waveguide array. (*d*) Protruding dielectric waveguide array. (*e*) Dual-frequency array.

solutions are published. Included in the figure are flush mounted arrays of rectangular [48] and circular [49] elements, ridge loaded elements [50, 51, 52], protruding dielectric (TEM solution) [53], and a dual-frequency dielectric loaded configuration [54]. Among other published solutions are numerous interlaced multiple-frequency configurations [55], examples of dielectric loading [56], iris loading [57] and fence [58], and corrugated plate [59] loadings for impedance match, as well as several very wide band configurations for waveguide [60] and stripline [61]. In addition there have been a number of infinite dipole array solutions published [62].

Conformal Arrays

The need for conformal or low profile arrays for aircraft and missile applications, and for ground-based arrays with 360° azimuth coverage or hemispherical coverage, has grown continually with requirements that emphasize maximum utilization of available space and minimum cost. The earliest and continuing stimulus for cylindrical and circular array development is the need for inexpensive systems with mechanical or electronic scanning with constant gain throughout the 360° coverage sector. There are also a number of spacecraft and aircraft applications requiring low profile or conformal arrays. Fig. 25 shows a possible configuration of an airborne array for satellite communication.

Array elements on curved bodies point in different directions, and so it is usually necessary to turn off those elements that radiate primarily away from the desired direction of radiation. For this reason also, one cannot factor an element pattern out of the total radiation pattern and therefore conformal array synthesis is very difficult. In addition, mutual coupling problems can be severe and difficult

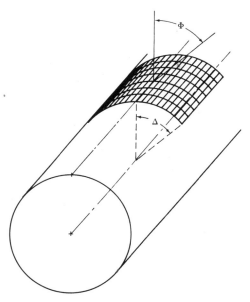

Fig. 25. Conformal array geometry for 2Δ arc array on cylinder.

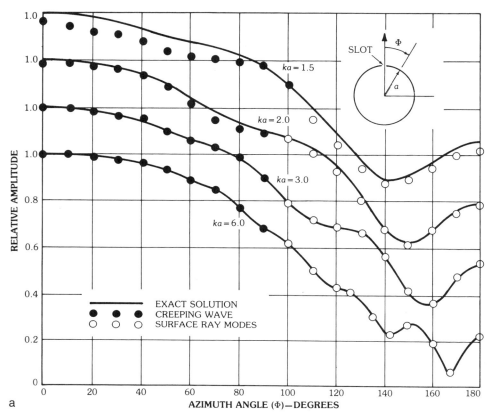

Fig. 26. Patterns of slots and arrays on cylinders. (*a*) Patterns of a thin axial slot in a perfectly conducting cylinder. (*After Pathak and Kouyoumjian [63], © 1974 IEEE*) (*b*) Element patterns for dipole arrays on a cylinder. (*After Herper et al. [68], © 1980 IEEE*)

to analyze because of the extreme asymmetry of structures like cones and because of multiple coupling paths between elements (for example, the clockwise and counterclockwise paths between two elements on a cylinder). Cross-polarization effects arise because of the different pointing directions for elements on curved surfaces causing the polarization vector projections to be nonaligned. There is also a need to use different collimating phase shifts in the azimuth plane of a cylindrical array scanned in elevation due to the fact that steering in azimuth and elevation planes is not separate. Another phenomenon related to mutual coupling is the evidence of ripples on the element patterns of cylindrical arrays. This phenomenon can be explained in terms of creeping-wave contributions.

The behavior of slot and dipole elements is altered by the presence of the curved surface. Pathak and Kouyoumjian [63] give a very convenient extension of the geometrical theory of diffraction (GTD) for apertures in curved surfaces. Fig. 26a shows the patterns of an axial slot element in perfectly conducting circular cylinders of various radii, as computed by Pathak and Kouyoumjian. The pattern compares the exact solution with that obtained using the appropriate GTD

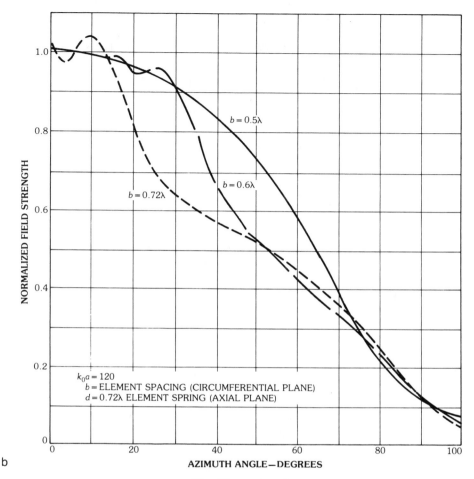

Fig. 26, continued.

expressions in several regions of space, and shows the GTD formalism to be highly accurate, even for relatively small cylinders. The figure also emphasizes the way in which the finite cylinder alters slot element patterns, for a similar slot in a flat ground plane would have a constant radiation pattern from $\phi = 0°$ to $180°$. A number of other references give the radiation patterns of slots and dipole elements on a variety of generalized surfaces [64, 65, 66].

Arrays of slots or dipoles on curved surfaces also behave differently from those on plane surfaces, and often have highly rippled element patterns [67] that make low side lobe synthesis impossible. Fig. 26b, however, shows that the rippled element pattern characteristic shown for a dipole array over a cylinder does not occur if the element spacing is restricted to about a half-wavelength [67]. The unwelcome rippled effect shown for larger spacings has been attributed [67] to interference between the grating lobe of the fast creeping wave and the direct ray.

At $\lambda/2$ element spacing this grating lobe does not radiate, and the resulting element patterns are free from ripples.

Cylindrical arrays can have low side lobe patterns, but it is important to maintain close element spacing and not to wrap the array too far around the cylinder. Fig. 27 shows the results attributable to Sureau and Hessel [67] that illustrate both of these effects. This figure shows that for the array with elements wrapped entirely around the cylinder, doubling the size of the excited sector of the array only increases the gain by 1 dB due to inefficient radiation of the edge elements. The narrowed array element patterns, also shown in the figure, emphasize the fact that edge elements are required to provide coverage at the outer limits of their active element patterns.

Array Errors and Phase Quantization

The ability of an array to create a desired antenna pattern in space is limited by diffraction effects resulting from finite antenna size, by element pattern ripple, and by random and correlated errors in the array illumination.

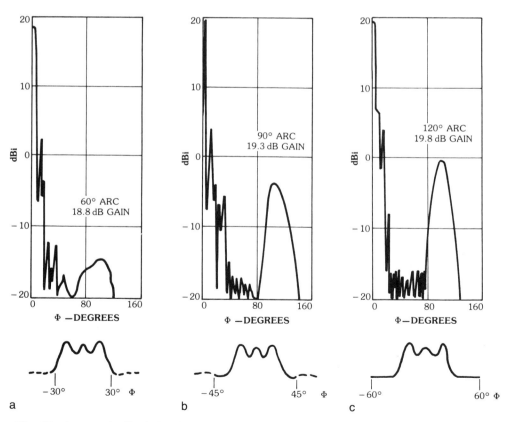

Fig. 27. Array gain (in dBi = dB relative to isotropic) and radiation patterns for 60°, 90°, and 120° arc arrays on a cylinder, where the lower curves show angular extent of element patterns used for given arc. (*After Sureau and Hessel [67], © 1972 Artech House*)

Periodic Arrays

In the case of an array with random phase and amplitude errors, and including randomly failed elements, the average side lobe level far from the beam peak is given by [69, 70]

$$\bar{\sigma}^2 = \frac{\bar{\varepsilon}^2}{P_e N \eta} = \frac{[(1 - P_e) + \bar{\Delta}^2 + P_e \bar{\delta}^2]}{P_e N \eta} \qquad (70)$$

where

$\bar{\Delta}^2$ = the amplitude error variance normalized to unity
$\bar{\delta}^2$ = the phase error variance
$\bar{\varepsilon}^2$ = the error variance
P_e = the probability of survival for any element in the array
η = array efficiency
N = total number of elements

This equation gives the normalized side lobe level relative to the average array gain. The failed elements in the array are assumed to be randomly located, and the average value of the phase and amplitude errors is assumed to be zero. The side lobe level above should be considered the average of a number of antenna patterns, not the average level of any one antenna.

If the broadside, no-error gain of an array with elements $\lambda/2$ apart is $\pi P_e N \eta$, the side lobe level is given from (70) as

$$\bar{\sigma}^2 = \pi \bar{\varepsilon}^2 / G \qquad (71)$$

in terms of the gain G and the error variance.

Peak side lobe levels are also given in the literature. A convenient result is obtained when the errors are sufficiently large compared to side lobes or null depths that structured minor-lobe radiation is negligible and the statistics of the field intensity pattern are described by a Rayleigh density function. In this case the probability $P(v > v_0)$ that a particular side lobe level v_0^2 is exceeded at any point is [70]

$$P(v > v_0) = e^{-v_0^2/\bar{\sigma}^2} \qquad (72)$$

where $\bar{\sigma}^2$ is the average side lobe level of (70).

Starting with the expression above, valid at a particular point, Allen [69] derives the following rule of thumb for the error $\bar{\varepsilon}^2$ allowable for an array with gain G, in terms of the far side lobe level $1/R$, assuming 99-percent probability that all side lobes are below the given level

$$\bar{\varepsilon}^2 \leq \frac{1}{10\pi} \frac{G}{R} \qquad (73)$$

which results in an allowable phase error of about 10° when the side lobe level is numerically equal to the gain. This important relationship explains why it is not difficult to design arrays with side lobes at the isotropic level ($G = R$), but to maintain side lobes of 20 dB below the isotropic level would require 1° phase error, an extremely difficult goal and one barely within the present state of the art.

In this expression, and in all equations given in this section, when the parameter N is the number of elements in the array, the side lobe levels are those distributed throughout all real space; but when the errors are correlated in one plane, as they would be for power divider or phase shifter errors in the plane of scan for an array of columns, then N is the number of columns, and the side lobe level given by these equations is in the principal scan plane.

For the case of an array of columns of $\lambda/2$-spaced elements the gain of a column is equal to $\sqrt{\pi}\eta NP_e$, and (70) becomes

$$\bar{\sigma}^2 = \frac{\sqrt{\pi}\,\bar{\varepsilon}^2}{G_1} \tag{74}$$

where G_1 is the column gain, $\bar{\varepsilon}^2$ is the error variance between columns, and $\bar{\sigma}^2$ the average side lobe level in the plane perpendicular to the columns. For a square array this reduces to

$$\bar{\sigma}^2 = \frac{\sqrt{\pi}\,\bar{\varepsilon}^2}{\sqrt{G}} \tag{75}$$

where now G is the total array gain.

Fig. 28 shows the average side lobe level, (71), for a square array with random phase shift errors, and a square array of columns (75) with no errors in the plane

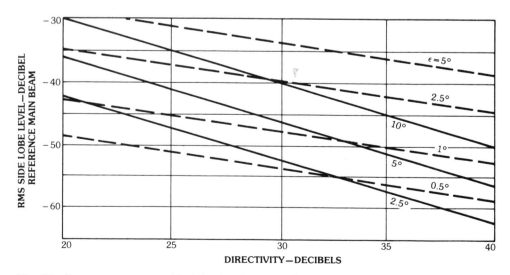

Fig. 28. Root-mean-square side lobe level versus directivity for square array with phase errors at elements (solid lines) or at columns (dashed lines, in plane of arrayed columns).

of the columns, and the phase error between columns randomly distributed in the scan plane. This figure illustrates that extreme precision is required for arrays organized into columns when low principal-plane side lobes are required.

The reduction in directivity due to these errors is given by

$$\frac{D}{D_0} = \frac{P_e}{1 + \bar{\Delta}^2 + \bar{\delta}^2} \qquad (76)$$

In the case of digitally controlled phase shifts, a p bit phase shifter has 2^p phase states separated by phase steps of $2\pi/(2^p)$. Miller [71] has analyzed the resulting peak and rms side lobe levels for this staircase approximation to the desired linear phase progression and has shown that the loss in array gain due to the triangular error distribution is

$$\Delta G = \frac{1}{3} \frac{\pi^2}{2^{2p}} \qquad (77)$$

which is on the order of 0.23 for a 3-bit phase shifter and 0.06 for a 4-bit phase shifter. More significant are the average side lobe levels which, based on an average array loss of 2 dB to account for illumination taper and scan degradation, are

$$\text{rms side lobes} = \frac{5}{N \times 2^{2p}} \qquad (78)$$

where N is the number of elements in the array. For a one-dimensionally scanned array, N is the number of phase controls, and the rms side lobe level above is measured in the plane of scan. The net result, as before, is to require extreme precision for unidimensional scanned arrays. Fig. 29 shows the side lobe level for various phase shifter bits p and N up to 10 000 elements. For -50-dB rms side lobes an array of 1000 elements requires 5-bit phase shifters, but an array of 10 000 elements can maintain 50-dB side lobes with only 3 phase bits.

Of greater significance to antenna design is that the phase errors have a periodic variation across the array and tend to collimate as individual side lobes, called *phase-quantization side lobes*, that are much larger than the rms levels. A detailed discussion of this phenomenon is given by Miller along with simple formulas for evaluating the resulting lobes. In the case of a perfectly triangular quantization error the quantization lobe level is $1/2^p$, which gives -30 dB for 5-bit phase shifters. Cheston and Frank [72] show that for discrete phase shifters the error is not triangular and that the maximum quantization lobe can be substantially larger.

One solution to the peak quantization lobe problem is to decorrelate the phase shifter errors. Decorrelation occurs naturally in space-fed arrays, where the phase shifters collimate the beam as well as steer it. In such arrays the phase error is distorted from the triangular shape and the quantization lobe is substantially reduced. Alternatively, in an array with in-phase power division one can introduce

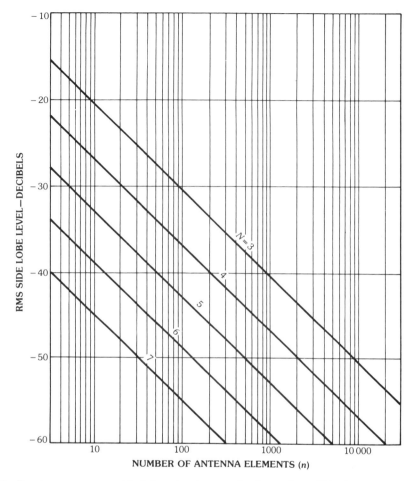

Fig. 29. Root-mean-square side lobes due to quantization, where N is the number of phase shifter bits. (*After Miller [71]*)

a phase error into each path and then program the phase shifter to remove the error in addition to steering the beam. Optimizing this error can reduce the peak side lobes very close to the rms side lobe level, but this is a consideration that must be carefully accounted for in the array design.

An entirely different solution to the quantization lobe problem is often achieved at the system level by recycling all the phase shifters between consecutive radar pulses or between transmit and receive. This process, called *beam dithering* [73], consists of adding a fixed phase shift to the phase command and recomputing phase shifts. The net result is to change all the phase states so that the quantization is made differently for each pulse (or between transmit and receive). If this procedure is compatible with other radar processing, one can use simple row-column steering but introduce randomness into the quantization steps to reduce the peak quantization lobes.

Array Elements

Array elements are usually some form of dipole or slot excited by a waveguide or other transmission line. Waveguide arrays, though heavy, tend to have low loss, good bandwidth, and relatively graceful scan degradation. They also have been the subject of numerous design studies, and so their behavior is well documented and predictable. Early examples of specific waveguide element designs are the studies of Wheeler [74, 75] in which matching networks were derived using waveguide transmission circuits like that shown in Fig. 30 consisting of dielectric slabs mounted in and above the waveguide. McGill and Wheeler [76] introduced the use of a dielectric sheet, often called a WAIM (wide-angle impedance-matching) sheet, to produce a susceptance variation with scan angle that partially cancels the scan mismatch of the array face. A significant development in impedance matching of waveguide arrays is the synthesis of double-tuned response characteristics achieved using dielectric loading and a cutoff waveguide section [77]. Fig. 31 shows a loaded rectangular waveguide (phase shifter), a transformer to circular guide, two dielectric disks, and an unloaded section of guide that is below cutoff at the operating frequencies.

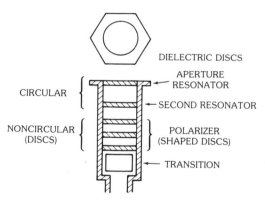

Fig. 30. Circular element for triangular grid array. (*After Wheeler [74, 75], © 1968 IEEE*)

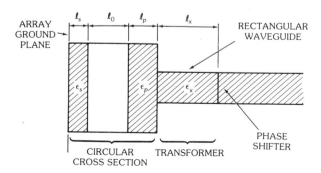

Fig. 31. Doubly tuned waveguide array element. (*After Lewis, Kaplan, and Hanfling [77], © 1974 IEEE*)

Dipole arrays have also received substantial attention and have generally graceful scan properties when properly designed. Fig. 32 shows several common varieties of stripline printed dipoles. One design, attributable to Wilkinson [78], uses metallization on two sides of a microstrip line to produce a complete dipole fed by a two-wire line in the plane of the dipoles. This dipole and printed circuit distribution network is fabricated by two photographic exposures using a two-sided printed-circuit board, and so is an example of low-cost technology. The array was mounted a quarter wavelength above a ground plane and uniformally illuminated by a reactive power divider to form a pencil beam. Another convenient circuit for dipole design, shown in Fig. 32, is described in a report by Hanley and Perini [79] and is a printed-stripline folded dipole with a Schiffman balun. One major advantage of this element is that it is printed in a single process, all on one side of a circuit board and so is relatively inexpensive to produce.

Fig. 33 shows one means of exciting flush-mounted stripline slot antennas. Most often these elements are isolated from the rest of the stripline medium by the

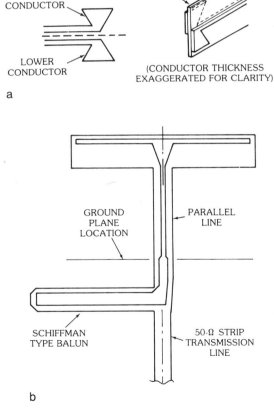

Fig. 32. Printed-circuit dipole configurations. (*a*) Conductors and dielectric. (*b*) Printed-stripline folded dipole and balun.

Periodic Arrays

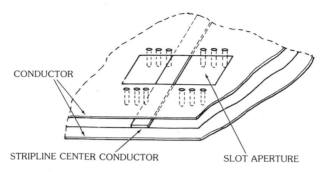

Fig. 33. Stripline slot radiator.

use of plated through holes or rivets that form a cavity as shown in the figure, which also serves to suppress higher-order modes.

Several very broadband elements have been described in the literature. Among these the broadest-band element is the flared-notch antenna (Fig. 34) studied by Lewis and others [61], which exhibits up to an octave band when used in an array, but needs careful design for any given array configuration because of the possibility of array blindness effects for critical frequencies and scan angles.

The microstrip patch element and its variations are inexpensive and lightweight, and have found increasing use in a variety of array applications. The basic path element (Fig. 35) is narrow-band, with percentage bandwidth given approximately by

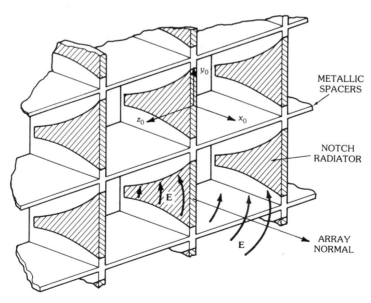

Fig. 34. Flared-notch array elements. (*After Lewis, Kaplan, and Hanfling [77]*, © 1974 *IEEE*)

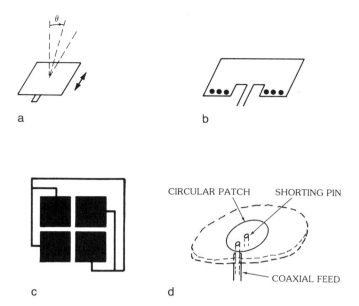

Fig. 35. Useful microstrip radiator types. (*a*) Patch antenna elements. (*b*) Shorted patch. (*c*) Crossed slot of four shorted patches. (*d*) Circular disk.

$$100 \frac{\Delta f}{f_0} = 5tf_0 \tag{79}$$

for a thickness t in centimeters for the air-loaded patch. A number of very creative microstrip elements have been developed. Fig. 35 shows a square patch design for radiating circular polarization, a shorted rectangular patch for producing a much wider element pattern in the scan plane, and a combination of shorted patches that radiates circular polarization and is the microstrip equivalent of crossed slot radiations [80]. Many other microstrip radiators have found practical application. Of these, the most significant is the circular disk radiator of Howell [81] (Fig. 35d), which can be excited by a microstrip line but which also is very suitably excited from below the ground plane, as shown in the figure. Still other transmission line media offer advantages for array use in a variety of applications.

Passive Components for Arrays: Polarizers and Power Dividers

The trends toward greater flexibility, more accurate pattern control, and lightweight, compact array structures are also having an impact on the type of components used in arrays. This section lists some of the passive components used to control radiated polarization and to provide power distribution networks for precise pattern synthesis.

Waveguide polarizers using a probe or obstacle to excite both polarizations and some variation of a quarter-wave plate to produce the requisite ±90° delay for circular polarization have long been used as phased array components.

Waveguide polarizers in current use include the tapered septum polarizer [82] for converting linear to circular polarization, which operates over about a 2- to 5-percent bandwidth, and the stepped-septum polarizer of Chen and Tsandoulas, a three-port device that allows polarization agility over at least 20-percent bandwidth, with isolation greater than 26 dB [83].

Often it is less expensive to insert a polarizer in front of the whole array. The earliest polarizer of that type is the use of a grid of quarter-wave plates, but a much more popular recent solution that is compatible with wide-angle electronic scanning is the use of meander line polarizers following the work of Young's group [84].

Power divider networks for array feeds need to be extremely precise to synthesize low side lobe radiation patterns. Other requirements often impose extreme high-power specifications and still others demand very lightweight or compact construction practices. Waveguide and coaxial-line corporate feed networks are most often used for high-power arrays but the increasing ease and quality of stripline construction has made that the medium of choice for many new array developments. Often it is convenient to produce hybrid combinations of waveguide, coaxial line, and stripline to take advantage of inexpensive stripline network techniques for lower-power sections of the array while using waveguide or coaxial line at the high-power regions of the feed network.

Precise feed synthesis requires both equal and unequal power dividers. Among the various developments in stripline components the most commonly used are reactive tee power dividers, branch line couplers, and parallel coupled line and in-line power dividers. Although they are most simple and inexpensive to construct, reactive tee power divider networks have no isolated port and hence offer serious mismatch and isolation problems when used to feed mismatched elements. Reactive corporate-feed networks are therefore useful mainly for fixed beam arrays or for power division in the unscanned plane of arrays with one plane of scan. Single-section [85] branch line couplers occupy an area approximately $\lambda/4$ square and are most useful for coupling ranges from 3 to 9 dB. These couplers are easily fabricated using a conventional stripline by machining or etching the center conductor. Parallel coupled stripline power dividers for loose coupling (greater than 10 dB) can also be designed from conventional stripline using side coupled parallel lines, but tighter coupling requires the use of three-layer stripline for broadside coupled lines (3 to 6 dB coupling) or variable overlap couplers for intermediate values. Single-section parallel coupled power dividers are $\lambda/4$ long but occupy less area than branch line hybrids. Another likely choice for array feed networks is the Wilkinson [86] in-line power divider or its impedance-compensated derivative split-tee power dividers [87].

Single-section Wilkinson power dividers are $\lambda/4$ long for equal power division and $\lambda/2$ long for unequal power division. Split-tee power dividers have an extra stage of impedance matching and so are longer by approximately $\lambda/4$, although they have the advantage of wider bandwidth.

In either case the in-line power dividers have excellent broadband characteristics in comparison with branch and coupled line hybrids because the coupling ratio is determined by relative impedance ratios, not line length. Similarly, in-line hybrids are in-phase power dividers and so there is little phase error introduced with frequency change. The output ports of branch and coupled line hybrids have substantially different phases ($\pi/2$ for equal power division) and, although this can be compensated at center frequency, networks of these hybrids tend to be very narrow band relative to in-line hybrids. Typical bandwidths for individual in-line hybrids can reach an octave. The selection of power division networks is critical to array design and the choice can vary substantially with the application.

Array Phase Control

A discussion of time-delay devices for arrays is omitted because the primary components used to date are switched transmission lines and are governed by the same critical components as phase shifters.

Phasing networks are most often implemented at the rf operating frequency because this is usually the most efficient process. Notable exceptions use phase shifters at intermediate frequencies and up-convert to rf with amplifiers to improve efficiency. In addition, many ingenious intermediate-frequency phase scanning systems have used harmonically derived phased shifts or frequency displaced signals across an array to produce time-varying beam positions. Systems of this type are described in the literature and their operation is beyond the scope of this chapter. In general, diode phase shifters dominate the frequency range below 2 GHz, and ferrites are usually selected above 5 GHz for high-power applications. Diode phase shifters are compact and very lightweight compared with ferrite devices and so are gaining popularity in lightweight array configurations through 40 GHz, often in combination with low-cost monolithic microstrip antenna circuits. The rapid development of solid-state amplifier modules for arrays also favors the use of diode phase shifters at all frequencies.

At present the most popular types of diode phase shifter designs for arrays are the hybrid coupled, switched-line, and loaded-line phase shifters using pin diodes. These three fundamental networks are shown in Fig. 36. Hybrid and loaded-line (transmission) phase shifters require two diodes per bit, and switched-line phase shifters require four per bit. Switched-line hybrids also have greater insertion loss and an undesirable phase dependence with frequency that usually makes them unsuitable for low side lobe array control. They do possess distinct advantages in weight and compactness, however, and have been used successfully in monolithic microstrip antennas for many years. Hybrid and transmission phase shifters have lower loss and better bandwidth performance. An *S*-band stripline hybrid phase shifter is reported by White [88] to have an average phase loss of only 3° over 20 percent bandwidth. This device had about 0.8 dB average loss from 3.0 to 3.5 GHz and was tested to high-power burnout at 4 kW peak with 0.1-ms pulses and 0.05 duty cycle. Insertion loss for *X*-band and K_u-band phase shifters was about 2 and 3 dB.

Broadband low–side-lobe array designs are possible using Schiffman phase shifters because this phase shifter produces a nearly constant phase shift over extremely wide frequency ranges. White gives data for a 90° bit over a frequency ratio of 2.27:1.

A precise low–side-lobe array developed by Tsandoulas [89] used six-bit diode phase shifters with phase tolerance limits of less than 0.0° rms for the 90° and 180° bits, 0.4° for 45°, 22.5°, and 11.25° bits, and 0.2° for the 5.625°. These remarkable results were achieved in a practical testbed array described in the reference.

Ferrite phase shifters have been built to operate up to 60 GHz and possess

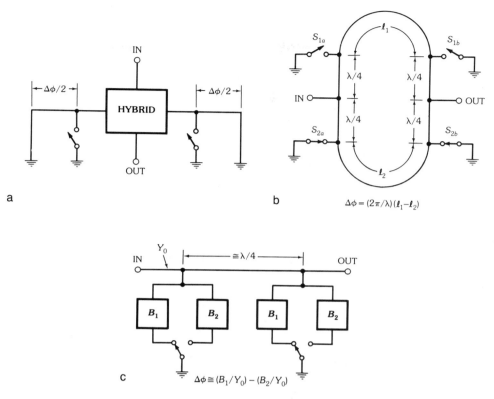

Fig. 36. Diode phase shifter circuits. (*a*) Hybrid coupled reflection phase shifter. (*b*) Switched-line phase shifter. (*c*) Loaded-line phase shifter.

excellent characteristics for many phased array applications. Several recent survey articles and an annotated bibliography [90, 91, 92] summarize progress in this field and list numerous references to devices and to the fundamental theory of ferrite phasor operation. Nonreciprocal ferrite phase shifters include early twin-slab designs (Fig. 37) that require a transverse-switched external magnetic field and the well-known toroid designs that use a longitudinal wire to drive the ferrite magnetization to saturation as in a latching phase shifter, or to various points on the magnetization curve with flux drive circuitry. Typical digital latching phase shifters

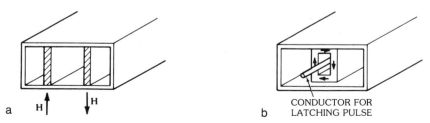

Fig. 37. Dual-slab and toroid ferrite phase shifters. (*a*) Dual-slab. (*b*) Toroid.

can have bandwidths in excess of 10 percent and insertion losses between 0.5 and 1 dB. Power levels supported by these devices can vary from 1 kW to as much as 150 kW peak with average power levels to 400 W. Latching phase shifters can be switched in about 1 µs and have become standard throughout the industry. Flux drive circuits with toroidal phase shifters are analog devices and not restricted to specific phase bits. Their major disadvantage is the need to reset them between transmit and receive functions for radar applications.

Among several varieties of reciprocal ferrite phase shifters the dual-mode phase shifters have replaced Reggia-Spencer phase shifters, which have been found to have low values of phase shift per wavelength. Dual-mode phasors are Faraday rotation devices in which a linearly polarized incident waveguide mode is converted to circularly polarized energy by a nonreciprocal ferrite quarter-wave plate, phase-shifted by Faraday rotation, and then converted back to linear polarization. A signal from the opposite direction is converted to circular polarization of the opposite sense, but since the directions of propagation and polarization are both opposite it incurs the same, reciprocal phase shift. Dual-mode phase shifters are very competitive with toroid phase shifters and have average power levels up to 1.5 kW at S-band and peak powers to 150 kW. Insertion loss can be 0.6 dB through X-band. Switching speed can be on the order of tens of microseconds for a latched design, depending on the application. The reference by Ince and Temme [90] compares specific phasor examples for S-band through K_a-band.

One other important reciprocal phase shifter is the analog rotary-field phase shifter of Boyd [93]. This phase shifter is based on the principle of the commercial rotary-vane phase shifters after Fox [94] and uses a ferrite rod of circular cross section fitted with a slotted stator in which are wound two sets of coils each generating a four-pole field. In comparison with the well-known rotating half-wave plate of the Fox phase shifter, the stator windings produce a rotating four-pole field distribution with the orientation of the principal axes proportional to the coil driving current values. The dc distribution in the ferrite serves to rotate a virtual half-wave plate that is converted to a phase shift just as is the mechanical half-wave plate rotation of the Fox phase shifter. This circuit has the disadvantages of requiring substantial drive power and having relatively long switching times (200 to 500 µs). Its advantages for many applications far outweigh these disadvantages because it has nearly dispersionless phase shift that can be maintained within a degree or two over substantial bandwidths, has insertion loss well under 1 dB, and can handle very high peak and average power levels. An S-band model operates at 90 kW peak, 3 kW average power, 0.5 dB insertion loss, and phase tracking within $\pm 1.5°$ over about 9 percent bandwidth.

5. References

[1] R. S. Elliott, "The theory of antenna arrays," in Chapter I, in *Microwave Scanning Antennas*, vol. II, ed. by R. C. Hansen, New York: Academic Press, 1966, pp. 29–31.

[2] C. T. Tai, "The optimum directivity of uniformly spaced broadside arrays of dipoles," *IEEE Trans. Antennas Propag.*, vol. AP-12, pp. 447–454, July 1964.

[3] C. H. Walter, *Traveling-Wave Antennas*, New York: McGraw-Hill Book Co., 1965, pp. 121–122 and 322–325.

[4] S. A. Schelkunoff, "A mathematical theory of linear arrays," *Bell Syst. Tech. J.*,

vol. 22, pp. 80–107, 1943.

[5] R. S. Elliott, "Design of line source antennas for sum patterns with side lobes of individually arbitrary heights," *IEEE Trans. Antennas Propag.*, vol. AP-24, pp. 76–83, June 1976.

[6] R. S. Elliott, "Design of line source antennas for difference patterns with side lobes of individually arbitrary heights," *IEEE Trans. Antennas Propag.*, vol. AP-24, pp. 310–316, June 1976.

[7] W. L. Stutzman and G. A. Thiele, *Antenna Theory and Design*, New York: John Wiley and Sons, 1966.

[8] P. M. Woodward, "A method of calculating the field over a plane aperture required to produce a given polar diagram," *J. IEE* (London), pt. IIIA, vol. 93, pp. 1554–1558, 1947.

[9] P. M. Woodward and J. P. Lawson, "The theoretical precision with which an arbitrary radiation pattern may be obtained from a source of finite size," *J. AIEE*, vol. 95, P1, pp. 362–370, September 1948.

[10] C. L. Dolph, "A current distribution for broadside arrays which optimizes the relationship between beamwidth and side lobe level," *Proc. IRE*, vol. 34, pp. 335–345, June 1946.

[11] H. J. Riblett, "Discussion of Dolph's paper," *Proc. IRE*, vol. 35, pp. 489–492, May 1947.

[12] D. Barbiere, "A method for calculating the current distribution of Tschebyscheff arrays," *Proc. IRE*, vol. 40, pp. 78–82, January 1952.

[13] R. J. Stegen, "Excitation coefficients and beamwidths of Tschebyscheff arrays," *Proc. IRE*, vol. 41, pp. 1671–1674, November 1953.

[14] L. B. Brown and G. A. Scharp, "Tschebyscheff antenna distribution, beamwidth, and gain tables," *NAVORD Rep. 4629 (NOLC Rep. 383)*, Naval Ordnance Lab., Corona, California, February 1958.

[15] C. J. Drane, Jr., "Useful approximations for the directivity and beamwidth of large scanning Dolph-Chebyshev arrays," *Proc. IEEE*, vol. 56, pp. 1779–1787, November 1968.

[16] T. T. Taylor, "Design of line source antennas for narrow beamwidth and low side lobes," *IEEE Trans. Antennas Propag.*, vol. AP-3, pp. 16–28, January 1955.

[17] R. C. Hansen, ed., *Microwave Scanning Antennas, Volume 1*, app. 1, New York: Academic Press, 1966.

[18] E. T. Bayliss, "Design of monopulse antenna difference patterns with low side lobes," *Bell Syst. Tech. J.*, vol. 47, pp. 623–640, 1968.

[19] R. S. Elliott, "Design of line source antennas for difference patterns with side lobes of individually arbitrary heights," *IEEE Trans. Antennas Propag.*, vol. AP-24, no. 3, pp. 310–316, May 1976.

[20] T. T. Taylor, "Design of circular apertures for narrow beamwidth and low side lobes," *IRE Trans. Antennas Propag.*, vol. AP-8, pp. 17–22, January 1960.

[21] F. I. Tseng and D. K. Cheng, "Optimum scannable planar arrays with an invariant side lobe level," *Proc. IEEE*, vol. 46, pp. 1771–1778, 1968.

[22] R. Tang, "Survey of time-delay beam-steering techniques," *Phased Array Antennas: Proceedings of the 1970 Phased Array Antenna Symposium*, Dedham: Artech House, pp. 254–260, 1972.

[23] V. Borgiotti, "An antenna for limited scan in one plane: design criteria and numerical simulation," *IEEE Trans. Antennas Propag.*, vol. AP-25, pp. 232–243, March 1977.

[24] R. L. Fante, "Systems study of overlapped subarrayed scanning antennas," *IEEE Trans. Antennas Propag.*, vol. AP-28, pp. 668–679, September 1980.

[25] P. D. Hrycak, "The theoretical and experimental investigation of a constrained-feed totally overlapped subarray antenna system," *IEEE 1982 AP-S Symp. Dig.*, pp. 695–698, May 24–28, 1982.

[26] R. J. Mailloux, "An overlapped subarray for limited scan application," *IEEE Trans. Antennas Propag.*, vol. AP-22, pp. 487–489, May 1974.

[27] E. C. Dufort, "Constrained feeds for limited scan arrays," *IEEE Trans. Antennas*

Propag., vol. AP-26, pp. 407–413, May 1978.

[28] W. Rotman and P. Franchi, "Cylindrical microwave lens antenna for wideband scanning application," *IEEE Intl. Symp. Dig.*, vol. AP-S, pp. 564–567, June 1980.

[29] J. H. Richmond, "Digital computer solutions of the rigorous equations for scattering problems," *Proc. IEEE*, vol. 53, pp. 796–804, August 1965.

[30] R. F. Harrington, *Field Computation by Moment Methods*, New York: Macmillan Co., 1968.

[31] R. W. P. King, *The Theory of Linear Antennas*, Cambridge: Harvard University Press, 1956, p. 79.

[32] G. H. Brown and R. King, "High-frequency models in antenna investigations," *Proc. IRE*, vol. 22, pp. 457–480, April 1934.

[33] P. S. Carter, "Circuit relations in radiating systems and applications to antenna problems," *Proc. IRE*, vol. 20, pp. 1004, 1041, June 1932.

[34] C. T. Tai, "Coupled antennas," *Proc. IRE*, vol. 36, pp. 487–500, April 1948.

[35] R. W. P. King, R. B. Mack, and S. S. Sandler, *Arrays of Cylindrical Dipoles*, London: Cambridge University Press, 1968, pp. 282–283.

[36] V. W. H. Chang and R. W. P. King, "Theoretical study of dipole arrays of *N* parallel elements," *Radio Sci.*, vol. 3 (New Series), no. 5, September–October 1968.

[37] N. Amitay, V. Galindo, and C. P. Wu, *Theory and Analysis of Phased Array Antennas*, New York: Wiley Interscience, 1972.

[38] R. J. Mailloux, "First-order solutions for mutual coupling between waveguides which propagate two orthogonal modes," *IEEE Trans. Antennas Propag.*, vol. AP-17, no. 6, pp. 740–746, November 1969.

[39] K. E. Golden et al., "Approximation techniques for the mutual admittance of slot antennas in metallic cones," *IEEE Trans. Antennas Propag.*, vol. AP-22, pp. 44, 48, 1979.

[40] H. Steyskal, "Analysis of circular waveguide arrays on cylinders," *IEEE Trans. Antennas Propag.*, vol. AP-25, pp. 610–616, 1977.

[41] C. P. Wu, "Analysis of finite parallel-plate waveguide arrays," *IEEE Trans. Antennas Propag.*, vol. AP-18, no. 3, pp. 328–334, May 1970.

[42] G. F. Farrell, Jr., and D. H. Kuhn, "Mutual coupling effects of triangular grid arrays by modal analysis," *IEEE Trans. Antennas Propag.*, vol. AP-14, pp. 652–654, September 1966.

[43] G. F. Farrell, Jr., and D. H. Kuhn, "Mutual coupling in infinite planar arrays of rectangular waveguide horns," *IEEE Trans. Antennas Propag.*, vol. AP-16, pp. 405–414, July 1968.

[44] J. C. Herper, C. J. Esposito, C. Rottenberg, and A. Hessel, "Surface resonances in a radome covered dipole array," *1977 IEEE AP-S Intl. Symp. Dig.*, pp. 198–201.

[45] G. H. Knittel, A. Hessel, and A. A. Oliner, "Element pattern nulls in phased arrays and their relation to guided waves," *Proc. IEEE*, vol. 56, no. 11, pp. 1822–1836, November 1968.

[46] V. W. H. Chang, "Infinite phased dipole array," *Proc. IEEE*, vol. 56, no. 11, pp. 1892–1900, November 1968.

[47] E. Mayer and A. Hessel, "Feed region modes in dipole phased arrays," *IEEE Trans. Antennas Propag.*, to be published.

[48] G. N. Tsandoulas and G. H. Knittel, "The analysis and design of dual-polarization square waveguide phased arrays," *IEEE Trans. Antennas Propag.*, vol. AP-21, pp. 796–808, November 1973.

[49] N. Amitay and V. Galindo, "The analysis of circular waveguide phased arrays," *Bell Syst. Tech. J.*, vol. 47, pp. 1903–1931, November 1968.

[50] M. H. Chen and G. N. Tsandoulas, "Bandwidth properties of quadruple ridge circular and square waveguide radiators," *IEEE AP-S Intl. Symp. Rec.*, pp. 391–394, June 1973.

[51] J. P. Montgomery, "Ridged waveguide phased array elements," *IEEE Trans. Antennas Propag.*, vol. AP-24, no. 1, pp. 46–53, January 1976.

[52] S. S. Wang and A. Hessell, "Aperture performance of a double-ridge rectangular

waveguide in a phased array," *IEEE Trans. Antennas Propag.*, vol. AP-26, pp. 204–214, March 1978.

[53] L. R. Lewis, A. Hessell, and G. H. Knittel, "Performance of a protruding-dielectric waveguide element in a phased array," *IEEE Trans. Antennas Propag.*, vol. AP-20, pp. 712–722, November 1972.

[54] R. J. Mailloux and H. Steyskal, "Analysis of a dual frequency array technique," *IEEE Trans. Antennas Propag.*, vol. AP-27, no. 2, pp. 130–134, March 1979.

[55] J. K. Hsiao, "Computer aided impedance matching of an interleaved waveguide phased array," *IEEE Trans. Antennas Propag.*, vol. AP-20, pp. 505–506, July 1972.

[56] V. Galindo and C. P. Wu, "Dielectric loaded and covered rectangular waveguide phased arrays," *Bell Syst. Tech. J.*, vol. 47, pp. 93–116, January 1978.

[57] S. W. Lee and W. R. Jones, "On the suppression of radiation nulls and broadband impedance matching of rectangular waveguide phased arrays," *IEEE Trans. Antennas Propag.*, vol. AP-19, pp. 41–51, May 1971.

[58] R. J. Mailloux, "Surface waves and anomalous wave radiation null phased arrays of TEM waveguides with fences," *IEEE Trans. Antennas Propag.*, vol. AP-20, pp. 160–166, January 1972.

[59] E. C. Dufort, "Design of corrugated plates for phased array matching," *IEEE Trans. Antennas Propag.*, vol. AP-16, pp. 37–46, January 1968.

[60] C. C. Chen, "Octave band waveguide radiators for wave-angle scan phased arrays," *IEEE AP-S Intl. Symp. Rec.*, pp. 376–377, June 1972.

[61] L. R. Lewis, M. Fassett, and J. Hunt, "A broadband stripline array element," *IEEE AP-S Intl. Symp. Dig.*, pp. 335–337, June 1974.

[62] W. H. Chang, "Infinite phased dipole array," *Proc. IEEE*, vol. 56, pp. 1892–1900, 1968.

[63] B. H. Pathak and R. G. Kouyoumjian, "An analysis of the radiation from apertures in curved surfaces by the geometrical theory of diffraction," *Proc. IEEE*, vol. 62, no. 11, pp. 1433–1447, November 1974.

[64] W. D. Burnside, R. J. Marhefka, and C. L. Yu, "Roll-plane analysis of on-aircraft antennas," *IEEE Trans. Antennas Propag.*, vol. AP-21, no. 6, pp. 780–786, November 1973.

[65] W. D. Burnside, M. C. Gilreath, R. J. Marhefka, and C. L. Yu, "A study of KC-135 aircraft antenna patterns," *IEEE Trans. Antennas Propag.*, vol. AP-23, no. 3, pp. 309–316, May 1975.

[66] B. H. Pathak, N. Wang, W. D. Burnside, and R. G. Kouyoumjian, "A uniform GTD solution for the radiation from sources on a convex surface," *IEEE Trans. Antennas Propag.*, vol. AP-29, pp. 609–622, July 1981.

[67] J. C. Sureau and A. S. Hessel, "Realized gain function for a cylindrical array of open-ended waveguides," *1970 Proc. Phased Array Antennas*, ed. by A. A. Oliner and G. H. Knittel, Dedham: Artech House, 1972, p. 283.

[68] J. C. Herper, C. Mandarino, R. Hessel, and B. Tomasic, "Performance of a dipole element in a cylindrical array—a modal approach," *IEEE AP-S Intl. Symp.*, pp. 162, 165, 1980.

[69] J. L. Allen, "The theory of array antennas," *Tech Rep. 323*, MIT Lincoln Labs, July 1963.

[70] R. E. Collin and F. J. Zucker, eds., *Antenna Theory*, New York: McGraw-Hill Book Co., 1969.

[71] C. J. Miller, "Minimizing the effects of phase quantization errors in an electronically scanned array," *Proc. 1964 Symp. on Electron. Scanned Array Tech. Appl.*, RADC TR-64-225, vol. 1, pp. 17–38, RADC, GAFB, NY.

[72] T. C. Cheston and J. Frank, *Radar Handbook*, Chapter 11, ed. by M. I. Skolnik, New York: McGraw-Hill Book Co., 1970, pp. 11–42.

[73] E. Brookner, ed., *Radar Technology*, Dedham: Artech House, 1977.

[74] H. A. Wheeler, "A systematic approach to the design of a radiator element for a phased array antenna," *Proc. IEEE*, vol. 56, pp. 1940–1951, November 1968.

[75] H. A. Wheeler, "A survey of the simulator techniques for designing a radiating

[76] E. G. McGill and H. A. Wheeler, "Wide-angle impedance matching of a planar array antenna by a dielectric sheet," *IEEE Trans. Antennas Propag.*, vol. AP-14, no. 1, pp. 49–53, January 1966.

[77] L. R. Lewis, L. J. Kaplan, and J. L. D. Hanfling, "Synthesis of a waveguide phased array element," *IEEE Trans. Antennas Propag.*, vol. AP-22, no. 4, pp. 536–540, July 1974.

[78] W. C. Wilkinson, "A class of printed circuit antennas," *IEEE AP-S Intl. Symp. Dig.*, pp. 270–273, 1974.

[79] G. R. Hanley and H. R. Perini, "Column network study for a planar array used with an unattended radar," *RADC TR-80*, final report, RADC, GAFB, NY, March 1980.

[80] G. Sanford and L. Klein, "Increasing the beamwidth of a microstrip radiating element," *Intl. Symp. Dig. of Antennas Propag. Soc.*, Univ. of Washington, pp. 126–129, June 1979.

[81] J. Q. Howell, "Microstrip antennas," *IEEE Trans. Antennas Propag.*, vol. AP-23, pp. 90–93, January 1975.

[82] D. Davis, O. J. Digiandomenico, and J. A. Kempic, "A new type of circularly polarized antenna element," *IEEE G-AP Symp. Dig.*, pp. 2–23, 1967.

[83] M. H. Chen and G. N. Tsandoulas, "A wideband square waveguide array polarizer," *IEEE Trans. Antennas Propag.*, vol. AP-21, pp. 389–391, May 1973.

[84] L. Young, L. A. Robinson, and C. A. Hacking, "Meander line polarizer," *IEEE Trans. Antennas Propag.*, vol. AP-21, pp. 376–378, May 1973.

[85] H. Howe, *Stripline Circuit Design*, Dedham: Artech House, 1974.

[86] E. Wilkinson, "An n-way hybrid power divider," *IEEE Trans. Microwave Theory Tech.*, vol. MTT-8, no. 1, pp. 116–118, January 1960.

[87] L. I. Parad and R. L. Moynihan, "Split-tee power divider," *IEEE Trans. Microwave Theory Tech.*, vol. MTT-13, pp. 91–95, January 1965.

[88] J. F. White, *Semiconductor Control*, Dedham: Artech House, 1977.

[89] G. N. Tsandoulas, "Unidimensionally scanned phased arrays," *IEEE Trans. Antennas Propag.*, vol. AP-28, no. 1, pp. 86–98, January 1980.

[90] W. J. Ince and D. H. Temme, "Phasors and time-delay elements," *Advances in Microwaves*, vol. 4, pp. 2–183, New York: Academic Press, 1969.

[91] L. R. Whicker and C. W. Young, "The evolution of ferrite control components," *Microwave J.*, vol. 21, pp. 33–37, November 1978.

[92] L. R. Whicker and D. M. Bolle, "Annotated literature survey of microwave ferrite control components and materials for 1968–1974," *IEEE Trans. Microwave Theory Tech.*, vol. MTT-23, no. 11, pp. 908–918, November 1975.

[93] C. R. Boyd, "Analog rotary-field ferrite phase shifters," *Microwave J.*, vol. 20, pp. 41–43, December 1977.

[94] A. G. Fox, "An adjustable waveguide phase changer," *Proc. IRE*, vol. 35, pp. 1489–1498, December 1947.

Chapter 14

Aperiodic Arrays

Y. T. Lo
University of Illinois

CONTENTS

1. Introduction — 14-3
2. A Brief Review — 14-4
3. Spaced-Tapered Arrays — 14-6
   Design Procedure for a Symmetrical Space-Tapered Array   14-6
4. Probabilistic Approach — 14-8
   Theoretical Results   14-9
   Illustrative Examples   14-18
   The Mutual Coupling Effect and Blind Angles   14-20
   The Holey Plate Experiment   14-27
   Other Remarks   14-32
5. References — 14-35

Yuen T. Lo is a professor and the director of the Electromagnetics Laboratory (formerly the Antenna Laboratory) in the Electrical and Computer Engineering Department, University of Illinois at Urbana-Champaign. He is a member of the National Academy of Engineering and a Fellow of IEEE, and a member of the International Union of Radio Science. He received the 1964 IEEE AP-S Bolljahn Memorial Award, the 1964 IEEE AP-S Best Paper Award, the 1979 IEEE AP-S Best Paper Award, the IEEE Centennial Medal, and the Halliburton Education Leadership award. He served as an AP-S AdCom member, the Chairman of the AP-S Education and Tutorial Papers Committee, and twice (1979–1982 and 1984–1987) as IEEE AP-S National Distinguished Lecturer. Dr. Lo is an honorary professor of the Northwest Telecommunication Engineering Institute and also the Northwestern Polytechnical University, both at Xian, China. He has published over a hundred technical articles in refereed journals covering a wide spectrum, from theoretical to experimental works. His works include large antenna arrays, radio telescopes, multiple-beam antennas, multiple scattering, antenna synthesis, antijamming antennas, antenna in plasmas, corrugated guides and horns, artificial dielectrics, and microstrip antennas. He designed the University of Illinois Radio Telescope, considered to be the world's largest antenna in the early 1960s.

1. Introduction

Arrays with uniformly spaced elements (often called *uniformly spaced arrays*) have been widely studied, mainly for two reasons: mathematical tractability in many cases and simplicity in fabrication. As discussed in Chapter 11, the pattern function of a uniformly spaced array, no matter how it is excited, is always periodic in the wave-vector **k**-space, or sometimes expressed in the so-called **u**-space. Thus a beam in the pattern function will repeat itself an infinite number of times. However, the visible region corresponds only to a finite region of this space. To ensure that only a single beam appears in the visible region (i.e., no grating lobes), adjacent element spacing must be kept sufficiently small, and as a result a large number of elements must be used to fill a given aperture, which is determined by the desired beamwidth and, sometimes, directivity also.

In contrast, arrays with incommensurable element spacings have aperiodic pattern functions, and are thus called *aperiodic arrays*. In general, they have no grating lobes, and, as a result, the required number of elements is not directly related to the grating lobe condition. Some nonuniformly spaced arrays, even with commensurable element spacings, may have very large periods in the pattern functions as compared with that of a uniformly spaced array and thus exhibit characteristics similar to those of aperiodic arrays. They may be regarded as pseudo-aperiodic arrays or, for all practical purposes, simply aperiodic arrays. This situation occurs actually in *all* practical arrays since all element spacings must be rounded up numerically to a finite number of the digits and are therefore divisible by a common unit.

Aperiodic arrays may be used in many different ways, but the most interesting applications are in array thinning, beamwidth narrowing, and element-interaction reduction. In regard to directivity it is shown in Chapter 11 that for element spacing greater than a half-wavelength the directivity for a uniform excitation is nearly maximum, approximately equal to the number of elements. This conclusion can also be deduced from the fact that, in general, the *mutual* radiation resistance becomes less significant in comparison with the self-radiation impedance. As an example, for half-wavelength dipoles, as the element spacing increases beyond a half-wavelength the total power radiated depends mainly on the self-radiation resistance and thus reaches approximately a constant value. From this point of view one can expect that the same conclusion can be drawn even for nonuniformly spaced arrays, as most element spacings, if not all, are greater than $\lambda/2$. On the other hand, if element spacings are allowed to assume any value, the result in Chapter 11 shows that maximum directivity can be reached when element spacings become vanishingly small. However, this is the so-called superdirectivity and cannot be realized in practice. From this argument one can therefore conclude that the optimization of element spacings for maximum directivity is a problem of little practical significance. Thus in this chapter we shall direct our attention to other aspects of the array.

One of the interesting array synthesis problems is to find an optimum set of element spacings and excitations that would minimize the highest side lobe level in the entire visible region. Many attempts have been made, including some analytic methods (using the Poisson summation formula), numerical methods (using the quadrature approximation and the so-called dynamic programming), and statistical and probabilistic approaches [1–21]. It is shown [22] that none of these methods can yield a solution even close to the true optimum. The difficulty of this problem may be attributed to the fact that the side lobe level depends on the element spacings in a highly nonlinear manner, and that, in general, there is no known analytical method to determine the highest side lobe level, or the angular direction where the highest side lobe may occur, even with all element positions given. Except for small arrays a numerical search with a modern computer is considered impractical. The reason is that the highest side lobe position does not change *continuously* with the element positions; thus there is no simple way to keep track of the highest side lobe as the element position changes unless an entire three-dimensional pattern is computed. With this understanding it is not surprising to find that so far all the attempts are not successful. In the following we shall make a brief review of these works and finally focus our discussion on the probabilistic approach because, first, the theory is more complete, second, it has been supported by various experiments, and, last, it can provide a useful practical solution which, though not optimum in the ordinary sense, is optimum in some probabilistic sense (see below). It will be seen that the array so obtained can have a performance much superior to that of conventional arrays in many respects.

2. A Brief Review

Unz [1] used the Fourier-Bessel expansion to relate the pattern function to the element positions. As pointed out by various authors it is difficult to make use of this expansion to yield useful numerical results. King, Packard, and Thomas [2] computed the pattern functions of a few sample arrays with preassigned spacings. Their computed results reveal some interesting properties of nonuniformly spaced arrays. At about the same time, nonuniformly spaced arrays were studied at the University of Illinois in connection with the feed of a radio telescope [3]. In that paper a certain optimization procedure and the method of relating the element spacings to the excitation function were proposed and applied. Maffett [4] independently proposed the same method of relating the element position function to the excitation function. Andreason [5] suggested a procedure for using a computer to optimize the element positions such that the side lobe level in the visible region is minimized. This method has been independently applied by Lo [3], who also stated that, at best, the solution so obtained is only optimum locally. For large arrays this procedure becomes tedious and time consuming, if not completely impossible. In his article Andreason gave an interesting lower bound of the side lobe level of widely spaced arrays by using the fact that the directivity of these arrays is proportional to the number of elements.

Ishimaru [6], Yen and Chow [7], and Ishimaru and Chen [8], using the Poisson summation formula, reduced the pattern function of a finite sum to an infinite sum. In general, this would make the computation more difficult; however, for small u, it is reasonable to assume that only the first term of the series is important. However, for large u or, equivalently, for very wide average element spacing, which, unfortunately, is the case of greater interest, other terms also become significant. Ishimaru and Chen in particular considered a spacing function of the type $x + (2A_1/\pi) \sin \pi x$ with $A_1 < \frac{1}{2}$, which is a perturbed uniform spacing, particularly when A_1 is small. As $A_1 = \frac{1}{2}$, this spacing function becomes identical with that considered earlier by Lo [3]. However, using this type of spacing function, Ishimaru expressed the pattern function in terms of the Anger function, and, for convenience, he and his associate compiled a table for this function [9]. In principle, for a general spacing function one may need to consider more terms of this type as indicated in their paper, but it is doubtful that the numerical convenience could be retained. Ishimaru and others claimed that the choice of a single term with $A_1 = \frac{1}{6}$ gave a pattern quite close to the optimum. As shown in [22], this is not so.

Skolnik, Nemhauser, and Sherman [11] realized that a straightforward search for an optimum set of element spacings, using a high-speed computer, is next to impossible, even for a moderately large array. As a result they proposed a systematical method using the so-called dynamic programming technique. Unfortunately, this method does not lead to a truly optimal solution since, as noted by the authors, the "principle of optimality" (which is essential in the successful use of the dynamic programming) does not apply to this problem [11]. In particular, the assumption that the optimal position of the first element depends *only* on the position of the second elements or in general that the $(n-1)$st element depends on the nth element is not valid for large arrays.

In addition, it is perhaps proper to cite the works of Harrington [14] and Baklanov, Pokrovshi, and Surdotovich [15], whose investigations are not concerned with reducing the number of elements but rather with achieving certain pattern characteristics by spacing weighting. Baklanov, Pokrovshi, and Surdotovich, in particular, used some complicated numerical techniques in an attempt to derive a spacing-weighted array from the Dolph-Chebyshev counterpart. From a practical point of view this may appear to have some advantage since it is much easier to achieve spacing weighting than excitation weighting. In doing so, however, some element spacings have to be smaller than usual, and the strong coupling effects between elements may offset this apparent advantage appreciably. Furthermore, the method seems to be too cumbersome to be applicable to large arrays.

Finally, we may mention the probabilistic and statistical approaches to this problem. Rabinowitz and Kolar [16] analyzed the case that the placement of elements over a uniform grid system in an aperture is determined by the outcome of a random experiment. They obtained only the *mean* side lobe level which, for the example studied in their paper, is about 7 dB lower than what was computed actually. Later, Skolnik, Sherman, and Ogg [17] considered the same problem and made a very similar analysis. They too obtained only the mean side lobe level; likewise, their examples showed that the predicted level is about 7 to 8 dB lower than the actually computed result in what they called "principal" planes.

Maher and Cheng [18] studied the problem of random removal of elements in a uniformly spaced array. Their assumption that the removals of elements are statistically independent events may result in the removal of the same element many times.

Starting with the probability distribution function of the elements in the aperture, Lo [12] obtained the distribution function of the antenna response at each observation angle, the half-power beamwidth, and the directivity, etc. He also obtained an approximate distribution of the side lobe level in any range of u. Later, these results have been verified by using the Monte Carlo simulation [13] and actual holey plate experiment [23]. Later Agrawal and Lo [24] refined the analysis, extended the theory for small arrays, and included the effect of element interactions. Clearly this method does not lead to an optimum design in the true sense.

The confusion of various theories finally prompted Lo and Lee [22] to conduct a comparison study with a few small arrays. They found that none of the theories yields the optimum solution; in fact, most are far from it. Similar to the so-called space-tapered arrays, some of these designs can control near-in side lobes only to some extent, and the far-out side lobe level not at all.

3. Spaced-Tapered Arrays

A spaced-tapered array is one in which the interelement spacing varies in some fashion across the array, usually increasing from the center to the edge of an array. This method is motivated by an attempt to remove the following two disadvantages of a conventional uniformly spaced array [3]:

(1) *Inefficiency*, in the sense that, in most low side lobe level designs, a large number of elements in the outer portion of the array serves only the purpose to cancel partially the high side lobes produced by the central portion of the array but contributes very little to the radiated power.

(2) *Feeding difficulty*, in that the design may require the element excitation power level over a range of several orders of magnitude.

As discussed in Chapter 11 a discrete array could be regarded as a sampling of an aperture antenna with "continuous" illumination functions. When the sampling is uniform one obtains a uniformly spaced array with element excitation weighted according to the illumination function. If the sampling interval is small enough, one expects a close similarity between the pattern functions of the discrete and the continuous array, such as $\sin(N\psi/2)/\sin(\psi/2)$ for the former against $\sin(L\psi/2)/(\psi/2)$ for the latter in the case of a uniform excitation with aperture length $L \cong Nd/\lambda$. On the other hand, if the sampling is made such that each interval contains the same amount of excitation power, one obtains a space-tapered array with all elements excited equally in power as illustrated in Fig. 1.

Design Procedure for a Symmetrical Space-Tapered Array

(a) Choose a proper illumination function $f(x)$ (see comments below).
(b) Define and compute

$$g(x) = \int_{-a/2}^{x} f(x)\,dx \tag{1}$$

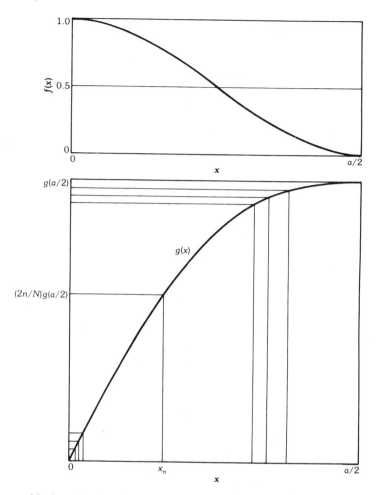

Fig. 1. A graphical method for determining element location of a space-tapered array.

where a = aperture dimension in wavelengths.

(c) Let N be the number of elements, which is approximately equal to the array directivity.

(d) Solve for x_n, the nth element location, in

$$g(x_n) = \frac{2g(a/2)}{N}n, \quad n = 1, 2, \ldots, N/2 \qquad (2)$$

which could easily be solved graphically as illustrated in Fig. 1.

Advantages—Usually a smaller number of elements are needed and all elements excited *equally*.

Disadvantages—With the exception of the region near the main beam, the array pattern may differ greatly from that of the corresponding aperture antenna; usually the side lobe level increases with observation angle.

Comments—The design is useful if only a low near-in side lobe is desired, or directive elements are used.

The set of element positions obtained above can be used as a starting set for an iteration procedure in which the element positions are moved one at a time in a direction where the side lobe level is decreasing. Of course, the final set so determined is at best an optimum locally [3].

The extension of this method to two- or even three-dimensional arrays is obvious, since it needs only to divide the illumination function in equal power per area or volume where an element will be placed.

4. Probabilistic Approach

Since the pattern function is an analytic function it is theoretically impossible to have contributions from all elements completely canceled for all observation angles except for a single direction in which the main beam maximum is intended. Therefore all patterns (except for the trivial case of a very wide beam) have side lobes. Consequently the best design one could hope for in many applications would be one having all side lobes equal in level. When a large number of elements is used the uniformly spaced Chebyshev-Dolph array has already provided the solution. But for economic reasons one may ask whether or not a similar performance could be achieved with substantially fewer elements. As discussed in the review section, so far all attempts have failed. In view of this fact it is natural to resort to a probabilistic approach so that contributions from all elements would be as incoherent as possible for all observation angles except in the main beam maximum direction where all contributions should be completely coherent. This can be achieved by using random element positions (or spacings, but with progressive phase excitation in case of scan). It will be seen later that, in so doing, the probability for the pattern to exceed a certain level at any observation angle outside the main beam region can be made the same as at any other angle. In the language of information theory this is a state of maximum entropy. It is only in this sense that the probabilistic design is optimum; in other words, all side lobes are equal in a probability sense. Therefore this design is, in a sense, a probabilistic approach to the Chebyshev-Dolph array.

For simplicity consider first a linear array, while the extension to the planar array is straightforward as will be shown later. In the following, all length dimensions are understood to be measured in wavelength λ. Let $g(X)$ = probability density function for placing an element at X, with $|X| \leq a/2$ and

$$\int_{-a/2}^{a/2} g(X) \, dX = 1 \tag{3}$$

If there are N equally excited elements each of which is to be placed within the array aperture $(-a/2, a/2)$ according to the same probability function $g(X)$ but independently of each other, then for each set of random samples $\{X\}$: $(X_1, X_2, \ldots, X_N)$ there is associated a sample pattern function

Aperiodic Arrays

$$F(u) = \frac{1}{N} \sum_{n=1}^{N} \exp[j2\pi(\sin\theta - \sin\alpha) X_n] = \frac{1}{N} \sum_{n=1}^{N} \exp(jux_n) \quad (4)$$

where the last expression is a normalized pattern function with

θ = any observation angle
α = main beam angle
$x_n = 2X_n/a$ = normalized element position
$u = a\pi(\sin\theta - \sin\alpha)$

As a result of the normalization in X, the aperture becomes $(-1, 1)$ and (3) can be rewritten as

$$g(x) \equiv 0 \quad \text{for } |x| > 1$$
$$\int_{-1}^{1} g(x)\,dx = 1 \quad (5)$$

The ensemble of all sample sets $\{x\}$ constitutes a corresponding ensemble of $\{F(u)\}$. Our objective is to study the probabilistic properties of $F(u)$, in particular, those related to the array performance. For later discussion it is noted that the array dimension a appears *only* in the parameter u. Thus the pattern of an array with any value a can be obtained simply from that of a normalized array. In other words, if a is increased by a factor of, say, two, one needs only to extend the computation of $F(u)$ for twice the range of u while $F(u)$ in the first half range remains the same except that the scale of u changes by a factor of two. This scaling is not important if one is only interested in the highest side lobe level in the visible region.

Theoretical Results

Let $E\{\cdot\}$ be a probability average operator. A few important results are summarized below:

(a) The mean of $F(u)$ is

$$\phi(u) \equiv E\{F(u)\} = \int_{-\infty}^{\infty} g(x) e^{jux}\,dx \quad (6)$$

which is simply the Fourier transform of $g(x)$, or in probability theory the characteristic function of the random variable x, or in antenna theory the pattern function of a "continuous" aperture antenna with illumination function $g(x)$. This implies that by choosing a proper $g(x)$, at least the mean pattern $\phi(u)$ can behave in some desired manner. It may also be noted that since $g(x)$ is an aperture-limited function, $\phi(u)$ is analytic.

(b) For *any given* u, the joined probability density function of the real and imaginary parts of $F(u)$, namely $F_1(u)$ and $F_2(u)$, is asymptotically normal:

$$f(F_1, F_2) = \frac{1}{2\pi\sigma_1\sigma_2} \exp\left\{-\left[\frac{1}{2}\frac{(F_1 - \phi)^2}{\sigma_1^2} + \frac{F_2^2}{\sigma_2^2}\right]\right\} \quad (7)$$

where the independent variable u has been suppressed, and $g(x)$, for simplicity, is assumed to be even, implying

$$\text{Im}\{\phi(u)\} = 0$$

$$\sigma_1^2(u) = \text{variance of } F_1(u) = E\{[F_1(u) - \phi(u)]^2\}$$

$$= \frac{1}{2N}[1 + \phi(2u)] - \frac{1}{N}\phi^2(u) \quad (8)$$

$$\sigma_2^2(u) = \text{variance of } F_2(u) = E\{F_2^2(u)\}$$

$$= \frac{1}{2N}[1 - \phi(2u)] \quad (9)$$

(c) Let $\Pr\{\cdot\}$ be the probability measure for the event in the curly braces; then

$$\Pr\{|F(u)| < r\} = \iint_{(F_1^2 + F_2^2) < r^2} f(F_1, F_2)\, dF_1 dF_2 \quad (10)$$

which is a generalized noncentral chi-square distribution with two degrees of freedom. Tables of percentiles in r for various values of the parameters are available [25]. When r is large compared with σ_1 and σ_2, an asymptotic solution is given by

$$\Pr\{|F| < r\} = \frac{1}{2}\text{erf}(sr)[\text{erf}(r + m) + \text{erf}(r - m)]$$
$$- \frac{1}{4\sqrt{\pi}s^2 r}\exp\left(-\delta^2\left\{1 + \frac{3\delta}{4s^2 r} + \left[\frac{3}{8s^2} + \frac{15(2\delta^2 - 1)}{48s^4}\right]\frac{1}{r^2}\right.\right.$$
$$\left.\left.+ \, 0(s^{-4}r^{-3}) + 0[\sqrt{s}kr\exp(-s^2k^2r^2)]\right\}\right) \quad (11)$$

where $m = \phi/\sigma_1$, $s = \sigma_1/\sigma_2$, $\delta = r - m$, and $0 < k < 1$.

For the special case when $\sigma_1 = \sigma_2$ (i.e., $s = 1$) which occurs for u outside the main beam region, the above expansion reduces to that given by Rice [26], who utilized the asymptotic expansion of the Bessel function in his derivation. For large $|u|$, $\phi(u) \cong 0$, and the distribution becomes simply Rayleigh. For the general case, however, it can approximately be computed by a method due to Patnaik [27] which approximates a noncentral chi-square distribution by a central one with different degrees of freedom. The latter can then be read from an incomplete gamma function table prepared by Pearson [28], which is shown as $I(v, p)$ in Fig. 2 with

$$\Pr\{|F| < r\} \cong I(v, p) \quad (12)$$

where

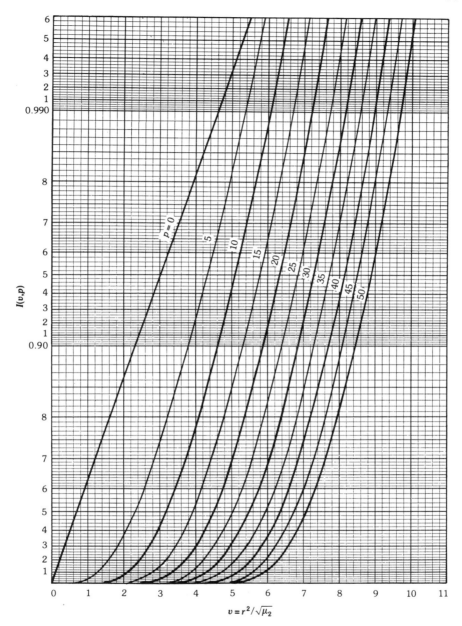

Fig. 2. Incomplete gamma function approximation for noncentral chi-square distribution. The curve for $p = 0$ is the Gausian distribution. (*After Lo [13]*, © *1964 by the American Geophysical Union*)

$$v = r^2/\sqrt{\mu_2}$$
$$p = \mu_1^2/\mu_2 - 1$$
$$\mu_1 = \sigma_1^2 + \sigma_2^2 + \phi^2$$
$$\mu_2 = 2(\sigma_1^4 + \sigma_2^4 + 2\phi^2\sigma_1^2)$$

In general, except for u in the main beam region, $\sigma_1^2 \cong \sigma_2^2 \cong 1/2N$, $\phi \cong 0$ and $p \cong 0$, and therefore the distribution becomes simply Rayleigh $I(Nr^2, 0)$ in Fig. 2 and independent of u. This implies that, although the pattern behavior in the main beam region is determined by $g(x)$, outside the main beam region it is determined *only* by N, the number of elements. Unless the near-in side lobe level is of interest it is advantageous to use the uniform density function for $g(x)$ so that a narrow beam is obtained.

As an example, consider

$$g(x) = \begin{cases} \cos^2(\pi x/2) & \text{for } |x| \leq 1 \\ 0 & \text{for } |x| > 1 \end{cases}$$

Fig. 3 shows the mean pattern $\phi(u)$ and variances $\sigma_1^2(u)$ and $\sigma_2^2(u)$ for each value of u/π. It is seen that in the neighborhood of main beam maximum the variances are nearly zero, indicating that in this region the beam is almost deterministic. Outside the main beam the variances quickly reach a constant value $1/2N$. Figs. 4 through 6 show the level curve as well as the mean pattern $|\phi(u)|$ for comparison, for $N = 10^2$, 10^4, and 10^5, respectively. A *p-percent* level curve is a plot of r_p

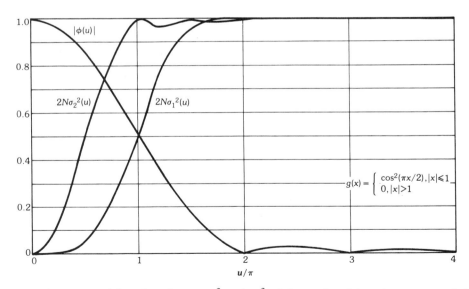

Fig. 3. The mean $\phi(u)$ and variances σ_1^2 and σ_2^2 of the real and imaginary parts of the random pattern function $|F(u)|$ as functions of u, for a cosine-square probability density function and uniform excitation. (*After Lo [12]*, © *1964 IEEE*)

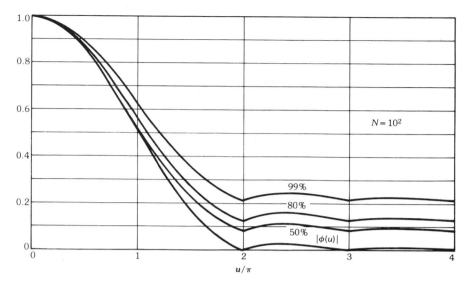

Fig. 4. Level curves of $|F(u)|$ for cumulative probability equal to 99 percent, 80 percent, and 50 percent with a cosine-square probability density function and uniform excitation, for $N = 10^2$. (*After Lo [12],* © *1964 IEEE*)

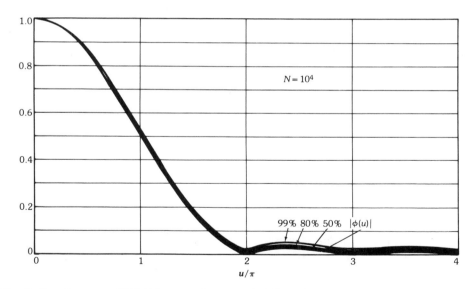

Fig. 5. Level curves of $|F(u)|$ for cumulative probability equal to 99 percent, 80 percent, and 50 percent with a cosine-square probability density function and uniform excitation, for $N = 10^4$. (*After Lo [12],* © *1964 IEEE*)

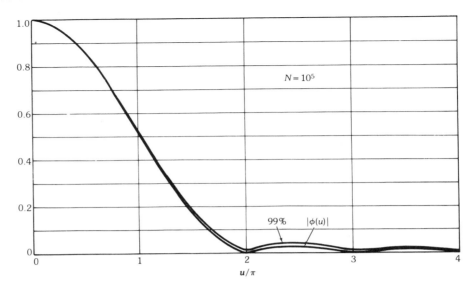

Fig. 6. Level curves of $|F(u)|$ for cumulative probability equal to 99 percent with a cosine-square probability density function and uniform excitation, for $N = 10^5$. (*After Lo [12]*, © *1964 IEEE*)

versus u with $\Pr\{|F(u)| < r_p\} = p$ percent. For $N \geqq 10^4$, $|F(u)|$ is almost equal to $|\phi(u)|$ with 99 percent probability. But it should be emphasized that these curves are meaningful only for *each* value of u. In other words, the probability of obtaining the *entire* 99 percent level curve for all *u*s in an interval is *not* 99 percent. In general, however, one is not interested in the exact pattern but rather the highest side lobe level. This is discussed next.

(d) The probability for a side lobe level less than r for $g(x) = 1/2$, $|x| < 1$, and u in the entire visible region (including the largest scan range) is

$$\Pr\{|F(u)| < r, |u| \in (u_f, 2\pi)\} \cong [1 - \exp(-Nr^2)] \\ \times \exp\{[-4\pi\sqrt{N}r\exp(-Nr^2)] \\ \times (a^2/12\pi)^{1/2}\} \qquad (13)$$

where u_f is the first null (or the main beam null) of $\phi(u)$. Computer simulations [24] have shown that this formula gives accurate results even for an array with as few as eleven elements over an aperture of 5λ to 10λ. For large numbers of elements the formula assumes a simpler form:

$$\Pr\{|F(u)| < r, |u| \in (u_f, 2\pi)\} \cong [1 - 10^{-0.4343Nr^2}]^{[4a]} \qquad (14)$$

where $[4a]$ is the integer part of $4a$. This formula shows that the probability for achieving a certain side lobe level r is nearly zero as N is below a certain value and increases sharply as N increases to a certain critical value. After that the probability increases very slowly. Thus to ensure a good probability of success a sufficient number of elements must be used for a desired side lobe level. For a 90 percent probability the required number N is plotted in Fig. 7 versus side lobe level in both ratio and decibel scale for various aperture dimensions $a = 10^q \lambda$. In this figure the graphs to the left of the dash-dot line are less accurate; here (13) should be used. It is of interest to see that for a 25-dB side lobe level the array needs about 4700 elements over $10^5 \lambda$ aperture with a half-power beamwidth of about 0.0005°, or $(5 \times 10^{-4})°$, and average element spacing of 20λ. If these same elements (4700) are spread over an aperture ten times larger, i.e., $10^6 \lambda$, the side lobe level increases only by about 0.5 dB but the beamwidth will be reduced by a factor of 10, i.e., to 0.00005°, or $(5 \times 10^{-5})°$. Now the average element spacing becomes 200λ, and for this spacing an ordinary equally spaced array would have hundreds of grating lobes. Thus it is not necessary to use a large number of elements in order to achieve a narrow beam. In the case of a planar array the saving in the number of elements is even more dramatic. However, as will be seen later, this does not imply that the directivity can be increased in this manner. The common notion of associating a narrow beamwidth with high directivity is not always correct.

(e) The random variable u_0 defined by the smallest positive root of the following equation determines the half-power beamwidth $2\theta_0$:

$$|F(u_0)| = 1/\sqrt{2} \qquad (15)$$

where $F(u)$ is the random pattern function given by (4) and $u_0 = \pi a(\sin \theta_0 - \sin \alpha)$. Even for the deterministic problem the solution u_0 to the above problem cannot be obtained analytically. Since in practice one is interested only in large arrays with narrow beams an approximate distribution for u_0 can be found from the joint probability of $F(u)$ and $dF(u)/du$ for u in the neighborhood of the mean pattern half-power value u_1, i.e., $|\phi(u_1)| = 1/\sqrt{2}$. Details can be found in Lo [12]. For example, when $g(x) = \cos^2(\pi x/2)$ and $N = 10^4$ the half-power angle θ_0 will fall in $0.7244/a < \sin \theta_0 - \sin \alpha < 0.7377/a$ with 90 percent probability, where α is the scan angle. This implies that the half-power beamwidth deviates from its mean no more than ± 0.91 percent with 90 percent probability; in other words, it is almost certain in practice that the half-power beamwidth is equal to that of the mean pattern. This conclusion is generally true for most functions $g(x)$ of interest and, in fact, verified by Monte Carlo simulations and holey plate experiments to be discussed later.

(f) Since the element locations are drawn from a collection of random numbers, for each sample set of these numbers there is associated a sample radiation pattern and thus a sample directivity. It is of interest to know how the directivity is distributed probabilistically for all possible sets of the element arrangement. Using the Karhunen and Loeve theorem for the expansion of a random function [29], one can obtain the following approximate distribution for the norm $\|F(u) - \phi(u)\|$:

$$\Pr\{\|P(u) - \phi(u)\| < k \|\phi(u)\|\} = \Phi\left[\{(k^2\|g(x)\|^2/d_{av}) - 1\}\sqrt{\frac{a}{2}}\Big/\|g(x)\|\right] \qquad (16)$$

where

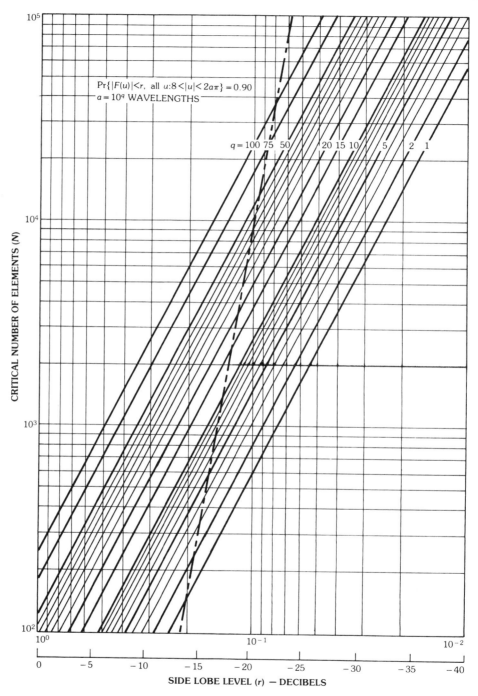

Fig. 7. Number of elements required as a function of the side lobe level for various values of $a = 10^q \lambda$ with a 90 percent probability. (*After Lo [12], © 1964 IEEE*)

$$\|F(u) - \phi(u)\|^2 = \int_{\substack{\text{visible} \\ \text{region}}} |F(u) - \phi(u)|^2 \, du$$

$$\|g(x)\|^2 = \int_{-1}^{1} |g(x)|^2 \, dx$$

d_{av} = average element spacing $\cong a/N$

k = a positive real number

$\Phi(\cdot)$ = standardized normal distribution function (i.e., one with zero mean and unit variance)

Because generally a is a very large number, if k^2 is slightly greater than $d_{av}/\|g\|^2$, the above probability is nearly unity, while if k^2 is slightly less than $d_{av}/\|g\|^2$, the above probability drops sharply to zero. From this result and the definition of directivity D associated with each sample pattern function, and the directivity D_0 associated with the mean pattern function $\phi(u)$, one can conclude that with probability nearly one

$$(D_0 - D) \, \text{dB} \leq 20 \log(1 + \sqrt{d_{av}}/\|g(x)\|) \tag{17}$$

In other words, it is practically certain that D is lower than D_0 by a quantity no greater than $20 \log(1 + \sqrt{d_{av}}/\|g\|)$ dB, which for large average element spacing equals $20 \log \sqrt{d_{av}}/\|g(x)\|$. The interpretation of this result can easily be understood if we consider two arrays of a different number of elements, say, N_1 and N_2. Let their corresponding directivities be D_1 and D_2, respectively; then the above results show that

$$(D_1 - D_2) \, \text{dB} \cong 10 \log(N_1/N_2) \tag{18}$$

with a probability nearly equal to 1. In other words it is practically certain that the directivity D is proportional to N, the number of elements. This result is in agreement with that obtained in Chapter 11. The reason why the probability is not exactly 100 percent is that there exist particular element arrangements, but with very small probabilities, for which D is not exactly equal to N, such as superdirectivity, as shown in Chapter 11.

(g) There is little difficulty in extending most of the above results to an array of higher dimensions. For example, consider a rectangular aperture of $ab\lambda^2$ in the xy plane with a probability density function $g(x, y)$ for the element locations which would produce a mean pattern with sufficiently low side lobe level as before. Then the relation between the total number of elements and the side lobe level is still approximately given by (14), except that [4a] is replaced by [16ab]. Assuming that $a = 10^q$ and $b = 10^p$, this relation is again shown by Fig. 7 except that q should be replaced by $(p + q)$, and 90 percent probability by 80 percent (approximately).

To give an example, the Benelux Cross antenna [30] is considered. This antenna consists of two perpendicular linear arrays each having an aperture of roughly $7 \times 10^3 \lambda \times 80\lambda$, or $ab \cong 10^6 \lambda^2$. From Fig. 7, with a -30-db side lobe level and 90 percent chance of success, each arm requires only about 1.9×10^4 elements. But according to Christiansen and Hogbem [30], each arm would require 2×10^6 uniformly spaced dipoles. This is about 100 times larger than the former estimate. Moreover, if elements were randomly spread over a square aperture $10^4\lambda \times 10^4\lambda$, a total of only 2.3×10^4 elements would be needed to achieve a -30-dB side lobe level. This is very significant since the cross type telescope, being a multiplicative antenna system, will have a side lobe level of -15 dB in the two principal planes even if each arm is designed to have a -30-dB side lobe level. Of course, as already noted before, the directive gain of the randomly spaced array is perhaps 10 or 15 dB lower than that of the uniformly spaced array. Since the former has uniform amplitude weighting function while a uniformly spaced array must have a strongly tapered illumination in order to achieve a -30-dB side lobe level, then from a practical point of view (considering the feeding system and the aperture efficiency) the loss in directive gain may not be as much as indicated above.

Illustrative Examples

Suppose that it is desired to design a linear array whose array pattern will have approximately a half-power beamwidth of 1 minute of arc and a side lobe level of -18 dB. If $g(x)$ is chosen to be uniform, the near-in side lobe level would likely be -13 dB. To ensure that it is below -18 dB, one may, although not necessarily, let

$$g(x) = \cos^2(\pi x/2) \qquad (19)$$

For this function the half-power beamwidth in degrees is approximately $83/a$ and, therefore, $a \cong 83 \times 60 \cong 5 \times 10^3(\lambda)$. From Fig. 7, N (the number of elements needed for a 90 percent probability of success) is approximately 800. The next step is to generate 800 random numbers, between $(0, 1)$ under the probability density function $g(x)$ given above, and these numbers, after scaled by the factor a, determine the actual element locations in the aperture. According to the theory this array when excited uniformly will have a 90 percent probability to yield a pattern with 1 minute of arc in half-power beamwidths and -18-dB side lobe level. Of course, before an investment is made for constructing this array it is prudent to first compute the pattern with the given set of element positions just found, since there is still a 10 percent chance of failure. In a sense the theory predicts that in about nine out of ten trials one should obtain the desired pattern. Hundreds of computer simulations have been made for many different arrays, and in almost all cases the desired properties as predicted by the theory were obtained in the very first trial. Since the *same* basic array theory is used for computing this pattern as for any conventional equally spaced array, there is no reason to doubt this array performance cannot be realized. In fact in this case, all elements being widely spaced, the mutual coupling will have a much less deleterious effect on the pattern. After the pattern properties are confirmed by the computations the array can then be constructed. From that step on, the array is as *deterministic* as any conventional array. This will be demonstrated in a holey plate experiment later.

For simplicity in pattern computation a symmetrical array has been considered for the example stated above. In this case the required number of elements becomes about 1000, somewhat more than that shown in Fig. 7, since there are in effect only 500 random numbers [13]. These numbers were taken directly from Owen's table [31]. Since these numbers are uniformly distributed, the probability integral transformation [32] is then used to convert the set of numbers into another set which obeys the cosine-square law prescribed in (19). This is done by solving the following equation for x_n:

$$y_n = \int_0^{x_n} g(x)\, dx \qquad (20)$$

where y_n is the uniformly distributed random number between 0 and 1, and x_n is the $g(x)$ distributed random number. This method is similar to that used to convert a set of uniform element spacings to a set of nonuniform element spacings in Fig. 1. With this transformation the distribution of the 500 random numbers taken from Owen's table is shown as dotted curves in Fig. 8. For comparison the exact cosine-square distribution is shown in solid line. The computed pattern for $0 \leq u/\pi \leq 19$ is shown as the solid curve in Fig. 9. Also shown are three other cases with $N = 100$, 300, and 600. It is clear that all four cases have almost identical main beams, but the side lobe level, as expected, increases as N decreases even for the small region of angle of observation shown in the figure. It is impractical to show the pattern for the entire visible range. Instead, a statistical distribution of $|F(u_i)/\sigma|^2$ for roughly 4×10^4 values of u_i uniformly spaced in the visible region is made for the case $N = 1000$ shown as the stepped sample curve in Fig. 10. The theory developed before predicts an χ^2 distribution which is also shown in the figure in a normalized scale. The agreement is nearly perfect. This implies that the theory, although it cannot predict the exact shape of the pattern curve, can predict quite accurately the *distribution* of the nearly 4×10^4 numbers of $\{|F(u_i)|\}$ actually computed. Unfortunately, a very accurate prediction of this distribution tells practically nothing about the side lobe level since if *one* of the numbers in the set $\{|F(u_i)|\}$ is changed to a very large value, it will have no effect on the distribution yet it alone will determine a high side lobe level. It is seen once again that the determination of side lobe level is a much more difficult problem. Using the approximate theory given earlier, however, one can predict the side lobe level as the aperture size a or visible region increases. This is shown in Fig. 11 for the four different cases along with the actually computed values for the sample arrays. In this figure the side lobe level in decibels is plotted against both a in wavelengths and the half-power beamwidth in minutes of arc. It is seen that even with as few as 300 uniformly excited elements one can obtain a half-power beamwidth of about 1 minute with a side lobe level of -13 dB. If the conventional design with a half-wavelength spacing is used, it would require 10^4 elements to produce the same beamwidth and side lobe level. The reduction in the number of elements is at a ratio of 100:3; in other words, 97 percent of the elements are saved. The directivity, however, being roughly proportional to N, cannot be obtained without paying the correct price. On the other hand, for the same number of elements as required for the directivity one can achieve a much narrower beam and lower side lobe level with a simple uniform excitation. As will be seen below, there are other significant advantages (such as absence of blind angles and a smaller deleterious effect due to phase errors in a digitalized phase system) for using this type of array.

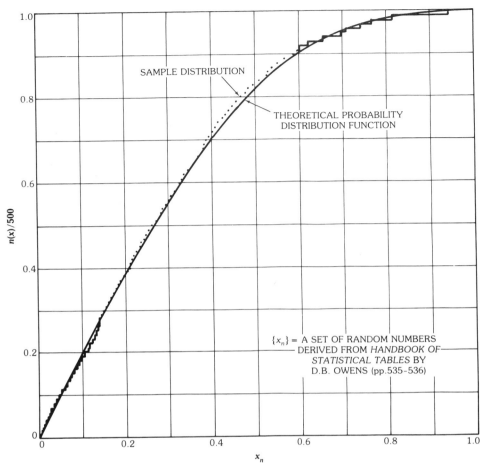

Fig. 8. Distribution of element positions $\{x_n\}$ for an assumed cosine-square density function. (*After Lo [13]*, © 1964 by the American Geophysical Union)

The Mutual Coupling Effect and Blind Angles

In general, the mutual coupling in a phased array with randomly spaced elements is not a serious problem since for such an array elements are commonly many wavelengths apart from each other. However, it may be shown that even if array thinning is not the objective and elements are not spaced far apart, random spacings can still be used advantageously in some applications.

Rigorously speaking, the mutual coupling effect in a practical array is a difficult problem to analyze. The approximate methods commonly used, as discussed in other chapters, are two. The first is to consider a practical array as a truncated portion of a corresponding infinite array [33]. Only when the array is a periodic structure is this method applicable. For arrays with unequal or random element spacings one must resort to the second approach in which the mutual impedances between two elements at various spacings are determined first and their effect in an

Aperiodic Arrays

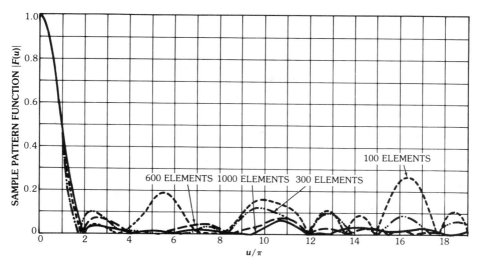

Fig. 9. Sample pattern functions for $0 \leq u/\pi \leq 19$ of four symmetrical arrays with $N = 100$, 300, 600, and 1000 elements, respectively. (*After Lo [13],* © *1964 by the American Geophysical Union*)

array is then studied with a circuit treatment. In this approach it is assumed that the current or aperture field distributions on the two elements remain unchanged whether or not all other elements are present. This seems generally to be a valid assumption for elements of many types at not too close a spacing.

Let I_n be the actual feed current of the nth element in an array and $I_n^{(0)}$ be the current when all mutual couplings are absent. Then by circuit analysis,

$$I_n = I_n^{(0)} - \sum_{\substack{m=1 \\ m \neq n}}^{N} \frac{Z_{mn}}{Z_{11} + Z_0} I_m \qquad (21)$$

where

Z_{mn} = mutual impedance between the mth and nth elements
Z_{11} = self-impedance of the element at the feed terminals
Z_0 = generator or the feed-line impedance

For all the elements, (21) can be written as

$$\mathbf{I} = \mathbf{I}^{(0)} - \mathbf{CI} \quad \text{or} \quad \mathbf{I} = (\mathbf{E} + \mathbf{C})^{-1} \mathbf{I}^{(0)} \qquad (22)$$

where

$$\mathbf{I} = \begin{bmatrix} I_1 \\ \vdots \\ I_N \end{bmatrix}, \qquad \mathbf{I}^{(0)} = \begin{bmatrix} I_1^{(0)} \\ \vdots \\ I_N^{(0)} \end{bmatrix}$$

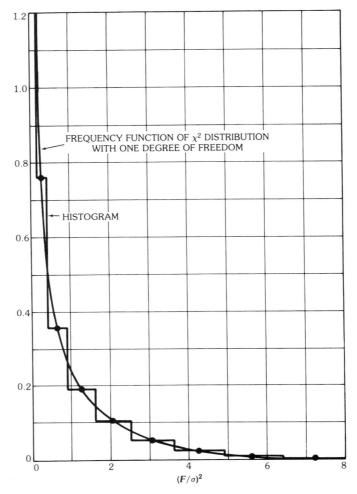

Fig. 10. Sample distribution of 4×10^4 points of the normalized pattern function squared $|F/\sigma|^2$, equally spaced in u/π from 0 to 10^4 (corresponding to $a = 10^4 \lambda$) for $N = 1000$ as compared with the theoretical density function, namely the chi-square density function with one degree of freedom. (*After Lo [13], © 1964 by the American Geophysical Union*)

C is an $N \times N$ matrix with elements

$$C_{mn} = \frac{Z_{mn}}{Z_{11} + Z_0} \quad \text{and} \quad C_{mm} = 0$$

E = identity matrix

For uniform cophasal excitation $I_n^{(0)} = \exp(-jvx_n)$, the pattern function

$$F(u, v) = \frac{1}{N} \sum_{n=1}^{N} I_n \exp(jux_n) \tag{23}$$

Aperiodic Arrays

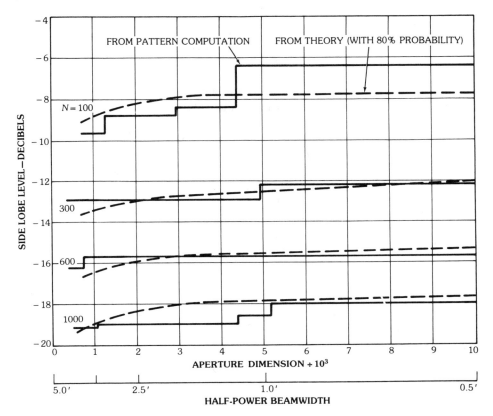

Fig. 11. Side lobe level for four sample symmetrical arrays with $N = 100$, 300, 600, and 1000 elements as a function of the aperture dimension in wavelengths, or the half-power beamwidth in minutes of arc. (*After Lo [13], © 1964 by the American Geophysical Union*)

where

$v = 2\pi \sin \alpha$

α = beam scan angle

$u = 2\pi \sin \theta$

θ = observation angle

x_n = nth element position in wavelengths

Except for the uninteresting case of close element spacing one may use the following approximation for (22),

$$\mathbf{I} \cong (\mathbf{E} - \mathbf{C})\mathbf{I}^{(0)} \qquad (24)$$

Then

$$F(u, v) = \frac{1}{N} \sum_{n=1}^{N} \exp[j(u - v)x_n] - \frac{1}{N} \sum_{n=1}^{N} \sum_{\substack{m=1 \\ m \neq n}}^{N} C_{mn} \exp[j(ux_n - vx_m)] \quad (25)$$

The first summation gives the pattern in the absence of mutual coupling and has been studied thus far, whereas the second, double summation is due to the coupling. The determination of Z_{mn}, or C_{mn}, is a boundary value problem depending on a particular type of antenna, which is not the subject matter of this chapter. At the present it is assumed that Z_{mn}, as a function of spacing, is known either analytically or experimentally. Its effect on the side lobe level distribution can then be found approximately by formulating the problem in terms of a diffusion equation for the probability function of the up-crossing of $|F(u, v)|$ at any given level η for u in any given region [24]. The equation is then solved numerically. To see the effect of mutual coupling on the side lobe level distribution, use has been made of the coupling coefficient

$$C_{mn} = \frac{0.11}{0.22 + |x_{mn}|} \exp(-j2.22\pi|x_{mn}|) \quad (26)$$

where $x_{mn} = x_m - x_n$, as determined experimentally by Lechtreck for two horn elements [34].

Fig. 12 shows the side lobe distributions for an array of 50 such elements uniformly distributed at random with three different average spacings: $d_{av} = 0.5\lambda$, 1.0λ, and 2.5λ. When the coupling is considered, the distributions are shown in solid curves, and when ignored, in dashed curves. As expected the coupling results in a higher side lobe level, but to a lesser degree for larger average spacing. Fig. 13 shows more clearly how the increase in side lobe level depends on the average spacings. In this case, $a = 500\lambda$ and $d_{av} = 0.5\lambda$, 1.0λ, 2.5λ, 5.0λ, and 10λ are considered. It is seen that when $d_{av} \geq 2.5\lambda$ the coupling may be ignored. It may be noted that for $d_{av} = 2.5\lambda$ and $N = 200$, the increase in side lobe level is less than that shown in Fig. 12 for the same d_{av} and for $N = 50$. Thus even for the same average spacing, the effect of mutual coupling on side lobe level becomes less as N increases. This is also expected as the double summation in (25), consist of terms in random phase, contributes less and less as N increases, for the same reason that the side lobe level decreases when N increases even in the case of no coupling as shown in (14).

If $u = v$ in (25), one obtains the main beam maximum. For most antenna elements, if not spaced very closely, the phase of C_{mn} is nearly linear in spacing as shown in (26). Let

$$C_{mn} = |C_{mn}| \exp(j\beta|x_{mn}|), \quad x_{mn} = x_m - x_n$$

Then the main beam magnitude as a function of scan angle α, or its corresponding value v, sometimes called the *array scan characteristics*, is

Aperiodic Arrays

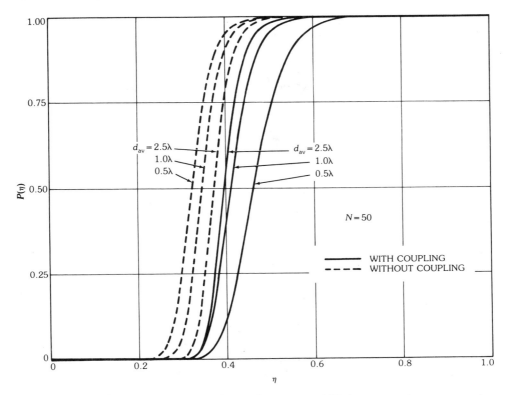

Fig. 12. Distribution of side lobe level for random arrays of 50 elements and average spacing $d_{av} = 0.5\lambda$, 1.0λ, and 2.5λ with and without mutual coupling. (*After Agrawal and Lo [24]*, © *1972 IEEE*)

$$F(v, v) = \frac{1}{N} \sum_{n=1}^{N} \left[1 - \sum_{\substack{n=1 \\ n \neq m}}^{N} |C_{mn}| \exp(-jvx_{mn} + j\beta|x_{mn}|) \right] \quad (27)$$

The quantity in the square brackets is exactly the pattern function of the nth element with all other elements terminated in the match load Z_0. It may be called the *array element pattern*, namely, the pattern of an element in an array environment, which can be vastly different from that of an isolated element. The scan characteristic is therefore the average of all array element patterns. It is interesting to see that if $x_n = nd$ (i.e., uniform spacing), and if

$$(v \pm \beta)d = 2\pi p, \quad p = 0, 1, 2, \ldots$$

the terms in the second sum of (27) will add up in phase to cause a sharp decrease in $F(v, v)$. For a very large array this sum could be as large as unity, thus resulting in the so-called blind angle phenomenon; namely, in that scan direction, v, the main beam maximum and therefore the total radiated power drop to nearly zero. In other words the array ceases to function as an antenna and all the power sent by

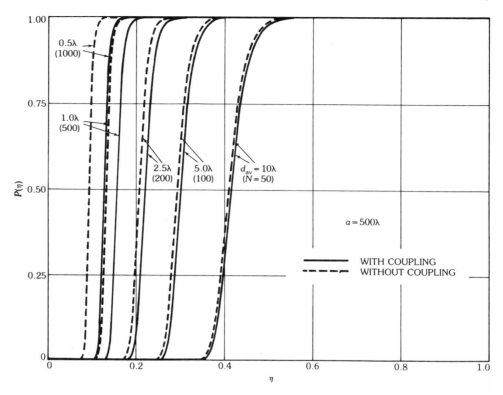

Fig. 13. Distribution of side lobe level for a 500λ aperture and 50, 100, 200, 500, and 1000 elements with and without mutual coupling. (*After Agrawal and Lo [24]*, © *1972 IEEE*)

the exciting generator will be reflected, or the element impedance in the array becomes reactive. Now if $x_n \neq nd$ and $\{x_n\}$ is a set of random numbers, the in-phase condition for all the terms in the summation will be unlikely to occur. In fact, the larger the array (i.e., more elements), the smaller will be the sum and thus the smaller the drop of main beam magnitude. This conclusion is in exact opposition to that of a uniformly spaced array, a very gratifying result indeed.

The probability distribution of the main beam magnitude $F(v,v)$ can be determined approximately once the coupling coefficient C_{mn} for a given type of element to be used and the probability density function $g(x)$ for placing the elements are known. The readers may refer to Agrawal and Lo [24] for details. In the following only some typical results are given.

Let the *fluctuation* of the main beam magnitude be defined as

$$r = \max_{v \in [0, 2\pi]} F(v,v) - \min_{v \in [0, 2\pi]} F(v,v) \tag{28}$$

When $g(x) = 1/2$ for $|x| \leq 1$ and 0 otherwise, C_{mn} is given by (26), $N = 50$, and $d_{av} = 0.5\lambda$, the probability for $r < \xi$ versus ξ can be computed and is plotted in Fig. 14. Also shown is the "experimental" result for 50 random sample arrays simulated by using the Monte Carlo method. The close agreement between the two results

Aperiodic Arrays

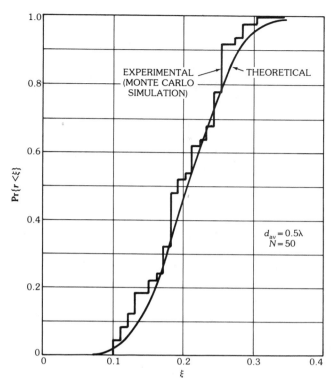

Fig. 14. Distribution of main beam amplitude fluctuation r over the entire scanning range for 50 elements and average spacing 0.5λ. (*After Agrawal and Lo [24], © 1972 IEEE*)

provides some confidence in the approximate theory. Fig. 15 shows the theoretical results for the same array except that $d_{av} = 0.5, 1.0, 2.5, 5,$ and 10λ. As expected, the larger the average spacing d_{av}, the smaller the coupling effect and the smaller the fluctuation of main beam magnitude over the scanning range.

The Holey Plate Experiment

Side Lobe Level and Half-Power Beamwidth—A physical experiment for a large array is a rather costly enterprise. Fortunately, the holey plate method as described by Stone at optical frequency [35] and by Skolnik at microwave frequency [36] provides a simple modeling technique. In this method the array is modeled by a large conducting plate perforated with small circular holes, each simulating an antenna element. The "holey" plate is illuminated by an incident plane wave from one side, and the field is measured from the other side for all directions. The pattern so obtained would be nearly the same as that of the actual array. The sketch in Fig. 16 shows the major components used. The holey plate is placed against a microwave lens which converts a spherical wave emitted from a reduced open waveguide at the focus into a plane wave. The assembly of these components is enclosed in a box made of absorbing material, and placed over a turntable.

In the actual setup the holey plate is used as a receiving array. Fig. 17 shows the

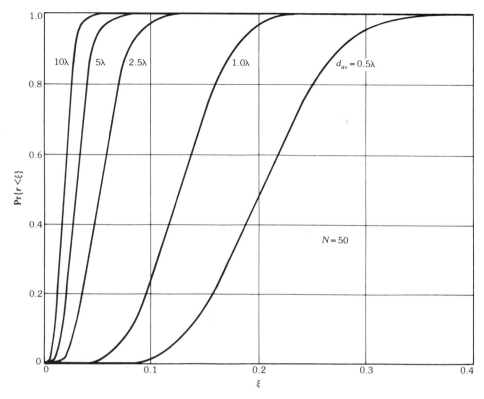

Fig. 15. Distribution of main beam amplitude fluctuation r over the entire scanning range for 50-element arrays with 0.5λ, 1.0λ, 2.5λ, 5λ, and 10λ average spacing. (*After Agrawal and Lo [24], © 1972 IEEE*)

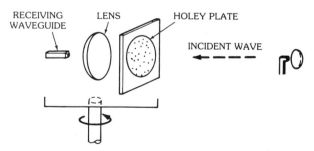

Fig. 16. Major components for the holey plate experiment. (*After Lo and Simcoe [23], © 1967 IEEE*)

physical experimental setup at 4 mm. The paraboloidal reflector at the far right is the transmitting antenna. The incident wave is diffracted by the holey plate and then focused on an open-ended waveguide on the left by a polystyrene lens. The receiving assembly, consisting of the plate, lens, and the open-ended waveguide, is

Aperiodic Arrays

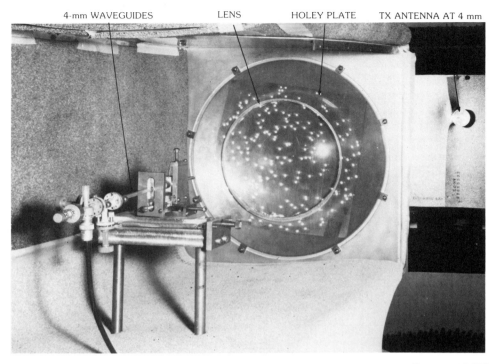

Fig. 17. Experimental setup for 4-mm waves. (*After Lo and Simcoe [23]*, © *1967 IEEE*)

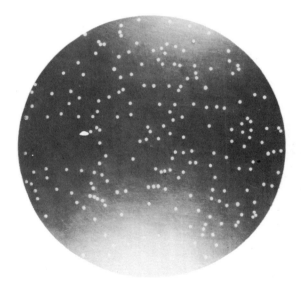

Fig. 18. Photo of the holey plate sample array, where the coordinates of elements are obtained from a set of uniformly distributed random numbers. (*After Lo and Simcoe [23]*, © *1967 IEEE*)

placed over a turntable. Except for the plate the assembly is enclosed with microwave-absorbing materials which are partially removed in order to show the details inside. In this picture the range has been reduced so that the transmitting antenna can also be shown. The holey plate contains 210 elements uniformly distributed at random over a circular aperture of about 56λ in diameter as shown in Fig. 18. Each element is a hole of about $\lambda/4$ in diameter. As is clear from this figure there exists no plane of symmetry and thus no principal plane. Therefore the measured patterns in a few planes cannot be used to infer the overall performance. It would be desirable to measure a three-dimensional pattern, but in this experiment a total of 90 cuts was taken (roughly twice as many as the highest spatial frequencies). A typical pattern with expanded main beam is shown in Fig. 19. The measured beams in all cuts are almost identical, with a half-power beamwidth ranging from 0.9° to 1.25° and a statistical mean of 1.04°. The difference between this value and the theoretical mean of 1.05° is well within the experimental error.

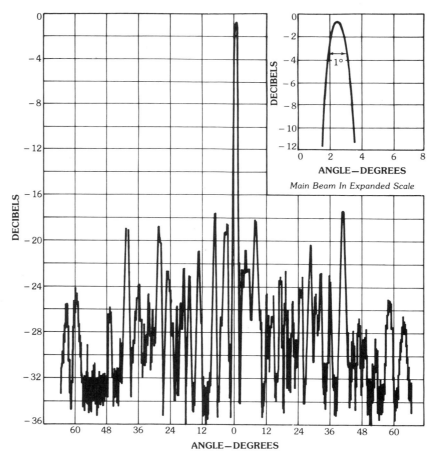

Fig. 19. A typical pattern in a plane cutting through the sample array as shown in Fig. 18. (*After Lo and Simcoe [23]*, © *1967 IEEE*)

Aperiodic Arrays

The 90 measured patterns are statistically "identical" in the sense that all have nearly the same beam and a few high side lobes, ranging approximately from -18 dB to -13 dB. However, no two patterns are really identical in details. Fig. 20a shows the measured as well as the theoretically predicted distributions of the highest side lobe levels associated with the 90 patterns. The agreement is remarkable. To safeguard against the luck in this experiment, a second completely independently designed holey plate was also constructed and tested. The same close agreement was obtained as shown in Fig. 20b. Finally, it is worth noting again that if a conventional array design (i.e., using uniformly spaced elements) is used to produce the same beam and side lobe level, 10^4 elements would be required. This number is about 50 times what was used in the experiment. Obviously the

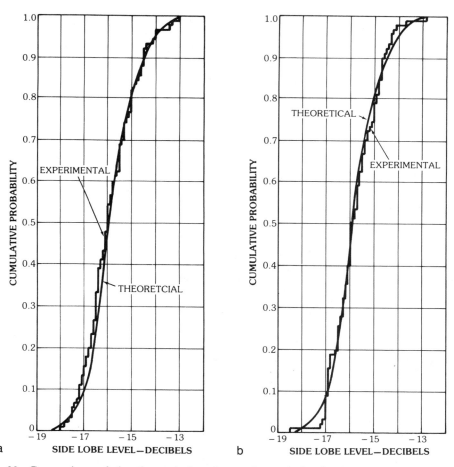

Fig. 20. Comparison of the theoretical and experimental distributions for the side lobe levels of linear arrays. (*a*) Experimental distribution obtained from a set of 90 pseudo-random linear arrays generated by the planar array 1. (*b*) Experimental distribution obtained from a set of 90 pseudo-random linear arrays generated by the planar array 2. (*After Lo and Simcoe [23],* © *1967 IEEE*)

difference is by no means insignificant. Of course, the two designs would not have the same directivity. On the other hand, suppose that the desired directivity is less than 10^4, say 10^3. Then the random spacing approach will provide a design which uses not only 1/10 of the elements but also can yield a lower side lobe level and/or narrower beam. In other words, in this random design the required number of elements is determined not directly by the aperture size but rather by the desired directivity and side lobe level.

Scanning Characteristics—It is a simple matter to obtain cophasal excitation in a holey plate experiment. If the holey plate is rotated at an angle α with respect to the lens as shown in Fig. 21, all elements will automatically be excited cophasally to produce a beam at an angle α from the broadside. In Fig. 21 a linear uniformly spaced slot array (or grating), etched from a copper-clad substrate with relative permittivity $\epsilon_r = 4.25$, is placed in front of a lens. In this case if the receiving waveguide and the lens are held fixed while only the grating rotates, one obtains the scanning characteristics, namely, the main beam magnitude variation (or the scanning pattern) with the scan angle α. The measured scanning pattern for a linear array of 61 uniformly spaced slots when **E** is perpendicular to the slots is shown in Fig. 22. It is clearly shown that at about ±18° the main beam magnitude drops sharply to −20 dB. These are the blind angles for this particular array. When the same 61 slots were rearranged to have random spacings but the same average spacing $d_{av} = 0.6\lambda$, the blind angles disappeared, as clearly seen in Fig. 22. Fig. 23 shows the superimposed measured main beams for another linear array of 45 slots, both uniformly spaced (Fig. 23a) and randomly spaced (Fig. 23b), all with the same average spacing $d_{av} = 0.6\lambda$. For this case the **E** vector is parallel to the slots and the blind angles appear at approximately ±30° as shown in Fig. 23a when the slots are spaced uniformly and become absent in Fig. 23b when they are spaced at random.

Figs. 24a through 24c show the computed patterns for a single element in an array environment of 21 elements, i.e., $F(v, v)$ in (27), for the case of perpendicular polarization and uniform spacing. With the exception of the few edge elements, they all are almost identical with the scanning pattern shown in Fig. 22 so that the sharp drop at about ±18° becomes cumulative, resulting in the blind angle [37]. From this, one expects that the larger the array, the deeper the drop. When spacings are random, however, the array element pattern is completely different from one element to another as shown in Fig. 25, thus resulting in a scanning pattern free from blind angles. For details and the theoretical analysis, readers are referred to Agrawal and Lo [24, 37].

Other Remarks

In most phased arrays quantized phases are used. For uniformly spaced elements the phase errors incurred in the elements are systematic and, as a result, not only the main beam magnitude may be reduced but the side lobe level raised also. In case of randomly spaced elements such deleterious effects can be reduced particularly for large arrays. Another method that can be used to achieve this goal in a similar manner is to deliberately add a small amount of random phase to each quantized level. However, this method is effective only when a large number of levels (or bits) is used in the system.

Aperiodic Arrays

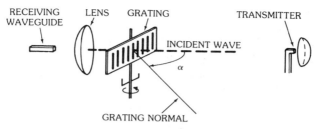

Fig. 21. Sketch of the experimental setup for measuring main beam magnitude fluctuation. (*After Agrawal and Lo [37]*)

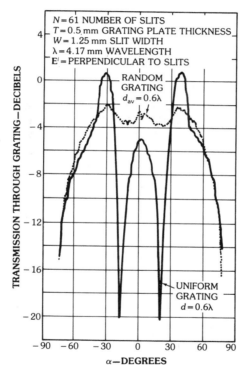

Fig. 22. Measured transmitted power through uniform and random gratings versus orientation angle α (perpendicular polarization). (*After Agrawal and Lo [37]*)

For many conformal arrays, such as cylindrical and spherical arrays, even if the elements are placed in a regular manner (for example, equiangular spacings), their pattern functions cannot be simply expressed in terms of polynomials as for linear arrays. Thus there is, in general, no analytic advantage to considering a particular element arrangement. Commonly the patterns are computed numerically, and clearly for large arrays this brute force method is very costly. However, the contributions from all elements to a pattern in a plane have phases proportional to

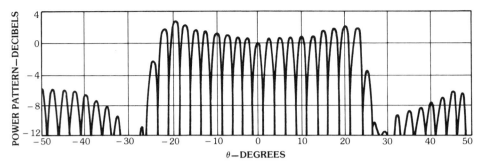

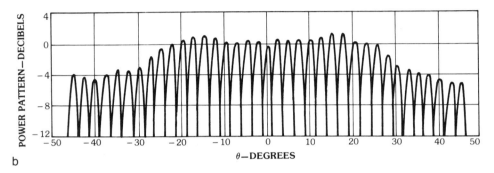

Fig. 23. Measured main beam patterns, superimposed for various scan angles α (parallel polarization). (*a*) Uniform array with $d = 0.6\lambda$, $N = 45$. (*b*) Random array with $d_{av} = 0.6\lambda$, $N = 45$. (*After Agrawal and Lo [37]*)

the projections of the element positions on that plane which are in general pseudo-random. For this reason a probabilistic approach to large cylindrical and spherical arrays has been studied [38].

With the advent of log-periodic antennas, extremely wide band antennas become a reality. This outstanding family of antennas, however, has its own limitations, namely, low directivity and wide beamwidth, since for each frequency only a limited portion of the antenna is active. To form an array with these elements for high directivity, one will be confronted with the difficulty that the physical element spacing increases with frequency, causing a serious grating lobe problem. To alleviate this difficulty random spacings of log-periodic elements, particularly for large arrays, should be used since such an array can tolerate large element spacings and is inherently frequency insensitive, as shown in this chapter.

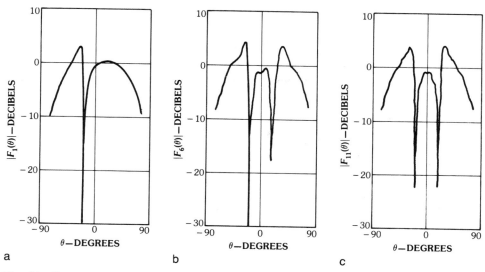

Fig. 24. Computed single-slit pattern $|F_k(\theta)|$ when only the kth element of a uniformly spaced array with 21 elements is excited, where $d = 0.6\lambda$, $\epsilon_r = 4.25$, $T = 0.5$ mm, $W = 1.25$ m, $\lambda = 4.17$ mm, and **E** is perpendicular to slit. (*a*) First-element pattern. (*b*) Sixth-element pattern. (*c*) Central-element pattern. (*After Agrawal and Lo [37]*)

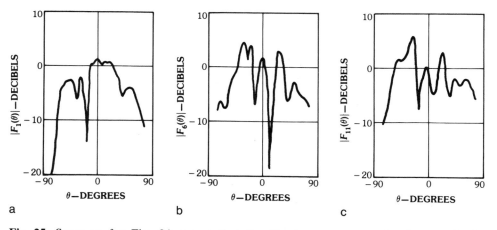

Fig. 25. Same as for Fig. 24 except that the 21 elements are randomly spaced with $d_{av} = 0.6\lambda$. (*a*) First-element pattern. (*b*) Sixth-element pattern. (*c*) Central-element pattern. (*After Agrawal and Lo [37]*)

5. References

[1] H. Unz, "Linear arrays with arbitrarily distributed elements," *IRE Trans. Antennas Propag.*, vol. AP-8, pp. 222–223, March 1960.

[2] D. D. King, R. F. Packard, and R. K. Thomas, "Unequally spaced broadband antenna arrays," *IRE Trans. Antennas Propag.*, vol. AP-8, pp. 380–385, July 1960.

[3] Y. T. Lo, "A nonuniform linear array system for the radio telescope at the University of Illinois," 1960 Spring URSI and IRE Joint Meeting, Washington, D.C., subsequently published in the *IRE Trans. Antennas Propag.*, vol. AP-9, pp. 9–16, January 1961, under the title, "The University of Illinois radio telescope," by G. W. Swenson and Y. T. Lo.

[4] A. L. Maffett, "Array factors with nonuniform spacing parameter," *IRE Trans. Antennas Propag.*, vol. AP-10, pp. 131–136, March 1962.

[5] M. G. Andreason, "Linear arrays with variable interelement spacings," *IRE Trans. Antennas Propag.*, vol. AP-10, pp. 137–143, March 1962.

[6] A. Ishimaru, "Theory of unequally spaced arrays," *IRE Trans. Antennas Propag.*, vol. AP-10, pp. 691–702, November 1962.

[7] J. L. Yen and Y. L. Chow, "On large nonuniformly spaced arrays," *Can. J. Phys.*, vol. 41, p. 1, January 1963.

[8] A. Ishimaru and Y. S. Chen, "Thinning and broadbanding antenna arrays by unequal spacings," *IEEE Trans. Antennas Propag.*, vol. AP-13, pp. 34–42, January 1965.

[9] G. Bernard and A. Ishimaru, *Tables on the Anger and Lommel-Weber Functions*, Seattle: University of Washington Press, 1962.

[10] Y. L. Chow and J. L. Yen, "Grating plateaux of planar nonuniformly spaced arrays," presented at the 1964 Fall URSI Meeting, Urbana, Il.

[11] M. I. Skolnik, G. Nemhauser, and J. W. Sherman, III, "Dynamic programming applied to unequally spaced arrays," *IEEE Trans. Antennas Propag.*, vol. AP-12, pp. 35–43, January 1964.

[12] Y. T. Lo, "On the theory of randomly spaced antenna arrays," *Tech. Rep. 1*, Antenna Lab., Dept. of Electr. Eng., Univ. of Illinois, Urbana, under NSF G 14894, 1962. Part of this report appeared in "A mathematical theory of antenna arrays with randomly spaced elements," *IEEE Trans. Antennas Propag.*, vol. AP-12, pp. 257–268, May 1964.

[13] Y. T. Lo, "A probabilistic approach to the problem of large antenna arrays," *Radio Sci.*, vol. 68D, pp. 1011–1019, September 1964.

[14] R. F. Harrington, "Side lobe reduction by nonuniform element spacing," *IRE Trans. Antennas Propag.*, vol. AP-9, pp. 187–192, March 1961.

[15] Y. V. Baklanov, V. L. Pokrovshi, and G. I. Surdotovich, "A theory of linear antennas with unequal spacing," *Radio Eng. Electron. Phys.*, no. 6, pp. 905–913, June 1962.

[16] S. J. Rabinowitz and R. F. Kolar, "Statistical design of space-tapered arrays," presented at the 1962 Twelfth Annual Symposium on USAF Antenna Res. and Dev. Prog., Univ. of Illinois, Urbana.

[17] M. I. Skolnik, J. W. Sherman, III, and F. C. Ogg, Jr., "Statistically designed density-tapered arrays," *IEEE Trans. Antennas Propag.*, vol. AP-12, pp. 408–417, July 1964.

[18] T. M. Maher and D. K. Cheng, "Random removal of radiators from large linear arrays," *IEEE Trans. Antennas Propag.*, vol. AP-11, pp. 106–111, March 1963.

[19] Y. T. Lo, "Side lobe level in nonuniformly spaced antenna arrays," *IEEE Trans. Antennas Propag. (Communications)*, vol. AP-11, p. 511, July 1963.

[20] Y. T. Lo, "High-resolution antenna arrays with elements at quantized random spacings," presented at the 1964 International Conference on Microwaves, Circuit Theory, and Information Theory, Tokyo.

[21] Y. L. Chow, "On grating plateaus of nonuniformly spaced arrays," *IEEE Trans. Antennas Propag.*, vol. AP-13, pp. 208–215, March 1965.

[22] Y. T. Lo and S. W. Lee, "A study of spaced-tapered arrays," *IEEE Trans. Antennas Propag.*, vol. AP-14, pp. 22–30, January 1966.

[23] Y. T. Lo and R. J. Simcoe, "An experiment on antenna arrays with randomly spaced elements," *IEEE Trans. Antennas Propag.*, vol. AP-15, pp. 231–235, March 1967.

[24] V. D. Agrawal and Y. T. Lo, "Mutual coupling in phased arrays of randomly spaced antennas," *IEEE Trans. Antennas Propag.*, vol. AP-20, pp. 288–295, May 1972.

[25] A. R. DiDonato and M. P. Jarnagin, "Integration of the general bivariate Gaussian distribution over an offset ellipse," *NWL Report 1710*, Naval Weapons Laboratory, Dahlgren, Va., 1960.

[26] S. O. Rice, "Mathematical analysis of random noise," *Bell Syst. Tech. J.*, vol. 24, p. 46, 1945.
[27] P. B. Patnaik, "The noncentral χ^2 and F distributions and their application," *Biometrica*, vol. 36, pp. 202–232, 1949.
[28] K. Pearson, *Tables of Incomplete Gamma Functions*, Department of Scientific and Industrial Research, Cambridge: Cambridge University Press, 1934.
[29] K. Karhunen, "Über lineare methoden in der wahrscheinlichkeits rechnung," *Ann. Acad. Sci. Fennicae*, ser. A, no. I, pp. 37–79, 1947.
[30] W. N. Christiansen and J. H. Hogbem, "A design for the Benelux Cross antenna," *BCAP Tech. Rep. No. 3*, Benelux Cross Antenna Project, Sterrewacht, Leiden, Netherlands.
[31] D. B. Owen, *Handbook of Statistical Tables*, pp. 535–536, Reading: Addison-Wesley Pub. Co., 1962.
[32] E. Parzen, *Modern Probability Theory and Its Applications*, New York: John Wiley & Sons, 1960.
[33] N. Amitay, V. Galindo, and C. P. Wu, *Theory and Analysis of Phased Array Antennas*, New York: Wiley-Interscience, 1972.
[34] L. W. Lechtreck, "Effects of coupling accumulation in antenna arrays," *IEEE Trans. Antennas Propag.*, vol. AP-16, pp. 31–37, January 1968.
[35] J. M. Stone, *Radiation and Optics*, New York: McGraw-Hill Book Co., pp. 146–152, 1963.
[36] M. I. Skolnik, "A method of modeling array antennas," *IEEE Trans. Antennas Propag.*, vol. AP-11, pp. 97–98, January 1963.
[37] V. D. Agrawal and Y. T. Lo, "Anomalies of dielectric-coated gratings," *Appl. Opt.*, vol. 11, pp. 1946–1951, September 1972.
[38] A. R. Panicali and Y. T. Lo, "A probabilistic approach to large circular and spherical arrays," *IEEE Trans. Antennas Propag.*, vol. AP-17, pp. 514–522, July 1969.

Chapter 15

Reflector Antennas

Y. Rahmat-Samii
Jet Propulsion Laboratory

CONTENTS

1. Introduction   15-5
2. Basic Formulations for Reflector Antenna Analysis   15-6
      Physical Optics Analysis   15-8
      Geometrical Theory of Diffraction Analysis   15-11
      Aperture Field Method   15-13
3. Simple Formulas for Far Fields of Tapered-Aperture Distributions   15-15
      Two-Parameter Model   15-15
      One-Parameter Model   15-18
      Near, Fresnel, and Far Fields   15-21
4. Some Important Geometrical Features of Conic-Section–Generated Reflector Antennas   15-23
      Conic Sections   15-23
      Reflector Surfaces   15-26
      Intersection with a Circular Cone   15-26
      Special Cases   15-29
      Intersection with a Circular Cylinder   15-30
5. Offset (Symmetric) Parabolic Reflectors   15-31
      Geometrical Parameters   15-31
      Idealized Feed Patterns   15-34
      Edge and Feed Tapers   15-36
      Reflector Pattern Characteristics for On-Focus Feeds   15-37
      Reflector Pattern Characteristics for Off-Focus Feeds   15-49
6. Dual-Reflector Antennas   15-61
      General Parameters   15-67
      Performance Evaluation   15-69
      Cross-Polarization Reduction   15-69
      Scan Performance   15-71
      Shaped Reflectors   15-73

Dr. Yahya Rahmat-Samii is a Professor of Electrical Engineering at the University of California Los Angeles (UCLA). He has been a Senior Research Scientist at NASA's Jet Propulsion Laboratory/California Institute of Technology since 1978, where he contributed significantly to advance antenna technology for space programs, satellite communications, large space systems, and microwave holography. Dr. Rahmat-Samii received the M.S. and Ph.D. degrees in Electrical Engineering from the University of Illinois, Champaign-Urbana.

Dr. Rahmat-Samii is a Fellow of IEEE (1985), a Fellow of IAE (1986), and the 1984 recipient of the Henry Booker Award of URSI, awarded triennially to the most outstanding young radio scientist in North America. He was appointed an IEEE Antennas and Propagation Society Distinguished Lecturer and presented lectures internationally. He was an elected IEEE AP-S AdCom member for the second term and has been an Associate Editor of the IEEE Antennas and Propagation Transactions and Magazine. He was the Chairman of the IEEE Antennas and Propagation Society of Los Angeles in 1987/88/89. In 1989, his chapter received the best Antennas and Propagation chapter award from the AP Society. He is one of the three International Editors of the IEE book series on Electromagnetics and Antennas. He is also one of the Directors of AMTA (Antenna Measurement Techniques Association) and the Electromagnetic Society. Dr. Rahmat-Samii has organized many short courses and presented numerous invited seminars. He is listed in Who's Who in America, and Who's Who in Frontiers of Science and Technology.

Dr. Rahmat-Samii has authored or co-authored, over 210 technical journal articles and conference papers and has written chapters in eight books. He has pioneered/contributed significantly to the developments of the Spectral Theory of Diffraction (STD) and GTD, Near-field plane-polar and bi-polar antenna measurements, Microwave holographic diagnostics, mobile satellite communication antennas, reflector surface compensation, multi-reflector antenna diffraction analysis and synthesis, scattering and radiation from complex objects and RCS computations, singularity in dyadic Green's function, high power microwave (HPM) antennas, EMP and aperture penetration, etc.

For his contributions, Dr. Rahmat-Samii has received numerous NASA Certificates of Recognition and recently earned the JPL Team NASA's Distinguished Group Achievement Award. In 1992, he was the recipient of the Best Application Paper Award (Wheeler Award) for his paper published in the IEEE Antennas and Propagation Transactions in 1991. Dr. Rahmat-Samii is a member of Commissions A, B and J of USNC/URSI, Sigma Xi, Eta Kappa Nu, AMTA and Electromagnetics Academy.

7. Contour Beam Reflectors — 15-80
8. Feeds for Reflectors — 15-84
   Radiation Patterns of Simple Feeds 15-89
   Complex Feeds 15-94
   $\cos^q(\theta)$ Type Patterns 15-99
9. Effects of Random Surface Errors — 15-105
10. Appendix: Coordinate Transformations for Antenna Applications — 15-115
    Cartesian and Spherical Components 15-115
    Eulerian Angles 15-116
    Feed and Reflector Coordinates 15-117
    Determination of Eulerian Angles 15-120
11. References — 15-120

1. Introduction

The fine art of synthesizing, analyzing, and designing reflector antennas of many various geometries did not really advance until the days of World War II, when numerous radar applications evolved to satisfy diverse technical demands. Subsequent demands for reflectors for use in radio astronomy, microwave communications, satellite communications and tracking, and the like have resulted in both the development of sophisticated reflector configurations and analytical and experimental design techniques. Reflector antennas may take many configurations, some of the most popular ones being plane, corner, curved reflectors, and so on. In this chapter only the curved reflectors, such as parabolic and Cassegrain, will be discussed. The reader is referred to references [1, 2, 3, 4] for other shapes.

The reflector antennas (curved) can be classified in a variety of ways, and one of the more recently suggested models shown in Fig. 1 is obtained from reference [5]. Fig. 1 identifies reflectors according to pattern type, reflector type, and feed type. In Fig. 1, pencil-beam reflectors are very popular and are commonly used in point-to-point microwave communications, since their patterns yield the maximum boresight gain and typically their beam directions are fixed at the time of antenna installation. In satellite communication systems the uplink beam of these pencil-beam reflectors may be either fully steerable by reflector movements, as in INTELSAT ground stations, or capable of limited steering, as in the Canadian domestic systems. Many new generations of satellite reflectors have produced other popular types of pattern classifications: contour (shaped) beams and multiple beams. These applications demand reflectors with improved off-axis beam characteristics, which result in more sophisticated configurations as suggested in Fig. 1. Many microwave communication antennas operate with one sense of polarization at a given frequency and require only reasonable discrimination between orthogonal polarizations. However, many of the current generation of microwave communication antennas operate with dual polarizations at the same frequency to enhance their so-called frequency reuse capabilities. Such requirements impose stringent demands on the polarization performance of reflectors and could be used as factors in pattern classification (Fig. 1).

Finally, the performance of reflectors cannot be properly examined without knowledge of the feed configurations of the reflectors. For this reason column three of Fig. 1 suggests a classification based on feed types. Horn or waveguide feeds operating in a single pure mode have been the most popular feeds for reflectors. However, in order to meet radio astronomy, earth station, and satellite antenna requirements considerable efforts have been focused on the development of new feeds to efficiently illuminate either the main reflector or the subreflector of the antenna. For example, hybrid-mode feeds (combining TE and TM fields) are used to match efficiently the feed distribution with the desired focal field distribution of the reflector, which also results in an ideal feed for reducing cross

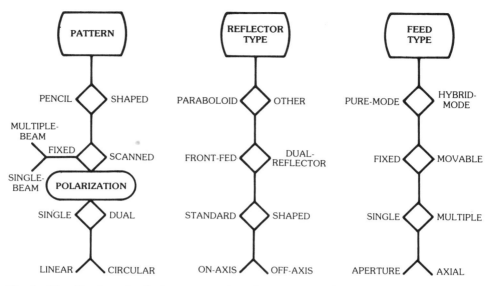

Fig. 1. Classification of reflector antennas based on pattern, reflector, and feed types. (*After Clarricoats and Poulton [5]*, © *1977 IEEE*)

polarization. Among the hybrid-mode feeds are feeds with corrugated metal walls. Another important consideration is the complexity of the feed system in terms of the number of feed elements used, for instance, to create contour or multiple beams. The proximity of feed elements necessitates an understanding of mutual coupling and the overcoming of difficulties in designing an acceptable beam-forming network. The reader may refer to references [6, 7] for a detailed discussion of the feeds.

The objective of this chapter is to provide a summarized description of the performances of many modern reflector configurations as discussed above and of advanced analytical techniques for analyzing reflectors. The emphasis is to present useful design data that complement information given in other reflector handbooks [1, 3, 7, 8]. In particular a considerable amount of data is provided for offset reflector configurations which are used extensively in today's antenna design. The are a few recently published review papers and collections of papers that the reader is strongly advised to study [1, 5, 9, 10]. The additional references assembled in these papers are not repeated here and only the most recent and relevant ones are referred to in this chapter.

2. Basic Formulations for Reflector Antenna Analysis

Since analytical/numerical techniques are used very extensively in the performance evaluation of reflector antennas and since there are numerous successful checks of the results of these techniques against measured data, this section describes application of some of the most currently applied techniques. Dual-offset reflector antennas are used as the base configuration for the description of these techniques.

The geometry of a dual-offset reflector with a feed arbitrarily positioned is shown in Fig. 2. This is the most general configuration and all other cases, such as single reflectors, symmetric configurations, Cassegrains, etc., can be regarded as special cases. Reflectors with more than one subreflector (beam waveguide systems) may also be handled by replacing the subreflector of Fig. 2 with the appropriate beam waveguide systems. Three coordinate systems are erected to define the main reflector, the subreflector, and the feed position (or array of feeds). The position and field vectors of these coordinate systems can be interrelated using the Eulerian angles construction [11] (see Appendix). For instance, the fields of feed can be expressed in feed coordinates (x_F, y_F, z_F) and then transformed into subreflector coordinates (x_S, y_S, z_S) to determine the scattered field from the subreflector and then transformed again into main reflector coordinates (x_M, y_M, z_M) to finally obtain the radiated field of the main reflector.

There are many different analytical/numerical techniques to determine the radiation characteristics of reflectors. Among them one may refer to physical optics (PO), aperture field (AF), geometrical optics (GO), geometrical theory of diffraction (GTD), method of moments (MOM), or any combination of these. All of these techniques have advantages or disadvantages, depending on the particular reflector configuration, far-field pattern domain, polarization, computation time, accuracy, and so on. For example, MOM gives the most accurate result but it is impossible to use it economically for reflectors larger than approximately 5 wavelengths. The aperture field method is not very accurate for offset configuration with displaced feeds when the edge diffracted rays are not included in the construction of the aperture fields. The GTD method is not easily applied in the caustics regions of pencil beams. The PO method can take an excessive amount of computation time for large reflectors, in particular, in the dual-reflector configuration, etc. Many advances have been reported in the last decade that improve both the accuracy and the computation time of analysis by applying a variety of techniques. One of the most appealing techniques has been the application of GTD for the subreflector and physical optics for the main reflector, in conjunction with

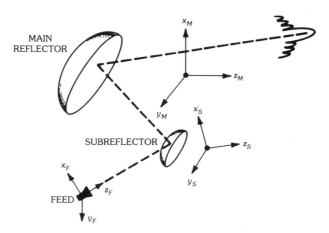

Fig. 2. Geometry of a dual-offset reflector antenna.

efficient expansions, such as Jacobi-Bessel, sampling theorem, etc. It is these last procedures which are summarized in this paper.

Physical Optics Analysis

The foundation of physical optics, or PO, rests on the assumption that the induced current on the reflector surface is given by (for a perfect conductor)

$$\mathbf{J} = \begin{cases} 2\hat{\mathbf{n}} \times \mathbf{H}^i & \text{illuminated region} \\ 0 & \text{otherwise} \end{cases}$$

where $\hat{\mathbf{n}}$ is the unit normal to the surface and $\mathbf{H}^i$ is the incident magnetic field. This incident field may emanate directly from the source or be scattered from the subreflector. Although PO current is an approximation for the true current on the reflector surface (obtainable using MOM), it nevertheless gives very accurate results for far fields of reflectors as small as approximately 5 wavelengths in diameter. The PO current displays certain errors, particularly near the edge of the reflector, which can be augmented by the incorporation of the fringe current [12]. Unfortunately the incorporation of the fringe current is not in general an easy task, although it may be required to accurately determine the far fields for a very wide range of observation angles. The radiated $\mathbf{E}$ field can be constructed using

$$\mathbf{E}(\mathbf{r}) = -\frac{j}{\omega\epsilon} \int_\Sigma [\mathbf{J} \cdot \nabla) \nabla + k^2 \mathbf{J}] g(R) \, dS' \quad (1)$$

which is applicable for both the near- and far-field zones [4, 13]. The above expression can be easily derived using the concept of vector potentials. In (1), $\mathbf{r}'$, $\mathbf{r}$, and $\mathbf{R}$ are described in Fig. 3, $k = \omega\sqrt{\mu\epsilon}$, and

$$\begin{aligned} g(R) &= \frac{e^{-jkR}}{4\pi R} \\ \nabla g(R) &= \left(jk + \frac{1}{R} \right) g(R) \hat{\mathbf{R}} \\ (\mathbf{J} \cdot \nabla)\nabla g(R) &= \left[-k^2 (\mathbf{J} \cdot \hat{\mathbf{R}})\hat{\mathbf{R}} + \frac{3}{R}\left(jk + \frac{1}{R} \right) \right. \\ &\quad \left. \times (\mathbf{J} \cdot \hat{\mathbf{R}})\hat{\mathbf{R}} - \frac{1}{R}\left(jk + \frac{1}{R} \right)\mathbf{J} \right] g(R) \end{aligned} \quad (2)$$

Equation 1 can be further simplified to give the following well-known expression for the radiated field in the far-field zone:

$$\mathbf{E}(\mathbf{r}) = -\frac{jk^2}{\omega\epsilon} g(r)(\bar{\bar{\mathbf{I}}} - \hat{\mathbf{r}}\hat{\mathbf{r}}) \cdot \mathbf{T}(\theta, \phi) \quad (3)$$

where

Reflector Antennas

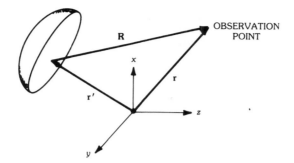

Fig. 3. Reflector geometry and integration and observation parameters for constructing the physical optics (PO) formulation.

$$\mathbf{T} = \int_{\Sigma} \mathbf{J} e^{jk\hat{\mathbf{r}}\cdot\mathbf{r}'} d\Sigma' \tag{4}$$

In deriving the above expression the standard far-field approximations are used, namely,

$$R = |\mathbf{r} - \mathbf{r}'| \cong r - \hat{\mathbf{r}}\cdot\mathbf{r}'$$

A similar but slightly more complicated expression can also be developed for the field in the Fresnel zone [13].

The surface integration in (4) is performed on the curved surface Σ. This integration, however, can be recast in terms of integration variables defined in a planar aperture by properly using the surface projection transformation to obtain

$$\mathbf{T} = \int_{\sigma} \mathbf{J}_{\text{eff}} e^{jk\hat{\mathbf{r}}\cdot\mathbf{r}'} dS' \tag{5}$$

where

$$\mathbf{J}_{\text{eff}} = \mathbf{J}\sqrt{1 + \left(\frac{\partial f}{\partial x'}\right)^2 + \left(\frac{\partial f}{\partial y'}\right)^2}$$

where σ is the projected aperture in this plane (for example, the xy plane), and $z' = f(x', y')$ defines the description of the reflector surface. One can show that this integral may be represented in terms of a summation series of many two-dimensional Fourier integrals [14, 15]. Typically one needs the first few terms of this series [14] to achieve a converging solution.

The PO radiation integral may be evaluated in many different ways and its efficient evaluation has been a challenging problem. A successful numerical method has been used based on the small patch constant phase technique [16] which can be applied to both the subreflector and main reflector geometries. Efficient numerical methods, which are particularly applicable to the main reflector with circular projected apertures, have been developed in [14, 17] and used frequently. One of these methods, which is based on the Jacobi-Bessel expansion, is particularly effective when it is used to evaluate (5). It can be shown that, based on this expansion, the radiation integral can be expressed as [14, 15]

$$T \propto \sum_{p=0}^{P \to \infty} \frac{1}{p} (jk)^p (\cos\theta - \cos\theta_B)^p \sum_{n=0}^{N \to \infty} \sum_{m=0}^{M \to \infty} J^n[{}_pC_{nm}\cos n\phi + {}_pD_{nm}\sin n\phi]$$
$$\times \sqrt{2(n + 2m + 1)} \left(\frac{J_{n+2m+1}[kaB]}{kaB} \right) \quad (6)$$

In the previous equation, a is the radius of the geometrically projected circular aperture, J is the Bessel function, and

$$B = \sqrt{(\sin\theta\sin\phi - \sin\theta_B\sin\phi_B)^2 + (\sin\theta\cos\phi - \sin\theta_B\cos\phi_B)^2}$$
$$\phi = \tan^{-1}\left(\frac{\sin\theta\sin\phi - \sin\theta_B\sin\phi_B}{\sin\theta\cos\phi - \sin\theta_B\cos\phi_B} \right) \quad (7)$$

where (θ_B, ϕ_B) is the direction of the anticipated beam maximum (or its vicinity). Note that when $B = 0$, i.e., in the boresight direction, only the first term of the series contributes. This first term is the Airy disk function, a typical far-field pattern of a uniform aperture. Another important feature of (6) is the fact that the coefficients **C** and **D** are independent of the observation points. These coefficients take the following form:

$$\begin{Bmatrix} {}_pC_{nm} \\ {}_pD_{nm} \end{Bmatrix} = \frac{\varepsilon_n}{2\pi} \int_0^{2\pi} \int_0^1 Q_p \begin{Bmatrix} \cos n\phi' \\ \sin n\phi' \end{Bmatrix} F_m{}^n(s') s'\, d\phi'\, ds' \quad (8)$$

where ε_n is the Neumann constant ($\varepsilon_n = 1$ for $n = 1$ and $\varepsilon_n = 2$ otherwise), $F_m{}^n$ is the modified Jacobi polynomial [15], and Q_p is directly related to the physical optics induced current. The exact mathematical form of Q_p can be found in [14, 15].

Another useful representation of the far field is obtained by applying the sampling theorem, which allows one to express the far fields from a knowledge of field at a limited number of sampling points [18], viz.,

$$\mathbf{E} = \sum \sum \mathbf{E}_{nm} \frac{\sin(u - n\pi)}{u - n\pi} \frac{\sin(v - m\pi)}{v - m\pi} \quad (9)$$

where $u = ka'\sin\theta\cos\phi$ and $v = ka'\sin\theta\sin\phi$ (a' is the radius of an enlarged aperture [18] in front of the reflector) and the $\mathbf{E}_{nm}$ are the fields at the sampling points dictated by the Wittaker-Shannon theorem with spacing $\Delta u \leq \pi/a'$ and $\Delta v \leq \pi/a'$. This far-field representation is particularly useful when one is evaluating the fields at many observation points. A generalization of this procedure can also be performed using nonuniform sampling techniques [19].

Reflector Antennas

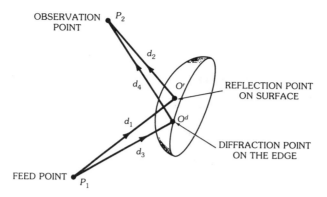

Fig. 4. Reflector geometry and reflection and diffraction parameters for constructing the geometrical theory of diffraction (GTD) formulation.

Geometrical Theory of Diffraction Analysis

Since the fundamentals of the geometrical theory of diffraction, or GTD, are discussed in another chapter, here only its application to reflectors is summarized. The essence of GTD lies in its concepts of localization and geometrical rays satisfying Fermat's principle. These rays may be classified as the geometrical and edge-diffracted rays emanating from surface reflection and edge-diffracted points in accordance with Snell's law and Keller's theory [20]. For the reflector of Fig. 4 the total scattered **H** field can be expressed as

$$\mathbf{H}^t = \theta^i \mathbf{H}^i + \sum_i \theta_i^r \mathbf{H}_i^r + \sum_j \mathbf{H}_j^d \tag{10}$$

where the superscripts i, r, and d designate the incident, reflected, and diffracted fields, respectively, and the subscripts i and j refer to the number of surface reflection and edge diffraction specular points, respectively, for a given feed and observation location. Furthermore, θ^i and θ^r designate the incident and reflected shadow indicators (i.e., $\theta = 0$ if the ray is blocked, and $\theta = 1$ if the ray path is unblocked) [21,22,23]. The **E** field can also be constructed in a straightforward manner using the fact that on each ray the **E** and **H** fields assume a plane wave relationship, namely,

$$\mathbf{E} = Z\mathbf{H} \times \hat{\mathbf{r}}$$

where Z is the intrinsic impedance of the medium ($Z = 120\pi$ ohms for the free space) and $\hat{\mathbf{r}}$ is the unit vector in the direction of each ray.

To apply the ray construction it is necessary to associate a ray field with the radiation field of the feed. For a single feed with a well-defined phase center one typically associates rays emanating from the phase center. For more distributed sources (arrays of feeds) one may have to associate many phase centers (one per feed element) and repeat the GTD construction for each of them. For a given feed position and specified observation point (this can be a point on the main reflector) one must first determine the GO reflection and edge-diffracted points (there can be more than one point for each type). There are different techniques for this determination; however, all of them are based on Fermat's principle, which requires that the optical path be optimal (stationary). Mathematically speaking, it is required that the functional

$$d = d_{1(3)} + d_{2(4)} \tag{11}$$

as defined in Fig. 4 be stationary with the constraint that (11) gives a solution(s) on the reflector surface and the edge boundary. Once reflection and diffraction points are obtained from the solution of the above nonlinear equations, one can then implement the field construction of GTD to determine both the reflected and diffracted fields. The details of this construction may be found in [22] and other references [23].

The geometrical optics field (reflected field) transported by the reflected rays can be formulated as

$$\mathbf{H}^r(P_2) = (\mathrm{DF}) e^{-jkd_2} \{ \mathbf{H}^i(O^r) - 2[\mathbf{H}^i(O^r) \cdot \hat{\mathbf{n}}] \hat{\mathbf{n}} \} \tag{12}$$

which is given in terms of the incident field $\mathbf{H}^i$ at the reflection point O^r, the unit surface normal $\hat{\mathbf{n}}$ at O^r, and a divergence factor (DF). This factor can be expressed as

$$\mathrm{DF} = \frac{1}{\sqrt{1 + (d_2/R_1^r)}} \frac{1}{\sqrt{1 + (d_2/R_2^r)}} \tag{13}$$

where the square roots take positive real, negative imaginary, or zero values (so that DF is positive real, positive imaginary, or infinite). The terms R_1^r and R_2^r are the radii of principal curvature of the reflected wavefront passing through O^r. Their computation can be found in [22] or in the GTD chapter of this handbook.

Corresponding to each diffraction point O^d of Fig. 4 there is a contribution to the diffracted field $\mathbf{H}^d$ in (10). Following the formulation of [22], the diffracted field can be expressed as

$$\mathbf{H}^d(P_2) = \frac{1}{2\sqrt{2\pi d_4}} e^{-j(d_4 + \pi/4)} \frac{1}{\sqrt{1 + (d_4/R_1)}} \frac{1}{\sin\beta} [\boldsymbol{\beta} D^h H_\beta^i + \boldsymbol{\alpha} D^s H_\alpha^i] \tag{14}$$

where

R_1 = the radius of curvature of the diffracted wavefront passing through O^d
β = the angle between the tangent to the edge and $\mathbf{d}_4$
$\hat{\boldsymbol{\alpha}}, \hat{\boldsymbol{\beta}}$ = the unit vectors of the diffracted-ray coordinates
D^h, D^s = the soft and hard diffraction coefficients
H_α^i, H_β^i = the projections of the $\mathbf{H}^i$ incident field on the ray coordinates at O^d

The detailed mathematical construction of these parameters can be found in [22]. It is worthwhile to mention that Keller's standard diffraction coefficients diverge at the incident and reflected shadow boundaries, which can be remedied by applying a uniform diffraction coefficient. There are different uniform expressions for the diffraction coefficients and the reader may refer to [24, 25, 26, 27] for a detailed comparison.

Aperture Field Method

Another widely used reflector analysis technique is based on the application of the aperture field, or AF, method. In this method first a hypothetical planar aperture is erected in front of the reflector and then the tangential fields are determined in this aperture using the GO and GTD constructions. The aperture is typically truncated to the reflector projected aperture size when one deals with well-focused pencil-beam antennas and only the GO field construction is used. For cases where the feed is defocused or for fields at wide observation angles, however, one must use larger apertures to properly incorporate the contribution of the edge-diffracted fields. (Alternatively one may use an aperture which caps the reflector's rim.) Once the tangential **E** and **H** fields in the aperture are determined, the far fields can then be obtained.

There are many different representations for the construction of far fields which use either the aperture tangential **E** or **H** fields separately or a combination of them. These different representations are customarily referred to as different Kirchhoff approximations [2]. One of the representations which uses both the aperture tangential **E** and **H** fields and results in the far-field radiated **E** field can be formulated as

$$\mathbf{E} = \frac{-jk^2}{\omega\epsilon}g(r)\int_A \left\{ \hat{\mathbf{n}} \times \mathbf{H}^A - [(\hat{\mathbf{n}} \times \mathbf{H}^A)\cdot\hat{\mathbf{r}}]\hat{\mathbf{r}} - \frac{1}{Z}[(\hat{\mathbf{n}} \times \mathbf{E}^A) \times \hat{\mathbf{r}}] \right\} e^{jk\hat{\mathbf{r}}\cdot\mathbf{r}'} dA \quad (15)$$

where $\mathbf{E}^A$ and $\mathbf{H}^A$ are the **E** and **H** fields in the aperture with aperture size A and $\hat{\mathbf{n}}$ is the normal to this planar aperture as shown in Fig. 5. Furthermore, $g(r)$ is defined

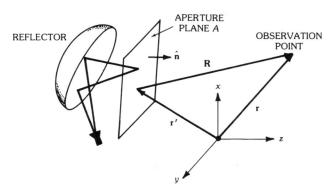

Fig. 5. Reflector geometry and integration and observation parameters for constructing the aperture field (AF) formulation.

in (2). For a planar aperture, (15) is a two-dimensional Fourier transform and can be evaluated in many different ways, one of which includes the use of the fast Fourier transform (FFT) algorithm. The reader is referred to [1, 28, 29] for different computational methods and comparisons among them (see also Fig. 6).

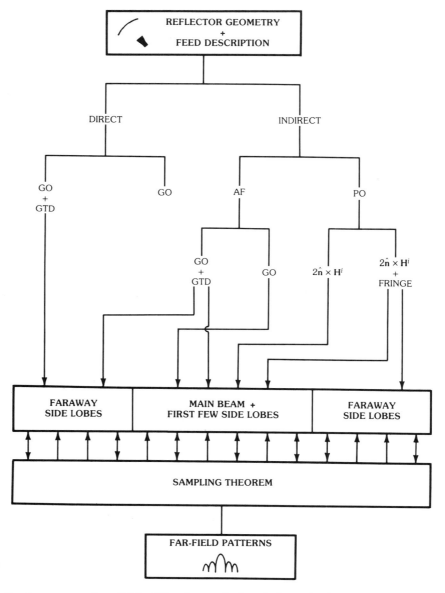

Fig. 6. A summary of available diffraction analysis techniques for focusing reflectors, where PO = physical optics, AF = aperture field, GO = geometrical optics, and GTD = geometrical theory of diffraction. (*After Rahmat-Samii [29], © 1984 IEEE*)

Reflector Antennas

3. Simple Formulas for Far Fields of Tapered-Aperture Distributions

For many applications it is difficult to employ the formulations of PO or GTD (without the use of efficient computer programs) and arrive at simple expressions for the fields or to initiate a preliminary design. For reflectors with pencil-beam radiation characteristics, such as parabolic reflectors, there are simple models which predict their performances. Two such models are discussed in this section.

Two-Parameter Model

A useful and simple analytical model for characterizing the far-field patterns of aperturelike antennas with uniform phase distributions involves the application of the following amplitude distribution across the aperture:

$$Q(\varrho) = C + (1 - C)[1 - (\varrho/a)^2]^P \qquad (16)$$

where P and C are parameters used to control the shape of the amplitude distribution and a is the radius of the circular aperture. In particular, C can be related to the aperture edge taper ET by

$$\text{ET} = 20 \log C \qquad (17)$$

Figs. 7 and 8 show the plot of (16) for values of $P = 1$ and $P = 2$ and different edge tapers. The far-field pattern of this aperture distribution is simply proportional to its Fourier transform. For the aperture of radius a and the blockage radius of a_0, one can arrive at the following integral expression for the far-field pattern:

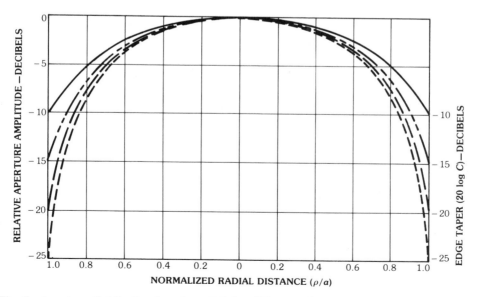

Fig. 7. Aperture distribution based on (16) for different edge tapers (-10, -15, -20, and -25 dB) and $P = 1$.

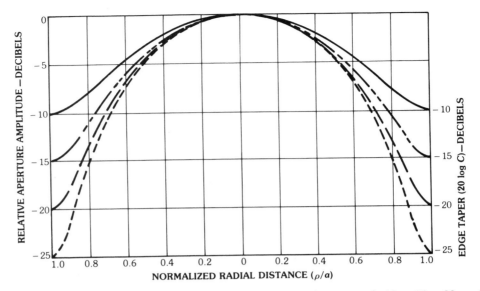

Fig. 8. Aperture distribution based on (16) for different edge tapers (-10, -15, -20, and -25 dB) and $P = 2$.

$$T = \int_{a_0}^{a} \int_0^{2\pi} Q(\varrho') e^{jk\varrho' \sin\theta \cos(\phi - \phi')} \varrho' d\varrho' d\phi' \tag{18}$$

For the cases of no blockage, i.e., $a_0 = 0$, (18) can be evaluated in closed forms for all values of P to give the following expressions for the pattern and the peak value:

$$\begin{aligned} T &= \pi a^2 [C \Lambda_1(ka \sin\theta) + (1 - C) \Lambda_{P+1}(ka \sin\theta)] \\ T(0°) &= \pi a^2 [C + (1 - C)/(P + 1)] \end{aligned} \tag{19}$$

where

$$\Lambda_{P+1}(\zeta) = 2^{P+1} \Gamma(P + 1) \frac{J_{P+1}(\zeta)}{\zeta^{P+1}} \tag{20}$$

in which Γ and J are the gamma and Bessel functions, respectively. Furthermore, by defining the aperture taper (illumination) efficiency η_t as a measure of the gain loss between the tapered and uniform aperture distributions, one obtains

$$\eta_t = \frac{[C + (1 - C)/(P + 1)]^2}{C^2 + 2C(1 - C)/(P + 1) + (1 - C)^2/(2P + 1)} \tag{21}$$

It can readily be observed how the beamwidth and side lobe levels vary as a function of edge tapers and aperture parameter P. For $P = 1$ and 2 cases Table 1 contains the values of pertinent parameters for quick reference.

Table 1. Characteristics of Tapered Circular-Aperture Distribution Based on Two-Parameter Model

Edge Illumination		P = 1			P = 2		
ET (dB)	C	Half-Power Beamwidth (rad)	Side Lobe Level (dB)	η_t	Half-Power Beamwidth (rad)	Side Lobe Level (dB)	η_t
0	1	$1.01\lambda/2a$	-17.6	1	$1.01\lambda/2a$	-17.6	1
-8	0.398	$1.12\lambda/2a$	-21.5	0.942	$1.14\lambda/2a$	-24.7	0.918
-10	0.316	$1.14\lambda/2a$	-22.3	0.917	$1.17\lambda/2a$	-27.0	0.877
-12	0.251	$1.16\lambda/2a$	-22.9	0.893	$1.20\lambda/2a$	-29.5	0.834
-14	0.200	$1.17\lambda/2a$	-23.4	0.871	$1.23\lambda/2a$	-31.7	0.792
-16	0.158	$1.19\lambda/2a$	-23.8	0.850	$1.26\lambda/2a$	-33.5	0.754
-18	0.126	$1.20\lambda/2a$	-24.1	0.833	$1.29\lambda/2a$	-34.5	0.719
-20	0.100	$1.21\lambda/2a$	-24.3	0.817	$1.32\lambda/2a$	-34.7	0.690

For apertures with central blockage ($a_0 \neq 0$), (18) can only be evaluated in closed forms for integer values of P. It can be shown that the following expressions hold [30] for $P = 1$ and $P = 2$:

$$T = T_1 - T_0 \tag{22}$$

for $P = 1$:

$$T_n = \pi a_n \left[C \frac{2}{u_n} J_1(u_n) + (1 - C) \left\{ \frac{2}{u_n} J_1(u_n) - \frac{a_n^2}{a^2} \left[\frac{2}{u_n} J_1(u_n) - \left(\frac{2}{u_n}\right)^2 J_2(u_n) \right] \right\} \right] \tag{23}$$

$$T_n(0°) = \pi a_n^2 \left[C + (1 - C)\left(1 - \frac{a_n^2}{2a^2}\right) \right]$$

and for $P = 2$:

$$T_n = \pi a_n^2 \left[C \frac{2}{u_n} J_1(u_n) + (1 - C) \left\{ \frac{2}{u_n} J_1(u_n) \right. \right.$$
$$- \frac{2a_n^2}{a^2} \left[\frac{2}{u_n} J_1(u_n) - \frac{4}{u_n^2} J_2(u_n) \right]$$
$$+ \frac{a_n^4}{a^4} \left[\frac{2}{u_n} J_1(u_n) - 2\left(\frac{2}{u_n}\right)^2 J_2(u_n) \right.$$
$$\left. \left. \left. + 2\left(\frac{2}{u_n}\right)^3 J_3(u_n) \right] \right\} \right] \tag{24}$$

$$T_n(0°) = \pi a_n^2 \left[C + (1 - C)\left(1 - \frac{a_n^2}{a^2} + \frac{a_n^4}{3a^4}\right) \right]$$

where, in the above equations, the following holds:

$$u_n = ka_n \sin\theta; \quad a_1 = a, \quad a_0 = a_0 \qquad (25)$$

Obviously, for the cases $a_0 = 0$, (23) and (24) are directly obtainable from (19).

Next, numerical results for apertures with different blockage values are presented. First, let us define the blockage parameter b as the ratio of the blockage diameter to the aperture diameter, viz.,

$$b = \frac{D_0}{D} = \frac{2a_0}{2a}$$

For different values of parameter b, the far-field patterns are shown in Figs. 9 and 10. Again, these far-field patterns are normalized with respect to the no-blockage case. The reader can develop a good understanding by comparing these results with the no-blockage cases, in order to identify the effects of central blockage. Also shown in Figs. 11 and 12 are the boresight gain loss and first side lobe levels as a function of the blockage parameter b for different edge tapers and cases $P = 1$ and $P = 2$.

Plots of Figs. 7 through 12 should provide a good basis for predicting the performance of most aperturelike antennas. The values of C and P should be chosen such that they provide similar edge tapers and amplitude slopes at the edge as compared with the actual feed/reflector configuration. In most cases the desired values of P are $1 \leqq P \leqq 2$.

One-Parameter Model

A one-parameter simple expression is also available for modeling the radiated field of a circular aperture and the reader may refer to [31] for details. In this model, which applies to the no-blockage circular aperture case, the far-field pattern is proportional to

$$\begin{aligned} T &= \frac{2J_1[\sqrt{u^2 - H^2}]}{\sqrt{u^2 - H^2}}, \quad u \geqq H \\ T &= \frac{2I_1[\sqrt{H^2 - u^2}]}{\sqrt{H^2 - u^2}}, \quad H \geqq u \end{aligned} \qquad (26)$$

where $u = ka \sin\theta$, J_1 and I_1 are Bessel and modified Bessel functions of the first kind of the first order, and the constant H is the single parameter. The J_1 part of the pattern provides the side lobe structure plus part of the main beam, and the remainder of the main beam is the I_1 part. The transition occurs at $u = H$, when the pattern function is unity. For this model the first side lobe level (SL) is at

$$\text{SL} = -17.57 \text{ dB} - 20\log(2I_1(\pi H)/\pi H) \qquad (27)$$

Furthermore, the inverse Fourier transform of (26) can be obtained analytically [31] to result in the following expression for the aperture distribution, viz.,

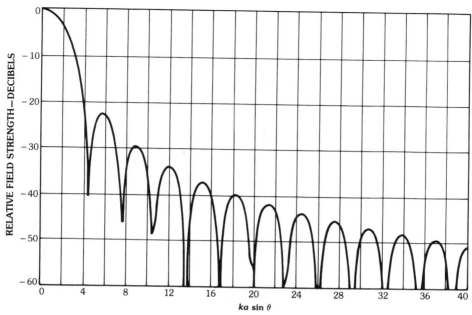

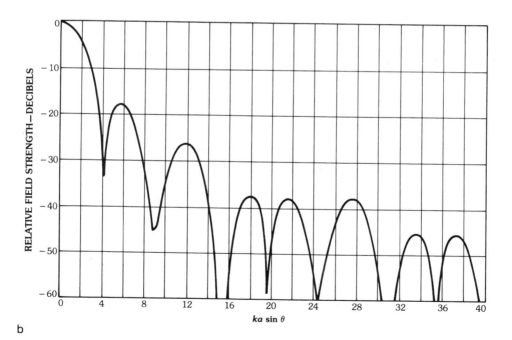

Fig. 9. Far-field patterns based on the aperture distribution (16), with ET = −10 dB and $P = 1$, for different values of $b = D_0/D$. (*a*) For central blockage ratio $b = 0$. (*b*) For central blockage ratio $b = 0.2$.

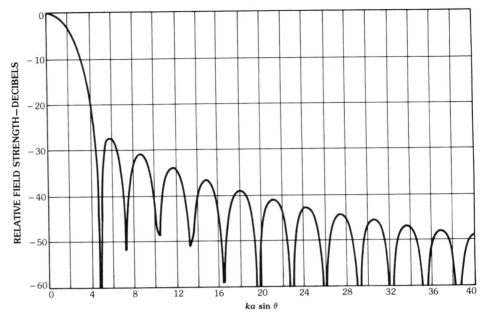

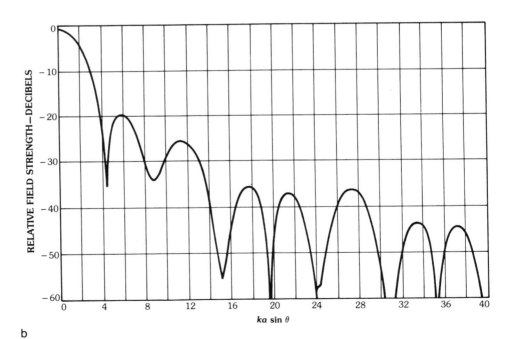

Fig. 10. Far-field patterns based on the aperture distribution (16), with ET = −10 dB and $P = 2$, for different values of $b = D_0/D$. (*a*) For central blockage ratio $b = 0$. (*b*) For central blockage ratio $b = 0.2$.

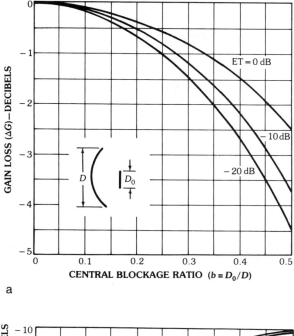

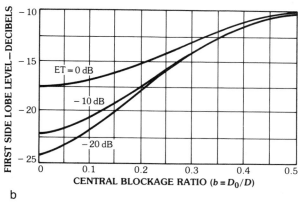

Fig. 11. Gain loss and first side lobe level as a function of $b = D_0/D$ for different edge tapers when (16) with $P = 1$ is the aperture distribution. (a) Gain loss. (b) First side lobe level.

$$Q(\varrho) = I_0[H\sqrt{1 - (\varrho/a)^2}] \qquad (28)$$

where I_0 is the modified Bessel function of the zeroth order. By using expressions (26)–(28) different characteristics of the pattern can be determined, as detailed in Table 2 [31].

Near, Fresnel, and Far Fields

Before closing this subsection it is worthwhile to present data on the near, Fresnel, and far fields of aperturelike antennas. This information is useful for

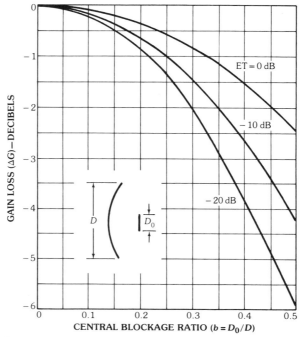

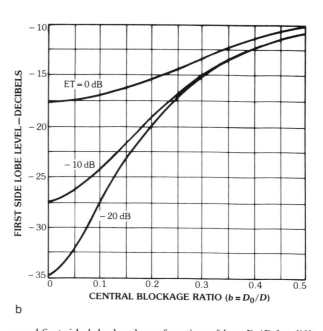

Fig. 12. Gain loss and first side lobe level as a function of $b = D_0/D$ for different edge tapers when (16) with $P = 2$ is the aperture distribution (*a*) Gain loss. (*b*) First side lobe level.

Table 2. Characteristics of Tapered Circular-Aperture Distribution Based on One-Parameter Model (*After Hansen [31]*, © *1975 IEE*)

First Side Lobe Level	H/π	Edge Taper (dB)	Half-Power Beamwidth*	Efficiency η_t
−17.57	0	0	1.0000	1.0000
−20	0.4872	−4.49	1.0483	0.9786
−25	0.8899	−12.35	1.1408	0.8711
−30	1.1977	−19.29	1.2252	0.7595
−35	1.4708	−25.78	1.3025	0.6683
−40	1.7254	−31.98	1.3741	0.5964

*Relative to a uniform distribution.

developing an understanding of the evolution of radiated fields away from the reflector. For the sake of simplicity, only the uniform aperture distribution is considered and its radiated fields at many different distances from the aperture are constructed by numerically evaluating (1). Results are shown in Fig. 13, where the reader can readily observe how the far-field pattern is formed at distances $R > 2D^2/\lambda$. For very low side-lobe reflectors, distances much greater than $2D^2/\lambda$ are required to obtain an accurate description of inner side lobes.

4. Some Important Geometrical Features of Conic-Section–Generated Reflector Antennas

Both single- and dual-offset reflector antennas are being applied more frequently in the design of low side lobe antenna systems. Due to their very unique optical focusing characteristics, surfaces generated from conic sections are most commonly used in practice. Typically these surfaces are constructed as a result of a translation or rotation of the conic sections. Offset reflectors are carved-out portions of these surfaces, resulting from their intersections with cylinders or cones. The cylinders have their axes parallel to the axes of the parent reflector surfaces, and the cones have their tips at one of the foci of the reflectors. In this section geometrical characteristics of offset (symmetric) conic section reflectors are presented in a unified fashion and some of their important geometrical features are described. Only the final results are given here and the reader is referred to [32, 33] for details.

Conic Sections

Conic sections are basically second-degree planar curves which can be generated in many ways, including the intersection of a circular cone with a planar surface. Referring to Fig. 14, with z as the abscissa and x as the ordinate of a plane Cartesian coordinate system, the equation of a conic section can be expressed in the following general form (when the principal axes of a conic section are tilted with respect to the x and z axes a more general quadratic expression results):

$$\frac{(z-c)^2}{(f+c)^2} + \frac{x^2}{(f+c)^2 - c^2} = 1 \tag{29}$$

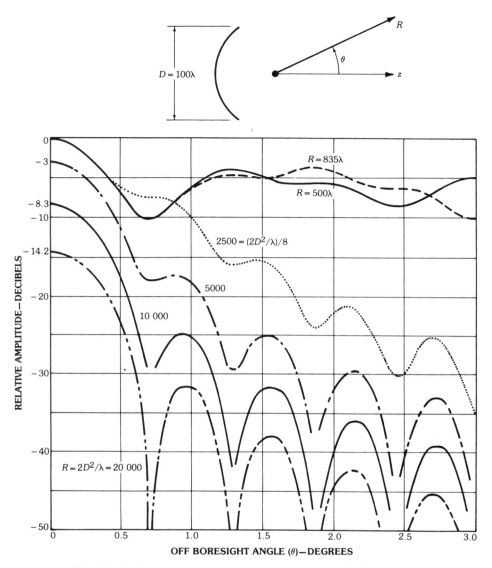

Fig. 13. Reflector antenna patterns as a function of distance R.

In polar coordinates this equation is written as

$$\varrho = \frac{f + 2c}{f + (1 - \cos\phi)c} f \qquad (30)$$

where in both of the previous equations the parameters have the following definitions: f is the focal length (distance from a focus to the nearest apex) having a positive or zero value, and $2c = F_1F_2 \cdot \hat{\mathbf{z}}$ is the distance between the two foci having

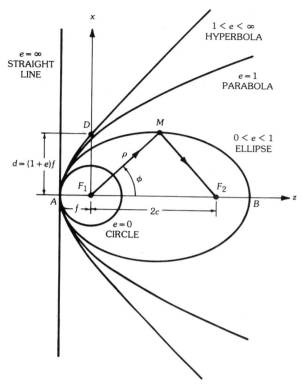

Fig. 14. Geometrical configuration of conic sections as a function of eccentricity e and for focal length f. (*After Jamnejad and Rahmat-Samii [32], © 1980 IEEE*)

an algebraic value which is positive if F_2 is to the right of F_1 (ellipse) and negative otherwise (hyperbola) where $\hat{z}$ is a unit vector along the positive z direction. One of the common properties of the conic sections is that either the sum or difference of distances from any point on the curve to the focal points is constant. Another manifestation of this property from the geometrical optics viewpoint is that any ray passing through one focus and crossing the curve at a given point is reflected along the line connecting the crossing point to the second focus.

Eccentricity e is defined as

$$e = \frac{c}{c+f} = \frac{1}{1+f/c} \qquad (31)$$

which is always a positive number. This parameter is a measure of the off-centeredness of the focal points for a given focal length. Note that the center is located at the midpoint of the foci. Thus, when the foci are at the center (a circle), the eccentricity is zero, and when they are infinitely apart, it is unity (a parabola), and so on.

In terms of this new parameter (29) and (30) can be rewritten as

$$\frac{[(z - ez - ef)/(1 - e)]^2}{f^2/(1 - e)^2} + \frac{x^2}{(1 - e^2)f^2/(1 - e)^2} = 1 \qquad (32)$$

and

$$\varrho = \frac{1 + e}{1 - e \cos \phi} f \qquad (33)$$

In order to better appreciate the characteristics of these conic section curves, the cases are outlined in Table 3. For the sake of comparison all the conic sections with one common focal point and an identical focal length are superimposed in Fig. 14. As the eccentricity e increases, the curves broaden, i.e., $F_1D = d$ and R, the radius of curvature at the apex A, increases accordingly, since it can be shown that

$$R = d = (1 + e)f \qquad (34)$$

Reflector Surfaces

Reflector surfaces are commonly generated either by translation of a conic section along the y axis, or by its rotation about the focal axis z, as in Fig. 15. In this presentation the rotationally generated surfaces are considered.

An offset reflector can be constructed by carving out a portion of the rotationally symmetric reflector. This is typically achieved by either intersecting the reflector with a circular or elliptical cylinder with its axis parallel to the reflector axis, or by intersecting the reflector with a circular or elliptical cone with its tip at the focal point. The following subsections discuss the geometrical characteristics of reflectors thus obtained.

Intersection with a Circular Cone

The cone of rays emanating from a source located at the focal point F_1 intersects the reflector surface as shown in Fig. 15. The equation of the reflector surface, produced by the revolution of a conic section about the z axis, is obtained by simply replacing x^2 by $x^2 + y^2$ in (32) to arrive at

Table 3. Classification of Conic Curves

Foci Separation/2	Eccentricity	Equation	Type
$c = 0$	$e = 0$	$\frac{z^2}{f^2} + \frac{x^2}{f^2} = 1$	Circle
$0 < c < \infty$	$0 < e < 1$	(32)	Ellipse
$c = \infty$	$e = 1$	$(z + f) - \frac{1}{4f}x^2 = 0$	Parabola
$-\infty < c < -2f$	$1 < e < \infty$	(32)	Hyperbola
$c = -f$	$e = \infty$	$z + f = 0$	Straight line

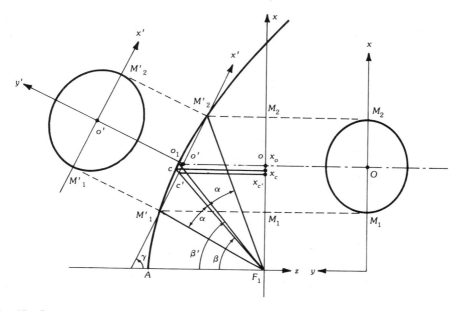

Fig. 15. Geometry of an offset reflector obtained from the intersection of a circular cone and a surface generated by rotation of a conic section about its focal axis. (*After Jamnejad and Rahmat-Samii [32]*, © *1980 IEEE*)

$$\frac{[(z - ez - ef)/(1 - e)]^2}{f^2/(1 - e)^2} + \frac{x^2 + y^2}{(1 - e^2)f^2/(1 - e)^2} = 1 \qquad (35)$$

The equation of a cone whose axis lies in the xz plane making an angle β with the negative z axis and having a half-angle α, as shown in Fig. 15, is

$$(x^2 + y^2 + z^2)\cos^2\alpha = (x \sin\beta - z \cos\beta)^2 \qquad (36)$$

The equation of the intersection curve of the surface of revolution and the cone is the solution of (35) and (36). Elimination of the variable y from these equations leads to the projection of the intersection curve on the xz plane with a final result of

$$x = \frac{\cos\beta + e \cos\alpha}{\sin\beta} z + \frac{(1 + e) \cos\alpha}{\sin\beta} f \qquad (37)$$

which is the equation of the projection of the intersection on the xz plane. Clearly this projection is a *straight* line, indicating that the intersection curve (rim) lies in a *plane* perpendicular to the xz plane. In other words it is shown that the intersection of a reflector surface, generated from the rotation of a conic section about its focal axis, and a circular cone, with its tip on a focal point of the reflector surface, is always a *planar curve*. In fact it will be shortly demonstrated that this curve is an ellipse.

By eliminating variable z in equations (35) or (36) and (37) the equation of the projection of the intersection curve (rim) on the xz plane is obtained. This elimination process is rather lengthy and the final result is

$$\frac{y^2}{a^2} + \frac{(x - x_o)^2}{b^2} = 1 \tag{38}$$

in which

$$x_o = \frac{(1 + e)(e \cos\beta + \cos\alpha) \sin\beta}{(1 - e^2) \sin^2\beta + (\cos\beta + e \cos\alpha)^2} f$$

$$a^2 = \frac{(1 + e)^2 \sin^2\alpha}{(1 - e^2) \sin^2\beta + (\cos\beta + e \cos\alpha)^2} f^2 \tag{39}$$

$$b^2 = \frac{(1 + e)^2 (\cos\beta + e \cos\alpha)^2 \sin^2\alpha}{[(1 - e^2) \sin^2\beta + (\cos\beta + e \cos\alpha)^2]^2} f^2$$

It is now an easy matter to determine the equation of the intersection curve (rim) in the plane of intersection. This is done by designating the $x'y'$ plane as the intersection plane, as shown in Fig. 15, and by choosing o' as the origin of this coordinate system to finally arrive at

$$\frac{y'^2}{a'^2} + \frac{x'^2}{b'^2} = 1 \tag{40}$$

where

$$a'^2 = a^2$$
$$b'^2 = \frac{(1 + e)^2 [\sin^2\beta + (\cos\beta + e \cos\alpha)^2] \sin^2\alpha}{[(1 - e^2) \sin^2\beta + (\cos\beta + e \cos\alpha)^2]^2} \tag{41}$$

in which a is given in (39).

It is of some interest to note that the point of intersection of the cone axis with the intersection plane, i.e., point c', does not coincide with the center of the intersection curve, i.e., point o' (Fig. 15). This is easily demonstrated by calculating the ordinate of point c' in the xz plane, which results in

$$x_{c'} = \frac{(1 + e) \cos\alpha \sin\beta}{1 + e \cos\alpha \cos\beta} f \tag{42}$$

which is clearly different from $x_{o'} = x_o$ given in (39). Neither do points c (intersection of the cone axis and the reflector surface) and o_1 (intersection of the central axis of the reflected cylinder, $o'o$, and the reflector surface) coincide. It can be shown that

$$x_c = x_o\bigg|_{\alpha=0} = \frac{(1+e)\sin\beta}{1+e\cos\beta}f \qquad (43)$$

which is different from $x_{o_1} = x_o$.

Special Cases

The equation of the intersection curve (rim) in the $x'y'$ plane and its projection on the xy plane as given by (40) and (38), respectively, and also the ordinate of some points of interest as given in (43), (42), and (39), are simplified in the following cases:

(a) spherical surface ($e = 0$)

$$\begin{aligned} a^2 &= f^2 \sin^2\alpha, & a'^2 &= a^2 \\ b^2 &= a^2 \cos^2\beta, & b'^2 &= a^2 \end{aligned} \qquad (44)$$

and

$$\begin{aligned} x_o &= f \sin\beta \cos\alpha \\ x_c &= x_o \\ x_{c'} &= f \sin\beta \end{aligned} \qquad (45)$$

In this case the intersection curve (rim) is a *circle* and its projection on the xy plane is an *ellipse* with its minor axis along the y axis.

(b) paraboloidal surface ($e = 1$)

$$\begin{aligned} a^2 &= \frac{4\sin^2\alpha}{(\cos\beta + \cos\alpha)^2}f^2, & a'^2 &= a^2 \\ b^2 &= a^2, & b'^2 &= a^2\left[1 + \frac{\sin^2\beta}{(\cos\beta + \cos\alpha)^2}\right] \end{aligned} \qquad (46)$$

and

$$\begin{aligned} x_o &= \frac{2\sin\beta}{\cos\beta + \cos\alpha}f \\ x_c &= \frac{2\sin\beta}{1 + \cos\beta}f \\ x_{c'} &= \frac{2\sin\beta\cos\alpha}{1 + \cos\alpha\cos\beta}f \end{aligned} \qquad (47)$$

In this case the intersection curve (rim) is an *ellipse* with its major axis along the x' axis, while its projection on the xy plane reduces to a *circle*.

It is also of interest in this case to determine the relationship between the angles β and β' as shown in Fig. 15. After some manipulations one arrives at

$$\tan\beta' = \frac{1}{1 - (\sin^2\alpha)/2(\cos^2\beta + \cos\beta\cos\alpha)}\tan\beta \tag{48}$$

(c) planar surface ($e = \infty$)

$$a^2 = \frac{\sin^2\alpha}{\cos^2\alpha - \sin^2\beta}f^2, \qquad a'^2 = a^2$$
$$b^2 = a^2\cos^2\alpha, \qquad\qquad b'^2 = b^2 \tag{49}$$

and

$$x_o = \frac{\cos\beta\sin\beta}{\cos^2\alpha - \sin^2\beta}f$$
$$x_c = f\tan\beta \tag{50}$$
$$x_{c'} = x_c$$

It is obvious that, in this case, the intersection curve (rim) and its projection on the xy plane are the same and display an *ellipse* elongated in the x direction.

As a concluding remark for this subsection one can say that, starting from zero eccentricity $e = 0$ (a sphere), where the intersection curve (rim) is a circle, as eccentricity increases, the rim curve becomes more oblong in the x' direction. On projection onto the xy plane, however, the curve becomes contracted in the x direction to the extent that for $e = 1$ (a paraboloid) the projected curve becomes a circle. For values of $e < 1$ (an ellipsoid) the projected curve has its minor axis in the x direction, while for values of $e > 1$ (a hyperboloid) it has its major axis in that direction.

Intersection with a Circular Cylinder

The equation of a circular cylinder with its axis in the yz plane and parallel to the z axis is given as

$$y^2 + (x - x_o)^2 = R^2 \tag{51}$$

where x_o and R are the offset height and the radius of the cylinder, respectively. The intersection of this cylinder with the reflector surface can be found from (51) and (22). The projection of the intersection curve (rim) on the yz plane is found by eliminating y in these two equations, which results in

$$(1 - e^2)z^2 - 2e(1 + e)fz + 2xx_o = (1 + e)^2f^2 + x_o^2 - R^2 \tag{52}$$

The above equation, in general, describes a parabola. It is therefore concluded that the rim lies on a *parabolic cylinder* and is *not* a planar curve. However, in the special case of $e = 1$ (paraboloid surface) the curve given by (52) is a line, which indicates that the rim is planar and, because it is the intersection of a circular cylinder and a plane, is indeed an ellipse.

Reflector Antennas

5. Offset (Symmetric) Parabolic Reflectors

In this section the radiation characteristics of offset (symmetric) parabolic reflector antennas illuminated by a single feed element (fixed phase center) are presented. As is typical in any reflector design, there are too many almost-independent parameters which may be varied to achieve a particular design goal. For instance, to design an offset parabolic reflector one must study the effects of such parameters as offset angles, illumination tapers, F/D ratios, locations and orientations of the feed, polarizations, etc., on such far-field pattern characteristics as scan loss, beamwidth, side lobe level, cross-polarization level, efficiency, and so on. Obviously it is not possible to perform a comprehensive study of all these parameters; rather, attempts are made to present the most important reflector characteristics based on a few key parameters. The results of this section can therefore be used as a guideline for an initial design, which can then be refined into greater detail by using computer programs and measured data.

Geometrical Parameters

Although some of the general geometrical features of conic sections were presented in Section 4, features which are particular to parabolic reflectors are presented in this section. The geometry of an offset parabolic reflector with focal length F, diameter D, and offset height H is shown in Fig. 16. There are also other parameters used to characterize offset parabolic reflectors which are defined below:

F = focal length

D = reflector diameter (diameter of the circular projected aperture)

H = offset height ($H = -D/2$ for symmetric reflectors)

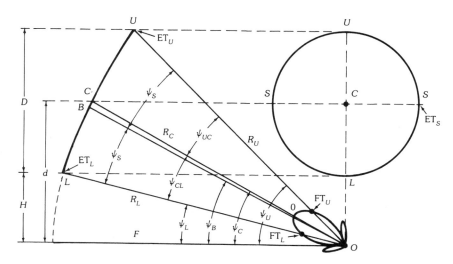

Fig. 16. Geometrical parameters of an offset parabolic reflector (for the symmetric case $H = -D/2$).

d = offset height of the circular projected aperture center
 = $D/2 + H$

ψ_U = angle subtended to the upper tip
 = $2 \tan^{-1}[(D + H)/2F]$

ψ_C = angle subtended to the center of the projected aperture
 = $2 \tan^{-1}[D/2 + H)/2F]$

ψ_L = angle subtended to the lower tip
 = $2 \tan^{-1}[H/2F]$

ψ_B = angle subtended in the bisect direction
 = $(\psi_U + \psi_L)/2$

ψ_S = half-angle subtended to the upper and lower tips
 = $(\psi_U - \psi_L)/2$

ψ_{SC} = half-angle subtended to the right and left sides
 = $\sin^{-1}\left(\dfrac{D}{2} \bigg/ \sqrt{\left(H + \dfrac{D}{2}\right)^2 + \left(\dfrac{D}{2}\right)^2 + \left[\dfrac{(D/2)^2 + (H + D/2)^2}{4F} - F\right]^2}\right)$

D_p = parent parabola diameter
 = $2(D + H)$ for $H \geqq -D/2$

In some cases F, ψ_B, and ψ_S are given, and from them D and H may be constructed. This can be done by using the following expressions:

$$D = 4F \sin \psi_S/(\cos \psi_B + \cos \psi_S)$$
$$H = 2F (\sin \psi_B - \sin \psi_S)/(\cos \psi_B + \cos \psi_S) \tag{53}$$

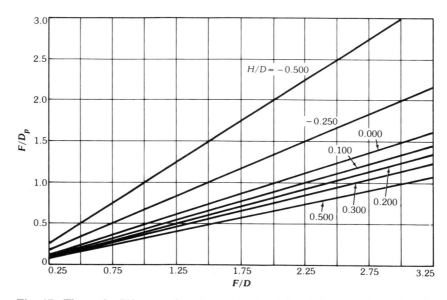

Fig. 17. The ratio F/D_p as a function of F/D and for different values of H/D.

Reflector Antennas

In the design of offset reflectors any combination of the previously mentioned parameters may be given and the others may be constructed. In this subsection the parameters D, F/D, and H/D are used as independent parameters and the others are obtained. For example, Figs. 17 and 18a through 18c show plots of F/D_p, ψ_{UC}, ψ_{CL}, ψ_{CS}, and $20\log(R_C/R_{U,L,S})$ as functions of F/D for different H/D values. It is clear for symmetric reflectors that $F/D_p = F/D$ and $\psi_{UC} = \psi_{CL} = \psi_{CS}$, because $H/D = -0.5$. In the following subsections it will become apparent that these parameters play important roles to the extent that the far-field characteristics of reflectors are concerned. For example, different path losses at different edges (tips) of the reflector can cause different side lobe levels in the far-field pattern cuts. For most offset parabolic antennas, H/D typically varies between $0.1 \leq H/D \leq 0.3$ to provide clearance for the feed assemblies.

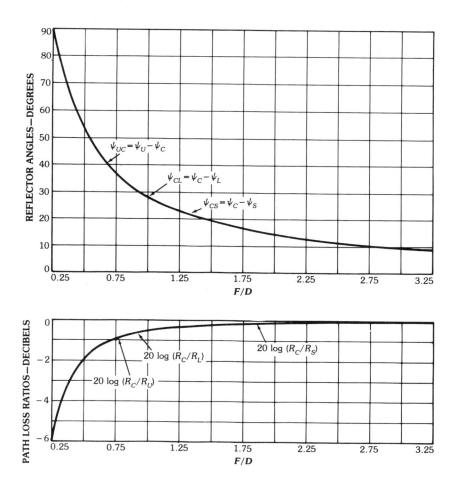

a

Fig. 18. Reflector angles and path loss ratios, as defined in Fig. 16, as a function of F/D. (a) For $H/D = -0.5$ (symmetric reflector case). (b) For $H/D = 0.10$. (c) For $H/D = 0.20$.

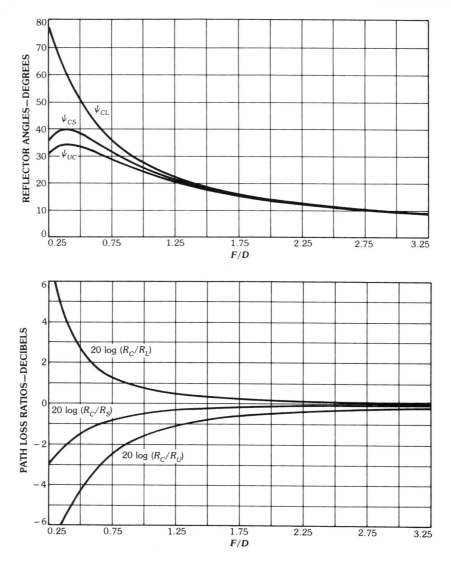

Fig. 18, *continued.*

Idealized Feed Patterns

As mentioned in the introduction a reflector's far-field pattern characteristics cannot be determined without a proper description of its feed patterns. A discussion of commonly used feed elements will be given in another section of this chapter. However, in order to present parametric results concerning the performance of parabolic reflectors, idealized feed patterns, which have proven to be very useful models, are used. For an idealized feed with a fixed phase center, its radiation pattern may be described as

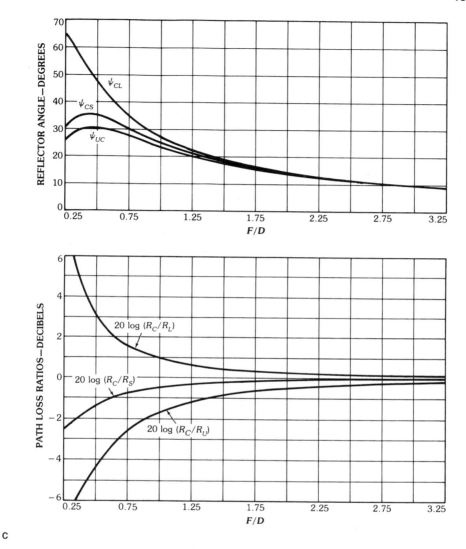

Fig. 18, *continued.*

$$\mathbf{E}(\mathbf{r}) = A_0 \begin{cases} \hat{\boldsymbol{\theta}} C_E(\theta) \cos\phi - \hat{\boldsymbol{\phi}} C_H(\theta) \sin\phi \\ \hat{\boldsymbol{\theta}} C_E(\theta) \sin\phi + \hat{\boldsymbol{\phi}} C_H(\theta) \cos\phi \end{cases} \frac{e^{-jkr}}{r}, \quad \begin{array}{l} \text{for } \hat{\mathbf{x}} \text{ polarized} \\ \text{for } \hat{\mathbf{y}} \text{ polarized} \end{array} \quad (54)$$

where A_0 is a complex constant and

$$\begin{aligned} C_E(\theta) &= (\cos\theta)^{q_E} = E\text{-plane pattern} \\ C_H(\theta) &= (\cos\theta)^{q_H} = H\text{-plane pattern} \end{aligned} \quad (55)$$

for $0 \leq \theta \leq \pi/2$ and zero otherwise.

Furthermore, in the above equations the coordinate variables are defined with respect to the feed coordinates, although this is not explicitly indicated. The shape of the pattern is controlled by q_E and q_H, which are determined by matching (55) to the given feed pattern (to be discussed further in later sections). A proper superposition of the x and y polarized components in (54) also generates a circularly polarized field, namely,

$$\mathbf{E}(\mathbf{r}) = A_0 e^{j\tau\phi}[\hat{\boldsymbol{\theta}} C_E(\theta) + \hat{\boldsymbol{\phi}} j\tau C_H(\theta)] \frac{e^{-jkr}}{r} \qquad (56)$$

where $\tau = +1$ for left-handed circular polarization, and $\tau = -1$ for right-handed circular polarization. Note that (56) represents a perfect circular polarized wave only in the main beam direction ($\theta = 0$). Away from this direction it is generally elliptically polarized unless $q_E = q_H$.

One of the attractive features of expression (54) is that it allows the directivity to be derived in closed form, viz.,

$$D(\theta = 0) = \frac{2(2q_E + 1)(2q_H + 1)}{q_E + q_H + 1} \qquad (57)$$

which consequently allows computation of the directivity of the reflector illuminated by this feed [34]. It must be noted that the back radiation of the feed pattern is ignored in deriving (57).

Edge and Feed Tapers

An important parameter which is widely employed to characterize the effects of the feed element pattern on the far-field pattern of the reflector is the "edge taper." In a broad sense this signifies the ratio of the field intensity at the reflector edge to the intensity at its center in decibels. Although this definition is unambiguous when it is applied to symmetric reflectors it can become ambiguous for offset reflectors. For this reason another definition, referred to as "feed taper," is introduced.

Referring to Fig. 16, the feed taper FT_U in the upper-tip direction is defined as

$$FT_U = 20 \log \left[\frac{C(\psi_U - \psi_C)}{C(0°)} \right] \qquad (58)$$

where $0°$ refers to the central direction (i.e., OC in Fig. 16) and C denotes the feed pattern as defined in (55). Similar definitions for FT_L and FT_S can also be given at the lower and side angles by using $\psi_C - \psi_L$ and ψ_S, respectively. The edge taper ET_U at the upper tip may now be expressed as

$$ET_U = FT_U + 20 \log (R_C/R_U) \qquad (59)$$

where R_C and R_U are the path lengths from the feed to the center and the upper tip of the reflector, respectively. The second term in (59) is also called the *path loss*

Reflector Antennas

term. Similar definitions can also be given for ET_L and ET_S. It is noted that, for symmetric reflectors, ET and FT only need to be defined at one edge (tip). For different values of F/D and H/D the path loss curves are shown in Figs. 18a through 18c. It is worthwhile to add that for most cases of interest ET_U, ET_L, and ET_S have nearly equal values. Since ET directly controls the reflector's aperture amplitude taper it has a more dominant effect on the far-field pattern than EF does. From the results of Section 3 it is noted that it is not only the taper level which controls the reflector pattern but also the overall shape of the illumination distribution. In particular, the slopes of the illumination pattern at the reflector's edge can affect the side lobe levels. The results of this section are primarily based on the $\cos^q(\theta)$ type illumination patterns.

Reflector Pattern Characteristics for On-Focus Feeds

In this section, reflector pattern characteristics are discussed for beams generated by on-focus feeds. Results are shown for beamwidths, side lobe levels, first-null positions, cross-polarization levels, directivity efficiency, etc., as functions of the edge taper ET (or feed taper FT) and the reflector geometries. To simplify the presentation and limit the number of graphs, cases are considered in which the path losses are small (less than 0.5 dB) and, therefore, no substantial differences may be observed for path losses at the different tips of the reflector. For these cases, $\psi_{UC} \cong \psi_{CL} \cong \psi_{CS} \cong \psi_S$ and FT $\cong$ ET. The reader should attempt to properly interpret the results when the path losses are substantial. This effect is clearly demonstrated in Fig. 19, which displays far-field patterns for symmetric reflectors ($H/D = -0.5$) for the cases of $F/D = 0.4$ and $F/D = 2.0$, with an edge taper of ET $= -10$ dB. From Fig. 18a it is concluded that the path losses are -2.86 dB and -0.13 dB for F/Ds of 0.4 and 2.0, respectively. Nevertheless, the patterns are very similar, which indicates that ET $= -10$ dB is the controlling factor. Note that for these values of F/D the feed taper FT takes the values of -7.14 dB and -9.87 dB, respectively. In the latter case, ET $\cong$ FT. Similar results are shown in Fig. 20, where the feed taper FT is kept constant at -10 dB. The patterns for cases in which $F/D = 0.4$ and 2.0 show differences of up to 2.8 dB at the first side lobe level which is obviously a manifestation of the effects of the path loss differences.

For offset parabolic reflectors the far-field patterns in different cuts are, in general, different even when the feed has a symmetric pattern. Also the pattern can be slightly asymmetric in the plane of offset (xz plane of Fig. 16) depending on the F/D ratio (see Fig. 21) [29]. However, for the results shown here, F/D is large in order to reduce the path loss effects and asymmetry of the pattern. Fig. 22 shows the half-beamwidth, first and second null positions, and first side lobe positions as functions of the edge taper ET. It is worthwhile to mention that for edge tapers beyond -20 dB the pattern characteristics depend heavily on the actual feed pattern description and results shown here are for $\cos^q(\theta)$ feed patterns. Also shown are the first side lobe levels and spillover and taper efficiencies as functions of edge tapers in Fig. 23. From Fig. 23 it is readily observed that the resulting efficiency $\eta = \eta_s \eta_t$ is maximized for edge tapers about -11 dB with a value of 81 percent. Some representative far-field patterns are shown in Figs. 24a through 24f for different edge tapers. Again, it is obvious that for edge tapers in the neighborhood of ET $= -20$ dB, the first side lobe starts to merge with the main beam,

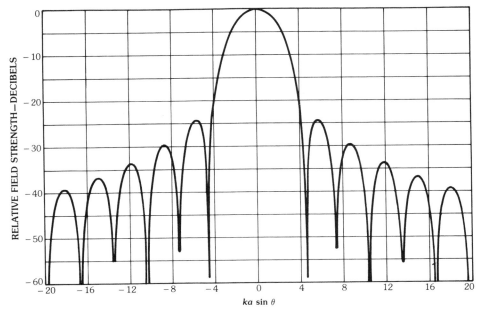

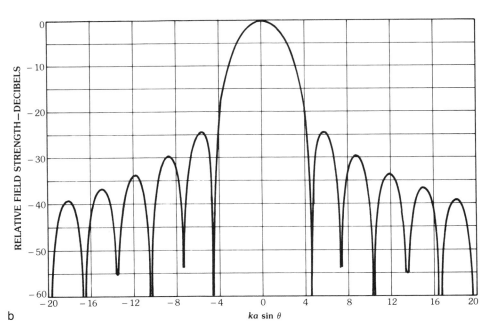

Fig. 19. Far-field patterns of a symmetric reflector antenna with an edge taper illumination of ET = -10 dB and different path losses (different F/D values). (*a*) For $F/D = 0.4$. (*b*) For $F/D = 2.0$.

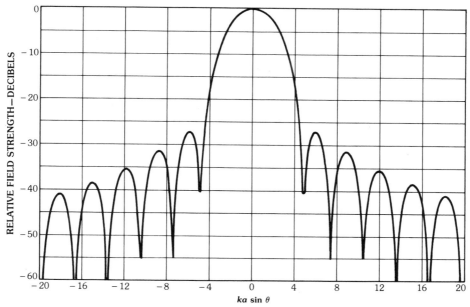

a

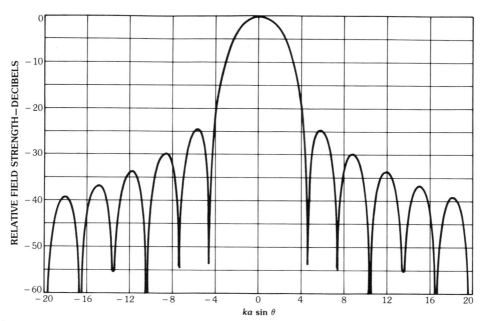

b

Fig. 20. Far-field patterns of a symmetric reflector antenna with a feed taper illumination of FT = −10 dB and different path losses (different F/D values). (a) For $F/D = 0.4$. (b) For $F/D = 2.0$.

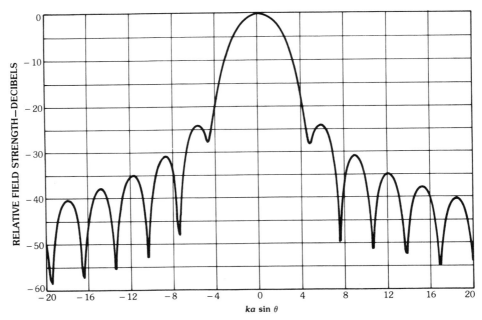

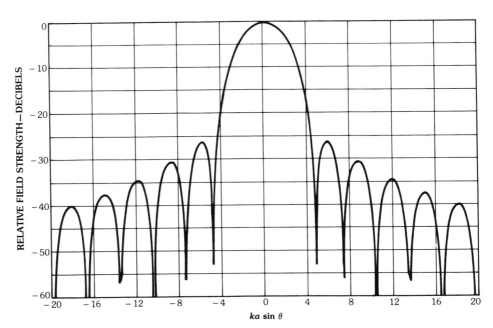

Fig. 21. Patterns of an offset reflector with $H/D = 0.5$ ($F/D_p = 0.25$) illuminated by a symmetric feed with a feed taper of FT = -10 dB. (*a*) For pattern in plane $\phi = 0°$ (plane of offset). (*b*) For pattern in plane $\phi = 90°$.

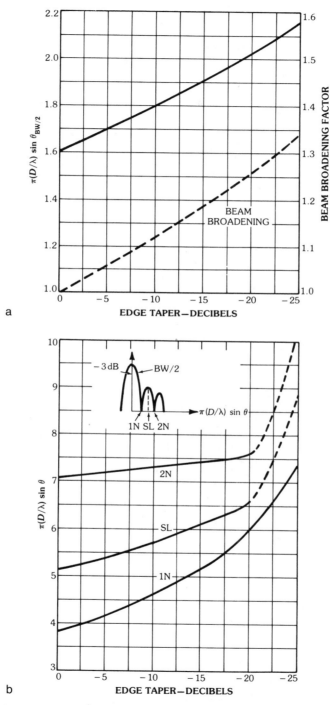

Fig. 22. Half-power beamwidth, first-null, second-null, and side lobe positions as functions of the edge taper, where D/λ is the diameter of the reflector in terms of the wavelength. (*a*) Half-power beamwidth. (*b*) First and second nulls and side lobe positions.

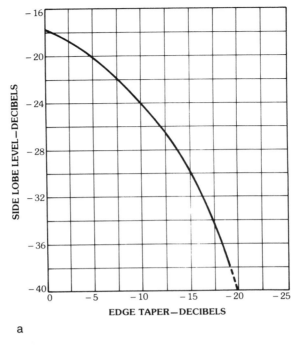

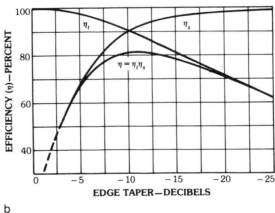

Fig. 23. Side lobe level, taper efficiency, spillover efficiency, and overall efficiency as functions of edge taper ET for $\cos^q(\theta)$ type feed patterns. (*a*) Side lobe level. (*b*) Taper, spillover, and overall efficiencies.

which results in a widened beam. For this level of edge taper the exact distribution of the feed pattern can have a significant effect on the pattern characteristics.

Another important reflector parameter is the level of generated cross-polarized field. This topic has been addressed in [1, 35, 36] for both symmetric and offset reflectors. Here, some representative cases are presented. There are many dif-

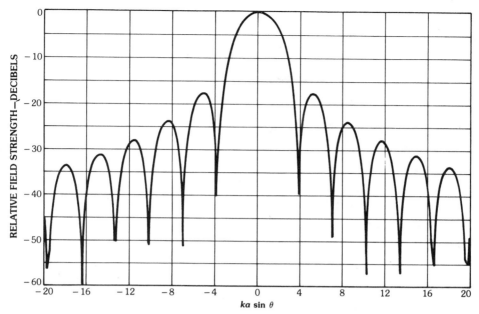

a

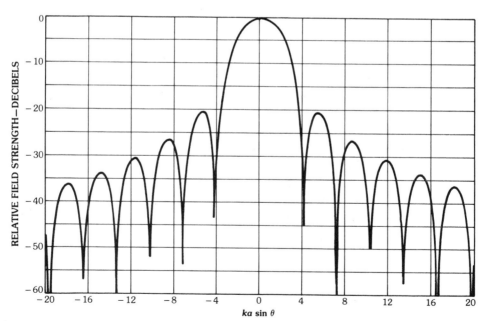

b

Fig. 24. Reflector far-field patterns for different edge tapers. (*a*) ET = 0 dB. (*b*) ET = −5 dB. (*c*) ET = −10 dB. (*d*) ET = −15 dB. (*e*) ET = −20 dB. (*f*) ET = −25 dB.

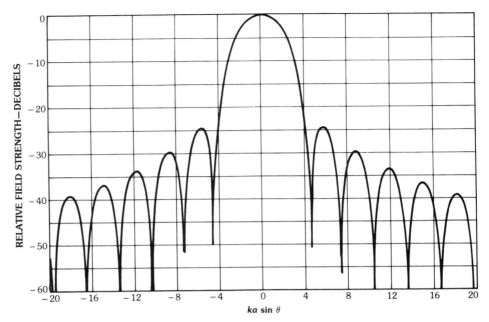

c

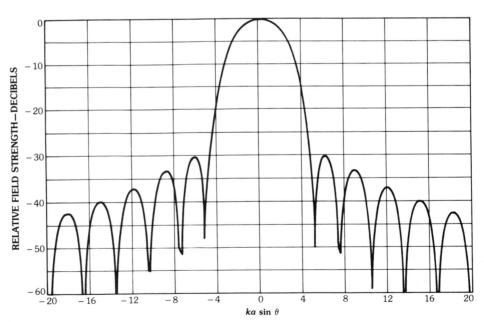

d

Fig. 24, *continued.*

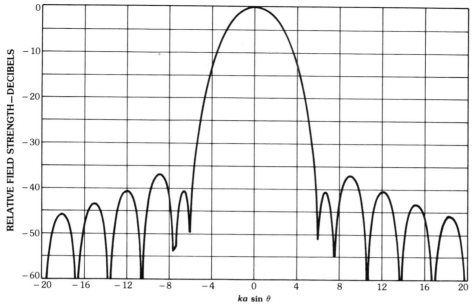

e

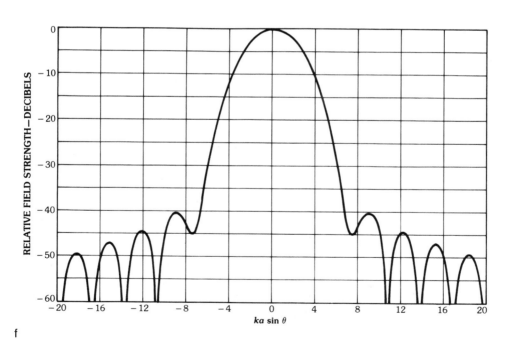

f

Fig. 24, *continued.*

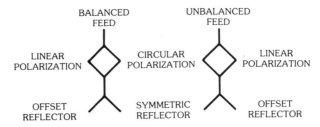

Fig. 25. Different cases of the generation of cross-polarized fields in reflector antennas.

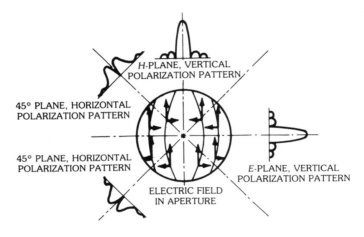

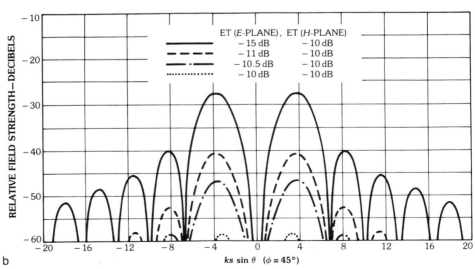

Fig. 26. The generation of cross-polarized fields and some typical patterns of symmetric reflectors. (*a*) Mechanism of generation of cross-polarized fields for symmetric parabolic reflectors. (*b*) Cross-polarized field (normalized with respect to the peak) of a symmetric reflector ($F/D = 0.3$) illuminated by a linearly polarized feed with unbalanced E- and H-patterns.

ferent cases which must be studied, summarized in Fig. 25. In this figure the unbalanced feed refers to a feed with different E- and H-plane patterns. For the symmetric reflector the mechanism of generation of the cross-polarized field is shown in Fig. 26a and some typical patterns are given in Fig. 26b. It should be noted that for a symmetric parabolic reflector illuminated by a linearly polarized feed the maximum of cross-polarized fields occurs in the plane $\phi = 45°$. As the feed becomes more balanced, the level of the cross-polarized field decreases substantially, as demonstrated by the results in Fig. 26. It is worth mentioning that these results are dependent on the F/D ratio and the edge taper.

Next, the cross-polarization characteristics of offset parabolic reflectors are considered. In contrast to symmetric reflectors, which have very low levels of cross-polarized fields for balanced and linearly polarized feeds, offset parabolic reflectors can have high levels of cross polarization. Even for balanced feeds located at the focal point, levels of cross-polarized fields can be high, depending on the tilt angle of the feed axis with respect to the reflector axis for linearly polarized feeds. For example, Fig. 27 shows the generation of cross-polarized fields for various feed-axis tilt angles for a fixed offset reflector configuration. In this example the feed pattern is chosen to be isotropic in order to show clearly the generation of the cross-polarized fields. It should be mentioned that for offset parabolic reflectors the cross-polarized field is predominantly observed in the plane $\phi = 90°$ (normal to the plane of the offset). Clearly, in practice, in order to reduce spillover the feed axis is always tilted toward the center of the reflector; hence a high level of cross-polarized field should be expected. The levels of cross-polarized fields for different values of

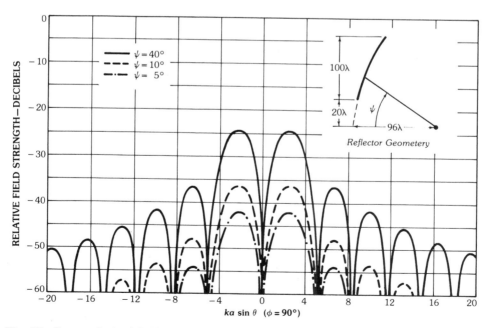

Fig. 27. Cross-polarized fields in the plane $\phi = 90°$ (normal to the offset plane) for an offset parabolic reflector for various feed-axis tilt angles ψ and illuminated by a balanced linearly polarized feed.

bisect angle ψ_B and the half-angle ψ_S (see Fig. 16) are shown in Fig. 28 [35].

When symmetric or offset parabolic reflectors are illuminated with balanced circularly polarized feeds very low levels of cross-polarized field result. For offset parabolic reflectors illuminated by circularly polarized feeds, however, an additional feature is observed, which is referred to as the *beam squint*. This means that the beam peak is shifted from the axis on the plane normal to the plane of offset (*yz* plane in Fig. 16). The amount of squint depends on the tilt angle of the feed axis and the reflector geometry. It can be shown that the following expression is a good approximation:

$$\sin \theta_s = \mp \frac{\sin \psi_B}{4\pi(F/\lambda)} \qquad (60)$$

where $\mp$ signs are for right and left circularly polarized cases and θ_s is the amount of squint. As an example, Fig. 29 shows the squinted patterns for right and left circularly polarized feeds located at the focal point. (Angles ψ_B and ψ_S are defined in Fig. 16.) The amount of squint obtained from experimental data and diffraction

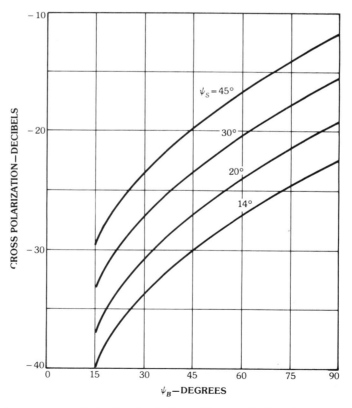

Fig. 28. Maximum cross-polarization level of an offset parabolic reflector illuminated by a balanced linearly polarized feed. (*After Chu and Turrin [35], © 1973 IEEE*))

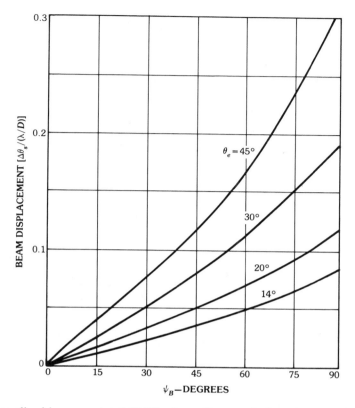

Fig. 29. Normalized beam squint $\theta_S/(\lambda/D)$ of an offset parabolic reflector illuminated by a circularly polarized feed located at the focal point. (*After Chu and Turrin [35], © 1973 IEEE*))

analysis [35] agrees well with the approximate formula. Additional results are shown in Fig. 30 [35], where again ψ_B and ψ_S are used to characterize the reflector.

Reflector Pattern Characteristics for Off-Focus Feeds

In many applications, such as the design of multiple and contour beam reflectors, it becomes necessary to illuminate the reflector with feeds positioned away from the reflector focal point. This feed displacement introduces phase aberration, which results in pattern distortion in terms of gain loss, side lobe degradations, etc. In this subsection some of the key distortion characteristics of reflectors are presented for both the symmetric and offset parabolic reflectors and for both the axial and lateral feed defocusings.

Axial Displacements—For the symmetric reflector the results of feed axial defocusing for gain loss and patterns are shown in Figs. 31 through 33. It is assumed that the feed provides a −10-dB edge taper when it is located at the focal point. Results are demonstrated for feed displacements toward and away from the reflector. A small asymmetry can be observed. These results are dependent on the

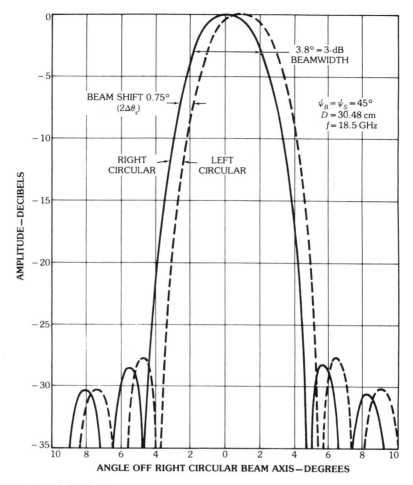

Fig. 30. Squinted far-field patterns in the plane $\phi = 90°$ of an offset parabolic reflector illuminated by right and left circularly polarized feeds located at its focal point. (*After Chu and Turrin [35], © 1973 IEEE*)

edge taper as discussed in [37]. Furthermore, for the cases where $\delta \ll F$ and the feed is an infinitesimal dipole with moment **p** polarized in the $\hat{y}$ direction, one can obtain the following approximate expression [37] for the field on axis, namely,

$$\mathbf{E}(R,0,0) = -\frac{jk^2 p}{4\pi\epsilon}(2kF)\hat{y}\frac{\exp(-jkR)}{R}\frac{1}{(4F/D)^2 + 1}$$
$$\exp\left[-jk\delta\frac{(4F/D)^2}{(4F/D)^2 + 1}\right]\frac{\sin\xi}{\xi} \qquad (61)$$

where

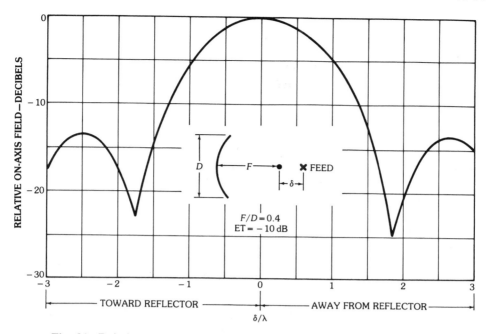

Fig. 31. Relative on-axis field as a function of axial feed displacement.

$$\xi = \frac{2\pi(\delta/\lambda)}{(4F/D)^2 + 1} \tag{62}$$

The dominant effect of the axial defocusing is the generation of the quadratic phase error across the reflector aperture. The axial field (61) becomes virtually zero for values of $\zeta = \pm\pi, \pm 2\pi, \pm 3\pi$, etc., which results in

$$\frac{\delta}{\lambda} = \pm\frac{n}{2}[(4F/D)^2 + 1], \qquad n = 1, 2, 3, \ldots \tag{63}$$

For these values the beam widens considerably and may also be bifurcated (see Figs. 32 and 33). The reader may have noticed a resemblance between these patterns and those resultant from the field of an aperture in the Fresnel zone.

Lateral Displacements—For a parabolic reflector, lateral feed displacements result in scanned beams. It is well known that these reflectors have limited scan capability which strongly depends on F/D and F/D_p ratios for symmetric and offset reflectors, respectively. First, results are presented for symmetric reflectors. An important parameter in dealing with scanned beams is the beam deviation factor (BDF) which is defined as

$$\text{BDF} = \frac{\theta_B}{\theta_F} \tag{64}$$

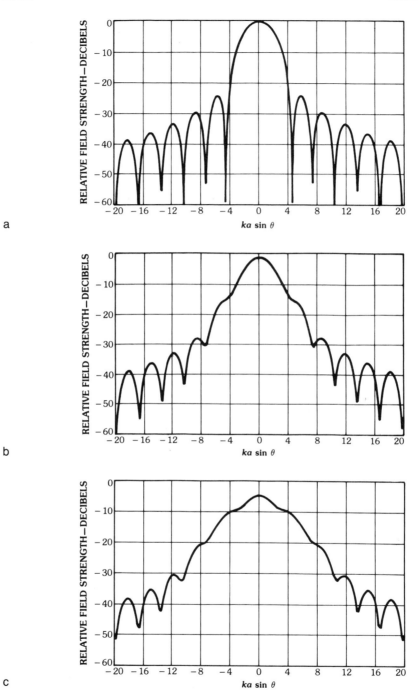

Fig. 32. Reflector far-field patterns for $\phi = 0°$ as a function of axial feed displacement away from the reflector (see Fig. 31). (*a*) For $\delta/\lambda = 0$. (*b*) For $\delta/\lambda = 0.5$. (*c*) For $\delta/\lambda = 1.0$. (*d*) For $\delta/\lambda = 2.0$. (*e*) For $\delta/\lambda = 3.0$.

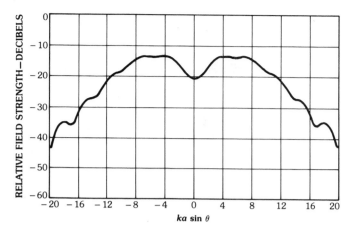

d

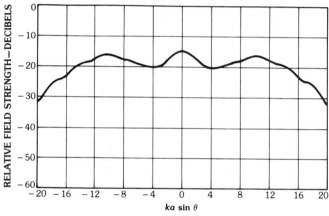

e

Fig. 32, *continued.*

where θ_B and θ_F are the beam scan angle and feed tilt angle, respectively, as shown in Fig. 34. The BDF has a strong dependence on F/D and weakly depends on the edge taper and δ/F (δ is the lateral feed displacement). A good approximation for BDF is [38]

$$\text{BDF} = \frac{\sin^{-1}(\delta/F)\{[1 + k(D/4F)^2]/[1 + (D/4F)^2]\}}{\tan^{-1}(\delta/F)} \quad (65)$$

where $0 < k < 1$ and $k = 0.36$ provides very accurate results when the BDF is compared with experimental and diffraction analysis data. For small δ/F, (65) may be simplified to give

$$\text{BDF}_0 = \frac{1 + k(D/4F)^2}{1 + (D/4F)^2} \quad (66)$$

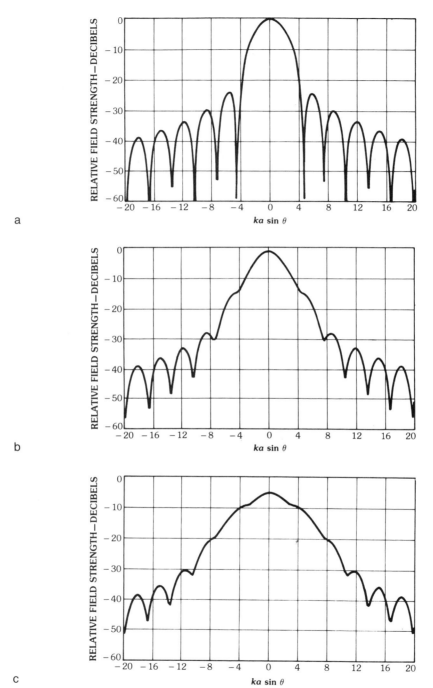

Fig. 33. Reflector far-field patterns for $\phi = 0°$ as a function of axial feed displacement toward the reflector (see Fig. 31). (*a*) For $\delta/\lambda = 0$. (*b*) For $\delta/\lambda = -0.5$. (*c*) For $\delta/\lambda = -1.0$. (*d*) For $\delta/\lambda = -2.0$. (*e*) For $\delta/\lambda = -3.0$.

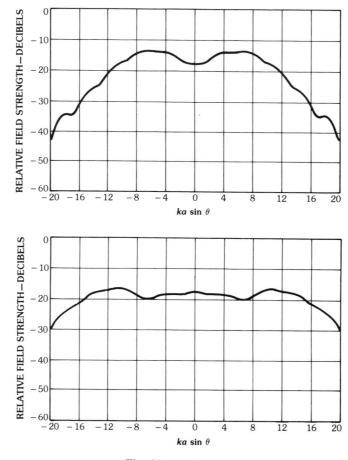

Fig. 33, *continued.*

which is independent of δ/F. Fig. 34 shows the BDF as a function of F/D and for $k = 0.36$. For large feed displacements one may have to resort to diffraction analysis results in order to obtain more accurate values for the BDF.

Due to the phase aberration introduced by defocused feeds the far-field pattern can be substantially degraded. This degradation depends very strongly on the F/D ratio and the angle of scan in terms of number of beamwidths scanned (bmws). There are some approximate formulas available for predicting the peak gain loss as a function of beamwidths scanned [39]. In many applications, however, the knowledge of peak gain loss is not sufficient and one must know the overall degradation effects on the far-field pattern. For this reason many selective but representative cases are given here to provide a clear picture of the pattern degradation. First, an $F/D = 0.4$ symmetric parabolic reflector illuminated by a $\cos^q(\theta)$ type feed with ET = -10 dB is considered. Far-field patterns for different numbers of beamwidths scanned are shown in Figs. 35b through 35e. It is important to note that the patterns are plotted versus the universal parameter $ka \sin \theta_p = \pi(D/\lambda) \sin \theta_p$

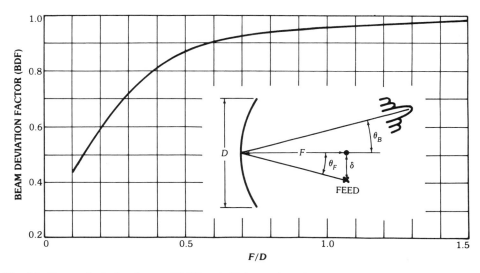

Fig. 34. Beam deviation factor (BDF = θ_B/θ_F) as a function of the F/D ratio for symmetric reflectors (ET = -10 dB).

in which θ_p is the angle measured from the axis passing through the peak of the beam which is directed in the $\theta_B = \text{BDF} \tan^{-1}(\delta/F)$ direction. The peak values of these patterns are normalized with respect to the peak value of the nonscanned beam. Note that for large scan angles, BDF = 0.82 (for F/D = 0.4) does not predict the location of the beam maximum accurately. Furthermore, the feed axis is kept parallel to the reflector axis and has not been tilted toward the reflector center. This condition is more practical when feeds are used in a planar array in contour and multiple beam applications. For directive feeds with large displacements, however, it may be necessary to tilt the feed toward the reflector center in order to reduce the amount of spillover. In this case the feeds may be arrayed on the spherical surface rather than the planar surface. Graphs of Figs. 35b through 35e clearly demonstrate how rapidly the reflector pattern can degrade for scan angles beyond two beamwidths scanned when F/D = 0.4. One simple way to improve this very limited scan capability is to increase the F/D ratio. Figs. 36a through 36d show the far-field patterns for F/D = 1.0 as a function of the beamwidths scanned. Note that for this value of F/D, beams with much larger scan angles can be generated with adequate characteristics. The prime drawbacks of employing large F/D values are the larger structural size and the need for more directive feeds to provide the required edge taper. Finally, Fig. 37 shows the gain loss curves for different values of F/D as a function of the beamwidths scanned (bmws). Although most of the results are shown for the case of ET = -10 dB, similar observations can be made for other edge tapers. For example, Fig. 38 shows the results of experimental data [40]. Excellent agreement has been observed as far as the effects of the feed displacements are concerned. There are, however, differences in the third side lobe levels due to the fact that the experimental feed patterns are not exactly modeled by the $\cos^q(\theta)$ type pattern.

The previously mentioned results for symmetric reflectors only demonstrate

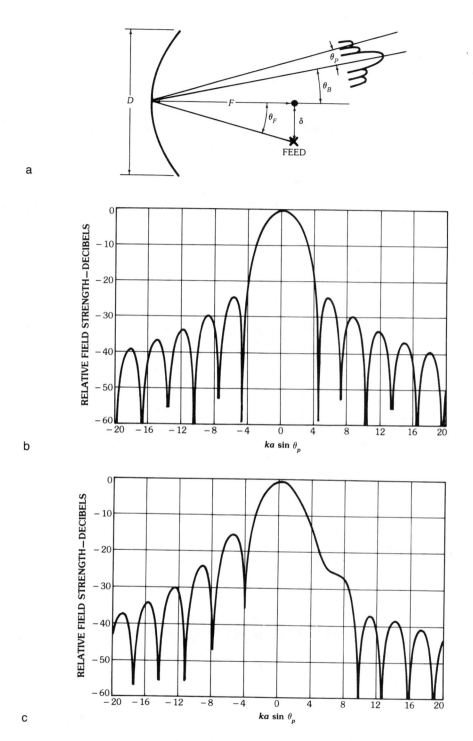

Fig. 35. Symmetric reflector far-field patterns for $\phi = 0°$ as a function of lateral feed displacement in terms of the number of beamwidths scanned ($F/D = 0.4$, ET = -10 dB, BDF = 0.82). (*a*) Reflector parameters. (*b*) 0 bmws. (*c*) 2 bmws. (*d*) 6 bmws. (*e*) 10 bmws.

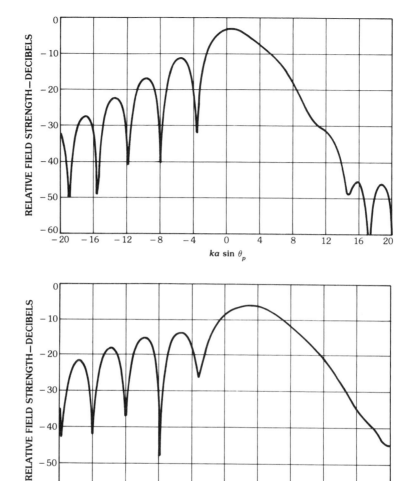

d

e

Fig. 35, *continued.*

the scan characteristics while ignoring the blockage effects, which could become very severe if large feed arrays are used to illuminate reflectors for multiple and contour beam applications. To overcome these blockage effects the designers of satellite communications systems frequently employ offset reflectors which have scan characteristics as discussed here. For offset reflectors the beam deviation factor is computed as shown in Fig. 39. Many studies have shown that the F/D_p ratio characterizes the offset reflector patterns better than the F/D ratio. Far-field patterns as a function of the number of beamwidths scanned are shown in Figs. 40a through 40d for $F/D_p = 0.4$ and ET $= -10$ dB. For this case $F/D = 0.96$ ($F = 96\lambda$, $D = 100\lambda$, $H/D = 0.2$), which is almost two and a half times larger than F/D_p.

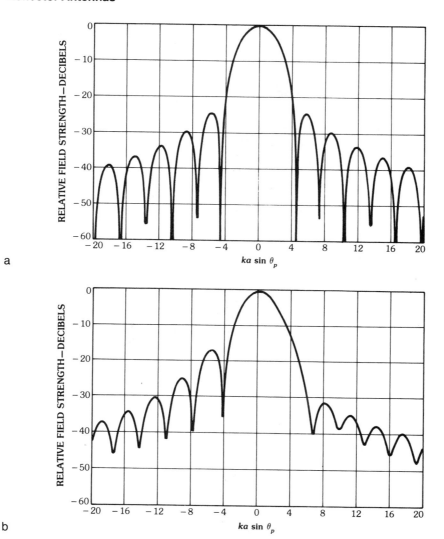

Fig. 36. Symmetric reflector far-field patterns for $\phi = 0°$ as a function of lateral feed displacement in terms of the number of beamwidths scanned ($F/D = 1.0$, ET = -10 dB, BDF = 0.966). (*a*) 0 bmws. (*b*) 8 bmws. (*c*) 16 bmws. (*d*) 20 bmws.

Notice the similarity between these patterns and those of the symmetric reflector with $F/D = 0.4$ (Fig. 35). Similar results are also shown for an offset reflector with $F/D_p = 1.0$ and $F/D = 2.4$. Figs. 41a through 41c show the patterns for the cases in which the feed direction is kept fixed as the feed is displaced in the plane orthogonal to the line joining the focal point with the reflector center. For this large value of F/D_p, the feed is very directive in order to provide the -10-dB edge taper and, therefore, for large scans the reflector would be very poorly illuminated. Results for the cases in which the feed is tilted toward the reflector center while it

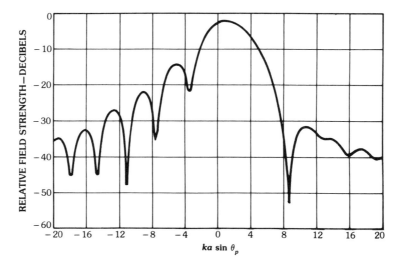

c

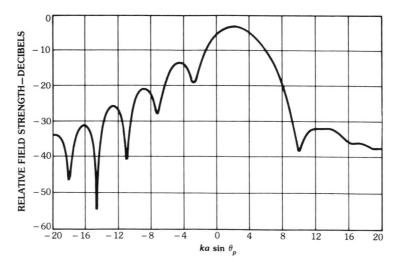

d

Fig. 36, *continued.*

is displaced in the focal plane are shown in Figs. 42a through 42f. Notice that a marked improvement can be observed in the patterns. Finally, the peak gain losses as a function of beamwidths scanned and for different values of F/D_p are plotted in Fig. 43. It is worthwhile to mention that there are a considerable number of ongoing attempts to improve the scan performance of single-reflector antennas by employing the concept of conjugate matched focal-plane feed arrays. This approach may also be utilized to overcome the deterministic reflector surface distortions.

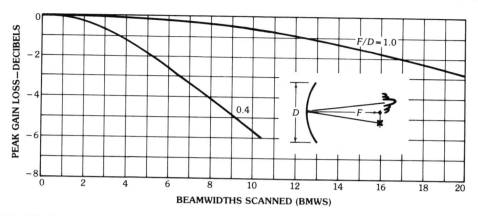

Fig. 37. Peak gain loss of symmetric parabolic reflectors as a function of beamwidths scanned for different F/D ratios.

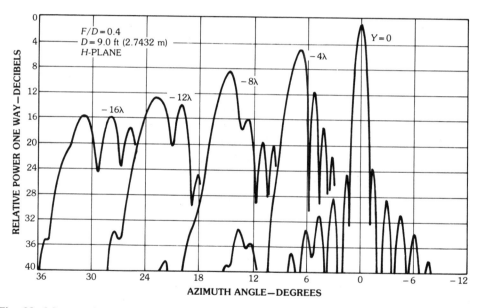

Fig. 38. Measured secondary patterns as a function of lateral primary feed displacement for a symmetric parabolic reflector. (*After Imbriale, Ingerson, and Wong [40], © 1974 IEEE*)

6. Dual-Reflector Antennas

The application of dual reflectors at microwave frequencies has evolved from their counterparts used in optical telescopes. Originally, most of the designs were symmetric configurations with large main reflectors to reduce the blockage effects of the subreflector. These designs include Cassegrain, Gregorian, and shaped. The reader may refer to references [1, 7, 33, 39, 41, 42], which discuss many aspects of these reflectors. Recent stringent performance requirements have increased the

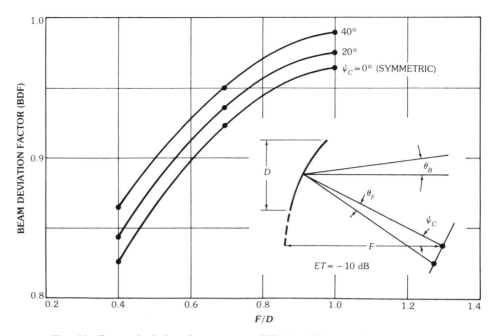

Fig. 39. Beam deviation factor versus F/D for different offset angles ψ_C.

application of offset dual-reflector antennas in both communication satellite and earth station terminals to achieve compact systems with reduced blockage effects. It is the purpose of this section to focus on offset dual reflectors with particular emphasis on offset Cassegrain reflectors.

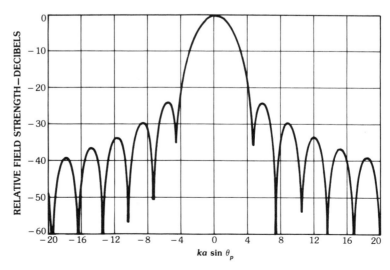

a

Fig. 40. Offset reflector far-field patterns as a function of feed displacement in terms of the number of beamwidths scanned ($F/D_p = 0.4$, $F/D = 0.96$, ET $= -10$ dB, BDF $= 0.983$, $\phi = 0°$). (a) 0 bmws. (b) 4 bmws. (c) 6 bmws. (d) 10 bmws.

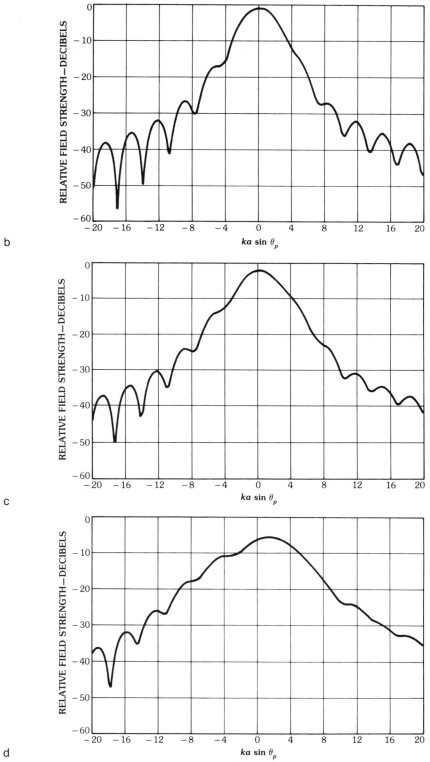

Fig. 40, *continued.*

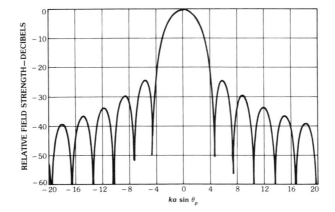

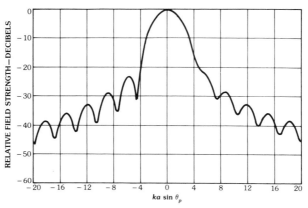

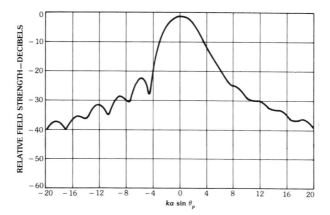

Fig. 41. Offset reflector far-field patterns as a function of feed displacement in terms of the number of beamwidths scanned ($F/D_p = 1.0$, $F/D = 2.4$, ET $= -10$ dB, BDF $= 0.998$, $\phi = 0°$). (*a*) 0 bmws. (*b*) 4 bmws. (*c*) 8 bmws.

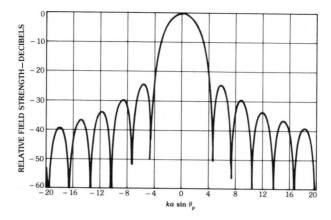

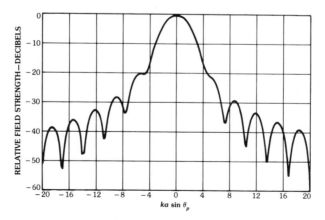

Fig. 42. Offset reflector far-field patterns for tilted feed as a function of feed displacement in terms of the number of beamwidths scanned ($F/D_p = 1.0$, $F/D = 2.4$, ET $= -10$ dB, BDF $= 0.998$, $\phi = 0°$). (*a*) 0 bmws. (*b*) 4 bmws. (*c*) 8 bmws. (*d*) 12 bmws. (*e*) 16 bmws. (*f*) 20 bmws.

d

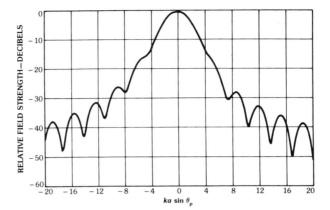

e

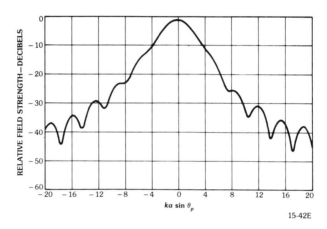

15-42E

f

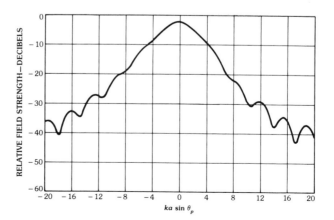

Fig. 42, *continued.*

Reflector Antennas

General Parameters

Geometries of an offset Cassegrain antenna and its equivalent paraboloid are shown in Fig. 44, and some of their important geometrical relationships among different parameters are listed below.

Main reflector:

$$\theta^B = \frac{\theta^U + \theta^L}{2}$$

$$\theta^C = \frac{\theta^U - \theta^L}{2}$$

$$D = \frac{4F \sin \theta^C}{\cos \theta^B + \cos \theta^C} \qquad (67)$$

$$d = \frac{2F \sin \theta^B}{\cos \theta^B + \cos \theta^C}$$

Subreflector:

$$e = \frac{\sin (\theta^U + \theta_s^U)/2}{\sin (\theta^U - \theta_s^U)/2}$$

$$M = \frac{e + 1}{e - 1}, \qquad e = \frac{M + 1}{M - 1}$$

$$\frac{z_0}{D'_s/2} = \frac{1}{\tan \theta^U} + \frac{1}{\tan \theta_s^U}$$

$$\frac{z_0}{D'_s/2 - D_s} = \frac{1}{\tan \theta^L} + \frac{1}{\tan \theta_s^L} \qquad (68)$$

$$\frac{2L}{z_0} = 1 - \frac{\sin (\theta^U - \theta_s^U)/2}{\sin (\theta^U + \theta_s^U)/2} = \frac{2}{1 + M}$$

$$\tan (\theta_s/2) = M^{-1} \tan (\theta/2)$$

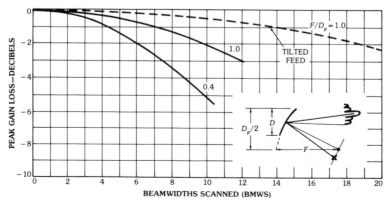

Fig. 43. Peak gain loss of offset parabolic reflectors as a function of the number of beamwidths scanned for different F/D_p ratios.

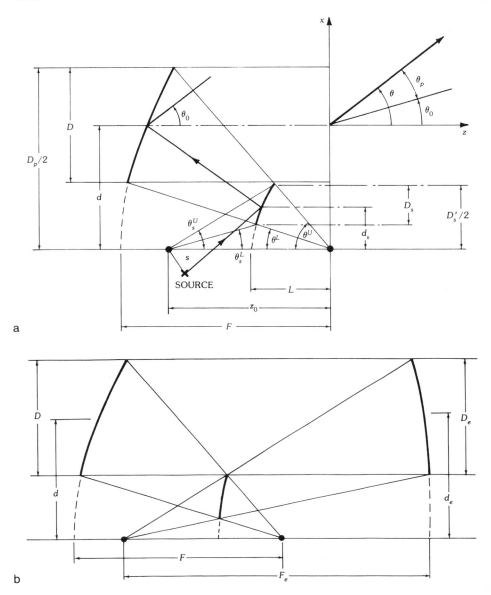

Fig. 44. Geometries of the offset Cassegrain reflector and its equivalent paraboloid. (*a*) Offset Cassegrain reflector. (*b*) Equivalent paraboloid.

Equivalent paraboloid:

$$\begin{aligned} D_e &= D \\ d_e &= d \\ F_e &= \mathbf{M} \cdot \mathbf{F} \end{aligned} \qquad (69)$$

Reflector Antennas

Performance Evaluation

Due to the appearance of the subreflector the analysis of dual reflectors is more complicated than for the single reflector. There are many different methods available based on techniques discussed in Section 2 and a comparative study has been done in [43]. Among the various techniques the applications of GTD from the subreflector and PO from the main reflector (GTD/PO), and PO from both the subreflector and the main reflector (PO/PO), have been most popular. Although in a majority of cases close agreement between the two approaches has been observed, there are situations, such as in small subreflectors (less than 10λ), where the results of GTD/PO do not agree with PO/PO, in particular, in the prediction of the cross-polarization level. As a matter of fact, in these cases PO/PO typically predicts a higher level of cross-polarized fields. It has been conjectured that for small subreflectors PO/PO provides a more accurate result.

For large subreflectors, however, PO/PO can become very time consuming and, for this reason, GTD/PO is used more frequently [1, 43]. One of the advantages of using this technique is that the effects of edge diffraction from the subreflector can be identified, when comparison is made between GO/PO and GTD/PO. In particular, it is known that, once the subreflector boundary is extended 1 to 2 wavelengths beyond the geometrical optical boundary, the diffraction effects can be significantly reduced.

In what follows, a few representative examples are shown. In all cases the offset Cassegrain of Fig. 44 is used, which is illuminated with a $\cos^q(\theta)$ type feed with different edge tapers. The results of the far-field patterns, each normalized to its own peak, for both GO/PO and GTD/PO are shown in Figs. 45 and 46. From these figures one can readily identify the contribution of edge diffracted rays. It is worthwhile to mention that for these cases subreflectors with optical extensions 0.5λ and 1λ beyond the optical extensions are used. Also shown in Fig. 47 is the variation of aperture efficiency as a function of feed edge taper for the cases of GO and GTD. It is interesting to note that for the optical extension case GTD predicts almost 8 percent less efficiency than GO due to the edge diffraction. However, in accordance with Fig. 47 considerable improvement can be achieved by extending the subreflector as small as 1λ beyond its optical edge.

Cross-Polarization Reduction

One of the important features of any dual-reflector system is to provide an additional degree of freedom to optimize a particular characteristic of these reflectors. For example, in a Cassegrain or Gregorian reflector system one can reduce considerably the level of cross-polarized field by simply tilting the axis of the subreflector with respect to the axis of the main reflector. Using the geometrical optics construction one can determine the amount of tilt to be [10, 44]

$$\tan(\alpha/2) = \frac{1}{M} \tan(\zeta/2) \qquad (70)$$

where M is the subreflector magnification, and angles ζ and α are angles shown in Fig. 48. As an example, for the offset Cassegrain reflector with dimensions as

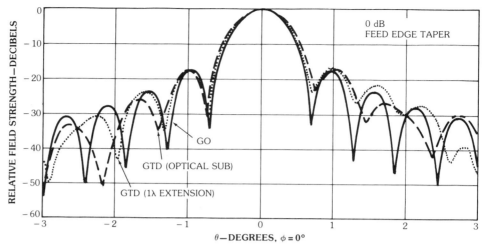

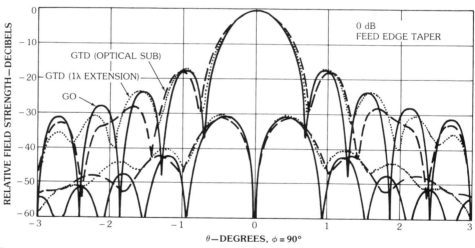

Fig. 45. A comparison between GO and GTD for a 0-dB tapered feed illuminating an offset Cassegrain antenna ($D = 100\lambda$, $d = 70\lambda$, $F = 96\lambda$, $Z_0 = 25\lambda$, $M = 1.5$). (*a*) $\phi = 0°$ cut (*xz* plane). (*b*) $\phi = 90°$ cut (*yz* plane). (*After Rahmat-Samii, "Subreflector extension for improved efficiencies in Cassegrain antennas—GTD/PO analysis,"* IEEE Trans. Antennas Propag., vol. AP-34, pp. 1266–1269, October 1986, © 1986 IEEE)

shown in Fig. 49, the optimal tilt angle is $\zeta = 9.26°$. For two cases of no tilt $\zeta = 0°$ and optimal tilt $\zeta = 9.26°$ the contour of the cross-polarized fields based on diffraction analysis is shown in Fig. 50, which clearly demonstrates how effectively the cross-polarization level can be reduced. Although the level of cross polarization can be reduced drastically for the feed at the focal point, the amount of reduction is not as drastic as for the case with a feed off the focal point.

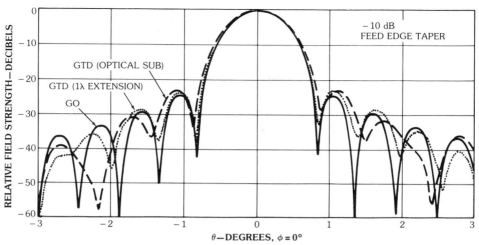

a

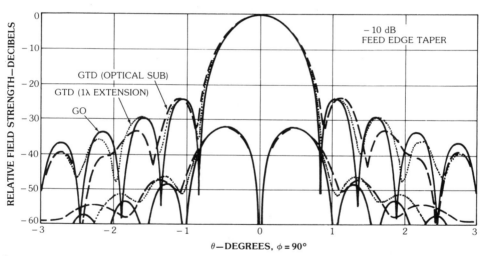

b

Fig. 46. A comparison between GO and GTD for a −10-dB tapered feed illuminating an offset Cassegrain antenna. (*a*) $\phi = 0°$ cut (*xz* plane). (*b*) $\phi = 90°$ cut (*yz* plane). (*After Rahmat-Samii, "Subreflector extension for improved efficiencies in Cassegrain antennas—GTD/PO analysis," IEEE Trans. Antennas Propag., vol. AP-34, pp. 1266–1269, October 1986, © 1986 IEEE*)

Scan Performance

Due to the fact that the equivalent paraboloid of a Cassegrain reflector has a longer focal length than the main reflector, it is expected to demonstrate a better scan performance than the single reflector. Earlier work reported in [33] on symmetric Cassegrain reflectors suggested that the scan performance of these

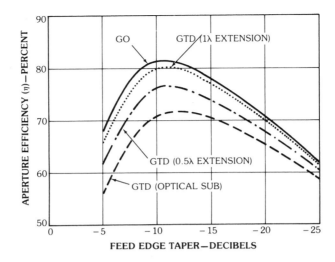

Fig. 47. A comparison between GO and GTD for reflector aperture efficiency. (*After Rahmat-Samii, "Subreflector extension for improved efficiencies in Cassegrain antennas—GTD/PO analysis," IEEE Trans. Antennas Propag., vol. AP-34, pp. 1266–1269, October 1986, © 1986 IEEE*)

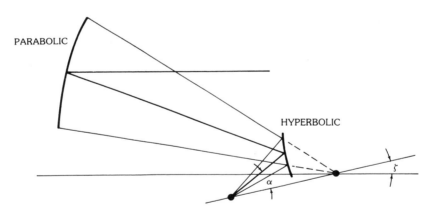

Fig. 48. A tilted subreflector for cross-polarization reduction.

reflectors may be predicted from their equivalent paraboloid counterparts. To this effect, results were presented in [45] based on the PO/PO diffraction analysis, which demonstrated close agreement between the Cassegrain and equivalent paraboloid. Fig. 51 shows the result for the scan performance. Similar results were also obtained using GTD/PO as shown in Fig. 52. More recent investigations have demonstrated that the concept of the equivalent paraboloid is only applicable for small scans and should be employed with care [43]. For example, for an offset Cassegrain with parameters as given in Fig. 53, comparisons are made between the main reflector, Cassegrain, and its equivalent paraboloid in Figs. 53–55. These

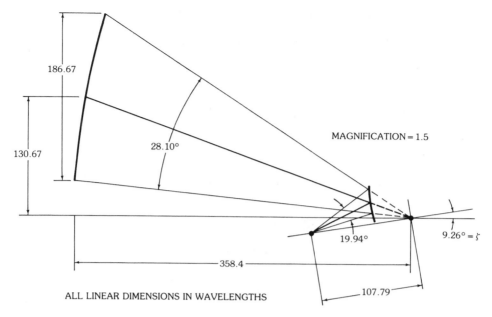

Fig. 49. An offset Cassegrain reflector with tilted subreflector.

results indicate that for small scans the performance of the Cassegrain is similar to the equivalent paraboloid, whereas for the large scan it deviates substantially. In all these cases an oversized subreflector was used for better illumination of the main reflector for large scans. Fig. 56 shows the gain loss as a function of the beamwidths scanned for all three cases. Additional improvement can be expected by locating the feeds on the optimal focal surface rather than the optimal focal plane tangent to this surface [46]. Similar observations can also be made for Gregorian reflectors [47].

Shaped Reflectors

An important class of dual reflectors is the shaped reflectors, with properly shaped subreflector and main reflector, which provide desired aperture amplitude and phase distributions. These reflectors are used particularly in the design of high-gain antennas and also extensively in the design of many ground and telemetry antennas. The foundations of these reflectors can be found in [1, 48, 49, 50, 51], where most of the designs have been for symmetric configurations. Recently, however, in order to further improve the overall efficiency, offset shaped reflectors have also been considered [52, 53, 54, 55, 56]. In particular, a recent evaluation [54] of this type of antenna has been reported which gives an overall efficiency of nearly 85 percent! A brief description of the concept behind these reflectors is given below [55].

With some exceptions reflector antennas are synthesized on the basis of geometrical optics (GO) with the assumptions that the wavelength is short relative to the overall dimensions and that the reflective surfaces have large radii of curvature.

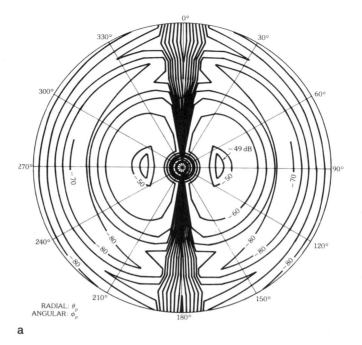
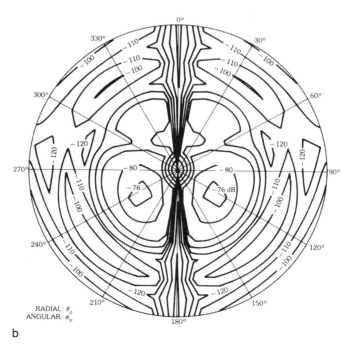

Fig. 50. A contour plot of the cross-polarized field for the Cassegrain reflector of Fig. 49. (*a*) With $\zeta = 0°$. (*b*) With $\zeta = 9.24°$.

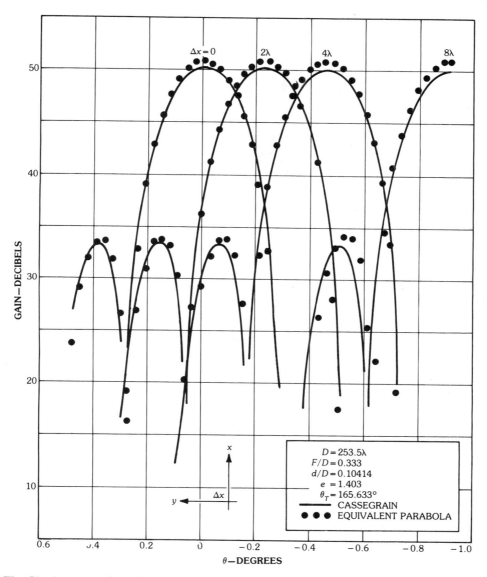

Fig. 51. A comparison between a symmetric Cassegrain reflector and its equivalent paraboloid as a function of feed displacement using PO/PO analysis. (*After Wong [45]*, © *1975 IEEE*)

Classically the paraboloid converts a spherical wave emanated from the source (feed) into a plane wave; hence one may state that the reflector transforms the shape of the *phase front*. Some single reflectors are also designed to convert the feed pattern into a given optical *energy* distribution, e.g., into a $\csc^2(\theta)$ type pattern [57]. It is interesting that a doubly curved noncylindrical reflector of this type has no exact solution. Dual reflectors have also been used to convert a spheri-

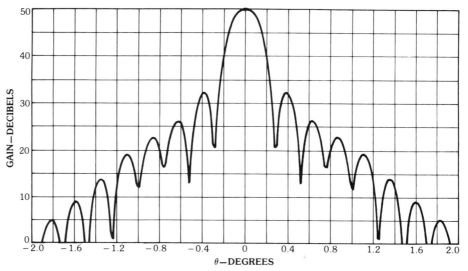

a

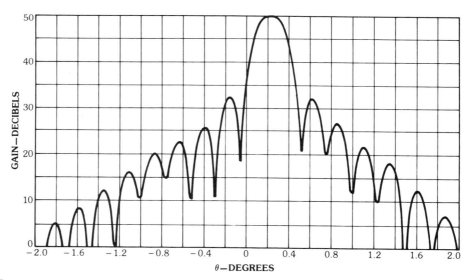

b

Fig. 52. Same as Fig. 51 except that GTD/PO diffraction analysis is used. (*a*) $\Delta = 0$. (*b*) $\Delta = -2\lambda$. (*c*) $\Delta = -4\lambda$. (*d*) $\Delta = -8\lambda$.

cal phase front into a plane phase front; for example, one may refer to the Cassegrain (paraboloid-hyperboloid) and Gregorian (paraboloid-ellipsoid) reflectors [33].

When two sequential reflections from the feed to the aperture are used for aperture magnification, it is possible to exercise control over both the resultant

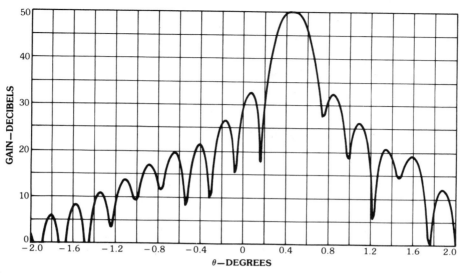

c

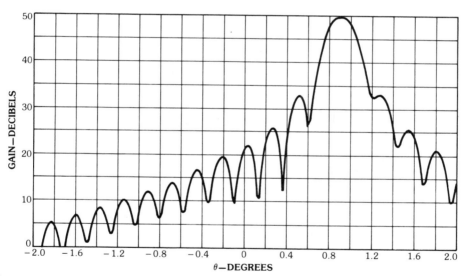

d

Fig. 52, *continued.*

aperture phase-front shape and the aperture energy distribution by appropriate shaping of the reflector surfaces. The basic objective of the GO dual-shaped synthesis is illustrated in Fig. 57. A bundle of rays radiated by the feed is a representation of the feed pattern in both amplitude and phase. This bundle, which has a well-defined periphery, is intercepted by the subreflector and then by the main reflector. The output bundle of rays, after two reflections, is required to have: (*a*)

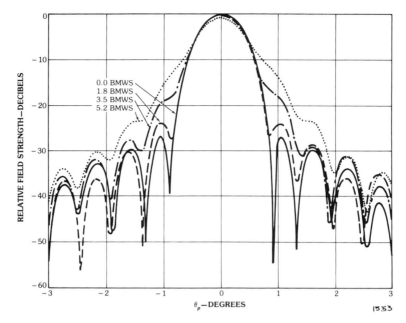

Fig. 53. Scan performance of an offset Cassegrain reflector ($D = 100\lambda$, $d = 70\lambda$, $F = 96\lambda$, $z_0 = 25\lambda$, $M = 1.5$, $D_s = 26\lambda$, oversized). (*After Rahmat-Samii and Galindo-Israel [43]*, © *1981, American Geophysical Union*)

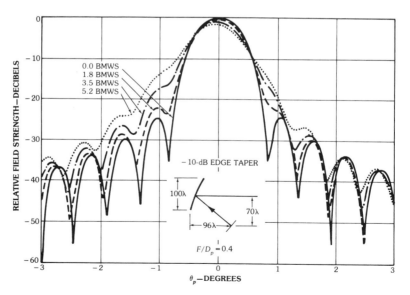

Fig. 54. Scan performance of an offset parabolic reflector with $F/D_p = 0.4$ illuminated by a linearly polarized incident field with -10-dB edge taper. (*After Rahmat-Samii and Galindo-Israel [43]*, © *1981, American Geophysical Union*)

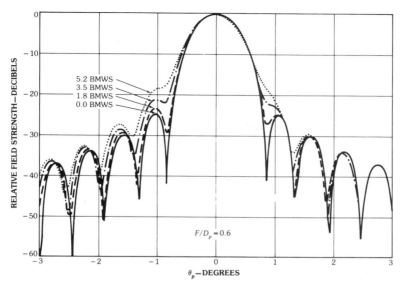

Fig. 55. Same as Fig. 54 except that $F = 144\lambda$ and $F/D_p = 0.6$. (*After Rahmat-Samii and Galindo-Israel [43], © 1981, American Geophysical Union*)

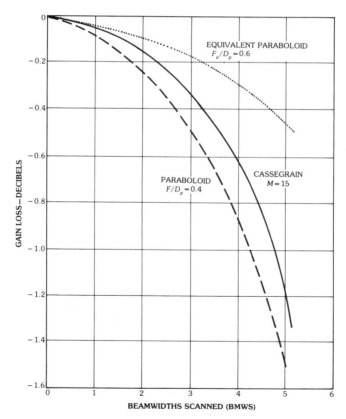

Fig. 56. Cassegrain gain loss performance with a comparison between the main-reflector paraboloid ($F/D_p = 0.4$) and its equivalent paraboloid ($F_e/D_p = 0.6$). (*After Rahmat-Samii and Galindo-Israel [43], © 1981, American Geophysical Union*)

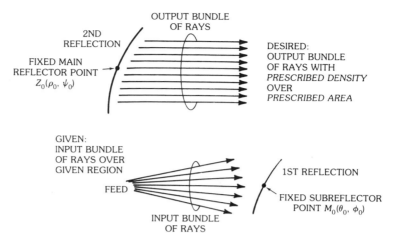

Fig. 57. Geometrical optics synthesis of a dual-shaped reflector—a statement of the problem.

a prescribed phase front, (b) a prescribed energy density distribution, (c) a prescribed periphery, (d) a prescribed fixed point on the subreflector (or a line), and (e) a prescribed fixed point on the main reflector (or a line) for a given feed pattern. These constraints define useful engineering problems of many types. It can be shown that the solution to these problems involves solving (numerically) two simultaneous, nonlinear, first-order, ordinary differential equations. Many numerical and approximate techniques have been developed to solve these equations and the reader is referred to [48–59] for details.

7. Contour Beam Reflectors

Another important class of reflectors is the combination of arrays with reflectors. This combination is used to generate multiple-beam [60, 61] improved scan performance [7, 62] and contour beam [7, 61, 63, 64] antennas. Among these applications the contour beam reflectors are widely used in communication satellite applications and some of the steps involved in designing them are briefly summarized in this section.

The major rf parameters in the design of contour beam reflectors are the coverage area, gain requirements, gain ripples, cross-polarization levels, overall losses, etc. To overcome the unwanted blockage effects of large arrays most contour beam reflectors use offset reflector configurations. So far, offset parabolic reflectors have been primarily used, although applications of offset Cassegrain and shaped reflectors are also becoming popular.

The principal design steps for an offset parabolic reflector with diameter D, focal length F, and offset angle ψ_c are summarized below [61]:

1. Plot a coverage map as viewed by the satellite from the synchronous orbit, or a composite coverage map including the pointing error and views from different orbit positions. Fig. 58 shows the typical azimuth and elevation coordinates

Reflector Antennas

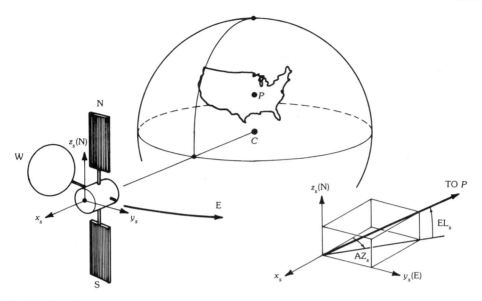

Fig. 58. An observation point P on earth as described by azimuth angle AZ_s and elevation angle EL_s in satellite coordinates.

used to define the coverage map and Fig. 59 depicts the United States from a synchronous satellite at 98° west longitude [61].

2. Select a set of contiguous (sometimes noncontiguous may also be used) squares or circles which represent the -3- to -4-dB beamwidth of the reflector illuminated by a single feed element. It takes a bit of experience to find the best arrangement. However, by setting the center of the outer elements close to the boundary of the desired coverage area, satisfactory results can be expected. Fig. 60 shows one such arrangement to cover the Eastern Time Zone (ETZ) of the United States.

3. Select an appropriate location for the antenna boresight and determine the proper scale factor between the lattice element centers and actual feed location in the feed plane. Beware of the beam deviation factor. For example, it has been found that the following expression provides a satisfactory estimate of the feed size:

$$\Delta = \tan\,(\theta_{\text{bmw}}/\text{BDF})\,\frac{2F}{1+\cos\psi_c} \qquad (71)$$

where ψ_c is the offset angle of the feed array, BDF is the beam deviation factor that can be approximated by

$$\text{BDF} \simeq \frac{1+k(D/4F)^2}{1+(D/4F)^2}, \qquad k=0.36, \qquad (72)$$

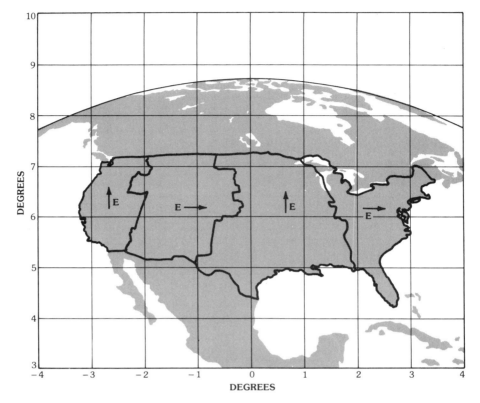

Fig. 59. Continental U.S. time zone map as viewed from a geosynchronous satellite at 98° west longitude. (*After Chen and Franklin [61]*)

and θ_{bmw} is the -3-dB beamwidth of the reflector with diameter D/λ (λ = wavelength) illuminated by a single feed, viz.,

$$\theta_{bmw} \cong 2\sin^{-1}\left(\frac{1.6 - 0.02\text{ET}}{\pi D/\lambda}\right) \quad (73)$$

In the above equation, ET is a single-feed edge taper which typically takes the value ET = -4 to -5 dB. The previous expression can be further approximated to give [61]

$$\Delta = \frac{1.06}{D/\lambda} \frac{2F}{1 + \cos\psi_c} \quad (74)$$

4. Select sample points on the map for gain optimization. Fig. 61 shows an example of the selection of sample points on ETZ. Typically the gain variations at a subset of these points are specified by the system requirements.
5. Determine the reflector far field due to each feed with unity excitation.
6. Determine the optimum amplitude and phase of each feed by numerical

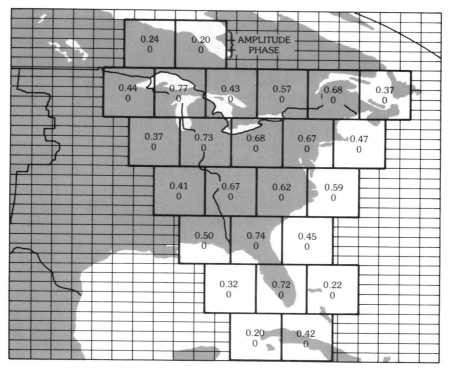

Fig. 60. Reflector image of ETZ feed cluster and its amplitude and phase distribution overlaid on the ETZ map. (*After Chen and Franklin [61]*)

optimization techniques. The proper selection of the optimization algorithm is a very crucial step in obtaining contours which satisfy prescribed constraints. Some algorithms are discussed in [7, 61]. For example, Fig. 62 shows a block diagram of the steps involved in an optimization. In this figure, P designates the total power radiated by the array for the directivity computation [34].

7. Construct the contour pattern of the reflector illuminated by the feed array with optimized coefficients. For example, the excitation coefficients of an array for the ETZ coverage is shown in Fig. 60 and the reflector contour map is depicted in Fig. 63 [61].
8. Finally, verify the results of the computer simulation with the experimental data. One such comparison is reported in [61] and the results are depicted in Fig. 64.

One of the difficult steps in accurately predicting the performance of a reflector illuminated by a densely packed feed array is the inclusion of mutual coupling effects. So far, a simple approach has not yet been developed for incorporating these coupling effects. This is typically done either by using the measured pattern of the feed elements in the array environment or by using numerical methods for simple feed array configurations [64, 65]. In a recent work [66], the planar near-field probing technique has provided a means to significantly reduce the effects of

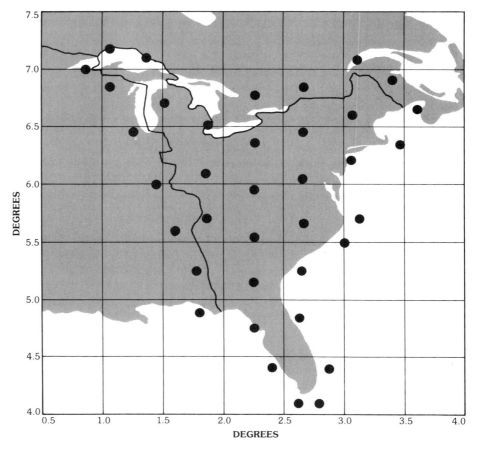

Fig. 61. Selected sample points for gain optimization, with satellite at 98° west longitude. (*After Chen and Franklin [61]*)

mutual coupling by the insertion of additional tuning into the feed lines of appropriate elements.

8. Feeds for Reflectors

Antenna feeds share a major role in the successful and efficient operation of any reflector system, independently of its optical setup. Unfortunately there is no such thing as an all-purpose universal feed, as in practice for a given reflector configuration and performance requirements the feed must be properly tailored. In the previous sections, independently of the actual feed elements, the feed patterns were approximated by $\cos^q(\theta)$ type patterns to allow for detailed parametric studies. In this section the characteristics of the most commonly used reflector feed elements are presented. These elements may take a variety of configurations including dipoles, log spirals, open-ended rectangular and circular waveguides, rectangular and circular horns, corrugated horns, dual- and hybrid-mode horns,

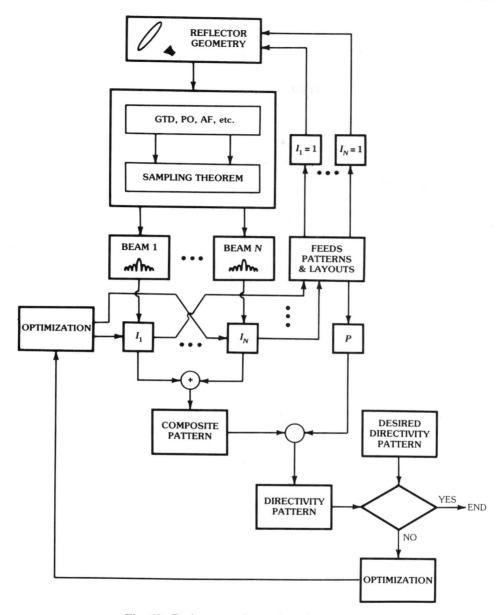

Fig. 62. Basic steps of a contour beam design.

microstrips, etc., or arrays of these elements. In this section only a few are considered and the reader may refer to [6, 7, 9, 67] for a more comprehensive treatment of the subject.

In the ideal situation, in which the reflector optics possess symmetric path losses (such as symmetric reflectors), the ideal feed element (although unachievable in reality) is one that has symmetric patterns, has a unique phase center, and

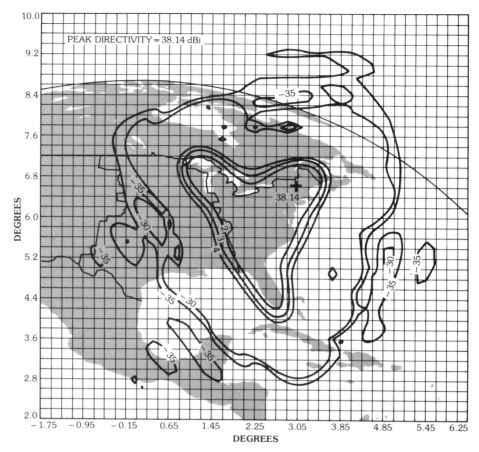

Fig. 63. Downlink copolarized beam isolation contours at 11.95 GHz. (*After Chen and Franklin [61]*)

confines its radiation only in a cone subtended to the reflector edge. For example, to obtain the highest gain in a symmetric parabolic reflector it is desirable to have feed patterns which both compensate for the path losses and provide a uniform aperture field. It is readily obtained that such a feed pattern is

$$C(\theta) = \sec^2(\theta/2) \tag{75}$$

where θ is the spherical angle from the feed axis which coincides with the reflector axis (see Section 5, under "Idealized Feed Patterns"). Note that, in terms of the power pattern, $\sec^4(\theta/2)$ will be used.

In the spherical coordinates of the feed as shown in Fig. 65, the radiated field may, in general, be presented as

$$\mathbf{E}(\mathbf{r}) = A_0[\hat{\boldsymbol{\theta}} C_\theta(\theta,\phi) + \hat{\boldsymbol{\phi}} C_\phi(\theta,\phi)] \frac{e^{-jkr}}{r} \tag{76}$$

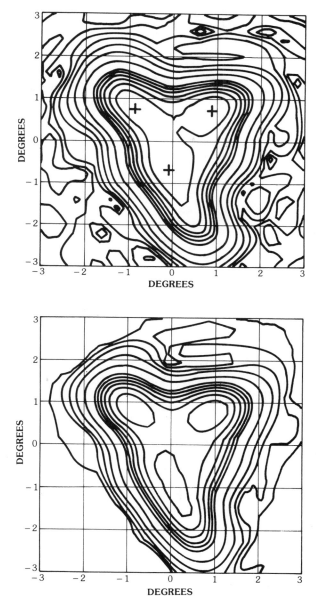

Fig. 64. Calculated and measured far-field gain contours. (*a*) Calculated gain contours from measured near-field feed data. (*b*) Measured far-field gain contours. (*After Chen and Franklin [61]*)

where A_0 is a complex constant. For a specified feed element its pattern can be measured in many ϕ = constant planes and then interpolated or expanded in terms of spherical harmonics to give the overall volumetric pattern. Or it may be numerically computed if the feed element possesses a mathematically tractable

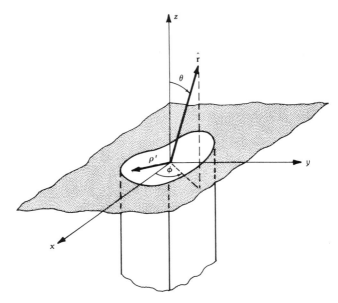

Fig. 65. Cartesian and spherical coordinates of a feed element with a planar aperture opening.

configuration. In practice, however, only the two principal plane patterns are used and then the overall volumetric pattern is approximated. These principal plane patterns are customarily referred to as the E- and H-plane patterns. For example, for an $\hat{\mathbf{x}}$ and $\hat{\mathbf{y}}$ polarized feed, the pattern may be described as

$$\mathbf{C}(\theta,\phi) = \begin{cases} \hat{\boldsymbol{\theta}} C_E(\theta)\cos\phi - \hat{\boldsymbol{\phi}} C_H(\theta)\sin\phi, & \text{for } \hat{\mathbf{x}} \text{ polarized} \\ \hat{\boldsymbol{\theta}} C_H(\theta)\sin\phi + \hat{\boldsymbol{\phi}} C_E(\theta)\cos\phi, & \text{for } \hat{\mathbf{y}} \text{ polarized} \end{cases} \quad (77)$$

For the case of a feed element with the above pattern illuminating a parabolic reflector (or equivalent parabola in dual reflectors) it is possible to construct the aperture field in a straightforward fashion, resulting in

$$\mathbf{E}_A = A_0[-\hat{\mathbf{x}}(C_\theta \cos\phi - C_\phi \sin\phi) + \hat{\mathbf{y}}(C_\theta \sin\phi + C_\phi \cos\phi)]\frac{e^{-j2kF}}{r} \quad (78)$$

where $r = \sec^2(\theta/2)$ and it is assumed that the z axis of the feed and reflector coordinates coincide (but lie in opposite directions) and that the aperture plane passes through the focal point with a focal length of F. The above expression for the aperture field is correct as long as it is further assumed that the incident and reflected fields have spherical and planar wavefronts, respectively. More complicated expressions result from general cases. It is worthwhile to emphasize that the reader should pay careful attention so as to clearly differentiate between feed and reflector coordinates, even if similar notations are used to describe them. For instance, in (78) $\hat{\mathbf{x}}$ and $\hat{\mathbf{y}}$ are referred to as the antenna coordinates, whereas in (77)

Reflector Antennas

they are referred to as feed coordinates. These coordinates can, in general, be related using Eulerian angles as described in the Appendix. Expression (78) allows one to clearly identify the generation of copolarized and cross-polarized fields in the aperture.

Radiation Patterns of Simple Feeds

There are different analytical approaches available to determine radiated fields of simple feeds. Among these methods the aperture E-field model has been used very successfully because of both its simplicity and the good comparison of its results with measured data. This model assumes that the tangential electric field is known in the aperture and is zero outside the aperture, which is the same as assuming that the aperture is in a ground plane. Denoting the feed's tangential electric field by $E^a(\varrho')$, one obtains the following Cartesian components for the radiated field:

$$C_x(\theta,\phi)\hat{x} + C_y(\theta,\phi)\hat{y} = 2\iint_{\substack{\text{feed}\\\text{aperture}}} E^a(\varrho')\, e^{jk\hat{r}\cdot\varrho'}\, dS' \tag{79}$$

where $\hat{r}$ and ϱ' are shown in Fig. 65. The far-field spherical components are (using the vector potential approach)

$$\begin{bmatrix} C_\theta \\ C_\phi \end{bmatrix} = \begin{bmatrix} \cos\phi & \sin\phi \\ -\cos\theta\sin\phi & \cos\theta\cos\phi \end{bmatrix} \begin{bmatrix} C_x \\ C_y \end{bmatrix} \tag{80}$$

If Ludwig's third definition [36] of cross polarization is used, the copolar C_p and cross-polar C_q components of the radiated field are

$$\begin{bmatrix} C_p \\ C_q \end{bmatrix} = \begin{bmatrix} \sin\phi & \cos\theta \\ \cos\phi & -\sin\phi \end{bmatrix} \begin{bmatrix} C_\theta \\ C_\phi \end{bmatrix} \tag{81}$$

Note that this assumes that the feed principal polarization lies along the y axis. If the principal polarization lies along the x axis, C_p and C_q must be interchanged. The above expression may further be combined with (80) to result in

$$\begin{bmatrix} C_p \\ C_q \end{bmatrix} = \cos^2(\theta/2) \begin{bmatrix} \sin 2\phi\,\tan^2(\theta/2) & 1 - \tan^2(\theta/2)\cos 2\phi \\ 1 + \tan^2(\theta/2)\cos 2\phi & \sin 2\phi\,\tan^2(\theta/2) \end{bmatrix} \begin{bmatrix} C_x \\ C_y \end{bmatrix} \tag{82}$$

which is a simple form for determining the copolar and cross-polar fields in planes $\phi = 0°$, $45°$, and $90°$.

Open-Ended Rectangular Waveguides—For an open-ended rectangular waveguide propagating the TE_{10} mode (y polarized) and with sides $x = 2a$ and $y = 2b$, the aperture field is (ignoring the end effects)

$$\mathbf{E}^a(x', y') = E_0 \cos(\pi x'/2a)\hat{\mathbf{y}} \tag{83}$$

The radiated field from (79) and (80) is

$$C_\theta(\theta, \phi) = 4\pi ab E_0 \sin\phi \, \frac{\cos U}{(\pi/2)^2 - U^2} \frac{\sin V}{V}$$

$$C_\phi(\theta, \phi) = 4\pi ab E_0 \cos\theta \cos\phi \, \frac{\cos U}{(\pi/2)^2 - U^2} \frac{\sin V}{V} \tag{84}$$

where

$$U = ka \sin\theta \cos\phi$$
$$V = kb \sin\theta \sin\phi \tag{85}$$

Figs. 66 and 67 show some important characteristics of the open-ended rectangular waveguides [68].

Open-Ended Circular Waveguides—For an open-ended circular waveguide propagating the TE_{11} mode (y polarized) and with radius a, the aperture field may be approximated by

$$\mathbf{E}^a(\varrho', \phi') = E_0 \frac{1}{\varrho'} J_1\left(\frac{\chi}{a}\varrho'\right) \sin\phi \, \hat{\boldsymbol{\varrho}} + E_0 \frac{\partial}{\partial \varrho'}\left[J_1\left(\frac{\chi}{a}\varrho'\right)\right] \cos\phi' \, \hat{\boldsymbol{\phi}} \tag{86}$$

where $\chi = 1.8411$. The radiated field from (79) and (80) is

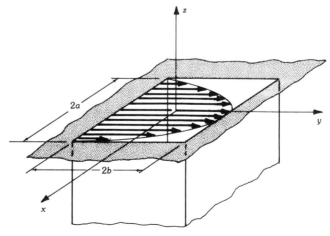

Fig. 66. A TE_{10}-mode open-ended rectangular waveguide feed with dimensions $2a$ and $2b$.

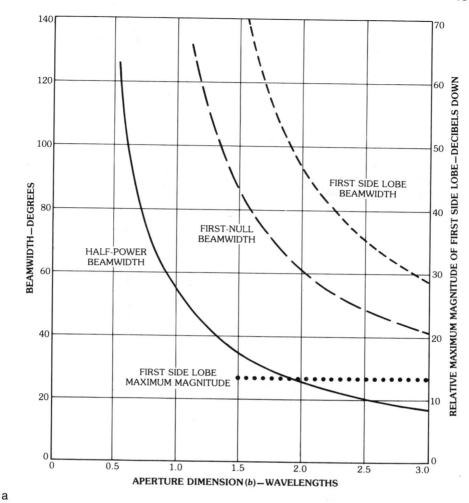

Fig. 67. Different beamwidths and first side lobe level for TE_{10}-mode open-ended rectangular waveguide on ground plane. (a) E-plane. (b) H-plane. (*After Balanis [68]*, © *1982 Harper & Row, Publishers, Inc.; reprinted with permission of the publishers*)

$$C_\theta(\theta,\phi) = 4\pi a E_0 J_1(\chi) \sin\theta \frac{J_1(U)}{U}$$
$$C_\phi(\theta,\phi) = 4\pi a E_0 J_1(\chi) \cos\theta \cos\phi \frac{J_1'(U)}{1 - (U/\chi)^2}$$
(87)

where

$$U = ka\sin\theta$$
$$J_1'(U) = J_0(U) - J_1(U)/U$$
(88)

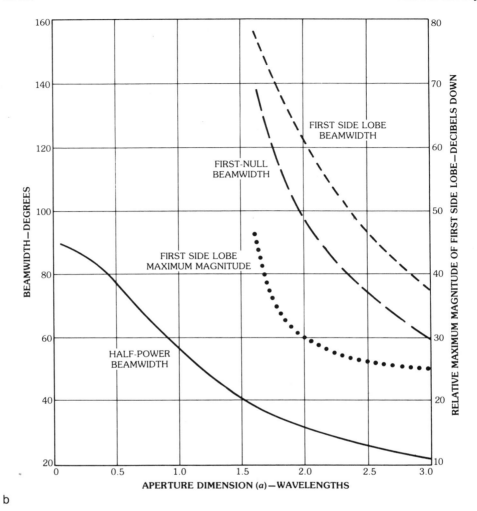

Fig. 67, *continued.*

Figs. 68 and 69 show some important characteristics of the open-ended circular waveguides [68].

Pyramidal Horns—Pyramidal horns may be classified as E-plane sectoral, H-plane sectoral, or pyramidal horns depending on to what plane its opening tapers. The type, direction, and amount of taper can have a significant effect on the overall performance of these horns. The radiation characteristics of these horns can be determined using the aperture field method or GTD construction [68, 69, 70]. The latter also provides the near-field characteristics of these horns. There are considerable amounts of design data available on these horns and the reader is referred to the just-mentioned references.

To apply the aperture field method the horn's aperture field is approximated by

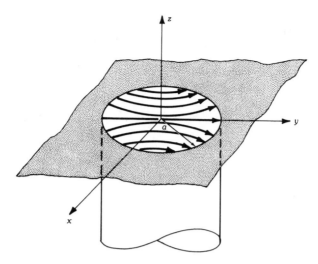

Fig. 68. A TE_{11}-mode circular waveguide feed with radius a.

$$\mathbf{E}^a(x', y') = E_0 \cos\left(\frac{\pi x'}{2a_1}\right) \exp\left\{-jk\left[\frac{(x')^2}{2L_x} + \frac{(y')^2}{2L_y}\right]\right\}\hat{\mathbf{y}} \qquad (89)$$

where L_x and L_y are the flare lengths for a pyramidal horn as shown in Fig. 70c. For E- and H-plane sectoral horns, only one of the terms in the exponent is used. The need for the quadratic phase term stems from the fact that the phase in the horn's aperture is not constant as it propagates from the throat to the opening. By substituting (89) into (79), the radiated field can finally be expressed in terms of the well-known sine and cosine Fresnel integrals [68]. Here only representative results are shown.

Figs. 71 and 72 are the plots of universal patterns for E- and H-plane sectoral horns as a function of their geometrical parameters. For pyramidal horns the geometrical parameters may be chosen to achieve the so-called optimum horn. Fig. 73 shows the relationship between the parameters to design an optimum horn for a specified gain (the overall efficiency of these horns is typically about 50 percent). Fig. 74 is the plot of the E- and H-plane patterns of such horns.

Conical Horns—The geometry of a circular horn is shown in Fig. 75. In contrast to pyramidal horns which are typically fed by rectangular waveguides, the circular horn is usually fed by circular waveguides. The aperture field of these horns can be constructed in a fashion similar to that of pyramidal horns by simply multiplying the aperture field of the circular waveguide by a quadratic phase term. The resulting integral for the computation of the radiated field can then be evaluated numerically. Fig. 75 gives the proper horn dimensions for constructing an "optimum" horn for a specified gain. These horns typically possess more symmetric E- and H-plane patterns than do their pyramidal horn counterparts. They can also be used more effectively to create circularly polarized fields.

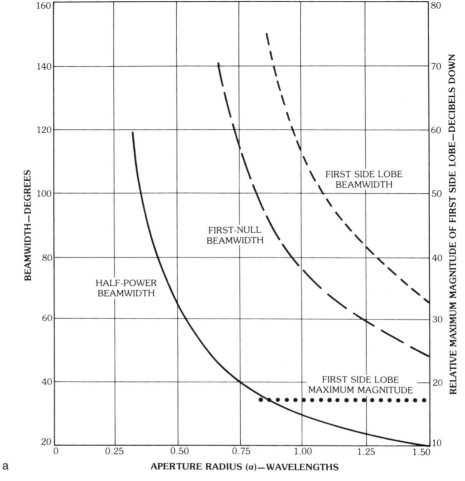

Fig. 69. Different beamwidths and first side lobe level for a TE_{11}-mode open-ended circular waveguide on ground plane. (a) E-plane. (b) H-plane. (*After Balanis [68], © 1982 Harper & Row, Publishers, Inc.; reprinted with permission of the publishers*)

Complex Feeds

So far the feed discussion has been concentrated around the single dominant-mode TE_{10} and TE_{11} structures. These feeds in general do not possess the needed ideal pattern symmetry in the subtended angle of the reflector, nor do they have good cross-polarization characteristics. Nevertheless it is possible to achieve an acceptable symmetry down to levels of 10 to 12 dB by properly adjusting the aperture dimensions. The prime advantage of using these single-mode feeds is their compactness which makes them in particular very attractive for contour and multiple-beam feed array designs.

The requirements for an optimum compromise between illumination efficiency, spillover, and cross polarization for a wide range of *F/D* values and frequencies

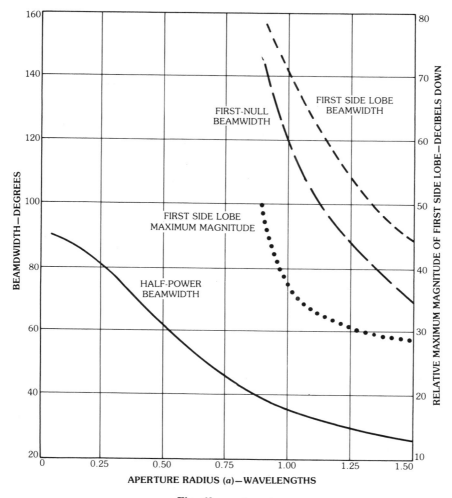

Fig. 69, *continued.*

have resulted in the generation of a new class of feeds. These feeds in particular are used in large reflectors for radioastronomy and tracking applications and offset reflectors. Among them one may refer to multimode, corrugated (hybrid-mode), matched, etc., feed horns [6, 7].

Multimode horns were invented to equalize the pattern asymmetry of single-mode horns. For instance, in the Potter horn [71] the TM_{11} mode is generated along with the dominant TE_{11} mode of a circular horn. Although this new TM_{11} mode does not have any appreciable effect on the H-plane radiation pattern, when it is properly phased and combined with the TE_{11} mode it can effectively alter the E-plane aperture distribution, which results in a symmetric radiation pattern. All these favorable features, however, are not properly realized until the feed aperture diameter exceeds about 1.3λ. For example, Fig. 76 shows the radiation pattern of these dual-mode horns as reported in [71]. As shown in this figure a partial con-

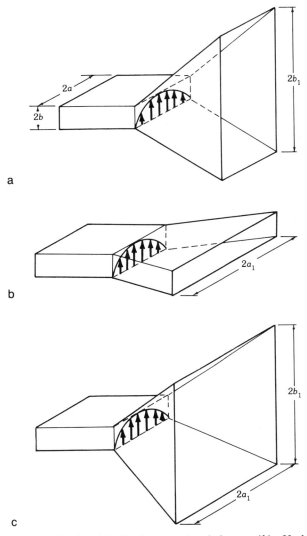

Fig. 70. Typical horn feeds. (*a*) *E*-plane sectoral horn. (*b*) *H*-plane sectoral horn. (*c*) Pyramidal horn.

version of the TE_{11} mode energy to a TM_{11} mode happens in the flared section of the horn while the straight section of length ℓ enforces the condition that both modes have the proper phase relationship at the aperture which can be maintained over a bandwidth of less than 10 percent. In general these feeds are very well suited for large *F/D* reflectors. There are also available other types of multimode horns, which result from a combination of modes such as TE_{10}, TE_{12}, and TM_{12} in a square-aperture pyramidal horn.

Corrugated horns are capable of creating similar boundary conditions at all polarizations which result in similar tapers in the aperture field distribution in all

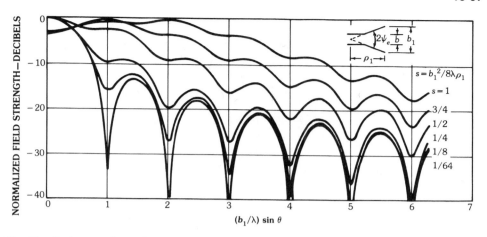

Fig. 71. *E*-plane universal patterns for *E*-plane sectoral and pyramidal horns. (*After Balanis [68]*, © 1982 Harper & Row, Publishers, Inc.; reprinted with permission of the publishers)

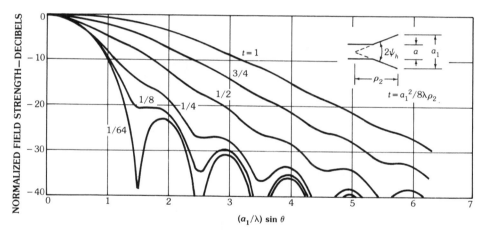

Fig. 72. *H*-plane universal patterns for *H*-plane sectoral and pyramidal horns. (*After Balanis [68]*, © 1982 Harper & Row, Publishers, Inc.; reprinted with permission of the publishers)

planes. Due to these boundary conditions symmetric radiation patterns can be obtained at levels as low as −25 dB in both the *E*- and *H*-planes. A corrugated horn can be realized by grooving the *E*-plane of a pyramidal horn or the entire wall of a circular horn with, typically, ten or more slots (corrugations) per wavelength. For circular corrugated horns Fig. 77 shows the plots of the pattern widths at different levels as a function of the opening angle. The appearance of the corrugations, especially near the waveguide-horn junction, affects the vswr of the horn. The usual practice is to begin the corrugations at a small distance from the junction. These horns are also classified as hybrid-mode horns because they support modes in

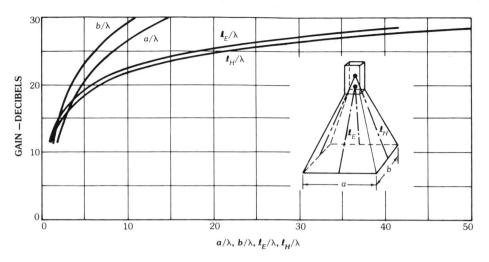

Fig. 73. Gain characteristics of pyramidal horns of optimum design.

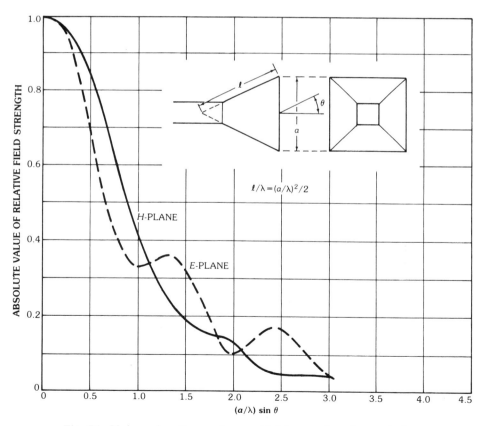

Fig. 74. Universal patterns of pyramidal horns of optimum design.

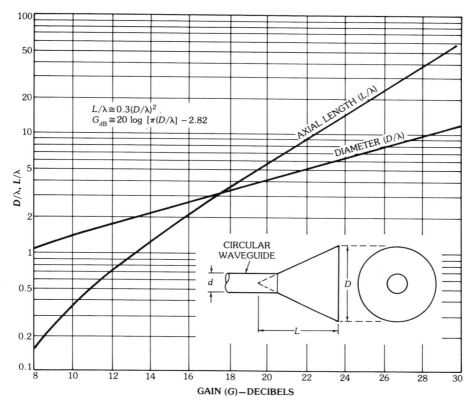

Fig. 75. Gain characteristics of conical horns of optimum design.

which both longitudinal **E**- and **H**-field components are present. In a circular corrugated horn a natural mixture of TE_{11} and TM_{11} results in the generation of a hybrid-mode HE_{11}. In contrast to dual-mode horns there is no need for a mode converter, and therefore the hybrid-mode horns have typically wider bandwidths than the dual-mode horns. In particular, one version of these hybrid horns (scalar horns), which has a large flare angle with a relatively short horn, has radiation characteristics with little dependence on frequency.

There are other classes of feeds which are properly tailored to specifically overcome some undesirable characteristics of reflectors. For example, matched feeds [72, 73] are used to significantly reduce the generation of unwanted cross-polarized fields in an offset parabolic reflector illuminated with a tilted feed from its axis. The basic concept behind these feeds is to match the feed aperture field distribution with the receiving focal plane distribution of the reflector as closely as possible.

$cos^q(\theta)$ *Type Patterns*

In the previous sections, independently of the actual feed elements, the feed patterns were approximated by $cos^q(\theta)$ type patterns to allow for detailed parametric studies of reflector characteristics. Some simple expressions are

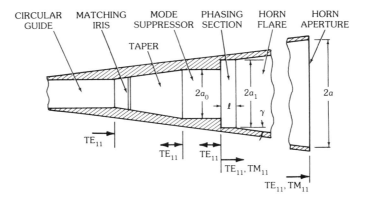

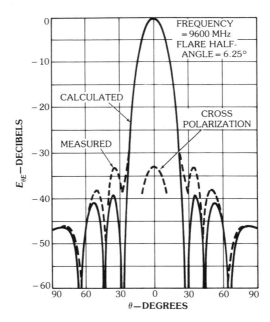

Fig. 76. Dual-mode conical horn of Potter [71]. (*a*) Horn structure. (*b*) Experimental data. (*c*) *E*-plane. (*d*) 45° plane. (*e*) *H*-plane. (*Reprinted with permission of* Microwave Journal, *from June 1963 issue,* © *1963 Horizon House–Microwave, Inc.*)

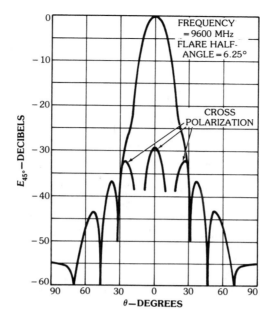

d

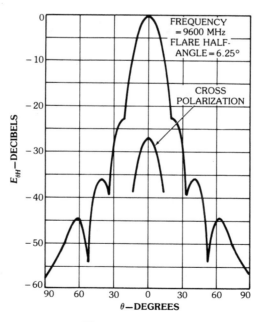

e

Fig. 76, *continued.*

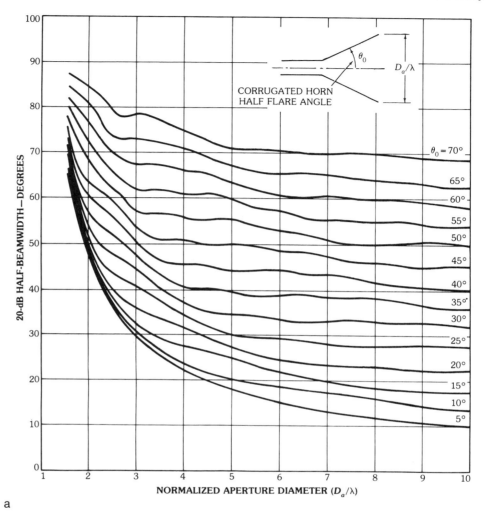

Fig. 77. Half-beamwidths of corrugated horns. (*a*) 20-dB. (*b*) 10-dB. (*c*) 3-dB.

presented in this subsection to relate the qs with the actual patterns of the most commonly used feed elements. Typically the patterns are matched in the principal planes and the values of q in these planes are designated by q_1 and q_2, respectively. There are several schemes to determine q, based on different matching requirements.

The $\cos^q(\theta)$ type pattern may be matched to the element pattern in the principal planes at both the peak and at a specified off-axis angle, which may coincide with the subtended angle of the reflector in these planes. In this case one obtains the same feed taper as the actual feed element at the reflector edge. Denoting the principal plane patterns by $C_E(\theta)$ and $C_H(\theta)$, and the subtended angles by θ_1 and θ_2, one obtains

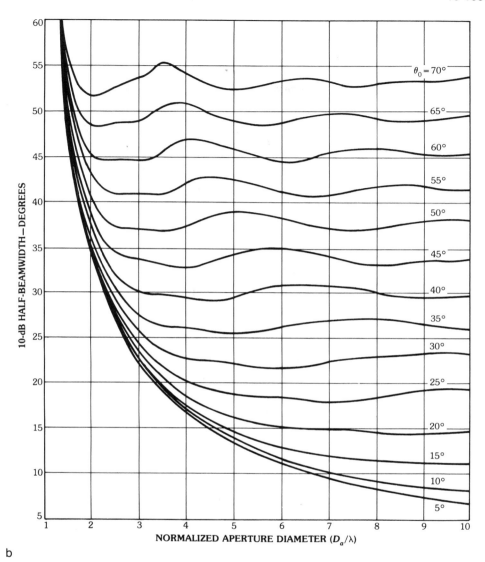

Fig. 77, *continued.*

$$q_{1,2} = \frac{\log C_{E,H}(\theta_{1,2})}{\log \cos \theta_{1,2}} \tag{90}$$

Patterns C_E and C_H may be given in either a closed form or in terms of measured data.

The match in (90) can be imposed at the 3-dB beamwidth level in the E- and H-planes. In this case, q is determined as

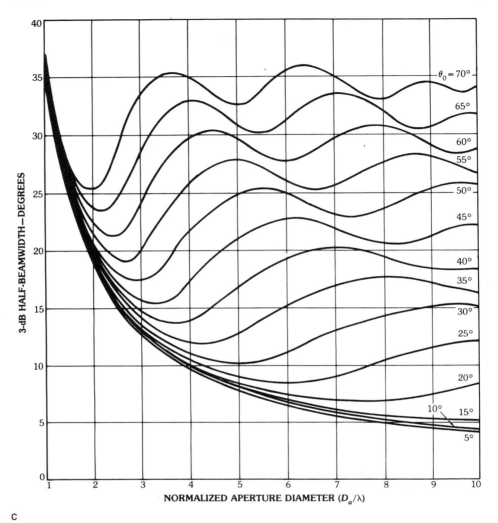

c

Fig. 77, *continued.*

$$q = \frac{-0.15}{\log \cos \theta_{1/2}} \qquad (91)$$

where $\theta_{1/2}$ is one-half of the half-power beamwidths in each of the principal planes. For instance, for an open-ended rectangular waveguide feed with sides $2a$ (H-plane) and $2b$ (E-plane) the following is obtained:

$$2\theta_{1/2} = \begin{cases} \dfrac{50.6°}{2b/\lambda}, & \text{for } E\text{-plane} \\ \dfrac{68.8°}{2a/\lambda}, & \text{for } H\text{-plane} \end{cases} \qquad (92)$$

Similarly, the following expressions are derived for the beamwidths of an open-ended circular waveguide feed with radius a:

$$2\theta_{1/2} = \begin{cases} \dfrac{29.2°}{a/\lambda} \\ \dfrac{37.0°}{a/\lambda} \end{cases} \quad (93)$$

Finally, values of q may be determined from directivity considerations. In this case the condition is imposed that the directivity of the feed element be the same as the directivity of the feed with $\cos^q(\theta)$ patterns. For the latter it has been shown that

$$D = \frac{2(2q_1 + 1)(2q_2 + 1)}{q_1 + q_2 + 1} \quad (94)$$

Furthermore, if it is assumed that the E- and H-plane patterns are the same, the result is

$$D = 2(2q + 1) \quad (95)$$

From (95), q may be obtained by making D equal to the directivity of the actual feed element. For example, for an open-ended rectangular waveguide the following is obtained:

$$D = \frac{32}{\pi} \frac{(2a)(2b)}{\lambda^2} \quad (96)$$

Similarly, for an open-ended circular waveguide the following expression is obtained:

$$D = 10.5\pi(a/\lambda)^2 \quad (97)$$

In some cases the value of q is approximated as an average value of all the qs obtained using the three previously mentioned methods.

9. Effects of Random Surface Errors

Due to imperfections of the manufacturing process, reflectors can only be constructed within tolerance ranges, which results in reflector surfaces that are randomly distorted from the ideal paraboloidal configuration. Random distortions cause radiation patterns to deteriorate by reducing the reflector aperture efficiency and increasing its side lobe envelope. A successful model has been suggested in [30, 74, 75] for the determination of these distortion effects. If it is assumed that the reflector's aperture phase error at a given point in the aperture has a zero mean with a Gaussian distribution and correlation radius c, it can then be shown that the distorted pattern is [74]

$$G(\theta, \phi) = G_0(\theta, \phi)e^{-\sigma^2} + \left(\frac{\pi D}{\lambda}\right)^2 e^{-\sigma^2}\left(\frac{2c}{D}\right)^2 f(\sigma, c\sin\theta) \tag{98}$$

where the "diffuse term" f is

$$f(\sigma, c\sin\theta) = \sum_{n=1}^{\infty} \frac{(\sigma^2)^n}{n!n} e^{-[(\pi c\sin\theta)/\lambda]^2/n} \tag{99}$$

In the above equations G_0 is the nondistorted pattern, D is the diameter of the circular aperture, and

$$\sigma \cong 2\varkappa(2\pi/\lambda)\epsilon_{\rm rms} \tag{100}$$

where $\epsilon_{\rm rms}$ is the rms surface error, the factor of 2 accounts for the two-way path incurred by the reflected ray, $\varkappa \leqq 1$ depends on the reflector geometry (F/D ratio), and $\varkappa = 1$ for shallow reflectors. As a rule of thumb the rms surface error is approximately one-fourth to one-third that of the surface peak error.

From (98) the peak gain reduction at $\theta = 0°$ becomes

$$G(0,0) = \eta\left(\frac{\pi D}{\lambda}\right)^2 e^{-\sigma^2}\left[1 + \frac{1}{\eta}\left(\frac{2c}{D}\right)^2 \sum_{n=1}^{\infty} \frac{(\sigma^2)^n}{n!n}\right] \tag{101}$$

where η is the aperture efficiency. For small correlation intervals one finally obtains the following well-known result:

$$G(0,0) \cong \eta\left(\frac{\pi D}{\lambda}\right)^2 e^{-(4\pi\varkappa\epsilon_{\rm rms}/\lambda)^2} \tag{102}$$

where values of $\varkappa$ may be estimated from [74]

$$\varkappa = \frac{4F}{D}\sqrt{\ln[1 + 1/(4F/D)^2]} \tag{103}$$

or can more accurately be determined [30] from Fig. 78.

From (102) and with the assumption of constant aperture efficiency, the peak gain first increases with the square of the frequency until the tolerance effect becomes significant. The shortest wavelength at which the gain is maximized can then be obtained by differentiating (102), which results in

$$\lambda_{\rm shortest} = 4\pi\varkappa\epsilon_{\rm rms} \tag{104}$$

At this wavelength the reflector gain is 4.3 dB (equal to 10 log e) below the error-free gain. Fig. 79 shows the antenna gain as a function of frequency for different surface roughness values [76]. In this figure $\varkappa = 1$ is used. It is worthwhile to observe that at $\lambda_{\rm shortest}$ ($\epsilon_{\rm rms}/\lambda \cong 1/16$), the gain is maximized; nevertheless the higher-order side lobes can be unacceptably distorted, as will be shown shortly.

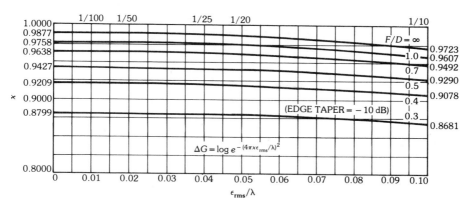

Fig. 78. The correction factor $\varkappa$ as a function of ϵ_{rms}/λ for different F/D values. (*After Rahmat-Samii [30], © 1983 IEEE*)

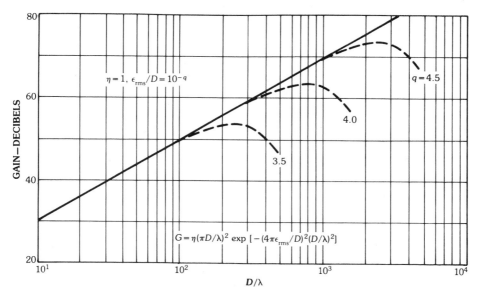

Fig. 79. The effect of reflector roughness on antenna gain. (*After Love [76], © 1976 American Geophysical Union*)

Although (98) determines the effects of surface tolerances on the reflector pattern it does not properly account for the tapered aperture distributions and nonuniform rms distortions. A simplified model has been developed in [30], which allows for the incorporation of the just-mentioned factors. The model assumes that in each prescribed annular region of the antenna the geometrical rms surface value is known and is designated by $\epsilon_{(rms)n}$ in the nth annular region. The rms aperture phase error can then be defined as

$$\sigma_n = \frac{4\pi}{\lambda} \epsilon_{(\text{rms})n} \cos \xi \tag{105}$$

where

$$\tan \xi = \frac{a_n + a_{n-1}}{4F} \tag{106}$$

and a_n and a_{n-1} are shown in Fig. 80. If an aperture taper distribution similar to (16) is used, viz.,

$$Q(\varrho') = C + B[1 - (\varrho'/a)^2]^P$$
$$B + C = 1 \tag{107}$$

one can then arrive at the following expression for the average power pattern [30]:

$$G(\theta, \phi) = \sum_{n=1}^{N} \sum_{m=1}^{N} E_{n,n-1} E^*_{m,m-1} e^{-0.5(\sigma_n^2 + \sigma_m^2)}$$
$$+ \sum_{n=1}^{N} E_{n,n-1} E^*_{n,n-1} (1 - e^{-\sigma_n^2}) \tag{108}$$

where N and * designate the total number of annular regions and the conjugate operator, respectively. Furthermore,

$$E_{n,n-1} = E_n - E_{n-1} \tag{109}$$

which takes the following expressions for $P = 1$ and 2.

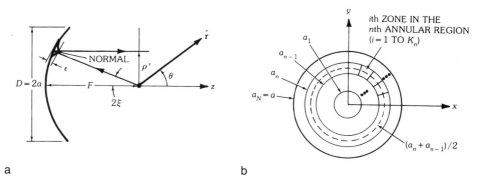

Fig. 80. Treatment of random surface errors. (a) Parabolic reflector with random surface errors. (b) Reflector aperture divided into N annular regions, which are further divided into K_n zones in the nth region ($n = 1$ to N).

$P = 1$:

$$E_n = \pi a_n^2 \left[C \frac{2}{u_n} J_1(u_n) + B \left\{ \frac{2}{u_n} J_1(u_n) - \frac{a_n^2}{a^2} \left[\frac{2}{u_n} J_1(u_n) - \left(\frac{2}{u_n}\right)^2 J_2(u_n) \right] \right\} \right] \quad (110)$$

$$E_n(0°) = \pi a_n^2 \left[C + B\left(1 - \frac{a_n^2}{2a^2}\right) \right]$$

$P = 2$:

$$E_n = \pi a_n^2 \left[C \frac{2}{u_n} J_1(u_n) + B \left\{ \frac{2}{u_n} J_1(u_n) - \frac{2a_n^2}{a^2} \left[\frac{2}{u_n} J_1(u_n) - \frac{4}{u_n^2} J_2(u_n) \right] \right.\right.$$
$$\left.\left. + \frac{a_n^4}{a^4} \left[\frac{2}{u_n} J_1(u_n) - 2\left(\frac{2}{u_n}\right)^2 J_2(u_n) + 2\left(\frac{2}{u_n}\right)^3 J_3(u_n) \right] \right\} \right] \quad (111)$$

$$E_n(0°) = \pi a_n^2 \left[C + B\left(1 - \frac{a_n^2}{a^2} + \frac{a_n^4}{3a^4}\right) \right]$$

In the previous equations,

$$u_n = k a_n \sin \theta$$

For the special case of the uniform aperture phase distortion

$$\sigma_n = \sigma_m = \sigma$$

(109) can be further simplified as

$$G(\theta, \phi) = e^{-\sigma^2} G_0(\theta) + (1 - e^{-\sigma^2}) \sum_{n=1}^{N} E_{n,n-1} E_{n,n-1}^* \quad (112)$$

where the undistorted gain function is proportional to

$$G_0(\theta) = \pi^2 a^4 \left[C \frac{2}{u} J_1(u) + B \frac{2^{P+1} P!}{u^{P+1}} J_{P+1}(u) \right]^2 \quad (113)$$

in which $u = ka \sin \theta$. Typically, for values of $N > 30$, insignificant changes are observed in the numerical results of (108).

Numerical results based on (108) for different values of edge tapers and ϵ_{rms}/λ are shown in Figs. 81 and 82. From these results one can estimate how the average power pattern becomes degraded. In particular, the variation of the peak and first three side lobes for the ET $= -10$-dB case are shown in Fig. 83. For these values of $P = 2$ and $F/D = 0.7$ the first, second, and third side lobe levels for the undistorted case are at -27.06, -30.83, and -33.85 dB, respectively.

The preceding results were constructed for a parabolic reflector. Similar results can also be derived for dual reflectors [77]. For example, the peak gain of a dual

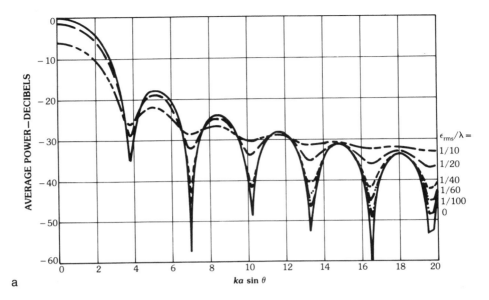

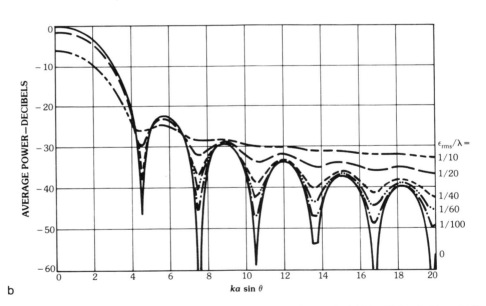

Fig. 81. Far-field patterns for various surface errors, with $P = 1$. (*a*) For ET = 0 dB and $F/D = 0.7$. (*b*) For ET = -10 dB and $F/D = 0.7$. (*c*) For ET = 0 dB and $F/D = 0.3$. (*d*) For ET = -10 dB and $F/D = 0.3$. (*After Rahmat-Samii [30], © 1983 IEEE*)

reflector may be estimated from

$$G(0,0) \cong \eta \left(\frac{\pi D}{\lambda}\right)^2 e^{-(4\pi\varkappa\epsilon_{\rm rms}/\lambda)^2} e^{-(4\pi\varkappa'\epsilon'_{\rm rms}/\lambda)^2} \qquad (114)$$

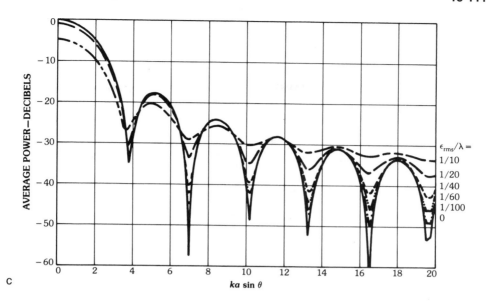

c

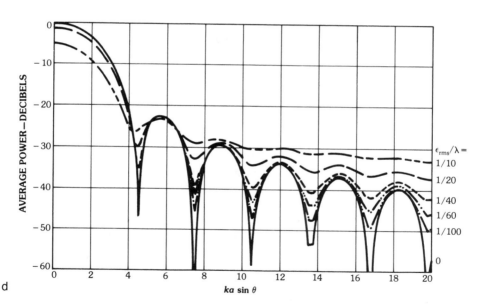

d

Fig. 81, *continued.*

where η is the aperture efficiency, D is the main parabolic reflector diameter, ϵ_{rms} and ϵ'_{rms} are, respectively, the main and subreflector surface rms, and $\varkappa$ and $\varkappa'$ are the correction factors for the main and subreflector, respectively, as shown in Figs. 78 and 84. The results of Fig. 84 are for uniform aperture illumination and do not vary substantially for other tapers. In this figure $\varkappa'^2$ is shown in terms of the paraboloid F/D ratio and the subreflector's eccentricity.

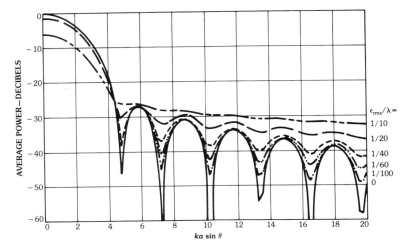

Fig. 82. Far-field patterns for various surface errors, for $P = 2$, $F/D = 0.7$, and $ET = -10$ dB. (*After Rahmat-Samii [30], © 1983 IEEE*)

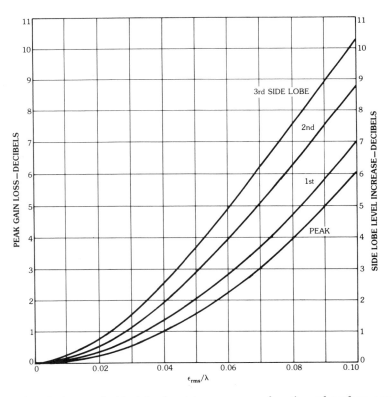

Fig. 83. Peak gain loss and side lobe level increase as a function of surface errors, with $P = 2$, $F/D = 0.7$, and $ET = -10$ dB. (*After Rahmat-Samii [30], © 1983 IEEE*)

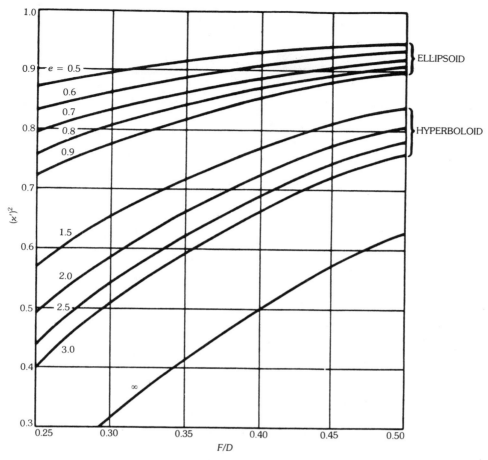

Fig. 84. The correction factor $(\varkappa')^2$ for subreflectors. (*After Rusch and Wolleben [77]*, © *1982 IEEE*)

So far results have only been shown for average power patterns. However, it is more meaningful to determine the probabilistic relationship between random reflector surface errors and side lobe levels [78]. For example, Figs. 85a and 85b provide the surface tolerance versus degradation of maximum side lobe level for different values of the probability of occurrence of the specified degradation. Clearly, as the required side lobe levels become lower, more stringent surface rms values would be needed for a desired probability. These kind of data should be of interest to system engineers in assessing the reliability of a particular design.

Additionally, one should incorporate the effects of other degradations, such as systematic surface distortions, mesh surface effects, strut blockages, etc., in the evaluation of the performance of reflectors. These evaluations are, in particular, important for the currently designed high-performance satellite and ground reflector antennas. The reader may refer to [79, 80] for many representative test cases.

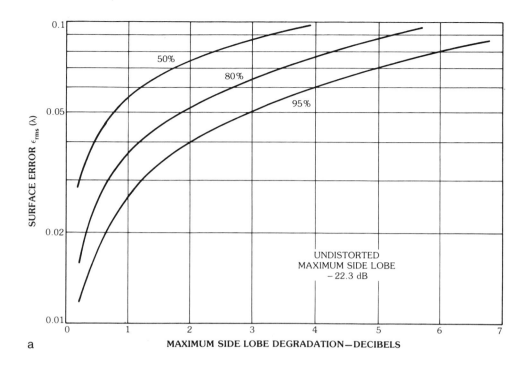

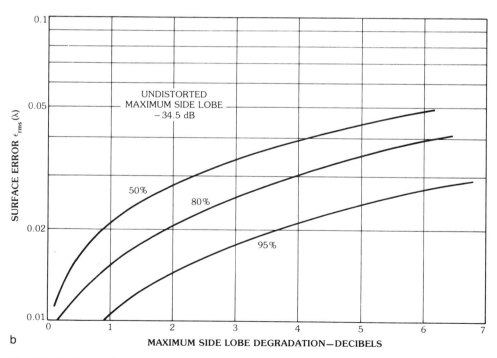

Fig. 85. Surface tolerance versus degradation of maximum side lobe level. (*a*) For a reflector with design side lobe level of −22.3 dB. (*b*) For a reflector with design side lobe level of −34.5 dB.

10. Appendix: Coordinate Transformations for Antenna Applications

Feed coordinates, far-field pattern coordinates, and reflector coordinates do not, in general, coincide. In many practical design and measurement applications it is necessary to have a general coordinate transformation which relates the Cartesian or spherical components of one coordinate system to the components of another system. For example, when a reflector antenna is illuminated by an array of feed horns one is required to find the illuminating field on the reflector in the reflector coordinates. The radiated field of each horn is known in its own system of coordinates which in general do not coincide with the reflector coordinates; therefore a coordinate transformation is needed to relate the two systems. Also, in many cases one does not want to compute the far-field pattern in the spherical coordinates of the antenna. Instead, it is desirable to compute them in the spherical coordinates which coincide with the main beam direction. In this appendix some basic transformations which are frequently applied to relate the Cartesian and spherical components of one coordinate system with those of another system are presented in a coherent and an easily usable format. In general, these two systems can be constructed with the use of some translations and rotations. For details the reader is referred to [11].

Cartesian and Spherical Components

Fig. A-1 illustrates Cartesian and spherical coordinate systems for defining points or vectors. A vector field **H** can be expressed by its Cartesian or spherical components as follows:

$$\mathbf{H} = \sum_{i=1}^{3} H_i^c \hat{\mathbf{c}}_i = \sum_{i=1}^{3} H_i^s \hat{\mathbf{s}}_i \tag{115}$$

where $\hat{\mathbf{c}}_i$ and $\hat{\mathbf{s}}_i$ are Cartesian ($\hat{\mathbf{c}}_1=\hat{\mathbf{x}}$, $\hat{\mathbf{c}}_2=\hat{\mathbf{y}}$, $\hat{\mathbf{c}}_3=\hat{\mathbf{z}}$) and spherical ($\hat{\mathbf{s}}_1=\hat{\mathbf{r}}$, $\hat{\mathbf{s}}_2=\hat{\boldsymbol{\theta}}$, $\hat{\mathbf{s}}_3=\hat{\boldsymbol{\phi}}$) unit vectors, respectively. The objective is to relate the $\{\hat{\mathbf{s}}\}$ to the $\{\hat{\mathbf{c}}\}$ and the $\{H^s\}$ to the $\{H^c\}$, where $\{\hat{\mathbf{s}}\} = \{\hat{\mathbf{s}}_1, \hat{\mathbf{s}}_2, \hat{\mathbf{s}}_3\}$. Note that $H_1 = H_r$, $H_2 = H_\theta$, and $H_3 = H_\phi$ and similarly for $\{H^c\}$. The desired connection can best be made by introducing the transformation matrix $[^sT^c]$ defined below:

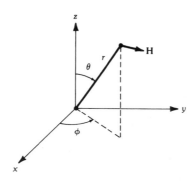

Fig. A-1. Cartesian and spherical coordinates. (*After Rahmat-Samii [11], © 1979 IEEE*)

$$[^sT^c] = \begin{bmatrix} \sin\theta\cos\phi & \sin\theta\sin\phi & \cos\theta \\ \cos\theta\cos\phi & \cos\theta\sin\phi & -\sin\theta \\ -\sin\phi & \cos\phi & 0 \end{bmatrix} \quad (116)$$

where superscripts s and c are used to denote transformations from Cartesian to spherical components. One can then easily demonstrate that for $(\hat{s}) = (\hat{s}_1, \hat{s}_2, \hat{s}_3)'$,

$$(\hat{s}) = [^sT^c](\hat{c}), \quad (H^c)_s = [^sT^c](H^c) \quad (117)$$

i.e., the same transformation holds for both the unit vectors and the vector components. Furthermore, it can be shown that

$$[^cT^s] = [^sT^c]^{-1} = [^sT^c]^t \quad (118)$$

where, in this case, $[^cT^s]$ defines a transformation from spherical to Cartesian components.

Eulerian Angles

Let us consider two Cartesian coordinate systems $\{\hat{c}\}$ and $\{\hat{c}'\}$ as shown in Fig. A-2. In the most general case these Cartesian systems can be aligned via three rotations. The angles of these rotations are known as *Eulerian angles*. Although one can find different definitions for these angles in the literature, the definition used in [11] is used. As displayed in Fig. A-2, angles α, β, and γ are employed to define the Eulerian angles. Angle α describes a counterclockwise rotation about

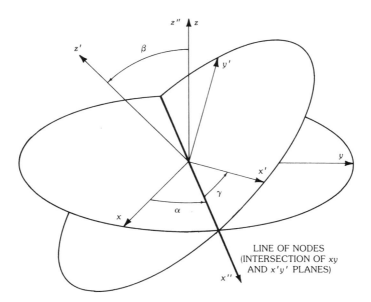

Fig. A-2. Eulerian angles relating primed and unprinted coordinates. (*After Rahmat-Samii [11], © 1979 IEEE*)

the z axis which brings the x axis to the x'' axis aligned with the line of nodes (line of intersection between xy and $x'y'$ planes), angle β defines a rotation about the line of nodes in a counterclockwise sense as indicated, bringing the z axis to z', and finally angle γ is another rotation about the z' axis which aligns the x'' axis with the x' axis in a counterclockwise sense.

By successively performing these three rotations one arrives at the following transformation equations among unit Cartesian components and Cartesian components of vectors, viz., for column vectors $(\hat{e}')$, $(\hat{e})$, $(H^{c'})$, and (H^c):

$$(\hat{e}') = [^{c'}A^c](\hat{e}), \qquad (H^{c'}) = [^{c'}A^c](H^c) \tag{119}$$

where $[^{c'}A^c]$ is the transformation matrix from Cartesian coordinates $(\hat{e})$ to $(\hat{e}')$ defined as

$$[^{c'}A^c] = \begin{bmatrix} \cos\gamma & \sin\gamma & 0 \\ -\sin\gamma & \cos\gamma & 0 \\ 0 & 0 & 1 \end{bmatrix} \begin{bmatrix} 1 & 0 & 0 \\ 0 & \cos\beta & \sin\beta \\ 0 & -\sin\beta & \cos\beta \end{bmatrix}$$

$$\times \begin{bmatrix} \cos\alpha & \sin\alpha & 0 \\ -\sin\alpha & \cos\alpha & 0 \\ 0 & 0 & 1 \end{bmatrix} \tag{120}$$

with components

$$\begin{aligned}
A_{11} &= \cos\gamma\cos\alpha - \sin\gamma\cos\beta\sin\alpha \\
A_{12} &= \cos\gamma\sin\alpha + \sin\gamma\cos\beta\cos\alpha \\
A_{13} &= \sin\gamma\sin\beta \\
A_{21} &= -\sin\gamma\cos\alpha - \cos\gamma\cos\beta\sin\alpha \\
A_{22} &= -\sin\gamma\sin\alpha + \cos\gamma\cos\beta\cos\alpha \\
A_{23} &= \cos\gamma\sin\beta \\
A_{31} &= \sin\beta\sin\alpha \\
A_{32} &= -\sin\beta\cos\alpha \\
A_{33} &= \cos\beta
\end{aligned} \tag{121}$$

The superscripts c and c' are deleted in the above expressions for simplicity, and the first and second subscripts, respectively, denote rows and columns of a 3×3 matrix. In addition, one can easily establish the following relations:

$$[^cA^{c'}] = [^{c'}A^c]^{-1} = [^{c'}A^c]^t \tag{122}$$

Feed and Reflector Coordinates

Having derived the transformation matrices (116) and (120) one can now solve an important antenna design and analysis problem. As depicted in Fig. A-3, the

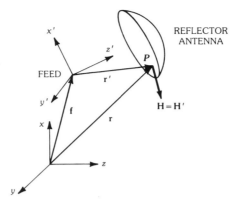

Fig. A-3. Feed and antenna coordinate systems. (*After Rahmat-Samii [11], 1979 IEEE*)

reflector coordinate system is $\{\hat{c}\}$ and the feed coordinate system is $\{\hat{c}'\}$. The radiated field of the feed is given in $\{\hat{c}'\}$ in terms of its spherical components. The aim is to determine this radiated field at a point on the reflector antenna in the reflector coordinate system $\{\hat{c}\}$ in terms of both the Cartesian and spherical components. Coordinates $\{\hat{c}\}$ and $\{\hat{c}'\}$ are, in general, related through both a translation and three rotations using Eulerian angles. This problem can be solved in two steps. First, the spherical coordinates of a point given on the reflector surface are determined in terms of the feed coordinate system $\{\hat{c}'\}$. Second, the field expression of the radiated field of the feed is derived in the reflector coordinate system $\{\hat{c}\}$.

The vectors $\mathbf{r}$ and $\mathbf{r}'$ are designated as the position vectors for point P on the antenna. Furthermore, as shown in Fig. A-3, the vector $\mathbf{f}$ is used to define the location of the feed in the reflector or unprimed system $\{\hat{c}\}$. Clearly, the following holds:

$$\mathbf{r}' = \mathbf{r} - \mathbf{f}$$

With $\mathbf{r}$ and $\mathbf{f}$ given, one first determines the Cartesian components of $\mathbf{r}'$ in $\{\hat{c}\}$ and then employs the transformation matrix to find its Cartesian components in $\{\hat{c}'\}$. Cartesian components of $\mathbf{r}'$ in $\{\hat{c}\}$ can be written as

$$\mathbf{r}'^c = (r'^c) = \{r\sin\theta\cos\phi - f_1, r\sin\theta\sin\phi - f_2, r\cos\theta - f_3\}^t \quad (123)$$

where f_1, f_2, and f_3 are the Cartesian components of $\mathbf{f}$ in $\{\hat{c}\}$ and t is the transpose operator. Using the second equation in (119), one can determine the Cartesian components of $\mathbf{r}'$ in $\{\hat{c}'\}$, namely

$$\mathbf{r}'^{c'} = (r'^{c'}) = [{}^{c'}A^c](r'^c) \quad (124)$$

Since

$$\mathbf{r}'^{c'} = (r'\sin\theta'\cos\phi', r'\sin\theta'\sin\phi', r'\cos\theta')^t \quad (125)$$

one can then easily determine r', θ', and ϕ' in terms of r, θ, and ϕ via the application of (123) and (124).

Once the coordinates of reflector point P in $\{\hat{\mathbf{e}}'\}$ are obtained, one next designates the spherical components of the radiated field of the feed at this point by $(H'^{s'})$. The task is now to describe the Cartesian and spherical components of $\mathbf{H}' = \mathbf{H}$, namely $\mathbf{H}^c$ and $\mathbf{H}^s$, in $\{\hat{\mathbf{e}}\}$. Using (118) in the primed coordinates one can find the Cartesian components of $\mathbf{H}'^{s'}$ from

$$\mathbf{H}' = (H'^{c'}) = [^{c'}T^{s'}](H'^{s'}) \tag{126}$$

The Cartesian components of this vector field in $\{\hat{\mathbf{e}}\}$ can then be found from (122), namely,

$$\mathbf{H} = (H^c) = [^c A^{c'}](H'^{c'}) = [^c A^{c'}][^{c'}T^{s'}](H'^{s'}) \tag{127}$$

It is worth mentioning that both $\mathbf{H}'$ and $\mathbf{H}$ represent the same vector and that the only difference lies in their functional dependence on primed and unprimed coordinates. With the use of (117), one can next derive the spherical components of $\mathbf{H}$ from (127) to arrive at

$$\mathbf{H} = (H^s) = [^s T^c](H^c) = [^s T^c][^c A^{c'}][^{c'}T^{s'}](H'^{s'}) \tag{128}$$

In summary, expressions (124) and (128) may finally be written more explicitly as follows:

$$\begin{bmatrix} r'\sin\theta'\cos\phi' \\ r'\sin\theta'\sin\phi' \\ r'\cos\theta' \end{bmatrix} = \begin{bmatrix} A_{11} & A_{12} & A_{13} \\ A_{21} & A_{22} & A_{23} \\ A_{31} & A_{32} & A_{33} \end{bmatrix} \begin{bmatrix} r\sin\theta\cos\phi - f_1 \\ r\sin\theta\sin\phi - f_2 \\ r\cos\theta - f_3 \end{bmatrix} \tag{129}$$

and

$$\begin{bmatrix} H_r(r,\theta,\phi) \\ H_\theta(r,\theta,\phi) \\ H_\phi(r,\theta,\phi) \end{bmatrix} = \begin{bmatrix} \sin\theta\cos\phi & \sin\theta\sin\phi & \cos\theta \\ \cos\theta\cos\phi & \cos\theta\sin\phi & -\sin\theta \\ -\sin\phi & \cos\phi & 0 \end{bmatrix}$$

$$\times \begin{bmatrix} A_{11} & A_{21} & A_{31} \\ A_{12} & A_{22} & A_{32} \\ A_{13} & A_{23} & A_{33} \end{bmatrix}$$

$$\times \begin{bmatrix} \sin\theta'\cos\phi' & \cos\theta'\cos\phi' & -\sin\phi' \\ \sin\theta'\sin\phi' & \cos\theta'\sin\phi' & \cos\phi' \\ \cos\theta' & -\sin\theta' & 0 \end{bmatrix}$$

$$\times \begin{bmatrix} H'_{r'}(r',\theta',\phi') \\ H'_{\theta'}(r',\theta',\phi') \\ H'_{\phi'}(r',\theta',\phi') \end{bmatrix} \tag{130}$$

In most cases one deals with the far field of the feed and therefore H_r is zero in (130). Expressions 129 and 130 are used in constructing the physical optics radiation integral for a reflector illuminated by an arbitrarily positioned feed source as described in Section 2.

Determination of Eulerian Angles

Eulerian angles can be determined by properly rotating the (x, y, z) system and aligning it with the (x', y', z') system, as shown in Fig. A-2, and then calculating the rotation angles. In many cases, however, it is desirable to have simple formulas which provide these angles. From Fig. A-2 the Eulerian angles α, β, and γ can be obtained by first assuming that the direction $\hat{z}'$ is known, which then allows the construction of x'' as

$$\hat{z}' = \hat{k}' = k_x' \hat{x} + k_y' \hat{y} + k_z' \hat{z}$$
$$\hat{x}'' = \hat{z} \times \hat{z}'/\Delta, \qquad \Delta^2 = k_x'^2 + k_y'^2 = 1 - k_z'^2 \tag{131}$$

Furthermore, by using the above expressions, α, β, and γ can be determined from the following equations:

$$\sin \alpha = (\hat{x} \times \hat{x}') \cdot \hat{z} = \hat{x} \cdot \hat{z}'/\Delta = k_x'/\Delta$$
$$\cos \alpha = \hat{x} \cdot \hat{x}'' = -\hat{y} \cdot \hat{z}'/\Delta = -k_y'/\Delta \tag{132}$$

$$\sin \beta = (\hat{z} \times \hat{z}') \cdot \hat{x}'' = \Delta$$
$$\cos \beta = \hat{z} \cdot \hat{z}' = k_z' \tag{133}$$

$$\sin \gamma = (\hat{x}'' \times \hat{x}') \cdot \hat{z}' = \hat{z} \cdot \hat{x}'/\Delta$$
$$\cos \gamma = \hat{x}'' \cdot \hat{x}' = \hat{z} \cdot \hat{y}'/\Delta \tag{134}$$

It is worthwhile to mention that α and β are determined once the direction of $\hat{z}'$ is identified. Angle γ defines a rotation about $\hat{z}'$, which is not fixed, but can be chosen, based on such considerations as polarization and the like. For example, once the $\hat{z}'$ axis of the feed coordinates in Fig. A-3 is known, the α and β angles can be constructed from (132) and (133), respectively. Then γ can be used, for instance, to adjust the polarization of the feed pattern with respect to the antenna coordinates.

11. References

[1] A. W. Love, ed., *Reflector Antennas*, New York: IEEE Press, 1978.
[2] E. C. Collin and F. J. Zucker, eds., *Antenna Theory, Parts 1 and 2*, New York: McGraw-Hill Book Co., 1969.
[3] H. Jasik, *Antenna Engineering Handbook*, New York: McGraw-Hill Book Co., 1961.
[4] S. Silver, ed., *Microwave Antenna Theory and Design*, chapter 12, New York: McGraw-Hill Book Co., 1949.
[5] P. J. B. Clarricoats and G. T. Poulton, "High-efficiency microwave reflector antennas—a review," *Proc. IEEE*, vol. 65, no. 10, pp. 1470–1504, October 1977.
[6] A. W. Love, *Electromagnetic Horn Antennas*, New York: IEEE Press, 1976.
[7] A. W. Rudge et al., eds., *The Handbook of Antenna Design, Volumes I and II*, IEE

Electromagnetic Waves Series 15, London: Peter Peregrinus Ltd., 1982.

[8] M. I. Skolnik, ed., *Radar Handbook*, chapter 9, New York: McGraw-Hill Book Co., 1970.

[9] P. J. B. Clarricoats, "Feeds for reflector antennas—a review," Part 1: Antennas, *Second Intl. Conf. Antennas Propag.*, April 13–16, 1981, York, England, pp. 309–317.

[10] A. W. Rudge and N. A. Adatia, "Offset-parabolic-reflector antennas: a review," *Proc. IEEE*, vol. 66, no. 12, pp. 1592–1618, December 1978.

[11] Y. Rahmat-Samii, "Useful coordinate transformations for antenna applications," *IEEE Trans. Antennas Propag.*, vol. AP-27, pp. 571–574, July 1979.

[12] Y. Hwang, C. H. Tsao, and C. C. Han, "Uniform analysis of reflector antenna for satellite application," *IEEE/AP-S Intl. Symp. Natl. Radio Sci. Mtg.*, Houston, Texas, May 23–26, 1983.

[13] V. Galindo-Israel and Y. Rahmat-Samii, "A new look at a Fresnel field computation using the Jacobi-Bessel series," *IEEE Trans. Antennas Propag.*, vol. AP-29, pp. 885–898, November 1981.

[14] Y. Rahmat-Samii and V. Galindo-Israel, "Shaped reflector antenna analysis using the Jacobi-Bessel series," *IEEE Trans. Antennas Propag.*, vol. AP-28, no. 4, pp. 425–435, July 1980.

[15] Y. Rahmat-Samii, R. Mittra, and V. Galindo-Israel, "Computation of Fresnel and Fraunhofer fields of planar apertures and reflector antennas by Jacobi-Bessel series—a review," *J. Electromagnetics*, vol. 1, no. 2, pp. 155–185, April–June 1981.

[16] A. C. Ludwig, "Calculation of scattered patterns from symmetrical reflectors," *Tech. Rep. No. 32-1430*, Jet Propulsion Laboratory, California Institute of Technology, Pasadena, February 1970.

[17] V. Galindo-Israel and R. Mittra, "A new series representation for the radiation integral with application to reflector antennas," *IEEE Trans. Antennas Propag.*, vol. AP-25, pp. 631–641, September 1977.

[18] G. Franceschetti, "Analytical techniques for reduction of computational effort in reflector antenna analysis," *IEE Intl. Symp. Antennas Propag. Dig.*, York, England, pp. 457–465, April 1981.

[19] Y. Rahmat-Samii and R. Chueng, "Nonuniform sampling techniques for antenna applications," *IEEE Trans. Antennas Propag.*, vol. AP-35, pp. 268–279, March 1987.

[20] J. B. Keller, "Geometrical theory of diffraction," *J. Opt. Soc. Am.*, vol. 52, pp. 116–130, 1962.

[21] S. W. Lee, "Uniform asymptotic theory of electromagnetic edge diffraction: a review," in *Electromagnetic Scattering*, New York: Academic Press, 1978, pp. 67–119.

[22] S. W. Lee, P. Cramer, Jr., K. Woo, and Y. Rahmat-Samii, "Diffraction by an arbitrary subreflector: GTD solution," *IEEE Trans. Antennas Propag.*, vol. AP-27, pp. 305–316, May 1979.

[23] R. G. Kouyoumjian, "The geometrical theory of diffraction and its application," in *Topics in Applied Physics, Volume 3: Numerical and Asymptotic Techniques in Electromagnetics*, Heidelberg: Springer-Verlag, 1975, pp. 165–215.

[24] S. W. Lee and G. A. Deschamps, "A uniform asymptotic theory of electromagnetic diffraction by a curved wedge," *IEEE Trans. Antennas Propag.*, vol. AP-24, no. 1, pp. 25–34, January 1976.

[25] R. G. Kouyoumjian and P. H. Pathak, "A uniform geometrical theory of diffraction for an edge in a perfectly conducting surface," *Proc. IEEE*, vol. 62, no. 11, pp. 1448–1461, November 1974.

[26] Y. Rahmat-Samii and R. Mittra, "Spectral analysis of high-frequency diffraction of an arbitrary incident field by a half-plane—comparison with four asymptotic techniques," *Radio Sci.*, vol. 13, pp. 31–48, January–February 1978.

[27] J. Boersma and Y. Rahmat-Samii, "Comparison of two leading uniform theories of edge diffraction with the exact uniform asymptotic solution," *Radio Sci.*, vol. 15, pp. 1179–1194, November–December 1980.

[28] J. F. Kauffman, W. F. Croswell, and L. J. Jowers, "Analysis of the radiation patterns

of reflector antennas," *IEEE Trans. Antennas Propag.*, vol. AP-24, pp. 53–65, January 1976.

[29] Y. Rahmat-Samii, "A comparison between GO/aperture field and physical optics methods for offset reflectors," *IEEE Trans. Antennas Propag.*, vol. AP-32, pp. 301–306, March 1984.

[30] Y. Rahmat-Samii, "An efficient computational method for characterizing the effects of random surface errors on the average power pattern of reflectors," *IEEE Trans. Antennas Propag.*, vol. AP-31, pp. 92–98, January 1983.

[31] R. C. Hansen, "Circular aperture distribution with one parameter," *Electron. Lett.*, vol. II, p. 184, April 1975.

[32] V. Jamnejad and Y. Rahmat-Samii, "Some important geometrical features of conic-section–generated offset reflector antennas," *IEEE Trans. Antennas Propag.*, vol. AP-28, pp. 952–957, November 1980.

[33] P. W. Hannan, "Microwave antennas derived from the Cassegrain telescope," *IRE Trans. Antennas Propag.*, vol. AP-9, pp. 140–153, March 1961.

[34] Y. Rahmat-Samii and S. W. Lee, "Directivity of planar array feeds for satellite reflector applications," *IEEE Trans. Antennas Propag.*, vol. AP-31, pp. 463–470, May 1983.

[35] T. S. Chu and R. H. Turrin, "Depolarization properties of offset reflector antennas," *IEEE Trans. Antennas Propag.*, vol. AP-21, pp. 339–345, May 1973.

[36] A. C. Ludwig, "The definition of cross polarization," *IEEE Trans. Antennas Propag.*, vol. AP-21, no. 1, pp. 116–119, January 1973.

[37] P. G. Ingerson and W. V. T. Rusch, "Radiation from a paraboloid with an axially defocused feed," *IEEE Trans. Antennas Propag.*, vol. AP-21, no. 1, pp. 104–106, January 1973.

[38] Y. T. Lo, "On the beam deviation factor of a parabolic reflector," *IRE Trans. Antennas Propag.*, vol. AP-8, pp. 347–349, May 1960.

[39] J. Ruze, "Lateral-feed displacement in a paraboloid," *IEEE Trans. Antennas Propag.*, vol. AP-13, pp. 660–665, September 1965.

[40] W. A. Imbriale, P. G. Ingerson, and W. C. Wong, "Large lateral feed displacements in a parabolic reflector," *IEEE Trans. Antennas Propag.*, vol. AP-22, pp. 742–745, November 1974.

[41] P. J. Woods, *Reflector Antenna Analysis and Design*, London: Peter Peregrinus, 1980.

[42] W. V. T. Rusch and P. D. Potter, *Analysis of Reflector Antennas*, New York: Academic Press, 1970.

[43] Y. Rahmat-Samii and V. Galindo-Israel, "Scan performance of dual offset reflector antennas for satellite communications," *Radio Sci.*, vol. 16, no. 6, pp. 1093–1099, November–December 1981.

[44] Y. Mitzugutch, M. Akagawa, and H. Yokoi, "Offset dual reflector antenna," *IEEE Intl. Symp.*, pp. 2–5, Amherst, Massachusetts, October 11–15, 1976.

[45] W. C. Wong, "On the equivalent parabola technique to predict the performance characteristics of a Cassegrainian system with an offset feed," *IEEE Trans. Antennas Propag.*, vol. AP-21, no. 3, pp. 335–339, May 1973.

[46] V. Krichevsky and D. F. DiFonzo, "Optimum feed locus for beam scanning in offset Cassegrain antennas," *IEEE/AP-S International Symposium*, Quebec, Canada, June 1980.

[47] V. Krichevsky, and D. F. DiFonzo, "Beam scanning in the offset Gregorian antenna," *Comsat Tech. Rev.*, vol. 12, pp. 251–269, 1982.

[48] B. Y. Kinber, "On two reflector antennas," *Radio Eng. Electron. Phys.*, vol. 6, June 1962.

[49] V. Galindo, "Design of dual reflector antennas with arbitrary phase and amplitude distribution," *IEEE Trans. Antennas Propag.*, vol. AP-12, pp. 403–408, July 1964.

[50] W. F. Williams, "High-efficiency antenna reflector," *Microwave J.*, vol. 8, p. 79, 1965.

[51] P. J. Woods, "Reflector profiles for the pencil-beam Cassegrain antenna," *Marconi Rev.*, pp. 121–138, 2nd quarter, 1972.

[52] V. Galindo-Israel, R. Mittra, and A. Cha, "Aperture amplitude and phase control of

offset dual reflectors," *IEEE Trans. Antennas Propag.*, vol. AP-27, no. 2, pp. 154–164, March 1979.

[53] J. J. Lee, L. I. Parad, and R. S. Chu, "A shaped offset-fed dual-reflector antenna," *IEEE Trans. Antennas Propag.*, vol. AP-27, no. 2, pp. 165–171, March 1979.

[54] A. Cha and D. A. Bathker, "Preliminary announcement of an 85-percent efficient reflector antenna," *IEEE Trans. Antennas Propag.*, vol. AP-31, no. 2, pp. 341–342, March 1983.

[55] V. Galindo-Israel, Y. Rahmat-Samii, W. Imbriale, and R. Mittra, "Recent advances in electromagnetic synthesis and analysis of dual-shaped reflector antennas," *SPIE*, vol. 294, pp. 98–112, 1981.

[56] P. Balling, "Wavefront synthesis of contoured beam antennas with low cross-polarization," *Rep. S-92-01*, under Grant 1112/69-78, from the Danish Space Board, TICRA, ApS, 1979.

[57] R. G. Spencer, *Microwave Antenna Theory*, Section 13.8, Radiation Lab Series No. 12, ed. by S. Silver, New York: McGraw-Hill Book Co., 1949, pp. 502–509.

[58] C. A. Smith, "A review of the state of the art in large spaceborne antenna technology," *Pub. 78-88*, Jet Propulsion Lab., Pasadena, California, November 15, 1978.

[59] R. S. Elliott, *Antenna Theory and Design*, Englewood Cliffs: Prentice-Hall, 1981.

[60] Y. Rahmat-Samii, "Chapter 3 of 30/20-GHz Lewis antenna support," *Final Report*, NASA No. 643-10-01-04-00, February 1981.

[61] C. C. Chen and C. F. Franklin, "$K\hat{u}$-band multiple-beam antenna," NASA *Contract Rep. 154364* for contract no. NAS 1-14814, NASA Langley Research Center, Hampton, Virginia 23665, December 1980.

[62] A. V. Mrstik, "Scan limits of off-axis fed parabolic reflectors," *IEEE Trans. Antennas Propag.*, vol. AP-27, no. 5, pp. 647–651, September 1979.

[63] D. T. Nakatani, F. A. Taormina, G. G. Kuhn, and D. K. McCarty, "Design aspects of commercial satellite antennas," Communications Satellite Antenna Technology Short Course, UCLA, March 1976.

[64] C. C. Han, A. E. Smoll, H. W. Bilenko, C. A. Chuang, and C. A. Klein, "A general beam shaping technique—multiple-feed offset reflector antenna system," AIAA Sixth Communications Satellite Systems Conference, Montreal, Canada, April 5–8, 1976.

[65] W. A. Imbriale, "Applications of the method of moments to thin-wire elements and Arrays," chapter 7 in *Topics in Applied Physics, Volume 3: Numerical and Asymptotic Techniques in Electromagnetics*, Heidelberg: Springer-Verlag, 1975.

[66] S. J. Hamada, P. G. Ingerson, and W. V. Rusch, "Reflector radiation from planar near-field measurements of array feed," *IEEE Trans. Antennas Propag.*, vol. AP-28, no. 4, pp. 436–442, July 1980.

[67] Y. Rahmat-Samii, P. Cramer, Jr., K. Woo, and S. W. Lee, "Realizable feed-element patterns for multibeam reflector antenna analysis," *IEEE Trans. Antennas Propag.*, vol. AP-29, pp. 961–963, November 1981.

[68] C. A. Balanis, *Antenna Theory Analysis and Design*, New York: Harper & Row, 1982.

[69] J. Huang, Y. Rahmat-Samii, and K. Woo, "A GTD study of pyramidal horns for offset reflector antenna applications," *IEEE Trans. Antennas Propag.*, vol. AP-31, no. 2, pp. 305–309, March 1983.

[70] W. L. Stutzman and G. A. Thiele, *Antenna Theory and Design*, New York: John Wiley & Sons, 1981.

[71] P. D. Potter, "A new horn antenna with suppressed side lobes and equal beamwidths," *Microwave J.*, vol. 6, pp. 71–78, June 1963.

[72] A. W. Rudge and N. A. Adatia, "Matched feeds for offset parabolic reflector antennas," *Proc. of the Sixth Eur. Microwave Conf.*, Rome, Italy, pp. 1–5, September 1976.

[73] A. W. Rudge and N. A. Adatia, "New class of primary-feed antennas for use with offset parabolic reflector antennas," *Electron. Lett.*, vol. 11, pp. 597–599, November 27, 1975.

[74] J. Ruze, "Antenna tolerance theory—a review," *Proc. IEEE*, vol. 54, no. 4, pp. 633–640, April 1966.
[75] T. B. Vu, "The effect of aperture errors on the antenna radiation pattern," *Proc. Inst. of Electr. Eng. (IEE)*, vol. 116, pp. 195–202, 1969.
[76] A. W. Love, "Some highlights in reflector antenna development," *Radio Sci.*, vol. 11, pp. 671–684, August–September 1976.
[77] W. V. T. Rusch and R. Wohlleben, "Surface tolerance loss for dual-reflector antennas," *IEEE Trans. Antennas Propag.*, vol. AP-30, no. 4, pp. 784–785, July 1982.
[78] H. Ling, Y. T. Lo, and Y. Rahmat-Samii, "Reflector sidelobe degradation due to random surface errors," *IEEE Trans. Antennas Propag.*, vol. AP-34, January 1986.
[79] Y. Rahmat-Samii and S. W. Lee, "Vector diffraction analysis of reflector antennas with mesh surfaces," *IEEE Trans. Antennas Propag.*, vol. AP-33, pp. 76–90, January 1985.
[80] Y. Rahmat-Samii, "Effects of deterministic surface distortions on reflector antenna performance," *Annales des Telecommunications*, France, vol. 40, pp. 350–360, August 1985.

Chapter 16
Lens Antennas

J. J. Lee
Hughes Aircraft Company

CONTENTS

1. Introduction — 16-5
2. Design Principles of the Dielectric Lens — 16-7
   - Rectangular Coordinates — 16-7
   - Polar Coordinates — 16-8
3. Simple Lenses with Analytic Surfaces — 16-9
   - Lens with a Flat Surface on S_2 — 16-9
   - Lens with a Flat Surface on S_1 — 16-10
   - Lens with a Spherical Surface on S_1 — 16-11
   - Lens with a Spherical Surface on S_2 — 16-12
4. Lens Aberrations and Tolerance Criteria — 16-12
5. Wide-Angle Dielectric Lenses — 16-19
   - Abbe Sine Condition — 16-19
   - Schmidt Corrector — 16-23
   - Spherical Thin Lens — 16-28
   - Bifocal Lenses — 16-30
6. Taper-Control Lenses — 16-33
7. Dielectric Lens Zoning — 16-38
8. Constrained Lenses — 16-41
   - Waveguide Lens and Zoning — 16-42
   - Equal Group Delay Lens — 16-45
   - Multifocal Bootlace Lens — 16-46
   - R-2R Lens — 16-48
   - Constrained Analog of Dielectric Lens — 16-49
9. Inhomogeneous Lenses — 16-51
   - Luneburg Lens — 16-52
   - Maxwell Fish-Eye Lens — 16-52
   - Hyperbolic Cosine Lens — 16-53

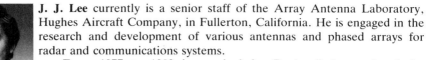

J. J. Lee currently is a senior staff of the Array Antenna Laboratory, Hughes Aircraft Company, in Fullerton, California. He is engaged in the research and development of various antennas and phased arrays for radar and communications systems.

From 1977 to 1982 he worked for Rockwell International, in Anaheim, California, where he was responsible for the design of antennas and adaptive phased arrays for satellite communications. Prior to that he was with GTE Sylvania, in Needham, Massachusetts, involved in the study of missile-site communications systems, ground-wave propagation, electromagnetic-pulse effects, and high-voltage simulation tests. From 1973 to 1974 he was associated with Cornell University in the Nuclear and Plasma Physics Laboratory.

Dr. Lee received his BSEE from National Taiwan University, R.O.C., in 1967 and his PhD in electrical engineering from the Case Institute of Technology, Cleveland, Ohio, in 1973. He is a member of the IEEE Antennas and Propagation Society. He has authored and coauthored 40 papers in various technical journals, and served as a reviewer for the *IEEE Transactions on Antennas and Propagation*. He has seven inventions, with several patents pending.

10. Bandwidth Limitation and Surface Mismatch 16-54
11. Summary 16-57
12. References 16-57

1. Introduction

The recent advent of millimeter-wave technology and the increasing demand for higher frequencies have stimulated a renewed interest in the design of lens antennas for radar and communication systems. For frequencies above K_u-band the penalty in weight and size associated with a lens is relatively small. In fact, for many ehf applications, lenses are selected for reasons of economy, performance, and reliability.

In general, lens antennas are easy to design and construct. They are mechanically rigid and are relatively tolerant of surface imperfections and load distortions. Compared with a dual-reflector design the lens antenna has a more desirable form factor due to its feedthrough characteristics. This unique feature eliminates the blockage problem and simplifies the mechanical design in the control of spillover with absorbing materials. It also offers greater design flexibility to meet various stringent requirements, such as low side lobes and wide scan coverage, by incorporating phase shifters or other phase compensation devices in the lens system.

Lenses, however, are not without drawbacks and problems. For high-gain antennas, especially those for spaceborne applications, the required aperture size, physical weight, and the transmission loss may prohibit the use of lenses. Edge diffraction, shadowing effect, and bandwidth limitation associated with lens zoning are also some of the common problems encountered in a lens design. In addition, impedance mismatch at the surface boundaries is always an important issue when very low side lobes are required.

Depending on their design, lenses can be broadly classified into two categories: dielectric lenses and constrained lenses. *Dielectric lens* refers to any focusing device of dielectric material which may or may not be homogeneous in its index of refraction. A *constrained lens* is defined as any optical transforming device in which the rays are guided and constrained to follow discrete paths that may have different propagation characteristics. The path lengths and the geometries of these guiding elements, which constitute the lens, are so designed that the exit rays produce the desired phase and amplitude distributions across the aperture. Constrained lenses include metal plate lenses, waveguide lenses, and other microwave lenses that consist of receiving and radiating elements with active or passive components for phase adjustments.

With computer controlled machines as readily available as they are today, a dielectric lens can be fabricated at very low cost. It can be easily contoured to any reasonable shape to produce a specified beam pattern. Recent advances in the development of low-loss dielectric materials make these lenses even more attractive to many antenna designers. Constrained lenses, on the other hand, are more expensive to construct. But normally they are lighter and more radiation resistant and therefore more suitable for space applications. Furthermore, they can be designed to meet very stringent requirements which cannot be fulfilled by dielectric lenses.

The design principles of lens antennas are very well known. Geometric optics is all that is required to formulate the problem. The lens can be considered as an optical transformer which, in most cases, transforms a spherical wavefront from a point source into a plane wave. In addition to this fundamental requirement the lens may be required to satisfy other conditions, depending on the application.

Sometimes one of the two surfaces is chosen to be flat for simplicity. For multibeam systems the lens may be designed to be coma free for a wide scan. For other applications the lens may be required to produce a special beam pattern and side lobe structure. Thus actual lens profiles depend on the specific conditions imposed in the design.

For uniform dielectric lenses Fermat's principle implies that Snell's law is automatically satisfied at each surface boundary of the lens. For constrained lenses, however, Snell's law is not necessarily satisfied at the boundary. Since Snell's law is not imposed at the boundaries of a constrained lens, other conditions must be specified to define the lens profiles in addition to the path length constraint. This provides much freedom in the design and for this reason many unique features can be achieved with a constrained lens.

Despite all the geometric differences, a dielectric lens is analogous to a dual-reflector system in many ways in an optical sense. The pickup surface of the lens (the one facing the feed element) is equivalent to the subreflector, which primarily controls the power distribution across the aperture. The second surface is equivalent to the main reflector, which chiefly corrects the phase. The space between the two reflectors can be viewed as a uniform dielectric medium which has a refraction index of -1. In this case, refraction is replaced by reflection at the boundary as a special case of Snell's law.

Since the lens design is based on geometric optics the aperture size of the lens must be sufficiently large, preferably more than 20 wavelengths, to make the geometric optics assumption valid. This consideration also limits the maximum number of zones that can be used for a given aperture size. Each zone must not be so small as to violate the geometric optics assumption. After the lens is designed by ray optics actual antenna characteristics must be determined by physical optics with diffraction taken into account. The far-field pattern is computed by standard methods, such as surface integration over the lens aperture as described in many antenna books.

The subject of lens antenna design has been treated in the past by many authors in considerable detail. Risser has summarized the early lens developments in Silver's book [1]. Brown also provides an excellent review on many lens designs in [2] and [3]. Detailed lens formulas with practical design considerations are given by Cohn [4] and Sengupta and Hiatt [5]. It is not the intent of this chapter to duplicate these works, but rather to present a complementary and systematic overview on various lenses as a design guide, with emphasis placed on more recently developed lens systems.

In the following sections, design formulas and performance characteristics of various dielectric and constrained lenses are presented. For brevity, discussions on theory and mathematical derivations are kept to a minimum. System applications are not stressed in this chapter, as these topics will be treated in more detail in Part C.

2. Design Principles of the Dielectric Lens

Fig. 1 shows a dielectric lens with its general contours S_1 and S_2 represented by (x_1, y_1) and (x_2, y_2), respectively. It is assumed that the lens is rotationally symmetric, and therefore only the cross section of the lens is of interest. Consider a lens of dielectric constant ϵ_r illuminated by a feed pattern emanating from a phase center located at the origin of the coordinates. The distance between the origin and the first surface of the lens is F. Let the central thickness of the lens be T and the diameter of the lens aperture be D.

Rectangular Coordinates

The most important condition to be imposed in the derivation of the lens profiles is the path length constraint. For phase reference an aperture plane is established at $x = S$, to the right of the lens for clarity. In general, the aperture phase distribution can be any rotationally symmetric function specified by $\varrho(\phi)$, where ϱ is the distance from the axis of the lens. For a nonuniform aperture phase front the lens profiles will be a function of S, but the dependence is minor since the

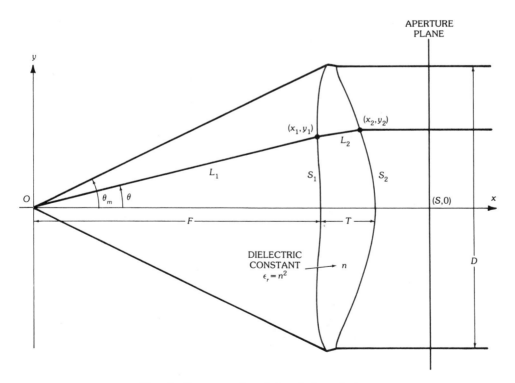

Fig. 1. Geometry for dielectric lens design.

aperture plane is usually very close to the lens. Examples of nonuniform phase distributions will be discussed in more detail later, but for the moment the aperture phase is assumed to be uniform. Mathematically the path length condition is given by

$$(x_1^2 + y_1^2)^{1/2} + n[(x_2 - x_1)^2 + (y_2 - y_1)^2]^{1/2} - x_2 = (n - 1)T \qquad (1)$$

where the central ray has been used as a path length reference and $n = \sqrt{\epsilon_r}$ is the refractive index of the lens, which is greater than unity for real dielectrics. It should be noted that the parameter S for the reference plane has been canceled out of both sides of (1), and the focal length F is implicitly imposed as an initial condition for x_1.

The second condition to be imposed is Snell's law derived from Fermat's principle. By taking the differential of y_1 with respect to x_1 in (1), the slope of the lens at (x_1, y_1) is obtained:

$$\frac{dy_1}{dx_1} = \frac{nL_1(x_2 - x_1) - L_2 x_1}{L_2 y_1 - nL_1(y_2 - y_1)} \qquad (2)$$

where

$$L_1 = (x_1^2 + y_1^2)^{1/2}$$
$$L_2 = [(x_2 - x_1)^2 + (y_2 - y_1)^2]^{1/2}$$

Similarly, the slope of the lens contour at (x_2, y_2) is given by

$$\frac{dy_2}{dx_2} = \frac{L_2 - n(x_2 - x_1)}{n(y_2 - y_1)} \qquad (3)$$

Polar Coordinates

Sometimes it is more convenient to formulate the problem in terms of polar coordinates using the phase center as the origin. Referring to Fig. 2, the path length constraint now becomes

$$r + n[R^2 + r^2 - 2Rr\cos(\theta - \phi)]^{1/2} - R\cos\phi = (n - 1)T \qquad (4)$$

Snell's law on surface 1 is given by

$$\frac{dr}{d\theta} = \frac{nRr\sin(\theta - \phi)}{n[R\cos(\theta - \phi) - r] - L_2} \qquad (5)$$

and Snell's law on surface 2 leads to

$$\frac{dR}{d\phi} = \frac{nRr\sin(\theta - \phi) - L_2 R\sin\phi}{n[R - r\cos(\theta - \phi)] - L_2\cos\phi} \qquad (6)$$

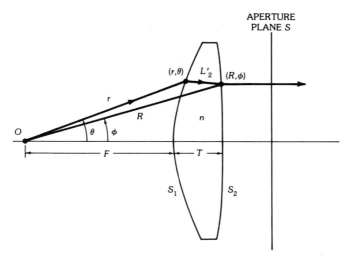

Fig. 2. Dielectric lens formulated in polar coordinates.

where

$$L_2 = n^{-1}[(n-1)T + R\cos\phi - r] = [R^2 + r^2 - 2Rr\cos(\theta - \phi)]^{1/2}$$

In the lens formulation there are four variables in (1)–(3): $x_1, y_1, x_2,$ and y_2. With one of these quantities considered to be the independent variable three independent equations are required to uniquely determine the lens contours. The details of how to apply and solve these equations are discussed in the following sections, where different types of lenses are presented.

3. Simple Lenses with Analytic Surfaces

A *simple lens* is defined to be a lens which can be described by an analytic expression for one or both surfaces. To be examined in this section are four commonly used designs.

Lens with a Flat Surface on S_2

The simplest case is the one that has a flat surface on S_2 as shown in Fig. 3. This lens transforms a spherical wavefront into a plane wave, or conversely focuses a beam from infinity onto one point. The conditions imposed to derive S_1 are $x_2 = T$, the slope on S_2 being infinity, and the equal path length constraint. It can be readily shown that S_1 is a hyperbolic surface given by [1]

$$y_1 = [(n^2 - 1)(x_1 - F)^2 + 2(n - 1)F(x_1 - F)]^{1/2} \qquad (7)$$

or equivalently

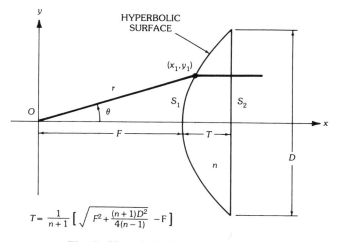

Fig. 3. Hyperbolic lens, planar on S_2.

$$r = \frac{(n-1)F}{n\cos\theta - 1} \qquad (8)$$

For a given diameter D, the central thickness of the lens is

$$T = \frac{1}{n+1}\left[\sqrt{F^2 + \frac{(n+1)D^2}{4(n-1)}} - F\right] \qquad (9)$$

Lens with a Flat Surface on S_1

Fig. 4 is a lens with a flat surface on S_1. By setting $x_1 = F$ and the slope on S_1 equal to infinity, a parametric solution for S_2 can be found from the path length constraint

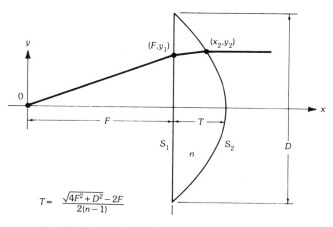

Fig. 4. Convex lens with planar surface on S_1.

$$x_2 = \frac{[(n-1)T - \sqrt{F^2 + y_1^2}]\sqrt{(n^2-1)y_1^2 + n^2F^2} + n^2F\sqrt{F^2 + y_1^2}}{n^2\sqrt{F^2 + y_1^2} - \sqrt{(n^2-1)y_1^2 + n^2F^2}} \quad (10)$$

$$y_2 = y_1\left[1 + \frac{x_2 - F}{\sqrt{(n^2-1)y_1^2 + n^2F^2}}\right] \quad (11)$$

In this case the central thickness is

$$T = \frac{\sqrt{4F^2 + D^2} - 2F}{2(n-1)} \quad (12)$$

Lens with a Spherical Surface on S_1

If S_1 is spherical, as shown in Fig. 5, the outer surface S_2 is an ellipse specified by

$$R = \frac{(n-1)(F+T)}{n - \cos\phi} \quad (13)$$

This is obtained by setting $\theta = \phi$ in (6) and integrating the differential equation. In rectangular coordinates the solution is

$$y_2 = \left[\left[\frac{x_2 + (n-1)(F+T)}{n}\right]^2 - x_2^2\right]^{1/2} \quad (14)$$

In this case the central thickness is

$$T = \frac{2F - \sqrt{4F^2 - D^2}}{2(n-1)} \quad (15)$$

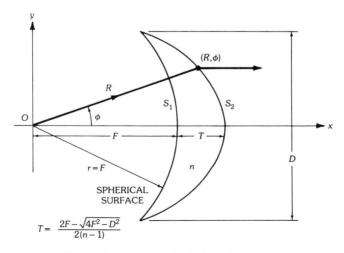

Fig. 5. Lens with spherical surface on S_1.

Lens with a Spherical Surface on S_2

Conversely, if S_2 is specified to be spherical, as shown in Fig. 6, the inner surface (r, θ) is defined by the following equations:

$$n^2[R^2 + r^2 - 2Rr\cos(\theta - \phi)] = [(n-1)T + R\cos\phi - r]^2 \qquad (16)$$

$$n^2 r \sin(\theta - \phi) = (\sin\phi)[(n-1)T + R\cos\phi - r] \qquad (17)$$

where $R = F + T$, and ϕ is the parametric variable. Equation 16 is derived from (4) and (17) is obtained by setting $dR/d\phi = 0$ in (6). For this lens the central thickness is related to D by

$$T = \sqrt{\frac{4(n-1)F^2 - (n-3)D^2}{4(n-1)(n-3)^2}} + \frac{F}{n-3} \qquad (18)$$

Shown in Fig. 7 are plots of the central thickness versus dielectric constant for these four designs with F/D as a parameter. It can be seen that for small F/D the lens with spherical surface on S_1 is much thicker than the other three designs. For larger F/D, however, all these lenses have about the same thickness for a given dielectric constant.

4. Lens Aberrations and Tolerance Criteria

Before proceeding to the discussion of other lens designs a brief review of various lens aberrations may be in order. The term "aberration" is primarily used by optical designers referring to any imperfection caused by a lens in reproducing the image of an object. For antenna engineers the more commonly used nomenclature is *phase error*. Phase error is defined as the deviation in phase of the

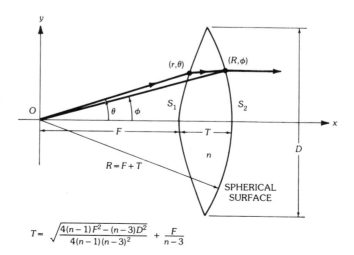

Fig. 6. Lens with spherical surface on S_2.

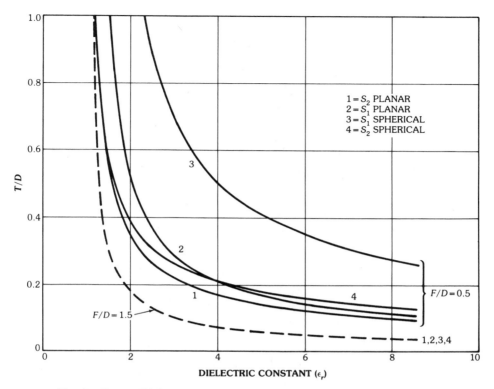

Fig. 7. Center thickness versus dielectric constant for different lenses.

wavefront from a prescribed reference, usually the aperture plane in front of the lens. Since amplitude errors generally do not occur in the lens design, they are therefore ignored here.

Phase errors may result from errors in the illumination pattern, displacement of the source from the focal point, or from the intrinsic errors of the lens. The phase error is normally expressed in terms of path length difference between a general ray and the reference central ray as a function of the exit point through the aperture. This aberration function, as defined in optics, can be represented as a surface above the aperture plane, defined by (ϱ, ϕ) in this case.

In general, degradations of the far-field pattern caused by a two-dimensional phase error distribution are difficult to visualize. However, the subject can be better understood if the error distribution is expanded into a power series in ϱ and $\cos\phi$, or equivalently in terms of the so-called Zernike cylindrical polynomials, as discussed in Born and Wolf [6]. The Zernike polynomials form an orthogonal set, which simplifies the evaluation of the diffraction integral over a circular aperture in the study of various aberration effects.

It has been shown that any aberration function of small amplitude can be approximated by the following series:

$$\Delta L(\varrho, \phi) = \alpha \varrho \cos\phi + \beta \varrho^2 + \gamma \varrho^2 \cos 2\phi + \delta \varrho^3 \cos\phi + \varepsilon \varrho^4 \qquad (19)$$

These are the five lowest-order terms of the power series, representing five primary aberrations which are better known as *Seidel aberrations*. They have been identified as distortion, curvature of field, astigmatism, coma, and spherical aberration, respectively. Note that the $\cos 2\phi$ term can also be represented by $\cos^2 \phi$ as they are similar in nature. The three terms with cosine factors account for the asymmetric part of the phase error in the azimuth plane. Fig. 8 illustrates the shapes of these aberrations.

Significant insights of the adverse effects caused by various aberrations can be obtained by simplifying the problem to a one-dimensional case. In one of the two principal planes defined by $\phi = 0$ or $\pi/2$, the five primary aberrations degenerate into four basic types of phase errors, namely linear, quadratic, cubic, and quartic (fourth-order) errors.

As is well known, a linear error does not perturb the antenna pattern; it merely shifts the beam by an angle given by

$$\theta_s = \sin^{-1}\left(\frac{a\lambda}{a\pi}\right)$$

where a is the linear dimension of the aperture. However, associated with this tilt angle the projected aperture dimension is reduced by a $\cos \theta_s$ factor.

The general effect of quadratic error is to raise both the side lobe level and the level of the minima. Since the phase error is symmetrical with respect to the center of the aperture the antenna pattern will always be symmetrical about the $\theta = 0$ axis. To gain further insights Silver's analysis [7] is followed. Let the far-field pattern be denoted by

$$g(u) = \frac{a}{2}\int_{-1}^{1} f(x) e^{j[ux - \beta x^2]} dx \tag{20}$$

where $u = (\pi a/\lambda)\sin\theta$, βx^2 stands for the quadratic phase error, and $f(x)$ is assumed to be an even function for the amplitude distribution over a normalized aperture. Then the power pattern for small β is given by

$$P(u) \cong \frac{a^2}{4}\{g_0^2(u) + \beta^2[g_0''(u)]^2\} \tag{21}$$

where $g_0(u)$ is the pattern in the absence of phase error ($\beta = 0$), and $g_0''(u)$ is the second derivative of $g_0(u)$.

Fig. 9 shows far-field patterns of a linear aperture with uniform amplitude distribution and quadratic phase error. It has been found that when β gets sufficiently large the main lobe becomes bifurcated, with maxima appearing on either side of the $\theta = 0$ axis. Fig. 10 exemplifies the effect of quadratic error on gain.

The cubic phase errors (δx^3) can be treated by the same technique. In this case the power pattern is

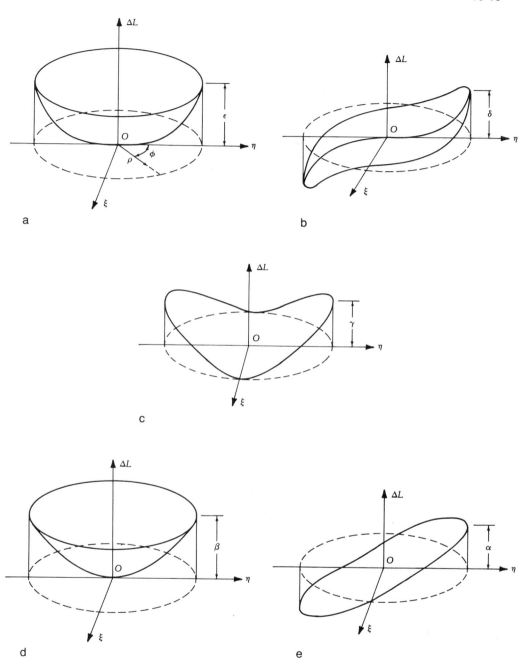

Fig. 8. Primary aberrations of a lens. (*a*) Spherical aberration: $\Delta L = \epsilon \varrho^4$. (*b*) Coma: $\Delta L = \delta \varrho^3 \cos\phi$. (*c*) Astigmatism: $\Delta L = \gamma \varrho^2 \cos^2\phi$ (similar to $\varrho^2 \cos 2\phi$ in nature). (*d*) Curvature of field: $\Delta L = \beta \varrho^2$. (*e*) Distortion: $\Delta L = \alpha \varrho \cos\phi$. (*After Born and Wolf [6]; reproduced by permission of M. Born and E. Wolf*, Principles of Optics, *Third Edition, New York and London: Pergamon Press, 1965*)

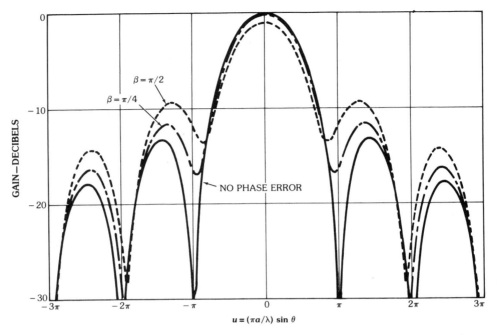

Fig. 9. Pattern degradation due to quadratic phase error (one-dimensional case).

$$P(u) \cong \frac{a^2}{4}[g_0(u) + \delta g_0'''(u)]^2 \quad (22)$$

The effect of the phase error is to tilt the beam as in the case of a linear error, and reduce the peak gain. In addition, the main lobe becomes asymmetrical, and the side lobes decrease on the side of the main lobe nearer $\theta = 0$ and increase on the other side of the main lobe as shown in Fig. 11.

Similarly, for quartic errors (εx^4) the corresponding power pattern becomes

$$P(u) \cong \frac{a^2}{4}\{g_0^2(u) + \varepsilon^2[g_0^{(4)}(u)]^2\} \quad (23)$$

A comparison of this with (21) indicates that the effect of a fourth-order error is similar to that of a quadratic one. As shown in Figs. 10 and 12, however, the effect is less pronounced because for the same peak error it has less total error field over the aperture.

Based on the simple analysis given above, tolerance criteria of various aberrations can be discussed. If gain degradation is a major concern in the design, a maximum allowable loss in the peak gain must be specified to establish a tolerance level for a particular phase error. For example, if a 1-dB drop in gain is allowed, a maximum deviation of $\lambda/4$ is tolerable for the quadratic error in a one-dimensional case. For a fourth-order error the tolerance level is relatively higher, but for cubic

Lens Antennas

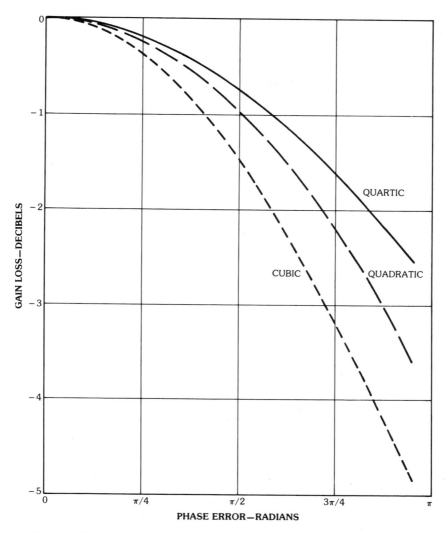

Fig. 10. Gain loss due to various phase errors (one-dimensional case).

error the tolerance for the same boresight gain loss is much tighter, as illustrated in Fig. 10.

On the other hand, if side lobe degradation is more of a concern, the criterion is taken to be the maximum error allowed for a peak side lobe level. Among the various aberrations the cubic error, which causes asymmetrical side lobe distortion, is the most undesirable one. In two-dimensional cases a contour plot of the diffraction pattern of a circular aperture with a cubic phase error (astigmatism) will show that concentric rings of equal intensity in the main lobe are shifted to one side and distorted into pear-shaped contours. This is commonly described as coma distortion in optics.

Usually quadratic errors can be corrected by displacing the feed from the focus

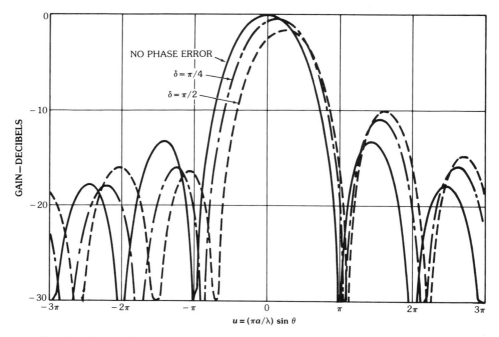

Fig. 11. Pattern degradation due to cubic phase error (one-dimensional case).

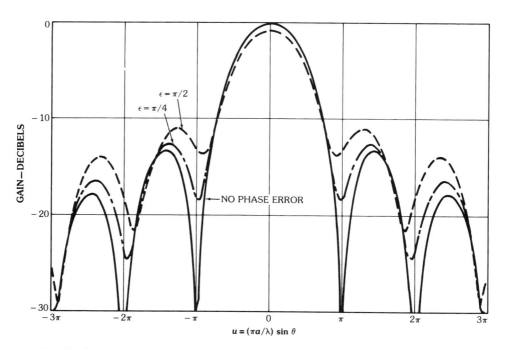

Fig. 12. Pattern degradation due to quartic phase error (one-dimensional case).

slightly along the axis. This defocusing technique cannot, however, compensate for the cubic or fourth-order errors. To minimize the coma aberration the system must be designed to satisfy the so-called Abbe sine condition, which will be discussed in the next section. For some applications the lens system is optimized to produce a minimum mean error instead of a particular type of error distribution. Thus the tolerance criterion is subject to system requirements and applications.

Although tolerance criteria for two-dimensional aberrations are more complex to analyze, the adverse effects are characterized by similar features derived in the one-dimensional case. If $\Delta L(\varrho, \phi) = a\varrho^n \cos m\phi$ is the distribution of the phase error, and $(1 - \varrho^2)^P$ is the amplitude taper over a normalized circular aperture, then the aperture efficiency (gain factor) is computed by

$$\eta = \frac{1}{\pi} \frac{\left[\int_0^1 \int_0^{2\pi} (1 - \varrho^2)^P e^{ja\varrho^n \cos m\phi} \varrho \, d\varrho \, d\phi\right]^2}{\int_0^1 \int_0^{2\pi} (1 - \varrho^2)^{2P} \varrho \, d\varrho \, d\phi}$$

$$= 2(4P + 2)\left[\int_0^1 (1 - \varrho^2)^P J_0(a\varrho^n) \varrho \, d\varrho\right]^2$$

Based on this formula the gain drop due to the phase error can be calculated. If gain loss is taken as the criterion for tolerance condition, then in general the tolerance of phase error for a circular aperture is not as tight as that for a one-dimensional case, and the tolerance increases with the power of ϱ. Shown in Table 1 are some numerical examples of gain loss as a function of the phase errors. Using 1-dB loss as a criterion, one can specify tolerance conditions for various errors as given in Table 2.

The phase errors discussed so far are strictly systematic errors. Random errors are not considered here. For a more detailed treatment on two-dimensional lens aberrations and random errors, readers are referred to Born and Wolf [6] and the collection of papers in [8] by Love.

5. Wide-Angle Dielectric Lenses

For multibeam or scanning applications a lens must be designed for wide-angle coverage. As the phase center is displaced from the axis to produce an off-axis beam, various lens aberrations will be generated. Among these aberrations the cubic phase error is the most undesirable effect as far as antenna performance is concerned. The resultant beam degradations are manifested by gain loss and coma lobe distortions [7]. Thus to maintain satisfactory scanning characteristics the cubic phase error due to an offset feed must be minimized. In this section several techniques to achieve relatively wide-angle coverage are introduced.

Abbe Sine Condition

As is well known in optics, a collimating lens can be designed to be coma free for a limited scan by imposing the Abbe sine condition [9]. Mathematically the Abbe sine condition requires that

Table 1. Boresight Gain Drop As a Function of Phase Error for Aperture Distribution $E(\varrho, \phi) = (1 - \varrho^2)^P \exp(ja\varrho^n \cos m\phi)$

	Gain Drop (dB)							
	$P = 0$				$P = 1$			
Peak Error	$n = 1$	$n = 2$	$n = 3$	$n = 4$	$n = 1$	$n = 2$	$n = 3$	$n = 4$
20°	0.13	0.09	0.07	0.05	0.09	0.04	0.02	0.02
40°	0.54	0.35	0.26	0.21	0.36	0.18	0.10	0.07
60°	1.22	0.80	0.59	0.47	0.80	0.40	0.23	0.15
80°	2.21	1.42	1.05	0.83	1.44	0.70	0.41	0.27
100°	3.55	2.23	1.63	1.28	2.28	1.09	0.64	0.42

Table 2. Primary Aberrations and Tolerance Conditions of a Circular

		Tolerance	
Type of Aberration	Representation	$P = 0$ (approximation)	$P = 1$
Spherical aberration	$\epsilon \varrho^4$	$\epsilon \leq 0.25\lambda$	$\epsilon \leq 0.43\lambda$
Coma	$\delta \varrho^3 \cos \phi$	$\delta \leq 0.22\lambda$	$\delta \leq 0.35\lambda$
Astigmatism	$\gamma \varrho^2 \cos 2\phi$	$\gamma \leq 0.19\lambda$	$\gamma \leq 0.27\lambda$
Curvature of field	$\beta \varrho^2$	$\beta \leq 0.19\lambda$	$\beta \leq 0.27\lambda$
Distortion	$\alpha \varrho \cos \phi$	$\alpha \leq 0.15\lambda$	$\alpha \leq 0.19\lambda$

$$y = F_e \sin \theta \qquad (24)$$

This condition is automatically fulfilled if the inner surface of a conventional waveguide lens is spherical. For a thin dielectric lens it is sufficient if the average shape of the lens is spherical [10]. The interpretation of this condition for a thick lens is that the initial and the final ray, when extended, intersect inside the lens on a circle of radius F_e. This is illustrated in Fig. 13, where F_e is the effective focal length, which is different from the F parameter defined previously.

Since the derivation of this condition is based on the approximation of paraxial rays, care should be taken not to apply the Abbe sine condition to systems with subtended feed angle much larger than 20°.

The first wide-angle dielectric lens based on the Abbe sine condition was developed by Friedlander [11] in 1946. He discovered that for a given focal length and thickness there is a family of lenses that satisfies the coma-free condition. But among these there is just one for which the aperture size is a maximum, characterized by the fact that the surfaces of the lenses meet at the edge. Furthermore, it is pointed out that if the dielectric constant is close to 2.6, the lens can be made to nearly satisfy the Abbe sine condition even with a flat inner surface. As reported in [11] a plano-convex lens of 50λ in diameter had been constructed

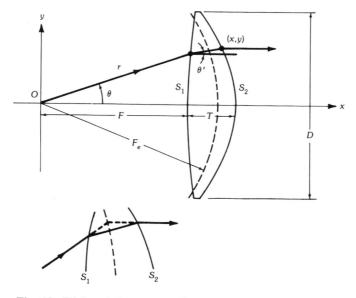

Fig. 13. Dielectric lens constrained by Abbe sine condition.

which produced a beam free of coma distortion over a scan coverage up to eight beamwidths.

In general, numerical integration is required to derive a coma-free lens. The design procedures are summarized in the following. When appropriate boundary conditions are specified, condition (24) and the phase constraint completely determine the lens contours. However, this does not imply that the solution can be analytically derived. It only means that a numerical solution can be obtained by step integration of the governing equations. To see this, first note that in reference to Fig. 13 the phase constraint can be written as

$$r + n[(y - r\sin\theta)^2 + (x - r\cos\theta)^2]^{1/2} - x = K \qquad (25)$$

where $K = (n - 1)T$ is a constant determined by the central ray as a boundary condition. Next, substitute (24) into (25) to eliminate y. After some manipulation a quadratic equation in x can be deduced:

$$Ax^2 + Bx + C = 0 \qquad (26)$$

where

$A = \epsilon_r - 1$
$B = 2(r - K) - 2\epsilon_r r \cos\theta$
$C = \epsilon_r r^2 \cos^2\theta + \epsilon_r (F_e - r)^2 \sin^2\theta - (r - K)^2$

The solution for x is simply

$$x = \frac{-B + (B^2 - 4AC)^{1/2}}{2A}$$

Thus x and y can be expressed in terms of r and θ. Now Snell's law is applied to get [12]

$$\frac{dr}{d\theta} = \frac{nr\sin(\theta - \theta')}{n\cos(\theta - \theta') - 1} \tag{27}$$

where

$$\theta' = \tan^{-1}\left[\frac{(F_e - r)\sin\theta}{x - r\cos\theta}\right]$$

It is clear from (27) that $dr/d\theta$ is a function of r and θ only, since x and y have been replaced by (24) and (26). Thus with the central ray as an initial condition, S_1 can be numerically solved [12] by step integration of (27).

Fig. 14 shows an example in which the lens was designed for satellite multibeam applications at 44 GHz. The Rexolite material has a dielectric constant of 2.54. The step size was 0.1° in θ. Other input parameters for the computer program include the aperture diameter D, the central thickness T, the distance F from the phase center to the vertex of the lens, and the effective focal length F_e.

The measured far-field patterns of the unmatched lens are shown in Fig. 15. The feed pattern is basically a $(\sin u)/u$ function of a square horn with a 20-dB edge taper. In this example the beam was only required to scan up to three beamwidths. Within this limited scan, as can be seen, there is virtually no coma distortion in the patterns. The first side lobe level of the lens is 23 dB below the peak. If the surface impedance is well matched [13], the side lobe level can be reduced by a few decibels.

It should be noted that not any combination of the input parameters can yield a

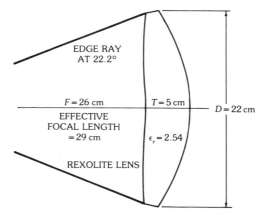

Fig. 14. Cross section of a coma-free dielectric lens.

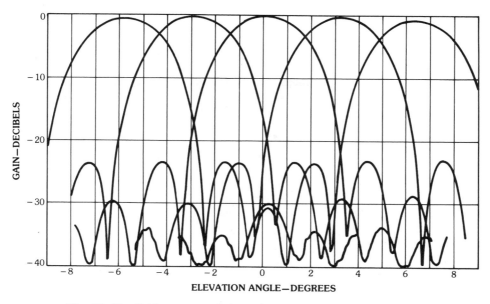

Fig. 15. Far-field patterns of the coma-free lens shown in Fig. 14.

complete lens. Some iterations are required to optimize a particular design. The numerical solution so derived for a given aperture size is certainly not unique. Any variation in the step size or the geometric parameters will lead to a different solution for the lens. Despite the fact that this is only an approximate solution, for all practical applications this numerical method provides versatile options and trade-offs for optimizing the performance.

It should be remarked that when the Abbe sine condition is imposed, the aperture power distribution can no longer be independently specified. In this case, as will be discussed in a later section, the aperture taper is mainly determined by the feed pattern. Hence a coma-free lens cannot provide very low side lobes if the feed pattern does not have enough illumination taper to begin with.

Schmidt Corrector

It is well known that a Luneburg lens is a truly wide-angle lens, because its radiation pattern is independent of the scan angle due to its inherent spherical symmetry and the property of possessing a perfect focus. Despite these unique features, however, the Luneburg lens has two basic drawbacks. One is that the peak side lobes cannot be held lower than about -17 dB in practice due to the inverse amplitude taper introduced by the lens [14]. The second disadvantage lies in the fact that the dielectric constant of the lens must vary as a function of radius, determined by $2 - (r/R)^2$, which makes it very difficult to fabricate. For these reasons a hemispherical or truncated spherical cap lens of uniform dielectric material, as shown in Fig. 16, has been used as an alternative for wide-scan applications.

A spherical cap lens does not have a well-defined focus, but the spherical aberrations can be minimized by properly adjusting the ratio of focal length to the

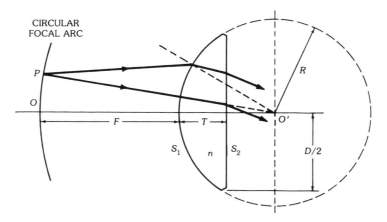

Fig. 16. Spherical cap lens.

radius of the lens for a given aperture diameter. Furthermore, the planar surface can be easily corrected to achieve a perfect focus if the aberration is small. Such corrected spherical cap lenses have been found to have excellent scanning characteristics. Other desirable features of these cap lenses are light weight, low volume, and simplicity in fabrication compared with the Luneburg lens. The only disadvantage of the cap lens is that the projected aperture decreases with scan angle as opposed to a constant aperture for a fully spherical lens.

Shown in Fig. 17 are some typical examples of the phase errors plotted across the aperture at various scan angles for two spherical cap lenses with different radii. It is seen that the phase error is almost symmetrical with respect to the central ray for scans up to $\pm 30°$. The maximum error is a function of the dielectric constant, radius of the inner surface, and the F/D ratio. If maximally flat phase over the central region is required, the F/R ratio [14] is $(n-1)^{-1}$. But the optimum condition is the one that produces a minimum average phase error over the whole aperture, similar to that shown in Fig. 17a. This kind of quadratic error can be easily corrected to yield perfect phase for the beam at boresight. Furthermore, the residual phase errors for other off-axis beams can be kept within an acceptable level.

A corrected spherical cap lens can be considered as a superposition of a cap lens and a Schmidt type corrector, which is commonly used in the optical design of a spherical mirror where spherical aberration is corrected.

To determine the corrector profile, basic lens equations presented in Section 2 are applied. Using x_1 as a running variable, the spherical pickup surface is defined by

$$y_1 = [R^2 - (x_1 - F - R)^2]^{1/2} \qquad (28)$$

With this given, the second surface can be found by solving (1) and (2). It can be shown that in terms of x_1 and y_1, the variable x_2 is given by

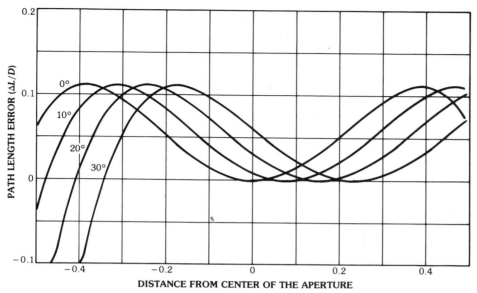

a

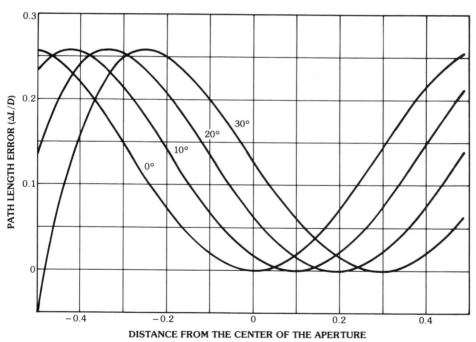

b

Fig. 17. Path length error for two spherical cap lenses across the aperture for different scan angles. (*a*) With $F/D = 1.0$, $R/D = 0.85$, $\epsilon_r = 2.54$. (*b*) With $F/D = 1.0$, $R/D = 1.0$, $\epsilon_r = 2.54$.

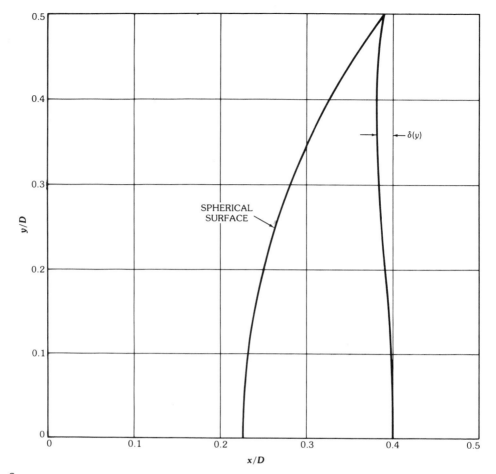

a

Fig. 18. Corrections for two spherical cap lenses with Schmidt correctors. (*a*) For $F/D = 1.0$, $R/D = 0.85$, $\epsilon_r = 2.54$. (*b*) For $F/D = 1.0$, $R/D = 1.0$, $\epsilon_r = 2.54$.

$$x_2 = \frac{-B + \sqrt{B^2 - 4AC}}{2A} \tag{29}$$

where

$$A = n^2(1 + P^2) - 1$$
$$B = 2n^2 P(Q - y_1) + 2\sqrt{x_1^2 + y_1^2} - 2n^2 x_1 - 2(n - 1)T$$
$$C = n^2(Q^2 - 2Qy_1 + x_1^2 + y_1^2) - [(n - 1)T - \sqrt{x_1^2 + y_1^2}]^2$$

and

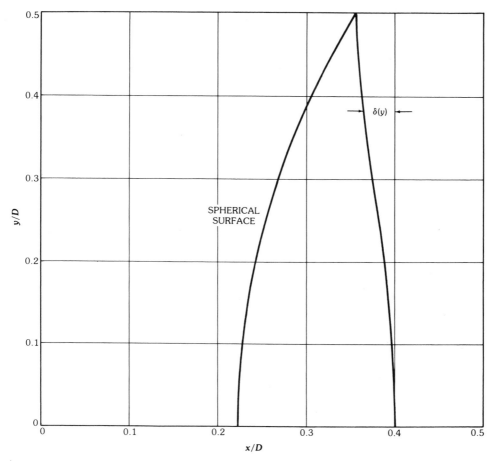

Fig. 18, *continued.*

$$P = \frac{F + R - n^2\sqrt{x_1^2 + y_1^2}}{n^2(F + R - x_1)} \frac{y_1}{\sqrt{x_1^2 + y_1^2}}$$

$$Q = (F + R) \frac{(n^2 - 1)\sqrt{x_1^2 + y_1^2} + (n - 1)T}{n^2(F + R - x_1)} \frac{y_1}{\sqrt{x_1^2 + y_1^2}}$$

The corresponding y_2 is computed by

$$y_2 = Px_2 + Q \tag{30}$$

If the lens is designed to have zero thickness at the edge, the central thickness must be chosen to be

$$T = \frac{\sqrt{D^2 + 4W^2} - 2W}{2(n - 1)} \quad (31)$$

where

$$W = F + R - \sqrt{R^2 - D^2/4}$$

With the second surface (x_2, y_2) determined, the deviation of S_2 from the planar surface at $x = T$ can be found. This deviation is the required correction of the cap lens to give a perfect wavefront, namely,

$$\delta(y_2) = F + T - x_2 \quad (32)$$

As examples the corrections for two spherical cap lenses are illustrated in Fig. 18. It is observed that a longer focal length will make the lens thinner, and a larger R/D ratio tends to make the correction thicker.

Spherical Thin Lens

As mentioned previously, if a lens is very thin and its average contour is very close to a spherical surface, the lens is a wide-angle lens. This has been shown by Shinn, who developed a thin lens for wide scan which is spherical on the outside and zoned on the inner surface [15]. Recently Rotman duplicated this design for satellite communication purposes with partial zoning and obtained remarkable scan performance as described below [10].

As depicted in Fig. 19 the outer surface is spherical with the center of curvature

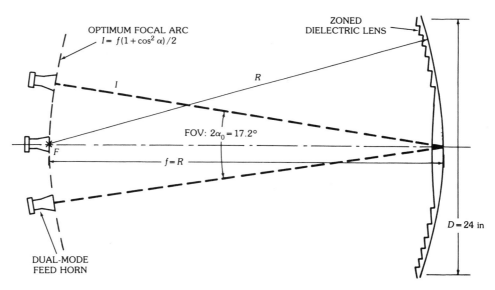

Fig. 19. Ehf aplanatic dielectric lens with multiple feeds. (*After Rotman [10]; reprinted with permission of Lincoln Laboratory, Massachusetts Institute of Technology, Lexington, Massachusetts*)

located at the focal point. The inner surface is determined by solving (16) and (17). Starting from the center the lens is stepped whenever the thickness reduces to a certain minimum value. For machining, the central and first outer zones of the inner surface of the lens are represented by a fourth-degree polynomial, while the outer five zones are approximated by straight line segments, determined by the points of discontinuity and corners on the contour. These approximations specify the inner surface of the lens to within 2 mils of the theoretical values. The lens was machined on a numerically controlled lathe in accordance with the numerical inputs.

As the beam is scanned and as the frequency changes, phase errors will occur across the radiating aperture. Since this lens is very thin, with its front surface radius R equal to its focal length, it obeys the Abbe sine condition and hence has minimum coma distortions. The only remaining significant phase error is the spherical aberration which, according to Shinn, is determined by the scanning locus (focal arc) and is independent of the shape of the lens.

The spherical aberration, measured as the path length error with respect to the central ray, is given by

$$\delta = \frac{1}{2}\frac{y^2}{f}\left(\frac{f}{\ell}\cos^2\alpha - 1\right) + \frac{1}{2}\frac{x^2}{f}\left(\frac{f}{\ell} - 1\right) \tag{33}$$

where the parameters are defined in Fig. 20. A compromised locus for the focal points, which balances the aberrations in both the yz and xz planes, is chosen to be [10]

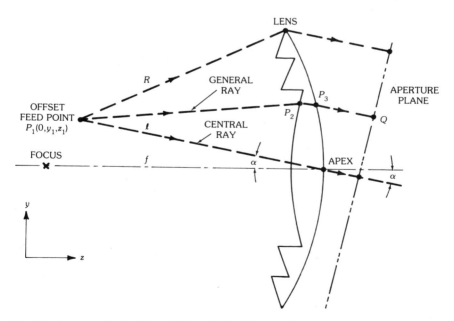

Fig. 20. Ray-trace analysis of zoned lens. (*After Rotman [10]; reprinted with permission of Lincoln Laboratory, Massachusetts Institute of Technology, Lexington, Massachusetts*)

$$\ell = \frac{1}{2}f(1 + \cos^2\alpha) \tag{34}$$

An experimental 90λ-diameter zoned lens operating at 44 GHz has been designed which generates a beam of 0.7° half-power beamwidth and 47 dBi gain, measured at the subsatellite point, over a 5-percent bandwidth. The beam can be steered over a total scan angle of 18° for full earth coverage with a scanning loss of less than 1 dB. Typical measured radiation patterns of the zoned lens are shown in Fig. 21.

Bifocal Lenses

Wide-scan capabilities can also be achieved by using bifocal systems, which are designed to have two perfect foci in the principal plane for two off-axis beams symmetrically displaced with respect to the axis. The aberrations of other beams that lie in between the limiting scans are relatively small compared with the cases where the system is designed for only one focal point on axis. Bifocal systems have been employed in both reflector [16, 17] and dielectric lens antennas [18]. It is, however, much more complex to design the bifocal dielectric lens because the algebra is complicated. For simplicity, only the design procedures are summarized below.

As shown in Fig. 22, let $(0, a)$ and $(0, -a)$ be the conjugate focal points of the

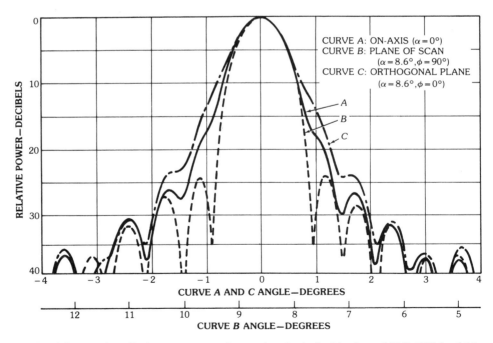

Fig. 21. Measured radiation patterns of zoned spherical thin lens (45.5 GHz). (*After Rotman [10]; reprinted with permission of Lincoln Laboratory, Massachusetts Institute of Technology, Lexington, Massachusetts*)

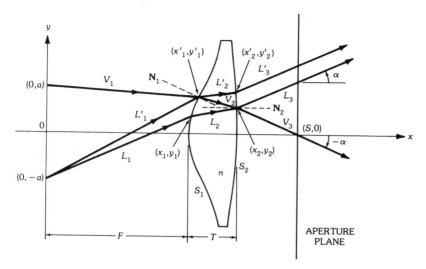

Fig. 22. Bifocal lens design.

lens for the two limiting scans at angles α and $-\alpha$ with respect to the boresight. The normal vector at (x_2, y_2) on S_2 can be denoted by

$$\mathbf{N}_2 = \hat{\mathbf{i}} + D_2 \hat{\mathbf{j}}$$

Then Snell's law requires that

$$n^2 \left[1 - \frac{[x_2 - x_1 + D_2(y_2 - y_1)]^2}{(1 + D_2^2)[(x_2 - x_1)^2 + (y_2 - y_1)^2]} \right] = 1 - \frac{(1 + D_2 \tan \alpha)^2}{(\sec^2 \alpha)(1 + D_2^2)} \quad (35)$$

which is derived from the angular relation between $\mathbf{L}_2$, $\mathbf{L}_3$, and $\mathbf{N}_2$ by taking the dot product of the vectors. From this equation D_2 can be solved as a function of (x_1, y_1) and (x_2, y_2), which are either given as initial conditions or previously determined in the step computations. Now the task is to compute (x'_1, y'_1) in terms of (x_1, y_1) and (x_2, y_2), where (x'_1, y'_1) is the point of incidence of the ray from the upper focal point which emerges at (x_2, y_2) with scan angle $-\alpha$. To this end, apply Snell's law at (x_2, y_2) for the ray with scan angle $-\alpha$, which leads to

$$n^2 \left[1 - \frac{[x_2 - x'_1 + D_2(y_2 - y'_1)]^2}{(1 + D_2^2)[(x_2 - x'_1)^2 + (y_2 - y'_1)^2]} \right] = 1 - \frac{(1 - D_2 \tan \alpha)^2}{(\sec^2 \alpha)(1 + D_2^2)} \quad (36)$$

This is defined by $\mathbf{V}_2$, $\mathbf{V}_3$, and $\mathbf{N}_2$. Next, impose the path length constraint for the ray along $\mathbf{V}_1$, $\mathbf{V}_2$, and $\mathbf{V}_3$,

$$[x'^2_1 + (y'_1 - a)^2]^{1/2} + n[(x_2 - x'_1)^2 + (y_2 - y'_1)^2]^{1/2}$$
$$+ (S - x_2) \sec \alpha + [y_2 - (S - x_2) \tan \alpha] \sin \alpha = K \quad (37)$$

where K is a constant determined by the central ray as a boundary condition. From (36) and (37) x'_1 and y'_1 can be solved as functions of (x_1, y_1) and (x_2, y_2). With (x'_1, y'_1) determined, surface 2 represented by (x'_2, y'_2) can be found by the following procedures.

First, note that the normal vector at (x'_1, y'_1), defined by

$$\mathbf{N}_1 = \hat{\mathbf{i}} + D_1 \hat{\mathbf{j}}$$

satisfies the relation

$$1 - \frac{[x'_1 + D_1(y'_1 - a)]^2}{(1 + D_1^2)[(x'_1)^2 + (y'_1 - a)^2]} = n^2 \left\{ 1 - \frac{[(x_2 - x'_1) + D_1(y_2 - y'_1)]^2}{(1 + D_1^2)[(x_2 - x'_1)^2 + (y_2 - y'_1)^2]} \right\} \tag{38}$$

This is derived from the refraction of V_1 and V_2. Thus D_1 is given in terms of (x'_1, y'_1) and (x_2, y_2). Similarly Snell's law for L'_1 and L'_2 leads to

$$1 - \frac{[x'_1 + D_1(y'_1 + a)]^2}{(1 + D_1^2)[x'_1{}^2 + (y'_1 + a)^2]} = n^2 \left\{ 1 - \frac{[(x'_2 - x'_1) + D_1(y'_2 - y'_1)]^2}{(1 + D_1^2)[(x'_2 - x'_1)^2 + (y'_2 - y'_1)^2]} \right\} \tag{39}$$

This is one of the two equations to be solved for x'_2 and y'_2. The other equation is obtained by imposing the path length constraint along $\mathbf{L}'_1$, $\mathbf{L}'_2$, and $\mathbf{L}'_3$, which is

$$[x'_1{}^2 + (y'_1 + a)^2]^{1/2} + n[(x'_2 - x'_1)^2 + (y'_2 - y'_1)^2]^{1/2}$$
$$+ (S - x'_2) \sec \alpha - [y'_2 + (S - x'_2) \tan \alpha] \sin \alpha = K \tag{40}$$

From (39) and (40) the two unknowns x'_2 and y'_2 can be uniquely determined. Repeating the process until the lens is completely shaped, a bifocal system is derived numerically. Clearly, the formulation is complex and extensive programming efforts are necessary; nevertheless, the design procedures are actually straightforward. Similar techniques have been applied to reflector systems with excellent results [16, 19].

As input parameters the focal points $(0, \pm a)$, the focal length F, the central thickness T, and the limiting scan angle α must be specified. The shaping starts with the central ray through P_1 and P_2 from the lower focal point. The exit point P_2 is given by $(F + T, b)$ where b is given by

$$\frac{a}{\sqrt{a^2 + F^2}} = n \frac{b}{\sqrt{b^2 + T^2}}$$

The step computations continue until a desirable aperture size is achieved. Due to the inherent symmetry of the lens in the principal plane, only one-half of the lens cross section is required in the computations. A complete lens is obtained by taking a figure of revolution of the contour so generated. This implies that a ring focus is obtained.

Lens Antennas

The shaping technique discussed in this section for dielectric lenses with bifocal points is different from those presented previously in that no step integration is involved and the step increments are relatively large. To completely define the surface points in between, a smoothing process of curve fitting is necessary. Due to the symmetry, only even power terms are needed. For most applications a fourth-order polynomial is sufficient. If, however, the geometry is such that the resultant step size is too large to warrant a smooth lens, this bifocal approach may not be acceptable. The other imperfection of this design is that there is a small amount of quadratic phase error in the orthogonal plane for any scan in the principal plane. This is due to the fact that the design is based on a two-dimensional analysis, whereas the actual lens is a figure of revolution of the contour generated.

6. Taper-Control Lenses

So far the treatment of lens design has been limited to the subjects of phase correction and wide-angle scanning. In addition to these lens there is another class of dielectric lens that is shaped to control the aperture taper for a special beam pattern and side lobe structure. The design principle for this type of lens is to explicitly impose the power conservation law in place of one of the constraints applied in the shaping formulation.

The significance of imposing the power law in the design can be illustrated by examining the aperture taper of a wide-angle lens constrained by the Abbe sine condition as discussed previously. Recall that when the Abbe sine condition is imposed on the lens design the aperture power distribution no longer can be independently specified. In this case the aperture power distribution is almost identical with that of the feed pattern. In other words, the lens performs no special transformation in the power distribution, a function that often is required in other applications. An examination of the power conservation law associated with the wide-angle lens verifies this last statement.

In differential form the power law requires that

$$f(\theta) \sin \theta \, d\theta = P(y) \, y \, dy$$

where $f(\theta)$ is the feed power pattern (in watts per unit solid angle) and $P(y)$ is the aperture power distribution (in watts per unit area). Using (24)

$$P(y) = \frac{f(\theta)}{F_e^2 \cos \theta}$$

For a feed angle (half-cone) less than 40°, the difference in edge taper between the feed pattern and the aperture distribution is less than 1 dB. Therefore, if taper control is desired, the power conservation law which specifies the aperture distribution must be imposed.

Referring to Fig. 23, consider a feed horn illuminating a dielectric lens with rotational symmetry. A reference plane for the aperture is taken at $x = S$. For nonuniform phase distribution, such as the one required for a special beam pattern [20], the lens shape is a function of the location of the reference plane, but the

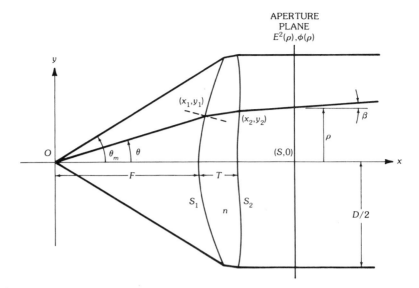

Fig. 23. Taper-control lens.

effect is inconsequential as long as the reference plane is not too far from the lens. Let the half-angle of the feed pattern subtended by the lens be θ_m. Then the power conservation law can be expressed as

$$\frac{\int_0^\theta g^2(\theta)\sin\theta\,d\theta}{\int_0^{\theta_m} g^2(\theta)\sin\theta\,d\theta} = \frac{\int_0^\varrho E^2(\varrho)\varrho\left[1-\left(\frac{1}{k}\frac{d\phi}{d\varrho}\right)^2\right]^{1/2}d\varrho}{\int_0^{D/2} E^2(\varrho)\varrho\left[1-\left(\frac{1}{k}\frac{d\phi}{d\varrho}\right)^2\right]^{1/2}d\varrho} \quad (41)$$

where $g^2(\theta)$ is the given feed pattern, watts per unit solid angle, and $E(\varrho)$ and $\phi(\varrho)$ are the specified aperture amplitude and phase distribution, respectively, all referred to the plane at $x = S$.

In (41), $k = 2\pi/\lambda$ is the wave number in free space, and the expression within the square brackets accounts for the cosine projection of the Poynting vector on the lens axis for a nonuniform wavefront. For the cases where a uniform phase is assumed, this factor reduces to unity. With $g(\theta)$, $E(\varrho)$, and $\phi(\varrho)$ given, a mapping relationship between θ and ϱ is established through (41).

The second condition to be imposed is Snell's law on surface 1, which was given in (2):

$$\frac{dy_1}{dx_1} = \frac{nL_1(x_2-x_1) - L_2 x_1}{L_2 y_1 - nL_1(y_2-y_1)} \quad (42)$$

The last constraint to be imposed is the path length condition specified by the phase distribution $\phi(\varrho)$, i.e.,

$$L_1 + nL_2 + [(S - x_2)^2 + (\varrho - y_2)^2]^{1/2} - \phi(\varrho)/k = K' \tag{43}$$

where $K' = S + (n - 1)T$.

To proceed, note that from the slope of the exit ray, y_2 is given by

$$y_2 = \varrho - (S - x_2)\tan\beta \tag{44}$$

The goal now is to solve (x_2, y_2) in terms of (x_1, y_1) and other related points on the same ray trajectory. This is accomplished by substituting (44) in (43). After a lengthy but straightforward algebraic manipulation one can show that [21]

$$x_2 = \frac{-B + (B^2 - 4AC)^{1/2}}{2A} \tag{45}$$

where

$$A = \frac{n^2 - 1}{n^2 \cos^2\beta}$$

$$B = 2\left(P\tan\beta - x_1 - \frac{Q}{n^2 \cos\beta}\right)$$

$$C = x_1^2 + P^2 - Q^2/n^2$$

and

$$\beta = \sin^{-1}\left(\frac{1}{k}\frac{d\phi(\varrho)}{d\varrho}\right)$$

$$P = \varrho - S\tan\beta - y_1$$

$$Q = K' - (x_1^2 + y_1^2)^{1/2} - S\sec\beta + \phi(\varrho)/k$$

To derive the contours of the lens the aperture radius is first divided into a large number of equal steps $\Delta\varrho$. The mapping then generates the corresponding delta increment $\Delta\theta$ for each $\Delta\varrho$. Using the central ray as an initial condition one can readily compute the slope at the starting point of surface 1. From (42), or by symmetry, the slope at the vertex is infinity. The intersection of this vertical tangent and the first emerging ray from the focal point with an angle $\Delta\theta$ is the first point of (x_1, y_1). The corresponding (x_2, y_2) on surface 2 can be found from (44) and (45). By repeating the same procedures until the whole aperture is covered in ϱ, a complete lens is shaped.

Geometrically (41) maps a cone of half-angle θ into a circle of radius ϱ on the aperture, a priori, without advance knowledge of the actual profiles of the lens. The design objective is to shape the lens to serve as an optical transformer, which transforms the given feed pattern into the desired aperture distribution.

In the step computation the process of generating the lookup table for ϱ and θ can be carried out in either direction. However, for those cases where the lens is shaped to converge the power density toward the axis for low side lobes, tracing backward to find $\Delta\theta$ with equal steps $\Delta\varrho$ is more convenient. The step size, $\Delta\varrho$, plays an important role in the accuracy of the numerical solution. Normally 200 points for a lens of 20λ in radius are sufficient. The accuracy is judged by the asymptotic behavior of the solution after iterations and how close Snell's law is satisfied on surface 2 as a check. For very rigorous solutions, of course, a higher-order approximation, such as the Runge-Kutta method, may be employed.

For most applications the specified aperture phase distribution is uniform. In this case the preceding formulation can be much simplified, because $\phi(\varrho)$ and β are set to zero. Also it should be remarked that for the computer program the feed pattern and the aperture distribution need not be given in closed analytic form. The (ϱ, θ) mapping can be generated with discrete input data points.

Based on the formulation just described, a general program can be developed. The inputs to the program are listed in the following:

feed location from the lens	F
feed power pattern (assumed axisymmetric)	$g^2(\theta)$
maximum feed angle	θ_m
aperture amplitude distribution	$E(\varrho)$
aperture phase distribution	$\phi(\varrho)$
aperture size	D
dielectric constant	ϵ_r
central thickness	T
reference plane for aperture	S

The outputs of the program specify the coordinates of the lens profiles for both sides, (x_1, y_1) and (x_2, y_2). The design procedure starts with a proper choice of the focal length F for the lens. An initial estimate of F can be made by finding the location of the lens at which the cross section of the feed cone is about the size of the aperture. For a given dielectric constant ϵ_r, the central thickness T must be large enough to warrant a complete solution. Too small a T sometimes does not give enough dielectric medium at the outer rim of the lens for the rays to converge and perform the power transformation. Some optimization efforts are required in the design.

Since no restriction on the shape of the aperture taper has been imposed in the formulation, any reasonable power distribution with rotational symmetry can be specified to suit a particular application. It may be a standard $(1 - \varrho^2)^P$ taper; or a special function, such as a truncated Bessel type distribution discussed by Love [22], where the antenna pattern has a dip at the center for better earth coverage from a geostationary satellite orbit. For satellite multibeam antennas it may be desirable for each beam to have a very broad beamwidth with a fast rolloff and low side lobes, so that a more uniform earth coverage can be achieved. To produce such shaped beams the generalized Taylor distributions with complex zeros [20, 23] are more appropriate.

As an example, Fig. 24 shows a lens shaped for an aperture distribution $E(\varrho) = [1 - (\varrho/1.05)^2]^3$ with a uniform phase designed for -35-dB side lobes. The feed pattern is a standard E-plane pattern of a square horn. The horn size was chosen to have an edge taper of 20 dB at $\theta_m = 20°$. The dielectric constant is 2.54 for Rexolite

Lens Antennas

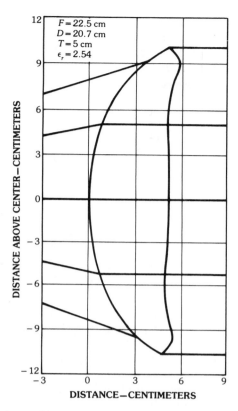

Fig. 24. A Rexolite lens with an aperture amplitude distribution of $[1 - (\varrho/1.05)^2]^3$.

material. The central thickness is 5 cm and the focal length is 22.5 cm. The aperture efficiency of this lens is 48 percent. The computed far-field pattern has a 3.2° beamwidth with a directivity of 36.4 dB for $D = 30.3\lambda$ (20.7 cm) at 44 GHz.

The transformation between the feed pattern and the aperture distribution is primarily accomplished by the first surface through refraction. The second surface is contoured mainly to satisfy the phase constraint. In general the solution of the lens is sensitive to the choice of F and T. This is usually true when the aperture distribution differs substantially from the feed pattern. When the required transformation is not too drastic, solutions exist over a wider range of F and T.

It should be noted that many lens solutions exist for the same power transformation if different input parameters, such as focal lengths, central thickness, etc., are specified. For example, a thinner lens can perform the same power transformation if a material of higher dielectric constant is used for the lens. Naturally, as the dielectric constant increases and the lens becomes thinner, surface mismatch becomes a limiting factor in choosing a very high dielectric constant. The F/D ratio is another factor that can be traded off with other design parameters. For a given D, a smaller F will make the subtended feed angle θ_m larger. If the horn size remains fixed, a larger θ_m means smaller spillover. In this case the lens will be less concave, since the illumination taper more closely matches the aperture taper, thus requiring less bending of the rays by the lens to achieve the power transformation. It should be noted that the actual lens profiles depend more on the shape of the feed pattern than just the taper level at the edge. When the input and output taper match closely, the lens is more or less flat.

7. Dielectric Lens Zoning

One of the inherent drawbacks of dielectric lenses is the physical volume and weight which may pose problems for certain applications. The penalty in weight for having a thick lens can be somewhat relaxed by classical zoning. That is, the lens can be designed to be very thin at the center, but each new zone is allowed to increase the thickness by $\Delta T = \lambda/(\sqrt{\epsilon_r} - 1)^{1/2}$, such that the overall phase difference is 2π.

Fig. 19 is an example in which the central thickness is kept to a minimum. Another example is shown in Fig. 25, where the lens accomplishes the same function as the one shown in Fig. 14, which is constrained by the Abbe sine condition. In this case the zoning is implemented on the outside of the lens, and the central thickness has been reduced to 3.5 cm, while the focal length F has been adjusted to 26.7 cm to maintain a relatively flat contour for the lens.

Reducing the physical weight of the lens is not the only merit of zoning. Another unique feature provided by zoning is to reduce coma aberrations in multibeam lens antennas. For satellite communications a lens antenna can be shaped to control the side lobe level and zoned to minimize the cubic phase errors for off-axis beams [21].

Based on the concept that if the average contour of a collimating lens is spherical the Abbe sine condition is approximately satisfied [24, 25], a special zoning technique for coma correction has been developed. This is accomplished by making the inner surface of the lens follow a circular arc on average. As shown in Fig. 26, when the inner surface of the shaped lens deviates from the circle by more than, in this case, 1λ, the lens is zoned to have the surface moved back by 2λ along

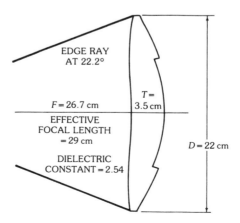

Fig. 25. Cross section of a coma-free zoned lens.

Lens Antennas

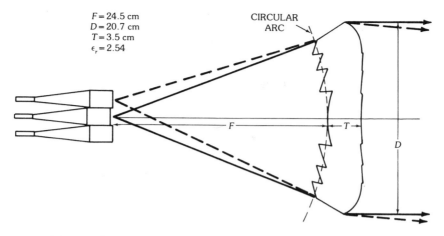

Fig. 26. Coma-corrected zoned lens for multibeam applications.

the ray path. Note that the zoning does not have to be 2λ per step. It can be any step size as long as the overall path length at the aperture satisfies the phase constraint. A smaller step, however, will result in a larger number of zones to conform to the spherical shape, which will cause more edge diffractions for off-axis beams and reduce the bandwidth to some extent.

The absolute phase, of course, can differ by any integral number of 2λ across each zone. In fact, to keep the second surface reasonably smooth in this example an additional wavelength is added to the total path length for each new zone to offset the effect of zoning on surface 1.

Shown in Fig. 26 is a zoned lens of Rexolite designed for the same aperture distribution and feed pattern as discussed in Fig. 24. The focal length is 24.5 cm with $T = 3.5$ cm and $D = 20.7$ cm.

It should be remarked that the zoning process does not significantly perturb the aperture distribution or the illumination function compared with the original design. The constraint of power conservation is unchanged; the additional step required in the shaping program is to give new boundary conditions as inputs whenever the inner surface deviates from the reference circle by a preset value. Effectively the lens is made up of zones of different lenses with different initial central thicknesses and focal lengths, but each zone is constrained to produce the corresponding portion of the same specified aperture distribution. Naturally, over an infinitesimal interval across the junction between zones the ray trajectory is not defined due to the discontinuity.

Since the edge of each zone is cut along the incident ray direction, very little scattering is expected for the beam on axis. For off-axis scans, however, a small amount of scattering and shadowing effect due to the steps occurs. The zoning technique presented here is similar to that proposed by Ronchi and Toraldo Di Francia [26] to correct the coma of a reflector, which was subsequently demonstrated by Provencher [27]. In 1961 Dasgupta and Lo presented a more detailed analysis on the same subject based on the diffraction theory [28]. Although this analysis does not apply to the case of lenses the approach is very similar.

Shown in Fig. 27 are measured antenna patterns of the zoned lens at 44 GHz. It can be seen that the lens collimates very well to form high-quality beams despite the zoning for coma reduction. The first side lobe level of the central beam is −30 dB, not as low as designed. This may be attributed to the surface mismatch and finite amount of scattering due to the zoning. For off-axis scans, coma distortion of the

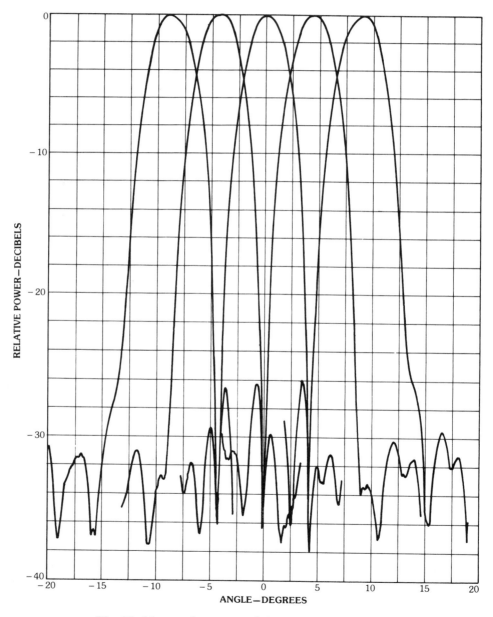

Fig. 27. Measured patterns of the coma-corrected lens.

main beam is negligible and the side lobe degradation is small. Thus it has been demonstrated that zoning and shaping can be combined to design a low side lobe, coma-corrected lens antenna.

8. Constrained Lenses

A constrained lens usually consists of an equal number of receiving and radiating elements on each surface of the lens, with corresponding elements connected by a transmission line for phase adjustment on a one-to-one basis. The lens transforms a divergent wavefront from a focal point into a collimated beam by providing the correct line lengths to compensate for the path differences. The term "transmission line" is used in a very broad sense here. Actually it can be any guiding structure such as waveguide, microstrip, coaxial line, etc., that serves as a phase or, ideally, time-delay medium.

The design principle of a constrained lens is quite simple. The only condition imposed in the design is the fundamental equal path length condition. As illustrated in Fig. 28, for a two-dimensional case this condition can be expressed as

$$[(x_1 - a)^2 + (y_1 - b)^2]^{1/2} + w + (T - x_2)\cos\alpha + y_2\sin\alpha = F + w_0 \qquad (46)$$

in which there are five variables, (x_1, y_1), (x_2, y_2), and w, that can be constrained. To uniquely determine the contour of either surface of a rotationally symmetric lens three more conditions must be specified. This freedom of choice in selecting additional constraints provides numerous options in design. For example, one of the two surfaces may be chosen to be planar or spherical. For some cases multiple focal points may be desired to enhance the scanning capabilities. Also, the question may be asked as to whether equal line lengths or unequal line lengths are to be used. Different considerations lead to different lens designs.

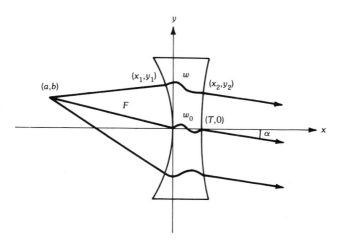

Fig. 28. Constrained lens design.

Waveguide Lens and Zoning

The use of waveguide lenses for microwave antenna applications has been reported by various authors [29, 30]. The waveguide lens is a special case of a constrained lens where straight sections of open-end waveguides are used as radiating elements. The radial positions of the elements on one side are identical with that of the corresponding ones on the opposite side of the lens. The simplest design is the one that has rotational symmetry with a single focal point on axis. The inner surface is a sphere of revolution of radius F centered at $(-F, 0)$, as shown in Fig. 29, while the outer surface is an ellipsoid centered at $[nF/(1-n), 0]$ given by

$$\frac{y_2^2}{F^2} + \left[\frac{x_2 - T - Fn/(1-n)}{Fn/(1-n)}\right]^2 = 1 \qquad (47)$$

where T is the central thickness and n is the index of refraction defined by λ/λ_g, with λ_g being the waveguide wavelength [31].

A two-dimensional waveguide lens exhibiting two focal points may also be designed. Following the treatment of Dion and Ricardi [32], let $(F\sin\alpha, -F\cos\alpha)$ be one of the two conjugate focal points of the lens in the principal plane, where F is the focal length and α the scan angle. Then the cross-section profile of the inner surface is

$$\frac{y_1^2}{F^2} + \frac{(x_1 + F\cos\alpha)^2}{F^2 \cos^2\alpha} = 1 \qquad (48)$$

i.e., an ellipse with foci at the two point sources, while the equation of the outer surface is given by

$$\frac{y_2^2}{F^2} + \left[\frac{x_2 - T - (nF\cos\alpha)/(\cos\alpha - n)}{(nF\cos\alpha)/(\cos\alpha - n)}\right]^2 = 1 \qquad (49)$$

which is also an ellipsoid.

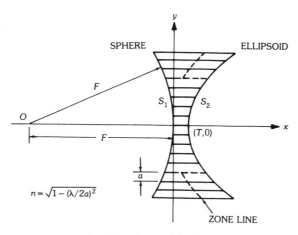

Fig. 29. Waveguide lens.

Lens Antennas

The thickness of the lens depends on the index of refraction; consequently the inside width of each waveguide determines the number of waveguides required for a given aperture size and the thickness of the lens, because n is a function of λ_g. For circular polarization, using square waveguides, the waveguide dimension a must be large enough to permit propagation of the dominant TE_{10} and TE_{01} modes and sufficiently small to suppress the higher-order modes. The latter requires that $a < 0.7\lambda$ and determines a maximum value of $n = 0.7$; usual values for n lie between 0.58 and 0.68. The lower limit is determined by the need to prevent an intolerable reflection loss at the surfaces. The F/D ratio determines the number of zones in the lens; increasing F/D decreases the number of zones, resulting in an increase in antenna efficiency.

In general it is necessary to zone the lens to reduce its thickness and to increase its bandwidth. Zoning a waveguide lens gives increased bandwidth because it reduces the dispersion in the waveguide. The discontinuity in the step is an integral multiple of $\lambda/(1 - n)$ so that the overall path length is increased by one or more wavelengths. The lens may be stepped on either or both of its surfaces. However, steps on the surface opposite the feed appear preferable in order to reduce shadowing effects.

As described in [33] by Dion, a number of zoning techniques can be applied to increase the bandwidth of a waveguide lens. One way to achieve this goal is to zone the lens for minimum thickness. In this design, as illustrated in Fig. 30, the location of the zones is obtained by making d equal to $d_0 + md_\lambda$, where $d_\lambda = \lambda/(1 - n)$, yielding $z_m = m\lambda$. At the start of each zone the waveguide length is the same as that of the central element, i.e., $d = d_0$ at $z = m\lambda$.

The phase error of a zoned lens across the aperture is given by

$$\Delta\phi = \frac{2\pi z}{\lambda} \frac{1 + n}{n} \left[1 - \frac{m\lambda}{(1 + n)z}\right] \frac{\Delta f}{f} \tag{50}$$

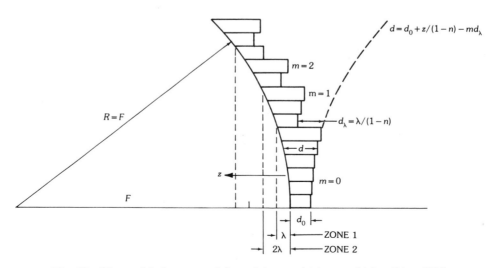

Fig. 30. Waveguide lens zoned for minimum thickness. (*After Dion [33]*)

where z is related to the radial distance r by $z = F - (F^2 - r^2)^{1/2}$, and $z \cong r^2/2F$ for large F/D. For the case of minimum-thickness lens the normalized phase error is a sawtooth curve superimposed with a linear slope, as plotted in Fig. 31 for $n = 0.6$.

On the other hand, a waveguide lens can be zoned to achieve minimum phase error at the expense of increased thickness at the center, as first reported by Colborn [34]. The basic concept is to ensure equal time delay at discrete points in addition to the constraint of equal phase delay for all rays. These discrete points correspond to the step locations of new zones. They can be found by first requiring that the time delay, based on group velocity, of a general ray from the focal point to an aperture plane is equal to that of the central ray. This condition leads to

$$(1 - n)(d - d_0) + nz = 0 \tag{51}$$

Next impose the equal phase constraint for a zoned lens to obtain

$$d = d_0 + \frac{z - m\lambda}{1 - n} \tag{52}$$

From (51) and (52) the step locations can be determined to be

$$z_m = \frac{m\lambda}{1 + n}, \quad m = 1, 2, \ldots \tag{53}$$

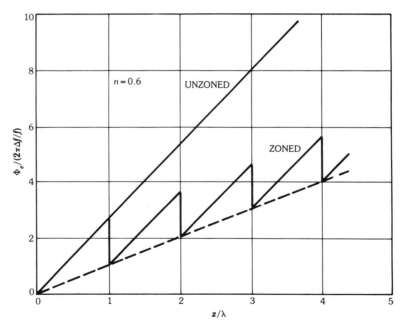

Fig. 31. Aperture phase error of the unzoned waveguide and minimum-thickness lens. (*After Dion [33]*)

Lens Antennas

This happens to be the condition that will reduce the phase error of the zoned lens to zero at the step locations, as can be proved from (50). A minimum phase error lens is illustrated in Fig. 32, and the resultant phase error distribution is shown in Fig. 33.

Equal Group Delay Lens

Waveguide lenses discussed so far are basically conventional designs that are based on the concept of equal phase delay. These lenses are intrinsically narrow-banded, because there is a large difference in time delay between the central and edge rays. Recently, Ajioka and Ramsey [35] developed an equal group delay waveguide lens which improves the bandwidth significantly. In this design all rays from the focal point to the aperture plane have equal time delay at the center frequency. The equality of time delay does not ensure equality of phase, however. The phase is then made equal by proper adjustment of a half-wave-plate phase

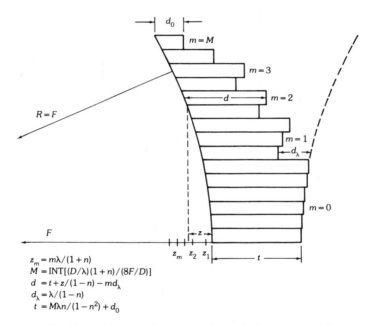

Fig. 32. Minimum phase error lens. (*After Dion [33]*)

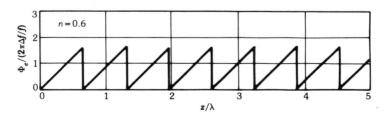

Fig. 33. Aperture phase error of the minimum phase error lens. (*After Dion [33]*)

shifter in each element. The half-wave plates add a constant length to all elements producing no phase error.

The design results in an aperture phase distribution which remains essentially constant over a much greater bandwidth than in the other lenses. The bandwidth of this group delay lens is only limited by the impedance mismatch between the lens radiating elements and free space, a limitation shared by all lenses, and the bandwidth of the half-wave plates.

As shown in Fig. 34 this lens consists of an array of waveguide elements that are uniformly spaced with various lengths. Each waveguide element contains a half-wave-plate phase shifter which orients in such a way that the phase of a circularly polarized wave incident on it is shifted by the proper amount to produce the correct phase at the aperture. The inner surface of the lens is spherical, for the benefit of the Abbe sine condition which ensures wide-angle performance as discussed previously. The outer lens surface is determined by the constraint of equal group delay for all rays. It can be shown that the outer surface is an ellipsoid.

The principle of the equal group delay lens also explains clearly why zoning enhances the bandwidth of a conventional waveguide lens. Since the index of refraction of a waveguide is less than unity the outer surface of the lens is concave in contour, with the longest waveguide element at the edge of the lens where the group delay is inherently longest. Zoning the lens diminishes the difference in time delay among the various rays and hence improves the bandwidth.

The disadvantages of the group delay lens are added complexity of the phase shifters and the inherent drawback of being in only one sense circularly polarized [35]. There will be 3-dB loss if the incident field is linearly polarized.

Multifocal Bootlace Lens

Using transmission lines for connecting the elements on the two surfaces of the lens, several bootlace designs have been described. Rotman and Turner have

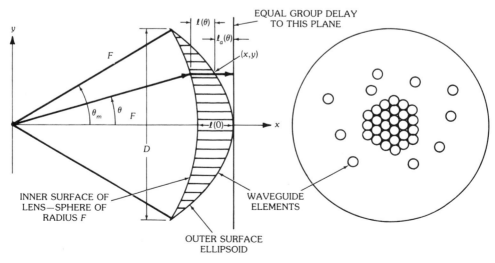

Fig. 34. Equal group delay lens. (*After Ajioka and Ramsey [35]*, © *1978 IEEE*)

derived a trifocal lens with a circular focal arc and a straight outer edge for a linear array [36]. Cornbleet also presents design details of quadrifocal lenses in [37]. Recently Rao [38] has investigated several bootlace lenses with simpler geometry as described in the following. As shown in Fig. 35 the bootlace lens in this case has a planar outer surface with focal points located on a vertical line. The shape of its inner surface is dependent on the condition of whether equal line length or unequal line length is imposed. If equal line length is used, a bifocal design can be deduced with the inner surface specified by

$$y_1^2 \cos^2\alpha + (x_1 + F_0)^2 = F_0^2 \tag{54}$$

and $y_1 = y_2$, $w = w_0$.

If the condition of unequal line length is selected, a trifocal system with the third focal point located on axis is available, where the inner surface is given by

$$y_1^2 + (x_1 + F_0)^2 = (F_0 - w + w_0)^2 \tag{55}$$

with

$$y_1 = y_2[1 - (w - w_0)(\cos\alpha)/F_0] \tag{56}$$

and

$$w - w_0 = (y_2^2/F_0)(\cos\alpha)\cos^2(\alpha/2) \tag{57}$$

Rao has also studied the phase error across the aperture when the feed is displaced from a focal point to scan the beam to other angles. With the feed confined to a straight line passing through the focal points, the maximum path length error versus scan angle for three multifocal systems plus the single focus case, with $F/D = 1$, is plotted in Fig. 36. It may be noted that, for a maximum path

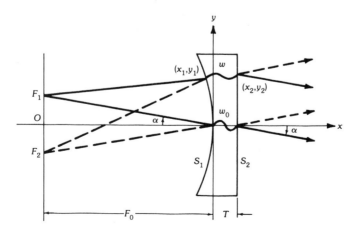

Fig. 35. Multifocal bootlace lens.

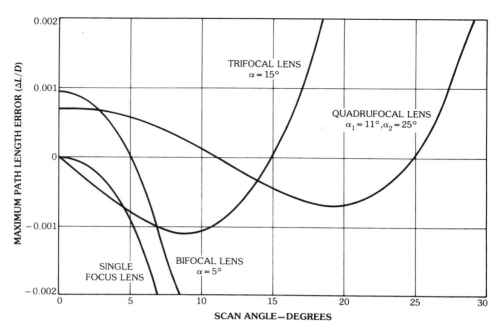

Fig. 36. Path length error versus scan angle for $F/D = 1$. (*After Rao [38]*, © *1979 IEEE*)

length error less than $0.001D$, a bifocal lens designed for $\alpha = 5°$ can be scanned up to $\pm 7°$, whereas a trifocal lens with $\alpha = 15°$ can be scanned up to $\pm 16.5°$. The comparison also suggests that if extremely wide scanning is required, a quadrifocal lens discussed by Cornbleet [37] is a better choice, since this lens minimizes the phase error over a much wider range in scan.

R-2R Lens

Another interesting design which is characterized by perfect focusing at any scan angle is commonly termed an *R-2R lens* [3]. This two-dimensional lens is constrained to have a variable focal length with the focal point following a circular path as the beam scans. The cross section of the inner surface is a circle, which coincides with the focal arc, as illustrated in Fig. 37. The outer surface is also a circle whose radius is twice that of the inner one, thus the appellation *R-2R*.

Another attribute of this lens is the fact that the relative positions of each element on the inner and outer surfaces can be determined by a very simple graphical method. Let an outgoing ray from the focal point on axis intercept the inner surface at P_1 and, when extended, the outer surface at P_2. If P_1 is the location of a receiving element on the inner surface, then P_2 is precisely the location of the corresponding radiating element on the outer one. The R-2R lens is characterized by using equal line lengths for the elements. Mathematically, the lens is described by the following conditions:

$$y_2^2 = F_0^2 - x_2^2 \tag{58}$$

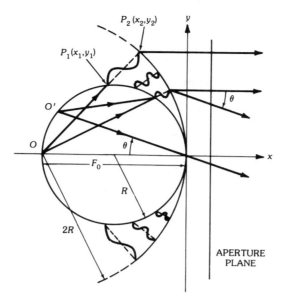

Fig. 37. R-$2R$ lens.

$$x_1 = F_0 - y_2^2/F_0 \tag{59}$$

$$y_1^2 = y_2^2 - y_2^4/F_0^2 \tag{60}$$

and $F = F_0 \cos\theta$.

In principle, there is no aberration with this lens at any scan angle. However, scanning is restricted to 60° in practice due to the fact that the effective aperture, and hence the peak gain, will drop significantly beyond that limit.

Constrained Analog of Dielectric Lens [39]

It is interesting to note that for any dielectric lens there is an equivalent constrained lens system that performs the same optical transformation as the dielectric lens. The converse, however, is not necessarily true. In other words, a dielectric lens can be replaced by a constrained lens system operating in free space without the dielectric medium, but a constrained lens cannot be substituted by a dielectric lens in general. Shown in Fig. 38 is a dielectric lens and its constrained analog that consists of two constrained lenses separated by a region of free space. The two lenses are identified as lens 1 and lens 2. Lens 1 is formed by two surfaces that are different from each other only by a scale factor. On the left the surface S_1 is identical with the inner surface of the dielectric lens. On the right the surface S_1' is an exact replica of S_1 except expanded by a factor n, where n is the refractive index of the dielectric lens. The lens has an equal number of radiating elements on each surface, with corresponding elements connected by equal line lengths on a one-to-one basis. The element spacing on S_1' is n times that on S_1 in all directions.

For lens 2 the above process is reversed. The surface on the right, S_2, is

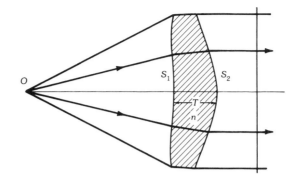

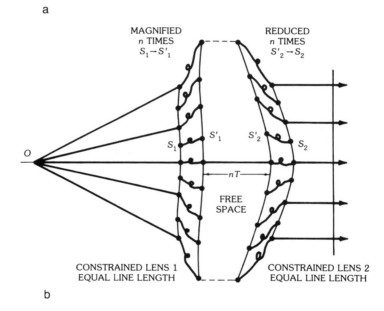

Fig. 38. Constrained analog of dielectric lens. (*a*) Dielectric lens. (*b*) Constrained analog of (*a*).

identical with the outer surface of the dielectric lens, whereas the surface on the left, S'_2, is an expanded version of S_2 with a scale factor of n. Similarly, equal line lengths are used to connect the corresponding elements on lens 2. The separation between S'_1 and S'_2 is n times the central thickness of the dielectric lens. Thus the region of free space confined by S'_1 and S'_2 has the same proportion as the dielectric lens in all directions.

To show that the constrained system accomplishes the same function as the dielectric lens the phase relationship illustrated in Fig. 39 is examined. Assume that the incidence angle of a ray at S_1 with respect to the local normal vector is θ_i and the exit angle at S'_1 is θ_r. Then the differential phase delay between two adjacent elements at S_1 is $kd \sin \theta_i$, and the corresponding phase difference at S'_1 is $nkd \sin \theta_r$.

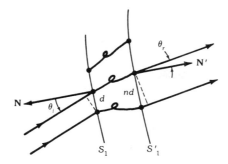

Fig. 39. Refraction through a constrained lens.

Since equal line lengths are used the phase distribution is preserved from S_1 to S_1'. This implies that

$$kd \sin \theta_i = nkd \sin \theta_r$$

or

$$\theta_r = \sin^{-1}\left(\frac{\sin \theta_i}{n}\right)$$

It can be recognized that this is Snell's law of refraction. Since S_1 and S_1' share the same form of surface contour, the normal vectors at the corresponding points on S_1 and S_2' are parallel to each other. Thus the ray refracts in the same way as it would across the inner surface of the dielectric lens. Likewise, the same kind of refraction takes place from S_2' to S_2. Furthermore, the time delay of each ray in the free space between S_1' and S_2' will be the same as in the dielectric lens, because the region is n times larger than the dielectric lens. It is evident that this constrained lens system behaves the same as the dielectric lens, and therefore the analogy between them is proved.

With this one-to-one correspondence established, a dielectric lens can be realized by a constrained lens, thereby eliminating such drawbacks of the dielectric lens as physical weight and complexity in machining. Moreover, all the design formulas presented in the sections on dielectric lenses can be applied to develop constrained lenses to meet special requirements.

9. Inhomogeneous Lenses

So far the refractive index of the lenses discussed has been assumed to be uniform. If, however, the lens medium is allowed to be inhomogeneous—as a function of radius, for example—a new class of lenses can be designed. These lenses have certain unique features that cannot be achieved with uniform lenses. The Luneburg lens and Maxwell fish-eye lens are two classical examples of such lenses.

Luneburg Lens

The Luneburg lens is a spherical lens with

$$n(r) = [2 - (r/a)^2]^{1/2} \tag{61}$$

It can be shown that the equation of the ray path which leaves the feed at an angle α to the lens axis is [3]

$$r^2[\sin^2\theta + \sin^2(\theta - \alpha)] = a^2 \sin^2\alpha \tag{62}$$

and all rays leaving a point P_1 on the surface emerge at the opposite end as a parallel beam, as shown in Fig. 40. Thus scanning is accomplished simply by moving the feed point around the surface. Due to the spherical symmetry the scanning performance is independent of the beam direction. The power distribution across the aperture is related to the feed power pattern by $P(y) = P(\theta) \sec\theta$, where $y = a\sin\theta$.

The Luneburg lens is difficult to construct due to the fact that its refractive index must vary continuously from $\sqrt{2}$ at the center to unity at the surface. However, satisfactory performance can be obtained by using a number of spherical shells of uniform dielectric constant to approximate a continuous gradient. Another drawback of this lens is the special feed required. It is difficult to design a useful feed which has a phase center that can be placed directly on the surface of the lens. This problem can be solved by moving the phase center away from the lens and requiring the lens to have a modified refractive index distribution. This has been investigated by Wolff [40] and Elliott [41] where numerical solutions of $n(r)$ are derived.

Maxwell Fish-Eye Lens

The Maxwell fish-eye lens is also a spherical lens which is characterized by

$$n(r) = \frac{n_0}{1 + (r/a)^2} \tag{63}$$

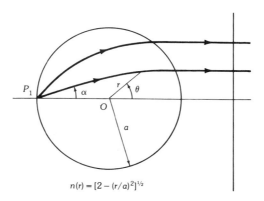

$n(r) = [2 - (r/a)^2]^{1/2}$

Fig. 40. Luneburg lens.

In this case, rays leaving point P_1 converge at P_2 diametrically opposite the point P_1, as shown in Fig. 41. The ray trajectory is given by

$$\sin\theta = \frac{C}{\sqrt{a^2 n_0^2 - 4C^2}} \frac{a^2 - r^2}{ar} \tag{64}$$

where C is a parametric constant determined by the product of $n(r)$ and r at Q, with the ray path normal to the line OQ. The ray paths are segments of circular arcs which form figures shaped like the eye of a fish.

To form a collimated beam the sphere in Fig. 41 has to be cut in half along OQ so that the lens becomes a hemisphere. This, however, may have a drastic effect on its scanning properties.

Hyperbolic Cosine Lens

A rectangular slab of dielectric material with index of refraction given by

$$n(y) = n_0 \operatorname{sech}(\pi y/2F) \tag{65}$$

also exhibits focusing effect. All rays leaving O, as in Fig. 42, emerge parallel to the x axis. If a line source at O radiates uniformly into each angular sector $d\theta$, the field intensity along the front surface follows the distribution $\cosh(\pi y/2F)$. The scanning performance of this lens is not known at this time.

Recently Brown has reviewed the subject and technology of gradient index optics [42]. A survey is given on the types of gradient available and the manufacturing processes being developed. When production techniques for inhomogeneous dielectrics become mature enough and cost effective, various inhomogeneous lenses such as those just discussed can be developed for special applications.

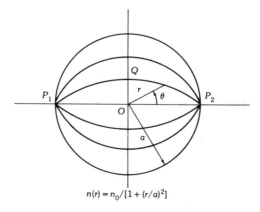

Fig. 41. Maxwell fish-eye lens.

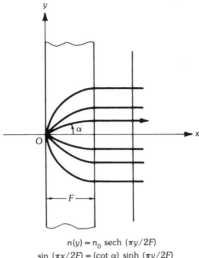

$n(y) = n_0 \text{ sech } (\pi y/2F)$
$\sin (\pi x/2F) = (\cot \alpha) \sinh (\pi y/2F)$

Fig. 42. Lens with hyperbolic cosine distribution.

10. Bandwidth Limitation and Surface Mismatch

A dielectric lens is a time-delay optical device which has virtually unlimited bandwidth for microwave frequencies if no zoning is introduced. However, the bandwidth of the dielectric lens is restricted once zoning is employed. Since zoning is implemented by reducing the lens thickness in steps by the amount of $\lambda/(n-1)$ with respect to the center frequency, error in the path length occurs as the frequency changes. If a total path length error of one-eighth wavelength from center to edge is tolerated, the bandwidth of a dielectric lens with M zones is given by [41]

$$B = \frac{25}{M-1} \text{ percent} \qquad (66)$$

It is clear that the more zones needed for a dielectric lens, the more narrowband the lens becomes. This is not true for a waveguide lens, which will be discussed later.

Mismatch in surface impedance is another problem that must be dealt with in a dielectric lens. As is well known in plane wave propagation, the vswr for a normally incident wave at the interface of free space and a semi-infinite dielectric medium is $\sqrt{\epsilon_r}$, where ϵ_r is the dielectric constant of the medium. Although the exact calculation of power reflected by a lens is complicated because of the lens curvature, a rough estimate may easily be found by calculating the mismatch loss corresponding to the normal incidence vswr. This mismatch loss is simply given in decibels by

$$\begin{aligned}\text{loss} &= -10\log(1-\Gamma^2) = -10\log\{1-[(\sqrt{\epsilon_r}-1)/(\sqrt{\epsilon_r}+1)]^2\} \\ &= 10\log[(\sqrt{\epsilon_r}+1)^2/4\sqrt{\epsilon_r}]\end{aligned}$$

Lens Antennas

It is therefore good practice to match the lens surface if the dielectric constant is large. A standard technique is to use a quarter-wave layer of intermediate refractive index. Other techniques involve cutting grooves or grids on the surface to simulate a region of intermediate dielectric constant. A detailed treatment on this subject has been given by Cohn in [4]. At high frequencies such as K_u-band or above, however, it becomes extremely difficult to implement these matching layers due to practical engineering problems. Thus a performance margin must be allowed to account for the gain drop and side lobe degradation in the design of a dielectric lens at high frequencies.

For constrained lenses the bandwidth is determined by many factors. The properties of the radiating elements, the connecting transmission lines, the array geometry, and the mutual coupling effects all play an important role in the bandwidth calculation. Many constrained lenses make use of phase shifting elements; therefore their frequency response affects the bandwidth as well. As a matter of fact, the criterion of bandwidth for a given lens must be carefully defined in terms of array performance. The definition of bandwidth from different points of view has been examined and summarized in [43] by Frank, in which the effects of aperture, feed, fill-up time, etc., on the bandwidth of a phased array are also discussed.

The bandwidth of a metal plate or waveguide lens is limited due to its dispersive characteristics. As discussed in [1], the bandwidth of an unzoned lens is

$$\text{bw} \cong 8.3 \frac{\lambda_0}{(1 - n_0) T} \text{ percent} \tag{67}$$

where λ_0 and n_0 are respectively the wavelength and refractive index of the lens at center frequency, and T is the difference between the thickness of the lens at the edge and at the center. Usually the bandwidth is only a few percent of the operating frequency. Unlike dielectric lenses, zoning actually improves the bandwidth of a waveguide lens because it reduces the dispersion of the waveguide. According to Risser in [1], the bandwidth of a zoned lens is

$$\text{bw} \cong 25 \frac{n_0}{1 + Kn_0} \text{ percent} \tag{68}$$

where K is the number of zones introduced. A comparison of equivalent zoned and unzoned lenses indicates that the bandwidth of a zoned lens is two to three times better than that of an unzoned one.

For a waveguide lens zoned for minimum thickness as discussed in [33], the bandwidth of the lens with a phase error less than $\pm\lambda/16$ is

$$\text{bw} \cong \frac{200n(F/D)}{(1 + n)(D/\lambda)} \frac{1}{1 - 8K(F/D)/[(1 + n)(D/\lambda)]} \text{ percent}$$

where K is the number of zones. If the lens is zoned for minimum phase error, the bandwidth is approximately equal to $25n$ percent.

Matching the surface impedance of a constrained lens is probably the most difficult part of the lens design. The inner surface facing the feed should be spherical, if possible, primarily for better impedance match due to normal incidence from a point source. A spherical surface also offers the desirable wide-scan features according to the Abbe sine condition. If planar surface is considered, the F/D ratio should be chosen as large as practical.

For wide-band and wide-scan applications internal matching of each element may be necessary. For waveguide elements this may involve impedance matching with use of dielectric plugs, stub chokes, and reactive irises at the apertures of the radiators. A dielectric sheet placed in front of the aperture is also one of the most common techniques used to achieve wide-angle impedance matching. The matching process can be best described by the equivalent circuit shown in Fig. 43. The dielectric sheet acts as an impedance transformer whose characteristic impedance and propagation constant change with scan [44]. For a thin sheet it behaves approximately as a variable shunt susceptance. By optimizing the dielec-

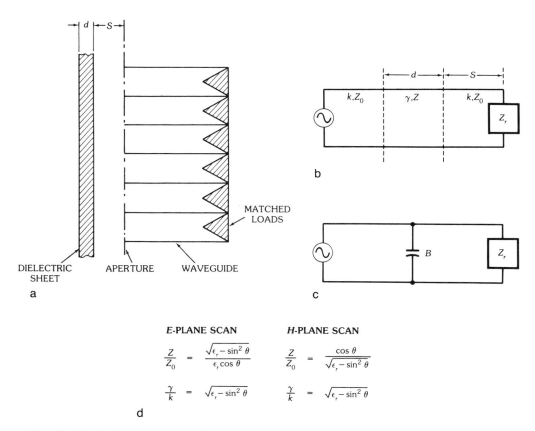

Fig. 43. Equivalent circuit of dielectric matching sheet. (a) Dielectric matching sheet. (b) Equivalent circuit. (c) Simplified equivalent circuit for thin sheet. (d) Mathematical relationships.

tric constant of the sheet and its thickness and distance from the elements through iterations, improved performance in matching can be obtained [45, 46].

11. Summary

Lenses have found many applications not only in optics but also in microwave antenna systems. The subject of lens design has been treated by numerous authors, and this field is relatively familiar to many antenna designers. In this chapter, only a very brief review has been offered due to space limitations. Care has been taken not to reproduce the design details that can be readily found in the existing reference literature. Emphasis has been placed on the more recently developed systems. For those who would like to explore more details of a particular design, the references listed below should be consulted.

Acknowledgments

Dr. Allan Love and Leni Parad's enlightening comments and corrections in proofreading this chapter are highly appreciated. Ms. Lynda Schenet's excellent typing of the manuscript is also gratefully acknowledged. Her patience and efficiency made writing this chapter a very pleasant task.

12. References

[1] S. Silver, ed., *Microwave Antenna Theory and Design*, chapter 11, New York: McGraw-Hill Book Co., 1949.
[2] J. Brown, *Microwave Lenses*, London: Methuen & Co., 1953.
[3] R. E. Collin and F. J. Zucker, eds., *Antenna Theory*, chapter 18, New York: McGraw-Hill Book Co., 1969.
[4] S. B. Cohn, "Lens-type radiators," chapter 14 in *Antenna Engineering Handbook*, ed. by H. Jasik, New York: McGraw-Hill Book Co., 1961.
[5] D. L. Sengupta and R. E. Hiatt, "Reflectors and lenses," chapter 10 in *Radar Handbook*, ed. by M. Skolnik, New York: McGraw-Hill Book Co., 1970.
[6] M. Born and E. Wolf, *Principles of Optics*, 3rd ed., New York: Pergamon Press, 1965, pp. 211–218 and pp. 468–484.
[7] S. Silver, ed., *Microwave Antenna Theory and Design*, section 6.7, New York: McGraw-Hill Book Co., 1949, p. 186.
[8] A. W. Love, *Reflector Antennas*, IEEE Press, 1978, pp. 291–336.
[9] F. A. Jenkins and H. E. White, *Fundamentals of Optics*, 4th ed., chapter 9, p. 173, New York: McGraw-Hill Book Co., 1973.
[10] W. Rotman, "Ehf dielectric lens antenna for satellite communication systems," *Report 620*, ESD-TR-82-102, Lincoln Labs, MIT, 1983.
[11] F. G. Friedlander, "A dielectric lens aerial for wide-angle beam scanning," *J. IEE*, vol. 93, pt. 3A, pp. 658–662, 1946.
[12] J. J. Lee, "Numerical methods make lens antennas practical," *Microwave*, pp. 81–84, September 1982.
[13] S. B. Cohn, "Lens-type radiators," chapter 14 in *Antenna Engineering Handbook*, ed. by H. Jasik, New York: McGraw-Hill Book Co., 1961.
[14] J. McFarland, D. Wesley, and R. Dell-Imagine, Private communication.
[15] D. H. Shinn, "The design of a zoned dielectric lens for wide-angle scanning," *Marconi Rev.*, no. 117, p. 37, 1955.
[16] J. B. L. Rao, "Bifocal dual-reflector antennas," *IEEE Trans. Antennas Propag.*, vol.

AP-22, pp. 711–714, 1974.
[17] K. C. Lang et al., "Gain, side lobe and cross-polarization performance of dual offset reflector antennas," *Proc. IEEE AP-S Symp.*, p. 269, 1982.
[18] R. M. Brown, "Dielectric bifocal lenses," *IRE Natl. Conv. Rec.*, pp. 180–187, 1956.
[19] H. Kumazawa and M. Kariokomi, "Multiple-beam antenna for domestic communication satellites," *IEEE Trans. Antennas Propag.*, vol. 21, p. 876, 1973.
[20] J. J. Lee, "A dielectric lens shaped for a generalized Taylor distribution," *Proc. IEEE AP-S Symp. vol. AP-5*, p. 124, 1982.
[21] J. J. Lee, "Dielectric lens shaping and coma correction zoning," *IEEE Trans. Antennas Propag.*, vol. AP-31, p. 211, 1983.
[22] A. W. Love, "Shaped-beam antennas for Earth coverage," Rockwell Space Div. internal report, *SD 74-SA-0074*, 1974.
[23] D. K. Waineo, "Low side lobe sector patterns for a circular aperture," *IEEE Antennas Propag. Symp. Dig.*, p. 527, 1977.
[24] K. S. Kelleher, "Microwave optics at NRL," *Proc. McGill Symp. Microwave Opt.*, AFCRC-TR59-118(I), p. 5, 1959.
[25] T. C. Cheston, "Microwave lenses," *Proc. McGill Symp. Microwave Opt.*, AFCRC-TR-59-118(I), p. 8, 1959.
[26] L. Ronchi and G. Toraldo Di Francia, "An application of parageometric optics to the design of microwave mirrors," *IRE Trans. Antennas Propag.*, vol. AP-6, p. 129, 1958.
[27] J. H. Provencher, "Experimental study of a diffraction reflector," *IRE Trans. Antennas Propag.*, vol. AP-8, p. 331, 1960.
[28] S. Dasgupta and Y. T. Lo, "A study of the coma-corrected zoned mirror by diffraction theory," *IRE Trans. Antennas Propag.*, vol. AP-9, p. 130, 1961.
[29] J. Ruze, "Wide-angle metal plate optics," *Proc. IRE*, vol. 38, pp. 53–59.
[30] S. Silver, ed., *Microwave Antenna Theory and Design*, chapter 11, New York: McGraw-Hill Book Co., 1949, pp. 401–412.
[31] R. W. Major and J. M. Devan, "Ehf multiple-beam antennas," *NOSC TR730*, San Diego 92152: Naval Ocean Systems Center, 1981.
[32] A. R. Dion and L. J. Ricardi, "A variable-coverage satellite antenna system," *Proc. IEEE*, vol. 59, p. 252, 1971.
[33] A. R. Dion, *Handbook of Antenna Design*, chapter 3.7, ed. by A. W. Rudge et al., Stevenage, Herts., UK: Peter Peregrinus, 1982, p. 293.
[34] C. B. Colborn, Jr., "Increased bandwidth waveguide lens antenna," *Aerospace Corp., Rep. No. TOR-0076(6403-01)-3*, December 8, 1975.
[35] J. S. Ajioka and V. W. Ramsey, "An equal group delay waveguide lens," *IEEE Trans. Antennas Propag.*, vol. AP-26, p. 519, 1978.
[36] W. Rotman and R. F. Turner, "Wide-angle microwave lens for line source applications," *IEEE Trans. Antennas Propag.*, vol. AP-11, pp. 623–632, 1963.
[37] S. Cornbleet, *Microwave Optics*, chapter 1, New York: Academic Press, 1976, pp. 73–77.
[38] J. B. L. Rao, "Multifocal three-dimensional bootlace lenses," *Proc. IEEE AP-S Symp.*, vol. AP-5(a)-2, p. 332, 1979.
[39] This analogy was pointed out by the late J. L. McFarland in private discussions.
[40] E. A. Wolff, *Antenna Analysis*, New York: John Wiley & Sons, 1966, p. 496.
[41] R. S. Elliott, *Antenna Theory and Design*, Englewood Cliffs: Prentice-Hall, 1981, p. 545.
[42] S. T. S. Brown, "An introduction to gradient index optics," *Marconi Rev.*, first quarter, p. 3, 1981.
[43] J. Frank, "Bandwidth criterion for phased array antennas," *Proc. 1970 Phased Array Antenna Symp.*, Dedham: Artech House, 1972.
[44] R. E. Collin, *Field Theory of Guided Waves*, New York: McGraw-Hill Book Co., 1960, pp. 87–94.
[45] S. W. Lee and W. R. Jones, "On the suppression of radiation nulls and broadband impedance matching of rectangular waveguide phased arrays," *IEEE Trans. Antennas Propag.*, vol. AP-19, pp. 41–51, 1971.

[46] C. C. Chen, "Wideband wide-angle impedance matching and polarization characteristics of circular waveguide phased arrays," *IEEE Trans. Antennas Propag.*, vol. AP-22, pp. 414–418, 1974.

Appendixes

CONTENTS

A. Physical Constants, International Units, Conversion of Units, and Metric Prefixes — A-3
B. The Frequency Spectrum — B-1
C. Electromagnetic Properties of Materials — C-1
D. Vector Analysis — D-1
E. VSWR Versus Reflection Coefficient and Mismatch Loss — E-1
F. Decibels Versus Voltage and Power Ratios — F-1

Appendix A

Physical Constants, International Units, Conversion of Units, and Metric Prefixes

Yi-Lin Chen
*University of Illinois**

Physical Constants

Quantity	Symbol	Value
Speed of light in vacuum	c	$2.997\,925 \times 10^8\,\text{ms}^{-1}$
Electron charge	e	$1.602\,192 \times 10^{-19}\,\text{C}$
Electron rest mass	m_e	$9.109\,558 \times 10^{-31}\,\text{kg}$
Boltzmann constant	k	$1.380\,622 \times 10^{-23}\,\text{JK}^{-1}$
Dielectric constant in vacuum	ϵ_0	$8.854\,185 \times 10^{-12}\,\text{Fm}^{-1}$ $\cong (36\pi \times 10^9)^{-1}\,\text{Fm}^{-1}$
Permeability in vacuum	μ_0	$4\pi \times 10^{-7}\,\text{Hm}^{-1}$

International System of Units (SI Units): Basic Units

Quantity	Symbol	Units
Length	ℓ	meters (m)
Mass	m	kilograms (kg)
Time	t	seconds (s)
Electric current	I	amperes (A)
Temperature	T	kelvins (K)
Luminous intensity	I	candelas (cd)

*On leave from the Chinese Aeronautical Laboratory, Beijing, China, during 1983.

Derived Units in Electromagnetics

Quantity	Symbol	Units
Electric-field strength	**E**	volts per meter (V/m)
Magnetic-field strength	**H**	amperes per meter (A/m)
Electric-flux density	**D**	coulombs per meter squared (C/m^2)
Magnetic-flux density	**B**	teslas (T) = Wb/m^2
Electric-current density	**J**	amperes per meter squared (A/m^2)
Magnetic-current density	**K**	volts per meter squared (V/m^2)
Electric-charge density	ϱ	coulombs per meter cubed (C/m^3)
Magnetic-charge density	ϱ_m	webers per meter cubed (Wb/m^3)
Voltage	V	volts (V)
Electric current	I	amperes (A)
Dielectric constant (permittivity)	ϵ	farads/meter (F/m)
Permeability	μ	henrys/meter (H/m)
Conductivity	σ	siemens per meter (S/m) = ℧/m
Resistance	R	ohms (Ω)
Inductance	L	henrys (H)
Capacitance	C	farads (F)
Impedance	Z	ohms (Ω)
Admittance	Y	siemens (S) or mhos (℧)
Power	P	watts (W)
Energy	W	joules (J)
Radiation intensity	I	watts per steradian (W/sr)
Frequency	f	hertz (Hz)
Angular frequency	ω	radians per second (rad/s)
Wavelength	λ	meters (m)
Wave number	k	1 per meter (m^{-1})
Phase shift constant	β	radians per meter (rad/m)
Attenuation factor	α	nepers per meter (Np/m)

Conversions of Units

Quantity	Symbol	SI Unit	Equivalent Number of CGS Electromagnetic Unit	Equivalent Number of CGS Electrostatic Unit
Electric charge	q	coulombs	10^{-1} abcoulomb	3×10^9 statcoulombs
Current	I	amperes	10^{-1} abampere	3×10^9 statamperes
Volume current density	$\mathbf{J}$	amperes/meter^2	10^{-5} abampere/centimeter^2	3×10^5 statamperes/centimeter^2
Voltage	V	volts	10^8 abvolts	$\frac{1}{3} \times 10^{-2}$ statvolt
Electric-field intensity	$\mathbf{E}$	volts/meter	10^6 abvolts/cm	$\frac{1}{3} \times 10^{-4}$ statvolt/centimeter
Electric-flux density	$\mathbf{D}$	coulombs/meter^2	$4\pi \times 10^{-5}$ abcoulomb/centimeter^2	$12\pi \times 10^5$ statcoulombs/centimeter^2
Magnetic-field intensity	$\mathbf{H}$	amperes/meter	$4\pi \times 10^{-3}$ oersted	$12\pi \times 10^7$ oersteds
Magnetic-flux intensity	$\mathbf{B}$	webers/meter^2	10^4 gausses	$\frac{1}{3} \times 10^{-6}$ gauss
Permittivity	ϵ	farads/meter	$4\pi \times 10^{-11}$ abfarad/centimeter	$36\pi \times 10^9$ statfarads/centimeter
Permeability	μ	henrys/meter	$\frac{1}{4\pi} \times 10^7$ gauss/oersted	$\frac{1}{36\pi} \times 10^{-13}$ gauss/oersted
Magnetic flux	Φ	webers	10^8 gilberts	$\frac{1}{3} \times 10^{-2}$ gilbert
Resistance	R	ohms	10^9 abohms	$\frac{1}{9} \times 10^{-11}$ statohm
Inductance	L	henrys	10^9 abhenrys	$\frac{1}{9} \times 10^{-11}$ stathenry
Capacitance	C	farads	10^{-9} abfarad	9×10^{11} statfarads
Conductivity	σ	siemens/meter	10^{-11} absiemen/centimeter	9×10^9 statsiemens/centimeter
Work	W	joules	10^7 ergs	10^7 ergs
Power	P	watts	10^7 ergs/second	10^7 ergs/second

Conversion of Length Units

Meters	Centimeters	Inches	Feet	Miles
1	100	39.37	3.281	6.214×10^{-4}
0.01	1	0.3937	3.281×10^{-2}	
0.0254	2.540	1	8.333×10^{-2}	
0.3048	30.48	12	1	1.894×10^{-4}
1609			5279	1

Metric Prefixes and Symbols*

Multiplication Factor	Prefix	Symbol
10^{18}	exa	E
10^{15}	peta	P
10^{12}	tera	T
10^{9}	giga	G
10^{6}	mega	M
10^{3}	kilo	k
10^{2}	hecto	h
10	deka	da
10^{-1}	deci	d
10^{-2}	centi	c
10^{-3}	milli	m
10^{-6}	micro	μ
10^{-9}	nano	n
10^{-12}	pico	p
10^{-15}	femto	f
10^{-18}	atto	a

*From *IEEE Standard Dictionary of Electrical and Electronics Terms*, p. 682, The Institute of Electrical and Electronics Engineers, Inc., 1984.

Appendix B

The Frequency Spectrum

Li-Yin Chen
*University of Illinois**

The wavelength of an electromagnetic wave in free space is $\lambda_0 = c/f$.

$$\lambda_0 = \frac{300\,000}{f(\text{kHz})}\,\text{m} = \frac{300}{f(\text{MHz})}\,\text{m} = \frac{30}{f(\text{GHz})}\,\text{cm}$$

$$= \frac{9.843 \times 10^5}{f(\text{kHz})}\,\text{ft} = \frac{9.843 \times 10^2}{f(\text{MHz})}\,\text{ft} = \frac{11.81}{f(\text{GHz})}\,\text{in}$$

The wave number of an electromagnetic wave in free space: $k_0 = \omega\sqrt{\mu_0 \epsilon_0} = 2\pi f/c$.

$$k_0 = f(\text{Hz}) \times 2.0944 \times 10^{-8}\,\text{m}^{-1} = f(\text{kHz}) \times 2.0944 \times 10^{-5}\,\text{m}^{-1}$$
$$= f(\text{MHz}) \times 2.0944 \times 10^{-2}\,\text{m}^{-1} = f(\text{GHz}) \times 20.944\,\text{m}^{-1}$$
$$= f(\text{Hz}) \times 6.383 \times 10^{-9}\,\text{ft}^{-1} = f(\text{kHz}) \times 6.383 \times 10^{-6}\,\text{ft}^{-1}$$
$$= f(\text{MHz}) \times 6.383 \times 10^{-3}\,\text{ft}^{-1} = f(\text{GHz}) \times 6.383\,\text{ft}^{-1}$$
$$= f(\text{GHz}) \times 0.532\,\text{in}^{-1}$$

---

*On leave from the Chinese Aeronautical Laboratory, Beijing, China, during 1983.

Nomenclature of Frequency Bands

Adjectival Designation	Frequency Range	Metric Subdivision	Wavelength Range
elf: Extremely low frequency	30 to 300 Hz	Megametric waves	10 000 to 1000 km
vf: Voice frequency	300 to 3000 Hz		1000 to 100 km
vlf: Very low frequency	3 to 30 kHz	Myriametric waves	100 to 10 km
lf: Low frequency	30 to 300 kHz	Kilometric waves	10 to 1 km
mf: Medium frequency	300 to 3000 kHz	Hectrometric waves	1000 to 100 m
hf: High frequency	3 to 30 MHz	Decametric waves	100 to 10 m
vhf: Very high frequency	30 to 300 MHz	Metric waves	10 to 1 m
uhf: Ultrahigh frequency	300 to 3000 MHz	Decimetric waves	100 to 10 cm
shf: Superhigh frequency	3 to 30 GHz	Centimetric waves	10 to 1 cm
ehf: Extremely high frequency	30 to 300 GHz	Millimetric waves	10 to 1 mm
	300 to 3000 GHz	Decimillimetric waves	1 to 0.1 mm

Standard Radar-Frequency Letter Bands*

Band Designation	Nominal Frequency Range
hf	3–30 MHz
vhf	30–300 MHz
uhf	300–1000 MHz
L	1000–2000 MHz
S	2000–4000 MHz
C	4000–8000 MHz
X	8000–12 000 MHz
K_u	12.0–18 GHz
K	18–27 GHz
K_a	27–40 GHz
Millimeter	40–300 GHz

*Reprinted from ANSI/IEEE Std. 100-1984, *IEEE Standard Dictionary of Electrical and Electronics Terms*, © 1984 by The Institute of Electrical and Electronics Engineers, Inc., by permission of the IEEE Standards Department.

Television Channel Frequencies*

Channel Number[†]	Band (MHz)	Channel Number[†]	Band (MHz)	Channel Number[†]	Band (MHz)
2	54–60	29	560–566	57	728–734
3	60–66	30	566–572	58	734–740
4	66–72	31	572–578	59	740–746
5	76–82	32	578–584	60	746–752
6	82–88	33	584–590	61	752–758
7	174–180	34	590–596	62	758–764
8	180–186	35	596–602	63	764–770
9	186–192	36	602–608	64	770–776
10	192–198	37	608–614	65	776–782
11	198–204	38	614–620	66	782–788
12	204–210	39	620–626	67	788–794
13	210–216	40	626–632	68	794–800
14	470–476	41	632–638	69	800–806
15	476–482	42	638–644	70	806–812
16	482–488	43	644–650	71	812–818
17	488–494	44	650–656	72	818–824
18	494–500	45	656–662	73	824–830
19	500–506	46	662–668	74	830–836
20	506–512	47	668–674	75	836–842
21	512–518	48	674–680	76	842–848
22	518–524	49	680–686	77	848–854
23	524–530	50	686–692	78	854–860
24	530–536	51	692–698	79	860–866
25	536–542	52	698–704	80	866–872
26	542–548	53	704–710	81	872–878
27	548–554	54	710–716	82	878–884
28	554–560	55	716–722	83	884–890
		56	722–728		

Note: The carrier frequency for the video portion is the lower frequency plus 1.25 MHz. The audio carrier frequency is the upper frequency minus 0.25 MHz. All channels have a 6-MHz bandwidth. For example, channel 2 video carrier is at 55.25 MHz and the audio carrier is at 59.75 MHz.

[†]Channels 2 through 13 are vhf; channels 14 through 83 are uhf. Channels 70 through 83 were withdrawn and reassigned to tv translator stations until licenses expire.

Appendix C

Electromagnetic Properties of Materials

Yi-Lin Chen
*University of Illinois**

Resistivities and Skin Depth of Metals and Alloys

Material	Resistivity* ($\mu\Omega$-cm)	Skin Depth[†] (μm at 1 GHz)
Aluminum	2.62	2.576
Brass (66% Cu, 34% Zn)	7.5	4.3586
Copper	1.7241	2.0898
Gold	2.44	2.4861
Iron	9.71	4.9594
Nickel	6.9	4.1807
Silver	1.62	2.0257
Steel (0.4–0.5% C, balance Fe)	13–22	5.7384–7.465
Steel, stainless (0.1% C, 18% Cr, 8% Ni, balance Fe)	90	15.0988
Tin	11.4	5.3737
Titanium	47.8	11.0036

*In solid form at 20°C; resistivity = (conductivity)$^{-1}$.
[†]Skin depth $\delta = (\pi\mu\sigma f)^{-1/2} = (20\pi)^{-1}[\sigma f(\text{GHz})]^{-1/2}$ m. The δs in the column are calculated at $f = 1$ GHz. For other frequencies, multiply them by $[f(\text{GHz})]^{-1/2}$.

---

*On leave from the Chinese Aeronautical Laboratory, Beijing, China, during 1983.

Characteristics of Insulating Materials*

Material Composition	T(°C)	Dielectric Constant† at (Frequency in Hertz)				Dissipation Factor† at (Frequency in Hertz)				Dielectric Strength in Volts/Mil at 25°C	DC Volume Resistivity in Ohm-cm at 25°C	Thermal Expansion (Linear) in Parts/°C	Softening Point in °C	Moisture Absorption in Percent
		10^4	10^6	3×10^9	2.5×10^{10}	10^4	10^6	3×10^9	2.5×10^{10}					
Ceramics:														
Aluminum oxide	25	8.80	8.80	8.79	—	0.00033	0.00030	0.0010	—	—	—	—	1400–1430	—
Barium titanate‡	26	1143	—	600	100	0.0105	—	0.30	0.60	75	10^{12}–10^{13}	—	—	0.1
Calcium titanate	25	167.7	167.7	165	—	0.0002	—	0.0023	—	100	10^{12}–10^{14}	—	1510	<0.1
Magnesium oxide	25	9.65	9.65	—	—	<0.0003	<0.0003	—	—	—	—	—	—	—
Magnesium silicate	25	5.97	5.96	5.90	—	0.0005	0.0004	0.0012	—	—	>10^{14}	9.2×10^{-6}	1350	0.1–1
Magnesium titanate	25	13.9	13.9	13.8	13.7	0.0004	0.0005	0.0017	0.0065	—	—	—	—	—
Oxides of aluminum, silicon, magnesium, calcium, barium	24	6.04	—	5.90	—	0.0011	—	0.0024	—	—	—	7.7×10^{-6}	1325	—
Porcelain (dry process)	25	5.08	5.04	—	—	0.0075	0.0078	—	—	—	—	—	—	—
Steatite 410	25	5.77	5.77	5.7	—	0.0007	0.0006	0.00089	—	—	—	—	1510	0.1
Strontium titanate	25	232	232	—	—	0.0002	0.0001	—	—	100	10^{12}–10^{14}	—	—	—
Titanium dioxide (rutile)	26	100	100	—	—	0.0003	0.00025	—	—	—	—	—	—	—
Glasses:														
Iron-sealing glass	24	8.30	8.20	7.99	7.84	0.0005	0.0009	0.00199	0.0112	—	10^{10} at 250°	132×10^{-7}	484	poor
Soda-borosilicate	25	4.84	4.84	4.82	4.65	0.0036	0.0030	0.0054	0.0090	—	7×10^7 at 250°	50×10^{-7}	693	—
100% silicon dioxide (fused quartz)	25	3.78	3.78	3.78	3.78	0.0001	0.0002	0.00006	0.00025	410 (0.25")	>10^{18}	5.7×10^{-7}	1667	—
Plastics:														
Alkyd resin	25	4.76	4.55	4.50	—	0.0149	0.0138	0.0108	—	—	—	—	—	—
Cellulose acetate-butyrate, plasticized	26	3.30	3.08	2.91	—	0.018	0.017	0.028	—	250–400 (0.125")	—	$11–17 \times 10^{-5}$	60–121	2.3
Cresylic acid–formaldehyde, 50% α-cellulose	25	4.51	3.85	3.43	3.21	0.036	0.055	0.051	0.038	1020 (0.033")	3×10^{12}	3×10^{-5}	>125	1.2
Cross-linked polystyrene	25	2.58	2.58	2.58	—	0.0016	0.0020	0.0019	—	—	—	—	—	—
Epoxy resin (Araldite CN-501)	25	3.62	3.35	3.09	—	0.019	0.034	0.027	—	405 (0.125")	>3.8×10^7	4.77×10^{-5}	109 (distortion)	0.14
Epoxy resin (Epon resin RN-48)	25	3.52	3.32	3.04	—	0.0142	0.0264	0.021	—	—	—	—	—	—
Foamed polystyrene, 0.25% filler	25	1.03	—	1.03	1.03	<0.0002	—	0.0001	—	—	—	—	85	low
Melamine—formaldehyde, α-cellulose	24	7.00	6.0	4.93	—	0.041	0.085	0.103	—	300–400	—	—	99 (stable)	0.4–0.6
Melamine—formaldehyde, 55% filler	26	5.75	5.5	—	—	0.0115	0.020	—	—	—	—	1.7×10^{-5}	—	0.6
Phenol—formaldehyde (Bakelite BM 120)	25	4.36	3.95	3.70	3.55	0.0280	0.0380	0.0438	0.0390	300 (0.125")	10^{11}	$30–40 \times 10^{-6}$	<135 (distortion)	<0.6
Phenol—formaldehyde, 50% paper laminate	26	4.60	4.04	3.57	—	0.034	0.057	0.060	—	—	—	—	—	—
Phenol—formaldehyde, 65% mica, 4% lubricants	24	4.78	4.72	4.71	—	0.0082	0.0115	0.0126	—	—	—	—	—	—
Polycarbonate	—	2.96	—	—	—	0.010	—	—	—	364 (0.125")	2×10^{16}	7×10^{-5}	135 (deflection)	—
Polychlorotrifluoroethylene	25	2.42	2.32	2.29	2.28	0.0082	—	0.0028	0.0053	—	10^{13}	—	—	—

Material	Temp (°C)												
Polyethylene	25	2.26	2.26	2.26	2.26	<0.0002	0.0002	0.00031	0.0006	1200 (0.033")	10^{17}	19×10^{-5} (varies)	95–105 (distortion)	0.03
Polyethylene-terephthalate	—	2.98	—	—	—	0.016	—	—	—	—	—	—	—	—
Polyethylmethacrylate	22	2.55	2.52	2.51	2.5	0.0090	—	0.0075	0.0083	4000 (0.002")	—	—	—	low
Polyhexamethylene-adipamide (nylon)	25	3.14	3.0	2.84	2.73	0.0218	0.0200	0.0117	0.0105	400 (0.125")	8×10^{14}	10.3×10^{-5}	60 (distortion) 65 (distortion)	1.5
Polyimide	—	3.4	—	—	—	0.003	—	—	—	570	—	—	—	low
Polyisobutylene	25	2.23	2.23	2.23	—	0.0001	0.0003	0.00047	—	600 (0.010")	—	—	25 (distortion)	low
Polymer of 95% vinyl-chloride, 5% vinyl-acetate	20	2.90	2.8	2.74	—	0.0150	0.0080	0.0059	—	—	—	—	—	—
Polymethyl methacrylate	27	2.76	—	2.60	—	0.0140	—	0.0057	—	990 (0.030")	$>5 \times 10^{16}$	$8–9 \times 10^{-5}$	70–75 (distortion)	0.3–0.6
Polyphenylene oxide	—	2.55	—	2.55	—	0.0007	—	0.0011	—	500 (0.125")	10^{17}	5.3×10^{-5}	195 (deflection)	—
Polypropylene	—	2.55	—	—	—	<0.0005	—	—	—	650 (0.125")	6×10^{16}	$6–8.5 \times 10^{-5}$	99–116 (deflection)	—
Polystyrene	25	2.56	2.55	2.55	2.54	0.00007	<0.0001	0.00033	0.0012	500–700 (0.125")	10^{18}	$6–8 \times 10^{-5}$	82 (distortion)	0.05
Polytetrafluoroethylene (Teflon)	22	2.1	2.1	2.1	2.08	<0.0002	<0.0002	0.00015	0.0006	1000–2000 (0.005"–0.012")	10^{17}	9.0×10^{-5}	66 (distortion) (stable to 300)	0.00
Polyvinylcyclohexane	24	2.25	2.25	2.25	—	<0.0002	0.0002	0.00018	—	860 (0.034")	$>5 \times 10^{16}$	7.7×10^{-5}	190	—
Polyvinyl formal	26	2.92	2.80	2.76	2.7	0.019	0.013	0.0113	0.0115	260 (0.125")	2×10^{14}	12×10^{-5}	148 (deflection)	1.3
Polyvinylidene fluoride	—	6.6	—	—	—	0.17	0.050	—	—	375 (0.085")	—	2.6×10^{-5}	152 (distortion)	2
Urea-formaldehyde, cellulose	27	5.65	5.1	4.57	—	0.027	—	0.0555	—	450–500 (0.125")	—	$10–20 \times 10^{-5}$	—	—
Urethane elastomer	—	6.5–7.1	—	—	—	0.057	0.0180	0.0072	—	300 (0.125")	2×10^{11}	15.8×10^{-5}	150	<0.1
Vinylidene–vinyl chloride copolymer	23	3.18	2.82	2.71	—						$10^{14}–10^{16}$			
100% aniline-formaldehyde (Dioctane-100)	25	3.58	3.50	3.44	—	0.0061	0.0033	0.0026	—	810 (0.068")	10^{16}	5.4×10^{-5}	125	0.06–0.08
100% phenol-formaldehyde	24	5.4	4.4	3.64	—	0.060	0.077	0.052	—	277 (0.125")	—	$8.3–13 \times 10^{-5}$	50 (distortion)	0.42
100% polyvinyl-chloride	20	2.88	2.85	2.84	—	0.0160	0.0081	0.0055	—	400 (0.125")	10^{14}	6.9×10^{-5}	54 (distortion)	0.05–0.15
Organic Liquids:														
Aviation gasoline (100 octane)	25	1.94	1.94	1.92	—	—	0.0001	0.0014	—	—	—	—	—	—
Benzene (pure, dried)	25	2.28	2.28	2.28	2.28	<0.0001	<0.0001	<0.0001	<0.0001	—	—	—	—	—
Carbon tetrachloride	25	2.17	2.17	2.17	—	<0.00004	<0.0002	0.0004	—	—	—	—	—	—
Ethyl alcohol (absolute)	25	24.5	23.7	6.5	—	0.090	0.062	0.250	—	—	—	—	—	—
Ethylene glycol	25	41	41	12	—	0.030	0.045	1.00	—	—	—	—	—	—
Jet fuel (JP-3)	25	2.08	2.08	2.04	—	0.0001	—	0.0055	—	—	—	—	—	—
Methyl alcohol (absolute analytical grade)	25	31	31.0	23.9	—	0.20	0.038	0.64	—	—	—	—	—	—
Methyl or ethyl siloxane polymer (1000 cs)	22	2.78	—	2.74	—	<0.0003	0.0001	0.0096	—	—	—	—	—	—
Monomeric styrene	22	2.40	2.40	2.40	—	<0.0003	—	0.0020	—	300 (0.100")	3×10^{12}	—	—	0.06
Transil oil	26	2.22	2.20	2.18	—	<0.0005	0.0048	0.0028	—	300 (0.100")	—	—	–40 (pour point)	—
Vaseline	25	2.16	2.16	2.16	—	<0.0001	<0.0004	0.00066	—	—	—	—	—	—

Characteristics of Insulating Materials* (cont'd.)

Material Composition	T (°C)	Dielectric Constant‡ at (Frequency in Hertz)				Dissipation Factor‡ at (Frequency in Hertz)				Dielectric Strength in Volts/Mil at 25°C	DC Volume Resistivity in Ohm-cm at 25°C	Thermal Expansion (Linear) in Parts/°C	Softening Point in °C	Moisture Absorption in Percent
		10^4	10^6	3×10^9	2.5×10^{10}	10^4	10^6	3×10^9	2.5×10^{10}					
Waxes:														
Beeswax, yellow	23	2.53	2.45	2.39	—	0.0092	0.0090	0.0075	—	—	—	—	45–64 (melts)	—
Dichloronaphthalenes	23	2.98	2.93	2.89	—	0.0003	0.0017	0.0037	—	—	—	—	35–63 (melts)	nil
Polybutene	25	2.34	2.30	2.27	—	0.00133	0.00133	0.0009	—	—	—	—	—	—
Vegetable and mineral waxes	25	2.3	2.3	2.25	—	0.0004	0.0004	0.00046	—	—	—	—	57	—
Rubbers:														
Butyl rubber	25	2.35	2.35	2.35	—	0.0010	0.0010	0.0009	—	—	—	—	—	—
GR-S rubber	25	2.90	2.82	2.75	—	0.0120	0.0080	0.0057	—	870 (0.040″)	2×10^{15}	—	—	—
Gutta-percha	25	2.53	2.47	2.40	—	0.0042	0.0120	0.0060	—		10^{15}	—	—	—
Hevea rubber (pale crepe)	25	2.4	2.4	2.15	—	0.0018	0.0050	0.0030	—	—	—	—	—	—
Hevea rubber, vulcanized (100 pts pale crepe, 6 pts sulfur)	27	2.74	2.42	2.36	—	0.0446	0.0180	0.0047	—	—	—	—	—	—
Neoprene rubber	24	6.26	4.5	4.00	4.0	0.038	0.090	0.034	0.025	300 (0.125″)	8×10^{12}	—	—	nil
Organic polysulfide, fillers	23	110	30	16	13.6	0.39	0.28	0.22	0.10	—	—	—	—	—
Silicone-rubber compound	25	3.20	3.16	3.13	—	0.0030	0.0032	0.0097	—	—	—	—	—	—
Woods:‡														
Balsa wood	26	1.37	1.30	1.22	—	0.0120	0.0135	0.100	—	—	—	—	—	—
Douglas fir	25	1.93	1.88	1.82	1.78	0.026	0.033	0.027	0.032	—	—	—	—	—
Douglas fir, plywood	25	1.90	—	—	1.6	0.0230	—	—	0.0220	—	—	—	—	—
Mahogany	25	2.25	2.07	1.88	1.6	0.025	0.032	0.025	0.020	—	—	—	—	—
Yellow birch	25	2.70	2.47	2.13	1.87	0.029	0.040	0.033	0.026	—	—	—	—	—
Yellow poplar	25	1.75	—	1.50	1.4	0.019	—	0.015	0.017	—	—	—	—	—
Miscellaneous:														
Amber (fossil resin)	25	2.65	—	2.6	—	0.0056	—	0.0090	—	2300 (0.125″)	Very high	—	200	—
DeKhotinsky cement	23	3.23	—	2.96	—	0.024	—	0.021	—	—	—	—	80–85	—
Gilsonite (99.9% natural bitumen)	26	2.58	2.56	—	—	0.0016	0.0011	—	—	—	—	9.8×10^{-5}	155 (melts)	—
Shellac (natural XL)	28	3.47	3.10	2.86	—	0.031	0.030	0.0254	—	—	10^{16}	—	80	low after baking
Mica, glass-bonded	25	7.39	—	—	—	0.0013	—	—	—	—	—	—	—	—
Mica, glass, titanium dioxide	24	9.0	—	—	—	0.0026	—	0.0040	—	3800–5600 (0.040″)	—	—	400	<0.5
Ruby mica	26	5.4	5.4	5.4	—	0.0003	0.0002	0.0003	—	202 (0.125″)	5×10^{13}	—	—	—
Paper, royalgrey	25	2.99	2.77	2.70	—	0.038	0.066	0.056	—	—	—	—	—	—
Selenium (amorphous)	25	6.00	6.00	6.00	6.00	<0.0003	<0.0002	0.00018	0.0013	—	—	—	—	—
Asbestos fiber–chrysotile paper	25	3.1	—	—	—	0.025	—	—	—	—	—	—	—	—
Sodium chloride (fresh crystals)	25	5.90	—	—	5.90	<0.0002	—	—	<0.0005	—	—	—	—	—

Material	Temp										
Soil, sandy dry	25	2.59	2.55	2.55	—	0.017	—	0.0062	—	—	—
Soil, loamy dry	25	2.53	2.48	2.44	—	0.018	—	0.0011	—	—	—
Ice (from pure distilled water)	−12	4.15	3.45	3.20	—	0.12	0.035	0.0009	—	—	—
Freshly fallen snow	−20	1.20	1.20	1.20	—	0.0215	—	0.00029	—	—	—
Hard-packed snow followed by light rain	−6	1.55	—	1.5	—	0.29	—	0.0009	—	—	—
Water (distilled)	25	78.2	78	76.7	34	0.040	0.005	0.157	0.2650	10^4	—

*Reproduced with permission of the publisher, Howard W. Sams & Company, Indianapolis, *Reference Data for Engineers: Radio, Electronics, Computer, and Communications*, 7th ed., by E. C. Jordan, ed., © 1985.

†The dissipation factor is defined as the ratio of the energy dissipated to the energy stored in the dielectric, or as the tangent of the loss angle. Dielectric constant and dissipation factor depend on electrical field strength.

‡Field perpendicular to grain.

Properties of Soft Magnetic Metals*

Name	Composition (%)	Permeability Initial	Permeability Maximum	Coercivity H_c (A/m)	Retentivity B_r (T)	B_{max} (T)	Resistivity (μΩ-cm)
Ingot iron	99.8 Fe	150	5 000	80	0.77	2.14	10
Low carbon steel	99.5 Fe	200	4 000	100	—	2.14	12
Silicon iron, unoriented	3 Si, bal Fe	270	8 000	60	—	2.01	47
Silicon iron, grain oriented	3 Si, bal Fe	1 400	50 000	7	1.20	2.01	50
4750 alloy	48 Ni, bal Fe	11 000	80 000	2	—	1.55	48
4-79 Permalloy	4 Mo, 79 Ni, bal Fe	40 000	200 000	1	—	0.80	58
Supermalloy	5 Mo, 80 Ni, bal Fe	80 000	450 000	0.4	—	0.78	65
2V-Permendur	2V, 49 Co, bal Fe	800	8 000	160	—	2.30	40
Supermendur	2V, 49 Co, bal Fe	—	100 000	16	2.00	2.30	26
Metglas[†] 2605SC	$Fe_{81}B_{13.5}Si_{3.5}C_2$	—	210 000	14	1.46	1.60	125
Metglas[†] 2605S-3	$Fe_{79}B_{16}Si_5$	—	30 000	8	0.30	1.58	125

*Reproduced with permission of the publisher, Howard W. Sams & Company, Indianapolis, *Reference Data for Engineers: Radio, Electronics, Computer, and Communications*, 7th ed., by E. C. Jordan, ed., © 1985.

[†]Metglas is Allied Corporation's registered trademark for amorphous alloys.

Appendix D

Vector Analysis

Yi-Lin Chen
*University of Illinois**

1. Change of Coordinate Systems

The transformations of the coordinate components of a vector **A** among the rectangular (x, y, z), cylindrical (θ, ϕ, z), and spherical (r, θ, ϕ) coordinates are given by the following relations (see Fig. 1):

$$A_x = A_\varrho \cos\phi - A_\phi \sin\phi = A_r \sin\theta \cos\phi + A_\theta \cos\theta \cos\phi - A_\phi \sin\phi$$
$$A_y = A_\varrho \sin\phi + A_\phi \cos\phi = A_r \sin\theta \sin\phi + A_\theta \cos\theta \sin\phi + A_\phi \cos\phi$$

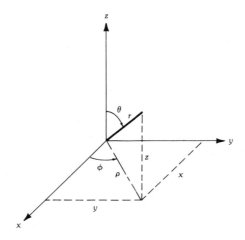

Fig. 1. Rectangular, cylindrical, and spherical coordinate systems.

---

*On leave from the Chinese Aeronautical Laboratory, Beijing, China, during 1983.

$$A_z = A_r \cos\theta - A_\theta \sin\theta$$
$$A_\varrho = A_x \cos\phi + A_y \sin\phi = A_r \sin\theta + A_\theta \cos\theta$$
$$A_\phi = -A_x \sin\phi + A_y \cos\phi$$
$$A_r = A_x \sin\theta \cos\phi + A_y \sin\theta \sin\phi + A_z \cos\theta = A_\varrho \sin\theta + A_z \cos\theta$$
$$A_\theta = A_x \cos\theta \cos\phi + A_y \cos\theta \sin\phi - A_z \sin\theta = A_\varrho \cos\theta - A_z \sin\theta$$

Differential element of volume:
$$dV = dx\,dy\,dz = \varrho\,d\varrho\,d\phi\,dz = r^2 \sin\theta\,dr\,d\theta\,d\phi$$

Differential element of vector area:
$$\mathbf{dS} = \hat{\mathbf{x}}\,dy\,dz + \hat{\mathbf{y}}\,dx\,dz + \hat{\mathbf{z}}\,dx\,dy$$
$$= \hat{\boldsymbol{\varrho}}\varrho\,d\phi\,dz + \hat{\boldsymbol{\phi}}\,d\varrho\,dz + \hat{\mathbf{z}}\varrho\,d\varrho\,d\phi$$
$$= \hat{\mathbf{r}}r^2 \sin\theta\,d\theta\,d\phi + \hat{\boldsymbol{\theta}}r \sin\theta\,dr\,d\phi + \hat{\boldsymbol{\phi}}r\,dr\,d\theta$$

Differential element of vector length:
$$d\boldsymbol{\ell} = \hat{\mathbf{x}}\,dx + \hat{\mathbf{y}}\,dy + \hat{\mathbf{z}}\,dz$$
$$= \hat{\boldsymbol{\varrho}}\,d\varrho + \hat{\boldsymbol{\phi}}\varrho\,d\phi + \hat{\mathbf{z}}\,dz$$
$$= \hat{\mathbf{r}}\,dr + \hat{\boldsymbol{\theta}}r\,d\theta + \hat{\boldsymbol{\phi}}r \sin\theta\,d\phi$$

2. ∇ Operator

In rectangular coordinates (x, y, z):

$$\nabla\Phi = \left(\hat{\mathbf{x}}\frac{\partial}{\partial x} + \hat{\mathbf{y}}\frac{\partial}{\partial y} + \hat{\mathbf{z}}\frac{\partial}{\partial z}\right)\Phi$$

$$\nabla\cdot\mathbf{A} = \frac{\partial A_x}{\partial x} + \frac{\partial A_y}{\partial y} + \frac{\partial A_z}{\partial z}$$

$$\nabla\times\mathbf{A} = \begin{vmatrix} \hat{\mathbf{x}} & \hat{\mathbf{y}} & \hat{\mathbf{z}} \\ \dfrac{\partial}{\partial x} & \dfrac{\partial}{\partial y} & \dfrac{\partial}{\partial z} \\ A_x & A_y & A_z \end{vmatrix}$$

$$\nabla^2\Phi = \nabla\cdot\nabla\Phi = \left(\frac{\partial^2}{\partial x^2} + \frac{\partial^2}{\partial y^2} + \frac{\partial^2}{\partial z^2}\right)\Phi$$

$$\nabla^2\mathbf{A} = \hat{\mathbf{x}}\nabla^2 A_x + \hat{\mathbf{y}}\nabla^2 A_y + \hat{\mathbf{z}}\nabla^2 A_z$$

In cylindrical coordinates (ϱ, ϕ, z):

$$\nabla\Phi = \left(\hat{\boldsymbol{\varrho}}\frac{\partial}{\partial \varrho} + \hat{\boldsymbol{\phi}}\frac{\partial}{\varrho\,\partial\phi} + \hat{\mathbf{z}}\frac{\partial}{\partial z}\right)\Phi$$

$$\nabla \cdot \mathbf{A} = \frac{1}{\varrho}\frac{\partial}{\partial \varrho}(\varrho A_\varrho) + \frac{1}{\varrho}\frac{\partial A_\phi}{\partial \phi} + \frac{\partial A_z}{\partial z}$$

$$\nabla \times \mathbf{A} = \frac{1}{\varrho}\begin{vmatrix} \hat{\varrho} & \varrho\hat{\phi} & \hat{z} \\ \frac{\partial}{\partial \varrho} & \frac{\partial}{\partial \phi} & \frac{\partial}{\partial z} \\ A_\varrho & \varrho A_\phi & A_z \end{vmatrix}$$

$$= \hat{\varrho}\left(\frac{1}{\varrho}\frac{\partial A_z}{\partial \phi} - \frac{\partial A_\phi}{\partial z}\right) + \hat{\phi}\left(\frac{\partial A_\varrho}{\partial z} - \frac{\partial A_z}{\partial \varrho}\right) + \hat{z}\left[\frac{1}{\varrho}\frac{\partial}{\partial \varrho}(\varrho A_\phi) - \frac{1}{\varrho}\frac{\partial A_\varrho}{\partial \phi}\right]$$

$$\nabla^2 \Phi = \frac{1}{\varrho}\frac{\partial}{\partial \varrho}\left(\varrho \frac{\partial \Phi}{\partial \varrho}\right) + \frac{1}{\varrho^2}\frac{\partial^2 \Phi}{\partial \phi^2} + \frac{\partial^2 \Phi}{\partial z^2}$$

$$\nabla^2 \mathbf{A} = \nabla\nabla\cdot\mathbf{A} - \nabla \times \nabla \times \mathbf{A} \neq \hat{\varrho}\nabla^2 A_\varrho + \hat{\phi}\nabla^2 A_\phi + \hat{z}\nabla^2 A_z$$

In spherical coordinates (r, θ, ϕ):

$$\nabla\Phi = \left(\hat{r}\frac{\partial}{\partial r} + \hat{\theta}\frac{1}{r}\frac{\partial}{\partial \theta} + \hat{\phi}\frac{1}{r\sin\theta}\frac{\partial}{\partial \phi}\right)\Phi$$

$$\nabla\cdot\mathbf{A} = \frac{1}{r}\frac{\partial}{\partial r}(r^2 A_r) + \frac{1}{r\sin\theta}\frac{\partial}{\partial \theta}(A_\theta \sin\theta) + \frac{1}{r\sin\theta}\frac{\partial A_\phi}{\partial \phi}$$

$$\nabla \times \mathbf{A} = \frac{1}{r^2\sin\theta}\begin{vmatrix} \hat{r} & r\hat{\theta} & (r\sin\theta)\hat{\phi} \\ \frac{\partial}{\partial r} & \frac{\partial}{\partial \theta} & \frac{\partial}{\partial \phi} \\ A_r & rA_\theta & (r\sin\theta)A_\phi \end{vmatrix}$$

$$= \hat{r}\frac{1}{r\sin\theta}\left[\frac{\partial}{\partial \theta}(A_\phi \sin\theta) - \frac{\partial A_\theta}{\partial \phi}\right] + \hat{\theta}\frac{1}{r}\left[\frac{1}{\sin\theta}\frac{\partial A_r}{\partial \phi} - \frac{\partial}{\partial r}(rA_\phi)\right]$$
$$+ \hat{\phi}\frac{1}{r}\left[\frac{\partial}{\partial r}(rA_\phi) - \frac{\partial A_r}{\partial \theta}\right]$$

$$\nabla^2\Phi = \frac{1}{r^2}\frac{\partial}{\partial r}\left(r^2\frac{\partial \Phi}{\partial r}\right) + \frac{1}{r^2\sin\theta}\frac{\partial}{\partial \theta}\left(\sin\theta\frac{\partial \Phi}{\partial \theta}\right) + \frac{1}{r^2\sin^2\theta}\frac{\partial^2\Phi}{\partial \phi^2}$$

$$\nabla^2\mathbf{A} = \nabla\nabla\cdot A - \nabla \times \nabla \times A \neq \hat{r}\nabla^2 A_r + \hat{\theta}\nabla^2 A_\theta + \hat{\phi}\nabla^2 A_\phi$$

3. Identities

$$\mathbf{a}\cdot\mathbf{b}\times\mathbf{c} = \mathbf{a}\times\mathbf{b}\cdot\mathbf{c} = \mathbf{b}\cdot\mathbf{c}\times\mathbf{a}$$

$$\mathbf{a}\times(\mathbf{b}\times\mathbf{c}) = (\mathbf{a}\cdot\mathbf{c})\mathbf{b} - (\mathbf{a}\cdot\mathbf{b})\mathbf{c}$$

$$(\mathbf{a}\times\mathbf{b})\cdot(\mathbf{c}\times\mathbf{d}) = \mathbf{a}\cdot\mathbf{b}\times(\mathbf{c}\times\mathbf{d}) = \mathbf{a}\cdot[(\mathbf{b}\cdot\mathbf{d})\mathbf{c} - (\mathbf{b}\cdot\mathbf{c})\mathbf{d}] = (\mathbf{a}\cdot\mathbf{c})(\mathbf{b}\cdot\mathbf{d}) - (\mathbf{a}\cdot\mathbf{d})(\mathbf{b}\cdot\mathbf{c})$$

$$(\mathbf{a}\times\mathbf{b})\times(\mathbf{c}\times\mathbf{d}) = (\mathbf{a}\times\mathbf{b}\cdot\mathbf{d})\mathbf{c} - (\mathbf{a}\times\mathbf{b}\cdot\mathbf{c})\mathbf{d}$$

$$\nabla(\Phi + \psi) = \nabla\Phi + \nabla\psi$$

$$\nabla(\Phi\psi) = \Phi\nabla\psi + \psi\nabla\Phi$$

$$\nabla\cdot(\mathbf{a} + \mathbf{b}) = \nabla\cdot\mathbf{a} + \nabla\cdot\mathbf{b}$$

$$\nabla \times (\mathbf{a}+\mathbf{b}) = \nabla \times \mathbf{a} + \nabla \times \mathbf{b}$$
$$\nabla \cdot (\Phi \mathbf{a}) = \mathbf{a} \cdot \nabla \Phi + \Phi \nabla \cdot \mathbf{a}$$
$$\nabla \times (\Phi \mathbf{a}) = \nabla \Phi \times \mathbf{a} + \Phi \nabla \times \mathbf{a}$$
$$\nabla (\mathbf{a} \cdot \mathbf{b}) = (\mathbf{a} \cdot \nabla)\mathbf{b} + (\mathbf{b} \cdot \nabla)\mathbf{a} + \mathbf{a} \times (\nabla \times \mathbf{b}) + \mathbf{b} \times (\nabla \times \mathbf{a})$$
$$\nabla \times (\mathbf{a} \times \mathbf{b}) = \mathbf{a} \nabla \cdot \mathbf{b} - \mathbf{b} \nabla \cdot \mathbf{a} + (\mathbf{b} \cdot \nabla)\mathbf{a} - (\mathbf{a} \cdot \nabla)\mathbf{b}$$
$$\nabla \cdot (\mathbf{a} \times \mathbf{b}) = \mathbf{b} \cdot \nabla \times \mathbf{a} - \mathbf{a} \cdot \nabla \times \mathbf{b}$$
$$\nabla \times \nabla \times \mathbf{a} = \nabla \nabla \cdot \mathbf{a} - \nabla^2 \mathbf{a}$$
$$\nabla \times \nabla \Phi \equiv 0$$
$$\nabla \cdot \nabla \times \mathbf{a} \equiv 0$$
$$\iiint_V \nabla \cdot \mathbf{a} \, dV = \oiint_S \mathbf{a} \cdot d\mathbf{S} \quad \text{(Gauss's theorem)}$$
$$\iint_S \nabla \times \mathbf{a} \cdot d\mathbf{S} = \oint_C \mathbf{a} \cdot d\boldsymbol{\ell} \quad \text{(Stokes's theorem)}$$

Green's first and second identities:

$$\iiint_V (\nabla \psi \cdot \nabla \Phi + \Phi \nabla^2 \psi) \, dV = \oiint_S \Phi \nabla \psi \cdot d\mathbf{S}$$
$$\iiint_V (\Phi \nabla^2 \psi - \psi \nabla^2 \Phi) \, dV = \oiint_S (\Phi \nabla \psi - \psi \nabla \Phi) \cdot d\mathbf{S}$$
$$\iiint_V (\nabla \times \mathbf{A} \cdot \nabla \times \mathbf{B} - \mathbf{A} \cdot \nabla \times \nabla \times \mathbf{B}) \, dV = \iiint_V (\nabla \cdot \mathbf{A} \times \nabla \times \mathbf{B}) \, dV$$
$$= \oiint_S \mathbf{A} \times \nabla \times \mathbf{B} \cdot d\mathbf{S}$$
$$\iiint_V (\mathbf{B} \cdot \nabla \times \nabla \times \mathbf{A} - \mathbf{A} \cdot \nabla \times \nabla \times \mathbf{B}) \, dV = \oiint_S (\mathbf{A} \times \nabla \times \mathbf{B} - \mathbf{B} \times \nabla \times \mathbf{A}) \cdot d\mathbf{S}$$

Appendix E

VSWR Versus Reflection Coefficient and Mismatch Loss

Yi-Lin Chen
*University of Illinois**

The following relations are used in the construction of the vswr table below.

$$\text{vswr} = \frac{1 + |\Gamma|}{1 - |\Gamma|}, \qquad |\Gamma| = \frac{\text{vswr} - 1}{\text{vswr} + 1}$$

$$\text{mismatch loss (dB)} = -10 \log_{10}(1 - |\Gamma|^2)$$

VSWR Versus Reflection Coefficient (Γ) and Mismatch Loss

VSWR	$|\Gamma|$	Mismatch Loss (dB)	VSWR	$|\Gamma|$	Mismatch Loss (dB)
1.01	.0050	.0001	1.12	.0566	.0139
1.02	.0099	.0004	1.13	.0610	.0162
1.03	.0148	.0009	1.14	.0654	.0186
1.04	.0196	.0017	1.15	.0698	.0212
1.05	.0244	.0026	1.16	.0741	.0239
1.06	.0291	.0037	1.17	.0783	.0267
1.07	.0338	.0050	1.18	.0826	.0297
1.08	.0385	.0064	1.19	.0868	.0328
1.09	.0431	.0081	1.20	.0909	.0360
1.10	.0476	.0099	1.21	.0950	.0394
1.11	.0521	.0118	1.22	.0991	.0429

*On leave from the Chinese Aeronautical Laboratory, Beijing, China, during 1983.

VSWR Versus Reflection Coefficient (Γ) and Mismatch Loss (cont'd.)

VSWR	|Γ|	Mismatch Loss (dB)	VSWR	|Γ|	Mismatch Loss (dB)
1.23	.1031	.0464	1.73	.2674	.3222
1.24	.1071	.0501	1.74	.2701	.3289
1.25	.1111	.0540	1.75	.2727	.3357
1.26	.1150	.0579	1.76	.2754	.3425
1.27	.1189	.0619	1.77	.2780	.3493
1.28	.1228	.0660	1.78	.2806	.3561
1.29	.1266	.0702	1.79	.2832	.3630
1.30	.1304	.0745	1.80	.2857	.3698
1.31	.1342	.0789	1.81	.2883	.3767
1.32	.1379	.0834	1.82	.2908	.3837
1.33	.1416	.0880	1.83	.2933	.3906
1.34	.1453	.0927	1.84	.2958	.3976
1.35	.1489	.0974	1.85	.2982	.4046
1.36	.1525	.1023	1.86	.3007	.4116
1.37	.1561	.1072	1.87	.3031	.4186
1.38	.1597	.1121	1.88	.3056	.4257
1.39	.1632	.1172	1.89	.3080	.4327
1.40	.1667	.1223	1.90	.3103	.4398
1.41	.1701	.1275	1.91	.3127	.4469
1.42	.1736	.1328	1.92	.3151	.4540
1.43	.1770	.1382	1.93	.3174	.4612
1.44	.1803	.1436	1.94	.3197	.4683
1.45	.1837	.1490	1.95	.3220	.4755
1.46	.1870	.1546	1.96	.3243	.4827
1.47	.1903	.1602	1.97	.3266	.4899
1.48	.1935	.1658	1.98	.3289	.4971
1.49	.1968	.1715	1.99	.3311	.5043
1.50	.2000	.1773	2.00	.3333	.5115
1.51	.2032	.1831	2.05	.3443	.5479
1.52	.2063	.1890	2.10	.3548	.5844
1.53	.2095	.1949	2.15	.3651	.6212
1.54	.2126	.2009	2.20	.3750	.6582
1.55	.2157	.2069	2.25	.3846	.6952
1.56	.2188	.2130	2.30	.3939	.7324
1.57	.2218	.2191	2.35	.4030	.7696
1.58	.2248	.2252	2.40	.4118	.8069
1.59	.2278	.2314	2.45	.4203	.8441
1.60	.2308	.2377	2.50	.4286	.8814
1.61	.2337	.2440	2.55	.4366	.9186
1.62	.2366	.2503	2.60	.4444	.9557
1.63	.2395	.2566	2.65	.4521	.9928
1.64	.2424	.2630	2.70	.4595	1.0298
1.65	.2453	.2695	2.75	.4667	1.0667
1.66	.2481	.2760	2.80	.4737	1.1035
1.67	.2509	.2825	2.85	.4805	1.1402
1.68	.2537	.2890	2.90	.4872	1.1767
1.69	.2565	.2956	2.95	.4937	1.2131
1.70	.2593	.3022	3.00	.5000	1.2494
1.71	.2620	.3088	3.10	.5122	1.3215
1.72	.2647	.3155	3.20	.5238	1.3929

Appendixes

VSWR Versus Reflection Coefficient (Γ) and Mismatch Loss

VSWR	|Γ|	Mismatch Loss (dB)	VSWR	|Γ|	Mismatch Loss (dB)
3.30	.5349	1.4636	8.10	.7802	4.0754
3.40	.5455	1.5337	8.20	.7826	4.1170
3.50	.5556	1.6030	8.30	.7849	4.1583
3.60	.5652	1.6715	8.40	.7872	4.1992
3.70	.5745	1.7393	8.50	.7895	4.2397
3.80	.5833	1.8064	8.60	.7917	4.2798
3.90	.5918	1.8727	8.70	.7938	4.3196
4.00	.6000	1.9382	8.80	.7959	4.3591
4.10	.6078	2.0030	8.90	.7980	4.3982
4.20	.6154	2.0670	9.00	.8000	4.4370
4.30	.6226	2.1302	9.10	.8020	4.4754
4.40	.6296	2.1927	9.20	.8039	4.5135
4.50	.6364	2.2545	9.30	.8058	4.5513
4.60	.6429	2.3156	9.40	.8077	4.5888
4.70	.6491	2.3759	9.50	.8095	4.6260
4.80	.6552	2.4355	9.60	.8113	4.6628
4.90	.6610	2.4945	9.70	.8131	4.6994
5.00	.6667	2.5527	9.80	.8148	4.7356
5.10	.6721	2.6103	9.90	.8165	4.7716
5.20	.6774	2.6672	10.00	.8182	4.8073
5.30	.6825	2.7235	11.00	.8333	5.1491
5.40	.6875	2.7791	12.00	.8462	5.4665
5.50	.6923	2.8340	13.00	.8571	4.7625
5.60	.6970	2.8884	14.00	.8667	6.0399
5.70	.7015	2.9421	15.00	.8750	6.3009
5.80	.7059	2.9953	16.00	.8824	6.5472
5.90	.7101	3.0479	17.00	.8889	6.7804
6.00	.7143	3.0998	18.00	.8947	7.0017
6.10	.7183	3.1513	19.00	.9000	7.2125
6.20	.7222	3.2021	20.00	.9048	7.4135
6.30	.7260	3.2525	30.00	.9355	9.0354
6.40	.7297	3.3022	40.00	.9512	10.2145
6.50	.7333	3.3515	50.00	.9608	11.1411
6.60	.7368	3.4002	60.00	.9672	11.9045
6.70	.7403	3.4485	70.00	.9718	12.5536
6.80	.7436	3.4962	80.00	.9753	13.1182
6.90	.7468	3.5435	90.00	.9780	13.6178
7.00	.7500	3.5902	100.00	.9802	14.0658
7.10	.7531	3.6365	200.00	.9900	17.0330
7.20	.7561	3.6824	300.00	.9934	18.7795
7.30	.7590	3.7277	400.00	.9950	20.0217
7.40	.7619	3.7727	500.00	.9960	20.9865
7.50	.7647	3.8172	600.00	.9967	21.7754
7.60	.7674	3.8612	700.00	.9971	22.4428
7.70	.7701	3.9049	800.00	.9975	23.0212
7.80	.7727	3.9481	900.00	.9978	23.5315
7.90	.7753	3.9909	1000.00	.9980	23.9881
8.00	.7778	4.0334			

Appendix F

Decibels Versus Voltage and Power Ratios*

Yi-Lin Chen
University of Illinois[†]

The decibel chart below indicates decibels for any ratio of voltage or power up to 100 dB. For voltage ratios greater than 10 (or power ratios greater than 100) the ratio can be broken down into two products, the decibels found for each separately, the two results then added. For example, to convert a voltage ratio of 200:1 to dB, a 200:1 voltage ratio equals the product of 100:1 and 2:1. Now, 100:1 equals 40 dB; 21:1 equals 6 dB. Therefore a 200:1 voltage ratio equals 40 dB + 6 dB, or 46 dB.

$$dB = 20\log_{10}(\text{voltage ratio}) = 10\log_{10}(\text{power ratio})$$

---

*Reprinted with permission of *Microwave Journal*, from *The Microwave Engineer's Handbook and Buyer's Guide*, 1966 issue, © 1966 Horizon House–Microwave, Inc.
[†]On leave from the Chinese Aeronautical Laboratory, Beijing, China, during 1983.

Decibels Versus Voltage and Power Ratios

Voltage Ratio	Power Ratio	−dB +	Voltage Ratio	Power Ratio	Voltage Ratio	Power Ratio	−dB +	Voltage Ratio	Power Ratio	Voltage Ratio	Power Ratio	−dB +	Voltage Ratio	Power Ratio
1.0000	1.0000	0	1.000	1.000	.5309	.2818	5.5	1.884	3.548	.2818	.07943	11.0	3.548	12.59
.9886	.9772	.1	1.012	1.023	.5248	.2754	5.6	1.905	3.631	.2876	.07762	11.1	3.589	12.88
.9772	.9550	.2	1.023	1.047	.5188	.2692	5.7	1.928	3.715	.2754	.07586	11.2	3.631	13.18
.9661	.9333	.3	1.035	1.072	.5129	.2630	5.8	1.950	3.802	.2723	.07413	11.3	3.673	13.49
.9550	.9120	.4	1.047	1.096	.5070	.2570	5.9	1.972	3.890	.2692	.07244	11.4	3.715	13.80
.9441	.8913	.5	1.059	1.122	.5012	.2512	6.0	1.995	3.981	.2661	.07079	11.5	3.758	14.13
.9333	.8710	.6	1.072	1.148	.4955	.2455	6.1	2.018	4.074	.2630	.06918	11.6	3.802	14.45
.9226	.8511	.7	1.084	1.175	.4898	.2399	6.2	2.042	4.169	.2600	.06761	11.7	3.846	14.79
.9120	.8318	.8	1.096	1.202	.4842	.2344	6.3	2.065	4.266	.2570	.06607	11.8	3.890	15.14
.9016	.8128	.9	1.109	1.230	.4786	.2291	6.4	2.089	4.365	.2541	.06457	11.9	3.936	15.49
.8913	.7943	1.0	1.122	1.259	.4732	.2239	6.5	2.113	4.467	.2512	.06310	12.0	3.981	15.85
.8810	.7762	1.1	1.135	1.288	.4677	.2188	6.6	2.138	4.571	.2483	.06166	12.1	4.027	16.22
.8710	.7586	1.2	1.148	1.318	.4624	.2138	6.7	2.163	4.677	.2455	.06026	12.2	4.074	16.60
.8610	.7413	1.3	1.161	1.349	.4571	.2089	6.8	2.188	4.786	.2427	.05888	12.3	4.121	16.98
.8511	.7244	1.4	1.175	1.380	.4519	.2042	6.9	2.213	4.898	.2399	.05754	12.4	4.169	17.38
.8414	.7079	1.5	1.189	1.413	.4467	.1995	7.0	2.239	5.012	.2371	.05623	12.5	4.217	17.78
.8318	.6918	1.6	1.202	1.445	.4416	.1950	7.1	2.265	5.129	.2344	.05495	12.6	4.266	18.20
.8222	.6761	1.7	1.216	1.479	.4365	.1905	7.2	2.291	5.248	.2317	.05370	12.7	4.315	18.62
.8128	.6607	1.8	1.230	1.514	.4315	.1862	7.3	2.317	5.370	.2291	.05248	12.8	4.365	19.05
.8035	.6457	1.9	1.245	1.549	.4266	.1820	7.4	2.344	5.495	.2265	.05129	12.9	4.416	19.50
.7943	.6310	2.0	1.259	1.585	.4217	.1778	7.5	2.371	5.623	.2239	.05012	13.0	4.467	19.95
.7852	.6166	2.1	1.274	1.622	.4169	.1738	7.6	2.399	5.754	.2213	.04898	13.1	4.519	20.42
.7762	.6026	2.2	1.288	1.660	.4121	.1698	7.7	2.427	5.888	.2188	.04786	13.2	4.571	20.89
.7674	.5888	2.3	1.303	1.698	.4074	.1660	7.8	2.455	6.026	.2163	.04677	13.3	4.624	21.38
.7586	.5754	2.4	1.318	1.738	.4027	.1622	7.9	2.483	6.166	.2138	.04571	13.4	4.677	21.88

.7499	.5623	2.5	1.334	1.778	.3981	.1585	8.0	2.512	6.310	.2113	.04467	13.5	4.732	22.39
.7413	.5495	2.6	1.349	1.820	.3936	.1549	8.1	2.541	6.457	.2089	.04365	13.6	4.786	22.91
.7328	.5370	2.7	1.365	1.862	.3890	.1514	8.2	2.570	6.607	.2065	.04266	13.7	4.842	23.44
.7244	.5248	2.8	1.380	1.905	.3846	.1479	8.3	2.600	6.761	.2042	.04169	13.8	4.898	23.99
.7161	.5129	2.9	1.396	1.950	.3802	.1445	8.4	2.630	6.918	.2018	.04074	13.9	4.955	24.55
.7079	.5012	3.0	1.413	1.995	.3758	.1413	8.5	2.661	7.079	.1995	.03981	14.0	5.012	25.12
.6998	.4898	3.1	1.429	2.042	.3715	.1380	8.6	2.692	7.244	.1972	.03890	14.1	5.070	25.70
.6918	.4786	3.2	1.445	2.089	.3673	.1349	8.7	2.723	7.413	.1950	.03802	14.2	5.129	26.30
.6839	.4677	3.3	1.462	2.138	.3631	.1318	8.8	2.754	7.586	.1928	.03715	14.3	5.188	26.92
.6761	.4571	3.4	1.479	2.188	.3589	.1288	8.9	2.786	7.762	.1905	.03631	14.4	5.248	27.54
.6683	.4467	3.5	1.496	2.239	.3548	.1259	9.0	2.818	7.943	.1884	.03548	14.5	5.309	28.18
.6607	.4365	3.6	1.514	2.291	.3508	.1230	9.1	2.851	8.128	.1862	.03467	14.6	5.370	28.84
.6531	.4266	3.7	1.531	2.344	.3467	.1202	9.2	2.884	8.318	.1841	.03388	14.7	5.433	29.51
.6457	.4169	3.8	1.549	2.399	.3428	.1175	9.3	2.917	8.511	.1820	.03311	14.8	5.495	30.20
.6383	.4074	3.9	1.567	2.455	.3388	.1148	9.4	2.951	8.710	.1799	.03236	14.9	5.559	30.90
.6310	.3981	4.0	1.585	2.512	.3350	.1122	9.5	2.985	8.913	.1778	.03162	15.0	5.623	31.62
.6237	.3890	4.1	1.603	2.570	.3311	.1096	9.6	3.020	9.120	.1758	.03090	15.1	5.689	32.36
.6166	.3802	4.2	1.622	2.630	.3273	.1072	9.7	3.055	9.333	.1738	.03020	15.2	5.754	33.11
.6095	.3715	4.3	1.641	2.692	.3236	.1047	9.8	3.090	9.550	.1718	.02951	15.3	5.821	33.88
.6026	.3631	4.4	1.660	2.754	.3199	.1023	9.9	3.126	9.772	.1698	.02884	15.4	5.888	34.67
.5957	.3548	4.5	1.679	2.818	.3162	.1000	10.0	3.162	10.000	.1679	.02818	15.5	5.957	35.48
.5888	.3467	4.6	1.698	2.884	.3126	.09772	10.1	3.199	10.23	.1660	.02754	15.6	6.026	36.31
.5821	.3388	4.7	1.718	2.951	.3090	.09550	10.2	3.236	10.47	.1641	.02692	15.7	6.095	37.15
.5754	.3311	4.8	1.738	3.020	.3055	.09333	10.3	3.273	10.72	.1622	.02630	15.8	6.166	38.02
.5689	.3236	4.9	1.758	3.090	.3020	.09120	10.4	3.311	10.96	.1603	.02570	15.9	6.237	38.90
.5623	.3162	5.0	1.778	3.162	.2985	.08913	10.5	3.350	11.22	.1585	.02512	16.0	6.310	39.81
.5559	.3090	5.1	1.799	3.236	.2951	.08710	10.6	3.388	11.48	.1567	.02455	16.1	6.383	40.74
.5495	.3020	5.2	1.820	3.311	.2917	.08511	10.7	3.428	11.75	.1549	.02399	16.2	6.457	41.69
.5433	.2951	5.3	1.841	3.388	.2884	.08318	10.8	3.467	12.02	.1531	.02344	16.3	6.531	42.66
.5370	.2884	5.4	1.862	3.467	.2851	.08128	10.9	3.508	12.30	.1514	.02291	16.4	6.607	43.65

Decibels Versus Voltage and Power Ratios, (cont'd.)

Voltage Ratio	Power Ratio	−dB +	Voltage Ratio	Power Ratio	Voltage Ratio	Power Ratio	−dB +	Voltage Ratio	Power Ratio	Voltage Ratio	Power Ratio	−dB +	Voltage Ratio	Power Ratio
.1496	.02239	16.5	6.683	44.67	.1259	.01585	18.0	7.943	63.10	.1059	.01122	19.5	9.441	89.13
.1479	.02188	16.6	6.761	45.71	.1245	.01549	18.1	8.035	64.57	.1047	.01096	19.6	9.550	91.20
.1462	.02138	16.7	6.839	46.77	.1230	.01514	18.2	8.128	66.07	.1035	.01072	19.7	9.661	93.33
.1445	.02089	16.8	6.918	47.86	.1216	.01479	18.3	8.222	67.61	.1023	.01047	19.8	9.772	95.50
.1429	.02042	16.9	6.998	48.98	.1202	.01445	18.4	8.318	69.18	.1012	.01023	19.9	9.886	97.72
.1413	.01995	17.0	7.079	50.12	.1189	.01413	18.5	8.414	70.79	.1000	.01000	20.0	10.000	100.00
.1396	.01950	17.1	7.161	51.29	.1175	.01380	18.6	8.511	72.44					
.1380	.01905	17.2	7.244	52.48	.1161	.01349	18.7	8.610	74.13			30		10^3
.1365	.01862	17.3	7.328	53.70	.1148	.01318	18.8	8.710	75.86	10^{-2}	10^{-3}	40	10^2	10^4
.1349	.01820	17.4	7.413	54.95	.1135	.01288	18.9	8.811	77.62		10^{-4}	50		10^5
.1334	.01778	17.5	7.499	56.23	.1122	.01259	19.0	8.913	79.43		10^{-5}	60		10^6
.1318	.01738	17.6	7.586	57.54	.1109	.01230	19.1	9.016	81.28	10^{-3}	10^{-6}	70	10^3	10^7
.1303	.01698	17.7	7.674	58.88	.1096	.01202	19.2	9.120	83.18	10^{-4}	10^{-7}	80	10^4	10^8
.1288	.01660	17.8	7.762	60.26	.1084	.01175	19.3	9.226	85.11	10^{-5}	10^{-8}	90		10^9
.1274	.01622	17.9	7.852	61.66	.1072	.01148	19.4	9.333	87.10		10^{-9}			
											10^{-10}	100	10^5	10^{10}

Index

Index

Abbe sine condition
  with lenses, 16-19 to 16-23, 16-29, 16-56, 21-18
  radiation pattern of lenses obeying, 19-97 to 19-99
Aberrations with lenses, 16-12 to 16-19, 21-23. *See also* Coma lens aberrations
Absolute gain measurements, 32-42 to 32-49
Absolute models, 32-69
Absolute polarization measurements, 32-60 to 32-61
Absorbers with UTD solutions, 20-17
Absorption
  of ionospheric propagation, 29-24
  of satellite-earth propagation, 29-8
ACS (attitude control systems), 22-46
Active element patterns
  of periodic arrays, 13-6
  with phased arrays, 21-8
Active region
  with log-periodic antennas
    dipole, 9-17 to 9-21, 9-24
    mono arrays, 9-65
    phasing of, 9-56
    zigzag wire, 9-35 to 9-36
  with log-spiral antennas, 9-75
    conical, 9-83, 9-84, 9-98 to 9-99
Adcock rotating antennas systems, 25-4, 25-9 to 25-10
  for in-rotating null patterns, 25-11 to 25-12
Admittance
  of edge slot arrays, 12-26
  of medium, 1-18
  mutual, between slots, 4-91
  of ridged TE/TM waveguides, 28-47
  of thin-wire antennas, 7-7 to 7-8, 7-14 to 7-15
  of transmission lines, 9-47 to 9-49, 28-5
Admittance matrices, 3-61
  and symmetries, 3-65, 3-67 to 3-68
Admittivity and dielectric modeling, 32-72
Advanced microwave sounding unit, 22-25, 22-48
Advanced Multifrequency Scanning Radiometer (AMSR), 22-48, 22-50, 22-51 to 22-52
AF method. *See* Aperture field method
Airborne Radiation Pattern code, 20-4, 20-70 to 20-85
Air coaxial transmission lines, 28-16
Aircraft and aircraft antennas
  blade antennas, 30-12, 30-16 to 30-19
  EMP responses for, 30-4, 30-6, 30-17 to 30-31
  interferometers mounted on, 25-21 to 25-23
  military, simulation of, 20-78 to 20-88
  modeling of, 32-74 to 32-76, 32-79 to 32-86
  monopoles, 20-62 to 20-68, 20-70, 20-73 to 20-75, 20-87 to 20-88
  numerical solutions for, 20-70 to 20-90
  radomes on, 31-3 to 31-4
  simulation of, 20-30, 20-37 to 20-42, 20-63 to 20-65
  structure of, as antenna, 20-61
Air gap with leaky-wave antennas, 17-97 to 17-98
Airport radar, antennas for, 19-19, 19-56
Air striplines, 21-58, 21-60
  for power divider elements, 21-50 to 21-51
  with satellite antennas, 21-7
Air traffic control antennas, 19-19, 19-56
Alignment of antenna ranges, 32-33 to 32-34
Aluminum alloys for satellite antennas, 21-28 to 21-29
AM antennas
  directional feeders for, 26-37 to 26-47
  ground systems for, 26-37

AM antennas (*cont.*)
  horizontal plane field strength in, 26-33 to 26-34
  patterns for
    augmented, 26-36
    size determination of, 26-22 to 26-33
    standard, 26-35 to 26-36
    theoretical, 26-34 to 26-35
    two-tower, 26-22
  power for system losses in, 26-36 to 26-37
  sky-wave propagation with, 29-25 to 29-27
  standard reference, 26-3 to 26-22
Ammeters, 2-30 to 2-32
Amplitude meters, 2-30 to 2-32
Amplitude patterns, measurement of, 32-39 to 32-41
Amplitude quantization errors, 21-12
Amplitude source, ideal, 2-29 to 2-30
AMSR, 22-48, 22-50, 22-51 to 22-52
AMSU (advanced microwave sounding unit), 22-25, 22-50, 22-51 to 22-52
Analytical formulations of modeling codes, 3-89
Analytical validation of computer code, 3-76
Anechoic chambers for indoor ranges, 32-25 to 32-28
  evaluation of, 32-38 to 32-39
Angles with log-periodic antennas, 9-12, 9-14, 9-15, 9-23
Annular patches for microstrip antennas, 10-43 to 10-45
  characteristics of, 10-17
  resonant frequency for, 10-47, 10-48
Annular phased array applicator for medical applications, 24-16 to 24-18
Annular-sector patches for microstrip antennas
  characteristics of, 10-17
  resonant frequency of, 10-47, 10-49
Annular slot antennas, 30-13
Antarctica, brightness temperature image of, 22-15
Antenna arrays. *See* Arrays
Antenna ranges. *See* Ranges, antenna
Antenna sampling systems for feeder systems, 26-41 to 26-44
Antipodal fin lines, 28-55
Aperiodic arrays, 14-3 to 14-6
  linear arrays as, 11-41
  optically fed, 19-57 to 19-59
  probabilistic approach to, 14-8 to 14-35
  space-tapered, 14-6 to 14-8
Aperture antennas, 5-5
  and discrete arrays, 11-29 to 11-30
  optimization of, 11-74
Aperture blockage, 5-16
  with Cassegrain feed systems, 19-62
  in compact ranges, 32-29, 33-23
  with HIHAT antennas, 19-87
  and lens antennas, 16-5, 21-16
  with millimeter-wave antennas, 17-28
  and offset parabolic antennas, 15-80
    dual-reflector, 15-61
    with off-focus feeds, 15-58
  with phased array feeds, 19-60 to 19-61
  with reflector antennas, 15-17 to 15-18
  and space-fed arrays, 19-54 to 19-55
Aperture couplers power division elements, 21-51 to 21-52
Aperture distributions. *See* Apertures and aperture distributions
Aperture efficiency
  of dielectric-loaded horn antennas, 8-73
  and effective area, 5-27 to 5-28
  of geodesic antennas, 17-32
  with lenses, 16-19, 16-20
  with offset phased-array feeds, 19-61
  and parallel feed networks, 19-6
  with reflector antennas, 21-24 to 21-25
  and reflector surface errors, 15-105
Aperture field method, 5-5
  for pyramidal horn fields, 15-92 to 15-93
  with reflector antennas, 8-72, 15-7, 15-13 to 15-15
Aperture fields, 15-88
  for circular arrays, 21-92
  for circumferential slots, 21-92
  of horn antennas
    conical, 15-93
    corrugated, 8-53
    E-plane, 8-5 to 8-7
    H-plane, 8-20
    pyramidal, 8-34, 8-36
Aperture-matched horn antennas, 8-50, 8-64
  bandwidths of, 8-65 to 8-66
  radiated fields of, 8-58, 8-66 to 8-67
  vswr of, 8-68
Apertures and aperture distributions, 11-13. *See also* Circular apertures; Planar apertures; Tapered aperture distributions; Uniform aperture distributions

Index

with Butler matrix, 19-8
on curved surfaces, 13-50 to 13-51
efficiency of. *See* Aperture efficiency
electric and magnetic fields in, 5-8
with feed systems
    radial transmission line, 19-28, 19-31, 19-33
    semiconstrained, 19-17
    unconstrained, 19-51
and gain, 5-26 to 5-27, 17-7, 17-8
illumination of, 13-30 to 13-32, 13-36 to 13-37
with lens antennas, 21-14, 21-16, 21-17
    phase distributions of, 16-7, 16-36
    power distributions of, 16-33, 16-36
of longitudinal array slots, 12-24
and parallel feed networks, 19-6
in perfectly conducting ground, 3-22
radiation from, 1-28
    circular, 5-20 to 5-25
    and equivalent currents, 5-8 to 5-10
    near-field, 5-24 to 5-26
    planar aperture distributions, 5-10 to 5-11
    and plane-wave spectra, 5-5 to 5-7
    rectangular aperture, 5-11 to 5-20
reflections of, 8-67
with reflector antennas, 15-15 to 15-23, 21-20
phase error of, 15-107 to 15-108
for satellite antennas, 21-5, 21-7, 21-14, 21-16, 21-17
size of
    and conical horn beamwidth, 8-61
    of E-plane horn antenna directivity, 8-16
    of lenses, 16-6
square, 5-32 to 5-33
with Taylor line source synthesis, 13-26
Aperture taper
    and radiometer beam efficiency, 22-32, 22-33
    with reflector antennas, 21-24
Apparent phase center, 8-75
Appleton-Hartee formula for ionospheric refractive index, 29-21
Applied excitation of periodic arrays, 13-6 to 13-7
Arabsat satellite antenna, 21-77
Arbitrary ray optical field, 4-8
Archimedean-spiral curves, 9-107 to 9-108, 9-110

Array blindness, 13-45 to 13-49
    with flared-notch antennas, 13-59
Array collimations, 13-7 to 13-10
Array geometry, linear transformations in, 11-23 to 11-25
Array pattern functions, 3-35, 11-8
    aperiodic, 14-4 to 14-5, 14-8 to 14-9, 14-15, 14-22, 14-35
    Dolph-Chebyshev, 11-17 to 11-18
    linear, 11-8, 11-9, 11-11 to 11-14
    with longitudinal slots, 12-17
    phased, 21-10
    transformation of, 11-23 to 11-48
    with UTD solutions, 20-6
Arrays, 3-33, 3-35 to 3-36, 3-38 to 3-43. *See also* Aperiodic arrays; Array theory; Broadside arrays; Circular arrays; Log-periodic arrays; Log-spiral antennas; Periodic arrays; Planar arrays; Slot arrays
    element patterns with, 14-25
    errors in, from phase quantization, 13-52 to 13-57
    horn antennas as elements in, 8-4
    large, design of, 12-28 to 12-34
    of microstrip antennas, for medical applications, 24-25
    millimeter-wave, 17-23 to 17-27
    phase control of, 13-62 to 13-64
    scan characteristics of, 14-24 to 14-25
    of tapered dielectric-rod antennas, 17-47
    thinning of, with aperiodic arrays, 14-3
Array theory
    directivity in, 11-63 to 11-76
    general formulation of, 11-5 to 11-8
    for linear arrays, 11-8 to 11-23
    linear transformations in, 11-23 to 11-48
    pattern synthesis in, 11-76 to 11-86
    for planar arrays, 11-48 to 11-58
    SNR in, 11-58 to 11-63, 11-70 to 11-76
Artificial-dielectric plates, 21-19
A-sandwich panels, 31-17 to 31-19
Aspect ratio
    of dielectric grating antennas, 17-77 to 17-78
    surface corrugations for, 17-64
    and leakage constants, 17-62, 17-63
    with log-periodic antennas, 9-12, 9-14
    and directivity, 9-21 to 9-24
Assistance with computer models, 3-68
Astigmatic lens aberrations, 16-15, 16-20, 21-13

Astigmatic ray tube, 4-8
Asymmetric aperture distributions, 21-17
Asymmetric strips, 17-84 to 17-87, 17-97 to 17-98
AT-536/ARN marker beacon antenna, 30-27 to 30-28
AT-1076 uhf antenna, 30-21
Atmosphere
  absorption by
    of millimeter-wave antennas, 17-5 to 17-6
    and satellite-earth propagation, 29-8
  noise from, 6-25, 29-50 to 29-51
  refraction of, and line-of-sight propagation, 29-32 to 29-33
  refractive index of, 29-30 to 29-32
  remote sensing of, 22-6 to 22-7, 22-13 to 22-14, 22-16, 22-21 to 22-22
  standard radio, 29-32
Attachment coefficients, 4-96
  for conducting cylinder, 21-97
  of microstrip line, 17-118
Attenuation and attenuation constants
  of atmosphere, 17-5 to 17-6, 29-8
  with biconical horns, 21-104
  for conducting cylinder, 21-97
  ground wave, 29-45
  with log-periodic antennas, 9-4 to 9-5
  with microstrip patch antennas, 17-110
  with periodic loads, 9-49 to 9-54
  from radiation, 7-12, 9-5
  by rain, 29-10 to 29-13
  of rf cables, 28-26
  of standard waveguides, 28-40 to 28-41
    circular, 1-49 to 1-50
    rectangular, 1-39, 1-43
    TE/TM, 28-37
  for transmission lines, 28-7
  with uniform leaky-wave antennas, 17-89, 17-92
Attitude control systems, 22-46
Augmented am antenna pattern, 26-36 to 26-37
Auroral blackout and ionospheric propagation, 29-25
Automated antenna ranges, 32-30 to 32-31
Axial current on thin-wire antennas, 7-5 to 7-6
Axial feed displacements, 15-49 to 15-55
Axial gain, near-field, 5-29 to 5-33
Axial power density, 5-32 to 5-33

Axial ratios
  with lens antenna feeds, 21-79, 21-80
  and polarization, 32-51
    with log-spiral antennas, 9-77 to 9-78, 9-90 to 9-93
    with polarization ellipse, 1-15 to 1-16
Axis slot on elliptical cylinder, 4-75, 4-77, 21-92, 21-95
Azimuth-over-elevation positioners, 32-12
Azimuthal symmetry, 1-33

Babinet principle, 2-13 to 2-16
  and EMP and slot antennas, 30-14 to 30-16
Backfire arrays, 9-64
Back lobes and horn antennas
  aperture-matched, 8-67
  corrugated, 8-55, 8-58
Baklanov-Tseng-Cheng design, 11-55 to 11-56
Balanced antennas
  dipole, 9-72 to 9-73
  log-periodic zigzag, 9-45
  slot, 9-72 to 9-73, 9-78 to 9-79
  spiral, 9-78
Balanced bifilar helix, 9-83
Balanced two- and four-wire transmission lines, 28-12, 28-13
Baluns
  for log-spiral antennas, 9-75
    conical, 9-95, 9-103, 9-105
  microstrip, 18-8
Bandwidth, 3-28 to 3-31
  of beam-forming networks, 18-21, 18-24, 18-26, 21-32
  and diode phase shifters, 13-62 to 13-64, 18-16
  of feed circuits
    array feeds, 13-61 to 13-62
    broadband array, 13-39 to 13-41
    parallel feed, 19-5 to 19-6
    space-fed beam-forming feeds, 18-18
    true-time-delay, 19-77 to 19-78
  of horn antennas
    aperture-matched, 8-64 to 8-65
    corrugated, 8-64
    hybrid-mode, 15-99
    millimeter-wave, 17-23
  of leaky-wave antennas, 17-98
  of lens antennas
    constrained, 16-55
    dielectric, 16-54 to 16-56

equal group delay, 16-45 to 16-46
millimeter-wave, 17-14
Rinehart-Luneberg, 19-44, 19-46 to 19-49
zoning of, 16-39, 16-43, 16-54 to 16-55, 21-16
of log-periodic dipole antennas, 9-17, 9-21
of log-spiral conical antennas, 9-98 to 9-99
of longitudinal-shunt-slot array, 17-33
of microstrip antennas, 10-6, 10-46, 17-106, 17-118, 17-122
   with circular polarization, 10-59, 10-61
   dipole, 17-118 to 17-120, 17-125 to 17-127
   and impedance matching, 10-50 to 10-52
of millimeter-wave antennas, 17-5
   biconical, 17-32
   dielectric grating, 17-74 to 17-75
   holographic, 17-130
   horn, 17-13
   lens, 17-14
   microstrip arrays, 17-106, 17-114, 17-116
   tapered dielectric-rod, 17-39 to 17-40, 17-44
of phased arrays, 13-19 to 13-20, 18-8, 21-12
and Q-factor, 6-22, 10-6
of receiving antennas, 6-22, 6-24
of satellite antennas, 21-4 to 21-5, 21-7
of self-complementary antennas, 9-10
with side-mount tv antennas, 27-21
of subarrays, 13-32 to 13-35, 13-39
and substrate height, 17-107
and temperature sounder sensitivity, 22-22
of TEM waveguides, 21-57
Bar-line feed network, 21-74, 21-78
Bar line transmission types, 21-58
Barn antennas, 21-84 to 21-85
Base impedance of am antennas, 26-8 to 26-17
Basic Scattering Code, 20-4
  for antennas on noncurved surfaces, 20-68
  for numerical simulations
   of aircraft, 20-70, 20-85 to 20-86, 20-90
   of ships, 20-90 to 20-97
Basis functions in modeling codes, 3-81, 3-83

Batwing tv antennas, 27-23 to 27-24
Bayliss line source pattern synthesis, 13-27 to 13-29
BCS (bistatic cross section), 2-22 to 2-23
BDF. *See* Beam deviation factor
Beacon antennas
  EMP responses for, 30-23, 30-27 to 30-28
  with satellite antennas, 21-84
Beads with coaxial lines, 28-21 to 28-22
Beam-broadening factor
  with linear arrays, 11-19 to 11-21
  with periodic arrays, 13-17 to 13-19
  with phased arrays, 21-10
Beam deviation factor
  with offset antennas, 15-81, 21-24
  with scanned beams, 15-51, 15-53, 15-55 to 15-59, 15-62
Beam dithering, 13-56
Beam efficiency
  in arrays, 11-74, 11-76
  for conical horn antennas, 8-62
  for microwave radiometers, 22-28 to 22-30, 22-32 to 22-33
Beam-forming feed networks
  constrained, 19-3 to 19-15
  cylindrical arrays, 19-98 to 19-119
  optical transform, 19-91 to 19-98
  for phased arrays, 18-17 to 18-26
  for satellite antennas, 21-5 to 21-6, 21-29 to 21-57
  semiconstrained, 19-15 to 19-49
  transmission lines for, 21-63
  unconstrained, 19-49 to 19-91
Beam-pointing errors, 18-27 to 18-28
Beam scanning, 14-25 to 14-26
  of arrays
   aperiodic, 14-32 to 14-35
   Dolph-Chebyshev, 11-52
   millimeter-wave, 17-25 to 17-26
  beam-forming feed networks for, 18-21, 18-24, 21-27 to 21-34, 21-37 to 21-49
  conical, 22-40 to 22-41
  with microwave radiometers, 22-33 to 22-43
  with millimeter-wave antennas, 17-9, 17-10, 17-24 to 17-26
   array, 17-25 to 17-26
   dielectric grating, 17-69 to 17-72
  with offset parabolic antennas, 15-51, 15-53

Beam scanning (*cont.*)
  with satellite antennas, 21-5
    phased-array, 21-71
    polar orbiting, 22-35 to 22-38
Beam-shaping efficiency, 21-5 to 21-6
Beam squint
  with constrained feeds, 18-19, 18-21 to 18-22
  with offset antennas, 15-48
  with phase-steered arrays, 13-20
  with reflector satellite antennas, 21-19
Beam steering
  computer for, in phased array design, 18-3 to 18-4, 18-26
  with lens antennas, 17-22
  with Wheeler Lab approach, 19-107
Beam tilt with tv antennas, 27-7, 27-26
Beamwidth, 1-20. *See also* Half-power beamwidths
  of arrays
    aperiodic, 14-3, 14-12
    linear, 11-15
    periodic, 13-10 to 13-12, 13-17 to 13-19
    phased, 19-61, 21-11 to 21-12
    planar, 11-53 to 11-55
    subarrays, 13-32
  and Dolph-Chebyshev pattern synthesis, 13-24 to 13-25
  of feed circuits
    limited scan, 19-57
    series feed networks, 19-5
  and gain, with tv antennas, 27-7
  of horn antennas
    diagonal, 8-70
    stepped-horn, 21-82 to 21-84
  and lens F/D ratio, 21-16
  of log-periodic antennas
    planar, 9-9 to 9-10
    zigzag, 9-39 to 9-40, 9-45 to 9-46
  of millimeter-wave antennas
    dielectric grating, 17-69 to 17-70, 17-74, 17-81 to 17-82
    geodesic, 17-32
    integrated, 17-134
    lens, 17-20
    metal grating, 17-69
    microstrip dipole, 17-119
    pillbox, 17-29
    reflector, 17-11, 17-13, 17-14
    slotted-shunt-slot array, 17-33
    spiral, 17-28
    tapered rod, 17-47
    uniform-waveguide leaky-wave, 17-85, 17-90, 17-98
  of multibeam antennas, 21-58
  and optimum rectangular aperture patterns, 5-18 to 5-20
  of reflector antennas
    offset, 15-37, 15-82
    tapered-aperture, 15-16 to 15-17
  of subarrays, 13-32
  and Taylor line source pattern synthesis, 13-26 to 13-27
  of tv antennas, 27-7
    side-mount tv antennas, 27-20
  of uhf antennas, 27-27
Beamwidths scanned and gain loss with reflector antennas
  dual-reflector, 15-73
  offset parabolic, 15-55, 15-58, 15-62 to 15-66
Benelux Cross antenna, 14-17 to 14-18
Beyond-the-horizon transmission, 29-21 to 29-30
Bhaskara-I and -II satellites, 22-17
Biconical antennas, 3-24 to 3-26, 3-29
  coaxial dipole, 3-30
  horn, 3-30
  millimeter-wave, 17-32
  for satellite antennas, 21-89, 21-100 to 21-109
Bifilar helix, 9-83
Bifilar zigzag wire antennas, 9-35, 9-37
Bifocal lenses, 16-30 to 16-33
Bilateral fin lines, 28-55
Binomial linear arrays, 11-10 to 11-11
Bistatic cross section, 2-22 to 2-23
Bistatic radar, 2-36 to 2-39
Blackbody radiation, 22-3 to 22-4
Blade antennas
  EMP responses for, 30-17 to 30-19
  equivalent circuits for, 30-16
  equivalent receiving area of, 30-12
Blass matrices
  with beam-forming feed networks, 18-25, 19-10 to 19-11
  with space-fed subarray systems, 19-68
  for switch beam circuits, 21-31 to 21-33
Blass tilt traveling-wave arrays, 19-80
Blind angles with arrays, 14-25 to 14-27, 14-32 to 14-35
Blindness, array, 13-45 to 13-49
  with flared-notch antennas, 13-59

Blind spots with horn antennas, 8-4
Blockage. *See* Aperture blockage
Block 5D satellite, 22-18
Body-stabilized satellites, 21-90
Boeing 707 aircraft and EMP, 30-4, 30-6
Boeing 737 aircraft
   simulation of, 20-64, 20-70 to 20-78
   slot antennas on, 20-85 to 20-86, 20-89, 20-90
Boeing 747 aircraft and EMP, 30-4, 30-6
Bombs, aircraft, simulation of, 20-79
Bone, human, phantom models for, 24-55
Booker-Gordon formula for scattering, 29-28
Boom length of log-periodic antennas
   dipole, 9-21, 9-23, 9-24, 9-26
   zigzag, 9-39
Bootlace lenses, 16-46 to 16-48
Boresight beam
   gain of, with reflector antennas, 15-5
   with offset parabolic antennas, 15-81
   of phased-array satellite antennas, 21-71
   and radomes, 31-3 to 31-4, 31-7 to 31-9, 31-27 to 31-29
Born-Rytov and computer solutions, 3-54
Boundary conditions, 1-8 to 1-9
   with corrugated horns, 8-50, 15-96 to 15-97
   and current distributions, 3-38 to 3-39
   and lens zoning, 16-39
   in longitudinal slots, 12-6
   numerical implementation of, 3-55
   numerical validation of, 3-79 to 3-80
   with patches, 10-16 to 10-17
   and perfect ground planes, 3-18 to 3-19
   for thin wires, 3-44 to 3-45
Bow-tie dipoles, 17-137 to 17-138
Bragg condition for diffraction, 29-28
Brain, human, phantom models for, 24-55
Bra-Ket notation and reciprocity, 2-16
Branch guide coupler power division elements, 21-51 to 21-52
Branch line couplers
   for array feeds, 13-61
   directional power division element, 21-50
Branch line guide for planar arrays, 12-21 to 12-23
Brewster angle phenomena
   modeling of, 32-79
   and radome design, 31-12 to 31-13

Brightness temperature
   of emitters, 2-40 to 2-41
   and humidity, 22-6 to 22-7
   of microwave radiometers, 22-28
   vs. physical temperature, 22-10
   and surface emissivity, 22-7 to 22-9, 22-15
Broadband array feeds, 13-39 to 13-41
Broadside arrays, 3-40, 11-11, 11-15
   maximum directivity of, 11-64 to 11-67
   radiation pattern of, 13-12 to 13-13
Broadside coupled transmission lines, 28-16, 28-17
Broadside radiation of microstrip arrays, 17-115
Broadwall millimeter-wave array antennas, 17-26
Broadwall shunt-slot radiators, 18-8
Broadwall slots
   for center-inclined arrays, 12-24
   longitudinal, 12-4
BSC. *See* Basic Scattering Code
Bulge, earth, and line-of-sight propagation 29-34
Butler matrices
   with arrays
      multimode circular, 25-18 to 25-19
      multiple-beam, 19-80
   with beam-forming feed networks, 18-23 to 18-24, 19-8 to 19-10, 19-12
   with cylindrical array feeds, 19-101 to 19-103
   as Fourier transformer, 19-95
   with reflector-lens limited-scan feed concept, 19-68
   with space-fed subarray systems, 19-68
   for switch beam circuits, 21-31, 21-33
   in transform feeds, 19-92 to 19-93

Cable effect with modeling, 32-84
Cabling with scanning ranges, 33-16
CAD (computer-aided design), 17-120 to 17-122
Calibration
   accuracy of, with microwave radiometers, 22-23
   of AMSU, 22-50
   of field probes, 24-54 to 24-57
   horn antennas as standard for, 8-3, 8-43, 8-69
   of microwave radiometers, 22-25 to 22-26, 22-48

Cameras, thermographic, 24-51
Cancer therapy. *See* Hyperthermia
Candelabras for multiple-antenna installations, 27-37
Capacitance of in-vivo probes, 24-33 to 24-34
Capacitor-plate hyperthermia applicators, 24-46 to 24-47
Capture area of receiving antenna, 6-6
Cardioid patterns
  with loop antennas, 25-6 to 25-7
  with slot antennas
    coaxial, 27-28
    waveguide, 27-30
  from slotted cylinder reflectors, 21-110
  for T T & C, 21-89 to 21-90
Carrier-to-noise ratio, 29-48
Cartesian coordinate systems, 15-115 to 15-120
Cassegrain feed systems
  near-field, 19-62 to 19-63
  for reflector antennas
    millimeter-wave, 17-9, 17-13 to 17-14
    satellite, 21-21 to 21-22
  with wide-angle systems, 19-84, 19-88
Cassegrain offset antennas
  cross polarization with, 15-69 to 15-70, 15-72 to 15-74
  dual-mode horn antennas for, 8-72
  parameters of, 15-67 to 15-68
  performance evaluation of, 15-69 to 15-72
  scan performance of, 15-71 to 15-73, 15-78 to 15-79
Caustic distances, 4-10
  edge-diffracted, 4-27
  and GO reflected field, 4-18
Caustics, 4-6, 8-76
  with diffracted rays, 4-5, 4-6
    surface-diffracted, 4-83
  of diffracted waves, 4-96 to 4-102
  and GO representation, 4-16
  and lens antennas, 17-16
  matching functions for, 4-103
Cavity-backed radiators
  in panel fm antennas, 27-35 to 27-36
  in side-mount tv antennas, 27-20 to 27-23
Cavity-backed slots with log-periodic arrays, 9-68 to 9-71
Cavity model for microstrip antennas, 10-10 to 10-21
Centered broad wall slots, 12-4

Center-fed antennas
  dipole
    microstrip, 17-123 to 17-124
    transient response of, 7-22
  slot, for tv, 27-29 to 27-30
Center-fed arrays, 17-119 to 17-120
Center-fed dual series feeds, 18-20
Center-fed reflectarrays, 19-55
Center-inclined slot arrays, 12-24
Channel-diplexing, 22-46
Channel guide mode for leaky-wave antennas, 17-98
Characteristic impedance, 3-28 to 3-31
  of dipoles, 7-6, 9-28, 9-30
  of fin lines, 28-57, 28-58
  of holographic antennas, 17-130
  of log-periodic zigzag antennas, 9-42 to 9-44, 9-48
  of microstrip arrays, 17-109 to 17-111
  of slot lines, 28-53 to 28-54
  of transmission lines, 28-7
    biconical, 3-24 to 3-25
    coaxial, 28-20
    microstrip planar, 28-30 to 28-31
    TEM, 28-10 to 28-18
    triplate stripline, 28-22, 28-28 to 28-29
    two-wire, 7-39, 28-11, 28-19
    waveguide planar, 28-33
  of waveguides
    rectangular, 2-28
    ridged, 24-14 to 24-16
    TE/TM, 28-36, 28-37
Chebyshev polynomials, 11-15 to 11-17. *See also* Dolph-Chebyshev arrays
Check cases for validating computer code, 3-80
Chi-square distribution with aperiodic arrays, 14-10 to 14-12
Chokes
  with coaxial sleeve antennas, 7-31 to 7-33, 7-36
  with microstrip antennas, 10-68
Circles, 15-25 to 15-26
  and intersection curve, 15-29
Circuit characteristics
  measurements for, 32-4
  models for, 10-26 to 10-28
Circular apertures
  Bayliss line source pattern synthesis for, 13-29
  distribution of
    axial field, 5-32
    with reflector antennas, 15-17, 15-23

efficiency of, 5-29, 5-30
radiation patterns of, 5-20 to 5-24
near-field, 5-25
Circular arrays
for direction-finding antennas, 25-17 to 25-20
of probes, with transmission line feeds, 19-26 to 19-30
sections of, 13-8 to 13-10
transformations with, 11-33 to 11-36, 11-41 to 11-47
for T T & C, 21-89 to 21-100
Circular coaxial lines, 28-20 to 28-22
Circular cones
geometry for, 4-60
intersection of reflector surface with, 15-26 to 15-29
Circular corrugated horns, 15-97, 15-99
Circular cylinders
and intersection curve, 15-30
radiation patterns of, 4-75, 4-77 to 4-79
Circular-disk microstrip patches
characteristics of, 10-16, 10-25
with CP microstrip antennas, 10-61, 10-62
elements for, 13-60
principal-plane pattern of, 10-22
resonant frequency of, 10-47
Circularity, of multiple antennas, 27-38 to 27-40
Circular log-periodic antenna, 9-3, 9-4
Circular loop antennas
for geophysical applications, 23-19 to 23-21
as magnetic-field probes, 24-40 to 24-43
radiation resistance of, 6-19
thin-wire, 7-42
Circular open-ended waveguides
radiators for, with phased arrays, 18-6
with reflector antennas, 21-74
for reflector feeds, 15-90 to 15-93
Circular patches, 10-41 to 10-44
Circular-phase fields, 9-109
Circular pillbox feeds, 19-19, 19-21 to 19-22
Circular polarization, 1-15 to 1-16, 1-20, 1-28 to 1-29
of arrays, 11-7
with constrained lenses, 16-43
conversion of linear to, 13-60 to 13-61
with dual-mode converters, 21-78, 21-80
with fm antennas, 27-33 to 27-34
gain measurements for, 32-49 to 32-51

with horn antennas
biconical, 21-107 to 21-108
corrugated, 8-51, 8-64
diagonal, 8-70
with image-guide-fed slot array antennas, 17-50
of log-spiral conical antennas, 9-90 to 9-91, 9-99
measurements of, 32-52, 32-58, 32-60
with microstrip antennas, 10-57 to 10-63, 10-69 to 10-70
microstrip radiator types for, 13-60
with phased arrays, 18-6
for power dividers, 21-78, 21-80
with reflector antennas
offset parabolic, 15-36, 15-48 to 15-50
for satellites, 21-19, 21-73, 21-74, 21-78 to 21-80
for satellite antennas, 21-7, 21-29, 21-89
reflector, 21-19, 21-73, 21-74, 21-78 to 21-80
for shipboard simulation, 20-91 to 20-93
with spiral antennas, 17-27
for T T & C, 21-89
with tv antennas, 27-8 to 27-9, 27-38 to 27-40
helical, 27-15 to 27-19
side-mount, 27-20 to 27-23
skewed-dipole, 27-11 to 27-12, 27-18
vee-dipole, 27-9 to 27-11
Circular-sector patches, 10-17, 10-44
resonant frequency of, 10-47, 10-48
Circular tapered dielectric-rod antennas, 17-37 to 17-38
Circular waveguides, 1-44 to 1-50
array of, with infinite array solutions, 13-48 to 13-49
for conical horn antennas, 8-46, 15-93
dielectric, 28-50 to 28-51
open-ended, 15-90 to 15-93, 15-105
TE/TM, 28-42 to 28-45
transmission type, 21-60
Circumferential slots
on cones, 4-92
on cylinders, 4-75, 4-77, 21-92 to 21-93, 21-96
CLAD (controlled liquid artificial dielectrics), 17-22
Classical antennas, 3-26 to 3-31
Cluttered environments, simulation of, 20-90 to 20-97
Cmn, explicit expressions for, 2-11

Coaxial cable. *See also* Sleeve antennas
  with implantable antennas, 24-26 to 24-28
  for in-vivo measurements, 24-32 to 24-35
  for log-periodic zigzag antennas, 9-44 to 9-45
  with log-spiral antennas, 9-75, 9-80
  with medical applications, 24-24 to 24-25
  for microstrip antennas, 10-51, 17-120
  with TEM waveguides, 21-57
Coaxial current loops for hyperthermia, 24-45 to 24-46
Coaxial dipole antennas, 3-30
Coaxial hybrids as baluns, 9-95
Coaxial-line corporate feed networks, 13-61
Coaxial line transmission type, 21-57, 21-60
  for TEM transmission lines, 28-23 to 28-25
    circular, 28-20 to 28-22
    impedance of, 28-15, 28-16, 28-18
Coaxial-probe feeds, 10-28, 10-29
Coaxial radiators
  dipole, for phased arrays, 18-7
  excited-disk, 18-8 to 18-9
Coaxial slot antennas, 27-28 to 27-29
COBRA modeling code, 3-88 to 3-91
Codes, computer modeling, 3-80 to 3-96
Collimations, array, 13-7 to 13-10
Collinear arrays, 3-43
Collocation moments method, 3-60
Column arrays
  collimation of, 13-8
  of dipoles, coupling between, 13-44
Coma lens aberrations, 16-15, 16-17, 21-13 to 21-14
  and Abbe sine condition, 16-19 to 16-21
  and millimeter-wave antennas, 17-19 to 17-21
  with Ruze lenses, 19-39
  with spherical thin lenses, 16-29
  and zoning, 16-38 to 16-41
Communication satellite antennas, 21-4 to 21-6
  beam-forming networks for, 21-29 to 21-57
  design of, 21-6 to 21-8, 21-68 to 21-80
  feed arrays for, 21-27 to 21-29
  multibeam, 21-57 to 21-68
  types of, 21-8 to 21-26

Compact antenna ranges, 32-28 to 32-30
  for field measurements, 33-22 to 33-24
Compensation theorem and computer solutions, 3-54
Complementary planes, 2-14 to 2-15
Completely overlapped space-fed subarray system, 19-68 to 19-76
Complex environments
  airborne antenna patterns, 20-70 to 20-90
  antennas for, 20-3 to 20-4
  far fields with, 20-53 to 20-54, 20-61 to 20-70
  numerical simulations
    for antennas, 20-5 to 20-7
    for environment, 20-7 to 20-53
  shipboard antenna patterns, 20-90 to 20-97
Complex feeds for reflectors, 15-94 to 15-99
Complex pattern function for arrays, 11-7
Complex polarization ratios, 32-52
Complex poles of antennas, 10-11
Complex power, 1-7
Component errors and phased-array performance, 18-26 to 18-28
Compound distributions for rectangular apertures, 5-16 to 5-17
Computation with computer models, 3-54 to 3-55, 3-68 to 3-72
Computer-aided design, 17-120 to 17-122
Computers and computer programs
  for beam-forming network topology, 21-57
  for beam steering, 18-3 to 18-4, 18-26
  for complex environment simulations, 20-3 to 20-4, 20-18
  for dipole-dipole arrays, 23-7
  for Gregorian feed systems, 19-65
  for integral equations, 3-52 to 3-55
    codes for, 3-80 to 3-96
    computation with, 3-68 to 3-72
    numerical implementation of, 3-55 to 3-68
    validation of, 3-72 to 3-80
  for ionospheric propagation, 29-25
  modeling with, 32-65
  for rf personnel dosimeters, 24-43
  for spherical scanning, 33-15
Computer storage and time
  with frequency-domain solutions, 3-61 to 3-62, 3-65

using NEC and TWTD modeling codes, 3-94
and symmetry, 3-65, 3-67
with time-domain solutions, 3-63 to 3-65
and unknowns, 3-57
Concentric lenses, 19-80
Concentric ring arrays, 19-57 to 19-58
Conceptualization for computer models, 3-53 to 3-54
Conductance
  in inclined narrow wall slot arrays, 12-25 to 12-26
  with microstrip antennas, 10-7 to 10-8
  of resonant-length slots, 12-14
Conducting cones, fields of, 1-10
Conducting wedges, fields of, 1-9 to 1-10
Conduction losses, 1-7
  of microstrip patch antennas, 17-111
  of transmission lines
    circular coaxial, 28-21
    coplanar waveguide planar, 28-34
    microstrip planar, 28-32
    triplate stripline, 28-28
  of waveguides, 1-39
    rectangular TE/TM, 28-38, 28-42
Conductivity of troposphere, 29-31
Conductors, modeling of, 32-70 to 32-71
Conductor surfaces, linear current density in, 12-3
C-141 aircraft, simulation of, 20-82 to 20-84
Cone-tip diffraction, 4-93 to 4-94
Cones
  circular, 4-60
  conducting, 1-10
  coupling coefficient of slots on, 4-92
  radial slots in, 4-78 to 4-80
  surface-ray paths on, 4-75 to 4-76
Conformal arrays, 13-49 to 13-52
  nonplanar, 13-41
  probabilistic approach to, 14-33 to 14-34
  for T T & C, 21-90
Conical cuts, 32-6
Conical horn antennas, 8-4. *See also* Corrugated horn antennas, conical
  design procedures for, 8-48 to 8-49
  directivity of, 8-46 to 8-47
  dual-mode, 8-71 to 8-72, 15-95, 15-100 to 15-101
  for earth coverage satellite antennas, 21-82

  for feeds for reflectors, 15-93 to 15-94, 15-99 to 15-101
  gain of, 8-48
  millimeter-wave, 17-23, 17-25
  phase center of, 8-76, 8-78 to 8-80
  radiated fields of, 8-46
Conical log-spiral antennas, 9-3, 9-4
  active region of, 9-83, 9-84, 9-98 to 9-99
  axial ratios with, 9-90 to 9-93
  current wave of, 9-83 to 9-84
  directivity of, 9-87 to 9-90
  feeds for, 9-95 to 9-96, 9-99 to 9-104, 9-111 to 9-112
  front-to-back ratios with, 9-89, 9-92
  half-power beamwidth of, 9-86, 9-87, 9-89, 9-97 to 9-98
  impedance of, 9-95 to 9-97, 9-99, 9-107
  phase center of, 9-83, 9-91 to 9-92, 9-94
  polarization of, 9-86, 9-90 to 9-91, 9-99
  propagation constant with, 9-80 to 9-84
  radiation fields for, 9-85 to 9-86, 9-89 to 9-92, 9-99 to 9-109
  transmission lines with, 9-95
  truncation with, 9-79, 9-84, 9-89 to 9-90, 9-95 to 9-97
Conical scanning, 22-38 to 22-43, 22-48
Conic-section-generated reflector antennas, 15-23 to 15-31
Constant-K lenses, 17-15 to 17-16, 17-19
Constituent waves with radomes, 31-5
Constrained feeds, 18-19 to 18-26, 21-34
  multimode element array technique, 19-12 to 19-15
  multiple-beam matrix, 19-8 to 19-12
  for overlapped subarrays, 13-36
  parallel feed networks, 19-5 to 19-6
  series feed networks, 19-3 to 19-5
  true time-delay, 19-7 to 19-8
Constrained lenses, 16-5 to 16-6, 16-41 to 16-48
  analog of dielectric lenses for, 16-49 to 16-51
  bandwidth of, 16-55
  wide-angle multiple-beam, 19-80 to 19-85
Constrained variables and radome performance, 31-4
Contiguous subarrays, 13-32 to 13-35
Continuity equations, 1-6. *See also* Discontinuities
  for electromagnetic fields, 3-6
  for field distributions, 5-18 to 5-19

Continuous metal strips for leaky-wave antennas, 17-84 to 17-86
Continuous scans, 22-35
  compared to step scans, 22-43
Contour beam reflectors, 15-5, 15-80 to 15-87
Contours and tv antenna strength, 27-6
Controlled liquid artificial dielectrics, 17-22
Convergence
  in large arrays, 12-28
  measures of, and modeling errors, 3-72 to 3-75
Convex cylinder, surface-ray paths on, 4-75 to 4-76
Convex surfaces
  GTD for, 4-28 to 4-30
  UTD for
    and mutual coupling, 4-84 to 4-96
    and radiation, 4-63 to 4-84
    and scattering, 4-50 to 4-63
Coordinate systems
  for antenna ranges, 32-8 to 32-14
  feed vs. reflector, 15-88 to 15-89
  for lenses, 16-7 to 16-9
  for numerical aircraft solutions, 20-76
  for probe antennas, 32-6 to 32-7
  with reflector antennas, 15-7
  transformations of, 15-115 to 15-120
Cophasal arrays
  Dolph-Chebyshev, 11-19
  end-fire, 11-69, 11-70
  uniform, 11-64 to 11-65
  circular, linear transformations with, 11-46 to 11-47
Coplanar waveguide transmission types, 21-59, 21-61
  applicators for medical applications, 24-24 to 24-25
  planar quasi-TEM, 28-33 to 28-34
  strip planar quasi-TEM, 28-35
Copolarization. See Reference polarization
Copper for satellite antenna feeds, 21-28 to 21-29
Corner-diffracted fields
  coefficient for, 4-44 to 4-48
  and UTD solutions, 20-12 to 20-13
Corporate feed networks, 19-5 to 19-6
  for arrays, 13-61
  for beam-forming, 18-24
Correction factors
  for propagation, 29-5
  with reflector surface errors, 15-106 to 15-107, 15-113
Corrugated horn antennas, 8-4, 8-50 to 8-51
  circular, 15-97, 15-99
  conical, 8-59 to 8-63
    cross-polarized pattern for, 8-69
    millimeter-wave, 17-23
    radiated fields of, 8-64, 8-65
  efficiency of, 8-50
  millimeter-wave, 17-23
  pyramidal, 8-51 to 8-52
    aperture fields of, 8-53
    half-power beamwidth of, 8-55
    radiated fields of, 8-53 to 8-59
  for reflector antennas, 15-96, 15-102 to 15-104
  for satellite antennas, 21-82
Corrugations
  depth of, and EDC, 17-55 to 17-56
  with periodic dielectric antennas, 17-48
Cosine lenses, hyperbolic, 16-53 to 16-54
Cosine representations of far fields, 1-28 to 1-29
Cosinusoidal aperture distributions
  and corrugated horns, 8-53
  near-field reduction factors for, 5-30 to 5-32
Cosmos-243 satellite, 22-17
Cosmos-384 satellite, 22-17
Cos type patterns in reflector feeds, 15-99, 15-102 to 15-105
Coupler/phase-shifter variable power dividers, 21-43 to 21-45
Couplers
  for array feeds, 13-61
  for series feed networks, 19-4 to 19-5
Coupling, 6-9. See also Mutual coupling
  of beam-forming feed networks, 18-25, 18-26
  with dipole arrays, 13-44
  microstrip antenna to waveguide, 17-119 to 17-121
  and phase shift, 9-58
  of transmitting and receiving antennas, 20-68
  with UTD solutions, 20-6 to 20-7
  for curved surfaces, 20-18
  of waveguides for planar arrays, 12-21 to 12-22
Coupling coefficient, 4-92
  with aperiodic arrays, 14-24, 14-26

Coupling equation
    for planar scanning, 33-8 to 33-9
    for spherical scanning, 33-13
Coverage area. *See also* Footprints
    of satellite antennas, 21-4 to 21-6, 21-73, 21-77
        maps for, 15-80 to 15-81
    of tv antennas, 27-4 to 27-6
Creeping waves
    with conformal arrays, 13-50 to 13-51
    with UTD solutions
        for curved surfaces, 20-20, 20-22
        for noncurved surfaces, 20-45
Critical frequencies of ionospheric propagation, 29-23
Crops, remote sensing for, 22-15
Cross dipole antennas, 1-24
Crossed yagis phased arrays, 21-12
Cross-flux of fields, 2-16, 2-18
Crossover
    of beam-forming feed networks, 18-25
    with pillbox feeds, 19-22 to 19-23
Cross polarization, 1-23 to 1-24
    with arrays
        center-inclined broad wall slot, 12-24
        conformal, 13-50
        inclined narrow wall slot, 12-24
        microstrip patch resonator, 17-108
        phased, 18-8
    with beam-forming networks, 21-57
    with edge-slot array antennas, 17-27
    with horn antennas
        aperture-matched, 8-68
        corrugated, 8-64, 8-69
        millimeter-wave, 17-23
    and matched feeds, 15-99
    with microwave antennas, 33-20 to 33-21
    with microwave radiometers, 22-30, 22-45
    with millimeter-wave antennas
        horn, 17-23
        reflector, 17-9
    with probe antennas, 32-8
    with reflector antennas, 15-5 to 15-6
        dual-reflector, 15-69 to 15-70, 15-72 to 15-74
        millimeter-wave, 17-9
        offset parabolic, 15-42, 15-46 to 15-48
        satellite, 21-19 to 21-20, 21-23
    with satellite antennas, 6-26, 6-27, 21-5, 21-7
    of side lobes, slots for, 8-63
Cross sections
    for leaky-wave antennas, 17-88
    receiving. *See* Receiving cross section
    with scattering, 2-22 to 2-25
    of tapered dielectric-rod antennas, 17-38 to 17-40
    of transmission lines, 21-63
Cross-track scanning, 33-7 to 33-10
Cubic phase errors with lenses, 16-14, 16-16 to 16-19, 16-38
    and coma aberrations, 17-19
Current and current distribution. *See also* Magnetic current distributions; Surface current
    with arrays, 3-33, 3-35
    in biconical transmission lines, 3-24
    and boundary-condition mismatch, 3-80
    with cavities, 10-12
    with coaxial sleeve antennas, 7-33, 7-35 to 7-37
    density of
        with microstrip antennas, 10-12
        PO, 4-19 to 4-21
        for thin wires, 3-44 to 3-45
    with dipole antennas
        arrays, 13-43
        folded, 7-37 to 7-39
        linear, 7-11
        loaded, 7-9, 7-19
        short, 6-14 to 6-15
        unloaded linear, 7-12 to 7-14, 7-18 to 7-19
    edge, 10-55 to 10-57
    of far fields, 1-21 to 1-22
    and imperfect ground planes, 3-27, 3-48
    with Kirchhoff approximations, 2-21
    measurement of, 2-31, 33-5 to 33-6
    of perfect conductors, 3-38 to 3-39
    in rectangular waveguides, 2-28 to 2-29, 12-3 to 12-4
    of reflected waves, 7-14
    short-circuit, 2-32, 6-5
    sinusoidal, 3-14, 3-17
    with sleeve antennas, 7-23 to 7-33
    with straight wire antennas, 3-49 to 3-50
    with thin-wire antennas, 7-5 to 7-6
        loops, 7-42 to 7-48
    for traveling-wave antennas, 3-16
    for Tseng-Cheng pattern, 11-56
    for UTD antenna solutions, 20-5
Current elements, 26-8
Current reflection coefficient, 2-28

Current source, ideal, 2-29 to 2-30
Current waves
  with log-spiral antennas, 9-74 to 9-75
    conical, 9-83 to 9-84
  in transmission lines, 28-6, 28-7
Curvature of field with lenses, 16-15, 16-20
Curved surfaces
  effects of, on conformal arrays, 13-50
  for diffraction, 8-64
  UTD solutions with, 20-4 to 20-5, 20-18 to 20-37
Cutoff frequency of waveguides, 28-40 to 28-41
  circular dielectric, 1-49, 28-50
  rectangular, 1-42
Cutoff wavelength
  of transmission lines, circular coaxial TEM, 28-20
  of waveguides, 28-40 to 28-41
    circular TE/TM, 28-42, 28-44
    rectangular TE/TM, 28-38
    ridged, 24-14 to 24-15
Cylinders and cylindrical antennas
  circular, 4-75, 4-77 to 4-79
  conducting, 21-92 to 21-100
  current distribution of, 3-14, 3-16
  dipole, 3-28
  dipole arrays on, 13-50 to 13-52
  elliptical, 4-75, 4-77
  illumination of, 4-12 to 4-13
  for intersection curve, 15-30
  for multiple-antenna analyses, 27-37 to 27-38
  radiation pattern for, 4-62 to 4-63
  strips mounted on, 20-23 to 20-24
  thin-wire, 7-10 to 7-11
Cylinder-to-cylinder interactions, 20-37, 20-44 to 20-53
Cylindrical array feeds, 19-98 to 19-101
  matrix-fed, 19-101 to 19-119
Cylindrical scanning for field measurements, 33-10 to 33-12

Data transmission, millimeter-wave antennas for, 17-5
Decoupling with coaxial sleeve antennas, 7-31 to 7-32
Defense meteorological satellites, 22-14
Defocusing techniques for field measurements, 33-24 to 33-26
Deicing of tv antennas, 27-9, 27-26
Delay lenses, equal group, 16-45 to 16-46

Design procedures and parameters
  for antenna ranges, 32-14 to 32-19
  for arrays
    Dolph-Chebyshev, 11-15 to 11-23
    feed, for satellite antennas, 21-27 to 21-29
    large, 12-28 to 12-34
    of longitudinal slots, 12-10 to 12-24
    phased, 18-3 to 18-28
    space-tapered, 14-6 to 14-8
    of wall slots, 12-24 to 12-28
  for dielectric grating antennas, 17-62 to 17-69
  for dielectric lenses, 16-8 to 16-9
  for horn antennas
    conical, 8-48 to 8-49
    E-plane, 8-19 to 8-20
    H-plane, 8-33 to 8-34
    pyramidal, 8-43 to 8-45
  for log-periodic antennas
    dipole, 9-20 to 9-32
    zigzag, 9-37 to 9-46
  for offset parabolic reflectors, 15-80 to 15-84
  and radome performance, 31-4, 31-8 to 31-10
  for satellite antennas, 21-6 to 21-8, 21-20 to 21-26
    feed arrays for, 21-27 to 21-29
Detour parameters, 4-42
Diagnostics, medical. *See* Medical applications
Diagonal horn antennas, 8-70 to 8-71
Diameter of offset parabolic reflectors, 15-31
Diathermy applicators, 24-46 to 24-48
Dicke radiometers, 22-24 to 22-25
Dielectric antennas
  bifocal lens, 16-30 to 16-33
  in integrated antennas, 17-132
  periodic, 17-34 to 17-35, 17-48 to 17-82
Dielectric constants, 1-6 to 1-7, 10-5. *See also* Effective dielectric constants
  and Abbe sine condition, 16-20
  and cross sections, 17-39
  of microstrip antennas, 10-7, 10-23 to 10-24
  vs. thickness, with lenses, 16-12 to 16-13, 16-23
Dielectric grating antennas, 17-50 to 17-53, 17-80 to 17-82

Dielectric horn antennas, 17-23, 17-25
Dielectric lenses, 16-5, 21-16
  constrained analog of, 16-49 to 16-51
  design principles of, 16-7 to 16-9
  and Snell's law, 16-6
  taper-control, 16-33 to 16-38
  wide-angle, 16-19 to 16-33
  zoning of, 16-38 to 16-41
Dielectric loading
  with horn antennas, 8-73
    for satellite antennas, 21-82, 21-84 to 21-86
  with lenses for cylindrical array feeds, 19-111, 19-114
  with pillbox feeds, 19-21
  with waveguides, 13-48 to 13-49
  for medical applications, 24-9
Dielectric logging for rock conductivity, 23-25
Dielectric loss
  in constant-K lenses, 17-19
  tangent for, 1-6
  with transmission lines, 28-9
    coplanar waveguide planar quasi-TEM, 28-34
    microstrip planar quasi-TEM, 28-31
  with waveguides
    rectangular, 1-39
    TE/TM, 28-37
Dielectric matching sheets, 16-56
Dielectric materials
  with cavities, 10-21
  and feed reactance, with microstrip antennas, 10-33
  and phase shift, 9-64
  properties of, 28-10
Dielectric millimeter-wave antennas, 17-48 to 17-82
Dielectric probes
  open-ended coaxial cable as, 24-32 to 24-34
  short monopoles as, 24-30 to 24-32
Dielectric resonator antennas, 17-35 to 17-36
Dielectric-rod antennas, 17-34 to 17-48, 17-133 to 17-134
Dielectrics
  and modeling, 32-70 to 32-75
  and radiation patterns, 32-81 to 32-82
Dielectric sheets, plane wave propagation through, 31-10 to 31-18, 31-20 to 31-25
Dielectric-slab polarizers, 21-30
Dielectric substrate in microstrip antennas, 10-5, 10-6
Dielectric waveguide transmission types, 21-59, 21-61
  circular, 28-50 to 28-51
  for fiber optics, 28-47
  rectangular, 28-51, 28-53
Dielectric weight of lens antennas, 17-15
Dielguides, 8-73
Difference beams with beam-forming feed networks, 18-25 to 18-26
Differential formulations, compared to integral, 3-55 to 3-58, 3-82
Diffracted rays, 4-3 to 4-5
  caustic regions with, 4-5
Diffractions and diffracted fields. *See also* Edges and edge-diffracted fields; High-frequency techniques
  coefficients of
    with reflector antennas, 15-13
    for UTD solutions, 20-13, 20-43, 20-44
  and curved surfaces, 8-64
  diffracted-diffracted, 20-28, 20-29, 20-34
  diffracted-reflected
    for curved surfaces, 20-24, 20-29, 20-34
    for noncurved surfaces, 20-37, 20-45, 20-46, 20-56
  and Fresnel ellipsoid, 29-36
  horn antennas to reduce, 8-4, 8-50
  propagation by, 29-30, 29-36, 29-41 to 29-44
  and radomes, 31-8, 31-10
  with UTD solutions, 20-13, 20-16, 20-43, 20-44
    correction of, 4-105 to 4-106
    for curved surfaces, 20-26, 20-29, 20-32, 20-34
    for noncurved surfaces, 20-49 to 20-50, 20-52, 20-59 to 20-60
Digital communications and rain attenuation, 29-13
Digital phase shifters
  for periodic arrays, 13-63 to 13-64
  for phased arrays, 18-12 to 18-13, 18-15, 18-18
  quantization errors wtih, 21-12
  in true-time-delay systems, 19-80
Dimensions of standard waveguides, 28-40 to 28-41

Diode detectors for rf radiation, 24-37
Diode-loaded circular loops, 24-40 to 24-43
Diode phase shifters
  for periodic arrays, 13-62 to 13-64
  for phased arrays, 18-12 to 18-17
  variable, for beam-forming networks, 21-38, 21-40 to 21-42
Diodes. *See* Pin diodes; Schottky diodes; Varactor diodes
Diode switches for beam-forming networks, 21-7, 21-37 to 21-38
Diode variable power dividers, 21-46 to 21-48
Diplexer/circulators, 19-110 to 19-113
Diplexers
  with fixed beam-forming networks, 21-7, 21-53, 21-54
  orthomode transducers as, 22-46
Dipole antennas, 2-27, 3-30. *See also* Dipole arrays; Electric dipole antennas; Log-periodic dipole antennas; Short antennas, dipole
  balanced, 9-72 to 9-73
  bow-tie, with integrated antennas, 17-137 to 17-138
  coaxial, 3-30
  compared to monopole, 3-20
  coupling between, 20-68
  for direction-finding antennas, 25-9 to 25-10
  as E-field probes, 24-38 to 24-39
  and EMP, 30-9 to 30-14
  in feed guides, for circular polarization, 8-64
  folded, 3-29, 7-36 to 7-40
  with implantable antennas, 24-27 to 24-28
  linear, 3-13 to 3-14, 7-6, 7-11 to 7-23
  magnetic, 4-59, 30-9
  microstrip, 17-106, 17-118 to 17-119, 17-122 to 17-126
  millimeter-wave, 17-32 to 17-33
  near-field radiation pattern of, 20-69
  with side-mount tv antennas, 27-20 to 27-21
  skewed, 27-11 to 27-18, 27-34 to 27-35
  sleeve, 7-26 to 7-29
  small, 3-28, 30-9
  spherical, 30-14
  thick, 3-28
  for vehicular-mounted interferometers, 25-23

Dipole arrays, 13-43 to 13-44, 13-58
  blindness in, 13-47 to 13-48
  on cylinders, 13-50 to 13-52
  mutual coupling with, 13-41 to 13-42
Dipole-dipole arrays
  for earth resistivity, 23-6 to 23-7
  for E-field probes, 24-51
Dipole radiators for phased arrays, 18-6 to 18-7, 18-10
Direct-contact waveguide applicators, 24-9
Direct-current mode with microstrip antennas, 10-25
Direct-current resistivity of earth, 23-4 to 23-8
Direct-ray method with radome analysis, 31-20 to 31-24
Direction-finding antennas and systems, 25-3
  with am antennas, 26-37 to 26-47
  with conical log-spiral antennas, 9-109 to 9-110
  interferometers, 25-21 to 25-23
  multimode circular arrays, 25-17 to 25-20
  multiple-signal, 25-24 to 25-25
  rotating antenna patterns, 25-4 to 25-17
Directive gain, 1-25, 3-12, 5-26, 5-27
  of arrays, 3-37
  of biconical antennas, 3-25
  of lens satellite antennas, 21-74
  of millimeter-wave antennas
    dielectric grating, 17-73 to 17-74, 17-76
    geodesic, 17-32
    integrated, 17-135 to 17-136
    lens, 17-18
    reflector, 17-12
    tapered dielectric-rod, 17-41, 17-44, 17-46 to 17-47
  of receiving antennas, 6-7
Directivity, 1-29, 3-12. *See also* Directive gain
  and aperture area, 17-7, 17-8
  of arrays, 11-5, 11-6, 11-58 to 11-76, 11-78
    aperiodic, 14-3, 14-15, 14-17 to 14-19, 14-32
    backfire, 9-64
    Dolph-Chebyshev, 11-19 to 11-21
    millimeter-wave, 17-33
    multimode element, 19-12

Index

periodic, 13-10 to 13-11, 13-54 to 13-55
planar, 9-10, 11-63 to 11-65
scanned in one plane, 13-12
uniform linear, 11-11 to 11-13
of batwing tv antennas, 27-24
of beam-forming feed networks, 18-25
of conical log-spiral antennas, 9-87 to 9-90
for dielectric grating antennas, 17-79
and Dolph-Chebyshev pattern synthesis, 13-24 to 13-25
with DuFort-Uyeda lenses, 19-49
of ferrite loop antennas, 6-18, 6-20
and gain, 1-25
of horn antennas
  conical, 8-46 to 8-47
  E-plane, 8-14 to 8-18
  H-plane, 8-29 to 8-33
  pyramidal, 8-37, 8-39 to 8-42
of lens antennas, 17-18
of log-periodic antennas, 9-10
  dipole arrays, 9-21 to 9-24, 9-27, 9-61, 9-64, 9-66
  zigzag, 9-39 to 9-40
measurement of, 32-39 to 32-41
  with modeling, 32-85
of microwave antennas, 33-20
of millimeter-wave antennas, 17-5, 17-33
  lens, 17-18
  reflector, 17-12 to 17-14
  tapered dielectric-rod, 17-38 to 17-39
of monopole on aircraft, 20-76
of offset parabolic reflectors, 15-36
of open-ended circular waveguide feeds, 15-105
of short dipoles, 6-10 to 6-11
of side-mount tv antennas, 27-22
with spherical scanning, 33-12
Disc-cone antennas, 3-30
  millimeter-wave, 17-32
Discontinuities and UTD solutions, 20-10, 20-12
  with curved surfaces, 20-23 to 20-24, 20-30
  with noncurved surfaces, 20-37, 20-48, 20-52 to 20-53
Discrete arrays and aperture antennas, 11-29 to 11-30
Disk radiators for phased arrays, 18-6
Dispersion characterics
  of dielectric grating antenna, 17-66
  of dielectric rod, 17-39
  of rectangular dielectric waveguides, 28-54
Displaced rectangular aperture distributions, 5-17 to 5-18
Dissipation factor of triplate stripline TEM transmission lines, 28-28, 28-29
Distortion
  from beam-forming networks, 21-57
  of integrated antennas, 17-135
  and lateral feed displacement, 21-20
  from lens aberrations, 16-15, 16-20, 21-13
  with microstrip patch resonator arrays, 17-109
  with offset parabolic antennas, 15-49
  from pin diodes, 21-38
  from reflector surface errors, 15-105 to 15-114
  with satellite antennas, 21-5
  scanning, 11-36 to 11-38
  with tapered dielectric-rod antennas, 17-44
  thermal, 21-28 to 21-29, 21-57
  from transformations, 11-57
  of wavefronts, from radomes, 31-7, 31-9
Distribution function with aperiodic arrays, 14-6, 14-10 to 14-12
Divergence factor and ocean reflection coefficient, 29-39 to 29-40
DMS (defense meteorological satellites), 22-14
DMSP Block 5D satellite, 22-17
DNA Wideband Satellite Experiment, 29-19
Documentation of computer models, 3-68, 3-70
Dolph-Chebyshev arrays
  linear, 11-13 to 11-23
  planar, two-dimensional, 11-49 to 11-52
  probabilistic approach to, 14-8
Dolph-Chebyshev pattern synthesis, 13-24 to 13-25
Doppler direction-finding antennas, 25-16 to 25-17
Doppler effects and mobile communication, 29-30
Dosimeters, personnel, rf, 24-43
Double-braid rf cable, 28-23
Double-ridged waveguides
  for medical applications, 24-12 to 24-16
  TE/TM waveguides, 28-45 to 28-47, 28-49

Doubly tuned waveguide array elements, 13-57
Downlink signals for satellite tv, 6-26
Driven-ferrite variable polarizer power dividers, 21-43 to 21-44
Driving-point impedance
  with directional antenna feeder systems, 26-43 to 26-45
  of microstrip antennas, 10-20, 10-34, 10-53
Drought conditions, remote sensing of, 22-15
Dual-band microstrip elements, 10-63 to 10-67
Dual fields, 2-5 to 2-6
  with Maxwell equations, 3-9, 3-11
Dual-frequency array with infinite array solutions, 13-48 to 13-49
Dual-hybrid variable power dividers, 21-43
Dual-lens limited-scan feed concept, 19-66 to 19-67
Dual-mode converters, 21-80
Dual-mode ferrite phase shifters, 13-64, 18-14 to 18-15, 21-38 to 21-40, 21-42
Dual-mode horn antennas
  conical, 8-71 to 8-72,
    radiation pattern of, 15-95, 15-100 to 15-101
  for satellite antennas, 21-82, 21-84 to 21-89
Dual-offset reflector antennas, 15-6 to 15-15
Dual polarizations
  circular, 21-73
  microstrip arrays with, 17-109
  reflector antennas with, 15-5
  satellite, 21-73 to 21-74, 21-78 to 21-80
Dual-reflector antennas, 15-61 to 15-62
  cross-polarization with, 15-69 to 15-71
  lens antennas as, 16-6
  millimeter-wave antennas, 17-9 to 17-11
  parameters of, 15-67 to 15-68
  for satellite antennas, 21-21
  scan performance of, 15-71 to 15-73
  shaped reflectors for, 15-73 to 15-80
  surface errors with, 15-109 to 15-111
Dual substrates with microstrip dipoles, 17-129
Dual-toroid ferrite phase shifters, 21-40 to 21-41
DuFort optical technique, 19-73 to 19-76

DuFort-Uyeda lenses, 19-49 to 19-53
Dummy cables
  for conical log-spiral antennas, 9-95, 9-103
  with log-spiral antennas, 9-75, 9-80
Dyadic coefficients
  edge-diffraction, 4-27, 4-101 to 4-105
    uniform, 4-32 to 4-33
  for surface reflection and diffraction, 4-52
Dyadic transfer function, 4-30
Dynamic programming technique, 14-5

Earth
  bulge of, and line-of-sight propagation, 29-34
  modeling of, 32-76 to 32-77
  resistivity of, 23-4 to 23-8
Earth coverage satellite antennas, 21-80 to 21-81
  horns
    dielectric-loaded, 21-84
    multistepped dual-mode, 21-84 to 21-89
    stepped, 21-82 to 21-84
    shaped beam, 21-89
Earth receiving antennas, 6-25 to 6-32
Earth-substended angles for satellites, 21-3
Eccentricity
  of conic sections, 15-25 to 15-26
  and intersection curve, 15-30
Eccentric line TEM transmission lines, 28-15
ECM. *See* Equivalent current method
EDC. *See* Effective dielectric constant method
Edge-coupled transmission lines, 28-16, 28-17
Edge-of-coverage gain with satellite antennas, 21-5
Edges and edge-diffracted fields, 4-32 to 4-33, 4-38
  and boundaries, 15-12
  corner-diffracted, 4-45 to 4-47
  currents on, 4-100 to 4-101
    magnetic, 10-55 to 10-57
  and ECM, 4-97 to 4-102
  elements with, and conformal arrays, 13-52
  GTD for, 4-25 to 4-28, 15-69
  and horn antenna fields, 8-3 to 8-4, 8-50

Index

and image-plane ranges, 32-23
and Maxwell equations, 1-9
PO for, 4-22
PTD for, 4-104 to 4-108
slope diffraction by, 4-38 to 4-40
UAT for, 4-40 to 4-43
UTD solutions for, 4-32 to 4-38, 20-10 to 20-12
  curved surfaces, 20-21 to 20-23
  noncurved surfaces, 20-37
Edge-slot array antennas, 17-26 to 17-27
Edge slots for slot arrays, 12-26 to 12-28
Edge tapers with reflector antennas, 15-36 to 15-38, 15-41 to 15-45
  and cross-polarization, 15-47
  with off-focus feeds, 15-53
  and surface errors, 15-109
  tapered-aperture, 15-16 to 15-18, 15-21, 15-23
Effective area of receiving antennas, 22-4
  and aperture efficiency, 5-27 to 5-29
Effective dielectric constants
  of fin lines, 28-56, 28-58
  for microstrip patch antennas, 24-19 to 24-21
  of transmission lines
    coplanar strip planar quasi-TEM, 28-35
    coplanar waveguide planar quasi-TEM, 28-33
Effective dielectric constant (EDC) method
  for dielectric grating antennas, 17-65, 17-68
  for phase constants, 17-53 to 17-57
Effective heights. *See also* Vector effective height
  of hf antennas, 30-30
  of L-band antennas, 30-19
  of uhf communication antennas, 30-21
Effective isotropically radiated power, 1-27
Effective length, 2-34
Effective loss tangent with cavities, 10-20
Effective phase center of log-spiral antennas, 9-91 to 9-92, 9-94
Effective radiated power
  of fm antennas, 27-4, 27-6
  of tv antennas, 27-3 to 27-5
Effective receiving area and EMP, 30-10

Effective relative dielectric constant, 10-7
Efficiency. *See also* Aperture efficiency; Beam efficiency; Radiation efficiency
  of arrays, 11-74, 11-76
    planar, 11-65
    spaced-tapered, 14-6 to 14-7
  beam-shaping, with satellite antennas, 21-5 to 21-6
  of corrugated horn antennas, 8-50, 8-62
  of dielectric grating antennas, 17-63
  and inductive loading, 6-13
  of integrated antennas, 17-135
  of loop antennas, 25-8
  of microstrip antennas, 10-6, 10-49 to 10-50
    dipole, 17-124
    and matching filters, 10-52
  of microstrip patch resonator arrays, 17-109 to 17-115
  of microwave radiometers, 22-28 to 22-30, 22-32 to 22-33
  polarization, 1-26, 2-36, 2-39, 32-42, 32-56
  and Q-factor, 6-22 to 6-23, 10-6
  of receiving antennas, 6-22 to 6-24
  of reflector antennas
    dual-reflector, 15-73
    offset parabolic, 15-42
    tapered-aperture, 15-16, 15-23
  of small transmitting antennas, 6-9 to 6-10
E-4 aircraft, antennas on, 30-29, 30-30
Ehf (extra high frequencies), lens antennas for, 16-5, 16-28
EIRP (effective isotropically radiated power), 1-27
Electrically rotating patterns
  high-gain, 25-14 to 25-16
  null, 25-11 to 25-12
Electrically scanning microwave radiometer, 22-10
  calibration of, 22-26
  as conical scanning device, 22-40
  modulating frequency of, 22-24
  as planar scanning device, 22-37
Electrical scanning, 21-5
Electrical well logging, 23-24 to 23-25
Electric dipole antennas, 3-13 to 3-14
  and EMP, 30-9
  far-field pattern of, 1-29, 4-60

Electric fields. *See also* Electric-field strength
 with arrays, 3-36
 and cavities, 10-12
 conical-wave, 4-99
 distribution of, 12-4, 12-6 to 12-10, 12-35
 and equivalent currents, 5-8 to 5-10
  edge, 4-101
 in far zone, 6-3 to 6-4
 GO for, 4-13, 4-17
 Green's dyads for, 3-46, 3-48
 GTD for, 4-23
  for convex surfaces, 4-28
  for edged bodies, 4-25
  for vertices, 4-30
 integral equations for, 3-44, 3-46 to 3-47, 3-51
 of isotropic medium, 4-9
 with plane-wave spectra, 1-32, 5-5 to 5-8
 PO, 4-19 to 4-21
 probes for, 24-37 to 24-41
  calibration of, 24-54 to 24-57
  implantable, 24-51 to 24-54
 PTD for, for edged bodies, 4-106 to 4-107
 of rectangular apertures, 5-11 to 5-12
 reflection coefficient for, 2-28
 scattering. *See* Scattering
 of short monopole, 4-65
 of small dipoles, 6-13
 of small loops, 6-18
 of spherical wave, 4-6 to 4-7
 of surface fields, 4-84
 of transmitting antenna, 6-3, 20-6
 UAT for, for edged bodies, 4-39
 in unbounded space, 2-8 to 2-9
 UTD for
  corner-diffracted, 4-44
  for edged bodies, 4-32, 4-37, 4-83
 vector for, 1-14, 7-46 to 7-47
 in waveguides, 12-4
Electric-field strength
 of hemispherical radiators, 26-31 to 26-32
 horizontal, with two towers, 26-30 to 26-34
 of isotropic radiators, 26-26
 of uniform hemispherical radiators, 26-27 to 26-29
 with UTD solutions, 20-6
Electric line dipole for UTD solutions, 20-19 to 20-20

Electric surface current, 2-26 to 2-27
Electric vector potential, 3-6 to 3-7, 3-22, 5-8 to 5-9
 of rectangular-aperture antennas, 3-22
Electrode arrays for resistivity measurements, 23-4 to 23-8
Electroformed nickel for satellite antennas, 21-28 to 21-29
Electromagnetic fields
 computer model for, 3-53
 equations for, 3-6 to 3-8
 integral vs. differential formulation of, 3-55 to 3-58
 models for
  materials for, 32-69 to 32-70
  theory of, 32-65 to 32-69
Electromagnetic pulses, 30-3
 and aircraft antennas, 30-17 to 30-31
 analyzing effects of, 30-8 to 30-16
 and mounting structures, 30-4 to 30-7
 and principal elements, 30-7 to 30-8
Electromagnetic radiation monitors, 24-37
Electromechanical measurements, 32-4
Electromechanical phase shifters, 21-38, 21-41 to 21-43
Electromechanical power dividers, 21-48 to 21-49
Electromechanical switches, 21-37 to 21-38
Electronically Scanning Airborne Intercept Radar Antenna, 19-55
Electronic scanning, 21-5, 22-36 to 22-37
Elements, array
 and EMP, 30-7 to 30-8
 for feed arrays for satellite antennas, 21-27
 interaction of, with aperiodic arrays, 14-3
 mutual impedance between, 11-5
 for periodic arrays, 13-57 to 13-60
 patterns of, 13-6
 in phased arrays, 21-12
 in planar arrays, 11-53
 spacing of, 3-36, 14-4 to 14-5
Elevated antenna ranges, 32-19 to 32-21
Elevated H Adcock arrays, 25-9 to 25-10
Elevation and noise in tvro antennas, 6-31
Elevation-over-azimuth positioners, 32-12
Elf. *See* Extremely low frequencies
Ellipses, 15-25 to 15-26
 and intersection curve, 15-27, 15-29 to 15-30
 polarization, 1-14 to 1-16

Elliptical arrays, transformations with, 11-33 to 11-36, 11-48
  uniformly excited, 11-44 to 11-45
Elliptical cylinders, 4-75, 4-77
Elliptical loops, 6-19
Elliptical patches, 10-18
Elliptical polarization, 1-15 to 1-16
  gain measurements for, 32-49 to 32-51
  with log-spiral antennas, 9-74
EM Scattering modeling code, 3-88 to 3-91
Emissivity in state, 2-40
Emissivity of surfaces, 22-7 to 22-9
EMP. *See* Electromagnetic pulses
End-correction network for input admittance, 7-7
End-fire arrays, 3-41, 11-12, 11-15
  maximum directivity of, 11-65 to 11-71
  radiation pattern for, 11-76, 11-77
Engines, aircraft, model of, 20-37
Environments. *See also* Complex environments
  measurements for, 32-4
  numerical simulation of, 20-7 to 20-53
E-plane gain correction factor for biconical horns, 21-104
E-plane horn antennas, 8-4
  aperture fields of, 8-5 to 8-7
  design procedure for, 8-19 to 8-20
  directivity of, 8-14 to 8-18
  field intensity of, 8-13 to 8-14
  gain of, 8-18 to 8-19
  half-power beamwidth of, 8-14 to 8-15
  phase center of, 8-76 to 8-77
  radiated fields of, 8-7 to 8-10
  for reflector feeds, 15-92 to 15-93, 15-96 to 15-97
  universal curves of, 8-10, 8-13 to 8-14
E-plane radiation patterns, 1-28, 5-12
  of apertures
    circular, 5-22
    rectangular, 5-12
  of dielectric grating antennas, 17-74, 17-78
  of DuFort-Uyeda lenses, 19-51
  of horn antennas
    corrugated, 8-55 to 8-59
    E-plane, 8-9 to 8-12
    H-plane, 8-22 to 8-25
    pyramidal, 8-37 to 8-40
    square, 17-20
  for log-periodic antennas
    diode arrays, 9-64 to 9-66
    metal-arm, 9-11 to 9-12
    zigzag, 9-39 to 9-42, 9-45, 9-46
  for microstrip antennas, 10-24
  and UTD solutions, 20-13 to 20-16
Equal group delay lenses, 16-45 to 16-46
Equiangular spirals, 9-72, 9-110
Equivalent current method, 4-5, 4-96 to 4-102
  and surface-diffracted rays, 4-83
Equivalent currents for surface fields, 5-8
Equivalent cylinder techniques, 27-37 to 27-38
Equivalent radii for thin-wire antennas, 7-10 to 7-11
Equivalent receiving area
  for aircraft blade antenna, 30-12
  for annular slot antenna, 30-13
  for center-fed ellipsoid antenna, 30-10 to 30-12
  for spherical dipole antenna, 30-14
ERP. *See* Effective radiated power
Errors. *See also* Aberrations; Path length constraint and errors; Phase errors; Random errors; Systematic errors
  in antenna ranges, 32-34 to 32-35
  array, 13-52 to 13-57
  boresight, 31-3 to 31-4, 31-7 to 31-9, 31-27 to 31-29
  component, for phased arrays, 18-26 to 18-28
  with far-field measurements, 32-5
  with microwave radiometers, 22-23, 22-25 to 22-28
  modeling, 3-69
  pattern, 21-65
  quantization, 13-52 to 13-57, 21-12
  with scanning ranges, 33-15 to 33-18
ESAIRA (Electronically Scanning Airborne Intercept Radar Antenna), 19-55
ESMR. *See* Electrically scanning microwave radiometer
Eulerian angles, 15-116 to 15-117, 15-120
Evaluation of antenna ranges, 32-33 to 32-39
Even distributions in rectangular apertures, 5-14
Excitations
  array element, 11-23, 13-6 to 13-7
  linear transformations on, 11-39 to 11-41
  nonuniform, 11-29 to 11-30

Excitations (*cont.*)
  of overlapped subarrays, 13-37
  in planar arrays, 11-52 to 11-53
  with TEM transmission lines, 28-11, 28-19
Excited-patch radiators, 18-8 to 18-9
Experimentation
  and modeling errors, 3-69
  and validation of computer code, 3-72 to 3-76
External fields in longitudinal slots, 12-7
External networks and EMP, 30-8
External noise, 6-24 to 6-25, 29-50
External numerical validation of computer code, 3-77 to 3-78
External quality factor with cavities, 10-20
External validation of computer codes, 3-77 to 3-78
Extra high frequencies
  lens antennas for, 16-5
    aplanatic dielectric, 16-28
Extrapolation techniques for field measurements, 33-26 to 33-28
  of gain, 32-47 to 32-48
Extraterrestrial noise, 29-50 to 29-51
Extremely low frequencies
  with geophysical applications, 23-3
  ionospheric propagation of, 29-27 to 29-28

Fade coherence time and ionospheric scintillations, 29-18
Faraday rotation
  depolarization from, 29-16
  with phase shifters, 13-64
  variable power dividers using, 21-44 to 21-46
Far-field diagnostics for slot array design, 12-36 to 12-37
Far fields and far-field radiation patterns, 1-13, 1-17 to 1-18, 3-10, 3-22 to 3-24, 6-3 to 6-4. *See also* Radiation fields and patterns
  of antennas vs. structures, 20-53 to 20-54, 20-61 to 20-70
  of aperture antennas, 3-22 to 3-24, 15-21 to 15-24
    circular, 5-20 to 5-24
    rectangular, 3-22 to 3-24, 5-11 to 5-12, 5-14 to 5-15
    sinusoidal, 3-24
  of arrays, 11-5 to 11-8
    circular, 21-91 to 21-92, 21-94
    cylindrical, feeds for, 19-99 to 19-100, 19-104
    periodic, 13-5
    phased, 21-9 to 21-10
  calculations of, 1-21 to 1-23
  with cavities, 10-15, 10-19 to 10-20
  components of, 8-73, 8-75
  coordinates of, transformations with, 15-115, 15-117 to 15-120
  of dipole antennas
    folded, 7-41
    half-wave, 6-10 to 6-11
    log-periodic, 9-19 to 9-20
    short, 6-10 to 6-11, 20-8
    sleeve, 7-27 to 7-28
  of E-plane horn antennas, 8-7 to 8-10
  and equivalent currents, 5-10
  expressions for, 3-10
  Friis transmission formula for, 2-38 to 2-39
  integral representation for, 3-8 to 3-9
  with lens antennas, 16-6, 16-13 to 16-14
    coma-free, 16-22 to 16-23
    obeying Abbe sine condition, 19-97 to 19-99
    reflector, limited-scan feed concept, 19-70 to 19-71
  measurements of. *See* Measurements of radiation characteristics
  with microstrip antennas, 10-10, 10-22, 10-24
  with multiple element feeds, 19-13 to 19-16
  with plane-wave spectra, 5-7
  polarization of, 1-23 to 1-24
  power density in, 6-7
  with radial transmission line feeds, 19-24
  with radomes, 31-3
  with reflector antennas, 15-8 to 15-10, 15-13 to 15-14
    with axial feed displacement, 15-52 to 15-55
    with lateral feed displacement, 15-56 to 15-66
    offset parabolic, 15-34 to 15-35, 15-37 to 15-39, 15-43 to 15-45
    surface errors on, 15-110 to 15-112

with tapered apertures, 15-15 to 15-23
with reflector-lens limited-scan feed
  concept, 19-70 to 19-71
representation of, 1-18 to 1-21, 1-28 to
  1-29
scattered, and PO, 4-20
with single-feed CP microstrip antennas,
  10-58 to 10-59
for sleeve monopoles, 7-31
for slot array design, 12-36 to 12-37
of space-fed subarray systems, 19-69
from thin-wire antennas, 7-8, 7-15 to
  7-17
  loops, 7-44 to 7-46
for triangular-aperture antennas, 3-24
with UTD solutions, 20-4 to 20-6, 20-13
  to 20-14
Fat, human
  phantom models for, 24-55
  power absorption by, 24-8, 24-9, 24-47
F/D ratios. *See* Focus length to diameter
  ratio
FDTD modeling code, 3-88 to 3-91
Feed blockage with offset reflector, 21-19,
  21-23. *See also* Aperture blockage
Feed circuits. *See also* Beam-forming feed
  networks; Cassegrain feed systems;
  Constrained feeds; Optical feeds;
  Semiconstrained feeds; Unconstrained
  feeds
  for am antennas, 26-37 to 26-47
  for arrays
    dipole, 9-15 to 9-18, 9-21, 9-27, 9-29 to
      9-31, 9-61 to 9-63
    monopole, 9-66 to 9-67
    periodic, synthesis of, 13-61
    phased, design of, 18-3 to 18-4, 18-6,
      18-17 to 18-26
    power dividers for, 13-61
    spaced-tapered, 14-6 to 14-7
    subarrays, 13-39 to 13-41
    of tapered dielectric-rod antennas,
      17-47 to 17-48
  for conical horn antennas, 8-46
    log-spiral, 9-95 to 9-96, 9-99 to 9-104,
      9-111 to 9-112
  cylindrical array, 19-98 to 19-119
  displacements of, with offset parabolic
    antennas, 15-49 to 15-61
  for log-periodic dipole arrays, 9-15 to
    9-18, 9-21, 9-27, 9-29 to 9-31, 9-61
  for log-spiral antennas, 9-75 to 9-78,
    9-80
    dipole arrays, 9-62 to 9-63
    mono arrays, 9-66 to 9-67
  for microstrip antennas, 10-5, 10-6,
    10-28, 17-108 to 17-119
    and impedance matching, 10-51 to
      10-52
  for millimeter-wave antennas, 17-12,
    17-20, 17-47 to 17-48
  offset, 17-9, 19-54, 19-63 to 19-66
  reactance of, with microstrip antennas,
    10-31 to 10-34
  for reflector antennas, 15-5 to 15-6,
    15-13, 15-84 to 15-89
    complex, 15-94 to 15-99
    cos type patterns of, 15-99 to 15-105
    radiation patterns of, 15-89 to 15-94
    rays with, 15-11 to 15-12, 21-21 to
      21-22
  for satellite antennas, 21-5 to 21-7, 21-27
    to 21-29
  for subarrays, 13-39 to 13-41
Feed coordinates
  with reflector antennas, 15-7, 15-36,
    15-88 to 15-89
  transformations with, 15-115, 15-117 to
    15-120
Feed elements
  horn antennas as, 8-3
  location of, for feed arrays for satellite
    antennas, 21-27
Feed gaps with log-spiral antennas, 9-75 to
  9-77
Feed patterns
  for reflector antennas
    dual-reflector, 15-77, 15-80
    offset parabolic, 15-34 to 15-36, 15-47
  transformation of, with taper-control
    lenses, 16-37
Feed tapers, 15-36 to 15-37, 15-39 to
  15-40, 17-40
Feedthrough with lens antennas, 16-5
Feed transmission media for beam-forming
  networks, 21-56 to 21-63
Feed types of reflector antennas, 15-5 to
  15-6
Fence guide transmission type, 21-62
Fermat's principle and rays, 4-8 to 4-9
Ferrite absorbers with UTD solutions,
  20-17

Ferrite loop antennas, 6-18, 6-20 to 6-22
  radiation resistance of, 6-19
Ferrite phase shifters
  for arrays
    periodic, 13-62 to 13-64
    phased, 18-12, 18-14 to 18-17
  with planar scanning, 22-37
  variable, for beam-forming networks, 21-38 to 21-40, 21-42
Ferrite switches for beam-forming networks, 21-7, 21-37 to 21-38
Ferrite variable power dividers, 21-43 to 21-46
F-4 fighter aircraft, simulation of, 20-78 to 20-82
Fiber optics
  dielectric waveguides for, 28-47
  with implantable temperature probes, 24-50
Fictitious isotropic radiator, 2-35
Field curvature lens aberrations, 21-13
Field distributions
  for circular dielectric waveguides, 28-50
  measurements of, with radomes, 31-6 to 31-7
Field equations, electromagnetic, 3-6 to 3-8
Field equation pairs, 3-7
Field probes for evaluating antenna ranges, 32-33
Fields. *See also* Aperture fields; Electric fields; Incident fields; Magnetic fields; Reactive fields
  with arrays, 3-33, 3-35 to 3-36
  circular-phase, 9-109
  dual, 2-5 to 2-6, 3-9, 3-11
  and Huygen's principle, 2-18 to 2-19
  using integration to determine, 3-56 to 3-57
  and Kirchhoff approximation, 2-19 to 2-21
  in longitudinal slots, 12-7
  of perfect conductors, 3-39
  of periodic structures, 1-33 to 1-37
  plane-wave spectrum representations of, 1-31 to 1-33
  in radome-bounded regions, 31-5 to 31-6
  of rectangular-aperture antennas, 3-21 to 3-24
  with short dipole, 3-34
  surface, 4-84 to 4-89, 5-8
  tangential, 5-7 to 5-9, 7-5 to 7-6
  TE and TM, 1-30 to 1-31
  time-harmonic, 1-5 to 1-8, 3-6 to 3-8
  of transmission lines, 2-27 to 2-29
  transmitting, 2-19 to 2-21, 2-34 to 2-35
  of waveguides
    circular, 1-44 to 1-50
    circular TE/TM, 28-43
    rectangular, 1-37 to 1-44
    rectangular TE/TM, 28-39
Field strength of am antennas, 26-8
Figure of merit
  EIRP as, 1-27
  for horn antennas, 8-14
  for tvro antenna, 6-30 to 6-31
Filamentary electric current with cavities, 10-20 to 10-21
Filter type diplexers, 21-54
Fin line transmission types, 21-59, 21-61, 28-55 to 28-57
Finite sources, far fields due to, 1-32
Finite waveguide arrays and coupling, 13-44 to 13-45
Five-wire transmission lines, 28-13
Fixed beam-forming networks, 21-34, 21-37, 21-49 to 21-56
Fixed line-of-sight ranges, 32-9 to 32-11
Fixed-wire antennas, EMP responses for, 30-23 to 30-27
Flared-notch antennas, 13-59
Flaring with horn antennas
  conical, 8-46
  corrugated, 8-60, 8-64
  E-plane, 8-9
  H-plane, 8-20, 8-29 to 8-30
  and phase center, 8-76 to 8-77
Flat dielectric sheets, propagation through, 31-10 to 31-18
Flat plate array antennas, 17-26
Flat-plate geometry for simulation of environments, 20-8, 20-18
Flat-plate Luneburg lenses, 19-44, 19-46 to 19-47
Flat surface lenses, 16-9 to 16-11
Flexible coaxial transmission lines, 28-22
Flood predicting with remote sensing, 22-15
Floquet space harmonics
  and scattered fields, 1-34 to 1-35
  and UP structures, 9-32
Fluctuation of aperiodic array beams, 14-26 to 14-28

FM broadcast antennas, 27-3 to 27-8, 27-32 to 27-40
Focal length
  of conic sections, 15-24 to 15-26
  of offset parabolic reflectors, 15-31
  for reflector satellite antennas, 21-22 to 21-24
  with taper-control lenses, 16-36
Focal points. *See* Caustics
Focal region of reflector satellite antennas, 21-19
Fock radiation functions, 4-68 to 4-69, 4-73
  for mutual coupling, 4-95
  for surface fields, 4-88 to 4-89
  for surface-reflection functions, 4-52 to 4-53, 4-66
Fock-type Airy functions, 4-53, 4-69
Fock type transition functions, 4-50n
Focus length to diameter (F/D) ratio
  with lens antennas, 16-12
    and dielectric constants, 16-24
    planar surface, 16-56
    for satellites, 21-16
    taper-control, 16-37
    and zoning, 16-43
  with microwave radiometers, 22-30
  and pyramidal horn antenna dimensions, 8-45
  with reflector satellites
    and cross-polarization, 15-47
    with off-focus feeds, 15-51, 15-53, 15-55 to 15-56, 15-58, 15-62
    offset parabolic reflectors, 15-32 to 15-33, 15-37
    for satellites, 21-22 to 21-23
  and wide-angle scans, 19-34 to 19-35
Fog, scintillation due to, 29-36
Folded antennas
  dipole, 3-29, 7-36 to 7-40
  slot, and Babinet principle, 30-15 to 30-16
  unipole, 3-30
Folded Luneburg lenses, 19-49
Folded pillbox feeds, 19-17 to 19-23
Folded-T waveguide power division elements, 21-50 to 21-52
Footprints. *See also* Coverage area
  of microwave radiometers, 22-30 to 22-32
  of tvro signal, 6-26, 6-27
Formulation for computer models, 3-53 to 3-54, 3-81

Four-electrode arrays for earth resistivity, 23-4 to 23-8
Fourier transformers, 19-95
Four-port hybrid junctions, 19-5 to 19-6
Four-tower systems, 26-33
Four-wire transmission lines, 28-12
Fox phase shifter, 13-64
Fraunhofer region, 1-18
Free space, resistance of, 26-25 to 26-26
Free-space loss, 2-39, 29-5 to 29-6
  of satellite-earth path, 29-8
Free space vswr evaluation method for anechoic chambers, 32-38 to 32-39
Frequencies
  and active region
    with conical log-spiral antennas, 9-83
    of log-periodic dipole antennas, 9-17 to 9-19
  allocation of
    for satellite antennas, 21-6
    for tv, 6-28, 27-3
  and beamwidth, 9-87
  and computer time, 3-62
  dependence on, by reflector antennas, 17-21 to 17-13
  and image impedance, 9-55
  and input impedance, 10-9
  and noise temperature, 2-43
  and phase center, 9-92
  and reactance, 10-30, 10-65
  and space harmonics, 9-35
  of standard waveguides, 28-40 to 28-41
    circular dielectric, 1-49, 28-50
    rectangular, 1-42
Frequency-agile elements, 10-66 to 10-69
Frequency-domain
  compared to time-domain, 3-65 to 3-66, 3-82
  and EMP, 30-4
  method of moments in, 3-57 to 3-62, 3-64
Frequency-independent antennas, 9-3 to 9-12. *See also* Log-periodic dipole antennas; Log-periodic zigzag antennas; Log-spiral antennas; Periodic structures
  millimeter-wave, 17-27 to 17-28
Frequency-modulation broadcast antennas, 27-3 to 27-8, 27-32 to 27-40
Frequency-scanned array antennas, 17-26 to 17-27

Frequency selective surface diplexer, 21-53
Fresnel ellipsoid and line-of-sight propagation, 29-36
Fresnel fields, 1-18. *See also* Near fields
Fresnel integrals with near-field patterns, 5-25
Fresnel-Kirchhoff knife-edge diffraction, 29-42 to 29-43
Fresnel reflection coefficients, 3-27, 3-33, 3-48
Fresnel zone-plate lens antennas, 17-20
Friis transmission formulas, 6-8 to 6-9
   for far fields, 2-38 to 2-39
Front ends, integrated, 17-134
Front-to-back ratios
   for log-periodic dipole arrays, 9-60 to 9-61
   for log-spiral conical antennas, 9-89, 9-92
F-16 fighter aircraft, simulation of, 20-82, 20-85 to 20-88
Fuel tanks, aircraft, simulation of, 20-79, 20-82
Fundamental waves and periodic structures, 9-33 to 9-35, 9-55 to 9-56
Fuselages of aircrafts
   and EMP, 30-4, 30-6
   simulation of, 20-70, 20-78

Gain, 1-24 to 1-27, 3-12 to 3-13, 3-28 to 3-31. *See also* Directive gain
   aperture, 5-26 to 5-34
      circular, 5-23
      rectangular, 5-29
      uniform fields, 5-14
   of arrays, 3-36, 3-38 to 3-43, 13-10 to 13-11
      array columns, 13-54
      edge-slot, 17-27
      feed, for satellite antennas, 21-27
      multimode element, 19-12
      phased, 18-11, 19-59, 21-11 to 21-12
   axial near-field, 5-32
   and bandwidth, with tv antennas, 27-7
   with DuFort optical technique, 19-76
   and effective area, 5-27 to 5-28
   of horn antennas
      biconical, 21-105 to 21-106, 21-108 to 21-109
      conical, 8-48
      correction factors with, 21-104
      E-plane, 8-18 to 8-19
      pyramidal, 8-43 to 8-44, 15-98
   of lens antennas, 17-18
   and lens F/D ratio, 21-16
   with limited scan feeds, 19-57
   of log-periodic zigzag antennas, 9-45 to 9-47
   measurement of, 32-42 to 32-51
   of microwave antennas, 33-20 to 33-21
   of millimeter-wave antennas
      lens, 17-18
      maximum-gain surface-wave, 17-40 to 17-41
      reflector, 17-12 to 17-14
      tapered dielectric rod, 17-40 to 17-41, 17-47
   reduction factor for, near-field, 5-30 to 5-32
   of reflector antennas
      boresight, 15-81
      millimeter-wave, 17-12 to 17-14
      surface errors on, 15-106 to 15-107
   ripple of, with feed arrays, 21-27
   with satellite antennas, 21-5 to 21-7, 21-12
      lateral feed displacement with, 21-20
      tvro, 6-30 to 6-31
   of short dipole, 6-10 to 6-11
   of small loops, 6-13, 6-17
Gain-comparison gain measurements, 32-49 to 32-51
Gain-to-noise ratio of tvro antenna, 6-31
Gain-transfer gain measurements, 32-49 to 32-51
Galactic noise, 6-25, 29-50
Galerkin's method
   with methods of moments, 3-60
   as weight function, 3-83
Gallium arsenide
   in integrated antennas, 17-132, 17-137 to 17-138, 17-140
   in sensors with implantable temperature probes, 24-50
Gap capacitance and input admittance, 7-8, 7-15
Gaseous absorption
   and line-of-sight propagation, 29-34 to 29-35
   of millimeter waves, 17-5
   of satellite-earth propagation, 29-8, 29-10

Gaseous emissions, noise from, 29-50 to 29-51
GEMACS modeling code, 3-88 to 3-91
Generalized equation for standard reference am antennas, 26-20 to 26-22
Geodesic lens antennas, 17-29 to 17-32
  matrix-fed, 19-108 to 19-119
  for phased array feeds, 19-17, 19-41, 19-44
Geodesic paths, 20-19
Geometrical models, 32-69
Geometrical optics, 4-3
  and computer solutions, 3-54
  fields with, 4-8 to 4-9
    incident, 4-10 to 4-13
    reflected, 4-13 to 4-18
  for lens antennas, 16-6
  and PTD, 4-103
  with reflector antennas, 15-7, 15-13, 15-14
    dual-reflector, 15-69 to 15-73, 15-77, 15-80
  for simulation of environment, 20-10
  with UTD solutions
    for curved surfaces, 20-35 to 20-36
    for noncurved surfaces, 20-45 to 20-46
Geometrical theory of diffraction, 4-3, 4-23 to 4-24. *See also* Uniform geometrical theory of diffraction
  for apertures on curved surfaces, 13-50 to 13-51
  and computer solutions, 3-54
  for convex surfaces, 4-28 to 4-30
  and ECM, 4-97 to 4-98
  for edges, 4-25 to 4-28
  for pyramidal horn fields, 15-92
  with reflector antennas, 15-7 to 15-8, 15-11 to 15-14
    dual-reflector, 15-69 to 15-72, 15-76 to 15-77
  for vertices, 4-30 to 4-31
Geophysical applications, antennas for, 23-3, 23-23 to 23-25
  electrode arrays, 23-4 to 23-8
  grounded wire, 23-8 to 23-17
  loop, 23-17 to 23-23
Geostationary Operational Environmental Satellite, 22-34, 22-43
Geosynchronous satellite antennas, 6-26
  microwave radiometers for, 22-43 to 22-46
G for microstrip antennas, 10-34 to 10-41

Gimbaling
  angles of, and radome field measurements, 31-6
  with satellite antennas, 21-5
Glass plates for microstrip arrays, 17-108
GOES (Geostationary Operational Environmental Satellite), 22-34, 22-43
GO methods. *See* Geometrical optics
Goniometers, rotating, 25-15 to 25-16
Good conductors, modeling of, 32-70 to 32-71
Good dielectrics, modeling of, 32-70 to 32-75
Goubau antenna, 6-24
Gradient synthesis method, 21-65
Graphite epoxy material for satellite antenna feeds, 21-28 to 21-29, 22-46
Grating antennas, 17-34 to 17-35, 17-48 to 17-82
Grating lobes
  with arrays
    aperiodic, 14-3, 19-57
    blindness in, 13-47
    Dolph-Chebyshev, 11-18 to 11-19
    for feeding satellite antennas, 21-27
    periodic, 11-32, 13-10
    phased, 18-9 to 18-10, 21-10 to 21-12
    scanned in one plane, 13-13 to 13-14
    scanned in two planes, 13-15 to 13-16
    space-fed, 19-69
    and subarrays, 13-32, 13-35, 19-69
  with multiple element feeds, 19-15
  with radial transmission line feeds, 19-31
  and space harmonics, 1-35
Grating period
  and leakage constants, 17-62
  and radiation angle in leaky modes, 17-56 to 17-58
Great-circuit cuts, 32-6
Green's dyads, 3-46 to 3-48
Green's function
  for field propagation, 3-56
  in unbounded space, 2-6 to 2-11
  with UTD solutions, 20-4 to 20-5
Gregorian feed systems
  offset-fed, 19-63 to 19-66
  with reflector satellite antennas, 21-21 to 21-22
Gregorian reflector systems
  cross polarization in, 15-69 to 15-70
  gain loss in, 15-73
Gridded reflector satellite antennas, 21-21

Groove depths
    of dielectric grating antennas, 17-77 to 17-78
    and leakage constants, 17-60 to 17-61
Groove guide transmission type, 17-35, 21-59, 21-62
    for leaky-wave antennas, 17-82 to 17-92
Ground-based hf antennas, modeling of, 32-76 to 32-77
Grounded wire antennas, 23-8 to 23-17
Ground-plane antenna ranges, 32-19
Ground planes
    and dipole arrays, 13-42
    imperfect, 3-27, 3-32 to 3-33, 3-47 to 3-50
    infinite, in slot array design, 12-34 to 12-35
    with microstrip antennas, 10-5, 10-6, 10-23 to 10-24
        circular polarization, 10-59
    perfect, 3-18 to 3-20, 3-47 to 3-48
    in UTD solutions, 20-7
Ground-reflection antenna ranges, 32-21 to 32-23
    gain measurement on, 32-48 to 32-49
Ground systems for am antennas, 26-37, 26-38
Ground-wave attenuation factor, 29-45
Group delay and satellite-earth propagation, 29-15 to 29-16
GTD. *See* Geometrical theory of diffraction
Guided waves from radomes, 31-5
Guidelines with modeling codes, 3-87, 3-95
Gyros for satellite antennas, 22-46
Gysel hybrid ring power division element, 21-50, 21-51

HAAT. *See* Height above average terrain
Half-power beamwidths, 1-20, 1-29
    of aperture antennas, 8-67
        circular, 5-22 to 5-24
        rectangular, 5-13, 5-16 to 5-17, 5-20, 15-91
    of arrays
        aperiodic, 14-15, 14-18, 14-23, 14-27 to 14-32
        Dolph-Chebyshev, 11-19
        linear, 11-15
        scanned in one plane, 13-12 to 13-15
    of classical antennas, 3-28 to 3-31
    and directivity, 32-41

    of horn antennas
        corrugated, 8-55, 8-59, 15-102 to 15-104
        E-plane, 8-14 to 8-15
        H-plane, 8-29 to 8-30
    of log-periodic zigzag antennas, 9-39 to 9-46
    of log-spiral conical antennas, 9-86, 9-87, 9-89, 9-97 to 9-98, 9-104
    and microwave radiometer beam efficiency, 22-27
    of millimeter-wave antennas
        dielectric grating, 17-79 to 17-80
        maximum-gain surface-wave, 17-40 to 17-42
        tapered dielectric-rod, 17-39, 17-47
    of reflector antennas
        lateral feed displacement, 21-20
        offset parabolic, 15-37, 15-41
        tapered-aperture, 15-16 to 15-17, 15-23
    of waveguide feeds, open-ended
        circular, 15-105
        rectangular, 15-104
Half-wave antennas, 26-8
    dipole, 3-17, 6-14 to 6-15
        radiation pattern of, 6-11
        receiving current of, 7-18 to 7-19
Half-wave radomes, 31-17
Hallen antenna, 6-24
Hansen-Woodyard end-fire linear array, 11-15
HAPDAR radar, 19-56
Harmonic distortion from pin diodes, 21-38
Harvesting of crops, remote sensing for, 22-15
H/D (height-to-diameter ratio), 15-32 to 15-33, 15-37, 15-40
Health care. *See* Medical applications
Heat sinking with integrated antennas, 17-133
Heating patterns, phantoms for checking, 24-54 to 24-57. *See also* Hyperthermia
Height. *See also* Vector effective height
    effective, 30-19, 30-21, 30-30
    for grating antennas
        dielectric, 17-64
        metal, 17-68 to 17-69
    and leakage constants, 17-58 to 17-59
    of log-spiral conical antennas, 9-101
    offset, of parabolic reflectors, 15-31 to 15-32

of substrates, 17-107, 17-109, 17-111, 17-124, 17-126
of test antennas in elevated antenna ranges, 32-20
Height above average terrain
and fm antenna ERP, 27-4, 27-6
and tv antenna ERP, 27-3 to 27-5
Height-to-diameter ratio, 15-32 to 15-33, 15-37, 15-40
Height-to-radius ratio
and impedance of dipoles, 9-61 to 9-62
of log-periodic antennas, and directivity, 9-27, 9-30
Helical antennas, 3-31
circularly polarized tv, 27-15 to 27-19
Helical-coil hyperthermia applicators, 24-48 to 24-49
Helical geodesic surface-ray paths, 4-75 to 4-76
Helix phased arrays, 21-12, 21-68, 21-70 to 21-71
Hemispherical lenses, 19-81 to 19-82
Hemispherical radiation pattern of am antennas, 26-5
Hemispherical radiators, 26-9, 26-27 to 26-29, 26-32
Hemispherical reflector antennas
high-resolution, 19-85, 19-87, 19-89 to 19-91
with wide-angle systems, 19-80 to 19-81
HEMP. *See* Electromagnetic pulses
Hertzian source, characteristics of, 3-13 to 3-14
Hexagonal arrays, 11-30 to 11-31
H-fields. *See* Magnetic fields
H-guide transmission type, 21-59, 21-62
modification of, for leaky-wave antennas, 17-93 to 17-94
High-attenuation rf cable, 28-25
High-delay rf cable, 28-25
Higher-order modes
junction coupling and scattering, 12-35 to 12-36
for leaky-wave antennas, 17-87 to 17-92, 17-98 to 17-103
with microstrip antennas, 10-25
High frequencies, ionosphere propagation of, 29-21 to 29-25
High-frequency antennas
fixed-wire, 30-23 to 30-27, 30-29 to 30-30
modeling of, 32-64, 32-76 to 32-77

High-frequency limit, 9-3
of log-spiral antennas, 9-75
High frequency techniques, 4-3 to 4-6
equivalent current method, 4-96 to 4-102
geometrical optics fields, 4-6 to 4-8
geometrical theory of diffraction, 4-23 to 4-96
physical optics field, 4-8 to 4-23
physical theory of diffraction, 4-102 to 4-115
High-gain antennas
direction finding, 25-14 to 25-16
dual-reflector, 15-73
lens, 16-5
limited scan, 19-57
microstrip dipole, 17-122 to 17-126
millimeter-wave antennas, 17-9 to 17-27
multimode element array, 19-12
rotating, 25-10 to 25-11
High-resolution antennas, 19-85, 19-87, 19-89 to 19-91
High-temperature rf cable, 28-24
HIHAT (high-resolution hemispherical reflector antennas), 19-85, 19-87, 19-89, to 19-91
Holey plate experiment, 14-27 to 14-35
Hollow-tube waveguides, 28-35 to 28-36
Holographic antennas, 17-106, 17-129 to 17-131
Homology with millimeter-wave antennas, 17-14
Horizon, radio, 29-34
Horizontal field strength with two towers, 26-30 to 26-34
Horizontal optimization, 11-82 to 11-83
Horizontal polarization
with leaky-wave antennas, 17-97
with loop fm antennas, 27-33
for microwave radiometers, 22-44 to 22-46
and surface emissivity, 22-8
with tv antennas, 27-23
Horizontal profiling for earth resistivity, 23-4
Horizontal scanners, 33-19
Horizontal stabilizers, simulation of, 20-82 to 20-84
Horizontal transmitting loop antennas, 23-17

Horn antennas, 3-31, 8-5. *See also* Aperture-matched horn antennas; Conical horn antennas; Corrugated horn antennas; Dielectric loading, with horn antennas; E-plane horn antennas; H-plane horn antennas; Pyramidal horn antennas
  as array elements, 8-4
  as calibration standard, 8-3
  millimeter-wave, 17-23 to 17-24
  multimode, 8-69 to 8-73, 15-95
  phase center, 8-73 to 8-84
  with satellite antennas, 21-7, 21-80, 21-82 to 21-84
Horn feeds, 15-5
Horn launchers, 17-46
Horn waveguides, 2-27
H-plane folded and septum T waveguide power division elements, 21-51 to 21-52
H-plane gain correction factor for biconical horns, 21-104
H-plane horn antennas, 8-4
  aperture fields of, 8-20
  design of, 8-23 to 8-34
  directivity of, 8-29 to 8-33
  half-power beamwidth of, 8-29 to 8-30
  phase center of, 8-76 to 8-77
  radiated fields of, 8-20 to 8-26, 8-28
  for reflector feeds, 15-92 to 15-93, 15-96 to 15-97
  universal curves of, 8-23, 8-27 to 8-29
H-plane radiation patterns, 1-28, 5-12
  of apertures
    circular, 5-22
    rectanguar, 5-12
  with cavity-backed slot arrays, 9-68 to 9-71
  of dielectric grating antennas, 17-73 to 17-74, 17-78 to 17-79
  of DuFort-Uyeda lenses, 19-52
  of horn antennas
    corrugated, 8-53 to 8-59
    E-plane, 8-9, 8-10
    H-plane, 8-23 to 8-26, 8-28
    pyramidal, 8-37 to 8-40
  for log-periodic antennas
    diode arrays, 9-63 to 9-66
    zigzag, 9-39 to 9-42, 9-45, 9-46
  with microstrip antennas, 10-25
  of monofilar zigzag antenna, 9-36
  and UTD solutions, 20-13 to 20-14

Hughes approaches with cylindrical array feeds
  matrix-fed Meyer geodesic lens, 19-108 to 19-119
  phased lens, 19-107 to 19-108
Human-made noise, 29-51 to 29-52
Humidity
  and brightness temperature, 22-10
  remote sensing of, 22-6 to 22-8
  and scintillations, 29-20
  sounders for, 22-14, 22-22
Huygens-Fresnel principle, 1-32, to 1-33
Huygen's principle and source, 2-18 to 2-20
Hybrid antennas with reflector antennas, 17-11
Hybrid beam-forming networks, 21-34
Hybrid-coupled phase shifters, 13-62 to 13-63, 18-13
Hybrid-mode horn feeds, 15-5, 15-97, 15-99
Hybrid-mode waveguides, 28-47, 28-50 to 28-58
Hybrid-ring coupler power division element, 21-50, 21-52
Hydrology, remote sensing for, 22-15 to 22-16
Hyperbolas, 15-25 to 15-26
Hyperbolic lenses, 16-10, 16-53 to 16-54
Hyperbolic reflectors, feeds with, 19-61 to 19-62
Hyperthermia for medical therapy, 24-5 to 24-7, 24-9
  applicators for, 24-43 to 24-49
  heating patterns with, 24-50 to 24-57
  microstrip antennas for, 24-18 to 24-25
  waveguide antennas for, 24-9 to 24-18
Hypodermic monopole radiators, 24-26 to 24-27

Ice, 22-9
  remote sensing of, 22-14 to 21-15
  and sidefire helical tv antennas, 27-26
Illuminating fields and imperfect ground planes, 3-48
Illumination control of apertures, 13-30 to 13-32
Image-guide-fed slot array antenna, 17-50
Image guides, 21-59, 21-61, 28-50 to 28-53

Image impedances
  for log-periodic mono arrays, 9-66
  with periodic loads, 9-50, 9-55
Image-plane antenna ranges, 32-11, 32-19, 32-22 to 32-24
Images
  and imperfect ground planes, 3-27, 3-32, 3-48
  and perfect ground planes, 3-20, 3-47
  theory of, 2-11 to 2-13
Imaging arrays, 17-137 to 17-141
Impedance. *See also* Characteristic impedance; Input impedance; Mean impedance; Wave impedance
  base, with am antennas, 26-8 to 26-17
  of bow-tie dipoles, 17-138
  of cavities, 10-20 to 10-21
  of complementary planar antennas, 2-15 to 2-16
  of dipoles, 9-56
    and height-to-radius ratio, 9-61 to 9-62
    skewed, 27-18
  driving point
    with directional antenna feeder systems, 26-43 to 26-45
    with microstrip antennas, 10-20, 10-34, 10-53
  and EMP, 30-4
  feeder, with log-periodic diode arrays, 9-65 to 9-66
  of ideal sources, 2-29 to 2-30
  of log-periodic planar antennas, 9-10
  of log-spiral conical antennas, 9-107
  for medical applications, 24-49 to 24-50
  of microstrip antennas, 10-26 to 10-44
  resonant, 10-28, 10-30 to 10-31
  of scale models, 32-64, 32-68, 32-76 to 32-77
  of short monopole probe, 24-31 to 24-32
  surface, 3-54, 32-8
  of transmission lines, 21-63, 28-5
    coplanar, 28-35
    triplate stripline TEM, 28-22, 28-28 to 28-29
    two-wire, 28-11, 28-19
  and UTD solutions, 20-5
  waveguide, 2-28
Impedance function of cavities, 10-11
Impedance-loaded antennas
  dipole, 7-19 to 7-20
  monopole, 7-20 to 7-21

Impedance matching
  with directional antenna feeder systems, 26-40 to 26-41
  and gain measurements, 32-42
  and lens antennas, 16-5
    transformers for, 21-19
  of microstrip antennas, 10-50 to 10-52
  with receiving antennas, 6-6 to 6-7
  of split-tee power dividers, 13-61
  of waveguide arrays, 13-57
Impedance matrices, 3-61
  and rotational symmetry, 3-67 to 3-68
Imperfect ground planes, 3-27, 3-32 to 3-33
  integral equations with, 3-47 to 3-50
Implantable antennas for cancer treatment, 24-25 to 24-29
  E-field probes, 24-51 to 24-54
  temperature probes, 24-50 to 24-51
Incident fields
  GO, 4-10 to 4-13
  reciprocity of, 2-32 to 2-33
  and UTD solutions, 20-8 to 20-9
Incident powers and gain, 1-24 to 1-25
Incident shadow boundary, 4-10 to 4-11, 4-17
Incident wavefronts, 4-15, 4-16
Inclined edge-slot radiators, 18-8
Inclined-slot array antennas, 17-26
Inclined slots, 12-4, 12-24
  narrow wall arrays, 12-25 to 12-28
Incremental conductance technique, 12-26, 12-34
Index of refraction. *See* Refractive index
Indoor antenna ranges, 32-26 to 32-30
Inductive loading of short dipole, 6-13
Inductive susceptance of microstrip antennas, 10-26
Infinite array solutions, 13-48 to 13-49
Infinite baluns, 9-103, 9-105
Infinite ground plane assumptions in slot array design, 12-34 to 12-35
Inhomogeneous lenses, 16-51 to 16-54
In-line power dividers for array feeds, 13-61
Input features in modeling codes, 3-84 to 3-85
Input impedance, 2-29
  of array elements, 11-5
  of biconical antennas, 3-26
  of coaxial sleeve antennas, 7-32 to 7-33, 7-36

Input impedance (cont.)
  of dipole antennas
    folded, 7-36, 7-39 to 7-41
    linear, 7-14 to 7-17
    microstrip, 17-125
    short, 6-11 to 6-12
    sleeve, 7-26 to 7-29
  of L-band antennas, 30-18
  of loaded microstrip elements, 10-54 to 10-55
  of log-periodic antennas
    dipole arrays, 9-20 to 9-21, 9-24, 9-27 to 9-30, 9-58, 9-62
    zigzag, 9-39, 9-42 to 9-44
  of log-spiral antennas, 9-78 to 9-79
    conical, 9-95 to 9-97, 9-99
  of loop antennas
    ferrite, 6-21 to 6-22
    small, 6-17 to 6-18
  of microstrip antennas, 10-9, 10-26 to 10-28, 10-31, 10-65
  of models, 32-85 to 32-86
  of open-ended coaxial cable, 24-35
  of self-complementary antennas, 9-5, 9-7
  of sleeve monopoles, 7-30 to 7-31, 7-34
  of thin-wire antennas, 7-7 to 7-8, 7-44, 7-47 to 7-48
  of transmission lines, 28-7
    two-wire, 7-40
  of uhf communication antennas, 30-21
  validation of computer code for, 3-74 to 3-75, 3-77
  of vhf antennas
    communication, 30-24
    localizer, 30-26
    marker beacon, 30-28
Input susceptance of linear thin-wire antenna, 7-15 to 7-17
Insertion delay phase and dielectric sheets, 31-14
Insertion loss
  and beam-forming network switches, 21-37
  of phase shifters, 18-15 to 18-17, 18-21, 21-41
  and power division elements, 21-51
    diode varible, 21-47
  of transmission lines, 21-63
  of waveguides, 21-7
Insertion-loss gain measurement method, 32-44 to 32-46

Instantaneous direction-finding patterns, 25-12 to 25-14
Instantaneous field of view, 22-30 to 22-32
Instrumentation for antenna ranges, 32-30 to 32-33
Insular waveguide transmission types, 21-59, 21-61
  for microstrip arrays, 17-117 to 17-118
Integral equations, 3-36, 3-38 to 3-39, 3-44 to 3-52
  codes for, 3-80 to 3-96
  compared to differential, 3-55 to 3-58, 3-82
  computation with, 3-68 to 3-72
  for far fields, 3-8 to 3-9
  numerical implementation of, 3-52 to 3-68
  for unbounded space, 2-9 to 2-11
  validation of, 3-72 to 3-80
Integrated antennas, 17-105 to 17-106, 17-131 to 17-141
Integrated optics, 28-47
Interaction terms and computational effort, 3-64
Interfaces, between media, 1-8 to 1-9
  reflections at, 3-32, 24-7, 31-12 to 31-13
Interference
  in antenna ranges, 32-19
  with satellite antennas, 21-5
Interferometers, 25-21 to 25-23
Intermodulation distortion from pin diodes, 21-38
Internal networks and EMP, 30-8
Internal noise, 6-24, 29-50
Internal validation of computer code, 3-78 to 3-80
Interpolation of measured patterns, 20-6
Interrupt ability of modeling codes, 3-86
Intersection curve of reflector surface, 15-27 to 15-30
Inverted strip guide transmission type, 21-62
  dielectric, 17-49 to 17-50
In-vivo measurement, antennas for, 24-30 to 24-35
Ionosondes, 29-24
Ionosphere
  propagation via, 29-21 to 29-30
    satellite-earth, 29-14 to 29-15
  scintillation caused by, 29-16 to 29-21

Index

Irrigation control, remote sensing for, 22-15
ISB (incident shadow boundary), 4-10 to 4-11, 4-17
Isolation
  with ferrite switches, 21-38
  for satellite antennas, 21-5 to 21-7
Isosceles beams, 19-11 to 19-12
Isosceles triangular lattices, 1-35 to 1-36
Isotropic radiators
  characteristics of, 3-28
  electric-field strength of, 26-26
  radiation intensity of, 1-27
  receiving cross section of, 2-35
Isotropic response for E-field probes, 24-39 to 24-40

Jamming signals, arrays for, 11-10
Johnson compact antenna range, 32-29 to 32-30
Junction coupling in slot array design, 12-36
Junction effect with sleeve antennas, 7-23 to 7-26

KC-135 aircraft, patterns for, 20-65 to 20-67
Keller cone, 4-33
Keller's diffraction coefficients, 15-13
Kirchhoff approximation, 2-19 to 2-21
  and physical optics approximation, 2-27
Knife-edge diffraction, 29-42 to 29-44
Kolmogroff turbulence, 29-20
Ku-band systems, 29-8

Land
  emissivity of, 22-10
  modeling of, 32-74 to 32-77
  remote sensing of, 22-15 to 22-16
Large arrays, design of, 12-28 to 12-34
LASMN3 modeling code, 3-88 to 3-91
Latching Faraday rotator variable power dividers, 21-44 to 21-46
Latching phase shifters, 13-63 to 13-64
Lateral feed displacement, 15-51 to 15-61, 21-20
Lateral width of dielectric grating antennas, 17-64
Lattices
  field representation of, 1-35
  for phased-array radiators, 18-9 to 18-10
  spacing of, for feed arrays, 21-27

Launching coefficients
  for conducting cylinder, 21-97
  surface-ray, 4-70
Law of edge diffraction, 4-33
L-band blade antennas, 30-17 to 30-19
Leakage constants
  for dielectric grating antennas, 17-65 to 17-69, 17-74 to 17-79, 18-82
  with leaky mode antennas, 17-53 to 17-63, 17-85, 17-87 to 17-90, 17-95 to 17-97
Leakage radiation in medical applications, 24-5, 24-50
Leaky modes with periodic dielectric antennas, 17-51 to 17-62
Leaky-wave antennas, 17-34 to 17-35
  uniform-waveguide, 17-82 to 17-103
Least squares moment method, 3-60
Left-hand circular polarization, 1-15, 1-20
  of arrays, 11-7
  dual-mode converters for, 21-78, 21-80
  far-field pattern for, 1-29
  for microstrip antennas, 10-59, 10-69
  with offset parabolic reflectors, 15-36, 15-48 to 15-50
  power dividers for, 21-78, 21-80
Left-hand elliptical polarization, 1-16, 9-74
Lenses and lens antennas, 16-5 to 16-6. *See also* Constrained lenses; Optical feeds; Semiconstrained feeds
  aberrations and tolerance criteria of, 16-12 to 16-19
  with analytic surfaces, 16-9 to 16-12
  in compact ranges, 32-29
  design principles of, 16-7 to 16-9
  with horn antennas
    conical, 8-46
    diagonal, 8-70
    H-plane, 8-23
  inhomogeneous, 16-51 to 16-54
  millimeter-wave, 17-14 to 17-22
  for satellite antennas, 21-7, 21-8, 21-12 to 21-19, 21-71 to 21-77
  surface mismatches of, 16-54 to 16-57
  taper-control, 16-33 to 16-38
  wide-angle, 16-19 to 16-33
  zoning of, 16-38 to 16-41
LEO (low earth orbiting) satellites, 22-35 to 22-43
Level curves, 14-12 to 14-14
Lightning and tv antennas, 27-8, 27-9

Limited-scan unconstrained feeds, 19-12, 19-52, 19-56 to 19-76
Lindenblad tv antenna, 27-12
Linear arrays, 11-8 to 11-23
   broadside, maximum directivity of, 11-64 to 11-68
   cophasal end-fire, 11-69, 11-70
   of longitudinal slots, 12-13 to 12-16
   millimeter-wave, 17-33
     uniform leaky-wave, 17-35
Linear dipole antennas, 3-13 to 3-14, 7-6, 7-11 to 7-23
Linear errors with lenses, 16-14
Linear polarization, 1-14 to 1-16, 1-20, 1-28 to 1-29
   conversion of, from circular, 13-60 to 13-61
   measurement of, 32-6, 32-8, 32-53, 32-59
   for microwave radiometers, 22-44 to 22-46
   with phased arrays, 18-6
   with satellite antennas, 21-7
     gridded reflector, 21-21
Linear transformations in arrays, 11-23 to 11-48
Linear wire antenna, polarization of, 1-24
Line-of-sight propagation, 29-30, 29-32 to 29-36
   with millimeter-wave antennas, 17-5
Line-of-sight ranges, 32-9 to 32-11
Line source illumination, 4-12 to 4-13
Line source patterns, 13-13
Link calculations for satellite tv, 6-27, 6-29 to 6-31
Liquid artificial dielectric, 17-22, 17-72
Lit zone, 4-75
   Fock parameter for, 4-73
   terms for, 4-71 to 4-72
Load current of receiving antennas, 6-3, 6-4
Loaded-line diode phase shifters, 13-62 to 13-63, 18-13
Loaded-lines, 9-46 to 9-62
Loaded loop antennas, 7-47 to 7-48
Loaded microstrip elements, 10-52 to 10-56
Loaded thin-wire antennas, 7-9
Lobes, 1-28. *See also* Grating lobes; Side lobes
   with log-spiral antennas, 9-75, 9-77
   shoulder, 32-16
Localizer antennas, 30-23, 30-25 to 30-26

Logarithmic spiral antennas, 17-28
Logarithmic spiral curve, 9-72
Log-periodic arrays, 9-3 to 9-4, 9-8 to 9-11
   periodic structure theory designs for, 9-62 to 9-71
   wire diameter of, 32-76
Log-periodic dipole antennas
   active regions in, 9-17 to 9-21, 9-24
   bandwidth of, 9-17, 9-21, 9-24
   circular, 9-3, 9-4
   design of, 9-20 to 9-32
   directivity of, 9-21 to 9-24, 9-27, 9-61, 9-66
   feeder circuits for, 9-16 to 9-18
   front-to-back ratios of, 9-60 to 9-61
   impedance of, 9-21, 9-24, 9-27 to 9-30, 9-58
   parameters for, 9-12, 9-14, 9-15, 9-24
   radiation fields for, and Q, 9-19 to 9-20, 9-62 to 9-66
   swr of, 9-58 to 9-59
   transposition of conductors in, 9-14 to 9-16
Log-periodic mono arrays, 9-64 to 9-67
Log-periodic zigzag antennas
   active region of, 9-35 to 9-36
   design procedures for, 9-37 to 9-46
   directivity of, 9-39 to 9-40
   half-power beamwidths for, 9-39 to 9-46
   radiation fields of, 9-39 to 9-40
Log-spiral antennas. *See also* Conical log-spiral antennas; Logarithmic spiral antennas
   feeds for, 9-75 to 9-78, 9-80
   geometry of, 9-8
   input impedance of, 9-78 to 9-79
   polarization of, 9-74, 9-77
   radiated fields of, 9-75
   truncation of, 9-72 to 9-73
Long-slot array antennas, 17-26
Longitude amplitude taper in antenna ranges, 32-15
Longitudinal broadwall slots, 12-4
Longitudinal currents in waveguides, 12-3
Longitudinal section electric mode, 8-73
Longitudinal slot arrays
   aperture distribution of, 12-24
   design equations for, 12-10 to 12-13
   E-field distribution in, 12-6 to 12-10, 12-35
   mutual coupling with, 12-30 to 12-31
   nonresonantly spaced, 12-16 to 12-20
   planar, 12-20 to 12-24

Index

resonantly spaced, 12-13 to 12-16
shunt-slot antennas, 17-33
Look angles for tvro, 6-29
Loop antennas
    coaxial current, 24-45 to 24-46
    ferrite, 6-18 to 6-22
    fm, 27-33 to 27-34
    for geophysical applications, 23-17 to 23-23
    as magnetic-field probes, 24-40 to 24-43
    rotating systems, 25-4 to 25-9
    small, 3-16, 3-19, 6-13, 6-17 to 6-19, 7-46 to 7-47
    spaced, 25-9 to 25-10
    thin-wire, 7-40 to 7-48
    vector effective area of, 6-5 to 6-6
Loop mutual impedance, 26-12 to 26-13
Looper radiators for medical applications, 24-20 to 24-22
Lorentz reciprocity theorem, 2-16 to 2-18
    and planar scanning, 33-7
Losses. *See also* Conduction losses; Path losses
    with conical horn antennas, 8-46
    free space, 2-39, 29-5 to 29-6, 29-8
    with microwave radiometers, 22-32 to 22-33
    system, power for, with am antennas, 26-36 to 26-37
    tangents for, 1-6
Lossy dielectrics, modeling of, 32-71 to 32-75
Low capacitance rf cable, 28-25
Low earth orbiting satellites, 22-35 to 22-43
Low frequencies, surface propagation by, 29-44
Low-frequency limit, 9-3
    of log-periodic planar antennas, 9-8 to 9-9
    of log-spiral antennas, 9-75
Low-frequency techniques, 3-5
    for classical antennas, 3-13 to 3-36
    for EMP analysis, 30-9 to 30-14
    integral equations for. *See* Integral equations
    theory with, 3-6 to 3-13
Low-frequency trailing-wire antennas, 30-31
Low-noise amplifiers for satellite tv, 6-26
Low-noise detectors, 17-136
Low-profile arrays. *See* Conformal arrays

Low-Q magnetic-field probes, 24-43, 24-44
Lowest usable frequency of ionospheric propagation, 29-25
LSE (longitudinal section electric) mode, 8-73
L-section power dividing circuits, 26-39 to 26-40
Ludwig's definition for polarization, 1-23 to 1-24
Lumped impedance for thin-wire antennas, 7-9
Luneburg lenses, 16-23
    for feeding arrays, 19-18, 19-44, 19-46 to 19-49
    as inhomogeneous lenses, 16-51 to 16-52
    vs. two-layer lenses, 17-18n
    with wide-angle multiple-beam, 19-80

Magic-T power division elements, 21-50 to 21-52
Magnetic cores. *See* Ferrite loop antennas
Magnetic current distributions
    and Kirchhoff approximation, 2-20 to 2-21
    for microstrip antennas, 10-22 to 10-25
    for dual-band elements of, 10-66
    edge of, in, 10-55 to 10-57
Magnetic dipoles
    and EMP, 30-9
    radiation pattern of, 4-59
Magnetic fields
    with arrays, 3-36
    conical-wave, 4-99
    and edge conditions, 1-9, 4-27 to 4-28, 4-101
    and equivalent currents, 4-101, 5-8 to 5-10
    in far zone, 6-3 to 6-4
    of GO reflected ray, 4-14
    Green's dyads for, 3-46, 3-48
    GTD for, 4-23
    for edge conditions, 4-27 to 4-28
    integral equation for, 3-46 to 3-47, 3-51
    with plane-wave spectra, 1-32, 5-7 to 5-8
    PO, 4-19 to 4-20
    probes for, 24-37 to 24-38, 24-40 to 24-44
    calibration of, 24-54 to 24-57
    reflection coefficient of, 2-28
    of small dipoles, 6-13
    of small loops, 6-18

Magnetic fields (*cont.*)
  surface, 4-84
    with cavities, 10-12 to 10-13
    for physical optics approximation, 2-26 to 2-27
  tangential, 1-32, 5-7
  with UTD, 4-83, 4-84, 4-95
  for vertices, 4-31
Magnetic loss tangent, 1-6
Magnetic vector potential, 3-6 to 3-7
  with arrays, 3-33, 3-35, 11-6
Magnetometers, 23-23 to 23-24
Magnetrode for hyperthermia, 24-45
Magnitude pattern for linear arrays, 11-9, 11-11
Main beam, 1-20
  with aperiodic arrays, 14-24, 14-26 to 14-34
  in array visible region, 11-31 to 11-32
  of corrugated horns, 8-55
  directivity of, 1-29
  with rectangular apertures, 5-12 to 5-14, 5-19
  and space harmonics, 1-35
Main line guide for planar arrays, 12-21 to 12-23
Main Scattering Program modeling code, 3-88 to 3-91
Maintenance of computer models, 3-68
Mangin Mirror with wide-angle multiple-beam systems, 19-85, 19-89
Mapping transformation, 11-25
Marine observation satellite (MOS-1), 22-47
Mariner-2 satellite, 22-17, 22-18
Marker beacon antennas, 30-23, 30-27 to 30-28
Masts, shipboard, simulation of, 20-94 to 20-96
Matched feeds and cross-polarization, 15-99
Matched four-port hybrid junctions, 19-5 to 19-6
Matched loads calibration method, 22-25 to 22-27
Matching. *See* Impedance matching
Matching filters, 10-52
Matching transformers for lenses, 21-19
Mathematical models, 32-65
Matrices. *See also* Admittance matrices; Blass matrices; Butler matrices
  impedance, 3-61, 3-67 to 3-68
  scattering, 12-31, 33-7
  transformation, 15-117, 15-119
Matrix-fed cylindrical arrays, 19-101 to 19-106
  with conventional lens approach, 19-107 to 19-108
  with Meyer geodesic lens, 19-108 to 19-119
Maximum effective height
  of short dipoles, 6-12, 6-14 to 6-15
  of small loops, 6-18
Maximum-gain antennas, 17-40 to 17-41
Maximum receiving cross section
  of short dipoles, 6-12 to 6-15
  of small loops, 6-18
Maximum usable frequency of ionospheric propagation, 29-22 to 29-23
Maxson tilt traveling-wave arrays, 19-80
Maxwell equations, 1-5 to 1-7
  for electromagnetic fields, 3-6
  using MOM to solve, 3-55
  in unbounded space, 1-9
Maxwell fish-eye lenses, 16-51 to 16-53
Meanderline polarizers, 21-29
Mean effective permeability of ferrite loops, 6-20 to 6-21
Mean impedance
  of log-periodic antennas
    dipole, 9-27
    zigzag, 9-46
  with self-complementary antennas, 9-10 to 9-11, 9-13
  of spiral-log antennas, 9-78
Mean resistance of log-periodic antennas
  dipole, 9-28
  zigzag, 9-40
Mean spacing parameter for log-periodic dipole antennas, 9-28
Measurement of radiation characteristics, 32-3 to 32-5, 32-39 to 32-63, 33-3 to 33-5
  compact ranges for, 33-22 to 33-24
  current distributions for, 33-5 to 33-6
  cylindrical scanning for, 33-10 to 33-12
  defocusing techniques for, 33-24 to 33-26
  errors in, 33-15 to 33-17
  extrapolation techniques for, 33-26 to 33-28
  interpolation of, 20-6

Index

modeling for, 32-74 to 32-86
planar scanning for, 33-7 to 33-10
plane-wave synthesis for, 33-21 to 33-22
and radiation cuts, 32-6 to 32-8
ranges for, 32-8 to 32-39, 33-15 to 33-21
for slot array design, 12-26 to 12-37
spherical scanning for, 33-12 to 33-15
Mechanical scanning
   conical, 22-39 to 22-40
   with millimeter-wave antennas, 17-9, 17-10, 17-24 to 17-26
   with satellite antennas, 21-5
Mechanical switches with satellite antennas, 21-7
Medical applications, 24-5 to 24-9
   characterization of antennas used in, 24-49 to 24-57
   hyperthermia applicators for, 24-43 to 24-49
   implantable antennas for, 24-25 to 24-29
   in-vivo antennas for, 24-30 to 24-35
   microstrip antennas for, 24-18 to 24-25
   monitoring rf radiation in, 24-36 to 24-43
   waveguide- and radiation-type antennas for, 24-9 to 24-18
Medium frequencies, propagation of
   ionospheric, 29-25 to 29-27
   surface, 29-44
Medium-gain millimeter-wave antennas, 17-9 to 17-27
Medium propagation loss, 29-6
Menzel's antennas, 17-102 to 17-104
Metal grating antennas, 17-49, 17-51, 17-68 to 17-69
   bandwidth with, 17-74
Metal waveguides, 17-117 to 17-118
Meteorology and remote sensing, 22-13 to 22-15
Meteors, ionization trails from, 29-30
Meteor satellite, 22-17
Meters for receiving antennas, 2-30 to 2-32
Method of moments
   for complex environment simulations, 20-4
   for coupling, 20-68
   frequency-domain, 3-57 to 3-62, 3-64
   with Maxwell's equations, 3-55
   radomes, 31-27 to 31-29
   with reflector antennas, 15-7
   time-domain, 3-62 to 3-64

Meyer lenses for feeds, 19-19, 19-32 to 19-37
   matrix-fed, 19-108 to 19-119
MFIE modeling code, 3-88 to 3-91
Microbolometers, 17-133, 17-137 to 17-138, 17-141
Microstrip antennas, 10-5 to 10-6
   applications with, 10-56 to 10-70
   arrays of, for medical applications, 24-25
   circular polarization with, 10-57 to 10-63
   coupling of, to waveguides, 17-199 to 17-121
   efficiency of, 10-49 to 10-50
   elements for
      dual-band, 10-63 to 10-66
      frequency-agile, 10-66 to 10-69
      loaded, 10-52 to 10-56
      polarization-agile, 10-69 to 10-70
   feeds for, 10-28, 10-51
   impedance of, 10-28 to 10-44
   leaky modes with, 17-35
   loop radiators, 24-20 to 24-22
   matching of, 10-50 to 10-52
   for medical applications, 24-18 to 24-20
   models of
      circuit, 10-26 to 10-28
      physical, 10-7 to 10-21
   radiation patterns of, 10-21 to 10-26
   resonant frequency of, 10-44 to 10-49
   slot, for medical applications, 24-22 to 24-24
Microstrip array antennas, 17-25
Microstrip baluns, 18-8
Microstrip dipole antennas, 17-106, 17-118 to 17-119, 17-122 to 17-126
Microstrip excited-patch radiators, 18-8 to 18-9
Microstrip lines, 13-58, 21-59, 21-60
   for leaky-wave antennas, 17-82 to 17-83, 17-98 to 17-103
   for planar transmission lines, 28-30 to 28-33
Microstrip patch resonator arrays, 17-106 to 17-118
Microstrip radiator types, 13-60
   dipole, for phased arrays, 18-7
Microstrip resonator millimeter-wave antennas, 17-103 to 17-106
   with thick substrates, 17-122 to 17-129
   with thin substrates, 17-107 to 17-122

Microstrip techniques for integrated antennas, 17-131, 17-134 to 17-137
Microstrip waveguides for microstrip arrays, 17-117 to 17-118
Microwave-absorbing material for indoor ranges, 32-25
Microwave antennas
 constrained lenses for, 16-42
 measurements for, 33-20
Microwave diode switches, 21-38
Microwave fading, 29-40 to 29-41
Microwave radiation
 antennas to monitor, 24-36 to 24-43
 transfer of, 22-4 to 22-7
Microwave radiometers and radiometry
 antenna requirements, 22-16 to 22-46
 fundamentals of, 22-3 to 22-9, 22-22 to 22-23
 future of, 22-47 to 22-52
 for remote sensing, 22-9 to 22-16
 spacecraft constraints with, 22-46 to 22-47
Microwave sounder units, 22-14
 modulation frequency of, 22-24 to 22-25
 onboard reference targets for, 22-26
 for planar scan, 22-37
 as step scan type, 22-43
 on TIROS-N, 22-37
Microwave transmissions, 29-32
Military aircraft, simulation of, 20-78 to 20-88
Military applications, antennas for, 17-6 to 17-7, 17-15
Millimeter-wave antennas, 17-5 to 17-8
 fan-shaped beam, 17-28 to 17-32
 high-gain and medium-gain, 17-9 to 17-27
 holographic, 17-129 to 17-131
 integrated, 17-131 to 17-141
 microstrip
  monolithic, 17-134 to 17-137
  with thick substrates, 17-122 to 17-129
  with thin substrates, 17-107 to 17-122
 near-millimeter-wave imaging array, 17-137 to 17-141
 omnidirectional, 17-32 to 17-34
 periodic dielectric, 17-48 to 17-82
 printed-circuit, 17-103 to 17-131
 spiral, 17-27 to 17-28
 tapered-dielectric rod, 17-36 to 17-48, 17-133 to 17-134
 uniform-waveguide, leaky-wave, 17-82 to 17-103
Milimeter-wave Sounders, 22-47, 22-50 to 22-51
Mine communications, 23-3
 grounded wire antennas for, 23-8 to 23-17
 loop antennas for, 23-17 to 23-23
Minimax synthesis method, 21-65
Mismatches
 with lenses, 16-54 to 16-55
 and signal-to-noise ratio, 6-25
 transmission line and antenna, 1-24
Missile racks, aircraft, simulation of, 20-79 to 20-80, 20-82
Mixer diodes for integrated antennas, 17-133 to 17-134, 17-138 to 17-139, 17-141
MMIC (monolithic microwave integrated circuit technology), 17-12
Modal field distributions of loaded elements, 10-53
Modal transverse field distributions
 in circular waveguides, 1-47 to 1-48
 in rectangular waveguides, 1-40 to 1-41
Mode converters, odd/even, 21-54 to 21-56
Mode distributions of patches, 10-16 to 10-19
Modeling. *See also* Scaling
 for antenna ranges, 32-63 to 32-86
 with computers
  codes for, 3-80 to 3-96
  computation with, 3-68 to 3-72
  validation of, 3-72 to 3-80
Modes, patch, 10-7 to 10-21, 10-24 to 10-26, 10-35 to 10-37
Mode voltage in common waveguides, 12-13, 12-16
Modulating radiometers, 22-24
MOM. *See* Method of moments
Moment Method Code for antenna currents, 20-91
Monitors
 for directional antenna feeder systems, 26-41 to 26-43
 of rf radiation, 24-36 to 24-43
Monofilar zigzag wire antennas, 9-34, 9-36
Monolithic microstrip antennas, 17-134 to 17-137
Monolithic microwave integrated circuit (MMIC) technology, 17-12

Monopole antennas
  on aircraft, 20-62 to 20-68
    Boeing 737, 20-70, 20-73 to 20-75
    F-16 fighter, 20-87 to 20-88
  bandwidth of, 6-24
  hypodermic, 24-26 to 24-27
  images with, 3-20
  impedance-loaded, 7-20 to 7-21
  for in-vivo measurements, 24-30 to 24-32
  millimeter-wave, 17-32
  radiation of, on convex surface, 4-64 to 4-66, 4-71, 4-83 to 4-85
  short, 4-94, 6-12
  sleeve, 7-29 to 7-31
Monopulse pyramidal horn antennas, 8-72 to 8-75
MOS (marine observation satellite), 22-47
Mounting structures, 30-4 to 30-7
Movable line-of-sight ranges, 32-9 to 32-11
MSU. See Microwave sounder units
Multiarm self-complementary antennas, 9-6 to 9-8
Multibeam applications, lens antennas for, 16-6, 17-21
  wide-angle dielectric, 16-19
Multibeam satellite antennas, 15-5, 15-80 to 15-87, 21-57 to 21-58, 21-62, 21-64 to 21-67
  zoning in, 16-38
Multidetectors with integrated antennas, 17-137
Multifocal bootlace lenses, 16-46 to 16-48
Multifrequency Imaging Microwave Radiometer (MIMR), 22-47, 22-48 to 22-25
Multimode circular arrays, 25-17 to 25-20
Multimode element array beam feed technique, 19-12 to 19-15
Multimode feeds, 9-110, 9-112
  radial transmission line, 19-23 to 19-32
Multimode generators, Butler matrix as, 19-10
Multimode horn antennas, 8-69 to 8-73
  for reflector antennas, 15-95
Multipath propagation, 29-30, 29-37 to 29-41
Multipath reflections with scanning ranges, 33-16, 33-18
Multiple-access phased arrays, 21-68 to 21-71
Multiple-amplitude component polarization measurement method, 32-61 to 32-62

Multiple-antenna installations, 27-36 to 27-40
Multiple-beam antennas, 15-80
  reflector antennas with, 15-5
Multiple-beam constrained lenses, 19-79 to 19-85
Multiple-beam-forming networks, 19-23
Multiple-beam matrix feeds, 19-8 to 19-12
Multiple-feed circular polarization, 10-61 to 10-63
Multiple quarter-wavelength impedance matching transformers, 21-19
Multiple reflections
  with compact ranges, 33-27
  with thin-wire transmitting antennas, 7-13
Multiple-signal direction finding, 25-24 to 25-25
Multiport impedance parameters, 10-34, 10-44
Multiprobe launcher polarizers, 21-30
Multistepped dual-mode horns, 21-82, 21-84 to 21-89
Muscle, human
  phantom models for, 24-55
  power absorption by, 24-8, 24-9, 24-24, 24-47
Mutual admittance between slots, 4-91
Mutual base impedance of am antennas, 26-12 to 26-17
Mutual conductance with microstrip antennas, 10-9
Mutual coupling, 4-49
  and array transformations, 11-23
  of arrays
    aperiodic, 14-20 to 14-27
    circular, on cylinder, 21-100
    conformal, 13-49 to 13-52
    inclined narrow wall slot, 12-25
    large, 12-28 to 12-33
    periodic, 13-7, 13-41 to 13-46
    phased, 21-8 to 21-10, 21-12
    planar, of edge slots, 12-27 to 12-28
    scanned in one plane, 13-12
    slots in, 12-5
  of tapered dielectric-rod antennas, 17-47
  of convex surfaces, UTD for, 4-84 to 4-96
  with integrated antennas, 17-137
  with microstrip antennas, 10-7

Mutual coupling (*cont.*)
  of millimeter-wave antennas
    dielectric grating, 17-79
    dipoles, 17-128 to 17-129
    tapered dielectric-rod, 17-47
  nulls from, 13-45 to 13-46
  and open-periodic structures, 9-56
  and phase shift, 9-58
  with reflector antennas, 15-6
    offset parabolic, 15-83 to 15-84
  of slots
    longitudinal, 12-15, 12-18
    radiating, 12-35 to 12-36
Mutual impedance, 6-8 to 6-9
  with am antennas, 26-12 to 26-13
  with aperiodic arrays, 14-3, 14-20 to 14-21
  of array elements, 11-5
  and dipole spacing, 9-58
  of microstrip antennas, 10-34
  of monopoles, 4-94
  and radiation resistance, 9-56, 14-3

Narrowband conical horn antennas, 8-61 to 8-62
Narrow wall coupler waveguide power division elements, 21-51 to 21-52
Narrow wall slots, inclined, arrays of, 12-25 to 12-28
National Oceanic and Atmospheric Administration (NOAA)—series satellites, 22-51
Natural modes for dipole transient response, 7-21 to 7-22
Navigation systems, ionospheric propagation for, 29-27
Near-degenerate modes, 10-28, 10-29, 10-60
Near fields and near-field radiation patterns, 1-16 to 1-18
  of aperture antennas, 5-24 to 5-26, 15-21 to 15-24
    axial gain of, 5-29 to 5-33
    with Cassegrain feed systems, 19-62 to 19-63
  of dipole arrays, 13-44
  of dipoles, 20-69
  experimental validation of computer code for, 3-76
  gain reduction factor for, 5-30 to 5-32
  measurement of. *See* Measurement of radiation characteristics
  in medical applications, 24-5 to 24-8
  for monopole on aircraft, 20-62 to 20-65
  in periodic structure theory, 9-33
  with pyramidal horns, 15-92
  with reflector antennas, 15-8, 15-21 to 15-23
  transformation of, to far-field, 20-61 to 20-62
  with UTD solutions, 20-4, 20-6, 20-13 to 20-14
  in zigzag wire antennas, 9-33
Near-millimeter-wave imaging arrays, 17-137 to 17-141
NEC (Numerical Electromagnetic Code) modeling code, 3-87 to 3-95
Needle radiators, 24-26 to 24-27
Nickel for satellite antenna feeds, 21-28 to 21-29
Nimbus-5 satellite, 22-17
  planar scanning on, 22-37
Nimbus-6 satellite, 22-17 to 22-18
Nimbus-7 satellite, 22-17
  SMMR on, 22-22
NNBW (null-to-null beamwidths), 22-28
Nodal curves with frequency-agile elements, 10-68 to 10-69
Nodal planes with circular polarization, 10-62
Noise and noise power
  with arrays, 11-59
  emitters of, 2-40 to 2-42
  and propagation, 29-47 to 29-52
  and receiving antennas, 6-24 to 6-25
  at terminal of receiver, 2-42 to 2-43
  at tvro antenna, 6-26, 6-30
Noise temperature, 2-39 to 2-43
  effective, 22-27
  and figure of merit, 1-27, 6-31
  and microwave radiometers, 22-22 to 22-23
Noncentral chi-square distribution for array design, 14-10 to 14-12
Nonconverged solutions and computer solutions, 3-54
Nondirectional antennas for T T & C, 21-89
Nondirective couplers, 19-4 to 19-5
Nonoverlapping switch beam-forming networks, 21-30 to 21-31
Nonperiodic configuration with multibeam antennas, 21-66
Nonradiative dielectric guides, 17-93 to 17-98
Nonreciprocal ferrite phase shifters, 13-63, 18-14 to 18-15, 21-38 to 21-39

Nonresonantly spaced longitudinal slots, 12-16 to 12-20
Nontrue time-delay feeds, 19-54 to 19-56
Nonuniform component of current, 4-103
Nonuniform excitation of array elements, 11-29 to 11-30, 11-47 to 11-48
Normal congruence of rays, 4-6
Normalized aperture distributions
　circular, 5-21 to 5-22
　rectangular, 5-19 to 5-20
Normalized intensity of far shields, 1-20 to 1-21
Normalized modal cutoff frequencies
　for circular waveguides, 1-46, 1-49
　for rectangular waveguides, 1-42
Normalized pattern functions, 11-11 to 11-13
Normalized resonant modes of microstrip antennas, 10-13, 10-15
　frequency of, 10-46
Norton's theorem for receiving antenna loads, 6-3, 6-4
Nose section of aircraft, simulation of, 20-78
N-port analogy, 3-64, 3-85
NRD guides, 17-35, 17-82 to 17-83, 17-92 to 17-98
Nuclear detonations. *See* Electromagnetic pulses
Null-to-null beamwidths, 22-28
Nulls
　with apertures
　　circular, 5-22 to 5-23
　　rectangular, 5-13 to 5-15, 5-19
　in Bayliss line source pattern synthesis, 13-28
　electrically rotating, 25-11 to 25-12
　filling in of, with tv antennas, 27-7
　with linear arrays, 11-15, 11-40
　with log-spiral conical antennas, 9-102
　with loop antennas, 25-5 to 25-6
　with microstrip antennas, 10-25 to 10-26
　with multiple element feeds, 19-13
　from mutual coupling, 13-45 to 13-46
　with offset reflector antennas, 15-37, 15-41
　prescribed, arrays with, 11-9 to 11-10
　with two-tower antennas, 26-22 to 26-24
Numerical Electromagnetic Code, 3-87 to 3-95
Numerical implementation for electromagnetic field problems, 3-53 to 3-68

Numerical modeling errors, 3-69 to 3-71
Numerical simulations for antennas, 20-5 to 20-7
Numerical solutions
　for aircraft antennas, 20-70 to 20-90
　for ship antennas, 20-90 to 20-97
Numerical treatment in modeling codes, 3-81, 3-83 to 3-84, 3-90
Numerical validation of computer code, 3-76 to 3-80

Obstacles, scattering by, 2-21 to 2-27
Ocean
　emissivity of, 22-10 to 22-11
　modeling of, 32-79
　reflection coefficient over, 29-38 to 29-40
　remote sensing of, 22-14 to 22-15
　and remote sensing of humidity, 22-6 to 22-7
　simulation of, 20-91
Odd distribution in rectangular apertures, 5-15
Odd/even converters, 21-7, 21-54 to 21-56
Odd-mode amplitude control, 19-14 to 19-15
Off-focus feeds with reflectors, 15-49 to 15-61
Offset distance with reflector antennas, 21-23
Offset feeds
　for Gregorian feed systems, 19-63 to 19-66
　for reflectarrays, 19-54
　for reflector antennas, 17-9
Offset height of offset reflectors, 15-31 to 15-32
Offset long-slot array antennas, 17-26
Offset parabolic reflectors
　edge and feed tapers with, 15-36 to 15-37
　feed patterns for, 15-34 to 15-36
　geometrical parameters for, 15-31 to 15-34
　off-focus feeds for, 15-49 to 15-61
　on-focus feeds for, 15-37 to 15-49
Offset phased array feeds reflector, 19-61 to 19-62
Offset reflector antennas, 15-6, 15-23. *See also* Cassegrain offset antennas

Offset reflector antennas (cont.)
  dual-reflector, 15-6 to 15-15, 15-73, 17-11
  satellite, 21-19 to 21-26
Ohmic losses with remote-sensing microwave radiometers, 22-32
Omega navigation systems, 29-27
Omnidirectional antennas
  dielectric grating, 17-80 to 17-82
  millimeter-wave, 17-32 to 17-34
  for satellite T T & C, 21-90
  tv, 27-5
On-axis gain of horn antennas, 8-3 to 8-4
One-parameter model for reflector antennas, 15-18 to 15-21
One plane, arrays scanned in, 13-12 to 13-15
On-focus feeds with reflectors, 15-37 to 15-49
Open-circuit voltage, 2-32
  with UTD solutions, 20-7
  and vector effective height, 6-5
Open-ended coaxial cable, 24-32 to 24-35
Open-ended waveguides
  arrays of, 13-44
  circular
    for reflector feeds, 15-90 to 15-93
    with reflector satellite antennas, 21-74
    half-power beamwidths of, 15-105
  dielectric-loaded, for medical applications, 24-9
  radiators of, for phased arrays, 18-6 to 18-7
  rectangular
    electric-field distribution of, 5-14 to 5-15
    half-power beamwidths of, 15-104
    for reflector feeds, 15-89 to 15-92
  TEM, for medical applications, 24-9 to 24-13
Open-periodic structures, 9-55 to 9-56
Open-wire transmission lines, 28-11
  two-wire, impedance of, 28-12, 28-13
Optical devices as Fourier transformers, 19-95 to 19-98
Optical feeds. *See also* Unconstrained feeds
  for aperiodic arrays, 19-57 to 19-59
  corporate, for beam-forming feed networks, 18-22
  transform, 19-91 to 19-98
Optics fields. *See* Geometrical optics; Physical optics

Optimization array problems, 11-81 to 11-86
Optimum pattern distributions, 5-18 to 5-20
Optimum working frequency and ionospheric propagation, 29-25
Orbits of satellite antennas, 6-26
Orthogonal beams, pattern synthesis with, 13-22 to 13-23
Orthogonal polarization, 21-7
Orthomode junctions, 21-7
Orthomode transducers, 22-45
Out-of-band characteristics and EMP, 30-8
Outdoor antenna ranges, 32-19 to 32-24, 33-28
Outer boundaries of radiating near-field regions, 1-18
Output features in modeling codes, 3-86
Overfeeding of power with am antennas, 26-36 to 26-37
Overlapped subarrays, 13-35 to 13-39
  space-fed system, 19-68 to 19-76
Overlapping switch beam-forming networks, 21-31 to 21-32
Overreach propagation, 29-41
Oversized waveguide transmission type, 21-60
Owf (optimum working frequency), 29-25
Oxygen
  absorption by, 29-35
  remote sensing of, 22-12 to 22-13
  temperature sounders for, 22-16, 22-21

Pancake coils, 24-47 to 24-48
Panel fm antennas, 27-35 to 27-36
Parabolas, 15-25 to 15-26
Parabolic patches, 10-18
Parabolic pillbox feeds, 19-19 to 19-20
Parabolic reflector antennas, 3-31. *See also* Offset parabolic reflectors
  far-field formulas for, 15-15 to 15-23
  near-field radiation pattern of, 4-109
Paraboloidal lenses, 21-18
Paraboloidal surfaces and intersection curve, 15-29 to 15-30
Paraboloid reflectors, 15-76
  with compact antenna ranges, 32-28 to 32-29
  phased array feeds with, 19-58 to 19-61
Parallel feed networks, 18-19, 18-21 to 18-22, 18-26, 19-5 to 19-6
Parallel plate optics. *See* Semiconstrained feeds

Index

Parallel polarization, 31-11 to 31-15
Parallel-resonant power dividing circuits, 26-39 to 26-40
Parasitic arrays, 3-42
Parasitic reflectors, 9-64 to 9-65
Partial gain, 1-26
Partial time-delay systems, 19-52, 19-68
Partial zoning, lenses with, 16-28
Passive components for periodic arrays, 13-60 to 13-62
Patches and patch antennas. *See also* Microstrip antennas
   elements for, 10-5 to 10-7, 13-60
   medical applications of, 24-18 to 24-20
   parameters for, 10-16 to 10-17
   resonant frequency of, 10-47 to 10-49
Patch radiators for phased arrays, 18-6
Path length constraint and errors with lens antennas, 16-7 to 16-10, 16-21, 21-18
   bootlace, 16-47 to 16-48
   constrained, 16-41 to 16-51, 19-82 to 19-83
   geodesic, 19-115, 19-119
   microwave, 19-43
   and phase errors, 16-13
   with pillbox feeds, 19-22
   spherical cap, 16-25
   taper-control, 16-34 to 16-35
Path losses
   from ionospheric propagation, 29-25
   and line-of-sight propagation, 29-34 to 29-35
   with offset parabolic reflectors, 15-36 to 15-37
   of tvro downlink, 6-31
Pattern cuts and radiation patterns, 32-6 to 32-8
Pattern error with multibeam antennas, 21-65
Pattern footprint of tvro signal, 6-26, 6-27
Pattern functions. *See* Array pattern functions
Patterns. *See also* Far fields; Fields; Near fields; Radiation fields and patterns
   of arrays
      aperiodic, 14-30 to 14-32
      multiplication of, 3-36, 11-8
      periodic, 13-12 to 13-29
      synthesis of, 11-76 to 11-86
   distortion of. *See* Distortion
   with microstrip antennas, 10-21 to 10-26
   periodic, 13-20 to 13-29
Peak gains, 1-26 to 1-27

Pekeris functions, 4-52 to 4-54
Pencil-beam reflector antennas, 15-5 to 15-6
   AF method with, 15-13
   with direction-finding antennas, 25-20
   far-field formulas for, 15-15 to 15-23
   for satellites, 21-5
Perfect conductors, 3-36, 3-38 to 3-39, 3-44, 3-46
   time-domain analyses with, 3-51
Perfect ground planes, 3-18 to 3-20
   Green's functions for, 3-47 to 3-48
Periodic arrays, 13-5 to 13-11. *See also* Phased arrays
   linear transformations with, 11-31 to 11-33
   organization of, 13-23 to 13-29
   patterns of, 13-12 to 13-23
   practical, 13-30 to 13-64
Periodic configuration with multibeam antennas, 21-66
Periodic dielectric antennas, 17-34 to 17-35, 17-48 to 17-82
Periodic structures, 1-33 to 1-37
   theory of, 9-32 to 9-37
      log-periodic designs based on, 9-62 to 9-71
      and periodically loaded lines, 9-37 to 9-46, 9-46 to 9-62
Period of surface corrugations, 17-64
Permeability, 1-6
   of core in ferrite loop, 6-20
Permittivity
   of corrugation regions, 17-53 to 17-54
   and dielectric modeling, 32-72 to 32-73
   of human tissue, 24-12, 24-30 to 24-35, 24-47
   and leakage constants, 17-61 to 17-63
   of microstrip patch antennas, 17-109
   of open-ended coaxial cable, 24-33
   of substrates, 17-114, 17-124, 17-127
   of troposphere, 29-31
Perpendicular polarization, 31-11 to 31-12, 31-14 to 31-20
Personnel dosimeters, rf, 24-43
Phantoms for heating patterns of antennas, 24-54 to 24-57
Phase-amplitude polarization measurement methods, 32-57 to 32-60
Phase angle of ocean surface, 29-38
Phase center
   with horn antennas, 8-73 to 8-75
      conical, 8-61, 8-76, 8-78 to 8-80

Phase center (*cont.*)
  E-plane, 8-76 to 8-77
  H-plane, 8-76 to 8-77
  of log-periodic dipole antennas, 9-24
  of log-spiral conical antennas, 9-83, 9-91 to 9-92, 9-94
  technique to measure, 8-80 to 8-84
  testing for, 32-9
Phase constants
  for dielectric grating antennas, 17-68, 17-71
  with leaky mode antennas, 17-51, 17-53 to 17-62
  for leaky-wave antennas, 17-84 to 17-85, 17-89 to 17-92, 17-95 to 17-96, 17-100
  of TE/TM waveguides, 28-37
  for transmission lines, 28-7
Phase constraint with dielectric lenses, 16-21
Phase control, array, 13-62 to 13-64
Phased arrays. *See also* Beam-forming feed networks; Periodic arrays
  bandwidth of, 13-19 to 13-20
  conical scanning by, 22-39
  design of, 18-3 to 18-5
    and component errors, 18-26 to 18-28
    feed network selection in, 18-17 to 18-26
    phase shifter selection in, 18-12 to 18-17
    radiator selection in, 18-6 to 18-12
  feeds for, 19-58 to 19-62
  horn antennas in, 8-3
  with integrated antennas, 17-134 to 17-137
  for medical applications, 24-16 to 24-18
  for satellite antennas, 21-7 to 21-12, 21-68 to 21-71
Phase delay
  with equal group delay lenses, 16-45
  with series feed networks, 19-3
Phased lens approach, 19-107 to 19-108
Phase differences
  and direction-finding antennas, 25-3
  with interferometers, 25-21 to 25-22
Phase errors
  in antenna ranges, 32-16 to 32-18
  of array feeds, 13-61
  with arrays
    aperiodic, 14-32
    microstrip patch resonator, 17-109
    periodic, 13-55
    phased, 18-26 to 18-28, 21-12
  with beam-forming feed networks, 18-25
  with horn antennas
    conical, 8-46
    H-plane, 8-23
    pyramidal, 8-39 to 8-40
  with lenses, 16-12 to 16-20, 16-33, 21-13 to 21-14
    bootlace, 16-47 to 16-48
    constrained, 19-82
    equal group delay, 16-46
    geodesic, 19-115
    spherical, 16-24 to 16-25, 16-29
    surface tolerance, 21-18
    zone constrained, 16-43 to 16-45
  with offset parabolic antennas, 15-49, 15-55
  with pillbox feeds, 19-21
  from reflector surface errors, 15-107 to 15-108
  with satellite antennas, 21-5
Phase fronts with reflector antennas, 15-75 to 15-76, 15-80
Phase pattern functions, 11-7
Phase progression with longitudinal slots, 12-17 to 12-18
Phase quantization and array errors, 13-52 to 13-57, 21-12
Phase shift
  and mutual coupling, 9-58
  with periodic loads, 9-49 to 9-54
Phase-shifted aperture distributions, 5-17 to 5-18
Phase shifters
  with arrays
    periodic, 13-20, 13-60 to 13-64
    phased, 18-3 to 18-4, 18-6, 18-12 to 18-17
  with beam-forming networks
    scanned, 21-7, 21-38 to 21-43
    switched, 21-32
  with Butler matrix, 19-9
  with feed systems
    broadband array, 13-40 to 13-41
    directional antenna, 26-40 to 26-43, 26-45 to 26-47
    for gain ripple, 21-27
    parallel, 19-6
    series, 19-3 to 19-4
  with planar scanning, 22-37
  quantization errors with, 21-12

Index

with subarrays
  contiguous, 13-32 to 13-35
  space-fed, 19-68
  in true-time-delay systems, 19-7, 19-80
  with Wheeler Lab approach, 19-105, 19-107
Phase squint with subarrays, 13-35
Phase steering, 13-7 to 13-8, 13-10
Phase taper in antenna ranges, 32-15 to 32-19
Phase term for aperture fields, 8-5
Phase velocity of TEM transmission lines, 28-10, 28-22
Phasors, Maxwell equations for, 1-5 to 1-7
Physically rotating antenna systems, 25-4 to 25-11
Physical models
  errors with, 3-69 to 3-71
  for microstrip antennas, 10-7 to 10-21
Physical optics and physical optics method, 4-5, 4-18 to 4-23
  and computer solutions, 3-54
  and PTD, 4-103
  with reflector antennas, 15-7 to 15-10, 15-14
    dual-reflector, 15-69, 15-72, 15-75 to 15-77
  for scattering, 2-25 to 2-27
Physical theory of diffraction, 4-5, 4-102 to 4-103
  for edged bodies, 4-104 to 4-108
Pillbox antennas, 17-28 to 17-30
Pillbox feeds, 18-22, 19-17 to 19-23
Pinched-guide polarizers, 21-30
Pin diodes
  for beam-forming network switches, 21-38
  for dielectric grating antennas, 17-71 to 17-72
  with integrated antennas, 17-135
  with microstrip antennas, 10-69
  in phase shifters, 13-62, 18-13
    variable, 21-40
  with variable power dividers, 21-46 to 21-47
Pin polarizers, 21-30
Pitch angle of log-periodic antennas, 9-39
Planar antennas
  impedance of, 2-15 to 2-16
  log-periodic antennas, 9-10 to 9-12
    radiated fields of, 9-8 to 9-9
    truncation with, 9-8

log-spiral antennas, 9-72
  with polar orbiting satellites, 22-35 to 22-38
  power radiated by, 5-26 to 5-27
Planar apertures
  feed elements with, 15-88
  and plane-wave spectra, 5-5
  radiation patterns of, 5-10 to 5-26
  for reflector antennas, 15-14
Planar arrays, 11-48 to 11-58
  directivity of, 11-63 to 11-65, 13-11
  millimeter-wave antennas, 17-26
  mutual coupling in, 12-32 to 12-33
  optimization of, 11-63 to 11-65
  periodic, transformations with, 11-26 to 11-29
  rectangular, patterns of, 13-15 to 13-17
  slot, 12-4 to 12-5
    edge, mutual coupling of, 12-27 to 12-28
    longitudinal, 12-4, 12-20 to 12-24
  triangular, grating lobes of, 13-15 to 13-17
  of vertical dipoles, geometry of, 11-85
  waveguide, 17-26
Planar curves
  conic sections as, 15-23 to 15-24
  for reflector surfaces, 15-27
Planar lenses, 21-18
Planar quartz substrates, 17-137
Planar scanning for field measurements,

Planar transmission lines, 28-30 to 28-35
Plane symmetry and computer time and storage, 3-65, 3-67
Plane-wave illumination
  for distance parameters, 4-56
  and GO incident fields, 4-11 to 4-12
Plane wave propagation
  and dielectric sheets, 31-20 to 31-25
  flat, 31-10 to 31-18
Plane waves
  polarization of, 1-13 to 1-16
  propagation direction of, 1-16
  reciprocity of, 2-33 to 2-34
  from spherical wavefronts, 16-6, 16-9 to 16-10
  synthesis of, for field measurements, 33-21 to 33-22
Plane-wave spectra, 5-5 to 5-8
  representation of, 1-31 to 1-33
Plano-convex lenses, 16-20 to 16-21

Plate-scattered fields, 20-23
PO. *See* Physical optics
Pockington's equation for dipole arrays, 13-43
Poincare sphere, 32-54 to 32-56
Point-current source, 3-13 to 3-14
Point matching
  with methods of moments, 3-60
  weight function with, 3-83
Point-source illumination
  for distance parameter, with UTD, 4-56
  and GO incident field, 4-11
Polar cap absorptions and ionospheric propagation, 29-25
Polar coordinates with lens antennas, 16-8 to 16-9
Polarization, 1-14 to 1-16, 1-19 to 1-21, 1-28 to 1-29, 3-28 to 3-31. *See also* Circular polarization; Cross polarization; Horizontal polarization; Linear polarization; Vertical polarization
  for am antennas, 26-3
  of arrays, 11-7
    phased, 18-6, 21-12
  and diagonal horns, 8-70
  for earth coverage satellite antennas, 21-82
  efficiency of, 1-26, 2-36, 32-56
    and gain measurements, 32-42
    and mismatch, 2-39
  and Faraday rotation, 29-16
  and Fresnel reflection coefficients, 3-27
  and gain, 1-25 to 1-26
  of leaky-wave antennas, 17-90, 17-94, 17-97
  of log-periodic antennas, 9-8
  with log-spiral antennas, 9-74, 9-77
    conical, 9-86, 9-90 to 9-91, 9-99
  measurements of, 32-51 to 32-64
  of microstrip antennas, 10-57 to 10-63, 10-69 to 10-70
  for microwave radiometers, 22-43 to 22-45
  parallel, 31-11 to 31-15
  perpendicular, 31-11 to 31-12, 31-14 to 31-20
  positioners for, 32-12, 32-14
  and radome interface reflections, 31-11 to 31-13
  and rain, 29-14
  reference, 1-23
  of reflector antennas, 15-5 to 15-6, 15-89, 21-19 to 21-20, 21-73
  for satellite antennas, 21-6 to 21-7, 21-80, 21-82
  and surface emissivity, 22-7 to 22-8
Polarization-agile elements, 10-69 to 10-70
Polarization ellipses, 1-14 to 1-16
Polarization-matching factor, 2-36, 6-7 to 6-8
Polarization pattern, 32-6
  polarization measurement method, 32-62 to 32-63
Polarization-transfer methods, 32-56 to 32-57
Polarizers
  for satellite antennas, 21-7, 21-29, 21-30, 21-74, 21-79 to 21-80
  waveguide, 13-60 to 13-61
Polar orbiting satellites, 22-35 to 22-43
Poles of cavities, 10-11
Polystyrene antennas, 17-44 to 17-46
Polytetrafluoroethylene bulb for implantable antennas, 24-27, 24-29
Porous pots with earth resistivity measurements, 23-8
Positioning systems for antenna ranges, 32-8 to 32-14, 32-30 to 32-32
Potter horns
  earth coverage satellite antennas, 21-82
  for reflector antennas, 15-95, 15-100 to 15-101
Power
  accepted by antennas, 1-24, 1-25, 22-4
  complex, 1-7
  conservation of, numerical validation of, 3-79
  incident to antennas, 1-24, 1-25
  input to antenna, 6-6
  overfeeding of, with am antennas, 26-36 to 26-37
  received by antenna, 2-38, 32-42, 33-22
  reflected, 2-29
  reflection coefficient for, 1-24
  sources of, for millimeter-wave antennas, 17-11
  time-averaged, 1-7 to 1-8, 3-11
    for modal fields, 1-46
    in waveguides, 1-39
  transfer ratio for, 6-8 to 6-9
  transformation of, with taper-control lenses, 16-37
  with transmission lines, 28-8, 28-10

Power amplifiers, integration of, 17-136
Power conservation law, 16-33 to 16-34
Power density
  with arrays, 11-59
  with bistatic radar, 2-37 to 2-38
  near-field, 5-29 to 5-33
  of plane waves, 1-14
  of receiving antenna, 6-7
Power detectors, integration of, 17-136
Power dividers
  for array feeds, 13-61
  for beam-forming networks
    scanned, 21-7, 21-43 to 21-49
    switched, 21-32 to 21-33
  for directional antenna feeder systems, 26-39 to 26-40
  fixed beam-forming networks, 21-49 to 21-53
  for gain ripple with feed arrays, 21-27
  LHCP and RHCP, 21-80
  quantiziation errors with, 21-12
Power flow integration method for am antennas, 26-22 to 26-33
Power gain with multibeam antennas, 21-64
Power handling capability
  measurements for, 32-4
  of transmission lines, 21-63
    circular coaxial TEM, 28-21
    microstrip planar quasi-TEM, 28-32 to 28-33
    rectangular TE/TM, 28-39, 28-42
    triplate stripline TEM, 28-28 to 28-29
    two-wire TEM, 28-19
  of waveguides
    circular TE/TM, 28-44
Power law coefficients for rain attenuation, 29-12
Power lines, noise from, 29-51
Power radiated. *See* Radiated power
Power rating
  of rf cables, 28-27
  of standard waveguides, 28-40 to 28-41
  of tv antennas, 27-7 to 27-8
Power ratio with bistatic radar, 2-36 to 2-39
Power transmittance, 2-29
  of A-sandwich panels, 31-18 to 31-19
  and dielectric sheets, 31-14 to 31-16
  efficiency of, 2-43
  with radomes, 31-3, 31-7
Poynting theorem, 1-7 to 1-8
Poynting vector for radiated power, 3-11

P-percent level curves, 14-12 to 14-14
Precipitation distributions, remote sensing of, 22-14
Prescribed nulls, 11-9 to 11-10
Pressurized air with transmission lines, 28-10
Principal polarization. *See* Reference polarization
Principle of stationary phase, 4-6 to 4-7
Printed circuits
  dipoles, 13-58
    bandwidth of, 17-122 to 17-127
  and integrated antennas, 17-130 to 17-131
  millimeter-wave, 17-25, 17-103 to 17-131
  waveguides, 21-57
    ridged, 28-57 to 28-58
    TEM, 21-57
    zigzag, 24-39, 24-40
Probabilistic approach to aperiodic arrays, 14-5, 14-8 to 14-35
Probability mean, 11-81
Probes
  for alignment, 32-33
  E-field, 24-37 to 24-41
    calibration of, 24-54 to 24-57
    implantable, 24-51 to 24-54
  in-vivo, 24-30 to 24-35, 24-49 to 24-57
  magnetic field, 24-37 to 24-38, 24-40 to 24-44
    calibration of, 24-54 to 24-57
  and polarization, 32-6 to 32-8
  for rf radiation detection, 24-37 to 24-43
  temperature, 24-40 to 24-51
Prolate spheroid
  geometry of, 4-82
  rectangular slot in, 4-79 to 4-82
Propagation, 29-5 to 29-6
  in computer models, 3-55 to 3-56
  ionospheric, 29-21 to 29-30
  with millimeter-wave antennas, 17-5
  noise, 29-47 to 29-52
  and polarization, 1-13 to 1-16
  satellite-earth, 29-7 to 29-21
  tropospheric and surface, 29-40 to 29-47
Propagation constants
  with conical log-spiral antennas, 9-80 to 9-84
  for dielectric grating antennas, 17-65
  of microstrip patch resonator arrays, 17-110
  for plane waves, 29-6

Propagation constants (*cont.*)
  for transmission lines, 28-7
  and UP structures, 9-32
Protruding dielectric waveguide arrays, 13-48 to 13-49
PTD. *See* Physical theory of diffraction
Ptfe bulb for implantable antennas, 24-27, 24-29
Pulse rf cable, 28-24 to 28-25
Purcell-type array antennas, 17-27
Push-pull power dividing circuits, 26-39
Pyramidal horn antennas, 8-4
  aperture and radiated fields of, 8-34 to 8-40
  corrugated, 8-51 to 8-59
  design procedure for, 8-43 to 8-45
  directivity of, 8-37 to 8-42
  for feeds for reflectors, 15-92 to 15-93, 15-96 to 15-98
  gain of, 8-43 to 8-44
  as surface-wave launcher, 17-46
Pyramidal log-periodic antenna, 9-4

Q. *See* Quality factor
Quadratic phase errors
  in antenna ranges, 32-16 to 32-18
  and conical horn antennas, 8-46
  with lenses, 16-3, 16-14, 16-16 to 16-18, 16-33
Quadratic ray pencil, 4-8
Quadrature hybrids for circular polarization, 10-61
Quadrature power dividing circuits, 26-39 to 26-40
Quadrifocal bootlace lenses, 16-47 to 16-48
Quality factor
  of arrays, 11-61 to 11-62
    dipoles in, 9-61 to 9-63
    directivity in, 11-67 to 11-68, 11-76
    planar, 11-65
    and SNR, 11-70 to 11-72
  of cavities, 10-11 to 10-13, 10-20, 10-21
  of dipoles
    in arrays, 9-61 to 9-63
    and phase shift, 9-58
  of ferrite loop, 6-22
  with log-periodic structures, 9-55 to 9-56
    dipole arrays, radiation fields in, 9-62 to 9-63
  of microstrip antennas, 10-6, 10-9, 10-49, 10-53

  of receiving antennas, 6-22 to 6-23
  and stopband width, 9-49
Quantization errors with phased arrays, 13-52 to 13-57, 21-12
Quarter-wave matching layers for lenses, 21-19
Quarter-wave vertical antennas, 26-9
Quartic errors with lenses, 16-16, 16-18
Quartz substrate with integrated antennas, 17-137

Radar cross section, 2-23 to 2-24
  measurements of, 32-64
Radar equation, 2-36 to 2-39
Radar requirements and phased array design, 18-4 to 18-5
Radial slots in cone, 4-78 to 4-80
Radial transmission line feeds, 19-23 to 19-32
Radiated power, 1-25, 3-11 to 3-12, 5-26 to 5-27, 6-6 to 6-7
  with arrays, 11-60
  of biconical antennas, 3-25
  and cavities, 10-20
  and directive gain, 5-26
  of fm antennas, 27-4, 27-6
  into free space, 2-29
  of space harmonics, 17-52
  of tv antennas, 27-3 to 27-5
Radiating edges, 10-24
Radiating elements
  for feed arrays, 21-27 to 21-28
  for phased array design, 18-3 to 18-12
Radiating near-field region, 1-18
Radiation angle and grating period in leaky modes, 17-56 to 17-58
Radiation conditions and Maxwell equations, 1-9
Radiation efficiency, 1-25, 3-12
  of microstrip dipole antennas, 17-119, 17-124 to 17-127
  of microstrip patch resonator arrays, 17-108 to 17-109
  of receiving antennas, 6-22 to 6-24
Radiation fields and patterns, 1-11 to 1-13, 1-20 to 1-21. *See also* Apertures and aperture distributions, radiation from; Far fields, Near fields; Patterns; Reactive fields
  of am antennas, 26-5 to 26-9, 26-17, 26-22 to 26-33
  standard reference, 26-3 to 26-8

with arrays, 3-36, 3-37
   aperiodic, 14-15, 14-17, 14-19 to 14-21
   circular, on cylinder, 21-100
   conformal, 13-52
   Dolph-Chebyshev, 11-19 to 11-20
   end-fire, 11-76, 11-77
   periodic, 13-5, 13-8 to 13-10, 13-12 to 13-17, 13-45 to 13-48
   subarrays, 13-30 to 13-32, 13-35 to 13-39
of biconical antennas, 3-26, 21-107, 21-109
blackbody, 22-3 to 22-4
with circular loop antennas, 23-21
of convex surfaces, 4-63 to 4-84
of dipole antennas
   short, 20-8 to 20-9
   skewed, 27-14 to 27-18
with dual fields, 2-5 to 2-6, 3-9, 3-11
with DuFort optical technique, 19-76, 19-78
of feed systems
   radial transmission line, 19-31 to 19-32
   reflector-lens limited-scan concept, 19-72 to 19-73
   for reflectors, 15-11 to 15-12, 15-86 to 15-94
   semiconstrained, 19-49, 19-51 to 19-53
   simple, 15-89 to 15-94
with grounded wire antennas, 23-9 to 23-16
with ground planes
   imperfect, 3-27, 3-33
   perfect, 3-22
of horn antennas, 8-3
   aperture-matched, 8-58, 8-66 to 8-67
   biconical, 21-107, 21-109
   corrugated, 8-53 to 8-59, 8-64, 8-65
   dielectric-loaded, 21-84 to 21-86
   dual-mode, 15-95, 15-100 to 15-101
   E-plane, 8-7 to 8-10
   E-plane-flared biconical, 21-107
   H-plane, 8-20 to 8-26, 8-28
   H-plane-flared biconical, 21-107
   pyramidal, 8-34 to 8-39, 15-92 to 15-93
and Huygen's principle, 2-18 to 2-19
and image theory, 2-11 to 2-13
of implantable antennas, 24-27 to 24-29
of lenses
   constrained, 19-82, 19-86 to 19-87
   effect of aperture amplitude distributions on, 21-14
   microwave, 19-45
   modified Meyers, 19-115 to 19-118
   pillbox, 17-28
   for satellites, 21-73, 21-75 to 21-77
   spherical thin, 16-30
   Teflon sphere, 17-188
of log-periodic antennas
   dipole, 9-17, 9-18, 9-62 to 9-63
   dipole arrays, 9-64 to 9-66
   planar, 9-8 to 9-9
   zigzag antennas, 9-39 to 9-40
of log-spiral antennas, 9-75
   conical, 9-85 to 9-86, 9-89 to 9-92, 9-99 to 9-109
of loop antennas, 25-6 to 25-7
measurements for. *See* Measurement of radiation characteristics
of microstrip antennas, 10-10
of millimeter-wave antennas
   dielectric grating, 17-69 to 17-70, 17-72 to 17-81
   dipoles, 17-118 to 17-119, 17-127 to 17-128
   holographic, 17-132
   leaky-wave, 17-86
   spiral, 17-28 to 17-29
   tapered dielectric-rod, 17-41 to 17-43
with models, 32-64
of monofilar zigzag wire antennas, 9-34, 9-36
of monopole on aircraft, 20-64, 20-74 to 20-75, 20-86 to 20-88
of multibeam antennas, 21-64
and polarization, 1-20 to 1-21, 1-25 to 1-26
with radomes, 31-25 to 31-27
reciprocity of, 2-32 to 2-33
with reflector antennas, 15-8 to 15-10, 21-20 to 21-21
   random surface distortion on, 21-23
   slotted cylinder, 21-110 to 21-111
with satellite antennas, 21-5
   lens, 21-73, 21-75 to 21-77
   multibeam, 21-64
   slotted cylinder reflector, 21-110 to 21-111
of side-mount tv antennas, 27-20, 27-22
for slot antennas, 27-27
   axial, 21-95, 21-100
   circumferential, 21-92 to 21-93, 21-96
of small loops, 6-13, 6-17

Radiation fields and patterns (*cont.*)
  of subarrays, 13-30 to 13-32, 13-35 to 13-39
  for TEM waveguides, 24-11
  in tissue, 24-51 to 24-54
  of transmission lines, 21-63
    radial feeds for, 19-31 to 19-32
    TEM, 28-11
  with traveling-wave antennas, 3-16, 3-18
    slot, 27-27
  and UTD solutions, 20-5
    for convex surfaces, 4-63 to 4-84
    with curved surface, 20-18, 20-20, 20-25 to 20-26, 20-31 to 20-36
    for noncurved surfaces, 20-51
  of vee-dipole antennas, 27-11
  of waveguide-slot antennas, 27-30 to 27-31
Radiation resistance, 3-12
  of ferrite loop, 6-21 to 6-22
  of microstrip patch antennas, 17-107
  and mutual impedance, 9-56
  of short dipole, 6-11 to 6-12, 6-14 to 6-15
  of small loops, 6-19, 25-7 to 25-8
  of vertical antenna, 6-16
Radiation sphere, 1-19, 1-21, 32-6 to 32-7
Radiation-type antennas for medical applications, 24-9 to 24-18
Radiative transfer, microwave, 22-4 to 22-7
Radio-astronomy
  feeds for, 15-94
  and microwave remote sensing, 22-9
  millimeter-wave antennas for, 17-14, 17-16
Radio-frequency cables, list of, 28-23 to 28-25
Radio-frequency power absorption, 24-30 to 24-35
Radio-frequency radiation, monitoring of, 24-36 to 24-43
  personnel dosimeters for, 24-43
Radio horizon, 29-34
Radius vector with conical log-spiral antennas, 9-79, 9-98, 9-100
Radome Antenna and RF Circuitry, 19-54 to 19-55
Radomes, 31-3 to 31-5
  on aircraft, numerical solutions for, 20-74
  and boresight error, 31-27
  design of, 31-18 to 31-20
  and flat dielectric sheets, 31-10 to 31-18, 31-20 to 31-25
  materials for, 31-29 to 31-30
  modeling of, 32-82 to 32-83
  and moment method, 31-27 to 31-29
  with omnidirectional dielectric grating antennas, 17-81 to 17-82
  patterns with, 31-25 to 31-27
  physical effects of, 31-5 to 31-10
Rain
  line-of-sight propagation path loss from, 29-35
  and millimeter-wave antennas, 17-5
  and noise temperature, 29-49, 29-50
  remote sensing of, 22-12, 22-14
  and satellite-earth propagation
    attenuation of, 29-10 to 29-13
    depolarization of, 29-14
Random errors
  with arrays
    periodic, 13-53 to 13-57
    phased, 18-26 to 18-28
  with microwave radiometers, 22-23
  with reflector surfaces, 15-105 to 15-114, 21-23 to 21-24
Random numbers with aperiodic array design, 14-18 to 14-19
Range equation, 2-38
Ranges, antenna
  design criteria for, 32-14 to 32-19
  errors with, 33-15 to 33-18
  evaluation of, 32-33 to 32-39
  for field measurements, 33-15 to 33-24
  indoor, 32-25 to 32-30
  instrumentation, 32-30 to 32-33
  outdoor, 32-19 to 32-25
  positioners and coordinate systems for, 32-8 to 32-14
RAR (Reflect Array Radar), 19-55
RARF (Radome Antenna and RF Circuitry), 19-54 to 19-55
Rayleigh approximation and computer solutions, 3-54
Rayleigh criterion for smoothness of antenna ranges, 22-30 to 22-32, 32-22
Rayleigh-Jeans approximation with blackbody radiation, 22-3
Rays, 4-6 to 4-8
  caustics of
    distance of, 4-8
    reflected and transmitted, 4-15
  construction of, with reflector antennas, 15-11 to 15-12

paths of, with UTD solutions, 20-48 to 20-49, 20-51
RCS (radar cross-section), 2-23 to 2-24
  measurements of, 32-64
Reactive fields, 1-16 to 1-18
  of aperture and plane-wave spectra, 5-5
  coupling of, in antenna ranges, 32-14
Reactive loads with dual-band elements, 10-63 to 10-66
Realized gain, 1-25
Real poles of cavities, 10-11
Real space of arrays scanned in one plane, 13-14
Received power, 2-38, 32-42, 33-22
Receiving antennas
  bandwidth and efficiency of, 6-22 to 6-24
  effective area of, 22-4
  equivalent circuit of, 6-3
  ferrite loop, 6-18 to 6-22
  Friis transmission formula for, 6-8 to 6-9
  grounded-wave, 23-16 to 23-17
  impedance-matching factor of, 6-6 to 6-7
  linear dipole, 7-18 to 7-19
  loops, 23-22 to 23-24
  meters for, 2-30 to 2-32
  mutual impedance between, 6-9
  and noise, 2-42 to 2-43, 6-24 to 6-25
  polarization-matching factor of, 6-7 to 6-8
  power accepted by, 1-24, 1-25, 22-4
  receiving cross section of, 6-6
  reciprocity of, 2-32 to 2-36
  satellite earth stations, 6-25 to 6-32
  small. *See* Small receiving antennas
  thin-wire, 7-8 to 7-9
  thin-wire loop, 7-46 to 7-47
  vector effective height of, 6-3 to 6-6
Receiving cross section, 2-35 to 2-36
  of receiving antennas, 6-6
  of short dipoles, 6-12 to 6-15
  of small loops, 6-18
Receiving polarization, 32-53
Receiving systems for antenna ranges, 32-30 to 32-32
Receptacles and plugs with modeling, 32-83
Reciprocal bases in transformations, 11-25
Reciprocal ferrite phase shifters
  dual-mode, 18-14 to 18-15, 21-38 to 21-40
  for periodic arrays, 13-64

Reciprocity, 2-32 to 2-36
  numerical validation of, 3-79
  and reciprocity theorem, 2-16 to 2-18, 33-7
  with scattering, 2-24 to 2-25
Recording systems for antenna ranges, 32-30 to 32-32
Rectangular anechoic chambers, 32-25 to 32-26
Rectangular apertures
  antennas with, 3-21 to 3-24
  compound distributions with, 5-16 to 5-17
  directive gain of, 5-27
  displaced, phase-shifted distributions with, 5-17 to 5-18
  effective area of, 5-27 to 5-28
  efficiency of, 5-28 to 5-29
  gain of, 5-29
  near-field gain reduction factors with, 5-30 to 5-31
  near-field pattern with, 5-25 to 5-26
  optimum pattern distributions with, 5-18 to 5-20
  radiation fields of, 5-11 to 5-20, 5-27
  simple distributions with, 5-13 to 5-15
  and uniform aperture distribution, 5-11 to 5-13
Rectangular coaxial TEM transmission lines, 28-18
Rectangular conducting plane, RCS of, 2-23 to 2-24
Rectangular coordinates with lens antennas, 16-7 to 16-8
Rectangular lattices
  field representation for, 1-35
  for phased-array radiators, 18-9
Rectangular loops
  for mine communications, 23-20
  radiation resistance of, 6-19
Rectangular patch antennas, 10-41 to 10-45, 24-19
  characteristics of, 10-16, 10-21 to 10-26
  medical applications of, 24-19 to 24-20
  principle-plane patterns of, 10-22
  resonant frequency of, 10-46 to 10-47
Rectangular planar arrays, patterns of, 13-15 to 13-17
Rectangular slots
  in cones, mutual coupling of, 4-91 to 4-92
  in prolate spheroid, radiation pattern of, 4-79 to 4-82

Rectangular tapered dielectric-rod antennas, 17-37 to 17-38
Rectangular waveguides, 1-37 to 1-44, 21-60
 currents in, 12-3 to 12-4
 dielectric, 28-51 to 28-54
 impedance of, 2-28
 with infinite array solutions, 13-48 to 13-49
 mutual coupling with, 13-44 to 13-45
 open-ended
  feeds for, half-power beamwidths of, 15-104
  radiators of, for phased arrays, 18-6
  for reflectors, 15-89 to 15-92, 15-104
 radiation from, 1-28
 TE/TM, 28-38 to 28-42
 voltage and current in, 2-28 to 2-29
Rectangular XY-scanners, 33-19
Reduction factor for line-of-sight propagation, 29-36
Redundant computer operations and symmetry, 3-65
Reference polarization, 1-23
 of microwave antennas, 33-20
 with probe antennas, 32-8
Reflect Array Radar, 19-55
Reflectarrays, 19-54 to 19-55
Reflected fields. *See* Reflections and reflected fields
Reflected power, 2-29
Reflection boundaries and UTD, 4-34
Reflection coefficients
 approximation of, for imperfect grounds, 3-48
 with coaxial sleeve antennas, 7-36
 for dielectric grating antennas, 17-71, 17-81
 and imperfect ground planes, 3-27
 with linear dipole antennas, 7-23
 of ocean surface, 29-38 to 29-40
 with tapered dielectric-rod antennas, 17-44 to 17-46
 with transmission lines, 2-28, 28-7, 28-8
 with unloaded transmitting antennas, 7-13
 with UTD noncurved surface solutions, 20-43, 20-44
Reflections and reflected fields
 from aircraft wings, 20-78
 in antenna ranges, 32-34 to 32-37
 errors from, in scanning ranges, 33-16

and Fresnel ellipsoid, 29-36
GO, 4-13 to 4-18
by human tissue, 24-7
and imperfect ground planes, 3-27, 3-33
lobes from, 19-5
and multipath propagation, 29-37
and plane boundaries, 31-10 to 31-12
with radomes, 31-5, 31-6, 31-10 to 31-17
reflected-diffracted, 20-29, 20-33, 20-37, 20-43, 20-45, 20-55
reflected-reflected, 20-29, 20-33, 20-45
in scattering problems, 2-13 to 2-14
shadow boundary, 4-13
space-fed arrays with, 21-34, 21-36
space feed types, 18-17 to 18-19
and spatial variations in antenna ranges, 32-19
with UTD solutions, 20-8 to 20-9, 20-15
 for curved surfaces, 20-21, 20-23 to 20-25, 20-29, 20-31
 for noncurved surfaces, 20-37, 20-42, 20-44, 20-49, 20-53 to 20-54, 20-59 to 20-60
Reflectivity of surfaces, 22-8 to 22-9
Reflector antennas, 15-5
 for antenna ranges, 32-28 to 32-30
 basic formulations for, 15-6 to 15-15
 bifocal lens, 16-30 to 16-33
 contour beam, 15-80 to 15-84
 and coordinate transformations, 15-115 to 15-120
 diameter of, 21-24
 dual. *See* Dual-reflector antennas
 far-field formulas for, 15-15 to 15-23
 with feed, rotating antenna systems, 25-4
 feeds for, 15-84 to 15-105
 generated, 15-23 to 15-31
 with implantable antennas, 24-27 to 24-28
 millimeter-wave, 17-9 to 17-14
 with off-focus feeds, 15-49 to 15-61
 offset parabolic, 15-31 to 15-61
 with on-focus feeds, 15-37 to 15-49
 phased arrays, 21-12
 random surface errors on, 15-105 to 15-114
 satellite, 21-7, 21-8, 21-19 to 21-26, 21-73 to 21-74, 21-77 to 21-81
 surfaces for, 15-26 to 15-29
 types of, 15-5 to 15-6, 21-20 to 21-22

Index

Reflector coordinates
  with reflector antennas, 15-7, 15-88 to 15-89
  transformations with, 15-115, 15-117 to 15-119
Reflector-lens limited-scan feed concept, 19-67 to 19-73
Refraction
  and line-of-sight propagation, 29-32 to 29-33
  and plane boundaries, 31-10 to 31-12
Refractive index, 29-7
  of atmosphere, 29-30 to 29-32
  and ionosphere, 29-14 to 29-15, 29-21 to 29-22
  lenses with varying, 16-51 to 16-54
  of short monopoles, 24-31
  of troposphere, 29-46
Regularization synthesis method, 21-65
Relabeling of bases in transformations, 11-25
Relative dielectric constant, 1-7
Relative permeability, 1-7
Remote sensing
  antenna requirements for, 22-16 to 22-46
  microwave radiometry for, 22-9 to 22-16
  sleeve dipole for, 7-26
Reradiated fields and imperfect ground planes, 3-48
Reradiative coupling in antenna ranges, 32-14
Resistance. *See also* Radiation resistance
  of free space, 26-25 to 26-26
  measurements of, electrode arrays for, 23-4 to 23-8
  surface, 1-43
Resonance
  in inclined narrow wall slot arrays, 12-26
  with planar arrays, 12-21 to 12-22
  in slot arrays, 12-14 to 12-15
  and thin-wire antennas, 7-10
  validation of computer code for, 3-73 to 3-76
Resonant arrays, 17-114, 17-118 to 17-119
Resonant frequency
  of am towers, 26-9 to 26-12
  for microstrip antennas, 10-9, 10-44 to 10-49
  of loaded elements, 10-53 to 10-54
Resonant impedance, 10-28, 10-30 to 10-31

Resonant loads with dual-band elements, 10-63 to 10-66
Resonant modes of cavities, 10-13, 10-15
Resonantly spaced longitudinal slots, 12-13 to 12-16
Resonator millimeter-wave antennas, 17-35 to 17-36
Rexolite with lenses, 16-22, 16-36 to 16-37
Rf. *See* Radio-frequency
RG-type cables, 28-22 to 28-25
Rhombic antennas, 3-31
  dielectric plate, 17-23, 17-25
Richmond formulation with radomes, 31-27 to 31-29
Ridged-waveguide antennas
  for medical applications, 24-12 to 24-16
  phased arrays for satellite antennas, 21-12
  TE/TM, 28-45 to 28-49
Ridge-loaded waveguide arrays, 13-48 to 13-49
Ridge waveguide feed transmissions, 21-56 to 21-58
Right-hand circular polarization, 1-15, 1-20
  of arrays, 11-7
  dual-mode converters for, 21-78, 21-80
  far-field pattern for, 1-29
  for microstrip antennas, 10-59, 10-69
  with offset parabolic reflectors, 15-36, 15-48 to 15-50
  power dividers for, 21-78, 21-80
Right-hand elliptical polarization, 1-16, 9-74
Rinehart-Luneburg lenses, 19-41 to 19-49
Ring focus
  with bifocal lenses, 16-32
  with DuFort-Uyeda lenses, 19-49
Ring-loaded slots, 8-64
Ripples
  with conformal arrays, 13-50 to 13-52
  with feed arrays, 21-27
Rock conductivity, dielectric logging for, 23-24 to 23-25
Rod antennas, effective receiving area of, 30-10
Roll-over-azimuth positioners, 32-12
Rotary-field phase shifters, 13-64
Rotary joints for beam-forming networks, 21-53, 21-55
Rotating antenna patterns, 25-4 to 25-11

Rotating reflector
  with conical scanning, 22-40
  with direction-finding antennas, 25-10
  with feed rotating antenna systems, 25-4 to 25-5
Rotating-source polarization measurement method, 32-63
Rotational symmetry
  and computer time and storage, 3-67 to 3-68
  dielectric grating antennas with, 17-80 to 17-81
Rotman and Turner line source lenses, 19-37 to 19-41
ROTSY modeling code, 3-88 to 3-91
Rounded-edge triplate striplines, 28-17
RSB (reflection shadow boundary), 4-13
R-3 singularity, integration involving, 2-9 to 2-10
R-2R lenses, 16-48 to 16-49
Run-time features in modeling codes, 3-86
Ruze lenses, 19-39

Saltwater, modeling of, 32-79, 32-80. *See also* Ocean
Sam-D Radar, 19-55
Sample directivity in aperiodic array design, 14-15, 14-17
Sample radiation patterns in aperiodic array design, 14-15, 14-17, 14-19 to 14-21
Sampling systems, 26-41 to 26-44
Sampling theorem, 15-10
SAR, 22-30
Satellite antennas and systems, 21-3
  communication. *See* Communication satellite antennas
  conformal array for, 13-49
  contour beam antennas for, 15-5, 15-80 to 15-87
  earth coverage, 21-80 to 21-89
  earth receiving antennas for, 6-25 to 6-32
  lenses for, 16-22
    millimeter-wave, 17-15, 17-19 to 17-20
    spherical-thin, 16-28
    taper-control, 16-36
    zoning of, 16-38
  limited scan antennas for, 19-56
  low earth orbiting 22-35 to 22-43
  for microwave remote sensing, 22-9 to 22-10
  millimeter-wave antennas for, 17-7, 17-11 to 17-12, 17-14
    lens, 17-15, 17-19 to 17-20
  modeling of, 32-86
  offset parabolic antennas for, 15-58
  polar orbiting, 22-34 to 22-42
  propagation for, 29-7 to 29-21
  reflector antennas for, 15-5
  testing of, 33-4
  tracking, telemetry, and command, 21-89 to 21-111
  weather, 22-14
S-band phased array satellite antennas, 21-68 to 21-71
Scalar conical horn antennas, 8-61
Scalar wave equation, 2-6 to 2-7
Scaling. *See also* Modeling
  with log-periodic dipole antennas, 9-12, 9-14
    and directivity, 9-20, 9-22 to 9-24
    and element length, 9-31 to 9-32
    and swr, 9-27, 9-28
  with log-periodic planar antennas, 9-9 to 9-10
  for millimeter-wave antennas, 17-7 to 17-9
  with modeling, 32-74 to 32-86
  with offset parabolic antennas, 15-81
  and principal-plane beamwidths, 9-9 to 9-10
SCAMS (scanning microwave spectrometer), 22-26, 22-37, 22-43
Scan angles vs. directivity, with arrays, 11-78
Scan characteristics, array, 14-24 to 14-25
Scanned beams
  beam-forming feed networks for, 18-21, 18-24, 21-27 to 21-34, 21-37 to 21-49
  with offset parabolic antennas, 15-51, 15-53
  with phased array satellite antennas, 21-71
Scanning. *See* Beam scanning
Scanning microwave spectrometers, 22-26, 22-37, 22-43
Scanning multichannel microwave radiometer, 22-22
  calibration of, 22-25 to 22-26
  compared to SSM/I, 22-48
  as conical scanning device, 22-40 to 22-42
  as continuous scan type, 22-43

Index

modulating frequency of, 22-24
momentum compensation devices in, 22-46 to 22-47
onboard reference targets for, 22-26
as step scan type, 22-43
Scanning ranges. *See* Ranges, antenna
Scanning thermographic cameras, 24-51
Scan performance
  of dual-reflector antennas, 15-71 to 15-73, 15-78 to 15-79
  of zoned lenses, 16-28
Scattering and scattered propagation fields, 2-21 to 2-27, 4-49
  and Babinet principle, 2-13 to 2-15
  cross section of, 2-22 to 2-25
  and dielectric thickness, 10-6
  with dual fields, 2-5 to 2-6
  field representation of, 1-33 to 1-37
  ionospheric, 29-28 to 29-30
  in longitudinal slots, 12-7, 12-10 to 12-11
  losses from, with remote-sensing microwave radiometers, 22-32, 22-33
  plate-scattered fields, 20-23
  and PO, 4-20, 4-22
  by radomes, 20-74, 31-5
  with reflector antennas, 15-7, 15-11
  with shipboard antennas, 20-90 to 20-97
  in slot array design, 12-35 to 12-36
  tropospheric, 29-46 to 29-47
  for UTD solutions
    centers of, 20-4 to 20-6
    for convex surfaces, 4-49 to 4-63
    for curved surfaces, 20-18
    for noncurved surfaaces, 20-47, 20-49, 20-51 to 20-52, 20-57 to 20-60
Scattering matrix for planar arrays, 12-21
Scattering matrix planar scanning, 33-7
Schelkunoff's induction theorem, 31-25
Schiffman phase shifters, 13-62
Schlumberger array, 23-5 to 23-7
Schmidt corrector with wide-angle lenses, 16-23 to 16-28
Schottky diodes for integrated antennas, 17-133 to 17-134, 17-138 to 17-139, 17-141
Schumaun resonance and ionospheric propagation, 29-27
Scintillation
  from fog and turbulence, 29-36
  index for, 29-16 to 29-17
  and satellite-earth propagation, 29-16 to 29-21

SCS (scattering cross section), 2-24 to 2-25
Sea. *See* Ocean
Seasat satellite, 22-17
  SMMR on, 22-22
Second-order effects in slot array design, 12-34 to 12-36
Second-order scattering, 20-49
Security, transmission, with millimeter-wave antennas, 17-6 to 17-7
Seidel lens aberrations, 16-14, 21-13
Self-admittance with microstrip antennas, 10-7
Self-baluns, 9-16, 9-17
Self base impedance of am antennas, 26-8 to 26-12
Self-complementary antennas, 9-5 to 9-7
  log-spiral, 9-98 to 9-99
Self-conductance with microstrip antennas, 10-7
Self-resistance of am antennas, 26-8
Semicircular arrays, 11-78 to 11-80
Semicircular-rod directional couplers, 21-50
Semiconstrained feeds, 19-15 to 19-19
  DuFort-Uyeda lens, 19-49
  Meyer lens, 19-32 to 19-37
  pillbox, 19-19 to 19-23
  radial transmission line, 19-23 to 19-32
  Rinehart-Luneburg lens, 19-41 to 19-49
  Rotman and Turner line source microwave lens, 19-37 to 19-41
Sense of rotation with polarization ellipse, 1-16
Sensing systems, millimeter-wave antennas for, 17-5. *See also* Remote sensing
Sensitivity and reception, 6-10
  with planar arrays, 11-65
SEO-I and -II satellites, 22-17
Septum polarizers, 21-30
  with reflector satellite antennas, 21-74, 21-79 to 21-80
  tapered, 13-60 to 13-61
Serial shift registers, 18-3 to 18-4
Series couplers, 19-4 to 19-5
Series-fed arrays
  for microstrip dipole antennas, 17-119
  microstrip patch resonator, 17-113 to 17-116
Series feed networks, 18-9, 18-21, 18-26, 19-3 to 19-5

Series impedance of transmission lines, 28-5
Series resonant magnetic field probes, 24-43, 24-44
Series-resonant power dividing circuits, 26-39 to 26-40
Series slot radiators with phased arrays, 18-8
Shadow boundaries
  and GO, 4-10, 4-17
  and PO, 4-22
  and PTD, 4-103 to 4-104
  and UTD, 4-32, 4-34, 4-36, 4-74, 20-8, 20-12
Shadow regions, 4-5, 4-75
  and diffracted rays, 4-3, 4-108
  Fock parameter for, 4-69
  and GO incident field, 4-10
  terms for, 4-67
  and UTD, 4-52, 4-63, 4-65
    with curved surfaces, 20-20 to 20-21
Shaped-beam antennas
  in antenna ranges, 32-19
  horns, 21-5, 21-82, 21-87 to 21-89
Shaped lenses, 21-18
Shaped reflectors, 15-73, 15-75 to 15-80
Shaped tapered dielectric-rod antennas, 17-37 to 17-38
Shapes for simulation of environments, 20-8
Shaping techniques for lens antennas, 17-19
Sheleg method, 19-103 to 19-105
Shielded-wire transmission lines, 28-11
  two-wire, impedance of, 28-15
Shields
  with loops, 25-9
  with scanning ranges, 33-16
Shift registers, 18-3 to 18-4
Ships and shipboard antennas
  models for, 32-74, 32-77 to 32-79
  numerical solutions for, 20-90 to 20-97
  simulation of, 20-7 to 20-8
Short antennas
  dipole, 3-13 to 3-15, 6-10 to 6-13, 6-15
    array of, 3-38 to 3-39
    compared to small loop antenna, 3-16
    fields with, 3-34
    UTD solutions for, 20-8 to 20-15
  monopole
    impedance of, 6-12

  for in-vivo measurements, 24-30 to 24-32
  mutual impedance of, 4-94
Short-circuit current, 2-32
  and vector effective height, 6-5
Shorted patch microstrip elements, 13-60
Shoulder lobes and phase taper, 32-16
Shunt admittance of transmission lines, 9-47 to 9-49, 28-5
Shunt couplers, 19-4
Shunt-slot array antennas, 17-26
Shunt slot radiators, 18-8
Side firing log-periodic diode arrays, 9-63 to 9-65
Side lobes, 1-20
  and antenna ranges
    compact, 33-23
    measurement of, 33-17
    phase errors in, 32-19
  and aperture field distributions, 21-14, 21-16
  with apertures
    circular, 5-22 to 5-23
    rectangular, 5-12 to 5-14, 5-16, 5-18 to 5-20
  with arrays, 11-15
    aperiodic, 14-4, 14-5, 14-8, 14-14 to 14-19, 14-23 to 14-32
    binomial, 11-11
    conformal, 13-52
    Dolph-Chebyshev, 11-13 to 11-14, 11-18, 11-21, 11-49 to 11-50
    edge-slot, 17-27
    feed, for satellite antennas, 21-27
    periodic, 9-44, 9-45, 13-10 to 13-11, 13-18
    phased, 18-3, 18-26 to 18-28, 19-60, 19-61
    from random errors, 13-53 to 13-55
    scanned in one plane, 13-12
  with beam-forming networks, 18-25, 21-57
  and beamwidth, 9-44, 9-45, 13-18
  with DuFort optical technique, 19-76
  with feed systems
    arrays for satellite antennas, 21-27
    constrained, 18-20
    Gregorian, 19-65
    parallel, 19-5
    transmission line, radial, 19-26, 19-31, 19-33

with horn antennas
  corrugated, 8-55
  multimode, 8-70
and lateral feed displacement, 21-20
with lens antennas, 16-5
  coma aberrations, 16-22 to 16-23, 16-41, 21-14
  distortion in, 16-17
  DuFort-Uyeda, 19-53
  millimeter-wave, 17-15, 17-19, 17-20
  obeying Abbe sine condition, 19-98
  power distribution of, 21-18
  quadratic errors in, 16-14
  taper-control, 16-33
  TEM, 21-16
and log-periodic zigzag antennas, 9-44, 9-45
measurement of, 33-17
and microwave radiometer beam efficiency, 22-28
with millimeter-wave antennas
  dielectric grating, 17-69, 17-75 to 17-77
  geodesic, 17-32
  lens, 17-15, 17-19, 17-20
  metal grating, 17-69
  microstrip patch resonator arrays, 17-115 to 17-116
  tapered dielectric-rod, 17-38 to 17-43, 17-47
with offset parabolic antennas, 15-33, 15-37, 15-41 to 15-42
with orthogonal beam synthesis, 13-22
with parabolic pillbox feeds, 19-20, 19-22
from radomes, 31-6
and reflections, 32-34 to 32-37
with reflector antennas
  millimeter-wave, 17-9
  satellite, 21-23
  surface errors on, 15-105 to 15-106, 15-109 to 15-110, 15-112 to 15-114, 21-26
  tapered-aperture, 15-16 to 15-17, 15-21 to 15-23
and reradiative coupling, 32-14
with satellite antennas, 21-5, 21-7
with Schiffman phase shifters, 13-62
and series feed networks, 19-5
slots for, 8-63
with space-fed arrays, 19-55
and subarrays, 13-30, 13-32, 13-39
with synthesized patterns, 13-23 to 13-29
with waveguides, open-ended
  circular, 15-94 to 15-95
  rectangular, 15-91 to 15-92
Side-mount antennas
  circularly polarized tv, 27-20 to 27-23
  for multiple-antenna installations, 27-37
Sidefire helical tv antennas, 27-24 to 27-26
Sidewall inclined-slot array antennas, 17-26
Signal-power to noise-power ratio for tvro link, 6-30
Signal-to-noise ratio and mismatches, 6-25
  with arrays, 11-58 to 11-63, 11-70 to 11-76
  with satellite antennas, 6-31
Significant height and ocean reflection coefficient, 29-38 to 29-40
Simple distributions
  with circular apertures, 5-23
  with rectangular apertures, 5-13 to 5-15
Simple lenses, 16-9 to 16-12
Simulation
  of antennas, 20-5 to 20-7
  of environment, 20-7 to 20-53
Simultaneous multi-beam systems, 19-79 to 19-80
Single-dielectric microstrip lines, 28-16
Single-feed circular polarization, 10-57 to 10-61
Single-wire transmission lines, 28-13 to 28-15
Sinusoidal-aperture antennas, 3-24
Sinusoidal current distribution, 3-14, 3-17
Skewed dipole antennas
  fm, 27-34 to 27-35
  tv, 27-11 to 27-18
Skin-effect resistance, 1-43
Skylab satellite, 22-17
Slant antenna ranges, 32-24 to 32-25
Sleeve antennas
  coaxial sleeve, 7-29 to 7-37
  dipole, 7-26 to 7-29
  with implantable antennas, 24-26 to 24-27
  junction effect with, 7-23 to 7-26
  monopole, 7-29 to 7-31
Slope of lenses, 16-9
Slope diffraction, 4-38 to 4-40, 4-108

Slot antennas, 3-31
  on aircraft wing, 20-85 to 20-86, 20-90
  annular, 30-13
  Babinet principle for, 30-14 to 30-16
  balanced, 9-72 to 9-73
  effective receiving area of, 30-10
  impedance of, 2-15 to 2-16, 9-78 to 9-79
  microstrip, for medical applications, 24-22 to 24-24
  radiation pattern of, 4-61
  stripline, 13-58 to 13-59
  tv, 27-26 to 27-31
Slot arrays
  cavity-backed, 9-68 to 9-72
  on curved surfaces, 13-51 to 13-52
  millimeter-wave, 17-26
  waveguide-fed, design of, 12-3 to 12-6
    aperture distribution in, 12-24
    and center-inclined broad wall slots, 12-24 to 12-25
    and E-field distribution, 12-6 to 12-10
    and equations for slots, 12-10 to 12-13
    far-field and near-field diagnostics for, 12-36 to 12-37
    and inclined narrow wall slots, 12-25 to 12-28
    for large arrays, 12-28 to 12-34
    and nonresonantly spaced slots, 12-16 to 12-20
    for planar arrays, 12-20 to 12-24
    and resonantly spaced slots, 12-13 to 12-16
    second-order effects in, 12-34 to 12-36
Slot line transmission types, 21-59, 21-61
Slot line waveguides, 28-53 to 28-57
Slot radiators
  for aperture antennas, 4-64
  with phased arrays, 18-8
Slots, 3-31
  axis, 4-75, 4-77, 21-92, 21-95
  in circular cylinder, 4-77 to 4-79
  mounted in plate-cylinder, 4-61
  for side lobes, 8-63
  in sphere, 4-77 to 4-79
  voltages in, 12-10 to 12-12
Slotted cylinder reflector antennas, 21-110 to 21-111
Slotted waveguide array antennas, 17-26, 17-33
Small dipole antennas, 3-28
  and EMP, 30-9
Small loop antennas, 3-16, 3-19, 6-13, 6-17 to 6-19, 7-46 to 7-47
Small receiving antennas, 6-9
  bandwidth of, 6-22 to 6-24
  short dipole, 6-10 to 6-16
  small loop, 3-16, 3-19, 6-13, 6-17 to 6-18, 7-46 to 7-47
SMMR. *See* Scanning multichannel microwave radiometer
Snell's law of refraction, 32-70
  and lenses, 21-18
    constrained, 16-51
    dielectric, 16-6, 16-8 to 16-9
    taper-control, 16-34
  with rays, 4-8 to 4-9
Snow and snowpack
  depolarization by, 29-14
  probing of, 23-25
  remote sensing of, 22-7, 22-15 to 22-16
SNR. *See* Signal-to-noise ratio
Soil
  emissivity of, 22-9 to 22-10
  probing of, 23-25
  remote sensing of moisture in, 22-13, 22-15 to 22-16
Solar radiation and ionospheric propagation, 29-14, 29-25
  noise from, 29-50
Solid bodies
  integral equations for, 3-46
  time-domain solutions with, 3-63
Solid-state components
  in integrated antennas, 17-131
  for millimeter-wave antennas, 17-11
Solid thin-wire antennas, 7-10 to 7-11
Sommerfield treatment, 3-49 to 3-50
Sounders. *See also* Microwave sounder units; Temperature sounders
  humidity, 22-14, 22-22
Source field patterns with UTD solutions, 20-15
  for curved surfaces, 20-25, 20-29, 20-31
  for noncurved surfaces, 20-37 to 20-42, 20-48
Sources
  radiation from, 1-11 to 1-13
  for transmitting antennas, 2-29 to 2-30
Space applications, lens antennas for, 16-5 to 16-6
Space-combination with millimeter-wave antennas, 17-11

Index

Space configuration with am antennas, 26-17 to 26-19
Spacecraft constraints for microwave radiometers, 22-46 to 22-47
Spaced loops for direction-finding antennas, 25-9 to 25-11
Spaced-tapered arrays, 14-6 to 14-8
Space factors, cylinder, 21-98 to 21-100
Space feeds
  for beam-forming feed networks, 18-17 to 18-19
  for arrays in, 21-34, 21-35 to 21-36
  for reflectarrays, 19-54 to 19-56
  for subarray systems, 13-26, 19-68 to 19-69, 19-70 to 19-76
Space harmonics
  for dielectric grating antennas, 17-72
  with periodic dielectric antennas, 17-51 to 17-53
  and periodic structure theory, 9-33 to 9-35, 9-55 to 9-56
  and scattered fields, 1-34 to 1-35
Space shuttle and remote sensing, 22-10
Space waves for leaky-wave antennas, 17-100 to 17-101
Spacing with arrays, 3-36
  aperiodic, 14-4 to 14-5
  with log-periodic dipole antennas, 9-28, 9-29, 9-31
S-parameter matrix, 10-34 to 10-37
Spatial coupling with UTD solutions, 20-6 to 20-7
Spatial diplexers, 21-54
Spatial resolution of microwave radiometers, 22-30 to 22-32
Spatial variations in antenna ranges, 32-19
Special sensor microwave/imager, 22-25
  momentum compensation devices in, 22-46 to 22-47
  polarization with, 22-48
  scanning by, 22-46 to 22-47
Special sensor microwave/temperature sounder, 22-14
Spectrum functions and aperture fields, 5-6
Specular reflections in indoor ranges, 32-27
Speed of phase propagation, 29-7
Sphere
  radiation, 1-19, 1-21, 32-6 to 32-7
  slot in, radiation pattern of, 4-78
Spherical aberrations with lenses, 16-15, 16-20
  spherical thin, 16-29
Spherical coordinates, 3-24
  positioning system for, 32-8 to 32-13
  transformations of, 15-115 to 15-120
Spherical dipole antennas, 30-14
Spherical lenses, 21-18. *See also* Luneburg lenses
  aberrations with, 21-13 to 21-14
  cap, 16-23 to 16-27
  Maxwell fish-eye, 16-52 to 16-53
  symmetrical, 17-15
  thin, 16-28 to 16-30
Spherical near-field test, 33-20
Spherical radiation pattern of am antennas, 26-4, 26-8 to 26-9
Spherical reflector systems
  with satellite antennas, 21-20 to 21-21
  with wide-angle multiple-beam systems, 19-80
Spherical scanning for field measurements, 33-12 to 33-15
Spherical surfaces
  and intersection curve, 15-29
  for lenses, 16-11 to 16-12
  for offset parabolic antennas, 15-56
Spherical-wave illumination
  for distance parameter, with UTD, 4-56
  and GO incident field, 4-11 to 4-12
Spherical waves
  and biconical antennas, 3-24 to 3-25
  transformation of, to plane waves, 16-6, 16-9 to 16-10
Spillover
  with DuFort optical technique, 19-76, 19-78
  and lens antennas, 16-5
  taper-control, 16-37
  with offset parabolic antennas, 15-37, 15-42, 15-47, 15-56
Spinning-diode circular polarization patterns, 10-60 to 10-62
Spinning geosynchronous satellites, 22-43
Spin-scan technique, efficiency of, 22-34
Spin-stabilized satellites, 21-90
Spiral angle with log-spiral antennas, 9-108
  and directivity, 9-89
Spiral antennas, 17-27 to 17-28. *See also* Log-spiral antennas
  balanced, 9-78

Spiral-phase fields, 9-109
Spiral-rate constant, 9-72, 9-102
Split-tee power dividers
  for array feeds, 13-61
  stripline, 21-50, 21-51
Spread-F irregularities
  and scintillations, 29-18 to 29-19
  and vhf ionospheric propagation, 29-29
Square apertures, 5-32 to 5-33
Square coaxial transmission lines, 28-18
Square log-periodic antennas, 9-3, 9-4
Square waveguides
  for circular polarization, 16-43
    normalized modal cutoff frequencies for, 1-42
SSB. *See* Surface shadow boundary
S-65-147 vhf antenna, 30-25 to 30-26
S-65-8262-2 uhf antenna, 30-22
SSM/I. *See* Special sensor microwave/imager
SSM/T (special sensor microwave/temperature sounder), 22-14
Stabilizers, aircraft, simulation of, 20-73, 20-78, 20-82 to 20-83
Staggered-slot array antennas, 17-26
Standard am antenna patterns, 26-35 to 26-36
Standard-gain horn, 8-43
Standard radio atmosphere, 29-32
Standard reference am antennas, 26-3 to 26-22
Standard waveguides, 28-40 to 28-41
Standing-wave fed arrays, 12-4, 12-13 to 12-16
Stationary phase point, 4-7
Steering
  with periodic arrays, 13-7 to 13-10
  with subarrays, 13-32 to 13-35
Step functions
  for convex surfaces, 4-28
  in scattering problem, 4-51
  for wedges, 4-25
Step scans compared to continuous scans, 22-42
Stepped-horn antennas, 21-82 to 21-84
Stepped-septum polarizer, 13-61
Stopbands
  with dielectric grating antennas, 17-71, 17-81
  with leaky-mode antennas, 17-52, 17-68
  for log-periodic dipole antennas, 9-61 to 9-62, 9-64 to 9-65
  with periodic loads, 9-49 to 9-50, 9-55, 9-58
Stored energy with cavities, 10-20
Straight lines and intersection curve, 15-27
Straight wire antennas, 3-49 to 3-50
Stratton-Chu formula, 2-21
Stray efficiency of microwave radiometers, 22-29
Strip antennas, impedance of, 2-15 to 2-16
Strip dielectric guide transmission type, 21-62
Strip edge-diffracted fields, 20-21 to 20-23
Stripline techniques, 17-106
  for arrays, 13-61
    blindness in, 13-47
  asymmetric, 17-84 to 17-87, 17-97 to 17-98
  for beam-forming networks, 20-50 to 20-51, 21-37
  for hybrid phase shifters, 13-62
  for printed dipoles, 13-58
  for slot antennas, 13-58 to 13-59
  for transmission types, 21-58, 21-60
    feeds for, 19-29 to 19-30
    for satellite antennas, 21-7
  TEM, 28-11, 28-16
  for waveguides, 21-57 to 21-58
Structural integrity, measurements for, 32-4
Structural stopbands, 9-49 to 9-50, 9-55
  with log-periodic arrays, 9-64 to 9-65
Structure bandwidth of log-periodic antennas, 9-17
Structures
  effect of, on radiation patterns, 20-53 to 20-54, 20-61 to 20-70
  and EMP, 30-4 to 30-7
  for simulation of environment, 20-8
Stub-loaded transmission lines
  attentuation curves for, 9-52, 9-54
  dispersion curves for, 9-52, 9-54, 9-57
Subarrays
  aperture illumination control with, 13-30 to 13-32
  overlapped, 13-35 to 13-39, 19-68 to 19-76
  time-delayed, 13-32 to 13-35
  with unconstrained feeds, 19-52
Subdomain procedures, 3-60
Submarine communications, 23-16

Subreflectors
- of Cassegrain offset antenna, 15-67, 15-69
- with reflector antennas, 15-7

Subsectional moment methods, 3-60

Substrates
- for integrated antennas, 17-131
- for microstrip antennas, 10-5, 17-107, 17-109 to 17-111, 17-114, 17-122
- dipole, 17-124 to 17-129

Substructures for simulation of environments, 20-8

Sum beams with beam-forming feed networks, 18-25

Sum-hybrid mode with microstrip antennas, 10-60 to 10-61

Sun
- and ionospheric propagation, 29-14, 29-25
- noise from, 29-51

Superdirective arrays, 11-61, 11-68 to 11-70
- aperiodic, 14-3

Supergain antennas, 5-27
- arrays, 11-61

Superstrates with microstrip dipoles, 17-127 to 17-128

Superturnstile tv antennas, 27-23 to 27-24

Surface corrugations
- for dielectric grating antennas, 17-64
- with periodic dielectric antennas, 17-48

Surface current
- decays of, with corrugated horn antennas, 8-52 to 8-54
- measurement of, 33-5 to 33-6
- and PO, 2-26 to 2-27
- and PTD, 4-103

Surface-diffracted waves with UTD, 4-57

Surface emissivity and brightness temperature, 22-7 to 22-9, 22-15

Surface errors of reflector antennas, 15-105 to 15-114
- and gain, 17-12

Surface fields
- equivalent currents for, 5-8
- with UTD, 4-84 to 4-89

Surface impedances
- and computer solutions, 3-54
- and modeling, 32-68

Surface integration method, 31-24 to 31-25

Surface mismatches with dielectric lenses, 16-54 to 16-56, 21-19

Surface Patch modeling code, 3-88 to 3-91

Surface perturbations for dielectric grating antennas, 17-70 to 17-71

Surface propagation, 29-30 to 29-47

Surface-ray field
- in GTD, 4-30
- in UTD, 4-58, 4-84 to 4-86

Surface-ray launching coefficients, 4-70

Surface representation of lens antennas, 21-13

Surface resistance, waveguide, 1-43

Surface roughness and conduction loss, 17-111, 22-31

Surface shadow boundary
- and GO fields, 4-13, 4-17
- and UTD, 4-50 to 4-51, 4-56
- transition regions, 4-58 to 4-59

Surface tolerances
- with lenses, 21-18 to 21-19
- with reflector satellite antennas, 21-23, 21-24, 21-26

Surface-wave antennas, 17-34
- periodic dielectric, 17-48 to 17-82
- tapered-rod, 17-35 to 17-48

Surface-wave launchers, 17-46

Surface waves
- with integrated antennas, 17-137
- for leaky-wave antennas, 17-100
- propagation with, 29-44 to 29-46
- from substrates, 17-122, 17-129

Susceptance
- with microstrip antennas, 10-7, 10-9
- with thin-wire antennas, 7-14 to 7-17

Suspended substrate transmission types, 21-59, 21-60

Switch beam-forming networks, 21-30 to 21-32

Switched-line phase shifters, 13-62 to 13-63, 18-13

Switches for beam-forming networks, 21-7, 21-37 to 21-38

Switching networks with Wheeler Lab approach, 19-105 to 19-106

Switching speeds of phase shifters, 18-16

Swr
- and feed cables, for log-spiral antennas, 9-78, 9-80
- and image impedance, 9-55

Swr (*cont.*)
   and log-periodic antennas, 9-27, 9-28
      dipole array, 9-58 to 9-59
      dipole scale factor, 9-27, 9-28
   and Q, in dipole arrays, 9-61
Symmetrical directional couplers, 21-50, 21-51
Symmetric parabolic reflectors. *See* Offset parabolic reflectors
Symmetry
   and computer time and storage, 3-65, 3-67
   with Dolph-Chebyshev arrays, 11-52
Synchronous satellites, 21-3
Synthesis methods
   for array patterns, 11-42 to 11-43, 11-76 to 11-86
   with multibeam antennas, 21-65
   for plane waves, 33-21 to 33-22
Synthetic aperture radar (SAR), 22-30
Systematic errors
   with microwave radiometers, 22-23, 22-25 to 22-28
   and phased array design, 18-26 to 18-28
System losses, power for, 26-36 to 26-37

TACOL structure, 19-56
Tangential fields, 5-7 to 5-9
   with thin-wire antennas, 7-5 to 7-6
Taper-control lenses, 16-33 to 16-38
Tapered anechoic chambers, 32-26 to 32-28
Tapered aperture distributions
   with parallel feed networks, 19-6
   with radial transmission line feeds, 19-31, 19-33
   for reflector antennas, 15-15 to 15-23
   with series feed networks, 19-5
Tapered-arm log-periodic zigzag antennas, 9-37 to 9-39
Tapered dielectric-rod antennas, 17-34 to 17-48
   for integrated antennas, 17-133 to 17-134
Tapered septum polarizer, 13-60 to 13-61
Taperline baluns, 9-95
Taylor line source pattern synthesis, 13-25 to 13-27
TDRSS (Tracking and Data Relay Satellite System), 21-67 to 21-71
TE field. *See* Transverse electric fields
TE/TM. *See* Transverse-electric/transverse-magnetic waveguides

TEC (total electron content), 29-15 to 29-16
Tee-bars for multiple-antenna installations, 27-37
Tee power dividers, 13-61, 26-40 to 26-41, 27-13, 27-15, 27-17
Teflon sphere lens antennas, 17-17 to 17-18
Television broadcast antennas, 27-3 to 27-8
   batwing, 27-23 to 27-24
   circularly polarized, 27-8 to 27-9
   helical, 27-15 to 27-19
      side-mount, 27-20 to 27-23
   horizontally polarized, 27-23
   multiple, 27-37 to 27-40
   sidefire helical, 27-24 to 27-26
   skewed dipole, 27-11 to 27-15
   slot
      coaxial, 27-28 to 27-29
      traveling-wave, 27-26 to 27-27
      waveguide, 27-29 to 27-31
   uhf, 27-27 to 27-28
   vee dipole array, 27-9 to 27-11
   zigzag, 27-31 to 27-32
TEM. *See* Transverse-electromagnetic fields
Temperature inversion layers and scintillations, 29-20
Temperature probes, implantable, 24-50 to 24-51
Temperature sensitivity of microwave radiometers, 22-21 to 22-22, 22-23
Temperature sounders
   and AMSU-A, 22-51
   for oxygen band, 22-16, 22-20
   for remote sensing, 22-13 to 22-14
Theoretical am antenna pattern, 26-34 to 26-36
Theory of uniform-periodic structures. *See* Periodic structure theory
Therapy, medical. *See* Hyperthermia
Thermal considerations with satellite antennas, 22-46
Thermal distortion, 21-28 to 21-29, 21-57
Thermal noise, 2-39 to 2-43, 29-50
Thermal temperature and sea surface emissivity, 22-14
Thermal therapy. *See* Hyperthermia
Thermistor detectors with implantable temperature probes, 24-50

Index

Thermocoupler detectors for rf radiation, 24-37
Thermographic cameras, 24-51
Thevenin's theorem for receiving antenna load, 6-3, 6-4
Thick dipole, 3-28
Thickness of lenses, 16-10 to 16-13. *See also* Zoning of lenses
    constrained, 16-43 to 16-44
    spherical, 16-27 to 16-29
    and surface tolerance, 21-18
    taper-control, 16-36
Thin dipole, 3-28
Thin-wall radome design, 31-17
Thin-wire antennas, 7-5 to 7-11
    and computer solutions, 3-54
    integral equations for, 3-44 to 3-46
    log-periodic zigzag, 9-45 to 9-47
    loop, 7-42 to 7-48
    receiving, 7-18 to 7-19
    time-domain analyses of, 3-51 to 3-52
    transmitting, 7-12 to 7-18
Thin-Wire Time Domain modeling code, 3-87 to 3-95
Three-antenna measurement methods
    for gain, 32-44, 32-46
    for polarization, 32-60 to 32-61
Three-axis stabilized satellites, 22-43 to 22-44
Three-wire transmission lines, 28-12
Through-the-earth communications. *See* Geophysical applications
Thunderstorms, noise from, 29-50
TID (traveling ionospheric disturbances), 29-25
Tilt angle and polarization, 32-51 to 32-53
    with polarization ellipse, 1-16
    with polarization pattern method, 32-62
    with reflector antennas, 15-48, 15-69 to 15-70, 15-72 to 15-73
Time-averaged power, 1-7 to 1-8, 3-11
    for modal fields, 1-46
    in rectangular waveguides, 1-39
Time delay
    with constrained analog of dielectric lenses, 16-51
    with constrained lenses, 16-44 to 16-46
    feed systems with
        matrix feed for beam-forming feed networks, 18-23, 18-25
        partial, 19-52, 19-68
        true, 19-7 to 19-8, 19-54, 19-76 to 19-91
    lenses for, 16-45 to 16-46, 16-54
    offset beams with, 13-39 to 13-41
    steering with
        with periodic arrays, 13-7 to 13-8
        with subarrays, 13-32 to 13-35
    with subarrays
        and contiguous subarrays, 13-32 to 13-35
    wideband characteristics of, 13-32
Time-domain
    compared to frequency-domain, 3-65 to 3-66, 3-82
    and EMP, 30-4
    integral equations in, 3-50 to 3-52
    method of moments in, 3-62 to 3-64
Time-harmonic excitation
    for grounded wire antennas, 23-8 to 23-12
    with loop antennas, 23-17 to 23-20
Time-harmonic fields, 1-5 to 1-8, 3-6 to 3-8
Tin-hat Rinehart lenses, 19-44, 19-46 to 19-47
TIROS-N satellite, 22-18
    planar scanning on, 22-36 to 22-37
    sounder units on, 22-14
Tissue, human
    permittivity of, 24-12, 24-30 to 24-35, 24-47
    phantoms for, 24-54 to 24-57
    power absorption by, 24-6 to 24-8
    radiation patterns in, 24-51 to 24-54
TM field. *See* Transverse magnetic fields
TMI microwave radiometer, 22-48 to 22-49
T-network as power dividers and phase shifters, 13-61, 26-40 to 26-41, 27-13, 27-15, 27-17
Toeplitz matrix, 3-68
Tolerance criteria for lenses, 16-16 to 16-20
Top loading of short dipole, 6-13
Topology with beam-forming networks, 21-57
Top-wall hybrid junction power division elements, 21-51 to 21-52
Toroidal beams for T T & C, 21-89
Toroidal ferrite phase shifters, 18-14, 21-40 to 21-42
Torsion factor, 4-87
Total electron content and satellite-earth propagation, 29-15 to 29-16

Towers for am antennas. *See also* Two-tower antenna patterns
  height of, and electric-field strength, 26-7
  mutual base impedance between, 26-12 to 26-17
  self-base impedance of, 26-8 to 26-12
  vertical radiation pattern of, 26-3 to 26-9
Tracking, telemetry, and command for satellite antennas, 21-89 to 21-111
Tracking and Data Relay Satellite System, 21-67 to 21-71
Trailing-wire antennas, EMP responses for, 30-27, 30-31
Transfer, microwave radiative, 22-4 to 22-7
Transformations
  of coordinates, 15-115 to 15-120
  distortion from, 11-57
  linear, 11-23 to 11-48
  of near-field data to far-field, 20-61 to 20-62
  for scale models, 32-67, 32-69
Transform feeds, 19-91 to 19-98
Transient conditions, 3-51
Transient excitation
  with grounded-wire antennas, 23-12 to 23-16
  with loop antennas, 23-20 to 23-21
Transient response
  of dipole antennas, 7-21 to 7-23
  of thin-wire antennas, 7-10
Transionospheric satellite-earth propagation, 29-14 to 29-15
Transition function with UTD solutions, 20-11
Transition regions, 4-5
  and PTD, 4-104
  and UTD, 4-32, 4-58 to 4-59
Translational symmetry and computer time and storage, 3-67 to 3-68
Transmission lines, 2-27 to 2-28. *See also* Constrained feeds
  analytical validation of computer code for, 3-76
  for beam-forming networks, 21-56 to 21-63
  with constrained lenses, 16-41
  current in, with folded dipoles, 7-38 to 7-39
  equations for, 28-5 to 28-10
  impedance of, with folded dipoles, 7-41
  and input admittance computations, 7-7
  with log-spiral conical antennas, 9-95
  for medical applications, 24-24 to 24-25
  for microstrip antennas, 10-7 to 10-10
  and mismatches, 1-24
  periodically loaded, 9-46 to 9-53, 9-57, 9-59 to 9-60
  planar quasi-TEM, 28-30 to 28-35
  radiator, for feeds, 19-23 to 19-32
  resonator for, with dual-band elements, 10-63, 10-67
  TEM, 28-10 to 28-29
Transmission space-fed arrays, 21-34, 21-35
Transmission-type space feeds, 18-17 to 18-19
Transmitting antennas
  linear dipole, 7-12 to 7-18
  reciprocity of, with receiving antennas, 2-32 to 2-36
  sources for, 2-29 to 2-30
  thin-wire, 7-8
    loop, 7-42 to 7-44
Transmitting fields
  and effective length, 2-34 to 2-35
  and Kirchhoff approximation, 2-19 to 2-21
Transmitting systems for antenna ranges, 32-30 to 32-32
Transposition of conductors with log-periodic antennas, 9-14 to 9-16, 9-59 to 9-60
Transverse amplitude taper in antenna ranges, 32-14
Transverse current in waveguides, 12-3
Transverse electric fields
  and circular waveguides, 1-44 to 1-50
  in rectangular waveguides, 1-37 to 1-41, 13-44
  representation of, 1-30 to 1-32
Transverse electric/transverse magnetic waveguides, 21-56 to 21-57, 28-35 to 28-47
  for medical applications, 24-9 to 24-12
Transverse electromagnetic fields
  and chambers for calibration of field probes, 24-54 to 24-57
  lenses for, 21-16
  transmission lines for, 28-10
    balanced, 28-12, 28-13
    circular coaxial, 28-20 to 28-22

Index

triplate stripline, 28-22 to 28-29
two-wire, 28-11, 28-19 to 28-20
and waveguide antennas
    for beam-forming networks, 21-56 to 21-58
    for medical applications, 24-9 to 24-13
Transverse magnetic fields
    and circular waveguides, 1-44 to 1-50
    in rectangular waveguides, 1-37 to 1-41
    representation of, 1-30 to 1-32
Transverse modes for leaky-wave antennas, 17-86, 17-88
Transverse slots, mutual coupling with, 12-30 to 12-31
Trapezoidal-tooth log-periodic antennas, 9-37 to 9-38
Trapped image-guide antenna, 17-49, 21-62
Trapped inverted microstrip transmission type, 21-59, 21-61
Trapped-miner problem, 23-23 to 23-24
Trapped surface waves, 29-45
Traveling ionospheric disturbances, 29-25
Traveling-wave antennas, 3-16, 3-18
    arrays
        for microstrip antennas, 17-118 to 17-119
        microstrip patch resonator, 17-103 to 17-106, 17-114 to 17-115
    television
        sidefire helical, 27-24 to 27-26
        slot, 27-26 to 27-27
Traveling-wave-fed arrays, 12-5
    linear array of edge slots, 12-26
    longitudinal slots, 12-16 to 12-20
    planar, 12-23 to 12-24
Triangular-aperture antennas, 3-24
Triangular lattices
    field representation of, 1-35
    for phased-array radiators, 18-9 to 18-10
Triangular patches, 10-18
Triangular planar arrays and grating lobes, 13-16 to 13-17
Triangular-tooth log-periodic antennas, 9-37 to 9-38
Trifocal bootlace lenses, 16-47
Triplate striplines, 28-22, 28-26 to 28-29
    impedance of, 28-17
Tropical Rainfall Measurement Mission (TRMM), 22-48 to 22-49
Troposphere and tropospheric propagation, 29-30 to 29-47
    permittivity and conductivity of, 29-31

scatter propagation by, 29-46 to 29-47
scintillations caused by, 29-20
True-time-delay feed systems, 19-7 to 19-8, 19-54, 19-76 to 19-91
Truncation
    of aperiodic arrays, 14-20
    with log-periodic antennas, 9-4 to 9-5, 9-8
    of log-spiral antennas, 9-72 to 9-73
        conical, 9-79, 9-84, 9-89 to 9-90, 9-95 to 9-97
    with microstrip antennas, 10-23 to 10-24
    and mutual coupling, 12-29, 12-34
    with optical devices, 19-98
    and scanning range errors, 33-17
T T & C (tracking, telemetry, and command), 21-89 to 21-111
Tumors. *See* Hyperthermia
Turbulence, scintillation due to, 29-36
Turnstile antennas, 3-29
    phased arrays for satellite antennas, 21-12
    with radial transmission line feeds, 19-24, 19-26 to 19-27
TV. *See* Television broadcast antennas
TVRO (television receiving only) antennas, 6-25 to 6-32
Twin-boom construction for log-periodic dipole antennas, 9-15 to 9-16
Twist reflector for millimeter-wave antennas, 17-11
Two-antenna gain measurement method, 32-43 to 32-44, 32-46
Two-dimensional arrays, 13-8, 13-9
    Dolph-Chebyshev planar, 11-49 to 11-52
Two-dimensional lenses
    aberrations in, 16-19
    waveguide, 16-42 to 16-43
Two-dimensional multiple-beam matrices, 19-11 to 19-12
Two-layer lenses, 17-15 to 17-18
Two-layer pillbox, 19-17 to 19-23
Two-parameter model for reflector antennas, 15-15 to 15-18
Two planes, arrays scanned in, 13-15 to 13-17
Two-point calibration method, 22-25 to 22-26
Two-terminal stopbands, 9-49 to 9-50
    with log-periodic diode arrays, 9-65
    phase shift in, 9-58
Two-tower antenna patterns, 26-22 to 26-24

Two-tower antenna patterns (*cont.*)
   horizontal electric-field strength with, 26-30 to 26-34
   mutual impedance of, 26-12 to 26-13
Two-wire transmission lines, 28-12
   field configuration of, 28-19
   impedance of, 7-39, 28-11, 28-14, 28-15, 28-20
   validation of computer code for, 3-76
TWTD (Thin-Wire Time Domain) modeling code, 3-87 to 3-95

UAT (uniform asymptotic theory), 4-5
   for edges, 4-32, 4-40 to 4-43
Uhf antennas, 27-3, 27-27 to 27-28
   bandwidth and gain of, 27-7
   EMP responses for, 30-17, 30-20 to 30-21
   and ionospheric propagation, 29-28 to 29-30
   slot, 27-29 to 27-31
Unconstrained feeds, 18-23 to 18-24, 19-49 to 19-54, 21-34
   limited scan, 19-56 to 19-76
   wide field of view, 19-54 to 19-56, 19-76 to 19-91
Underground communications, 23-3
   grounded-wire antennas for, 23-8 to 23-17
   loop antennas for, 23-17 to 23-23
Uniform aperture distributions, 5-11 to 5-13
   far-zone characteristics of, 3-23
   near-field reduction factors for, 5-30 to 5-31
   phase distortion with, 15-109
   radiation pattern with, 21-17
Uniform broadside linear arrays, 11-15
Uniform circular arrays, 11-46 to 11-47
Uniform dielectric layers, 17-65
Uniform diffracted fields, 20-44
Uniform dyadic edge-diffraction coefficient, 4-32 to 4-33
Uniform end-fire linear array, 11-15
Uniform geometrical theory of diffraction, 4-5
   for complex environment simulations, 20-4
   with convex surfaces
      mutual coupling, 4-84 to 4-96
      radiation, 4-63 to 4-84
      scattering, 4-50 to 4-63
   for edges, 4-32 to 4-38
   for numerical solutions. *See* Numerical solutions
   for vertices, 4-43 to 4-48
Uniform hemispherical radiators
   characteristics of, 26-9
   electric-field strength of, 26-27 to 26-29
Uniform linear arrays, 11-11 to 11-13
Uniformly perturbed millimeter waveguides, 17-35
Uniform reflected fields, 20-43
Uniform spherical radiators, 26-8
Uniform structures. *See* Periodic structures
Uniform-waveguide leaky-wave antennas, 17-35, 17-82 to 17-103
Unilateral fin lines, 28-55
Universal curves
   of E-plane horn antennas, 8-10, 8-13 to 8-14
   of H-plane horn antennas, 8-23, 8-27 to 8-29
Unknowns, and computer storage and time, 3-57
Unloaded linear dipole antennas
   receiving, 7-18 to 7-19
   transmitting, 7-12 to 7-18
Updating of computer models, 3-68
Uplink transmitters for satellite tv, 6-25 to 6-26
Upper frequency limit, 9-73
   of log-spiral antennas, 9-75
Use assistance with computer models, 3-68
UTD. *See* Uniform geometrical theory of diffraction

Validation of computer code, 3-55, 3-72 to 3-80
Varactor diodes
   with microstrip antennas, 10-67, 10-70
   in phase shifters, 18-13, 21-40
Variable amplitude networks, 21-33 to 21-34
Variable offset long-slot arrays, 17-26
Variable phase networks, 21-33 to 21-34
Variable phase shifters
   for beam-forming networks
      scanned, 21-38 to 21-43
      switched, 21-32
   for gain ripple with feed arrays, 21-27
   with Wheeler Lab approach, 19-105, 19-107

Index

Variable power dividers
  for beam-forming networks
    scanned 21-43 to 21-49
    switched, 21-32 to 21-33
  for gain ripple with feed arrays, 21-27
  quantization errors with, 21-12
Vector effective area, 6-5 to 6-6
Vector effective height, 6-3 to 6-6
  and power ratio, 6-8 to 6-9
  of short dipole, 6-12
  with UTD solutions, 20-6
Vector far-field patterns, 7-8
Vector wave equation, 2-7 to 2-8
Vee dipole tv antennas, 27-9 to 27-11
Vegetation and surface emissivity, 22-10, 22-15
Vehicular antennas, 32-9, 32-19
Vehicular-mounted interferometers, 25-21 to 25-23
Vertical antennas, radiation resistance of, 6-16
Vertical current element, 26-9
Vertical full-wave loops, 3-31
Vertical optimization, 11-82 to 11-83
Vertical polarization
  for am antennas, 26-3
  with leaky-wave antennas, 17-94
  for microwave radiometers, 22-44 to 22-46
  and surface emissivity, 22-8
  with surface-wave propagation, 29-44 to 29-45
Vertical radiation characteristics for am antennas, 26-3 to 26-8
Vertical sounding, 23-4, 23-18
Vertical stabilizers, simulation of, 20-73, 20-82, 20-84
Vertical transmitting loop antennas, 23-17
Vertices
  GTD for, 4-30 to 4-31
  UTD for, 4-43 to 4-48
Very high frequencies
  communication antennas for, and EMP, 30-17, 30-20 to 30-24
  and ionospheric propagation, 29-21, 29-28 to 29-30
  localizer antennas for, and EMP, 30-23, 30-25 to 30-26
  marker beacon antennas for, and EMP, 30-23, 30-27 to 30-28
Very low frequencies
  ionospheric propagation of, 29-27 to 29-28
  trailing-wire antennas for, and EMP, 30-27, 30-31
Viscometric thermometers, 24-50
Visible regions
  of linear arrays, 11-9
  and log-periodic arrays, 9-56
  with planar periodic arrays, 11-27
  single main beam in, 11-31 to 11-33
Vokurka compact antenna range, 32-30
Voltage-controlled variable power dividers, 21-46 to 21-47, 21-49
Voltage reflection coefficient, 2-28
  of surfaces, 22-8
  for transmission lines, 28-7
Voltages
  in biconical transmission line, 3-24
  diagrams of, with standard reference am antennas, 26-19 to 26-20
  mode, 12-13, 12-16
  in rectangular waveguide, 2-28 to 2-29
  slot, 12-10 to 12-12
  in transmission lines, 28-6 to 28-8
  with UTD solutions, 20-7
Voltage source, ideal, 2-29 to 2-30
Voltmeters for receiving antenna, 2-30 to 2-32
Vswr
  of corrugated horns, 15-97
  and couplers, 19-4 to 19-5
  and lens surface mismatch, 16-54 to 16-55
  with log-periodic antennas
    front-to-back ratios in, 9-60 to 9-61
    zigzag, 9-42 to 9-44
  and log-spiral conical antennas impedances, 9-97, 9-107
  with microstrip antennas, 17-120
  with phase shifters, 21-40 to 21-41
  for pyramidal horns, 8-68
  and stopbands, 17-52, 17-81
  with transmission lines, 28-8
  of tv broadcast antennas, 27-3

Wall currents in waveguides, 12-3, 12-4
Water
  modeling of, 32-74 to 32-75
  remote sensing of, 22-15
Water bolus for loop radiators in medical applications, 24-20, 24-22
Water vapor
  absorption by, 29-35
  noise from, 29-51
  remote sensing of, 22-12 to 22-14, 22-22

Watson-Watt direction-finding system, 25-12 to 25-14
Wavefronts
  aberrations in, and radomes, 31-7 to 31-9
  incident, 4-15, 4-16
  and lenses, 16-6, 16-9 to 16-10
  and rays, 4-6 to 4-8
Waveguide-fed slot arrays. *See* Slot arrays, waveguide-fed
Waveguide feeds
  for beam-forming networks, 21-56 to 21-58
  for reflector antennas, 15-5
Waveguide hybrid ring waveguide power division elements, 21-50 to 21-52
Waveguide lenses, 16-42 to 16-45, 21-16
  for satellite antennas, 21-71 to 21-77
Waveguide loss. *See also* Leaky-wave antennas; Surface wave antennas
  circular, 1-49
  rectangular, 1-39
  with reflector antennas, 17-11
Waveguide nonreciprocal ferrite phase shifters, 21-38 to 21-39
Waveguide phased arrays, 21-12
Waveguides, 2-27 to 2-29
  arrays of, 13-44 to 13-45, 13-57
  blindness in, 13-45, 13-47 to 13-49
  circular. *See* Circular waveguides
  coplanar, 28-30 to 28-35
  coupling of, to microstrip antennas, 17-119 to 17-121
  hybrid-mode, 28-47 to 28-58
  impedance of, 2-28 to 2-29
  for medical applications, 24-9 to 24-18
  for microstrip patch resonator arrays, 17-117 to 17-118
  open-ended, for reflector feeds, 15-89 to 15-93, 15-104 to 15-105
  polarizers for, 13-60 to 13-61
  power divider elements for, 21-50 to 21-52
  rectangular. *See* Rectangular waveguides
  ridged, 28-45 to 28-49
  standard, 28-40 to 28-41
  TE/TM, 28-35 to 28-47
  uniformly perturbed millimeter, 17-35
Waveguide slot radiators, 18-6, 18-10
Waveguide slot tv antennas, 27-29 to 27-31
Waveguide transmission lines, 21-37

Wave impedances, 1-6, 1-13 to 1-14, 1-18 to 1-19
  of hollow-tube waveguides, 28-36, 28-37
  and modeling, 32-68
Wavelength, 1-6 to 1-7. *See also* Cutoff wavelength
Wave number, 1-6 to 1-7
Weapon-locating radar, limited scan, 19-56
Weather forecasting, remote sensing for, 22-13 to 22-15
Wedge-diffraction and UTD solutions, 20-11, 20-17 to 20-18
  coefficient of, 20-46
Wedges, conducting, 1-10
Weight functions in modeling codes, 3-83 to 3-84
Well logging, 23-24 to 23-25
Wenner array for earth resistivity, 23-5, 23-7
Wheeler Lab approach to cylindrical array feeds, 19-105 to 19-107
Wide-angle antennas, 17-29 to 17-30
Wide-angle lenses
  dielectric, 16-19 to 16-33
  multiple-beam, 19-37, 19-80 to 19-85
Wide-angle scanning
  and F/D ratios, 19-34 to 19-35
  lens antennas for, 17-15
Wide antennas, leaky mode with, 17-53 to 17-62
Wideband conical horn antennas, 8-61 to 8-62
Wideband Satellite Experiment, 29-19
Wideband scanning with broadband array feeds, 13-40 to 13-41
Wide field of view unconstrained feeds, 19-52, 19-54 to 19-56, 19-76 to 19-91
Width
  of dielectric grating antennas, 17-64
  main beam, with rectangular apertures, 5-13
  of metal grating antennas, 17-68 to 17-69
  of microstrip antennas, 10-8
Wilkinson power dividers for array feeds, 13-61
Wind speed, remote sensing of, 22-14
Wind sway, effect of, 27-38
Wings, aircraft, simulation of, 20-37, 20-70 to 20-71, 20-78, 20-80
Winter anomaly period for ionospheric propagation, 29-25

Index

Wire-antennas, 7-5 to 7-11
   folded dipole, 7-37 to 7-40
   linear dipole, 7-11 to 7-23
   loop, 7-40 to 7-48
   sleeve, 7-23 to 7-37
Wire diameter and radiation patterns, 32-76, 32-79
Wire objects, computer storage and time for, analysis of, 3-57
Wire-outline log-periodic antennas, 9-13
Wire problems
   subdomain procedures for, 3-60
   time-domain, 3-63
Wullenweber arrays, 25-14 to 25-16

X-band horn antennas, 8-41 to 8-42, 8-44 to 8-45, 8-67
X-polarized antennas, 1-28
XY-scanners, 33-19

Yagi-Uda antennas, 32-76
Yardarm, shipboard, simulation of, 20-93 to 20-94

Yield of crops, forecasting of, 22-15
Y-polarized antennas, 1-28

Zernike cylindrical polynomials, 16-13
Zero-bias diodes, 24-43
Zeroes of the Airy function, 4-59
Zeroes of Bessel functions, 1-45
Zeroes of W, 4-89
Zigzag antennas. *See also* Log-periodic zigzag antennas
   dipoles, as E-field probes, 24-39 to 24-40
   tv, 27-31 to 27-32
   wire, 9-33 to 9-38
Zone-plate lens antennas, 17-20, 17-22
Zones, FCC allocation and assignment, 27-4
Zoning of lenses, 16-6, 16-28 to 16-30
   and bandwidth, 16-54 to 16-55, 21-16
   with constrained lenses, 16-43 to 16-45
   with dielectric lenses, 16-38 to 16-41
   with equal group delay lenses, 16-46
   with millimeter-wave antennas, 17-15, 17-19 to 17-21